Lecture Notes in Computer Science   9255

Commenced Publication in 1973
Founding and Former Series Editors:
Gerhard Goos, Juris Hartmanis, and Jan van Leeuwen

More information about this series at http://www.springer.com/series/7408

Gilles Pesant (Ed.)

Principles and Practice
of Constraint Programming

21st International Conference, CP 2015
Cork, Ireland, August 31 – September 4, 2015
Proceedings

 Springer

Editor
Gilles Pesant
École Polytechnique de Montréal
Montréal, Québec
Canada

ISSN 0302-9743   ISSN 1611-3349   (electronic)
Lecture Notes in Computer Science
ISBN 978-3-319-23218-8   ISBN 978-3-319-23219-5   (eBook)
DOI 10.1007/978-3-319-23219-5

Library of Congress Control Number: 2015946574

Springer Cham Heidelberg New York Dordrecht London

Springer International Publishing AG Switzerland is part of Springer Science+Business Media
(www.springer.com)

Preface

This volume contains the proceedings of the 21st International Conference on the Principles and Practice of Constraint Programming (CP 2015), which was held in Cork, Ireland, from August 31 to September 4, 2015. Detailed information about the conference is available at http://cp2015.a4cp.org. The CP conference is the annual international conference on constraint programming. It is concerned with all aspects of computing with constraints, including theory, algorithms, environments, languages, models, systems, and applications such as decision making, resource allocation, scheduling, configuration, and planning.

For the purpose of the conference's scientific programming, we invited submissions to the technical, application, and published journal tracks. We received 80, 25, and 14 submissions to these tracks respectively. Authors chose to submit either long (15 pages) or short (8 pages) papers to the technical and application tracks. The review process for the technical track relied on a two-level Program Committee and on additional reviewers recruited by Program Committee members. Each submission to the technical track was assigned to one member of the Senior Program Committee and three members of the track's Program Committee. Submissions to the application track were each assigned to three members of its Program Committee. Every paper received at least three reviews. Once the initial reviews were in, authors were given an opportunity to respond before a detailed discussion was undertaken at the level of the Program Committees, overseen by the Program Chair and the Senior Program Committee member or the Application Track Chair, as appropriate. For the first time this year the Senior Program Committee did not meet in person but deliberated by video conference instead. The published journal track gives an opportunity to discuss important results in the area of constraint programming that appeared recently in relevant journals, but had not been previously presented to the community at conferences. Submissions were evaluated by a separate Program Committee for relevance and significance.

At the end of the reviewing process, we accepted 39 papers from the technical track, 15 papers from the application track, and 10 papers from the published journal paper track. The Senior Program Committee awarded the Best Paper Prize to John N. Hooker for "Projection, Consistency, and George Boole" and the Best Student Paper Prize to Alexander Ivrii, Sharad Malik, Kuldeep S. Meel, and Moshe Vardi for "On Computing Minimal Independent Support and Its Applications to Sampling and Counting." The Application Track Program Committee awarded the Best Application Paper Prize to Tommaso Urli and Philip Kilby for "Long-Haul Fleet Mix and Routing Optimisation with Constraint Programming and Large Neighbourhood Search." The Program Chair and the *Constraints* journal editor-in-chief, Michela Milano, also invited six papers from the technical and application tracks for direct publication in that journal. These were presented at the conference like any other paper and they appear in the proceedings as a one-page abstract.

This edition of the conference was part of George Boole 200, a celebration of the life and work of George Boole who was born in 1815 and worked at the University College of Cork. It was also co-located with the 31st International Conference on Logic Programming (ICLP 2015). The conference program featured three invited talks by Claire Bagley, Gerhard Friedrich (joint with ICLP), and Douglas R. Smith. This volume includes one-page abstracts of their talks. The conference also featured four tutorials and six satellite workshops, whose topics are listed in this volume. The Doctoral Program gave PhD. students an opportunity to present their work to more senior researchers, to meet with an assigned mentor for advice on their research and early career, to attend special tutorials, and to interact with one another. The winners of the 2015 ACP Research Excellence Award and Doctoral Research Award presented their award talks. The results of the annual MiniZinc Challenge were announced and a joint ICLP/CP Programming Competition was held. For the first time an Industry Outreach effort, headed by Helmut Simonis, provided formal opportunities for CP researchers and representatives from industry to interact: The former were presented with industrial problems to solve in a friendly competition and the latter were presented with samples of industrial problems solved using CP technology. Following the conference, a CSPlib sprint event took place to update this library of constraint satisfaction problems. All these contributed to a very exciting program.

I am grateful to many people who made this conference such a success. First of all, to the authors who provided excellent material to select from. Then to the members of the Program Committees and additional reviewers who worked hard to provide constructive, high-quality reviews. To members of the Senior Program Committee who helped me ensure that each paper was adequately discussed, wrote meta-reviews for their assigned papers, and participated in live remote deliberations — for some, quite early or late in the day. To Christian Schulte in particular, who agreed to step in where I had a conflict of interest. Of course there is a whole team standing with me, who chaired various aspects of the conference: Ken Brown, Barry O'Sullivan (Conference Chairs) and their own local team (Barry Hurley as webmaster; Kathy Bunney who provided coordinating support; the technical and administrative support staff of the Insight Centre for Data Analytics at UCC), Louis-Martin Rousseau (Application Track Chair), Thomas Schiex (Published Journal Track Chair), David Bergman and Marie Pelleau (Doctoral Program Chairs), Willem-Jan van Hoeve (Workshop and Tutorial Chair), Helmut Simonis (Industry Outreach Chair), and Ian Miguel (Publicity Chair). Thank you for your dedication! I acknowledge and thank our sponsors for their generous support: they include, at the time of writing, the Association for Constraint Programming, the Association for Logic Programming, ECCAI — the European Coordinating Committee for Artificial Intelligence, the Insight Centre for Data Analytics, Science Foundation Ireland, Springer, and University College Cork. Finally, I thank the ACP Executive Committee for the trust they showed me in asking me to serve as Program Chair. It has been an honor.

June 2015   Gilles Pesant

Tutorials and Workshops

Tutorials

Constraints and Bioinformatics: Results and Challenges
Agostino Dovier
Lagrangian Relaxation for Domain Filtering
Hadrien Cambazard
Towards Embedded Answer Set Solving
Torsten Schaub
XCSP3
Frédéric Boussemart, Christophe Lecoutre, and Cédric Piette

Workshops

Workshop on Constraint-Based Methods in Bioinformatics (WCB 2015)
Alessandro Dal Palù and Agostino Dovier
6th International Workshop on Bin Packing and Placement Constraints (BPPC 2015)
Nicolas Beldiceanu and François Fages
*5th International Workshop on the Cross-Fertilization Between CSP and SAT
(CSPSAT 2015)*
Yael Ben-Haim, Valentin Mayer-Eichberger, and Yehuda Naveh
*14th International Workshop on Constraint Modelling and Reformulation
(ModRef 2015)*
Ozgur Akgun and Peter Nightingale
Workshop on Teaching Constraint Programming
Alan Frisch, Ciaran McCreesh, Karen Petrie, and Patrick Prosser
CP and Analytics
Youssef Hamadi and Willem-Jan van Hoeve

Conference Organization

Conference Chairs

Ken Brown   University College Cork, Ireland
Barry O'Sullivan   University College Cork, Ireland

Program Chair

Gilles Pesant   École Polytechnique de Montréal, Canada

Application Track Chair

Louis-Martin Rousseau   École Polytechnique de Montréal, Canada

Published Journal Track Chair

Thomas Schiex   INRA Toulouse, France

Doctoral Program Chairs

David Bergman   University of Connecticut, USA
Marie Pelleau   Université de Montréal, Canada

Workshop and Tutorial Chair

Willem-Jan van Hoeve   Carnegie Mellon University, USA

Industry Outreach Chair

Helmut Simonis   Insight Centre for Data Analytics, UCC, Ireland

Publicity Chair

Ian Miguel   University of St Andrews, UK

Senior Program Committee

Chris Beck   University of Toronto, Canada
Nicolas Beldiceanu   TASC (CNRS/Inria), Mines Nantes, France
Christian Bessiere   CNRS, France
Yves Deville   UCLouvain, Belgium

Pierre Flener Uppsala University, Sweden
John Hooker Carnegie Mellon University, USA
Peter Jeavons University of Oxford, UK
Christophe Lecoutre CRIL, University of Artois, France
Jimmy Lee The Chinese University of Hong Kong, Hong Kong,
    SAR China
Michela Milano DISI Università di Bologna, Italy
Jean-Charles Régin University of Nice Sophia Antipolis / CNRS, France
Christian Schulte KTH Royal Institute of Technology, Sweden
Peter J. Stuckey NICTA and the University of Melbourne, Australia
Pascal Van Hentenryck NICTA and ANU, Australia
Willem-Jan van Hoeve Carnegie Mellon University, USA
Toby Walsh NICTA and UNSW, Australia
Roland Yap National University of Singapore, Singapore

Technical Track Program Committee

Carlos Ansótegui Universitat de Lleida, Spain
Fahiem Bacchus University of Toronto, Canada
Pedro Barahona Universidade Nova de Lisboa, Portugal
Roman Bartak Charles University in Prague, Czech Republic
David Bergman University of Connecticut, USA
Hadrien Cambazard Grenoble INP, CNRS, Joseph Fourier University, France
Hubie Chen Universidad del País Vasco and Ikerbasque, Spain
Geoffrey Chu NICTA and the University of Melbourne, Australia
David Cohen Royal Holloway, University of London, UK
Remi Coletta University of Montpellier, France
Martin Cooper IRIT - Universitié Paul Sabatier, France
Sophie Demassey CMA, MINES ParisTech, France
François Fages Inria Paris-Rocquencourt, France
Alan Frisch University of York, UK
Maria Garcia De La Monash University, Australia
Banda
Arnaud Gotlieb SIMULA Research Laboratory, Norway
Stefano Gualandi Università di Pavia, Italy
Emmanuel Hebrard LAAS, CNRS, France
Philippe Jégou LSIS - UMR CNRS 7296 - Aix-Marseille University,
    France
George Katsirelos INRA, Toulouse, France
Zeynep Kiziltan Università di Bologna, Italy
Lars Kotthoff University of British Columbia, Canada
Philippe Laborie IBM, France
Michele Lombardi DISI Università di Bologna, Italy
Xavier Lorca Ecole des Mines de Nantes, France
Inês Lynce Inst. Superior Técnico INESC-ID Lisboa, Portugal
Arnaud Malapert University of Nice Sophia Antipolis / CNRS, France

Joao Marques-Silva	University College Dublin, Ireland
Ian Miguel	University of St. Andrews, UK
Justin Pearson	Uppsala University, Sweden
Justyna Petke	University College London, UK
Steve Prestwich	Insight, UCC, Ireland
Patrick Prosser	Glasgow University, UK
Claude-Guy Quimper	Université Laval, Canada
Andrea Rendl	NICTA and Monash University, Australia
Michel Rueher	University of Nice Sophia Antipolis / CNRS, France
Marius Silaghi	Florida Institute of Technology, USA
Stephen Smith	Carnegie Mellon University, USA
Christine Solnon	LIRIS CNRS UMR 5205 / INSA Lyon, France
Kostas Stergiou	University of Western Macedonia, Greece
Guido Tack	Monash University, Australia
Charlotte Truchet	LINA, UMR 6241, Université de Nantes, France
Nic Wilson	Insight, UCC, Ireland

Application Track Program Committee

Claire Bagley	Oracle Corporation, USA
Thierry Benoist	Innovation 24 - LocalSolver, France
Mats Carlsson	SICS, Sweden
Jean-Guillaume Fages	COSLING S.A.S., France
Carmen Gervet	Université de Savoie, LISTIC, France
Laurent Perron	Google, France
Ashish Sabharwal	AI2, USA
Pierre Schaus	UCLouvain, Belgium
Paul Shaw	IBM, France
Helmut Simonis	Insight, UCC, Ireland
Peter van Beek	University of Waterloo, Canada
Mark Wallace	Monash University, Australia
Tallys Yunes	University of Miami, USA

Published Journal Track Program Committee

Simon de Givry	MIAT-INRA, France
Yves Deville	UCLouvain, Belgium
John Hooker	Carnegie Mellon University, USA
Michela Milano	DEIS Università di Bologna, Italy
Nina Narodytska	Carnegie Mellon University, USA
Patrick Prosser	Glasgow University, UK
Francesca Rossi	University of Padova, Italy
Roland Yap Hock Chuan	National University of Singapore, Singapore

Invited Talks

Constraint-based Problems and Solutions in the Global Enterprise

Claire Bagley

Advanced Constraint Technology
Oracle Corporation, Burlington, MA 01803, USA
{claire.bagley@oracle.com}

Oracle is a large global technology organization, whose product offerings have grown organically and through the acquisition of first class companies. Its leadership has expanded to the entire technology stack, to span the full range of computer hardware, operating systems, programming languages, databases, enterprise software, collaboration management tools, and into the cloud.

With thousands of products and tools running the full imaginable breadth and depth of a technology stack, it is then not surprising that Oracle is exposed to a vast number of complex combinatorial problems, as well as the different types of technology to solve them. Indeed, many of the classical applications and variations of constraint problems are represented: planning, scheduling, rostering, vehicle routing, configuration, networking, grid optimization, logistics, analytics, and cloud management. As expected with the development of products and the acquisition of companies operating in these domains, a large number of technologies come into play including Constraint Programming (CP), Mathematical Programming (MP), local search, heuristics, knowledge-based reasoning, genetic algorithms, machine learning, and many more.

The Advanced Constraint Technology (ACT) group at Oracle is tasked with identifying, understanding and solving the complex combinatorial problems that arise in a diverse field of application environments. Our expertise and industry proven ACT products are available to assist Oracle development teams on how to best model and solve their problems using CP, MP, Heuristics, and Hybrid solutions.

In this talk we examine some of the successful solutions to constrained problems within such a large corporation. We discuss at a high level the many opportunities for further integration and unification of constraint technologies into more products and tools, including the challenge of modeling and solving in highly interactive scenarios. Most importantly, we open the discussion about various challenges faced by large front-end centered companies, with many degrees of separation between organizations, who must balance resources to focus over immediate and long-term deliverables.

Industrial Success Stories of ASP and CP: What's Still Open?

Gerhard Friedrich

Institut für Angewandte Informatik
Alpen-Adria Universität Klagenfurt, Austria
e-mail: gerhard.friedrich@aau.at

Abstract More than 25 years ago together with Siemens we started to investigate the possibility of substituting the classical rule-based configuration approach by model- based techniques. It turned out that in those days only constrained programming (CP) had any real chance of meeting the application demands. By exploiting CP we were able to significantly improve the productivity of highly trained employees (by more than 300 %) and to substantially reduce software development and maintenance costs (by more than 80 %) [4]. Consequently, CP has been our method of choice for problem solving in industrial projects since 1989 [3].

Some years ago, we started to investigate answer set programming (ASP) techniques [2], mainly because of the possibility to apply a very expressive logical first-order language for specifying problems. It emerged that, by using simply problem encoding, we were able to solve difficult real world problem instances witnessing the enormous improvements of logic programming over the last decades [1].

Although ASP and CP have proven their practical applicability, we will point out challenges of large problems of the electronic and the semiconductor industry. In particular, we will stress the power of problem-specific heuristics [5, 9] which turned out to be the key in many applications of problem solvers.

Looking at the famous equation "algorithm = logic + control" [6] most of the current work in the AI community assumes that control should be problem- independent and only the logical specification depends on the problem to be solved, i.e. "algorithm = logic(problem) + control". It is not surprising that for the current problem solving technology this is a practical approach up to a certain size of the problem instances, since we deal with NP-hard problems in many cases. However, it is observed (and examples are given [10, 7]) that problem-specific heuristics allow enormous run-time improvements. This success is based on problem-specific control, i.e. "algorithm = logic (problem) + *control(problem)*". Unfortunately, the design of such problem-specific heuristics is very time-consuming and redesigns are frequently required because of recurrent changes of the problem. Interestingly, humans are very successful at developing such problem-specific heuristics. Therefore, we argue that the automation of generating problem-specific heuristics with satisfying quality is still an important basic AI research goal with high practical impact that should be achievable [8].

References

1. Aschinger, M., Drescher, C., Friedrich, G., Gottlob, G., Jeavons, P., Ryabokon, A., Thorstensen, E.: Optimization methods for the partner units problem. In: Achterberg, T., Beck, J.C. (eds.) CPAIOR 2011. LNCS, vol. 6697, pp. 4–19. Springer, Heidelberg (2011). http://dx.doi.org/10.1007/978-3-642-21311-3

2. Brewka, G., Eiter, T., Truszczynski, M.: Answer set programming at a glance. Commun. ACM **54**(12), 92–103 (2011). http://doi.acm.org/10.1145/2043174.2043195

3. Falkner, A., Haselboeck, A., Schenner, G., Schreiner, H.: Benefits from three configurator generations. In: Blecker, T., Edwards, K., Friedrich, G., Hvam, L., Salvodor, F. (eds.) Innovative Processes and Products for Mass Customization, vol. 3, pp. 89–103 (2007)

4. Fleischanderl, G., Friedrich, G., Haselböck, A., Schreiner, H., Stumptner, M.: Configuring large systems using generative constraint satisfaction. IEEE Intell. Syst. **13**(4), 59–68 (1998)

5. Gebser, M., Kaufmann, B., Romero, J., Otero, R., Schaub, T., Wanko, P.: Domain-specific heuristics in answer set programming. In: desJardins, M., Littman, M.L. (eds.) Proceedings of the Twenty-Seventh AAAI Conference on Artificial Intelligence, July 14–18, 2013, Bellevue, Washington, USA. AAAI Press (2013). http://www.aaai.org/ocs/index.php/AAAI/AAAI13/paper/view/6278

6. Kowalski, R.A.: Algorithm = logic + control. Commun. ACM **22**(7), 424–436 (1979). http://doi.acm.org/10.1145/359131.359136

7. Mersheeva, V., Friedrich, G.: Multi-uav monitoring with priorities and limited energy resources. In: Brafman, R.I., Domshlak, C., Haslum, P., Zilberstein, S. (eds.) Proceedings of the Twenty-Fifth International Conference on Automated Planning and Scheduling, ICAPS 2015, Jerusalem, Israel, June 7-11, 2015, ICAPS 2015. pp. 347–356. AAAI Press (2015). http://www.aaai.org/ocs/index.php/ICAPS/ICAPS15/paper/view/10460

8. Pearl, J.: On the discovery and generation of certain heuristics. AI Mag. **4**(1), 23–33 (1983). http://www.aaai.org/ojs/index.php/aimagazine/article/view/385

9. Schrijvers, T., Tack, G., Wuille, P., Samulowitz, H., Stuckey, P.J.: Search combinators. Constraints **18**(2), 269–305 (2013). http://dx.doi.org/10.1007/s10601-012- 9137-8

10. Teppan, E.C., Friedrich, G., Falkner, A.A.: Quickpup: A heuristic backtracking algorithm for the partner units configuration problem. In: Fromherz, M.P.J., Muñoz- Avila, H. (eds.) Proceedings of the Twenty-Fourth Conference on Innovative Applications of Artificial Intelligence, July 22–26, 2012, Toronto, Ontario, Canada. AAAI (2012). http://www.aaai.org/ocs/index.php/IAAI/IAAI-12/paper/view/4793

Synthesis of Constraint Solvers

Douglas R. Smith and Stephen J. Westfold

Kestrel Institute, Palo Alto, CA 94034 USA
{smith,westfold}@kestrel.edu

In [2], we present a mathematical framework for specifying and formally designing high-performance constraint solving algorithms. The framework is based on concepts from abstract interpretation which generalizes earlier work on a Galois Connection-based model of Global Search algorithms. The main focus is on how to use the framework to automate the calculations necessary to construct correct, efficient problem-specific constraint solvers.

It is common practice in the constraint-solving community to solve a new problem P by building a reduction to a well-studied problem Q that has a well-engineered solver. One problem with this approach is that the reduction of P to Q often loses some key structure which cannot then be exploited by the Q-solver. Our thesis is that a native solver can always be generated for a constraint problem that outperforms a reduction to an existing solver.

This talk focuses on three main results from [2]:

1. *Algorithm theory* – We develop and prove an algorithm theory for constraint solving with propagation, conflict detection and analysis, backjumping, and learning that is parametric on the constraint logic.
2. *Design Method for Constraint Propagation* – We prove that Arc Consistency is a best-possible constraint propagation mechanism for arbitrary CSPs, and then showed how to calculate optimal code for propagation. From Arc Consistency formula schemes we calculate simple Definite Constraints that can be instantiated into the optimal Definite Constraint Solver scheme [1].
3. *Theory of Conflict Analysis* – There are several mathematical formalisms for generalizing conflict analysis to arbitrary logics. We present a general pattern for calculating resolution rules in a given logic, and prove how resolution can be iterated to soundly infer a new constraint for backjumping and learning purposes.

References

1. Rehof, J., Mogenson, T.: Tractable constraints finite semilattices. Sci. Comput. Program. **35**, 191–221 (1999)
2. Smith, D.R., Westfold, S.: Toward Synthesis Constraint Solvers. Tech. rep., Kestrel Institute (2013). http://www.kestrel.edu/home/people/smith/pub/CW- report.pdf

Contents

Application Track

Abstracts of Papers Fast Tracked to *Constraints* Journal

Abstracts of Published Journal Track Papers

Technical Track

Encoding Linear Constraints with Implication Chains to CNF

Ignasi Abío[1][✉], Valentin Mayer-Eichberger[1,2], and Peter J. Stuckey[1,3]

[1] NICTA, Canberra, Australia
{ignasi.abio,valentin.mayer-eichberger,peterj.stuckey}@nicta.com.au
[2] University of New South Wales, Sydney, Australia
[3] University of Melbourne, Melbourne, Australia

Abstract. Linear constraints are the most common constraints occurring in combinatorial problems. For some problems which combine linear constraints with highly combinatorial constraints, the best solving method is translation to SAT. Translation of a single linear constraint to SAT is a well studied problem, particularly for cardinality and pseudo-Boolean constraints. In this paper we describe how we can improve encodings of linear constraints by taking into account implication chains in the problem. The resulting encodings are smaller and can propagate more strongly than separate encodings. We illustrate benchmarks where the encoding improves performance.

1   Introduction

In this paper we study linear integer constraints (LI constraints), that is, constraints of the form $a_1 x_1 + \cdots + a_n x_n \ \# \ a_0$, where the a_i are integer given values, the x_i are finite-domain integer variables, and the relation operator $\#$ belongs to $\{<, >, \leq, \geq, =, \neq\}$. We will assume w.l.o.g that $\#$ is $\leq$, the a_i are positive and all the domains of the variables are $\{0, 1..d_i\}$, since other cases can be reduced to this one.[1] Special case of linear constraints are: pseudo-Boolean (PB) constraints where the domain of each variable is $\{0..1\}$, cardinality (CARD) constraints where additionally $a_i = 1, 1 \leq i \leq n$, and at-most-one (AMO) constraints where additionally $a_0 = 1$.

Linear integer constraints appear in many combinatorial problems such as scheduling, planning or software verification, and, therefore, many different SMT solvers [9,14] and encodings [4,6,11] have been suggested for handling them. There are two main approaches to encoding linear constraints: cardinality constraints are encoded as some variation of a sorting network [3]; multi-decision diagrams (MDDs) are used to encode more general linear constraints [5], which in the special case of pseudo-Boolean constraints collapse to binary decision diagrams (BDDs).

---

[1] See [5] for details. Note that propagation is severely hampered by replacing equalities by a pair of inequalities.

© Springer International Publishing Switzerland 2015
G. Pesant (Ed.): CP 2015, LNCS 9255, pp. 3–11, 2015.
DOI: 10.1007/978-3-319-23219-5_1

Any form of encoding linear constraints to SAT introduces many intermediate Boolean variables, and breaks the constraint up into many parts. This gives us the opportunity to improve the encoding if we can recognize other constraints in the problem that help tighten the encoding of some part of the linear constraint.

Example 1. Consider a pseudo-Boolean constraint $x_1 + 2x_2 + 2x_3 + 4x_4 + 5x_5 + 6x_6 + 8x_7 \leq 14$ where we also have that $x_2 + x_3 + x_5 \leq 1$ we can rewrite the constraints as $x_1 + 2x_{235} + 4x_4 + 3x_5 + 6x_6 + 8x_7 \leq 14$ and $x_{235} \equiv (x_2 + x_3 + x_5 = 1)$, where x_{235} is a new Boolean variable. Notice, that x_{235} can be used to encode the at-most-one constraint.   □

Example 2. Consider a pseudo-Boolean constraint $4x_1 + 2x_2 + 5x_3 + 4x_4 \leq 9$ and the implications $x_1 \leftarrow x_2$ (i.e. $x_1 \lor \neg x_2$) and $x_2 \leftarrow x_3$. Separately they do not propagate, but considered together we can immediately propagate $\neg x_3$.   □

In this paper we show how to encode pseudo-Boolean constraints taking into account implication chains, as seen in Example 2. The resulting encodings are no larger than the separate encoding, but result in strictly stronger propagation. The approach also allows us to encode general linear integer constraints, and is a strict generalization of the MDD encoding of linear integer constraints [5]. We show how these new combined encodings are effective in practice on a set of hard real-life sports scheduling examples, and that the combination of pseudo-Booleans with implication chains arises in a wide variety of models.

2   Preliminaries

2.1   SAT Solving

Let $\mathcal{X} = \{x_1, x_2, \ldots\}$ be a fixed set of propositional *variables*. If $x \in \mathcal{X}$ then x and $\neg x$ are *positive* and *negative literals*, respectively. The *negation* of a literal l, written $\neg l$, denotes $\neg x$ if l is x, and x if l is $\neg x$. A *clause* is a disjunction of literals $l_1 \lor \cdots \lor l_n$. An *implication* $x_1 \rightarrow x_2$ is notation for the clause $\neg x_1 \lor x_2$, similarly $x_1 \leftarrow x_2$ denotes $x_1 \lor \neg x_2$. A *CNF formula* is a conjunction of clauses.

A (partial) *assignment* A is a set of literals such that $\{x, \neg x\} \not\subseteq A$ for any $x \in \mathcal{X}$, i.e., no contradictory literals appear. A literal l is *true* ($\top$) in A if $l \in A$, is *false* ($\bot$) in A if $\neg l \in A$, and is *undefined* in A otherwise. A clause C is true in A if at least one of its literals is true in A. A formula F is true in A if all its clauses are true in A. In that case, A is a *model* of F. Systems that decide whether a formula F has any model are called SAT-solvers, and the main inference rule they implement is *unit propagation*: given a CNF F and an assignment A, find a clause in F such that all its literals are false in A except one, say l, which is undefined, add l to A and repeat the process until reaching a fix-point. A detailed explanation can be found in [7].

Let $[l..u]$ where l and u are integers represent the set $\{l, \ldots, u\}$. Let y be an integer variable with domain $[0..d]$. The *order encoding* introduces Boolean variables y_i for $1 \leq i \leq d$. A variable y_i is true iff $y < i$. The encoding also introduces the clauses $y_i \rightarrow y_{i+1}$ for $1 \leq i < d$.

2.2 Multi Decision Diagrams

A directed acyclic graph is called an *ordered Multi Decision Diagram* (MDD) if it satisfies the following properties:

- It has two terminal nodes, namely $\mathcal{T}$ (true) and $\mathcal{F}$ (false).
- Each non-terminal node is labeled by an array of Booleans $[x_{i1}, \ldots, x_{id_i}]$ representing the order encoding of integer variable y_i where y_i ranges from $[0..d_i]$. The variable y_i is called the *selector variable*.
- Every node labeled by y_i has $d_i + 1$ outgoing edges, labelled $x_{i1}, \neg x_{i1}, \ldots, \neg x_{id_i}$.
- Each edge goes from a node with selector y_i to a node with selector variable y_j has $i < j$.

The MDD is *quasi-reduced* if no isomorphic subgraphs exist. It is *reduced* if, moreover, no nodes with only one child exist. A *long edge* is an edge connecting two nodes with selector variables y_i and y_j such that $j > i + 1$. In the following we only consider quasi-reduced ordered MDDs without long edges, and we just refer to them as MDDs for simplicity. A *Binary Decision Diagram* (BDD) is an MDD where $\forall i, d_i = 1$.

An MDD represents a function $f : \{0, 1, \ldots, d_1\} \times \{0, 1, \ldots, d_2\} \times \cdots \times \{0, 1, \ldots, d_n\} \to \{\bot, \top\}$ in the obvious way. Moreover, given a fixed variable ordering, there is only one MDD representing that function. We refer to [17] for further details about MDDs.

A function f is *anti-monotonic* in argument y_i if $v_i \geq v_i'$ implies that $f(v_1, \ldots, v_{i-1}, v_i, v_{i+1}, \ldots, v_n) = \top \Rightarrow f(v_1, \ldots, v_{i-1}, v_i', v_{i+1}, \ldots, v_n) = \top$ for all values $v_1, \ldots, v_n, v_i'$. An MDD is *anti-monotonic* if it encodes a function f that is anti-monotonic in all arguments. We shall only consider anti-monotonic MDDs in this paper.

Given an anti-monotonic MDD $\mathcal{M}$, we can encode it into CNF by introducing a new Boolean variable b_o to represent each node o in the MDD $\mathcal{M}$; unary clauses $\{b_{\mathcal{T}}, \neg b_{\mathcal{F}}, b_r\}$ where r is the root node of the MDD; and clauses $\{\neg b_o \vee b_{o_0}\} \cup \{\neg b_o \vee x_{ij} \vee b_{o_j} \mid j \in [1..d_i]\}$ for each node o of the form $\mathsf{mdd}([x_{i1}, \ldots, x_{id_i}], [o_0, o_1, \ldots, o_{d_i}])$. See [5] for more details.

We can encode arbitrary MDDs to SAT using Tseitin transformation but the encoding is substantially more complicated.

3 Pseudo Boolean Constraints and Chains

A *chain* $x_1 \Leftarrow x_2 \Leftarrow \cdots \Leftarrow x_n$ is a constraint requiring $x_1 \leftarrow x_2$, $x_2 \leftarrow x_3$, $\ldots$, $x_{n-1} \leftarrow x_n$. A *unary chain* x_1 is the trivial case that imposes no constraint. A chain is *compatible* with an ordered list of Boolean variables $x_1, \ldots, x_n$ if the chain is of the form $x_l \Leftarrow x_{l+1} \Leftarrow \cdots \Leftarrow x_k, l \leq k$. Given an ordered list L of Boolean variables $x_1, \ldots, x_n$ a *chain coverage* S is a set of variable-disjoint compatible chains such that each variable appears in exactly one chain. We will sometimes treat a chain coverage S as a Boolean formula equivalent to the constraints of the chains appearing in S. Given a variable ordering and a set of disjoint compatible chains we can always construct a chain coverage by adding in unary chains.

Example 3. Given list L of variables $x_1, \ldots, x_9$ and chains $x_1 \Leftarrow x_2 \Leftarrow x_3$, $x_5 \Leftarrow x_6$, $x_7 \Leftarrow x_8$, then a chain coverage S of L is $\{x_1 \Leftarrow x_2 \Leftarrow x_3, x_4, x_5 \Leftarrow x_6, x_7 \Leftarrow x_8, x_9\}$. S represents the constraint $x_1 \leftarrow x_2 \wedge x_2 \leftarrow x_3 \wedge x_5 \leftarrow x_6 \wedge x_7 \leftarrow x_8$. $\square$

Given a Boolean variable ordering L, a PB constraint C and chain coverage S of L we will demonstrate how to build an MDD to encode the constraint C taking into account the chain constraints in S. First lets examine how chains arise in models.

At-most-one constraints. Given PB $C \equiv a_1x_1 + \ldots + a_nx_n \le a_0$ and AMO $A \equiv x_1 + x_2 + \ldots + x_k \le 1$. We can reformulate A using new variables x'_j where $x_1 + \ldots + x_k = x'_1 + \ldots + x'_k$ using the ladder encoding [12] which gives $x_i \rightarrow x'_i$ for $i = 1 \ldots k$, $x_i \rightarrow \neg x'_{i+1}$ and $x'_{i+1} \rightarrow x'_i$ for $i = 1 \ldots k - 1$. If $[a'_1, \ldots, a'_k]$ is the sorted array of coefficients $[a_1, \ldots, a_k]$ then C can be written as $a'_1x'_1 + (a'_2 - a'_1)x'_2 + \cdots + (a'_k - a'_{k-1})x'_k + a_{k+1}x_{k+1} + \cdots a_nx_n \le a_0$. The chain comes from the auxiliary variables: $x'_1 \Leftarrow x'_2 \Leftarrow \cdots \Leftarrow x'_k$.

General linear integer constraints. A general LI $C \equiv a_1y_1 + \cdots a_my_m \le a_0$ can be expressed as a PB with chains. We encode each integer y_i with domain $[0..d_i]$ by the order encoding $[x_{i1}, \ldots, x_{id_i}]$ and then C can be rewritten as $a_1(\neg x_{11}) + \cdots + a_1(\neg x_{1d_1}) + \cdots + a_m(\neg x_{m1}) + \cdots + a_m(\neg x_{md_m}) \le a_0$ with chains $\neg x_{i1} \Leftarrow \neg x_{i2} \Leftarrow \cdots \Leftarrow \neg x_{id_i}$.

Shared coefficients. Frequently, PB constraints contain a large number of coefficients that are the same. This structure can be exploited. A similar technique of grouping shared coefficients is described in [5] which in our context is restated using chains. Given a PB $C \equiv ax_1 + \cdots + ax_k + a_{k+1}x_{k+1} + \cdots + a_nx_n \le a_0$ where the first k variables share the same coefficient. We introduce new variables $x'_1, \ldots, x'_k$ to encode the sum $x_1 + \cdots + x_k$ so $x_1 + \ldots + x_k = x'_1 + \ldots + x'_k$ and encode this constraint (usually using some form of sorting network [3]). This ensures that $x'_1 \Leftarrow x'_2 \Leftarrow \cdots \Leftarrow x'_k$. Then C can be rewritten as $ax'_1 + \cdots + ax'_k + a_{k+1}x_{k+1} + \cdots + a_nx_n \le a_0$. There are several advantages with this rewritten version. The sorting network can be represented more compactly than the same logic in an MDD ($O(k \cdot \log^2 k)$ vs $O(k^2)$). Secondly, the introduced variables x'_j are meaningful for the constraint and are likely to be useful for branching and in conflict clause learning during the search. Moreover, the sorted variables may be reusable for rewriting other constraints.

Binary implicants. Finally, a more general method is to automatically extract chains from the global problem description. There are a number of methods to detect binary implicants of CNF encodings [10,13]. Given a set of binary implicants B and a PB constraint C we can search for a chain coverage S implied by B, and an ordering L of the variables in C with which S is compatible, and then encode the reordered constraint C making use of the chain coverage S.

In the experimental section of this paper we have only considered the first three types of chains.

4 Translating Through MDDs with Chains

The main algorithm in this section generalizes the construction of an MDD in [5]. We first restate definitions of the original algorithm and then show how to take advantage of chains in the new construction. The CNF decomposition has desirable properties, i.e. we show that the encoding is more compact and propagates stronger.

4.1 Preliminaries for the Construction

Let $\mathcal{M}$ be the MDD of pseudo-Boolean C and let ν be a node of $\mathcal{M}$ with selector variable x_i. We define the *interval* of ν as the set of values α such that the MDD rooted at ν represents the pseudo-Boolean constraint $a_i x_i + \cdots + a_n x_n \leq \alpha$. It is easy to see that this definition corresponds in fact to an interval. The key point in constructing the MDD is to label each node of the MDD with its interval $[\beta, \gamma]$.

In the following, for every $i \in \{1, 2, \ldots, n+1\}$, we use a set L_i consisting of pairs $([\beta, \gamma], \mathcal{M})$, where $\mathcal{M}$ is the MDD of the constraint $a_i x_i + \cdots + a_n x_n \leq a_0'$ for every $a_0' \in [\beta, \gamma]$ (i.e., $[\beta, \gamma]$ is the interval of $\mathcal{M}$). All these sets are kept in a tuple $\mathcal{L} = (L_1, L_2, \ldots, L_{n+1})$.

Note that by definition of the MDD's intervals, if both $([\beta_1, \gamma_1], \mathcal{M}_1)$ and $([\beta_2, \gamma_2], \mathcal{M}_2)$ belong to L_i then either $[\beta_1, \gamma_1] = [\beta_2, \gamma_2]$ or $[\beta_1, \gamma_1] \cap [\beta_2, \gamma_2] = \emptyset$. Moreover, the first case holds if and only if $\mathcal{M}_1 = \mathcal{M}_2$. Therefore, L_i can be represented with a *binary search tree-like* data structure, where insertions and searches can be done in logarithmic time. The function **search**(K, L_i) searches whether there exists a pair $([\beta, \gamma], \mathcal{M}) \in L_i$ with $K \in [\beta, \gamma]$. Such a tuple is returned if it exists, otherwise an empty interval is returned in the first component of the pair. Similarly, we also use function **insert**$(([\beta, \gamma], \mathcal{M}), L_i)$ for insertions.

4.2 Algorithm and Properties of the Construction

In this section we show how to translate a PB $C \equiv a_1 x_1 + \ldots a_n x_n \leq a_0$ and a chain coverage S for variable order $x_1, \ldots, x_n$. Algorithm 1 describes the construction of the MDD. The initial call is **MDDChain**$(1, C, S)$. The call **MDDChain**(i, C', S) recursively builds an MDD for $C' \wedge S$ by building the ith level. If the chain including x_i is $x_i \Leftarrow \cdots \Leftarrow x_k$ it builds an MDD node that has child nodes with selector x_{k+1}. If the chain for x_i is unary this is the usual MDD (BDD) construction.

Algorithm 1. Procedure **MDDChain**

Require: $i \in \{1, 2, \ldots, n+1\}$ and pseudo-Boolean constraint $C' : a_i x_i + \ldots + a_n x_n \leq$
    a'_0 and chain coverage S on $[x_1, \ldots, x_n]$
Ensure: returns $[\beta, \gamma]$ interval of C' and $\mathcal{M}$ its MDD
1: $([\beta, \gamma], \mathcal{M}) \leftarrow \mathbf{search}(a'_0, L_i)$.
2: **if** $[\beta, \gamma] \neq \emptyset$ **then**
3:    **return** $([\beta, \gamma], \mathcal{M})$.
4: **else**
5:    $\delta_0 \leftarrow 0$
6:    **let** $\{x_i \Leftarrow x_{i+1} \Leftarrow \cdots \Leftarrow x_k\} \in S$ % including unary chain x_i
7:    $u \leftarrow k - i + 1$
8:    **for all** j such that $0 \leq j \leq u$ **do**
9:        $([\beta_j, \gamma_j], \mathcal{M}_j) \leftarrow \mathbf{MDDChain}(k+1, a_{k+1}x_{k+1} + \cdots + a_n x_n \leq a'_0 - \delta_j, S)$.
10:       $\delta_{j+1} \leftarrow \delta_j + a_{i+j}$
11:   **end for**
12:   $\mathcal{M} \leftarrow \mathsf{mdd}([x_i, \ldots, x_k], \mathcal{M}_0, \ldots, \mathcal{M}_u)$
13:   $[\beta, \gamma] \leftarrow [\beta_0, \gamma_0] \cap [\beta_1 + \delta_1, \gamma_1 + \delta_1] \cap \cdots \cap [\beta_u + \delta_u, \gamma_u + \delta_u]$.
14:   $\mathbf{insert}(([\beta, \gamma], \mathcal{M}), L_i)$.
15:   **return** $([\beta, \gamma], \mathcal{M})$.
16: **end if**

Example 4. The MDDs that result from **MDDChain**$(1, C, S)$ where $C \equiv 4x_1 + 2x_2 + 5x_3 + 4x_4 \leq 9$ of Example 2 encoded with chain coverage (a) $S = \{x_1, x_2, x_3, x_4\}$ (no chains) and (b) $S = \{x_1 \Leftarrow x_2 \Leftarrow x_3, x_4\}$ are shown in Figure 1. The diagrams show $[\beta, \gamma]$ for each node with the remainder of the constraint at the left. Unit propagation of the CNF of (b) sets $x_3 = \bot$ immediately since $4x_4 \leq -1$ is $\bot$.

We can prove that the algorithm returns a correct MDD, that is no larger than the MDD (BDD) encoding of C, and that the resulting CNF encoding is domain consistent on the original variables $x_1, \ldots, x_n$. Proofs are available at [1].

Theorem 1. *Given a pseudo-Boolean constraint $C \equiv a_1 x_1 + \cdots + a_n x_n \leq a_0$ and chain coverage S on $[x_1, \ldots, x_n]$ then* **MDDChain**$(1, C, S)$ *returns an MDD $\mathcal{M}$ representing function f such that constraint $C \wedge S \models f$. The running time of the algorithm is $O(n \cdot a_0 \cdot \log a_0)$.* □

Theorem 2. *Given a pseudo-Boolean constraint $C \equiv a_1 x_1 + \cdots + a_n x_n \leq a_0$ and chain coverage S on $[x_1, \ldots, x_n]$ then the MDD* **MDDChain**$(1, C, S)$ *has no more nodes than* **MDDChain**$(1, C, \{x_1, \ldots, x_n\})$, *the BDD for C.* □

Theorem 3. *Given a pseudo-Boolean constraint $C \equiv a_1 x_1 + \cdots + a_n x_n \leq a_0$ and chain coverage S on $[x_1, \ldots, x_n]$ then unit propagation on the CNF encoding of* **MDDChain**$(1, C, S) \wedge S$ *enforces domain consistency of $C \wedge S$ on variables $x_1, \ldots, x_n$.* □

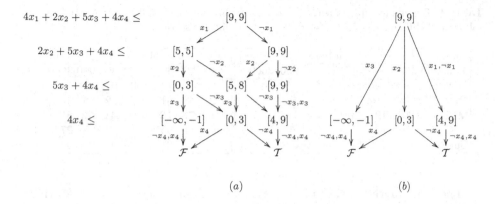

Fig. 1. The MDDs that result from $4x_1 + 2x_2 + 5x_3 + 4x_4 \leq 9$ encoded (a) without and (b) with the chain $x_1 \Leftarrow x_2 \Leftarrow x_3$.

5 Experiments

To illustrate the advantage of combined compilation we consider a challenging combinatorial optimization problem where both AMO and shared coefficients chains arise.

Sports league scheduling is a challenging combinatorial optimization problem. We consider scheduling a double round-robin sports league of N teams. All teams meet each other once in the first $N - 1$ weeks and again in the second $N - 1$ weeks, with exactly one match per team each week. A given pair of teams must play at the home of one team in one half, and at the home of the other in the other half, and such matches must be spaced at least a certain minimal number of weeks apart. Additional constraints include, e.g., that no team ever plays at home (or away) three times in a row, other (public order, sportive, TV revenues) constraints, blocking given matches on given days, etc.

Additionally, the different teams can propose a set of constraints with some importance (low, medium or high). We aim not only to maximize the number of these constraints satisfied, but also to assure that at least some of the constraints of every team are satisfied. More information can be found in [2].

Low-importance constraints are given a weight of 1; medium-importance, 5, and high-importance, 10. For every constraint proposed by a team i, a new Boolean variable $x_{i,j}$ is created. This variable is set to true if the constraint is violated. For every team, a pseudo-Boolean constraint $\sum_j w_{i,j}x_{i,j} \leq K_i$ is imposed. The objective function to minimize is $\sum_i \sum_j w_{i,j}x_{i,j}$. The data is based on real-life instances.

Desired constraints typically refer to critical weeks in the schedule, e.g. around Christmas, or other key dates, and preferences of different teams almost always clash. Double round-robin tournaments contain a lot of AMO and EO constraints (for instance, each week each team meets exactly one team). These AMO constraints can be used to simplify the desired constraints.

Table 1. Results for sports league scheduling, showing the number of runs that find a solution of different quality after different time limits (seconds).

Quality	Some solution			cost ≤ 30 + best			cost ≤ 20 + best			cost ≤ 10 + best		
Timelimit	300	900	3600	300	900	3600	300	900	3600	300	900	3600
MDD1	148	190	199	21	55	107	17	35	74	6	25	51
MDD2	151	**194**	199	27	59	115	19	38	81	12	25	43
MDD3	**160**	191	**200**	**56**	**107**	**162**	**45**	**72**	**121**	**41**	**52**	**87**
LCG	69	123	172	21	29	51	18	21	35	14	20	27
Gurobi	0	0	0	0	0	0	0	0	0	0	0	0

The benchmark consists of 10 instances and each method is run 20 times with different seeds, for 200 total runs. Compared methods are: MDD1, the usual MDD (in fact, BDD) method to encode PBs [4]; MDD2, the method of [5] using sorting networks for the identical coefficients and then using an MDD; MDD3, the method defined herein; LCG, using lazy clause generation [15]; and Gurobi, using the MIP solver Gurobi. *Barcelogic* [8] SAT Solver was used in methods MDD1, MDD2, MDD3 and LCG.

The results can be shown at Table 1. The number of times a solution has been found within the time limit can be found at columns 2-4. Columns 5-7 present the number of times (within the timelimit) a method finds a solution of cost at most *best* + 30, where *best* is the cost of the best solution found by any method. Similarly, columns 8-10 and 11-13 contain the number of times a solution of cost at most *best* + 20 and *best* + 10 has been found.

As we can see the new encoding substantially improves on previous encodings of the problem. For these sports leagues scheduling problems it is well known that other solving approaches do not compete with SAT encoding [2].

6   Conclusion and Future Work

We demonstrate a new method for encoding pseudo-Boolean constraints taking into account implications chains. The improved encoding is beneficial on hard benchmark problems. The approach is an extension on earlier work on encoding linear constraints to SAT [5]. The approach is related to the propagator for the `increasing_sum` constraint $y = a_1y_1 + \cdots + a_ny_n \wedge y_1 \leq y_2 \leq \cdots \leq y_n$ described in [16], which combines a linear constraint with a "chain" of integer inequalities. Interestingly, `increasing_sum` is not directly encodable as an MDD using the method herein, but it does suggest that the methods can be extended to arbitrary sets of chains all compatible with a global variable order. Another interesting direction for future work is to consider combining chains with the sorting networks encodings of linear constraints (e.g. [11]).

Acknowledgments. NICTA is funded by the Australian Government as represented by the Department of Broadband, Communications and the Digital Economy and the Australian Research Council through the ICT Centre of Excellence program.

References

1. Abio, I., Mayer-Eichberge, V., Stuckey, P.: Encoding linear constraints with implication chains to CNF. Tech. rep., University of Melbourne (2015). http://www.
   people.eng.unimelb.edu.au/pstuckey/papers/cp2015a.pdf
2. Abío, I.: Solving hard industrial combinatorial problems with SAT. Ph.D. thesis,
   Technical University of Catalonia (UPC) (2013)
3. Abío, I., Nieuwenhuis, R., Oliveras, A., Rodríguez-Carbonell, E.: A parametric
   approach for smaller and better encodings of cardinality constraints. In: Schulte,
   C. (ed.) CP 2013. LNCS, vol. 8124, pp. 80–96. Springer, Heidelberg (2013)
4. Abío, I., Nieuwenhuis, R., Oliveras, A., Rodríguez-Carbonell, E., Mayer-
   Eichberger, V.: A New Look at BDDs for Pseudo-Boolean Constraints. J. Artif.
   Intell. Res. (JAIR) **45**, 443–480 (2012)
5. Abío, I., Stuckey, P.J.: Encoding linear constraints into SAT. In: O'Sullivan, B.
   (ed.) CP 2014. LNCS, vol. 8656, pp. 75–91. Springer, Heidelberg (2014)
6. Bailleux, O., Boufkhad, Y., Roussel, O.: New encodings of pseudo-boolean constraints into CNF. In: Kullmann, O. (ed.) SAT 2009. LNCS, vol. 5584, pp. 181–194.
   Springer, Heidelberg (2009)
7. Biere, A., Heule, M., van Maaren, H., Walsh, T. (eds.): Handbook of Satisfiability,
   Frontiers in Artificial Intelligence and Applications, vol. 185. IOS Press (2009)
8. Bofill, M., Nieuwenhuis, R., Oliveras, A., Rodríguez-Carbonell, E., Rubio, A.: The
   barcelogic SMT solver. In: Gupta, A., Malik, S. (eds.) CAV 2008. LNCS, vol. 5123,
   pp. 294–298. Springer, Heidelberg (2008)
9. Dutertre, B., de Moura, L.: The YICES SMT Solver. Tech. rep., Computer Science
   Laboratory, SRI International (2006). http://yices.csl.sri.com
10. Eén, N., Biere, A.: Effective preprocessing in SAT through variable and clause elim-
    ination. In: Bacchus, F., Walsh, T. (eds.) SAT 2005. LNCS, vol. 3569, pp. 61–75.
    Springer, Heidelberg (2005)
11. Eén, N., Sörensson, N.: Translating Pseudo-Boolean Constraints into SAT. Journal
    on Satisfiability, Boolean Modeling and Computation **2**(1–4), 1–26 (2006)
12. Gent, I.P., Prosser, P., Smith, B.M.: A 0/1 encoding of the GACLex constraint for
    pairs of vectors. In: ECAI 2002 workshop W9: Modelling and Solving Problems
    with Constraints. University of Glasgow (2002)
13. Heule, M.J.H., Järvisalo, M., Biere, A.: Efficient CNF simplification based on
    binary implication graphs. In: Sakallah, K.A., Simon, L. (eds.) SAT 2011. LNCS,
    vol. 6695, pp. 201–215. Springer, Heidelberg (2011)
14. de Moura, L., Bjorner, N.: Z3: An Efficient SMT Solver. Tech. rep., Microsoft
    Research, Redmond (2007). http://research.microsoft.com/projects/z3
15. Ohrimenko, O., Stuckey, P., Codish, M.: Propagation via lazy clause generation.
    Constraints **14**(3), 357–391 (2009)
16. Petit, T., Régin, J.-C., Beldiceanu, N.: A $\Theta(n)$ bound-consistency algorithm for
    the increasing sum constraint. In: Lee, Jimmy (ed.) CP 2011. LNCS, vol. 6876, pp.
    721–728. Springer, Heidelberg (2011)
17. Srinivasan, A., Ham, T., Malik, S., Brayton, R.: Algorithms for discrete function
    manipulation. In: 1990 IEEE International Conference on Computer-Aided Design,
    ICCAD 1990, pp. 92–95. Digest of Technical Papers (1990)

Anytime Hybrid Best-First Search with Tree Decomposition for Weighted CSP

David Allouche, Simon de Givry$^{(\boxtimes)}$, George Katsirelos,
Thomas Schiex, and Matthias Zytnicki

MIAT, UR-875, INRA, F-31320 Castanet Tolosan, France
{david.allouche,simon.degivry,george.katsirelos,
thomas.schiex,matthias.zytnicki}@toulouse.inra.fr

Abstract. We propose Hybrid Best-First Search (HBFS), a search strategy for optimization problems that combines Best-First Search (BFS) and Depth-First Search (DFS). Like BFS, HBFS provides an anytime global lower bound on the optimum, while also providing anytime upper bounds, like DFS. Hence, it provides feedback on the progress of search and solution quality in the form of an optimality gap. In addition, it exhibits highly dynamic behavior that allows it to perform on par with methods like limited discrepancy search and frequent restarting in terms of quickly finding good solutions.

We also use the lower bounds reported by HBFS in problems with small treewidth, by integrating it into Backtracking with Tree Decomposition (BTD). BTD-HBFS exploits the lower bounds reported by HBFS in individual clusters to improve the anytime behavior and global pruning lower bound of BTD.

In an extensive empirical evaluation on optimization problems from a variety of application domains, we show that both HBFS and BTD-HBFS improve both anytime and overall performance compared to their counterparts.

Keywords: Combinatorial optimization · Anytime algorithm · Weighted constraint satisfaction problem · Cost function networks · Best-first search · Tree decomposition

1 Introduction

Branch and Bound search is a fundamental tool in exact combinatorial optimization. For minimization, in order to prune the search tree, all variants of Branch and Bound rely on a local lower bound on the cost of the best solution below a given node.

Depth-First Search (DFS) always develops a deepest unexplored node. When the gap between the local lower bound and a global upper bound on the cost of an optimal solution – usually provided by the best known solution – becomes empty, backtrack occurs. DFS is often used in Constraint Programming because it offers

© Springer International Publishing Switzerland 2015
G. Pesant (Ed.): CP 2015, LNCS 9255, pp. 12–29, 2015.
DOI: 10.1007/978-3-319-23219-5_2

polyspace complexity, it takes advantage of the incrementality of local consistencies and it has a reasonably good anytime behavior that can be further enhanced by branching heuristics. This anytime behavior is however largely destroyed in DFS variants targeted at solving problems with a reasonable treewidth such as BTD [7] or AND/OR search [6].

Best-First Search (BFS) instead always develops the node with the lowest lower bound first. It offers a running global lower bound and has been proved to never develop more nodes than DFS for the same lower bound [22]. But it has a worst-case exponential space complexity and the optimal solution is always the only solution produced.

An ideal Branch and Bound algorithm would combine the best of all approaches. It would have a bearable space complexity, benefit from the incrementality of local consistencies and offer both updated global upper and lower bounds as the problem is solved. It would also not loose all its anytime qualities when used in the context of treewidth sensitive algorithms such as BTD.

With updated global lower and upper bounds, it becomes possible to compute a current global optimality gap. This gap can serve as a meaningful indicator of search progress, providing a direct feedback in terms of the criteria being optimized. This gap also becomes of prime importance in the context of tree-decomposition based Branch and Bound algorithms such as BTD [7] as global bounds for each cluster can typically be used to enhance pruning in other clusters.

In this paper, we introduce HBFS, an hybrid, easy to implement, anyspace Branch and Bound algorithm combining the qualities of DFS and BFS. The only limitation of HBFS is that it may require to compromise the anytime updating of the global lower bound for space. This can be achieved dynamically during search. HBFS can also be combined with a tree-decomposition to define the more complex BTD-HBFS, a BTD variant offering anytime solutions and updated global optimality gap.

On a set of more than 3,000 benchmark problems from various sources (MaxCSP, WCSP, Markov Random Fields, Partial Weighted MaxSAT) including resource allocation, bioinformatics, image processing and uncertain reasoning problems, we observe that HBFS improves DFS in term of efficiency, while being able to quickly provide good solutions – on par with LDS and Luby restarts – and a global running optimality gap. Similarly, HBFS is able to improve the efficiency and anytime capacities of BTD.

2   Background

Our presentation is restricted to binary problems for simplicity. Our implementation does not have such restriction. A binary Cost Function Network (CFN) is a triplet (X, D, W). $X = \{1, \ldots, n\}$ is a set of n variables. Each variable $i \in X$ has a finite domain $D_i \in D$ of values than can be assigned to it. The maximum domain size is d. W is a set of cost functions. A binary cost function $w_{ij} \in W$ is a function $w_{ij} : D_i \times D_j \mapsto [0, k]$ where k is a given maximum integer cost corresponding to a completely forbidden assignment (expressing hard constraints).

If they do not exist, we add to W one unary cost function for every variable such that $w_i : D_i \mapsto [0, k]$ and a zero arity constraint $w_\varnothing$ (a constant cost payed by any assignment, defining a lower bound on the optimum). All these additional cost functions will have initial value 0, leaving the semantics of the problem unchanged.

The Weighted Constraint Satisfaction Problem (WCSP) is to find a minimum cost complete assignment: $\min_{(a_1,\ldots,a_n) \in \prod_i D_i} \{w_\varnothing + \sum_{i=1}^{n} w_i(a_i) + \sum_{w_{ij} \in W} w_{ij}(a_i, a_j)\}$, an optimization problem with an associated NP-complete decision problem.

The WCSP can be solved exactly using Branch and Bound maintaining some lower bound: at each node ν of a tree, we use the local non naive lower bound $\nu.lb = w_\varnothing$ provided by a given soft arc consistency [5]. Each node corresponds to a sequence of decisions $\nu.\delta$. The root node has an empty decision sequence. When a node is explored, an unassigned variable is chosen and a branching decision to either assign the variable to a chosen value (left branch, positive decision) or remove the value from the domain (right branch, negative decision) is taken. The number of decisions taken to reach a given node ν is the depth of the node, $\nu.depth$. A node of the search tree that corresponds to a complete assignment is called a leaf. At this point, $\nu.lb$ is assumed to be equal to the node cost (which is guaranteed by all soft arc consistencies).

The graph $G = (X, E)$ of a CFN has one vertex for each variable and one edge (i, j) for every binary cost function $w_{ij} \in W$. A tree decomposition of this graph is defined by a tree (C, T). The set of nodes of the tree is $C = \{C_1, \ldots, C_m\}$ where C_e is a set of variables ($C_e \subseteq X$) called a cluster. T is a set of edges connecting clusters and forming a tree (a connected acyclic graph). The set of clusters C must cover all the variables ($\bigcup_{C_e \in C} C_e = X$) and all the cost functions ($\forall \{i, j\} \in E, \exists C_e \in C$ s.t. $i, j \in C_e$). Furthermore, if a variable i appears in two clusters C_e and C_g, i must also appear in all the clusters C_f on the unique path from C_e to C_g in (C, T). If the cardinality of the largest cluster in a tree decomposition is $\omega + 1$ then the *width* of the decomposition is ω. The *treewidth* of a graph is the minimum width among all its decompositions [24].

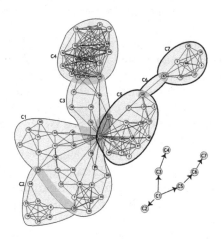

Fig. 1. A tree-decomposition of the CELAR06 radio frequency assignment problem, rooted in C_1 with subproblem P_5 highlighted.

3   Hybrid Best-First Search

Classical BFS explores the search tree by keeping a list *open* of open nodes representing unexplored subproblems. Initially, this list is reduced to the root node at depth 0. Iteratively, a best node is explored: the node is removed and replaced by its two left and right children with updated decisions, lower bound and depth. In this paper we always choose as best node a node with the smallest $\nu.lb$, breaking ties by selecting a node with maximum $\nu.depth$. The first leaf of the tree explored is then guaranteed to be an optimal solution [14,22]. The list *open* may reach a size in $O(d^n)$ and, if incrementality in the lower bound computation is sought, each node should hold the minimum data-structures required for soft arc consistency enforcing (in $O(ed)$ per node).

The pseudocode for Hybrid BFS is described as Algorithm 1. HBFS starts with the empty root node in the list of *open* nodes. It then iteratively picks a best node ν from the open list as above, replays all the decisions in $\nu.\delta$ leading to an assignment A_ν, while maintaining consistency. It then performs a depth-first search probe starting from that node for a limited number Z of backtracks. The DFS algorithm is a standard DFS algorithm except for the fact that, when the bound on the number of backtracks is reached, it places all the nodes corresponding to open right branches of its current search state in the *open* list (see Figure 2).

At the price of increased memory usage, this hybrid maintains the advantages of depth-first search. Since it spends a significant amount of its time in a DFS subroutine, it can exploit the incrementality of arc consistency filtering during DFS search without any extra space cost: nodes in the *open* list will just contain decisions δ and lower bound lb, avoiding the extra $O(ed)$ space required for incrementality during BFS. However, each time a node is picked up in *open*, the set of $\nu.depth$ decisions must be "replayed" and local consistency reinforced from the root node state, leading to redundant propagation. This cost can be mitigated to some degree by merging all decisions into one. Hence, a single fixpoint has to be computed rather than $\nu.depth$. Additionally, the cost can be further reduced using other techniques employed by copying solvers [26]. Regardless of these mitigation techniques, some redundancy is unavoidable, hence the number of backtracks performed at each DFS probe should be large enough to avoid excessive redundancy.

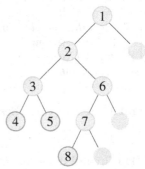

Fig. 2. A tree that is partially explored by DFS with backtrack limit = 3. Nodes with a bold border are leaves, nodes with no border are placed in the open list after the backtrack bound is exceeded. Nodes are numbered in the order they are visited.

Second, as it is allowed to perform Z backtracks in a depth-first manner before it picks a new node, it may find new and better incumbent solutions, thus it is anytime. The number of backtracks of each DFS probe should be sufficiently small to offer quick diversification: by exploring a new best node, we are offered the opportunity to reconsider early choices, similarly to what LDS [8] and Luby randomized restarts [18] may offer. Additionally, with early upper bounds, we can also prune the open node list and remove all nodes such that $\nu.lb \geq ub$.

To balance the conflicting objectives of reducing repeated propagation and diversification, we dynamically adjust the amount of backtracks Z that can be performed during one DFS probe by trying to keep the observed rate of redundantly propagated decisions between reasonable bounds (α and β). In all the algorithms here, we assume that the number of nodes (*Nodes*) and backtracks (*Backtracks*) are implicitly maintained during search.

Function HBFS(*clb,cub*) : pair(integer,integer)
  | $open := \nu(\delta = \varnothing, lb = clb)$;
  | **while** ($open \neq \varnothing$ **and** $clb < cub$) **do**
  | | $\nu :=$pop($open$) /* *Choose a node with minimum lower bound and maximum depth* */;
  | | Restore state $\nu.\delta$, leading to assignment A_ν, maintaining local consistency ;
  | | $NodesRecompute := NodesRecompute + \nu.depth$;
  | | $cub :=$DFS(A_ν,cub,Z)/* *puts all right open branches in open* */ ;
  | | $clb := \max(clb, lb(open))$;
  | | **if** ($NodesRecompute > 0$) **then**
  | | | **if** ($NodesRecompute/Nodes > \beta$ **and** $Z \leq N$) **then** $Z := 2 \times Z$;
  | | | **else if** ($NodesRecompute/Nodes < \alpha$ **and** $Z \geq 2$) **then** $Z := Z/2$;
  | **return** (clb, cub);

Algorithm 1. Hybrid Best-First Search. Initial call: HBFS($w_\varnothing$,k) with $Z = 1$.

This hybrid does not preserve the polyspace complexity of DFS. However, it can easily be made *anyspace*. If memory is exhausted (or a memory upper bound is reached, with the same effect), the algorithm can switch from bounded DFS to complete DFS. This means that for every node it picks from the open list, it explores the entire subtree under that node. Hence, it will not generate any new open nodes. It can continue in this mode of operation until memory pressure is relieved.

Finally, this method computes stronger global lower bounds than DFS, as the cost of a best node in the open list defines a global lower bound, as in BFS. DFS instead cannot improve on the global lower bound computed at the root until it finally visits the first right branch. In the context of a single instance this is only important in the sense that it provides a better estimation of the optimality gap. However, we will see that this can improve performance in decomposition-based methods.

3.1 Related Work

Alternate search space exploration schemes have been proposed in the field of heuristic search, as variations of A* search. These schemes can be applied to discrete optimization, yielding other variants of best-first search. However, depth-first search is not effective or even feasible in domains where A* search is used: for example, it is possible in planning to have exponentially long sequences of actions when short plans exist. Hence, methods like BRFSL(k) [27] can only do *bounded-depth* DFS probes. Also, in contrast to HBFS, they do not insert the open nodes of the DFS probes into the open list of BFS. Other methods like Weighted best-first search [23], ARA* [16] and ANA* [2] weigh future assignments more heavily in order to bias the search towards solutions. We do not need to modify the branching heuristic in any way in HBFS.

Stratification [1], which solves a weighted MaxSAT instance by iteratively considering larger subsets of its clauses, starting with those that have the highest weight, provides similar benefits to HBFS, as provides solutions quickly and produces lower bounds. This techniques, however, can be viewed as a wrapper over an optimization method and is therefore orthogonal to HBFS.

Alternate heuristics for choosing the next node to explore may yield different algorithms. When we can identify a preferred value to assign at each choice point, the *discrepancy* of a node ν is the number of right branches in the path from the root to ν. If we always open the node with the smallest discrepancy, set $Z = 1$ and disable the adaptive heuristic, HBFS is identical to Limited Discrepancy Search (LDS)[1] [8].

In ILP, a closely related approach is so-called BFS with *diving* heuristics [3]. Such heuristics perform a single depth-first probe trying to find a feasible solution. Although the idea is quite close to that of HBFS, it is typically restricted to a single branch, the open nodes it leaves are not added to the open node file and is treated separately from the rest of the search process. This is in part motivated by the fact that DFS is considered impractical in ILP [17] and by the fact that the lower bounding method (LP) used is not as lightweight as those used in WCSP.

4 Hybrid Best-First Search and Tree Decompositions

When the graph of a CFN has bounded treewidth, the $O(d^n)$ worst-case complexity of DFS can be improved using a tree decomposition of the CFN graph. We can trivially observe that the tree decomposition can be rooted by selecting a root cluster denoted C_1. The separator of a non root cluster C_e is $C_e \cap pa(C_e)$, where $pa(C_e)$ is the parent of C_e in T. Local consistency can be enforced on the problem and provide a cluster-localized lower-bound $w_\varnothing^e$ for each cluster C_e. The sum of these cluster-wise lower bounds is a lower bound for the complete problem. Beyond this trivial observation, Terrioux and Jégou [28] and de Givry et al. [7] have extended BTD [9] (which we call BTD-DFS here for clarity) from pure satisfaction problems to the case of optimization (WCSP), in a way similar to AND/OR search [19]. Next, we briefly describe BTD-DFS, as given by de Givry et al, as we base our own algorithm on this.

---

[1] In WCSP optimization, we always have a non naive value heuristic that selects a value (i, a) with minimum unary marginal cost $w_i(a)$ or better, the EAC support [13].

In BTD-DFS, by always assigning the variables of a cluster before the variables of its descendant clusters, it is possible to exploit the fact that assigning a cluster C_e separates all its child clusters $children(C_e)$. Each child cluster C_f is the root of a subproblem P_f defined by the subtree rooted in C_f which becomes independent of others. So, each subproblem P_f conditioned by the current assignment A_f of its separator, can be independently and recursively solved to optimality. If we memoize the optimum cost of every solved conditioned subproblem $P_e|A_e$ in a cache, then $P_e|A_e$ will never be solved again and an overall $O(nd^{\omega+1})$ time complexity can be guaranteed.

Although this simple strategy offers an attractive worst case theoretical bound, it may behave poorly in practice. Indeed, each conditioned subproblem $P_e|A_e$ is always solved from scratch to optimality. This ignores additional information that can be extracted from already solved clusters. Imagine C_e has been assigned and that we have an upper bound ub (a solution) for the problem $P_e|A_e$. Assume that C_e has two children C_f and $C_{f'}$ and that we have solved the first subproblem $P_f|A_f$ to optimality. By subtracting the lower bound $w_\varnothing^e$ and the optimum of $P_f|A_f$ from ub, we obtain the maximum cost that a solution of $P_{f'}|A_{f'}$ may have in order to be able to improve over ub. Instead of solving it from scratch, we can solve $P_{f'}|A_{f'}$ with this initial upper bound and either find an optimal solution – which can be cached – or fail. If we fail, we have proved a global lower bound on the cost of an optimal solution of $P_{f'}|A_{f'}$. This lower bound can be cached and prevent repeated search if $P_{f'}|A_{f'}$ is revisited with the same or a lower initial upper bound. Otherwise, the problem will be solved again and again either solved to optimality or fail and provide an improved global lower bound. This has been shown to improve search in practice while offering a theoretical bound on time complexity in $O(kn.d^{\omega+1})$ (each time a subproblem $P_f|A_f$ is solved again, the global lower bound increases at least by 1).

In practice, we therefore cache two values, $LB_{P_e|A_e}$ and $UB_{P_e|A_e}$, for every visited assignment A_e of the separator of every cluster C_e. We always assume caching is done implicitly: $LB_{P_e|A_e}$ is updated every time a stronger lower bound is proved for $P_e|A_e$ and $UB_{P_e|A_e}$ when an updated upper bound is found. When an optimal solution is found and proved to be optimal, we will therefore have $LB_{P_e|A_e} = UB_{P_e|A_e}$. Thanks to these cached bounds and to the cluster-wise local lower bounds $w_\varnothing^e$, an improved local lower bound $lb(P_e|A_e)$ for the subproblem $P_e|A_e$ can be computed by recursively summing the maximum of the cached and local bound (see [7]).

We show pseudocode for the resulting algorithm combining BTD and DFS in Algorithm 2. Grayed lines in this code are not needed for the DFS variant and should be ignored. The algorithm is called on root cluster C_1, with an assignment $A_1 = \varnothing$, a set of unassigned variables $V = C_1$ and initial lower and upper bound clb and cub set respectively to $lb(P_1|\varnothing)$ and k (the maximum cost). The last argument, $RecCall$ is a functional argument that denotes which function will be used to recurse inside BTD-DFS. Here, $RecCall$ will be initially equal to BTD-DFS itself. The algorithm always returns two identical values equal to the current

Function BTD-DFS($A,C_e,V,clb,cub,RecCall$) : pair(integer,integer)

> **if** ($V \neq \varnothing$) **then**
>> $i :=$pop(V) /* *Choose an unassigned variable in C_e* */ ;
>> $a :=$pop(D_i) /* *Choose a value* */ ;
>> Assign a to i, maintaining local consistency on subproblem $lb(P_e|A \cup \{(i = a)\})$;
>> $clb' := \max(clb, lb(P_e|A \cup \{(i = a)\}))$;
>> **if** ($clb' < cub$) **then**
>>> $(cub, cub) :=$ BTD-DFS($A \cup \{(i = a)\}$, C_e, $V - \{i\}$, clb', cub,$RecCall$);
>>
>> $C_e.backtracks := C_e.backtracks + 1$;
>> **if** ($\max(clb, lb(P_e|A)) < cub$) **then**
>>> Remove a from i, maintaining local consistency on subproblem
>>> $lb(P_e|A \cup \{(i \neq a)\})$;
>>> $clb' := \max(clb, lb(P_e|A \cup \{(i \neq a)\}))$;
>>> **if** ($clb' < cub$) **then**
>>>> **if** ($C_e.backtracks < C_e.limit$ **and** $Backtracks < P_e.limit$) **then**
>>>>> $(cub, cub) :=$ BTD-DFS($A \cup \{(i \neq a)\}$, C_e, V, clb', cub,$RecCall$);
>>>>
>>>> **else** /* *Stop depth-first search* */
>>>>> Push current search node in open list of $P_e|A$ at position clb' ;
>
> **else**
>> $S := Children(C_e)$;
>> /* *Solve all clusters with non-zero optimality gap and unchanged lb or ub* */ ;
>> **while** ($S \neq \varnothing$ **and** $lb(P_e|A) < cub$) **do**
>>> $C_f :=$pop(S) /* *Choose a child cluster* */ ;
>>> **if** ($LB_{P_f|A} < UB_{P_f|A}$) **then**
>>>> $cub' := \min(UB_{P_f|A}, cub - [lb(P_e|A) - lb(P_f|A_f)])$;
>>>> $(clb'', cub'') := RecCall$ (A, C_f, C_f, $lb(P_f|A_f)$, cub',$RecCall$);
>>>> Update $LB_{P_f|A}$ and $UB_{P_f|A}$ using clb'' and cub'';
>>
>> $cub := \min(cub, w_{\varnothing}^e + \sum_{C_f \in Children(C_e)} UB_{P_f|A})$;
>> **if** $\max(clb, lb(P_e|A)) < cub$ **then**
>>> Push current search node in open list of $P_e|A$ at position $\max(clb, lb(P_e|A))$;
>>> $C_e.limit := C_e.backtracks$ /* *Stop depth-first search* */ ;
>
> **return** (cub, cub)

Algorithm 2. BTD using depth-first search

upper bound.[2] Caches are initially empty and return naive values $LB_{P_e/A} = 0$ and $UB_{P_e/A} = k$ for all clusters and separator assignments.

4.1 Using HBFS in BTD

BTD-DFS has two main disadvantages: first, it has very poor anytime behavior, as it cannot produce a solution in a decomposition with k leaves until $k - 1$ leaf clusters have been completely solved. This affects the strength of pruning, as

---

[2] This is clearly redundant for BTD-DFS, but allows a more uniform presentation with BTD-HBFS.

values are only pruned if the current lower bound added to the marginal cost of the value exceeds the upper bound. Second, because child clusters are examined in order, only the lower bounds of siblings earlier than C_f in that order can contribute to pruning in C_f. For example, consider a cluster C_e with 3 child clusters $C_{f_1}, C_{f_2}, C_{f_3}$. Assume that $ub = 31$ and under an assignment A, $w_\varnothing^e$ has known cost 10 while $P_{f_1}|A_{f_1}$, $P_{f_2}|A_{f_2}$ and $P_{f_3}|A_{f_3}$ all have optimal cost 10, and $lb(P_{f_1}|A_{f_1}) = lb(P_{f_2}|A_{f_2}) = lb(P_{f_3}|A_{f_3}) = 0$. Clearly the subproblem under C_e cannot improve on the upper bound, but when we solve C_{f_1} and C_{f_2}, BTD-DFS does not reduce the effective upper bound at all. However, it may be relatively easy to prove a lower bound of 7 for each of the child clusters. If we had this information, we could backtrack.

HBFS has the ability to quickly provide good lower and upper bounds and interrupts itself as soon as the limit number of backtracks is reached. Using HBFS instead of DFS in BTD should allow to quickly probe each subproblem to obtain intermediate upper and lower bounds for each of them. The upper bounds can be used to quickly build a global solution, giving anytime behavior to BTD. The lower bounds of all subproblems can be used to improve pruning in all other clusters.

The pseudocode of BTD-HBFS is described as Algorithm 3. It takes the same arguments as BTD-DFS but ignores the last one (used only to pass information from BTD-HBFS to BTD-DFS). BTD-HBFS relies on BTD-DFS pseudocode, assuming that all grayed lines of BTD-DFS are active. These reactivated lines in Algorithm 2 impose per-cluster and per-subproblem backtrack limits. Every cluster C_e has a counter $C_e.backtracks$ for number of backtracks performed inside the cluster and an associated limit $C_e.limit$. Every subproblem P_e has a limit $P_e.limit$ on the number of backtracks N performed inside the subproblem. Initially, $C_e.limit = P_e.limit = \infty$ for all $C_e \in C$.

Every subproblem $P_e|A_e$ has its own list of open nodes $P_e|A_e.open$ for each upper bound which it is given. The value of the upper bound participates in the definition of the actual search space that needs to be explored. If the same subproblem is revisited later with a lower upper bound, then the search space shrinks and we can just copy the open list associated with the higher bound and prune all nodes ν such that $\nu.lb \geq ub$. But if the upper bound is more relaxed than any previous upper bound then we need to create a new open list starting with the root node.

Finally, the loop in line 1 is interrupted as soon as the optimality gap reduces or the number of backtracks reaches the subproblem limit, making the search more dynamic. If subproblems quickly update their bound, the remaining backtracks can be used in a higher cluster. However, the subproblem under each child cluster is guaranteed to get at least Z backtracks. The result is that we spend most of the time in leaf clusters. When one cluster exhausts its budget, the search quickly returns to the root cluster.

Example 1. Consider the example in figure 3. We have a CFN with the tree decomposition given in the box labeled (C, T) and the search in each cluster is shown in a box labeled by that cluster's name. Let $N = 2$ and $Z = 1$ in this example. The search visits nodes as they are numbered in the figure. When it reaches node 4, cluster C_1 is completely instantiated and hence it descends into C_2 and after node 7 it descends into C_4. After node 10, we have performed a backtrack in this cluster, and since $Z = 1$ we end this DFS probe and return control to BTD-HBFS. The limit on the number of backtracks in P_4 is still not exceeded, so we choose a new node from the open list, node 11, a conflict. Again we return control to BTD-HBFS and, having exceeded the backtrack limit on P_4, exit this cluster, but with an improved lower bound. Since C_4 exceeded its backtrack limit before improving its lower bound, no more search is allowed in parent clusters. The search is allowed, however, to visit sibling clusters, hence it explores C_5 (nodes 12–14), which it exits with an improved upper bound before exceeding its backtrack limit, and C_3 (nodes 15–18). Once it returns to node 7 after cluster C_5, that node is not closed, because one of the child clusters is not closed. It is instead put back on the open list. Similarly node 4 is put back on the open list of C_1. At that point, best-first search picks another node from the open list of C_1, node 19, and continues from there.   □

Function BTD-HBFS$(A,C_e,V,clb,cub,\_)$: pair(integer,integer)
  $open :=$ open list of $P_e|A(cub)$;
  if $(open = \varnothing)$ **then**
    | **if** exists minimum cub' s.t. $cub' > cub$ **and** $open(P_e|A(cub')) \neq \varnothing$ **then**
    | | $open = \{\nu \in open(P_e|A(cub')) \mid \nu.lb < cub\}$
    | **else**
    | | $open = \{\varnothing\}$ /* *Contains only the root node at position clb* */
  $P_e.limit := Backtracks + N$ /* *Set a global backtrack limit for the subproblem* */ ;
  $clb' := \max(clb, lb(open))$;
  $cub' := cub$;
1   **while** $(open \neq \varnothing$ **and** $clb' < cub'$ **and** $(C_e = C_1$ **or** $(clb' = clb$ **and** $cub' = cub$ **and** $Backtracks < P_e.limit)))$ **do**
    | $\nu :=$pop$(open)$ /* *Choose a node with minimum lower bound and maximum depth* */ ;
    | Restore state $\nu.\delta$, leading to assignment A_ν, maintaining local consistency ;
    | $NodesRecompute := NodesRecompute + \nu.depth$;
    | $C_e.limit := C_e.backtracks + Z$ /* *Set a depth-first search backtrack limit* */ ;
    | $(cub', cub') :=$BTD-DFS$(A_\nu,C_e,V_\nu,\max(clb', lb(\nu), lb(P_e|A_\nu)),cub',$BTD-HBFS$)$;
    | $clb' := \max(clb', lb(open))$;
    | **if** $(NodesRecompute > 0)$ **then**
    | | **if** $(NodesRecompute/Nodes > \beta$ **and** $Z \leq N)$ **then** $Z := 2 \times Z$;
    | | **else if** $(NodesRecompute/Nodes < \alpha$ **and** $Z \geq 2)$ **then** $Z := Z/2$;
  return (clb', cub') /* *invariant* $clb' \geq clb$ *and* $cub' \leq cub$ */ ;

Algorithm 3. Hybrid Best-First Search with Tree Decomposition.

BTD-HBFS addresses both the issues of BTD that we identified above. First, it is anytime, because as soon as $UB_{P_e|A_e} < k$ for all subproblems, we can combine the assignments that gave these upper bounds to produce a global solution. Second, it constantly updates lower bounds for all active subproblems, so the search effort in each subproblem immediately exploits all other lower bounds.

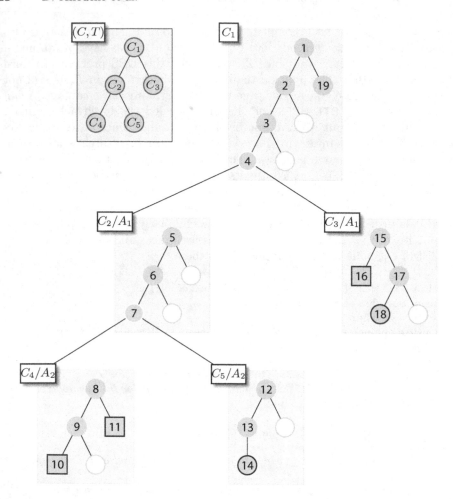

Fig. 3. An example of a run of BTD-HBFS. White nodes are open, nodes with a black border are conflicts (square) or solutions (circle). Grey nodes with a white border are explored but put back in the open list. Grey nodes with no border are closed.

Like HBFS, BTD-HBFS can be made *anyspace*, i.e., its memory usage can be limited to any amount beyond what is needed for BTD-DFS, including zero. The cache of bounds can also be limited, at the expense of additional subproblem recomputations, leading to worst-case complexity exponential in the tree decomposition height.

Theorem 1. *Given a CFN P with treewidth ω,* BTD-HBFS *computes the optimum in time $O(knd^{\omega+1})$ and space $O(knd^{2\omega})$.*

Proof (Sketch). For correctness, observe that BTD-DFS solves independent subproblems separately using DFS, hence using HBFS or any other solution method does not affect correctness. Each leaf node in an internal cluster is closed only when all child clusters are solved, hence all bounds for each subproblem and open node list are correct. Finally, exploration at the root cluster continues until the optimality gap is closed. Complexity stems from the complexity of BTD and the additional overhead of storing open lists for each separator assignment and upper bound.   □

We implemented a simpler version of this algorithm with better space complexity: each time BTD-HBFS is called on $P_e|A$ with a higher upper bound than previously stored, we wipe the open list and replace it with the root node of C_e. This removes theoretical guarantees on the performance of the algorithm, but does not hurt practical performance, as we will see.

4.2   Related Work

AND/OR Branch and Bound search has already been combined with BFS [20]. The resulting AOBF algorithm, has good worst-case time complexity similar to BTD-DFS, but otherwise has the space-intensive non anytime behavior of BFS.

The poor anytime ability of BTD has been addressed by breadth-rotating AND/OR search (BRAO) [21]. BRAO *interleaves* DFS on all components, so it can combine the incumbents of all components to produce a global solution. However, as it performs DFS on each component, it does not produce better lower bounds.

OR-decomposition [12] is an anytime method that exploits lower bounds produced by other clusters by performing DFS in which it interleaves variable choices from all components, and uses caching to achieve the same effect as BTD. However, the global lower bound it computes depends on the partial assignments of all components. Thus it may revisit the same partial assignment of one component many times. This may also inhibit its anytime behavior, as a high cost partial assignment in one component will prevent other components from reaching good solutions. Moreover, the local lower bound for each component is only updated by visiting the right branch at its root.

Russian Doll Search [25], uses DFS to solve each cluster of a rooted tree decomposition in topological order. This method is not anytime, as it cannot produce a solution until it starts solving the root cluster. Moreover, it computes lower bounds that are independent of the separator assignment, hence can be lower than their true value.

5   Experimental Results

We used benchmark instances including stochastic graphical models from the *UAI evaluation* in 2008 and 2010, the *Probabilistic Inference Challenge 2011*, the *Weighted Partial Max-SAT Evaluation 2013*, the MiniZinc Challenge 2012 and 2013, Computer Vision and Pattern Recognition problems from OpenGM2[3]

---

[3] http://hci.iwr.uni-heidelberg.de/opengm2/

and additional instances from the CostFunctionLib[4]. This is a total of more than 3,000 instances that can be encoded as Cost Function Networks, available at http://genoweb.toulouse.inra.fr/~degivry/evalgm, with domain sizes that range from $d = 2$ to 503, $n = 2$ to $903,884$ variables, and $e = 3$ to $2,912,880$ cost functions of arity from $r = 2$ to 580. We used `toulbar2` version 0.9.8.0-dev[5] on a cluster of 48-core Opteron 6176 nodes at 2.3 GHz with 378 GB RAM, with a limit of 24 simultaneous jobs per node.

In all cases, the local lower bound is provided by maintaining EDAC [13]. The variable ordering includes both weighted-degree [4] and last-conflict [15] heuristics. The value ordering is to select the EAC support value first [13]. All executions used a min-fill variable ordering for DAC preprocessing. For HBFS, we set the node recomputation parameters to $[\alpha, \beta] = [5\%, 10\%]$ and the backtrack limit N to $10,000$.

The methods based on BTD use a different value ordering heuristic: if a solution is known for a cluster, it keeps the same value if possible and if not uses EAC support values as the previous methods. A min-fill ordering is used for building a tree decomposition. Children of a cluster are statically sorted by minimum separator size first and smallest number of subproblem variables next.

Our aim is to determine whether HBFS is able to improve over DFS both in terms of number of problems solved (including the optimality proof) and in its anytime behavior. Similarly, we compare BTD-HBFS to BTD-DFS. We include in our comparison two methods that are known to significantly improve the upper bound anytime behavior of DFS: Limited Discrepancy Search [8] and DFS with Luby restarts [18].

We also include results from BRAO [21] using the `daoopt` solver[6] with static mini-bucket lower bounds of different strength (*i-bound* set to 15 and 35) and without local search nor iterative min-fill preprocessing. `daoopt` is restricted to cost functions expressed by complete tables, hence we exclude most MaxSAT families (except MIPLib and MaxClique) in tests where we use it.

5.1 Proving Optimality

We show in figure 4 a cactus plot comparing all the algorithms that do not use tree decompositions, but also include BTD-HBFS^{r_k} for reference (see below). We see that HBFS is the best performing decomposition-unaware algorithm. It outperforms DFS and DFS with Luby restarts significantly, and slightly outperforms LDS.

Although our benchmark set includes very large instances, and our HBFS implementation does not include automatic control of space usage, no instance required more than 32 GB. The median memory usage was 36.8 MB for DFS and 38.2 MB for hybrid BFS. The worst-case largest ratio between HBFS and

---

[4] https://mulcyber.toulouse.inra.fr/projects/costfunctionlib

[5] Available in the git repository at https://mulcyber.toulouse.inra.fr/projects/ toulbar2/ in the `bfs` branch.

[6] https://github.com/lotten/daoopt

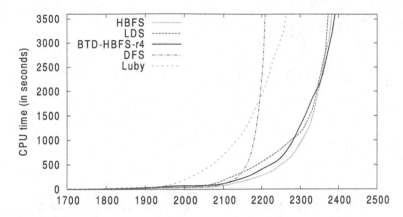

Fig. 4. Number of solved instances within a given time. Methods in the legend are sorted at time=20min.

DFS was $\frac{379.8MB}{12.1MB} = 31.4$ on MRF Grid instance `grid20x20.f15` (unsolved in one hour by both methods).

In figure 5, we compare algorithms exploiting a tree decomposition (BTD-like) or a pseudo-tree (BRAO).We see that BTD-HBFS slightly outperforms BTD-DFS, both outperforming BRAO. However, many of these instances have large treewidth and BTD-like methods are not ideal for these. Even for instances with small treewidth, the decomposition is often a deep tree in which each cluster shares all but one variables with its parent. In these cases, following the tree decomposition imposes a static variable ordering on BTD, while HBFS degrades to DFS. Finding good tree decompositions is not straightforward [10,11]. A simple way to improve one is to merge clusters until no separator has size greater than k, even if this increases width. We call the algorithms that apply this BTD-DFS^{r_k} and BTD-HBFS^{r_k}. Figure 5 includes results for BTD-DFS^{r_4} and BTD-HBFS^{r_4}. BTD-DFS^{r_4} is significantly better than BTD-DFS and BTD-HBFS^{r_4} outperforms BTD-DFS^{r_4} by an even greater margin. BTD-HBFS^{r_4} is also the overall best performer as shown in figure 4.

5.2 Anytime Behavior

To analyze the algorithms' anytime behavior, we first show in figure 6 the evolution of the lower and upper bounds for two instances: the SPOT5 404 (left) and the RLFAP CELAR06 instances (right). We solve both instances using DFS, LDS, DFS with Luby restarts, HBFS, BTD-DFS and BTD-HBFS. In both instances, we see that HBFS and BTD-HBFS improve significantly on the upper bound anytime ability of DFS and BTD-DFS, respectively. Moreover, the lower bound that they report increases quickly in the beginning and keeps increasing with time. For all other algorithms, the lower bound increases by small amounts and infrequently, when the left branch of the root node is closed. The HBFS variants are as fast as the base algorithms in proving optimality.

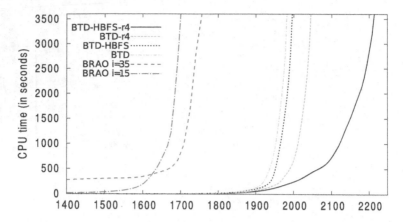

Fig. 5. Number of solved instances as time passes on a restricted benchmark set (without MaxSAT). Methods in the legend are sorted at time=20min.

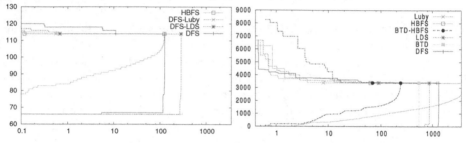

Fig. 6. Evolution of the lower and upper bounds (Y axis, in cost units) as time (X axis, in seconds) passes for HBFS, Luby restart, LDS, and DFS on SPOT5 404 instance (left) and also BTD, and BTD-HBFS for the RLFAP CELAR06 instance (right). Methods are sorted in increasing time to find the optimum. For each curve, the first point represents the time where the optimum is found and the second point the time (if any) of proof of optimality.

In figure 7, we summarize the evolution of lower and upper bounds for each algorithm over all instances that required more than 5 sec to be solved by DFS. Specifically, for each instance I we normalize all costs as follows: the initial lower bound produced by EDAC (which is common to all algorithms) is 0; the best – but potentially suboptimal – solution found by any algorithm is 1; the worst solution is 2. This normalization is invariant to translation and scaling. Additionally, we normalize time from 0 to 1 for each pair of algorithm A and instance I, so that preprocessing ends at time 0 and each run finishes at time 1. This time normalization is different for different instances and for different algorithms on the same instance. A point $\langle x, y \rangle$ on the lower bound line for algorithm A in figure 7 means that after normalized runtime x, algorithm A has proved on average over all instances a normalized lower bound of y and similarly for the upper bound. We show both the upper and lower bound curves for all

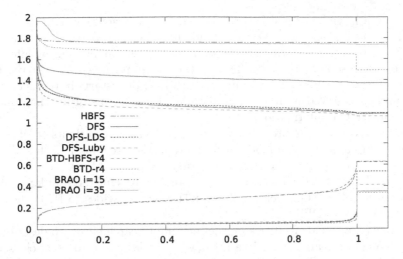

Fig. 7. Average evolution of normalized upper and lower bounds for each algorithm.

algorithms evaluated here. In order for the last point of each curve to be visible, we extend all curves horizontally after 1.0.

This figure mostly ignores absolute performance in order to illustrate the evolution of upper and lower bounds with each algorithm, hence cannot be interpreted without the additional information provided by the cactus plots in figures 4 and 5. It confirms that HBFS improves on DFS in terms of both upper and lower bound anytime behavior and similarly for BTD-HBFS^{r4} over BTD-DFS^{r4} and BRAO, with the latter being especially dramatic. The two HBFS variants are, as expected, significantly better than all other algorithms in terms of the lower bounds they produce. HBFS and BTD-HBFS^{r4} produce solutions of the same quality as LDS, while DFS-Luby is slightly better than this group on this restricted benchmark set (without MaxSAT).

Despite the fact that time to solve an instance is normalized away in figure 7, it does give some information that is absent from the cactus plots and that is the average normalized lower and upper bounds at time 1. Figure 7 tells us that DFS-Luby finds the best solution most often, as its upper bound curve is the lowest at time 1. It is followed closely by the HBFS variants and LDS, while DFS and BTD-DFS^{r4} are significantly worse. On the other hand, DFS-Luby is significantly worse than the HBFS variants in the cactus plot. HBFS and BTD-HBFS^{r4} give better lower bounds in those instances that they failed to solve, so their lower bound curves are higher at point 1.

6   Conclusions

Hybrid BFS is an easily implemented variant of the Branch and Bound algorithm combining advantages of BFS and DFS. While being a generic strategy, applicable to essentially any combinatorial optimization framework, we used it to improve Depth-First Branch and Bound maintaining soft arc consistency and tested it on a large benchmark set of problems from various formalisms, including Cost Function Networks, Markov Random Field, Partial Weighted MaxSAT and CP instances representing a variety of application domains in bioinformatics, planning, resource allocation, image processing and more. We showed that HBFS improves on DFS or DFS equipped with LDS or restarts in terms of number of problems solved within a deadline but also in terms of anytime quality and optimality gap information.

HBFS is also able to improve Tree Decomposition aware variants of DFS such as BTD, being able to solve more problems than the previous DFS based BTD on the same set of benchmarks. BTD is targeted at problems with relatively low treewidth and has been instrumental in solving difficult radio-link frequency assignment problems. On such problems, BTD-HBFS provides to BTD the same improvements as to DFS.

Its ability to provide feedback on the remaining search effort, to describe the current remaining search space in a list of open nodes and to decompose search in self-interrupted DFS probes makes it a very dynamic search method, very attractive for implementing multi-core search.

Acknowledgments. We are grateful to the Genotoul (Toulouse) Bioinformatic platform for providing us computational support for this work.

References

1. Ansótegui, C., Bonet, M.L., Gabàs, J., Levy, J.: Improving sat-based weighted maxsat solvers. In: Proc. of CP 2012, Québec City, Canada, pp. 86–101 (2012)
2. van den Berg, J., Shah, R., Huang, A., Goldberg, K.: ANA*: Anytime nonparametric A*. In: Proceedings of Twenty-Fifth AAAI Conference on Artificial Intelligence (AAAI 2011) (2011)
3. Berthold, T.: Primal heuristics for mixed integer programs. Master's thesis, Technischen Universität Berlin (2006). urn:nbn:de:0297-zib-10293
4. Boussemart, F., Hemery, F., Lecoutre, C., Sais, L.: Boosting systematic search by weighting constraints. In: ECAI, vol. 16, p. 146 (2004)
5. Cooper, M., de Givry, S., Sanchez, M., Schiex, T., Zytnicki, M., Werner, T.: Soft arc consistency revisited. Artificial Intelligence **174**(7), 449–478 (2010)
6. Dechter, R., Mateescu, R.: And/or search spaces for graphical models. Artificial Intelligence **171**(2), 73–106 (2007)
7. de Givry, S., Schiex, T., Verfaillie, G.: Exploiting tree decomposition and soft local consistency in weighted CSP. In: Proc. of the National Conference on Artificial Intelligence, AAAI 2006, pp. 22–27 (2006)
8. Harvey, W.D., Ginsberg, M.L.: Limited discrepency search. In: Proc. of the 14th IJCAI, Montréal, Canada (1995)
9. Jégou, P., Terrioux, C.: Hybrid backtracking bounded by tree-decomposition of constraint networks. Artif. Intell. **146**(1), 43–75 (2003)
10. Jégou, P., Terrioux, C.: Combining restarts, nogoods and decompositions for solving csps. In: Proc. of ECAI 2014, Prague, Czech Republic, pp. 465–470 (2014)
11. Jégou, P., Terrioux, C.: Tree-decompositions with connected clusters for solving constraint networks. In: Proc. of CP 2014, Lyon, France, pp. 407–423 (2014)

12. Kitching, M., Bacchus, F.: Exploiting decomposition in constraint optimization problems. In: Stuckey, P.J. (ed.) CP 2008. LNCS, vol. 5202, pp. 478–492. Springer, Heidelberg (2008)
13. Larrosa, J., de Givry, S., Heras, F., Zytnicki, M.: Existential arc consistency: getting closer to full arc consistency in weighted CSPs. In: Proc. of the 19th IJCAI, pp. 84–89, Edinburgh, Scotland (August 2005)
14. Lawler, E., Wood, D.: Branch-and-bound methods: A survey. Operations Research **14**(4), 699–719 (1966)
15. Lecoutre, C., Saïs, L., Tabary, S., Vidal, V.: Reasoning from last conflict(s) in constraint programming. Artificial Intelligence **173**, 1592–1614 (2009)
16. Likhachev, M., Gordon, G.J., Thrun, S.: ARA*: Anytime A* with provable bounds on sub-optimality. In: Advances in Neural Information Processing Systems, p. None (2003)
17. Linderoth, J.T., Savelsbergh, M.W.: A computational study of search strategies for mixed integer programming. INFORMS Journal on Computing **11**(2), 173–187 (1999)
18. Luby, M., Sinclair, A., Zuckerman, D.: Optimal speedup of las vegas algorithms. In: Proceedings of the 2nd Israel Symposium on the Theory and Computing Systems, pp. 128–133. IEEE (1993)
19. Marinescu, R., Dechter, R.: AND/OR branch-and-bound for graphical models. In: Proc. of IJCAI 2005, Edinburgh, Scotland, UK, pp. 224–229 (2005)
20. Marinescu, R., Dechter, R.: Best-first AND/OR search for graphical models. In: Proceedings of the National Conference on Artificial Intelligence, pp. 1171–1176. AAAI Press, MIT Press, Menlo Park, Cambridge (1999, 2007)
21. Otten, L., Dechter, R.: Anytime and/or depth-first search for combinatorial optimization. AI Communications **25**(3), 211–227 (2012)
22. Pearl, J.: Heuristics – Intelligent Search Strategies for Computer Problem Solving. Addison-Wesley Publishing Comp. (1985)
23. Pohl, I.: Heuristic search viewed as path finding in a graph. Artificial Intelligence **1**(3), 193–204 (1970)
24. Robertson, N., Seymour, P.D.: Graph minors. II. Algorithmic aspects of tree-width. Journal of Algorithms **7**(3), 309–322 (1986)
25. Sanchez, M., Allouche, D., de Givry, S., Schiex, T.: Russian doll search with tree decomposition. In: IJCAI, pp. 603–608 (2009)
26. Schulte, C.: Comparing trailing and copying for constraint programming. In: Logic Programming, Las Cruces, New Mexico, USA, pp. 275–289 (1999)
27. Stern, R., Kulberis, T., Felner, A., Holte, R.: Using lookaheads with optimal best-first search. In: AAAI (2010)
28. Terrioux, C., Jégou, P.: Bounded backtracking for the valued constraint satisfaction problems. In: Rossi, F. (ed.) CP 2003. LNCS, vol. 2833, pp. 709–723. Springer, Heidelberg (2003)

Improved Constraint Propagation via Lagrangian Decomposition

David Bergman[1], Andre A. Cire[2], and Willem-Jan van Hoeve[3](✉)

[1] School of Business, University of Connecticut, Mansfield, USA
david.bergman@business.uconn.edu
[2] University of Toronto Scarborough, Toronto, Canada
acire@utsc.utoronto.ca
[3] Tepper School of Business, Carnegie Mellon University, Pittsburgh, USA
vanhoeve@andrew.cmu.edu

Abstract. Constraint propagation is inherently restricted to the local information that is available to each propagator. We propose to improve the communication between constraints by introducing Lagrangian penalty costs between pairs of constraints, based on the Lagrangian decomposition scheme. The role of these penalties is to force variable assignments in each of the constraints to correspond to one another. We apply this approach to constraints that can be represented by decision diagrams, and show that propagating Lagrangian cost information can help improve the overall bound computation as well as the solution time.

1 Introduction

Modern finite-domain constraint programming (CP) solvers employ a constraint propagation process in which domain changes for the variables are propagated between constraints. To allow for more communication and knowledge sharing between constraints, several techniques have been proposed. One possibility is to propagate more structural information than variable domains, such as (relaxed) decision diagrams [1,10]. Another option, in the context of optimization problems, is to combine constraints with the objective function, and utilize mathematical programming relaxations for stronger cost-based filtering [7,15]. These approaches, however, have in common that consistency checks are done separately and independently for each constraint. Higher-order consistencies, such as pairwise consistency [12] can consider multiple constraints simultaneously, but may suffer from a relatively high computational cost.

We propose an alternative, and generic, approach to improve the propagation between constraints based on *Lagrangian decomposition* [9]. In Lagrangian decomposition, the constraint set of a given problem is partitioned into structured subproblems, each of which is defined on a duplicate copy of the original variables. To link the subproblems, constraints are added to ensure that each of the duplicates is equal to the original variable. These latter constraints are then relaxed with an associated Lagrangian multiplier, and moved to the objective. This results in independent subproblems that can be separately optimized.

© Springer International Publishing Switzerland 2015
G. Pesant (Ed.): CP 2015, LNCS 9255, pp. 30–38, 2015.
DOI: 10.1007/978-3-319-23219-5_3

Intuitively, the idea is to force the variables to take the same value in each constraint, via the Lagrangian penalties, which are iteratively updated. This will somehow synchronize the consistency checks for each of the constraints; instead of allowing each constraint to check its consistency w.r.t. an arbitrary tuple, we iteratively arrive at tuples with minimal disagreement.

Since constraint programming has been designed to work with (global) constraints that capture a specific structure of the problem, the application of Lagrangian decomposition in this context seems natural and promising. Indeed, we show that the Lagrangian decomposition is not only useful to improve the bound on the objective, but can also be applied for cost-based domain filtering.

The structure of the paper is as follows. We first provide an overview of the most relevant related work in Section 2. In Section 3 we recall the Lagrangian decomposition scheme. We apply this to constraint programming models in Section 4. Experimental results on instances with multiple `alldiff` constraints are given in Section 5, while Section 6 provides results on set covering problems. We conclude in Section 7.

2   Related Work

Lagrangian relaxations have been widely applied in operations research as well as constraint programming. One of the first applications in CP is the work by Benoist et al. [2] on the Traveling Tournament Problem. A formal treatment was provided by Sellmann [16] who showed that optimal Lagrangian multipliers may not result in the most effective domain filtering. Recently, [8] introduced a framework for automated Lagrangian relaxation in a constraint programming context. That work explicitly generalizes Lagrangian relaxations to CP problems using measures of constraint violations, or degrees of satisfiability.

Adapting weights for improving propagation has also been applied in the context of Valued Constraint Satisfaction Problems [6]. In that work, a linear programming model is proposed for computing Optimal Soft Arc Consistency, but Lagrangian relaxations are not used. Khemmoudj et al. [13] combine arc consistency with Lagrangian relaxation for filtering constraint satisfaction problems (CSPs). They consider binary CSPs (i.e., each constraint has at most two variables in its scope) in extensional form. Lastly, Bergman et al. [3] introduce Lagrangian relaxations in the context of propagating (relaxed) decision diagrams.

3   Lagrangian Decomposition

Lagrangian decomposition has been introduced to strengthen Lagrangian bounds for integer linear optimization problems [9]. Consider an integer linear program of the form:

$$(P) \quad \max\{fx \mid Ax \leq b, Cx \leq d, x \in X\},$$

for some feasible set X, where $x \in \mathbb{R}^n$ is a vector of variables, $f \in \mathbb{R}^n$ represents a 'weight' vector, A and C represent constraint coefficient matrices, and b and

c are constant right-hand size vectors. This is equivalent to the reformulated program

$$\max\{fx \mid Ax \leq b, Cx \leq d, x = y, x \in X, y \in Y\},$$

for any set Y containing X.

The Lagrangian decomposition of P consists in dualizing the *equality* constraints $x = y$ with Lagrangian multipliers $\lambda \in \mathbb{R}^n$:

$$L_P(\lambda) := \max\{fx + \lambda(y - x) \mid Cx \leq d, x \in X, Ay \leq b, y \in Y\}$$
$$= \max\{(f - \lambda)x \mid Cx \leq d, x \in X\} + \max\{\lambda y \mid Ay \leq by \in Y\}$$

The *Lagrangian dual* is to find those Lagrangian multipliers λ that provide the best bound:

$$\min_{\lambda} L_P(\lambda).$$

Guignard and Kim [9] show that the optimal bound obtained from this Lagrangian decomposition is at least as strong as the standard Lagrangian bounds from dualizing either $Ax \leq b$ or $Cx \leq d$. Lagrangian decomposition may be particularly useful when the problem is composed of several well-structured subproblems, such as those defined by (global) constraints in CP models.

4   Application to Constraint Programming

We apply Lagrangian decomposition to constraint optimization problems (COPs), which include constraint satisfaction problems (CSPs) as special case. It is important to note that this approach will transform each of the original constraints into an 'optimization constraint'; instead of representing a witness for feasibility the constraint now has to represent a witness for *optimality*, even if the constraint is not directly linked to the objective function, or in case of feasibility problems.

When the variables have numeric domains, the method from Section 3 can be directly applied. In general, however, domains need not be numeric, and we will therefore focus our discussion on this more general case. Consider a COP with variables $x_1, \ldots, x_n$ that have given finite domains $x_i \in D_i$:

$$\begin{aligned}
\max \ & f(x_1, \ldots, x_n) \\
\text{s.t.} \ & C_j(x_1, \ldots, x_n) \ j \in \{1, \ldots, m\} \\
& x_i \in D_i \qquad i \in \{1, \ldots, n\}
\end{aligned} \tag{1}$$

For simplicity we assume here that all variables appear in all constraints, but we can allow a different subset of variables for each constraint. Also, C_j may represent any substructure, for example a global constraint, a table constraint, or a collection of constraints. We introduce for each variable x_i and each constraint $j = 1, \ldots, m$ a duplicate variable y_i^j with domain D_i. We let the set y_i^1 represent our 'base' variables, to which we will compare the variables y_i^j for $j = 2, \ldots, m$. The reformulated COP is as follows:

$$
\begin{aligned}
\max\ & f(y_1^1, \ldots, y_n^1) \\
\text{s.t.}\ & C_j(y_1^j, \ldots, y_n^j)\ j \in \{1, \ldots, m\} \\
& y_i^1 = y_i^j \qquad i \in \{1, \ldots, n\}, j \in \{2, \ldots, m\} \\
& y_i^j \in D_i \qquad i \in \{1, \ldots, n\}, j \in \{1, \ldots, m\}
\end{aligned}
$$

To establish the Lagrangian decomposition, we relax the constraints $y_i^1 = y_i^j$ and move these into the objective as $y_i^1 \neq y_i^j$ with associated Lagrangian multipliers. To measure its violation, we propose to represent $y_i^1 \neq y_i^j$ with the set of constraints $((y_i^j = v) - (y_i^1 = v))$ for all $v \in D_i$, where $(y_i^j = v)$ is interpreted as a binary value representing the truth value of the expression. Lastly, we define a Lagrangian multiplier for each i, j $(j \geq 2)$ and each $v \in D(x_i)$ as a vector

$$
\overline{\lambda}_i^j := \lambda_i^j[v].
$$

The Lagrangian objective function can then be written as:

$$
\begin{aligned}
\max\ & f(y_1^1, \ldots, y_n^1) + \sum_{j=2}^{m} \sum_{i=1}^{n} \sum_{v \in D(x_i)} \lambda_i^j[v]((y_i^j = v) - (y_i^1 = v)) \\
& = f(y_1^1, \ldots, y_n^1) + \sum_{j=2}^{m} \sum_{i=1}^{n} \left(\lambda_i^j[y_i^j] - \lambda_i^j[y_i^1] \right)
\end{aligned}
$$

This leads to the following decomposition (for any given set of multipliers $\overline{\lambda}_i^j$):

$$
\begin{aligned}
\max\ & \left\{ f(y_1^1, \ldots, y_n^1) - \sum_{j=2}^{m} \sum_{i=1}^{n} \lambda_i^j[y_i^1] \mid C_1(y_1^1, \ldots, y_n^1) \right\} \\
& + \sum_{j=2}^{m} \left(\max \left\{ \sum_{i=1}^{n} \lambda_i^j[y_i^j] \mid C_j(y_1^j, \ldots, y_n^j) \right\} \right)
\end{aligned}
$$

which are m independent subproblems. Let z_j be the optimal objective value for subproblem $j \in \{1, \ldots, m\}$. Then $\sum_{j=1}^{m} z_j$ is a valid bound on $f(x_1, \ldots, x_n)$. Note that the duplicate variables have only been introduced for the formal description of the method. In practice, all constraints C_j use the original variables.

Design Choices. The Lagrangian decomposition scheme can be adapted by allocating parts of original objective to different subproblems. Moreover, we can introduce equality constraints between any pair of subproblems. We will illustrate the latter in the following example.

Example 1. Consider the following CSP:

C_1 : $\texttt{alldiff}(x_1, x_2, x_3)$ C_2 : $\texttt{alldiff}(x_2, x_4, x_5)$ C_3 : $\texttt{alldiff}(x_3, x_5)$
$x_1 \in \{a, b\}, x_2 \in \{b, c\}, x_3 \in \{a, c\}, x_4 \in \{a, b\}, x_5 \in \{a, b, c\}$

This CSP is domain consistent as well as pairwise consistent, and has one solution $(x_1, x_2, x_3, x_4, x_5) = (b, c, a, a, b)$.

We construct a Lagrangian decomposition based on the constraints C_1, C_2, C_3. To link these, we only need to introduce the constraints $y_2^2 = y_2^1$, $y_3^3 = y_3^1$, $y_5^3 = y_5^2$, and their associated multipliers. This yields the following three subproblems, with respective objective values z_1, z_2, z_3:

$$z_1 = \max \left\{ -\overline{\lambda}_2^2[y_2^1] - \overline{\lambda}_3^3[y_3^1] \mid \texttt{alldiff}(y_1^1, y_2^1, y_3^1) \right\}$$
$$z_2 = \max \left\{ \overline{\lambda}_2^2[y_2^2] - \overline{\lambda}_5^3[y_5^2] \mid \texttt{alldiff}(y_2^2, y_4^2, y_5^2) \right\}$$
$$z_3 = \max \left\{ \overline{\lambda}_3^3[y_3^3] + \overline{\lambda}_5^3[y_5^3] \mid \texttt{alldiff}(y_3^3, y_5^3) \right\}$$

This CSP can be considered as a COP with a zero-valued objective function so that the value $z_1 + z_2 + z_3$ is an upper bound on the satisfiability of this problem, for any Lagrangian multipliers; if the bound is below zero, the problem is unsatisfiable. And so, the optimal Lagrangian decomposition bound is 0.   □

Cost-Based Domain Filtering. In addition to pruning the search based on the overall bound $L_P(\lambda)$ and a given lower bound B, we can apply cost-based domain filtering. The difference with existing cost-based filtering methods is that the bounds from the different subproblems can all be conditioned on a specific variable/value pair. To this end, let $z_j|_{x_i=v}$ be the optimal objective value for subproblem $j \in \{1, \ldots, m\}$ in which $y_i^j = v$. We have the following result:

Proposition 1. *If $\sum_j z_j|_{x_i=v} < B$ then v can be removed from D_i.*

This result may be particularly effective when there is no single subproblem that collects all variables. We continue our example to give an illustration.

Example 2. Continuing Example 1, consider the following Lagrangian multipliers (all others are zero): $\lambda_2^2[b] = 0.5, \lambda_3^3[a] = 0.5, \lambda_5^3[a] = 0.5$. This yields $z_1 = -0.5$, $z_2 = 0.5$, $z_3 = 0.5$, and a total bound of 0.5. Even though this is not optimal, when we condition $x_2 = b$ or $x_5 = c$, the bound becomes -0.5 in both cases, and by Proposition 1 we can remove those values from their respective domains. We can similarly remove values a from D_1, b from D_2, and c from D_3 using the multipliers $\lambda_2^2[c] = -0.5, \lambda_3^3[c] = 0.5, \lambda_5^3[c] = 0.5$.   □

Example 2 implies the following result:

Proposition 2. *Cost-based filtering based on Lagrangian decomposition can be stronger than pairwise consistency.*

Implementation Issues. To apply Lagrangian propagation efficiently, it is important that each constraint is optimized efficiently. For many constraints optimization versions are already available [11], to which the Lagrangian costs can be immediately added. For example, in our experiments we represent constraints by decision diagrams, which permit to find the optimal solution quickly via a shortest (or longest) path calculation. Also, cost-based propagators are available that filter sub-optimal arcs from the decision diagram. Second, the search for optimal Lagrangian multipliers can be done with different methods [14]. Regardless, any set of multipliers results in a valid relaxation, and we do not necessarily need to solve the Lagrangian dual to optimality. In our implementation, we compute the multipliers once at the root node and reuse them during the CP search process.

5   Application: Multiple Alldifferent Constraints

As first application, we consider systems of multiple overlapping `alldiff` constraints, as in [1]. These are defined on a set $X = \{x_1, \ldots, x_n\}$ of variables with domain $\{1, \ldots, n\}$. Each `alldiff` constraint is defined on a subset of variables $S_j \subset X$, for $j = 1, \ldots, k$. We then consider the following COP:

$$\max\left\{ \sum_{i=1}^{n} w_i x_i \mid \texttt{alldiff}(S_j) \ \ \forall j \in \{1, \ldots, k\} \right\}$$

We generated instances with $n = 10, 11, 12, 13, 14$, $k = 4$, and $|S_j| = n - 2$ for all $j = 1, \ldots, 5$. For the Lagrangian decomposition, we partition the `alldiff` constraints into two arbitrary subsets of size two, and define one multi-valued decision diagram (MDD) for each subset. In other words, we apply MDD propagation to these subsets of `alldiff` constraints. The two MDDs thus formed are the basis for the Lagrangian decomposition, which follows the description in Section 4 (where the j-th MDD represents constraint set C_j).

We implemented the MDD propagation as well as the Lagrangian decomposition as a global constraint in IBM ILOG CPO 12.6, similar to [5]. The (near-)optimal Lagrangian multipliers were computed using the Kelly-Cheney-Goldstein method [14], using IBM ILOG CPLEX 12.6 as the linear programming solver. We fix the CP search to be lexicographic in the order of the variables, to ensure the search tree is the same across all instances. We compare the performance with and without Lagrangian multipliers.

In Figure 1.a we show the root node percent gap for the 25 instances (where the optimal value was obtained by formulating an integer linear program and solving the instances using CPLEX). The reduction in the gap can be substantial, in some case several orders of magnitude. This reduction in the optimality gap and additional cost-based filtering due to the Lagrangian multipliers enables more instances to be solved in shorter computational time, as depicted in Figure 1.b. Depending on the configuration of the `alldiff` systems the improvement can be marginal and in some cases negligible.

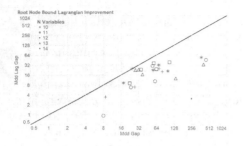
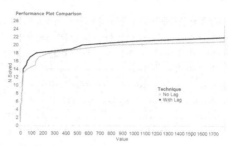

a. Root node gap comparison   b. Performance plot

Fig. 1. Evaluating the impact of the Lagrangian decomposition on systems of multiple `alldiff` constraints. (a) compares the root node gap obtained with Lagrangian decomposition (Mdd Lag Gap) and without (Mdd Gap) and (b) depicts a performance profile comparing the number of instances solved (N Solved) within a given time limit (horizontal axis) with Lagrangian decomposition (With Lag) and without (No Lag).

6 Application: Set Covering

The set covering problem is defined on a universe of n elements $U = \{1, \ldots, n\}$. Given a collection of subsets $C_1, \ldots, C_m \subseteq U$ and weights w_i ($i = 1, \ldots, n$), the problem is to find a set of elements $S \subset U$ of minimum total weight such that all $S \cap C_j$ is not empty for all $j = 1, \ldots, m$. Using a binary variable x_i to represent whether element i is in S, the problem can be formulated as the following COP:

$$\min \left\{ \sum_{i=1}^{n} w_i x_i \ \Big| \ \sum_{i \in C_j} x_i \geq 1 \ \ \forall j \in \{1, \ldots, m\}, x_i \in \{0, 1\} \ \ \forall i \in \{1, \ldots, n\} \right\}$$

Instead of defining a subproblem for each separate constraint, we create exact binary decision diagram (BDD) representations for collections of them. That is, using the construction method described in [4], we create a BDD by adding constraints one at the time, until the exact width exceeds a given limit (in our case 100 nodes on any layer). We then create the next BDD, and so forth. This forms a partition of the constraint set, each of which is represented by an exact BDD. For the instances we considered, we construct 10 or 11 BDDs per instance.

We also slightly modify the Lagrangian decomposition method by representing the original objective function in each of the BDDs, and dualizing constraints $x_i^j = x_i^{j'}$ for every pair (j, j'). Hence, the Lagrangian bound is no longer the sum of the bounds of the respective BDDs, rather the average over the objectives.

In previous work [4], it was shown that the bounds from BDDs were most effective when the constraint matrix has a relatively small bandwidth. We therefore used the same benchmark generator to evaluate the impact of the Lagrangian decomposition for increasing bandwidths. We generated instances with $n = 150$ variables, randomly generated costs, and uniform-randomly selected subsets C_j from within a given bandwidth of size 55 to 75 (five instances for each bandwidth). To generate the costs, we let $c(i)$ represent the number of subsets C_j that

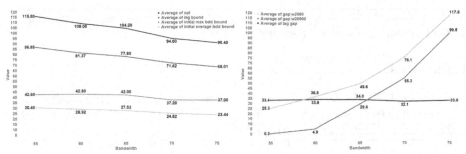

a. Impact of Lagrangian decomposition   b. Comparison with single relaxed BDD

Fig. 2. Evaluating the bound from Lagrangian decomposition for set covering problems of varying bandwidth.

contain element i. Then the cost for variable x_i is taken uniform randomly in $[0.75*c(i), 1.25*c(i)]$. The results are shown in Figure 2, showing the average over the five instances per bandwidth. Figure 2.a depicts four lines: the optimal solution (found by CPLEX), the average bound without using Lagrangian decomposition, the maximum bound without using Lagrangian decomposition, and lastly the average bound when using the Lagrangian decomposition. Lagrangian decomposition generates bounds of much better quality than the independent BDDs. For example, for bandwidth 65 the average bound of 27.53 is improved to 77.80 using the Lagrangian decomposition, on average.

We also compare the Lagrangian decomposition to the original BDD *relaxation* from [4] that represents all constraints in a single BDD respecting a given maximum width. A larger width leads to a stronger relaxation and better bounds. Figure 2.b compares the percent gap (between the lower bound and the optimal solution) of the BDD relaxation for maximum widths 2,000 and 20,000 with that of the Lagrangian decomposition. We note that the BDD relaxation with maximum width 2,000 has about the same memory requirements as the separate BDDs for the Lagrangian decomposition. As the bandwidth increases, the quality of BDD relaxation rapidly declines, while the Lagrangian decomposition is much more stable and outperforms the BDD relaxation (decreasing the gap from 117.6% to 33% for bandwidth 75 and maximum width 2,000). This demonstrates that Lagrangian decompositions can be used to improve BDD-based optimization when a single BDD relaxation can no longer provide sufficient power to represent the entire problem. We do note, however, that the Lagrangian decomposition takes more time to compute (on average 60s) compared to the single BDD relaxation (on average 1.4s for width 2,000 and 17s for width 20,000).

7   Conclusion

We have introduced Lagrangian decomposition in the context of constraint programming as a generic approach to improve the constraint propagation process. The key idea is that we penalize variables in different constraints to take different assignments. We have shown how this approach can be utilized for stronger cost-based domain filtering, and that it leads to improved bounds for systems of `alldiff` constraints and set covering problems.

References

1. Andersen, H.R., Hadzic, T., Hooker, J.N., Tiedemann, P.: A constraint store based on multivalued decision diagrams. In: Bessière, C. (ed.) CP 2007. LNCS, vol. 4741, pp. 118–132. Springer, Heidelberg (2007)
2. Benoist, T., Laburthe, F., Rottembourg, B.: Lagrange relaxation and constraint programming collaborative schemes for traveling tournament problems. In: Proceedings of the International Workshop on Integration of Artificial Intelligence and Operations Research Techniques in Constraint Programming for Combinatorial Optimization Problems (CPAIOR 2001) (2001)
3. Bergman, D., Cire, A.A., van Hoeve, W.J.: Lagrangian Bounds from Decision Diagrams. Constraints 20(3), 346–361 (2015)
4. Bergman, D., van Hoeve, W.-J., Hooker, J.N.: Manipulating MDD relaxations for combinatorial optimization. In: Achterberg, T., Beck, J.C. (eds.) CPAIOR 2011. LNCS, vol. 6697, pp. 20–35. Springer, Heidelberg (2011)
5. Cire, A.A., van Hoeve, W.J.: Multivalued Decision Diagrams for Sequencing Problems. Operations Research 61(6), 1411–1428 (2013)
6. Cooper, M.C., de Givry, S., Sanchez, M., Schiex, T., Zytnicki, M., Werner, T.: Soft arc consistency revisited. Artificial Intelligence 174(7–8), 449–478 (2010)
7. Focacci, F., Lodi, A., Milano, M.: Cost-based domain filtering. In: Jaffar, J. (ed.) CP 1999. LNCS, vol. 1713, pp. 189–203. Springer, Heidelberg (1999)
8. Fontaine, D., Michel, L., Van Hentenryck, P.: Constraint-based lagrangian relaxation. In: O'Sullivan, B. (ed.) CP 2014. LNCS, vol. 8656, pp. 324–339. Springer, Heidelberg (2014)
9. Guignard, M., Kim, S.: Lagrangian Decomposition: A Model Yielding Stronger Lagrangian Bounds. Mathematical Programming 39, 215–228 (1987)
10. Hoda, S., van Hoeve, W.-J., Hooker, J.N.: A systematic approach to MDD-based constraint programming. In: Cohen, D. (ed.) CP 2010. LNCS, vol. 6308, pp. 266–280. Springer, Heidelberg (2010)
11. van Hoeve, W.J., Katriel, I.: Global constraints. In: Handbook of Constraint Programming, pp. 169–208. Elsevier (2006)
12. Janssen, P., Jégou, P., Nouguier, B., Vilarem, M.C.: A filtering process for general constraint-satisfaction problems: achieving pairwise-consistency using an associated binary representation. In: IEEE International Workshop on Tools for Artificial Intelligence, Architectures, Languages and Algorithms, pp. 420–427. IEEE (1989)
13. Khemmoudj, M.O.I., Bennaceur, H., Nagih, A.: Combining arc-consistency and dual lagrangean relaxation for filtering CSPS. In: Barták, R., Milano, M. (eds.) CPAIOR 2005. LNCS, vol. 3524, pp. 258–272. Springer, Heidelberg (2005)
14. Lemaréchal, C.: Lagrangian relaxation. In: Jünger, M., Naddef, D. (eds.) Computational Combinatorial Optimization. LNCS, vol. 2241, pp. 112–156. Springer, Heidelberg (2001)
15. Régin, J.-C.: Arc consistency for global cardinality constraints with costs. In: Jaffar, J. (ed.) CP 1999. LNCS, vol. 1713, pp. 390–404. Springer, Heidelberg (1999)
16. Sellmann, M.: Theoretical foundations of cp-based lagrangian relaxation. In: Wallace, M. (ed.) CP 2004. LNCS, vol. 3258, pp. 634–647. Springer, Heidelberg (2004)

Strengthening Convex Relaxations with Bound Tightening for Power Network Optimization

Carleton Coffrin[1,2]([⊠]), Hassan L. Hijazi[1,2], and Pascal Van Hentenryck[1,2]

[1] NICTA - Optimisation Research Group, Canberra, Australia
[2] College of Engineering and Computer Science, Australian National University, Canberra, Australia
{carleton.coffrin,hassan.hijazi,pvh}@nicta.com.au

Abstract. Convexification is a fundamental technique in (mixed-integer) nonlinear optimization and many convex relaxations are parametrized by variable bounds, i.e., the tighter the bounds, the stronger the relaxations. This paper studies how bound tightening can improve convex relaxations for power network optimization. It adapts traditional constraint-programming concepts (e.g., minimal network and bound consistency) to a relaxation framework and shows how bound tightening can dramatically improve power network optimization. In particular, the paper shows that the Quadratic Convex relaxation of power flows, enhanced by bound tightening, almost always outperforms the state-of-the-art Semi-Definite Programming relaxation on the optimal power flow problem.

Keywords: Continuous constraint networks · Minimal network · Bound consistency · Convex relaxation · AC power flow · QC relaxation · AC optimal power flow

1 Introduction

In (mixed-integer) nonlinear optimization, convexification is used to obtain dual bounds, complementing primal heuristics. In many cases, these convex relaxations are parametrized by variable bounds and the tighter the bounds are, the stronger the relaxations. There is thus a strong potential for synergies between convex optimization and constraint programming. This paper explores these synergies in the context of power system optimization.

The power industry has been undergoing a fundamental transformation in recent years. Deregulation, the emergence of power markets, pressure for reduced capital investment, and the need to secure a clean sustainable energy supply all stress the importance of efficiency and reliability in the design and operation of power networks. As a result, optimization has become a critical component of

NICTA—NICTA is funded by the Australian Government through the Department of Communications and the Australian Research Council through the ICT Centre of Excellence Program.

© Springer International Publishing Switzerland 2015
G. Pesant (Ed.): CP 2015, LNCS 9255, pp. 39–57, 2015.
DOI: 10.1007/978-3-319-23219-5_4

the emerging *smart-grid* [28] and has resulted in millions of dollars in annual savings [32].

Power network applications range from long-term network design and investment tasks [7,12,21] to minute-by-minute operation tasks [14,16,17,19,23]. All of these optimization problems share a common core, the Alternating Current (AC) power flow equations, which model the steady-state physics of power flows. These equations form a system of continuous non-convex nonlinear equations that prove to be a significant challenge for existing general-purpose optimization tools. It is thus not surprising that, in the last decade, significant attention has been devoted to developing computationally efficient convex relaxations.

The main contribution of this paper is to show that constraint programming can substantially improve the quality of convex relaxations for power flow applications. To obtain this result, the paper defines the concept of constraint relaxation networks and generalizes traditional consistency notions to these networks, including minimal network and bound consistency. These concepts, and the associated algorithms, are then applied to optimal power flow applications with and without load uncertainty. The experimental results demonstrate the significant value of bound tightening for power flow applications. In particular,

1. Bound tightening reduces the domains of the variables by as much as 90% in many cases.
2. In over 90% of the test cases considered, propagation over the convex relaxation was sufficient to close the optimality gap within 1%. Only 4 of the test cases considered remain open.
3. The network consistency algorithm improves the quality of the Quadratic Convex (QC) relaxation [18] considerably. The QC relaxation now outperforms, in the vast majority of the cases, the established state-of-the-art Semi-Definite Programming (SDP) relaxation on the optimal power flow problem.
4. Parallelization can significantly reduce the runtime requirements of bound tightening, making the proposed algorithms highly practical.

The rest of the paper is organized as follows. Section 2 reviews the AC power flow feasibility problem and introduces the notations. Section 3 reviews the state-of-the-art QC power flow relaxation, which is essential for building efficient consistency algorithms. Section 4 formalizes the idea of constraint relaxation networks and Section 5 applies this formalism to AC power flows. Section 6 studies the quality of bound tightening in this application domain and Section 7 evaluates the proposed methods on the ubiquitous AC Optimal Power Flow problem. Section 8 illustrates the potential of the proposed methods on power flow applications incorporating uncertainty and Section 9 concludes the paper.

2   AC Power Flow

A power network is composed of a variety of components such as buses, lines, generators, and loads. The network can be interpreted as a graph (N, E) where the set of buses N represent the nodes and the set of lines E represent the edges.

Note that E is a set of directed arcs and E^R will be used to indicate those arcs in the reverse direction. To break numerical symmetries in the model and to allow easy comparison of solutions, a reference node $r \in N$ is also specified.

Every node $i \in N$ in the network has three properties, voltage $V_i = v_i \angle \theta_i$, power generation $S_i^g = p_i^g + iq_i^g$, and power consumption $S_i^d = p_i^d + iq_i^d$, all of which are complex numbers due to the oscillating nature of AC power. Each line $(i,j) \in E$ has an admittance $Y_{ij} = g_{ij} + ib_{ij}$, also a complex number. These network values are connected by two fundamental physical laws, Kirchhoff's Current Law (KCL),

$$S_i^g - S_i^d = \sum_{(i,j) \in E \cup E^R} S_{ij} \ \forall \, i \in N \qquad (1)$$

and Ohm's Law,

$$S_{ij} = Y_{ij}^*(V_iV_i^* - V_iV_j^*) \ \forall \, (i,j) \in E \cup E^R. \qquad (2)$$

Note that bold values indicate parameters that are constant in the classic AC power flow problem and non-bold values are the decision variables.

In addition to these physical laws, the following operational constraints are required in AC power flows. Generator output limitations on S^g,

$$S_i^{gl} \le S_i^g \le S_i^{gu} \ \forall i \in N. \qquad (3)$$

Line thermal limits on S_{ij},

$$|S_{ij}| \le s_{ij}^u \ \forall (i,j) \in E \cup E^R. \qquad (4)$$

Bus voltage limits on V_i,

$$v_i^l \le |V_i| \le v_i^u \ \forall i \in N \qquad (5)$$

and line phase angle difference limits on $V_iV_j^*$,

$$\theta_{ij}^{\Delta l} \le \angle(V_iV_j^*) \le \theta_{ij}^{\Delta u} \ \forall (i,j) \in E \qquad (6)$$

Note that power networks are designed and operated so that $-\pi/3 \le \theta^{\Delta l} \le \theta^{\Delta u} \le \pi/3$ [22] and values as low as $\pi/18$ are common in practice [33]. Additionally the values of v^l, v^u, s^u must be positive as they are bounds on the magnitudes of complex numbers.

Combining all of these constraints and expanding them into their real-number representation yields the AC Power Flow Feasibility Problem (AC-PF) presented in Model 1. The input data is indicated by bold values and a description of the decision variables is given in the model. Constraint (7a) sets the reference angle. Constraints (7b)–(7c) capture KCL and constraints (7d)–(7e) capture Ohm's Law. Constraints (7f) link the phase angle differences on the lines to the bus variables and constraints (7g) enforce the thermal limit on the lines. This particular formulation of AC-PF is advantageous as the auxiliary variables $\theta^\Delta, p,$

Model 1. The AC Power Flow Feasibility Problem (AC-PF)

---

variables:

$p_i^g \in (\boldsymbol{p}_i^{gl}, \boldsymbol{p}_i^{gu}) \ \forall i \in N$ - active power generation

$q_i^g \in (\boldsymbol{q}_i^{gl}, \boldsymbol{q}_i^{gu}) \ \forall i \in N$ - reactive power generation

$v_i \in (\boldsymbol{v}_i^l, \boldsymbol{v}_i^u) \ \forall i \in N$ - bus voltage magnitude

$\theta_i \in (-\infty, \infty) \ \forall i \in N$ - bus voltage angle

$\theta_{ij}^\Delta \in (\boldsymbol{\theta}_{ij}^{\Delta l}, \boldsymbol{\theta}_{ij}^{\Delta u}) \ \forall (i,j) \in E$ - angle difference on a line (aux.)

$p_{ij} \in (-\boldsymbol{s}_{ij}^u, \boldsymbol{s}_{ij}^u) \ \forall (i,j) \in E \cup E^R$ - active power flow on a line (aux.)

$q_{ij} \in (-\boldsymbol{s}_{ij}^u, \boldsymbol{s}_{ij}^u) \ \forall (i,j) \in E \cup E^R$ - reactive power flow on a line (aux.)

subject to:

$$\theta_r = 0 \tag{7a}$$

$$p_i^g - \boldsymbol{p}_i^d = \sum_{(i,j) \in E \cup E^R} p_{ij} \ \forall i \in N \tag{7b}$$

$$q_i^g - \boldsymbol{q}_i^d = \sum_{(i,j) \in E \cup E^R} q_{ij} \ \forall i \in N \tag{7c}$$

$$p_{ij} = \boldsymbol{g}_{ij} v_i^2 - \boldsymbol{g}_{ij} v_i v_j \cos(\theta_{ij}^\Delta) - \boldsymbol{b}_{ij} v_i v_j \sin(\theta_{ij}^\Delta) \ (i,j) \in E \cup E^R \tag{7d}$$

$$q_{ij} = -\boldsymbol{b}_{ij} v_i^2 + \boldsymbol{b}_{ij} v_i v_j \cos(\theta_{ij}^\Delta) - \boldsymbol{g}_{ij} v_i v_j \sin(\theta_{ij}^\Delta) \ (i,j) \in E \cup E^R \tag{7e}$$

$$\theta_{ij}^\Delta = \theta_i - \theta_j \ \forall (i,j) \in E \tag{7f}$$

$$p_{ij}^2 + q_{ij}^2 \leq (\boldsymbol{s}_{ij}^u)^2 \ \forall (i,j) \in E \cup E^R \tag{7g}$$

---

and q isolate the problem's non-convexities in constraints (7d)–(7e) and enable all but one of the operational constraints to be captured by the variable bounds. This continuous constraint satisfaction problem is NP-Hard in general [24,40] and forms a core sub-problem that underpins a wide variety of power network optimization tasks.

To address the computational difficulties of AC-PF, convex relaxations (i.e. polynomial time) have attracted significant interest in recent years. Such relaxations include the Semi-Definite Programming (SDP) [2], Second-Order Cone (SOC) [20], Convex-DistFlow (CDF) [13], and the recent Quadratic Convex (QC) [18] relaxations. To further improve these relaxations, this paper proposes consistency notions and associated propagation algorithms for AC power flows. A detailed evaluation on 57 AC transmission system test cases demonstrates that the propagation algorithms enable reliable and efficient methods for improving these relaxations on a wide variety of power network optimization tasks via industrial-strength convex optimization solvers (e.g., Gurobi, Cplex, Mosek). The next section reviews the QC relaxation in detail, which forms the core of the proposed propagation algorithms.

3   The Quadratic Convex (QC) Relaxation

The QC relaxation [18] was introduced to utilize the bounds on the voltage variables v and θ^Δ, which are ignored by the other relaxations. The key idea is to use the variable bounds to derive convex envelopes around the non-convex aspects of the AC-PF problem. The derivation begins by lifting the voltage product terms in to the higher dimensional W-space using the following equalities:

$$w_i = v_i^2 \quad i \in N \tag{8a}$$

$$w_{ij}^R = v_i v_j \cos(\theta_{ij}^\Delta) \quad \forall(i,j) \in E \tag{8b}$$

$$w_{ij}^I = v_i v_j \sin(\theta_{ij}^\Delta) \quad \forall(i,j) \in E \tag{8c}$$

When Model 1 is lifted into this W-space, all of the remaining constraints are convex. On its own, this lifted model is a weak relaxation but the QC relaxation strengthens it by developing convex relaxations of the nonlinear equations (8a)–(8c) for the operational bounds on variables v and θ^Δ. The convex envelopes for the square and bilinear functions are well-known [27], i.e.,

$$\langle x^2 \rangle^T \equiv \begin{cases} \check{x} \geq x^2 \\ \check{x} \leq (x^u + x^l)x - x^u x^l \end{cases} \tag{T-CONV}$$

$$\langle xy \rangle^M \equiv \begin{cases} \check{x}y \geq x^l y + y^l x - x^l y^l \\ \check{x}y \geq x^u y + y^u x - x^u y^u \\ \check{x}y \leq x^l y + y^u x - x^l y^u \\ \check{x}y \leq x^u y + y^l x - x^u y^l \end{cases} \tag{M-CONV}$$

Under the assumption that the phase angle difference bound is within $-\pi/2 \leq \theta^{\Delta l} \leq \theta^{\Delta u} \leq \pi/2$, relaxations for sine and cosine are given by

$$\langle \cos(x) \rangle^C \equiv \begin{cases} \check{c}x \leq \cos(x) \\ \check{c}x \geq \frac{\cos(x^l) - \cos(x^u)}{(x^l - x^u)}(x - x^l) + \cos(x^l) \end{cases} \tag{C-CONV}$$

$$\langle \sin(x) \rangle^S \equiv \begin{cases} \check{s}x \leq \cos\left(\frac{x^m}{2}\right)\left(x - \frac{x^m}{2}\right) + \sin\left(\frac{x^m}{2}\right) & \text{if } x^l < 0 \wedge x^u > 0 \\ \check{s}x \geq \cos\left(\frac{x^m}{2}\right)\left(x + \frac{x^m}{2}\right) - \sin\left(\frac{x^m}{2}\right) & \text{if } x^l < 0 \wedge x^u > 0 \\ \check{s}x \leq \sin(x) & \text{if } x^l \geq 0 \\ \check{s}x \geq \frac{\sin(x^l) - \sin(x^u)}{(x^l - x^u)}(x - x^l) + \sin(x^l) & \text{if } x^l \geq 0 \\ \check{s}x \leq \frac{\sin(x^l) - \sin(x^u)}{(x^l - x^u)}(x - x^l) + \sin(x^l) & \text{if } x^u \leq 0 \\ \check{s}x \geq \sin(x) & \text{if } x^u \leq 0 \end{cases} \tag{S-CONV}$$

where $x^m = \max(-x^l, x^u)$. These are a generalization of the relaxations proposed in [18] to support asymmetrical bounds on x. Utilizing these building

---

Model 2. The QC Power Flow Feasibility Problem (QC-PF)

---

variables: Variables of Model 1

$st_{ij} \in (-1, 1)$ $\forall (i,j) \in E$ - relaxation of the sine (aux.)

$ct_{ij} \in (0, 1)$ $\forall (i,j) \in E$ - relaxation of the cosine (aux.)

$vv_{ij} \in (\boldsymbol{v}_i^l \boldsymbol{v}_j^l, \boldsymbol{v}_i^u \boldsymbol{v}_j^u)$ $\forall (i,j) \in E$ - relaxation of the voltage product (aux.)

$w_i \in \left((\boldsymbol{v}_i^l)^2, (\boldsymbol{v}_i^u)^2 \right)$ $\forall i \in N$ - relaxation of the voltage square (aux.)

$w_{ij}^R \in (0, \infty)$ $\forall (i,j) \in E$ - relaxation of the voltage and cosine product (aux.)

$w_{ij}^I \in (-\infty, \infty)$ $\forall (i,j) \in E$ - relaxation of the voltage and sine product (aux.)

subject to: (7a)–(7c),(7f)–(7g)

$$\text{CONV}(w_i = v_i^2 \in (\boldsymbol{v}_i^l, \boldsymbol{v}_i^u)) \quad \forall i \in N \tag{10a}$$

$$\text{CONV}(ct_{ij} = \cos(\theta_{ij}^\Delta) \in (\boldsymbol{\theta}_{ij}^{\Delta l}, \boldsymbol{\theta}_{ij}^{\Delta u})) \quad \forall (i,j) \in E \tag{10b}$$

$$\text{CONV}(st_{ij} = \sin(\theta_{ij}^\Delta) \in (\boldsymbol{\theta}_{ij}^{\Delta l}, \boldsymbol{\theta}_{ij}^{\Delta u})) \quad \forall (i,j) \in E \tag{10c}$$

$$\text{CONV}(vv_{ij} = v_i v_j \in (\boldsymbol{v}_i^l, \boldsymbol{v}_i^u) \times (\boldsymbol{v}_j^l, \boldsymbol{v}_j^u)) \quad \forall (i,j) \in E \tag{10d}$$

$$\text{CONV}(w_{ij}^R = vv_{ij} ct_{ij} \in (\boldsymbol{vv}_{ij}^l, \boldsymbol{vv}_{ij}^u) \times (\boldsymbol{ct}_{ij}^l, \boldsymbol{ct}_{ij}^u)) \quad \forall (i,j) \in E \tag{10e}$$

$$\text{CONV}(w_{ij}^I = vv_{ij} st_{ij} \in (\boldsymbol{vv}_{ij}^l, \boldsymbol{vv}_{ij}^u) \times (\boldsymbol{st}_{ij}^l, \boldsymbol{st}_{ij}^u)) \quad \forall (i,j) \in E \tag{10f}$$

$$(w_{ij}^R)^2 + (w_{ij}^I)^2 \leq w_i w_j \quad \forall (i,j) \in E \tag{10g}$$

$$p_{ij} = \boldsymbol{g}_{ij} w_i - \boldsymbol{g}_{ij} w_{ij}^R - \boldsymbol{b}_{ij} w_{ij}^I \quad \forall (i,j) \in E \tag{10h}$$

$$q_{ij} = -\boldsymbol{b}_{ij} w_i + \boldsymbol{b}_{ij} w_{ij}^R - \boldsymbol{g}_{ij} w_{ij}^I \quad \forall (i,j) \in E \tag{10i}$$

$$p_{ji} = \boldsymbol{g}_{ij} w_j - \boldsymbol{g}_{ij} w_{ij}^R + \boldsymbol{b}_{ij} w_{ij}^I \quad \forall (i,j) \in E \tag{10j}$$

$$q_{ji} = -\boldsymbol{b}_{ij} w_j + \boldsymbol{b}_{ij} w_{ij}^R + \boldsymbol{g}_{ij} w_{ij}^I \quad \forall (i,j) \in E \tag{10k}$$

---

blocks, convex relaxations for equations (8a)–(8c) can be obtained by composing relaxations of the subexpressions, for example, $w_{ij}^R \equiv \langle\langle v_i v_j \rangle^M \langle \cos(\theta_i - \theta_j) \rangle^C \rangle^M$. Lastly, the QC relaxation proposes to strengthen these convex relaxations with a valid second-order cone constraint [11,18,20],

$$(w_{ij}^R)^2 + (w_{ij}^I)^2 \leq w_i w_j \quad \forall (i,j) \in E \tag{9}$$

The complete QC relaxation of the AC-PF problem is presented in Model 2 (QC-PF), which incorporates many of the components of Model 1. In the model, the constraint $\text{CONV}(y = f(x) \in D)$ is used to indicate that y lies in a convex relaxation of function f within the domain D. Constraints (10a)–(10f) implement the convex relaxations and constraints (10g) further strengthen these relaxations. Constraints (10h)–(10k) capture the line power flow in terms of the W-space variables.

The Impact of Tight Bounds in the QC Relaxation: Constraints (10a)–(10f) in Model 2 highlight the critical role that the bounds on v and θ^Δ play in

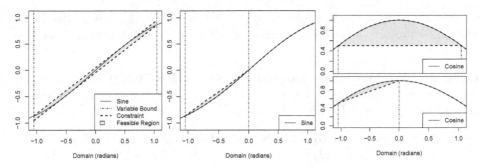

Fig. 1. The Impact of Variable Bounds on the Convex Relaxations.

the strength of the QC relaxation. Figure 1 illustrates this point by showing the convex relaxations for sine and cosine over the domains $\theta^\Delta \in (-\pi/3, \pi/3)$ and $\theta^\Delta \in (-\pi/3, 0)$. This figure indicates two key points: (1) Although the reduction in the size of the bound is 50% in this case, the area inside of the convex relaxations have been reduced even more significantly; (2) Both the sine and cosine functions are monotonic when the sign of θ^Δ is known, which produces tight convex relaxations.

4   Consistency of Constraint Relaxation Networks

As discussed in Section 2, the core of many applications in power network optimization is a continuous constraint network. Moreover, the QC relaxation of this continuous constraint network depends on the bounds of the variables. As a result, constraint propagation now has two benefits: On one hand, it reduces the domains of the variables while, on the other hand, it strengthens the relaxation. These two processes reinforce each other, since tighter constraints generate tighter bounds creating a virtuous cycle. This section generalizes traditional consistency notions to this new context through the concepts of constraint schemes and constraint relaxations. Since solutions to continuous constraint networks are real numbers and computer implementations typically rely on floating-point numbers, some care must be exercised in formalizing these notions. The formalization also assumes that only bound reasoning is of interest, since these are continuous constraint networks. However, the concepts generalize naturally to domain reasoning.

Continuous Constraint Networks: Constraint networks are defined in terms of a set of variables $X = \{x_1, \ldots, x_n\}$ ranging over intervals $I = \{I_1, \ldots, I_n\}$ and a set of constraints. An interval $I = [l, u]$ denotes the set of real numbers $\{r \in \Re \mid l \leq r \leq u\}$. This paper only considers floating-point intervals, i.e., intervals whose bounds are floating-point numbers. If r is a real number, $[r]$ denotes the smallest floating-point interval containing r, $[r]^-$ the largest floating-point number no greater than r, and $[r]^+$ the smallest floating-point number no smaller than r. A variable assignment assigns to each variable x_i a value from its interval I_i. A constraint is a function $(X \rightarrow \Re) \rightarrow Bool$ which, given a variable assignment, returns a truth value denoting whether the assignment satisfies the constraint.

Definition 1 (Continuous Constraint Network (CCN)). *A continuous constraint network is a triple (X, I, C) where $X = (x_1, \ldots, x_n)$ is a collection of variables ranging over $I = (I_1, \ldots, I_n)$ and C is a set of constraints.*

Definition 2 (Solution to a CCN). *A solution to a CCN $(X, \mathcal{I}, C)$, where $X = (x_1, \ldots, x_n)$ and $I = (I_1, \ldots, I_n)$, is an assignment $\sigma = \{x_1 \leftarrow v_1; \ldots; x_n \leftarrow v_n\}$ such that $v_i \in I_i$ and for all $c \in C$: $c(\sigma)$ holds. The set of solutions to a CCN $\mathcal{P}$ is denoted by $\Sigma(\mathcal{P})$.*

In the following we use $\max(x, \Sigma)$ to denote the maximum value of variable x in the assignments Σ, i.e., $\max(x, \Sigma) = \max_{\sigma \in \Sigma} \sigma(x)$, where $\sigma(x)$ denotes the value of variable x in assignment σ. The value $\min(x, \Sigma)$ is defined similarly. The following definition adapts the traditional concept of minimal constraint network [31] to continuous constraint networks.

Definition 3 (Minimal CCN). *A CCN $\mathcal{P} = (X, I, C)$, where $X = (x_1, \ldots, x_n)$ and $I = (I_1, \ldots, I_n)$, is minimal if, for each variable x_i, the interval $I_i = [l_i, u_i]$ satisfies $l_i = [\min(x_i, \Sigma(\mathcal{P}))]^- \wedge u_i = [\max(x_i, \Sigma(\mathcal{P}))]^+$.*

Note that the bounds are not necessarily solutions themselves but are as tight as the floating-point accuracy allows for. Given a CCN $\mathcal{P} = (X, I, C)$, its largest minimal network $\mathcal{P}' = (\mathcal{X}, \mathcal{I}_\Updownarrow, C)$ $(I_m \subseteq I)$ always exists and is unique since there are only finitely many floating-point intervals.

The concept of bound consistency [39] captures a relaxation of the minimal network: It only requires the variable bounds to be tight locally for each constraint.

Definition 4 (Bound Consistency for CCNs). *A CCN $\mathcal{P} = (X, I, C)$, where $X = (x_1, \ldots, x_n)$ and $I = (I_1, \ldots, I_n)$, is bound-consistent if each constraint c is bound-consistent with respect to I. A constraint c is bound-consistent with respect to I if the continuous constraint network $(X, I, \{c\})$ is minimal.*

Once again, given a CCN $\mathcal{P} = (X, I, C)$, its largest bound-consistent network $\mathcal{P} = (X, I_m, C)$ $(I_m \subseteq I)$ always exists and is unique. In the following, we use $minCCN(X, I, C)$ and $bcCCN(X, I, C)$ to denote these networks, i.e.,

$$\text{MINCCN}(X, I, C) = \max\{I_m \subseteq I \mid (X, I_m, C) \text{ is minimal}\},$$
$$\text{BCCCN}(X, I, C) = \max\{I_m \subseteq I \mid (X, I_m, C) \text{ is bound-consistent}\}.$$

Constraint Relaxation Networks: The convex relaxations used in the QC relaxation depend on the variable bounds, i.e., the stronger the bounds the stronger the relaxations. Since the relaxations change over time, it is necessary to introduce new consistency notions: constraint schemes and constraint relaxations.

Definition 5 (Continuous Constraint Scheme). *A constraint scheme r is a function* $\mathcal{I} \to (X \to \Re) \to Bool$ *which, given a collection of intervals, returns a constraint. Moreover, the scheme r satisfies the following monotonicity property:*

$$I \subseteq I' \Rightarrow (r(I')(\sigma) \Rightarrow r(I)(\sigma))$$

for all collections of intervals I and I', and variable assignment σ.

The monotonicity property ensures that tighter bounds produce tighter constraints. Traditional constraints are constraint schemes that just ignore the initial bounds. A constraint relaxation is a constraint scheme that preserves the solutions to the original constraint.

Definition 6 (Constraint Relaxation). *A constraint scheme r is a relaxation of constraint c if, for all assignment σ and bounds $I = ([\sigma(x_1)], \dots, [\sigma(x_n)])$, we have $r(I)(\sigma) \Rightarrow c(\sigma)$.*[1]

Example 1. Consider the constraint $c(x, y, z)$ which holds if $z = xy$. Given bounds $[x^l, x^u]$ and $[y^l, y^u]$ for variables x and y, the McCormick relaxation [27] is a constraint scheme specified by the collection of constraints in M-CONV. Note that this envelope ignores the bound on variable z. Additionally this constraint scheme is also a constraint relaxation of c because it is known to be the convex envelope of $z = xy$ for any bounds on x and y [27].

Definition 7 (Continuous Constraint Relaxation Network (CCRN)). *A constraint relaxation network is a triple (X, I, R) where X is a collection of variables ranging over I and R is a set of constraint relaxations.*

In the following, we use $R(I)$ to denote $\{r(I) \mid r \in R\}$ if R is a set of relaxations.

Consistency of Constraint Relaxation Networks: We now generalize the concepts of minimal and bound-consistent networks to CCRNs. The definitions capture the fact that no additional bound tightening is possible for the relaxations induced by the bounds.

Definition 8 (Minimal CCRN). *A CCRN $\mathcal{P} = (X, I, R)$, $X = (x_1, \dots, x_n)$ and $I = (I_1, \dots, I_n)$, is minimal if the CCN network $(X, I, R(I))$ is.*

Definition 9 (Bound-Consistent CCRN). *A CCRN $\mathcal{P} = (X, I, R)$, where $X = (x_1, \dots, x_n)$ and $I = (I_1, \dots, I_n)$, is bound-consistent if the CCN network $(X, I, R(I))$ is.*

Once again, the largest minimal or bound-consistent network of a CCRN exists and is unique by monotonicity of constraint relaxations. In the following, we use $minCCRN(X, I, C)$ and $bcCCRN(X, I, C)$ to denote these networks, i.e.,

$$\text{MINCCRN}(X, I, R) = \max\{I_m \subseteq I \mid (X, I_m, R) \text{ is minimal}\},$$
$$\text{BCCCRN}(X, I, R) = \max\{I_m \subseteq I \mid (X, I_m, R) \text{ is bound-consistent}\}.$$

---

[1] Note that some of the convex relaxations used in the QC relaxation are only valid within some bounds. This is easily captured by assuming that the constraint itself imposes these bounds.

MINCCRN(X, I, R)
  $I^n := I$;
  repeat
    $I^o := I^n$;
    $I^n :=$ MINCCN$(X, I^o, R(I^o))$;
  until $I^o = I^n$
  return I^n;

BCCCN(X, I, C)
  $I^n := I$;
  repeat
    $I^o := I^n$;
    for all $c \in C$
      $I^n_c :=$ MINCCN$(X, I^o, \{c\})$;
    $I^n := \bigcap_{c \in C} I^n_c$;
  until $I^o = I^n$
  return I^n;

Fig. 2. Computing the Minimal Continuous Constraint Relaxation Networks

Fig. 3. Computing the Largest Bound-Consistent Constraint Network.

The following property establishes the soundness of bound tightenings in CCRNs.

Proposition 1. *Let* (X, I, C) *be a CCN and let* (X, I, R) *be a CCRN such that* $R = \{r \mid c \in C \land r \text{ is a relaxation of } c\}$. *Then,*

$$\text{MINCCN}(X, I, C) \subseteq \text{MINCCRN}(X, I, R),$$

$$\text{BCCCN}(X, I, C) \subseteq \text{BCCCRN}(X, I, R).$$

The minimal and bound-consistent relaxation networks can be computed by a simple fixpoint algorithm that iterates the consistency algorithm over the increasingly tighter relaxation networks. Figure 2 depicts the algorithm for computing a minimal network. The algorithm is similar for bound consistency. Observe that the bound-consistency algorithm has a fixpoint algorithm embedded inside the top-level fixpoint.

4.1   Relation to Concepts in Global Optimization

The idea of bounds propagation for global optimization goes as far back as [6]: It was subsequently implemented in the NUMERICA system which also performs bound propagation on a linearization of the nonlinear constraints [37, 38]. The notion of using bound reductions for improving convex relaxations of non-convex programs was first widely recognized in the Branch-and-Reduce (BNR) algorithm [34]. BNR is a natural extension of Branch-and-Bound over continuous domains, which includes additional steps to reduce the domains of the variables at each search node. This line of work has developed into two core bound reduction ideas: (1) Feasibility-Based Range Reduction (FBRR), which is concerned with pruning techniques based on feasibility information and (2) Optimality Based Range Reduction (OBRR), which develops bound reductions based on Lagrangian-duality arguments [35]. A variety of methods have been developed for FBRR and OBRR with various pruning strength and computational time tradeoffs [5, 25, 34]. However, all these methods are non-global bound reduction techniques and may be iterated until a desired level of consistency is achieved.

CCRNs and the associated consistency notions (MINCCRN, BCCCRN) developed herein are examples of FBRR methods. The idea of computing MINCCRN is discussed informally in [4] for the special case where the relaxation is a system of linear or convex equations (note that the algorithm in Figure 2 applies for any kind of CSP). It is often noted in the FBRR literature that just one iteration of the MINCCRN is too costly to compute [4, 35], let alone the full fixpoint. The preferred approach is to perform some bound propagation (not always to the fixpoint) on linear relaxations of the non-convex problem [5, 25, 35]. In fact, specialized algorithms have been proposed for computing bound consistency on purely linear systems for this purpose [4]. The linear bound-consistency computations discussed in [4,5] are weaker forms of the BCCCRN notion considered here since it does not explicitly mention re-linearizing the relaxation after bound propagation is complete and re-computing bounds consistency. It is important to note that the algorithm in Figure 2 seamlessly hybridizes the FBRR ideas from global optimization to CP systems, which include arbitrary global constraints that are outside the scope of purely mathematical programs. This advantage is utilized in the next section.

5   Constraint Relaxation Networks for Power Flows

This section discusses how to compute the largest minimal and bound-consistent networks for the relaxation model (X, I, R) defined by Model 2. Observe first that the convex relaxations used in Model 2 are all monotonic.

Proposition 2. *The convex relaxations T-CONV, M-CONV, C-CONV, and S-CONV are monotonic.*

Minimal Network: The largest minimal network is computed by Algorithm QC-N which applies the fixpoint algorithm MINCCRN shown in Figure 2 to Model 2. The underlying MINCCN networks are computed by optimizing each variable independently, i.e.,

$$I_x^n := [\min_{\sigma : R(\sigma)} \sigma(x), \max_{\sigma : R(\sigma)} \sigma(x)];$$

Observe that this computation is inherently parallel, since all the optimizations are independent.

Bound-Consistent Network: The largest bound-consistent network is computed by Algorithm QC-B which applies the bound-consistency counterpart of algorithm MINCCRN to Model 2. The BCCCN networks needed in this fixpoint algorithm are computed by the algorithm shown in Figure 3. Algorithm BCCCN computes the intervals I_c^n that are bound-consistent for each constraint $c \in C$ before taking the intersection of these intervals. The process is iterated until a fixpoint is obtained. This algorithm was selected because the bound-consistency computations can be performed in parallel.

Observe also that the QC-B algorithm is applied to a version of Model 2 using a global constraint

$$\text{line\_power\_qc}(p_{ij}, q_{ij}, v_i, \theta_i, p_{ji}, q_{ji}, v_j, \theta_j)$$

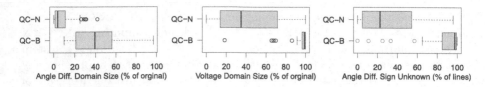

Fig. 4. QC Consistency Algorithms – Quality Analysis.

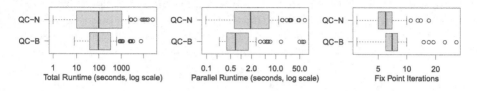

Fig. 5. QC Consistency Algorithms – Runtime Analysis.

that captures constraints (7f)–(7g), (10a)–(10k) for each line $(i,j) \in E$. The use of this global constraint means that QC-B computes a stronger form of bounds consistency than one based purely on Model 2. This stronger level of consistency is necessary to obtain reasonable bound tightenings. Note that all the optimizations in algorithms QC-N and QC-B are convex optimization problems which can be solved in polynomial time.

6   Strength and Performance of the Bound Tightening

This section evaluates the benefits of QC-N and QC-B on the general feasibility problem in Model 1. Algorithms QC-N and QC-B were implemented in AMPL [15] using IPOPT 3.12 [41] to solve the convex nonlinear programs. The propagation algorithms were executed on Dell PowerEdge R415 servers with Dual 2.8GHz AMD 6-Core Opteron 4184 CPUs and 64GB of memory with a convergence tolerance of $\epsilon = 0.001$. Their performance is evaluated on 57 transmission system test cases from the NESTA 0.3.0 archive [10] ranging from 3 to 300 buses.

Figure 4 summarizes the results of QC-B and QC-N on three key metrics: the phase angle difference domains (θ^Δ), the voltage domains (v), and the number of lines where the sign of θ^Δ is determined. Each plot summarizes the distribution of 57 values as a standard *box-and-whisker* plot, where the width of the box reflects the first and third quartiles, the black line inside the box is the median, and the whiskers reflect min and max values up to 1.5 IQR with the remaining data points plotted as outliers. In these plots values to the left are preferable. The domain reduction of the QC-N approach is substantial, typically pruning the domain θ^Δ by 90% and the domain of v by 30% and determining the sign of θ^Δ for about half of the lines. Across all of the metrics, it is clear that QC-N has significant benefits over QC-B.

Figure 5 summarizes the runtime performance of QC-B and QC-N on three key metrics: Total CPU time (T_1), fully parallel CPU wall-clock time (T_∞), and the number of fixpoint iterations. The total runtimes of the QC-B and QC-N algorithms vary widely based on the size of the network under consideration and can range from seconds to hours. Fortunately, regardless of the size of the network, the number of iterations in the fixpoint computation is small (often less than 10). As a result, the parallel runtime of the algorithms scale well with the size of the network and rarely exceeds 1 minute, which is well within the runtime requirements of the majority of network optimization applications.[2]

7   Application to AC Optimal Power Flow

This section assesses the benefits of QC-B and QC-N on the ubiquitous AC Optimal Power Flow problem (AC-OPF) [29,30]. The goal of the AC-OPF is to find the cheapest way to satisfy the loads given the network flow constraints and generator costs functions, which are typically quadratic. If c_{2i}, c_{1i}, c_{0i} are the cost coefficients for generating power at bus $i \in N$, the AC-OPF objective function is given by,

$$\textbf{minimize:} \sum_{i \in N} c_{2i}(p_i^g)^2 + c_{1i}(p_i^g) + c_{0i} \qquad (11)$$

The complete non-convex AC-OPF problem is Model 1 with objective (11) and the QC relaxation of this problem is Model 2 with objective (11).

The goal is to compare five AC-OPF relaxations for bounding primal AC-OPF solutions produced by IPOPT, which only guarantees local optimality. The five relaxations under consideration are as follows:

1. QC - as defined in Model 2.
2. QC-B - BCCCRN for Model 2.
3. QC-N - MINCCRN for Model 2.
4. SDP - a state-of-the-art relaxation based on semi-definite programming [26].
5. SDP-N - the SDP relaxation strengthened with bounds from QC-N.

There is no need to consider other existing relaxations as the QC and SDP dominate them [11]. The computational environment and test cases are those of Section 6. SDPT3 4.0 [36] was used to solve the SDP models.

Table 1 presents the detailed performance and runtime results on all 57 test cases. They can be summarized as follows: (1) The optimality gaps of the QC relaxation are significantly reduced by both QC-N and QC-B; (2) QC-N closes the AC-OPF optimality gap to below 1% in 90% of the cases considered and closes 10 open test cases; (3) QC-N almost always outperforms the SDP relaxation in quality with comparable parallel runtimes; (4) For the test cases with significant optimality gaps, QC-N outperforms the SDP relaxation most often, even when the SDP relaxation is strengthened with QC-N bounds (i.e., SDP-N).

---

[2] Dedicated high performance computational resources are commonplace in power system operation centers. The T_∞ runtime is realistic in these settings where high-level of reliability is critical.

Table 1. Quality and Runtime Results of Convex Relaxations on the AC-OPF Problem (**bold** - best in row (runtime used to break ties in quality), — - solving error)

Test Case	$/h AC	Optimality Gap (%)					T_∞ Runtime (sec.)					
		SDP-N	SDP	QC-N	QC-B	QC	AC	SDP-N	SDP	QC-N	QC-B	QC
case3_lmbd	5812	0.1	0.4	**0.1**	1.0	1.2	0.2	6.8	4.7	0.5	0.4	0.1
case4_gs	156	0.0	0.0	**0.0**	0.0	0.0	0.2	7.2	4.8	0.4	0.8	0.1
case5_pjm	17551	5.2	**5.2**	9.3	14.5	14.5	0.1	6.4	5.1	0.9	0.3	0.2
case6_c	23	0.0	0.0	**0.0**	0.3	0.3	0.0	6.9	5.4	1.3	0.4	0.1
case6_ww	3143	0.0	0.0	**0.0**	0.1	0.6	0.3	5.4	5.4	0.8	2.7	0.1
case9_wscc	5296	0.0	0.0	**0.0**	0.0	0.0	0.2	6.2	4.9	1.5	0.7	0.1
case14_ieee	244	0.0	0.0	**0.0**	0.1	0.1	0.1	4.8	5.2	2.0	0.4	0.1
case24_ieee_rts	63352	0.0	0.0	**0.0**	0.0	0.0	0.2	8.5	6.0	3.2	0.5	0.2
case29_edin	29895	0.0	**0.0**	0.0	0.1	0.1	0.4	8.2	7.8	15.8	1.4	1.1
case30_as	803	0.0	0.0	**0.0**	0.1	0.1	0.3	6.9	5.4	2.3	0.5	0.1
case30_fsr	575	0.0	**0.0**	0.1	0.3	0.4	0.2	5.5	6.1	2.2	1.0	0.2
case30_ieee	205	0.0	0.0	**0.0**	5.3	15.4	0.4	7.8	6.3	0.7	0.7	0.3
case39_epri	96505	0.0	0.0	**0.0**	0.0	0.0	0.2	6.5	7.1	2.1	0.6	0.2
case57_ieee	1143	0.0	0.0	**0.0**	0.1	0.1	0.1	11.4	9.1	5.1	0.9	0.4
case73_ieee_rts	189764	0.0	0.0	**0.0**	0.0	0.0	0.5	12.5	8.5	4.7	0.7	0.5
case118_ieee	3720	0.1	**0.1**	0.4	1.0	1.7	0.3	18.2	12.0	21.1	6.0	0.8
case162_ieee_dtc	4237	1.0	1.1	**0.7**	3.8	4.2	0.7	57.6	34.2	25.9	7.0	1.5
case189_edin	849	—	0.1	**0.1**	—	0.2	0.9	12.3	13.3	6.5	59.9	1.6
case300_ieee	16894	0.1	**0.1**	0.1	1.0	1.2	0.9	40.8	25.5	48.2	14.4	2.4
case3_lmbd__api	367	0.0	1.3	**0.0**	0.5	1.8	0.2	4.0	4.0	0.5	1.5	0.1
case4_gs__api	767	0.0	0.0	**0.0**	0.2	0.7	0.9	6.9	3.9	0.8	0.3	0.1
case5_pjm__api	2994	0.0	0.0	**0.0**	0.4	0.4	0.0	6.9	7.0	0.2	0.4	0.1
case6_c__api	807	0.0	0.0	**0.0**	0.5	0.5	0.6	5.3	5.4	0.3	0.4	0.1
case6_ww__api	273	—	0.0	**0.0**	2.1	13.1	0.2	4.5	15.0	0.4	0.4	0.1
case9_wscc__api	656	0.0	0.0	**0.0**	0.0	0.0	0.4	5.4	6.4	0.8	0.9	0.1
case14_ieee__api	323	0.0	**0.0**	0.2	1.3	1.3	0.1	6.4	4.7	0.5	0.4	0.1
case24_ieee_rts__api	6421	0.7	1.4	**0.3**	3.3	13.8	0.2	8.6	7.2	1.4	1.6	0.2
case29_edin__api	295764	—	—	**0.1**	0.4	0.4	0.3	12.6	7.8	28.4	1.1	3.2
case30_as__api	571	0.0	0.0	**0.0**	2.4	4.8	0.4	7.6	6.0	3.7	0.8	0.2
case30_fsr__api	372	3.6	11.1	**2.7**	42.8	46.0	0.2	7.9	6.7	1.4	0.4	0.2
case30_ieee__api	411	0.0	0.0	**0.0**	0.9	1.0	0.3	8.9	6.5	1.0	0.5	0.2
case39_epri__api	7466	0.0	0.0	**0.0**	0.8	3.0	0.1	9.2	6.5	4.9	1.9	0.2
case57_ieee__api	1430	0.0	0.1	**0.0**	0.2	0.2	0.4	8.8	8.1	3.2	0.6	0.4
case73_ieee_rts__api	20123	0.9	4.3	**0.1**	3.6	12.0	0.6	15.4	9.5	11.4	2.0	0.6
case118_ieee__api	10258	16.7	31.5	**11.8**	38.9	44.0	0.6	14.2	14.6	11.3	4.7	0.8
case162_ieee_dtc__api	6095	0.6	1.0	**0.1**	1.4	1.5	0.4	51.9	32.8	25.5	2.1	1.5
case189_edin__api	1971	—	0.1	**0.0**	—	5.6	0.3	14.3	13.5	8.3	67.1	1.1
case300_ieee__api	22825	0.0	**0.0**	0.2	0.6	0.8	0.9	47.5	28.5	71.1	3.6	2.6
case3_lmbd__sad	5992	0.1	2.1	**0.0**	0.2	1.2	0.1	5.1	4.2	0.2	0.9	0.1
case4_gs__sad	324	0.0	0.0	**0.0**	0.5	0.8	0.1	4.4	3.9	0.1	1.3	0.1
case5_pjm__sad	26423	0.0	0.0	**0.0**	0.7	1.1	0.1	5.7	5.3	0.2	0.4	0.1
case6_c__sad	24	0.0	0.0	**0.0**	0.4	0.4	0.1	6.7	4.6	0.2	0.3	0.1
case6_ww__sad	3149	0.0	0.0	**0.0**	0.1	0.3	0.1	5.9	5.4	0.2	0.2	0.1
case9_wscc__sad	5590	0.0	0.0	**0.0**	0.2	0.4	0.3	5.5	4.4	0.1	0.5	0.1
case14_ieee__sad	244	0.0	0.0	**0.0**	0.1	0.1	0.1	7.5	4.6	0.5	0.3	0.1
case24_ieee_rts__sad	79804	1.4	6.1	**0.1**	3.4	3.9	0.3	9.3	5.7	0.6	0.4	0.3
case29_edin__sad	46933	5.8	28.4	**0.9**	20.0	20.6	0.5	7.1	8.5	15.5	0.3	1.6
case30_as__sad	914	0.1	0.5	**0.0**	2.9	3.1	0.1	6.4	6.8	2.3	0.3	0.2
case30_fsr__sad	577	0.1	0.1	**0.1**	0.5	0.6	0.1	6.2	6.8	1.9	0.3	0.2
case30_ieee__sad	205	0.0	0.0	**0.0**	2.0	4.0	0.3	7.0	6.0	0.6	0.6	0.1
case39_epri__sad	97219	0.0	0.1	**0.0**	0.0	0.0	0.1	7.1	6.0	1.0	0.9	0.2
case57_ieee__sad	1143	0.0	0.0	**0.0**	0.1	0.1	0.4	8.8	7.6	1.9	0.8	0.3
case73_ieee_rts__sad	235241	2.4	4.1	**0.1**	3.1	3.5	0.3	9.7	8.4	3.6	0.6	0.8
case118_ieee__sad	4323	4.0	7.6	**1.4**	7.6	8.3	0.4	15.4	13.8	5.9	0.6	1.0
case162_ieee_dtc__sad	4368	1.7	3.6	**0.4**	5.9	6.9	0.9	46.8	37.7	27.3	2.1	1.4
case189_edin__sad	914	—	1.2	**0.5**	—	2.2	0.6	11.4	17.4	12.2	49.6	1.1
case300_ieee__sad	16912	0.1	**0.1**	0.1	0.8	1.2	0.9	25.2	30.8	45.6	5.5	2.4

Overall, these results clearly establish QC-N is the new state-of-the-art convex relaxation of the AC-OPF. General purpose global optimization solvers (e.g., Couenne 0.4 [3] and SCIP 3.1.1 [1,8]) were also considered for comparison. Preliminary results indicated that these general purpose solvers are much slower than the dedicated power flow relaxations considered here and cannot produce competitive lower bounds on these networks with in 10 hours of computation.

8   Propagation with Load Uncertainty

Loads in power systems are highly predictable. In transmission systems, it is commonplace for minute-by-minute load forecasts to be within 5% of the true values [9]. This high degree of predictability can be utilized by the bound tightening algorithms proposed here. Indeed, if the feasible set of Model 1 is increased to include a range of possible load values, determined by the forecast, then the algorithms compute a description of all possible future power flows. This section studies the power of bound propagation in this setting.

Model 3 presents an extension of Model 1 to incorporate load uncertainty. New decision variables for the possible load values are introduced (i.e., p^d, q^d) and their bounds come from the extreme values of the load forecasting model. The lower bounds on active power generation (p^g) are also increased to include 0, as generators may become inactive at some point in the future (e.g., due to scheduled maintenance or market operations). Constraints (12a)–(12b) incorporate the load variables into KCL. The other constraints remain the same as in Model 1. Because only the KCL constraints are modified in this formulation, the QC relaxation of Model 3 (QC-U) is similar to Model 1, as described in Section 3. For the experimental evaluation, the 57 deterministic test cases were extended into uncertain load cases by adopting a forecast model of ±5% of the deterministic load value.

Figure 6 compares the quality of MINCCRN on the QC-U model (QC-U-N) to MINCCRN in the deterministic case (QC-N) in order to illustrate the pruning

---

Model 3. The AC-PF Program with Load Uncertainty (AC-PF-U)

> **variables:** Variables of Model 1
>
> $p_i^d \in (p_i^{dl}, p_i^{du})$ $\forall i \in N$ - active power load interval
>
> $q_i^d \in (q_i^{dl}, q_i^{du})$ $\forall i \in N$ - reactive power load interval
>
> $p_i^g \in (0, p_i^{du})$ $\forall i \in N$ - active power generation interval
>
> **subject to:** (7a), (7d)–(7g)
>
> $$p_i^g - p_i^d = \sum_{(i,j)\in E\cup E^R} p_{ij} \ \forall i \in N \tag{12a}$$
>
> $$q_i^g - q_i^d = \sum_{(i,j)\in E\cup E^R} q_{ij} \ \forall i \in N \tag{12b}$$

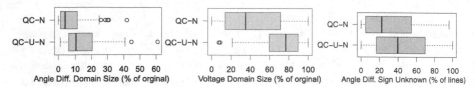

Fig. 6. QC Consistency Algorithms with Load Uncertainty – Quality Analysis.

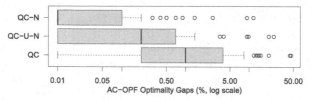

Fig. 7. Comparison of AC-OPF Bound Improvements of QC Variants.

loss due to uncertainty. The results indicate that, even when load uncertainty is incorporated, MINCCRN still prunes the variable domains significantly, typically reducing the voltage angle domains by 80% and the voltage magnitude domains by 10%, and determining the sign of θ^Δ for about 30% of the lines. The domain reduction on θ^Δ in QC-U-N is particularly significant.

Figure 7 considers the AC-OPF and summarizes the optimality gaps produced under load certainty and uncertainty. QC-U-N produces significant improvement in optimality gaps, moving from $< 5\%$ (QC) to less than $< 1\%$. Obviously, load certainty (QC-N) closes the remaining 1%.

9   Conclusion

This paper studied how bound tightening can improve convex relaxations by adapting traditional constraint-programming concepts (e.g., minimal network and bound consistency) to a relaxation framework. It showed that, on power flow applications, bound tightening over the QC relaxation can dramatically reduce variable domains. Moreover, on the ubiquitous AC-OPF problem, the QC relaxation, enhanced by bound tightening, almost always outperforms the state-of-the-art SDP relaxation on the optimal power flow problem. The paper also showed that bound tightening yields significant benefits under load uncertainty, demonstrating a breadth of applicability. These results highlight the significant potential synergies between constraint programming and convex optimization for complex engineering problems.

References

1. Achterberg, T.: Scip: solving constraint integer programs. Mathematical Programming Computation **1**(1), 1–41 (2009). http://dx.doi.org/10.1007/s12532-008-0001-1

2. Bai, X., Wei, H., Fujisawa, K., Wang, Y.: Semidefinite programming for optimal power flow problems. International Journal of Electrical Power & Energy Systems 30(67), 383–392 (2008)
3. Belotti, P.: Couenne: User manual (2009). https://projects.coin-or.org/Couenne/ (accessed April 10, 2015)
4. Belotti, P., Cafieri, S., Lee, J., Liberti, L.: On feasibility based bounds tightening (2012). http://www.optimization-online.org/DB_HTML/2012/01/3325.html
5. Belotti, P., Lee, J., Liberti, L., Margot, F., Wachter, A.: Branching and bounds tightening techniques for non-convex minlp. Optimization Methods Software 24(4–5), 597–634 (2009)
6. Benhamou, F., McAllester, D., Van Hentenryck, P.: Clp (intervals) revisited.Tech. rep., Brown University, Providence, RI, USA (1994)
7. Bent, R., Coffrin, C., Gumucio, R., Van Hentenryck, P.: Transmission network expansion planning: Bridging the gap between ac heuristics and dc approximations. In: Proceedings of the 18th Power Systems Computation Conference (PSCC 2014), Wroclaw, Poland (2014)
8. Berthold, T., Heinz, S., Vigerske, S.: Extending a cip framework to solvemiqcps. In: Lee, J., Leyffer, S. (eds.) Mixed Integer Nonlinear Programming. The IMA Volumes in Mathematics and its Applications, vol. 154, pp. 427–444. Springer New York (2012)
9. Chen, Y., Luh, P., Guan, C., Zhao, Y., Michel, L., Coolbeth, M., Friedland, P., Rourke, S.: Short-term load forecasting: Similar day-based wavelet neural networks. IEEE Transactions on Power Systems 25(1), 322–330 (2010)
10. Coffrin, C., Gordon, D., Scott, P.: NESTA, The NICTA Energy System Test Case Archive. CoRR abs/1411.0359 (2014). http://arxiv.org/abs/1411.0359
11. Coffrin, C., Hijazi, H., Van Hentenryck, P.: The QC Relaxation: Theoretical and Computational Results on Optimal Power Flow. CoRR abs/1502.07847 (2015). http://arxiv.org/abs/1502.07847
12. Coffrin, C., Van Hentenryck, P.: Transmission system restoration: Co-optimization of repairs, load pickups, and generation dispatch. International Journal of Electrical Power & Energy Systems (2015) (forthcoming)
13. Farivar, M., Clarke, C., Low, S., Chandy, K.: Inverter var control for distribution systems with renewables. In: 2011 IEEE International Conference on Smart Grid Communications (SmartGridComm), pp. 457–462, October 2011
14. Fisher, E., O'Neill, R., Ferris, M.: Optimal transmission switching. IEEE Transactions on Power Systems 23(3), 1346–1355 (2008)
15. Fourer, R., Gay, D.M., Kernighan, B.: AMPL: a mathematical programming language. In: Wallace, S.W. (ed.) Algorithms and Model Formulations in Mathematical Programming, pp. 150–151. Springer-Verlag New York Inc., New York (1989)
16. Fu, Y., Shahidehpour, M., Li, Z.: Security-constrained unit commitment with ac constraints*. IEEE Transactions on Power Systems 20(3), 1538–1550 (2005)
17. Hedman, K., Ferris, M., O'Neill, R., Fisher, E., Oren, S.: Co-optimization of generation unit commitment and transmission switching with n-1 reliability. In: 2010 IEEE Power and Energy Society General Meeting, pp. 1–1, July 2010
18. Hijazi, H., Coffrin, C., Van Hentenryck, P.: Convex quadratic relaxations of mixed-integer nonlinear programs in power systems (2013). http://www.optimization-online.org/DB_HTML/2013/09/4057.html

19. Hijazi, H., Thiebaux, S.: Optimal ac distribution systems reconfiguration. In: Proceedings of the 18th Power Systems Computation Conference (PSCC 2014), Wroclaw, Poland (2014)
20. Jabr, R.: Radial distribution load flow using conic programming. IEEE Transactions on Power Systems **21**(3), 1458–1459 (2006)
21. Jabr, R.: Optimization of ac transmission system planning. IEEE Transactions on Power Systems **28**(3), 2779–2787 (2013)
22. Kundur, P.: Power System Stability and Control. McGraw-Hill Professional (1994)
23. Lavaei, J., Low, S.: Zero duality gap in optimal power flow problem. IEEE Transactions on Power Systems **27**(1), 92–107 (2012)
24. Lehmann, K., Grastien, A., Van Hentenryck, P.: AC-Feasibility on Tree Networks is NP-Hard. IEEE Transactions on Power Systems (2015) (to appear)
25. Liberti, L.: Writing global optimization software. In: Liberti, L., Maculan, N. (eds.) Global Optimization, Nonconvex Optimization and Its Applications, vol. 84, pp. 211–262. Springer, US (2006). http://dx.doi.org/10.1007/0-387-30528-9_8
26. Madani, R., Ashraphijuo, M., Lavaei, J.: Promises of conic relaxation for contingency-constrained optimal power flow problem (2014). http://www.ee.columbia.edu/lavaei/SCOPF_2014.pdf (accessed February 22, 2015)
27. McCormick, G.: Computability of global solutions to factorable nonconvex programs: Part i convex underestimating problems. Mathematical Programming **10**, 146–175 (1976)
28. Miller, J.: Power system optimization smart grid, demand dispatch, and microgrids, September 2011. http://www.netl.doe.gov/smartgrid/referenceshelf/presentations/SE%20Dist%20Apparatus%20School_Final_082911_rev2.pdf (accessed April 22, 2012)
29. Momoh, J., Adapa, R., El-Hawary, M.: A review of selected optimal power flow literature to 1993. i. nonlinear and quadratic programming approaches. IEEE Transactions on Power Systems **14**(1), 96–104 (1999)
30. Momoh, J., El-Hawary, M., Adapa, R.: A review of selected optimal power flow literature to 1993. ii. newton, linear programming and interior point methods. IEEE Transactions on Power Systems **14**(1), 105–111 (1999)
31. Montanari, U.: Networks of Constraints : Fundamental Properties and Applications to Picture Processing. Information Science **7**(2), 95–132 (1974)
32. Ott, A.: Unit commitment in the pjm day-ahead and real-time markets, June 2010. http://www.ferc.gov/eventcalendar/Files/20100601131610-Ott,%20PJM.pdf (accessed April 22, 2012)
33. Purchala, K., Meeus, L., Van Dommelen, D., Belmans, R.: Usefulness of DC power flow for active power flow analysis. In: Power Engineering Society General Meeting, pp. 454–459 (2005)
34. Ryoo, H., Sahinidis, N.: A branch-and-reduce approach to global optimization. Journal of Global Optimization **8**(2), 107–138 (1996)
35. Sahinidis, N.: Global optimization and constraint satisfaction: the branch-and-reduce approach. In: Bliek, C., Jermann, C., Neumaier, A. (eds.) Global Optimization and Constraint Satisfaction. LNCS, vol. 2861, pp. 1–16. Springer, Heidelberg (2003)
36. Toh, K.C., Todd, M., Ttnc, R.H.: Sdpt3 - a matlab software package for semidefinite programming. Optimization Methods and Software **11**, 545–581 (1999)
37. Van Hentenryck, P., McAllister, D., Kapur, D.: Solving Polynomial Systems Using a Branch and Prune Approach. SIAM Journal on Numerical Analysis **34**(2) (1997)

38. Van Hentenryck, P., Michel, L., Deville, Y.: Numerica: a Modeling Language for Global Optimization. The MIT Press, Cambridge (1997)
39. Van Hentenryck, P., Saraswat, V., Deville, Y.: The design, implementation, and evaluation of the constraint language cc(FD). In: Podelski, A. (ed.) Constraint Programming: Basics and Trends. LNCS, vol. 910, pp. 293–316. Springer, Heidelberg (1995)
40. Verma, A.: Power grid security analysis: An optimization approach. Ph.D. thesis, Columbia University (2009)
41. Wächter, A., Biegler, L.T.: On the implementation of a primal-dual interior point filter line search algorithm for large-scale nonlinear programming. Mathematical Programming 106(1), 25–57 (2006)

Broken Triangles Revisited

Martin C. Cooper[1]([✉]), Aymeric Duchein[1], and Guillaume Escamocher[2]

[1] IRIT, University of Toulouse III, 31062 Toulouse, France
{cooper,Aymeric.Duchein}@irit.fr
[2] INSIGHT Centre for Data Analytics, University College Cork, Cork, Ireland
guillaume.escamocher@insight-centre.org

Abstract. A broken triangle is a pattern of (in)compatibilities between assignments in a binary CSP (constraint satisfaction problem). In the absence of certain broken triangles, satisfiability-preserving domain reductions are possible via merging of domain values. We investigate the possibility of maximising the number of domain reduction operations by the choice of the order in which they are applied, as well as their interaction with arc consistency operations. It turns out that it is NP-hard to choose the best order.

1 Introduction

The notion of broken triangle has generated a certain amount of interest in the constraints community: it has led to the definition of novel tractable classes [3,7], variable elimination rules [1] and domain reduction rules [4,5]. The merging of pairs of values in the same variable domain which do not belong to a broken triangle has been shown to lead to considerable reduction of search space size for certain benchmark instances of binary CSP [4]. The corresponding reduction operation, known as BTP-merging, is satisfiability-preserving and is therefore worthy of a deeper theoretical analysis as a potentially useful preprocessing operation. An obvious question is whether the order in which BTP-merging operations, and other domain-reduction operations such as arc consistency, are performed has an effect on the number of possible merges.

Definition 1. *A binary CSP instance I consists of*

- *a set X of n variables,*
- *a domain $\mathcal{D}(x)$ of possible values for each variable $x \in X$, with d the maximum domain size,*
- *a relation $R_{xy} \subseteq \mathcal{D}(x) \times \mathcal{D}(y)$, for each pair of distinct variables $x, y \in X$, which consists of the set of compatible pairs of values (a, b) for variables (x, y).*

A partial solution to I on $Y = \{y_1, \ldots, y_r\} \subseteq X$ is a set $\{\langle y_1, a_1 \rangle, \ldots, \langle y_r, a_r \rangle\}$ such that $\forall i, j \in [1, r]$, $(a_i, a_j) \in R_{y_i y_j}$. A solution to I is a partial solution on X.

M.C. Cooper—supported by ANR Project ANR-10-BLAN-0210 and EPSRC grant EP/L021226/1.

© Springer International Publishing Switzerland 2015
G. Pesant (Ed.): CP 2015, LNCS 9255, pp. 58–73, 2015.
DOI: 10.1007/978-3-319-23219-5_5

For simplicity of presentation, Definition 1 assumes that there is exactly one constraint relation for each pair of variables. An instance I is *arc consistent* if for each pair of distinct variables $x, y \in X$, for each value $a \in \mathcal{D}(x)$, there is a value $b \in \mathcal{D}(y)$ such that $(a, b) \in R_{xy}$.

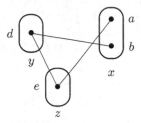

Fig. 1. A broken triangle on two values $a, b \in \mathcal{D}(x)$.

We now formally define the value-merging operation based on absence of certain broken triangles. A broken triangle on values a, b is shown in Figure 1. In all figures in this paper, the pairs of values joined by a solid line are exactly those belonging to the corresponding constraint relation.

Definition 2. *A broken triangle* on the pair of variable-value assignments $a, b \in \mathcal{D}(x)$ *consists of a pair of assignments* $d \in \mathcal{D}(y)$, $e \in \mathcal{D}(z)$ *to distinct variables* $y, z \in X \setminus \{x\}$ *such that* $(a, d) \notin R_{xy}$, $(b, d) \in R_{xy}$, $(a, e) \in R_{xz}$, $(b, e) \notin R_{xz}$ *and* $(d, e) \in R_{yz}$. *The pair of values* $a, b \in \mathcal{D}(x)$ *is BT-free if there is no broken triangle on* a, b.

BTP-merging values $a, b \in \mathcal{D}(x)$ *in a binary CSP consists in replacing* a, b *in* $\mathcal{D}(x)$ *by a new value* c *which is compatible with all variable-value assignments compatible with at least one of the assignments* $\langle x, a \rangle$ *or* $\langle x, b \rangle$ *(i.e.* $\forall y \in X \setminus \{x\}$, $\forall d \in \mathcal{D}(y)$, $(c, d) \in R_{xy}$ *iff* $(a, d) \in R_{xy}$ *or* $(b, d) \in R_{xy}$*)*

When $a, b \in \mathcal{D}(x)$ are BT-free in a binary CSP instance I, the instance I' obtained from I by merging $a, b \in \mathcal{D}(x)$ is satisfiable if and only if I is satisfiable. Furthermore, given a solution to the instance resulting from the merging of two values, we can find a solution to the original instance in linear time [4].

The paper is structured as follows. In Section 2 we investigate the interaction between arc consistency and BTP-merging. In Section 3 we show that finding the best order in which to apply BTP-mergings is NP-hard, even for arc-consistent instances. In Section 4 we prove that this remains true even if we only perform merges at a single variable. In Section 5 we take this line of work one step further by showing that it is also NP-hard to find the best sequence of merges by a weaker property combing virtual interchangeability and neighbourhood substitutability.

2   Mixing Arc Consistency and BTP-merging

BTP-merging can be seen as a generalisation of neighbourhood substitutability [6], since if $a \in \mathcal{D}(x)$ is neighbourhood substitutable for $b \in \mathcal{D}(x)$ then a, b

can be BTP-merged. The possible interactions between arc consistency (AC) and neighbourhood substitution (NS) are relatively simple and can be summarised as follows:

1. The fact that $a \in \mathcal{D}(x)$ is AC-supported or not at variable y remains invariant after the elimination of any other value b (in $\mathcal{D}(x) \setminus \{a\}$ or in the domain $\mathcal{D}(z)$ of any variable $z \neq x$) by neighbourhood substitution.
2. An arc-consistent value $a \in \mathcal{D}(x)$ that is neighbourhood substitutable remains neighbourhood substitutable after the elimination of any other value by arc consistency.
3. On the other hand, a value $a \in \mathcal{D}(x)$ may become neighbourhood substitutable after the elimination of a value $c \in \mathcal{D}(y)$ ($y \neq x$) by arc consistency.

Indeed, it has been shown that the maximum cumulated number of eliminations by arc consistency and neighbourhood substitution can be achieved by first establishing arc consistency and then applying any convergent sequence of NS eliminations (i.e. any valid sequence of eliminations by neighbourhood substitution until no more NS eliminations are possible) [2].

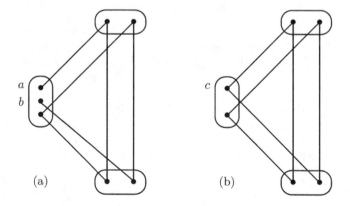

Fig. 2. (a) An instance in which applying AC leads to the elimination of all values (starting with the values a and b), but applying BTP merging leads to just one elimination, namely the merging of a with b (with the resulting instance shown in (b)).

The interaction between arc consistency and BTP-merging is not so simple and can be summarised as follows:

1. The fact that $a \in \mathcal{D}(x)$ is AC-supported or not at variable y remains invariant after the BTP-merging of any other pair of other values b, c (in $\mathcal{D}(x) \setminus \{a\}$ or in the domain $\mathcal{D}(z)$ of any variable $z \neq x$). However, after the BTP-merging of two arc-inconsistent values the resulting merged value may be arc consistent. An example is given in Figure 2(a). In this 3-variable instance, the two values $a, b \in \mathcal{D}(x)$ can be eliminated by arc consistency (which in turn leads to

the elimination of all values), or alternatively they can be BTP-merged (to produce the value c) resulting in the instance shown in Figure 2(b) in which no more eliminations are possible by AC or BTP-merging.

2. A single elimination by AC may prevent a sequence of several BTP-mergings. An example is given in Figure 3(a). In this 4-variable instance, if the value b is eliminated by AC, then no other eliminations are possible by AC or BTP-merging in the resulting instance (shown in Figure 3(b)), whereas if a and b are BTP-merged into a new value d (as shown in Figure 3(c)) this destroys a broken triangle thus allowing c to be BTP-merged with d (as shown in Figure 3(d)).

3. On the other hand, two values in the domain of a variable x may become BTP-mergeable after an elimination of a value $c \in \mathcal{D}(y)$ $(y \neq x)$ by arc consistency.

3   The Order of BTP-mergings

It is known that BTP-merging can both create and destroy broken triangles [4]. This implies that the choice of the order in which BTP-mergings are applied may affect the total number of merges that can be performed. Unfortunately, maximising the total number of merges in a binary CSP instance turns out to be NP-hard, even when bounding the maximum size of the domains d by a constant as small as 3. For simplicity of presentation, we first prove this for the case in which the instance is not necessarily arc consistent. We will then prove a tighter version, namely NP-hardness of maximising the total number of merges even in arc-consistent instances.

Theorem 1. *The problem of determining if it is possible to perform k BTP-mergings in a boolean binary CSP instance is NP-complete.*

Proof. For a given sequence of k BTP-mergings, verifying if this sequence is correct can be performed in $\mathcal{O}(kn^2d^2)$ time because looking for broken triangles for a given couple of values takes $\mathcal{O}(n^2d^2)$ [4]. As we can verify a solution in polynomial time, the problem of determining if it is possible to perform k BTP-mergings in a binary CSP instance is in NP. So to complete the proof of NP-completeness it suffices to give a polynomial-time reduction from the well-known 3-SAT problem. Let I_{3SAT} be an instance of 3-SAT (SAT in which each clause contains exactly 3 literals) with variables $X_1, \ldots, X_N$ and clauses $C_1, \ldots, C_M$. We will create a boolean binary CSP instance I_{CSP} which has a sequence of $k = 3 \times M$ mergings if and only if I_{3SAT} is satisfiable.

For each variable X_i of I_{3SAT}, we add a new variable z_i to I_{CSP}. For each occurrence of X_i in the clause C_j of I_{3SAT}, we add two more variables x_{ij} and y_{ij} to I_{CSP}. Each $\mathcal{D}(z_i)$ contains only one value c_i and each $\mathcal{D}(x_{ij})$ (resp. $\mathcal{D}(y_{ij})$) contains only two values a_i and b_i (resp. a_i' and b_i'). The roles of variables x_{ij} and y_{ij} are the following:

$$X_i = true \Leftrightarrow \forall j,\ a_i, b_i \ can\ be\ merged\ in\ \mathcal{D}(x_{ij}) \tag{1}$$

$$X_i = false \Leftrightarrow \forall j,\ a_i', b_i' \ can\ be\ merged\ in\ \mathcal{D}(y_{ij}) \tag{2}$$

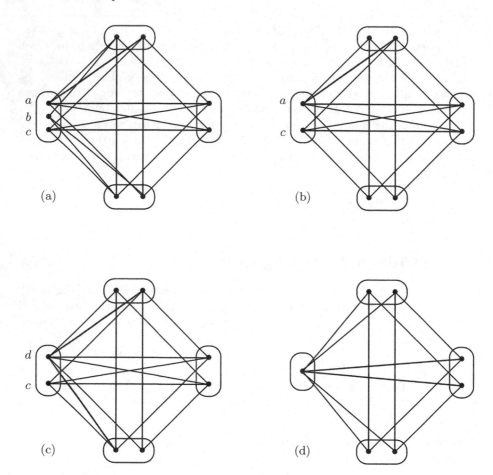

Fig. 3. (a) An instance in which applying AC leads to one elimination (the value b) (as shown in (b)), but applying BTP merging leads to two eliminations, namely a with b (shown in (c)) and then d with c (shown in (d)).

In order to prevent the possibility of merging both (a_i, b_i) and (a_i', b_i'), we define the following constraints for z_i, x_{ij} and y_{ij}: $\forall j$ $R_{x_{ij}z_i} = \{(b_i, c_i)\}$ and $R_{y_{ij}z_i} = \{(b_i', c_i)\}$; $\forall j$ $\forall$ k $R_{x_{ij}y_{ik}} = \{(a_i, a_i')\}$. These constraints are shown in Figure 4(a) for a single j (where a pair of points not joined by a solid line are incompatible). By this gadget, we create a broken triangle on each y_{ij} when merging values in the x_{ij} and vice versa.

Then for each clause $C_i = (X_j, X_k, X_l)$, we add the following constraints in order to have at least one of the literals X_j, X_k, X_l true: $R_{y_{ji}y_{ki}} = \{(a_j', b_k')\}$, $R_{y_{ki}y_{li}} = \{(a_k', b_l')\}$ and $R_{y_{li}y_{ji}} = \{(a_l', b_j')\}$. This construction, shown in Figure 4(b), is such that it allows two mergings on the variables y_{ji}, y_{ki}, y_{li} before a broken triangle is created. For example, merging a_j', b_j' and then a_k', b_k' creates a broken triangle on a_i', b_i'. So a third merging is not possible.

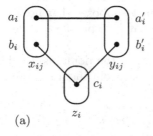 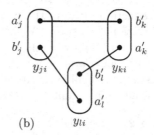

(a)   (b)

Fig. 4. (a) Representation of the variable X_i and its negation (by the possibility of performing a merge in $\mathcal{D}(x_{ij})$ or $\mathcal{D}(y_{ij})$, respectively, according to rules (1),(2)). (b) Representation of the clause $(X_j \vee X_k \vee X_l)$. Pairs of points joined by a solid line are compatible and incompatible otherwise.

If the clause C_i contains a negated literal $\overline{X_j}$ instead of X_j, it suffices to replace y_{ji} by x_{ji}. Indeed, Figure 5 shows the construction for the clause $(\overline{X_j} \vee X_k \vee \overline{X_l})$ together with the gadgets for each variable. The maximum number of mergings that can be performed are one per occurrence of each variable in a clause, which is exactly $3 \times M$. Given a sequence of $3 \times M$ mergings in the CSP instance, there is a corresponding solution to I_{3SAT} given by (1) and (2). The above reduction allows us to code I_{3SAT} as the problem of testing the existence of a sequence of $k = 3 \times M$ mergings in the corresponding instance I_{CSP}. This reduction being polynomial, we have proved the NP-completeness of the problem of determining whether k BTP merges are possible in a boolean binary CSP instance.

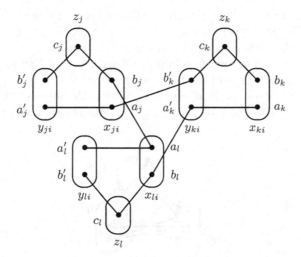

Fig. 5. Gadget representing the clause $(\overline{X_j} \vee X_k \vee \overline{X_l})$.

The reduction given in the proof of Theorem 1 supposes that no arc consistency operations are used. We will now show that it is possible to modify the reduction so as to prevent the elimination of any values in the instance I_{CSP} by arc consistency, even when the maximum size of the domains d is bounded by a constant as small as 3. Recall that an arc-consistent instance remains arc-consistent after any number of BTP-mergings.

Theorem 2. *The problem of determining if it is possible to perform k BTP-mergings in an arc-consistent binary CSP instance is NP-complete, even when only considering binary CSP instances where the size of the domains is bounded by 3.*

Proof. In order to ensure arc consistency of the instance I_{CSP}, we add a new value d_i to the domain of each of the variables x_{ij}, y_{ij}, z_i. However, we cannot simply make d_i compatible with all values in all other domains, because this would allow all values to be merged with d_i, destroying in the process the semantics of the reduction.

In the three binary constraints concerning the triple of variables x_{ij}, y_{ij}, z_i, we make d_i compatible with all values in the other two domains *except* d_i. In other words, we add the following tuples to constraint relations, as illustrated in Figure 6:

- $\forall i \forall j,\ (a_i, d_i),\ (b_i, d_i),\ (d_i, c_i) \in R_{x_{ij}z_i}$
- $\forall i \forall j,\ (a'_i, d_i),\ (b'_i, d_i),\ (d_i, c_i) \in R_{y_{ij}z_i}$
- $\forall i \forall j,\ (a_i, d_i),\ (b_i, d_i),\ (d_i, a'_i),\ (d_i, b'_i) \in R_{x_{ij}y_{ij}}$

This ensures arc consistency, without creating new broken triangles on a_i, b_i or a'_i, b'_i, while at the same time preventing BTP-merging with the new value d_i. It is important to note that even after BTP-merging of one of the pairs a_i, b_i or a'_i, b'_i, no BTP-merging is possible with d_i in $\mathcal{D}(x_{ij})$, $\mathcal{D}(y_{ij})$ or $\mathcal{D}(z_i)$ due to the presence of broken triangles on this triple of variables. For example, the pair of values $a_i, d_i \in \mathcal{D}(x_{ij})$ belongs to a broken triangle on $c_i \in \mathcal{D}(z_i)$ and $d_i \in \mathcal{D}(y_{ij})$, and this broken triangle still exists if the values $a'_i, b'_i \in \mathcal{D}(y_{ij})$ are merged. We can then simply make d_i compatible with all values in the domain of all variables outside this triple of variables. This ensures arc consistency, and does

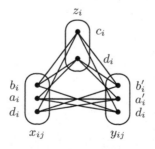

Fig. 6. Ensuring arc consistency between the variables z_i, y_{ij}, x_{ij} by addition of new values d_i.

not introduce any broken triangles on a_i, b_i or a_i', b_i'. With these constraints we ensure arc consistency without changing any of the properties of I_{CSP} used in the reduction from 3-SAT described in the proof of Theorem 1. For each pair of values $a_i, b_i \in \mathcal{D}(x_{ij})$ and $a_i', b_i' \in \mathcal{D}(y_{ij})$, no new broken triangle is created since these two values always have the same compatibility with all the new values d_k. As we have seen, the constraints shown in Figure 6 prevent any merging of the new values d_k.

Corollary 1. *The problem of determining if it is possible to perform k value eliminations by arc consistency and BTP-merging in a binary CSP instance is NP-complete, even when only considering binary CSP instances where the size of the domains is bounded by 3.*

From a practical point of view, an obvious question concerns the existence of a non-optimal but nevertheless useful heuristic for choosing the order of BTP-merges. Imagine that a heuristic concerning the order in which to apply BTP-merges involves first finding all possible merges before choosing between them. If this is the case, then once a merge has been chosen and performed, the list of possible merges has to be recalculated. This process is thus already an order of magnitude slower than the simple technique (applied in the experiments in [4]) consisting of performing a merge as soon as it is detected.

4 Optimal Sequence of BTP-mergings at a Single Variable

We now show that even when only considering a single domain, finding the optimal order of BTP-mergings is NP-Complete. For simplicity of presentation, we first prove this for the case in which the instance is not necessarily arc-consistent. We then prove a tighter version for arc-consistent instances.

Theorem 3. *The problem of determining if it is possible to perform k BTP-mergings within a same domain in a binary CSP instance is NP-Complete.*

Proof. For a given sequence of k BTP-mergings within a domain of a binary CSP instance I, verifying if this sequence is correct can be performed in $\mathcal{O}(kN^2d^2)$ time, with N being the number of variables in I and d being the size of the largest domain in I, because looking for broken triangles for a given couple of values takes $\mathcal{O}(N^2d^2)$. As we can verify a solution in polynomial time, the problem of determining if it is possible to perform k BTP-mergings within a single domain in a binary CSP instance is in NP. So to complete the proof of NP-Completeness, it suffices to give a polynomial-time reduction from the well-known SAT problem.

Let I_{SAT} be a SAT instance with n variables $X_1, X_2, \ldots, X_n$ and m clauses $C_1, C_2, \ldots, C_m$. We reduce I_{SAT} to a binary CSP instance I_{CSP} containing a variable v_0 such that I_{SAT} is satisfiable if and only if we can make $k = n + m$ BTP-mergings within $\mathcal{D}(v_0)$. I_{CSP} is defined in the following way:

1. I_{CSP} contains $1+n(2+4m+9(n-1))$ variables $v_0, v_1, v_2, \ldots, v_{n(2+4m+9(n-1))}$.

2. $\mathcal{D}(v_0)$ contains $3 \times n + m$ values with the following names: $x_1, x_2, \ldots, x_n, x_1 T,$
   $x_2 T, \ldots, x_n T, x_1 F, x_2 F, \ldots, x_n F, c_1, c_2, \ldots, c_m$. All other domains in I_{CSP}
   only contain one value.
3. $\forall i \in [1, n]$, $x_i T$ and $x_i F$ can never be BTP-merged. The idea here is to allow
   exactly one BTP-merging among the three values x_i, $x_i T$ and $x_i F$: x_i and
   $x_i T$ if X_i is assigned True in the SAT instance I_{SAT}, x_i and $x_i F$ if X_i is
   assigned False instead.
4. $\forall (i, j) \in [1, n] \times [1, m]$ such that C_j does not contain X_i (respectively $\overline{X_i}$),
   c_j can never be BTP-merged with $x_i T$ (respectively $x_i F$).
5. $\forall (i, j) \in [1, n] \times [1, m]$ such that C_j contains X_i (respectively $\overline{X_i}$), c_j can only
   be BTP-merged with $x_i T$ (respectively $x_i F$) if either c_j or $x_i T$ (respectively
   $x_i F$) has been previously BTP-merged with x_i.
6. $\forall i, j$ with $1 \leq i < j \leq n$, the nine following couples can never be BTP-
   merged: x_i and x_j, x_i and $x_j T$, x_i and $x_j F$, $x_i T$ and x_j, $x_i T$ and $x_j T$, $x_i T$
   and $x_j F$, $x_i F$ and x_j, $x_i F$ and $x_j T$, $x_i F$ and $x_j F$. The idea here is to prevent
   any BTP-merging between two CSP values corresponding to two different
   SAT variables.

When we say that two values a and b can never be BTP-merged, it means that
we add two variables v_a and v_b, with only one value a' in $\mathcal{D}(v_a)$, and only one
value b' in $\mathcal{D}(v_b)$, such that a is compatible with a' and incompatible with b', b
is compatible with b' and incompatible with a', a' is compatible with b' and all
other edges containing either a' or b' are incompatible. The purpose of making
a' (respectively b') incompatible will all values in the instance except a and b'
(respectively b and a') is twofold. First, it ensures that no future BTP-merging
can establish a compatibility between a' (respectively b') and b (respectively a)
and thus destroy the broken triangle. Second, it ensures that the only broken
triangle introduced by a' and b' is on a and b, so that the addition of a' and b'
does not prevent any other BTP-merging than the one between a and b.

   When we say that two values a and b can only be BTP-merged if either a
or b has been previously BTP-merged with some third value c, it means that
we add two variables v_a and v_b, with only one value a' in $\mathcal{D}(v_a)$, and only one
value b' in $\mathcal{D}(v_b)$, such that a is compatible with a' and incompatible with b',
b is compatible with b' and incompatible with a', c is compatible with both a'
and b', a' is compatible with b' and all other edges containing either a' or b' are
incompatible. Here again, the purpose of making a' (respectively b') incompatible
will all values in the instance except a, b' and c (respectively b, a' and c) is
twofold. First, it ensures that no future BTP-merging that does not include c
can establish a compatibility between a' (respectively b') and b (respectively a)
and thus destroy the broken triangle. Second, it ensures that the only broken
triangle introduced by a' and b' is on a and b, so that the addition of a' and b'
does not prevent any other BTP-merging than the one between a and b.

   For every couple of values that can never be BTP-merged, and for every
couple of values that can only be BTP-merged when one of them has been
previously BTP-merged with some third value, we add two new single-valued
variables to I_{CSP}. Therefore, the third point in the definition of I_{CSP} adds $2n$

variables to I_{CSP}, the fourth and fifth points in the definition of I_{CSP} add $4nm$ variables to I_{CSP} and the sixth point in the definition of I_{CSP} adds $9n(n-1)$ variables to I_{CSP}. Therefore, the total number of single-valued variables added to I_{CSP} is $n(2 + 4m + 9(n-1))$, as expected from the first point in the definition of I_{CSP}.

- The number of BTP-mergings is limited by $n + m$:
  From the third point in the definition of I_{CSP}, for all $i \in [1, n]$, we can BTP-merge at most once within the triple $\{x_i, x_iT, x_iF\}$. From the sixth point in the definition of I_{CSP}, we cannot BTP-merge two values within $\mathcal{D}(v_0)$ if they are associated to two different SAT variables X_i and X_j. Therefore, we have at most m BTP-mergings remaining, one for each c_j for $1 \leq j \leq m$.
- If we can BTP-merge $n + m$ times, then we have a solution for I_{SAT}:
  Since we have done the maximum number of BTP-mergings, we know that for all $i \in [1, n]$, x_i has been BTP-merged with either x_iT or x_iF, but not both. So we create the following solution for I_{SAT}: $\forall i \in [1, n]$, we assign $True$ to X_i if x_i and x_iT have been BTP-merged, and $False$ otherwise. From the fourth and fifth points in the definition of I_{CSP}, we know that for each j in $[1, m]$, C_j is satisfied by the literal associated with the value C_j has been BTP-merged with.
- If we have a solution for I_{SAT}, then we can BTP-merge $n + m$ times:
  $\forall i \in [1, n]$, we BTP-merge x_i with x_iT if $True$ has been assigned to X_i, with x_iF otherwise. $\forall (i, j) \in [1, n] \times [1, m]$, we BTP-merge c_j and the value that x_i has been BTP-merged with if X_i is satisfied in C_j and c_j has not been BTP-merged yet. From the fifth point in the definition of I_{CSP}, we know we can BTP-merge each c_j once.

Therefore I_{SAT} is satisfiable if and only if we can perform $k = n + m$ BTP-mergings within $\mathcal{D}(v_0)$, and we have the result.

We now generalise the result of Theorem 3 to arc-consistent binary CSP instances.

Theorem 4. *The problem of determining if it is possible to perform k BTP-mergings within a same domain in an arc-consistent binary CSP instance is NP-Complete.*

Proof. We transform the binary CSP instance I_{CSP} from the proof of Theorem 3 into an arc-consistent binary CSP instance I'_{CSP}. To do so, we add a new value d_i in $\mathcal{D}(v_i)$ for $1 \leq i \leq n(2 + 4m + 9(n-1))$ such that all d_i are incompatible with each other and compatible with all other points in I'_{CSP}. This ensures arc consistency. It remains to show that:

1. For any couple of values $(a, b) \in \mathcal{D}(v_0)$, adding the values d_i does not create the broken triangle on a and b, even if a or b is the result of a previous BTP-merging:
   Suppose that we have two values $a, b \in \mathcal{D}(v_0)$ such that adding the values d_i

creates a broken triangle on a and b. Let $a' \in \mathcal{D}(v_a)$ and $b' \in \mathcal{D}(v_b)$ be the other two values forming the broken triangle. Since it was the new values d_i that created this particular broken triangle, either a' or b' is one of the d_i. Without loss of generality, we assume that a' is one of the d_i. But since the d_i are compatible with all values from $\mathcal{D}(v_0)$, both a and b are compatible with a', even if a or b is the result of a previous BTP-merging. Therefore, there cannot be any broken triangle on a and b caused by the new values d_i.

2. For all $i \in [1, n(2 + 4m + 9(n - 1))]$, it is never possible to BTP-merge the two values in $\mathcal{D}(v_i)$:

We assume, for simplicity of presentation and without loss of generality, that the SAT instance I_{SAT} has more than one variable and that no clause contains both a variable and its negation. Let $i \in [1, n(2 + 4m + 9(n - 1))]$. Let a and d_i be the two points in $\mathcal{D}(v_i)$. From the proof of Theorem 3, we know that a is compatible with only one value from $\mathcal{D}(v_0)$. Let b be this value. If b is associated with one of the SAT variables from I_{SAT}, then from the sixth point in the definition of I_{CSP} in the proof of Theorem 3 we know that there is at least one value $c \in \mathcal{D}(v_0)$ that can never be BTP-merged with b, and therefore will always be incompatible with a. If on the other hand c is associated with one of the SAT clauses from I_{SAT}, then from the fourth point in the definition of I_{CSP} in the proof of Theorem 3 we know that there is at least one value $c \in \mathcal{D}(v_0)$ that can never be BTP-merged with b, and therefore will always be incompatible with a. Therefore, we have a value $c \in \mathcal{D}(v_0)$ that is always incompatible with a, even if c is the result of a previous BTP-merging. Let $j \in [1, n(2 + 4m + 9(n - 1))]$, such that $j \neq i$. Since the d_i are incompatible with each other, and compatible with all other values in I'_{CSP}, then d_j is compatible with both a and c, and d_i is compatible with c and incompatible with d_j. Therefore we have a broken triangle on a and d_i that can never be destroyed. Therefore a and d_i can never be BTP-merged and we have the result.

One motivation for studying the single-variable version of the problem was that if all values in $\mathcal{D}(x)$ can be BTP-merged, then the variable x can be eliminated since its domain becomes a singleton. Our proof of NP-hardness in the single-variable case relied on a large domain which was not actually reduced to a singleton. There remains therefore an interesting open question concerning the complexity of eliminating the largest number of variables by sequences of BTP-merging operations.

5   Virtual Interchangeability and Neighbourhood Substitution

Since testing whether two values $a, b \in \mathcal{D}(x)$ are BTP-mergeable requires testing all pairs of assignments to all pairs of distinct variables $y, z \neq x$, it is natural to investigate weaker versions which are less costly to test. Two such weaker versions are neighbourhood substitutability [6] and virtual interchangeability [8].

Given two values $a, b \in \mathcal{D}(x)$, a is neighbourhood substitutable by b (NS), and so can be merged with b, if for all variables $y \neq x$, $\forall c \in \mathcal{D}(y)$, $(a, c) \in R_{xy} \Rightarrow (b, c) \in R_{xy}$. Two values $a, b \in \mathcal{D}(x)$ are virtual interchangeable (VI), and so can be merged, if for all variables $y \neq x$ *except at most one*, $\forall c \in \mathcal{D}(y)$, $(a, c) \in R_{xy} \Leftrightarrow (b, c) \in R_{xy}$. Applying both VI and neighbourhood substitution (NS) operations until convergence provides a weaker and less time-costly alternative to applying BTP-merging operations until convergence. An interesting question is therefore whether it is possible to find in polynomial time an optimal (i.e. longest) sequence of VI and NS operations. The following theorem shows that this problem is in fact also NP-hard.

Theorem 5. *Determining whether there exists a sequence of VI and NS operations of length k that can be applied to a binary CSP instance is NP-complete.*

Proof. Since checking the validity of a sequence of VI and NS operations can be achieved in polynomial time, the problem is in NP. To complete the proof we demonstrate a polynomial reduction from 3SAT. Let I_{3SAT} be an instance of 3SAT. We will show how to construct a binary CSP instance I_{CSP} and a value k so that it is possible to perform k merges by VI and NS if and only if I_{3SAT} is satisfiable.

For each variable x in I_{3SAT}, we introduce Boolean variables called x and $\bar{x}$ in I_{CSP}: variable x in I_{3SAT} is assigned true (respectively, false) if and only if the two domain values in $\mathcal{D}(x)$ (respectively, $\mathcal{D}(\bar{x})$) are merged in the corresponding sequence of merges in I_{CSP}. Figure 7(a) shows the gadget for choosing the truth value for x. The variables x and $\bar{x}$ are both connected to another variable, not shown in this figure: this prevents the values in their domains being merged before the values in the domains of t or t' are merged. Initially, the only merges (by VI and NS) which can occur are in the domain of variable s: we can either merge values 1 and 2 or values 2 and 3. Once one of these merges is performed, the other it not possible. Figure 7 shows the propagation of merges which occurs in the second of these cases. Figure 7(b) shows the result of this merge in $\mathcal{D}(s)$. In Figure 7(b) the only merge that is possible is the merging of values 1 and 3 (by NS) in $\mathcal{D}(t)$. The result is shown in Figure 7(c). Now, the two values in the domain of x can be merged (by VI) since x is constrained by a single other variable (not shown in the figure). It is important to note that no other merges are possible. By a similar chain of propagations, if we had chosen to merge 1 and 2 in $\mathcal{D}(s)$, then the values 1 and 3 in $\mathcal{D}(t')$ would have been merged, and finally the two values in $\mathcal{D}(\bar{x})$. This gadget therefore allows us to choose a truth value for the corresponding variable x of I_{3SAT}.

In order to code the instance I_{3SAT}, we need to be able to have several copies of each variable. This is achieved by the gadget shown in Figure 8(a). The variables x_1 and x_2 are each assumed to be constrained by a single other variable not shown in the figure. The variable x is the variable in the gadget of Figure 7. If the values in $\mathcal{D}(x)$ are merged then this allows the merging (by VI) of the pair of values $0, 1 \in \mathcal{D}(u)$ and the merging of the pair of values 2,3. In the resulting instance, the two values in the domain of x_i ($i = 1, 2$) can be merged.

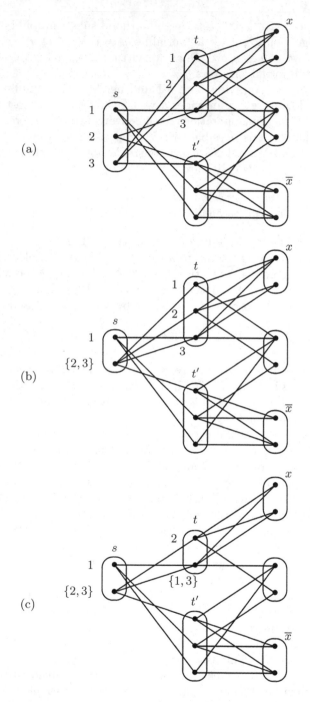

Fig. 7. (a) Gadget for choosing a truth value for x: true if the two values in $\mathcal{D}(x)$ are merged; false if the two values in $\mathcal{D}(\overline{x})$ are merged. This same gagdet (b) after merging the values 2 and 3 in $\mathcal{D}(s)$, then (c) after merging the values 1 and 3 in $\mathcal{D}(t)$.

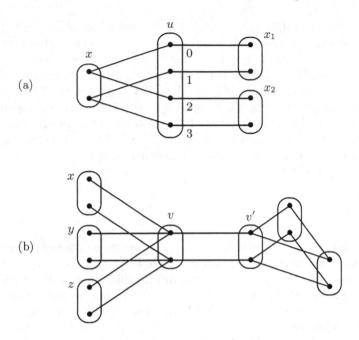

Fig. 8. (a) Gadget for making copies of a variable x: if the two values in $\mathcal{D}(x)$ can be merged, then the two values in $\mathcal{D}(x_1)$ and the two values in $\mathcal{D}(x_2)$ can be merged. (b) Gadget used in the simulation of a clause: the variable v is true (i.e. the two values in its domain can be merged) if and only if x, y, z are all true.

This gadget therefore allows us to make multiple copies of the variable x all with the same truth value.

To complete the proof, we now show how to code each clause c of I_{3SAT}. There are exactly seven assignments to the variables in c which satisfy this clause. For each of these assignments, we add a gadget of the form shown in Figure 8(b). The variables x, y, z are the output of the gadgets introduced above and correspond to the variables x, y, z occurring in the clause c in I_{3SAT}. In the example shown in the figure, the satisfying assignment is $x = y = z = true$. When the two values in each of the domains of these three variables can be merged (and only in this case), the values in the domain of v can also be merged. The triangle of variables to the right of v in Figure 8(b) prevents the merging of the values in $\mathcal{D}(v)$ when only two of the three variables x, y, z are assigned true.

In order to have the same number of copies of x and $\overline{x}$ in our construction, we also add a gadget similar to Figure 8(b) for the one non-satisfying assignment to the three variables of the clause c: in this case, the variable v is constrained by two other variables (as is the variable v' in Figure 8(b)) which prevents the merging of the values in $\mathcal{D}(v)$.

Suppose that there are n variables and m clauses in I_{3SAT}. The maximum total number of merges which can be performed in I_{CSP} is 3 per gadget shown in

Figure 7, 4 per gadget shown in Figure 8(a) and 1 per gadget shown in Figure 8(b) (provided the gadget corresponds to a truth assignment which satisfies the clause c). Each clause c requires four copies of each of the three variables x occurring in c (as well as four copies of $\bar{x}$). For each copy of each literal assigned the value true, there are 4 merges in the gadget of Figure 8(a). For the first occurrence of each variable, produced by the gadget of Figure 7, there is one less merge (3 instead of 4). Finally, for each satisfied clause there is one merge. This implies that we can perform a total of $k = 48m - n + m = 49m - n$ merges in I_{CSP} if and only if I_{3SAT} is satisfiable. Since this reduction is clearly polynomial, this completes the proof.

6   Conclusion

We have investigated the possibility of maximising the number of domain reduction operations in binary CSP instances by choosing an optimal order in which to apply them. Whereas for consistency and neighbourhood-substitution operations, the number of domain reduction operations can be maximised in polynomial time, the problem becomes NP-hard when we allow merging operations, such as virtual interchangeability or BTP-merging. We emphasise that this does not detract from the possible utility of such value-merging operations in practice, which is an independent question.

Different tractable subproblems of binary CSP have been defined based on the absence of certain broken triangles [3,5,7]. Instances can be solved by eliminating variables one by one and, in each case, the recognition of instances in the tractable class can be achieved in polynomial time by a greedy algorithm since the elimination of one variable cannot prevent the elimination of another variable. BTP-merging, on the other hand, performs reduction operations on a lower level than the elimination of variables. Given the NP-completeness results in this paper, recognizing those instances which can be reduced to a trivial problem with only singleton domains by some sequence of BTP-merges is unlikely to be tractable, but this remains an open problem.

References

1. Cohen, D.A., Cooper, M.C.: Guillaume Escamocher and Stanislav Živný, Variable and Value Elimination in Binary Constraint Satisfaction via Forbidden Patterns. J. Comp. Systems Science (2015). http://dx.doi.org/10.1016/j.jcss.2015.02.001
2. Martin, C.: Cooper, Fundamental Properties of Neighbourhood Substitution in Constraint Satisfaction Problems. Artif. Intell. **90**(1–2), 1–24 (1997)
3. Cooper, M.C., Jeavons, P.G., Salamon, A.Z.: Generalizing constraint satisfaction on trees: Hybrid tractability and variable elimination. Artif. Intell. **174**(9–10), 570–584 (2010)
4. Cooper, M.C., El Mouelhi, A., Terrioux, C., Zanuttini, B.: On broken triangles. In: O'Sullivan, B. (ed.) CP 2014. LNCS, vol. 8656, pp. 9–24. Springer, Heidelberg (2014)

5. Cooper, M.C.: Beyond consistency and substitutability. In: O'Sullivan, B. (ed.) CP 2014. LNCS, vol. 8656, pp. 256–271. Springer, Heidelberg (2014)
6. Freuder, E.C.: Eliminating interchangeable values in constraint satisfaction problems. In: Proceedings AAAI 1991, pp. 227–233 (1991)
7. Jégou, P., Terrioux, C.: The extendable-triple property: a new CSP tractable class beyond BTP. In: AAAI (2015)
8. Likitvivatanavong, C., Yap, R.H.C.: Eliminating redundancy in CSPs through merging and subsumption of domain values. ACM SIGAPP Applied Computing Review **13**(2) (2013)

A Microstructure-Based Family of Tractable Classes for CSPs

Martin C. Cooper[1][(⊠)], Philippe Jégou[2], and Cyril Terrioux[2]

[1] IRIT, University of Toulouse III, 31062 Toulouse, France
cooper@irit.fr

[2] Aix-Marseille Université, CNRS, ENSAM, Université de Toulon, LSIS UMR 7296, Avenue Escadrille Normandie-Niemen, 13397 Marseille Cedex 20, France
{philippe.jegou,cyril.terrioux}@lsis.org

Abstract. The study of tractable classes is an important issue in Artificial Intelligence, especially in Constraint Satisfaction Problems. In this context, the Broken Triangle Property (BTP) is a state-of-the-art microstructure-based tractable class which generalizes well-known and previously-defined tractable classes, notably the set of instances whose constraint graph is a tree. In this paper, we propose to extend and to generalize this class using a more general approach based on a parameter k which is a given constant. To this end, we introduce the k-BTP property (and the class of instances satisfying this property) such that we have 2-BTP = BTP, and for $k > 2$, k-BTP is a relaxation of BTP in the sense that k-BTP $\subsetneq (k+1)$-BTP. Moreover, we show that if k-TW is the class of instances having tree-width bounded by a constant k, then k-TW $\subsetneq (k+1)$-BTP. Concerning tractability, we show that instances satisfying k-BTP and which are strong k-consistent are tractable, that is, can be recognized and solved in polynomial time. We also study the relationship between k-BTP and the approach of Naanaa who proposed a set-theoretical tool, known as the directional rank, to extend tractable classes in a parameterized way. Finally we propose an experimental study of 3-BTP which shows the practical interest of this class, particularly w.r.t. the practical solving of instances satisfying 3-BTP and for other instances, w.r.t. to backdoors based on this tractable class.

1 Introduction

Finding islands of tractability, generally called *tractable classes* is an important issue in Artificial Intelligence, especially in Constraint Satisfaction Problems (CSPs [1]). Many studies have addressed this issue, from the very beginnings of Artificial Intelligence. These results are often theoretical in nature with, in certain cases, tractable classes which can be considered as somewhat artificial. But some tractable classes have actually been used in practice, such as the classes defined by constraint networks with bounded tree-width [2,3]. More recently, the concept of hybrid class has been defined, for example with the

Supported by ANR Project ANR-10-BLAN-0210 and EPSRC grant EP/L021226/1.

G. Pesant (Ed.): CP 2015, LNCS 9255, pp. 74–88, 2015.
DOI: 10.1007/978-3-319-23219-5_6

class *BTP* [4]. This class strictly contains both structural tractable classes (such as tree-structured CSPs) and tractable classes defined by language restrictions. One major advantage of this class, in addition to its generalization of already-known tractable classes, is related to its practical interest. Indeed, instances of this class can be solved in polynomial time using algorithms, such as MAC (Maintaining Arc-Consistency [5]) and RFL (Real Full Look-ahead [6]), implemented in efficient solvers which allows it to be used directly in practice. In addition, it may also help to explain theoretically the practical efficiency of solvers, even though the theoretical complexity of the algorithms employed by the solvers is exponential in the worst case.

In this paper, we return to this type of approach by generalizing the tractable class *BTP* which is defined by a property excluding certain patterns (called *Broken Triangles*) in the microstructure graph associated with a binary CSP instance. Very recent work in this same direction introduced the class *ETP* [7] which generalizes *BTP* by relaxing some of its conditions, since it tolerates some broken triangles which are forbidden by *BTP*. Here we propose a broader generalization called *k-BTP* which extends this previous work along two axes. First, in the spirit of *ETP*, the new class allows the presence of a larger number of broken triangles, generalizing strictly *ETP* (and thus *BTP*). Secondly, the class *k-BTP* is parameterized by a constant k, thus providing a generic version, which may prove of theoretical interest for general values of k, although in practice we consider the case $k = 3$ to be of most interest. Thus, while *BTP* is defined for sets of three variables and *ETP* for sets of four variables, *k-BTP* is defined on the basis of sets of $k + 1$ variables where k is a fixed constant. According to this approach, $BTP = 2\text{-}BTP$ while $ETP \subsetneq 3\text{-}BTP$. Thus, this approach makes it possible to strictly generalize these two classes. Furthermore, *k-BTP* retains some of their interesting properties and practical advantages mentioned above. Notably, we show that classical algorithms such as MAC or RFL can solve instances belonging to *k-BTP* in polynomial time, assuming that these instances verify *Strong k-Consistency* [8]. Moreover, we highlight the relationships of this class with known structural and hybrid classes. We show in particular that the class of constraint networks whose tree-width is bounded by k is strictly included in the class *k-BTP*. This result gives a first answer to a question recently asked by M. Vardi about the relationships between *ETP* and the tractable class induced by instances of bounded tree-width. We also highlight a recent but relatively unknown result that was proposed by Naanaa [9] whose relationships with *k-BTP* we investigate.

In Section 2 we recall the definitions of the tractable classes *BTP* and *ETP*. In Section 3 we define the new class *k-BTP* and show that instances from this class can be detected in polynomial time even when the variable order is not known in advance. Furthermore, we show that, under the extra hypothesis of strong k-consistency, such instances can be solved in polynomial time, and, in fact, standard algorithms will solve them. In Section 4 we investigate relationships between *k-BTP* and several known tractable classes and in Section 5 we report results of experimental trials on benchmark problems.

2   Background

Formally, a *constraint satisfaction problem* also called *constraint network* is a triple (X, D, C), where $X = \{x_1, \ldots, x_n\}$ is a set of n variables, $D = (D_{x_1}, \ldots, D_{x_n})$ is a list of finite domains of values, one per variable, and $C = \{c_1, \ldots, c_e\}$ is a finite set of e constraints. Each constraint c_i is a pair $(S(c_i), R(c_i))$, where $S(c_i) = \{x_{i_1}, \ldots, x_{i_k}\} \subseteq X$ is the *scope* of c_i, and $R(c_i) \subseteq D_{x_{i_1}} \times \cdots \times D_{x_{i_k}}$ is its *compatibility relation*. The *arity* of c_i is $|S(c_i)|$. In this paper, we only deal with the case of binary CSPs, that is CSPs for which all the constraints are of arity 2. Hence, we will denote by c_{ij} the constraints involving x_i and x_j. The structure of a constraint network is represented by a graph, called the *constraint graph*, whose vertices correspond to variables and edges to the constraint scopes. An assignment to a subset Y of X is said to be *consistent* if it does not violate any constraint whose scope is included in Y. We use the notation $R(c_{ij})[a]$ to represent the set of values in D_{x_j} compatible with $a \in D_{x_i}$. Thus, if there is a constraint with scope $\{i, j\}$, then $R(c_{ij})[a] = \{b \in D_{x_j} | (a, b) \in R(c_{ij})\}$; if there is no constraint with scope $\{i, j\}$, then, by default, $R(c_{ij})[a] = D_{x_j}$. We recall the BTP property presented in [4].

Definition (BTP). A binary CSP instance (X, D, C) satisfies the *Broken Triangle Property* (BTP) w.r.t. the variable ordering $<$ if, for all triples of variables (x_i, x_j, x_k) s.t. $i < j < k$, if $(v_i, v_j) \in R(c_{ij})$, $(v_i, v_k) \in R(c_{ik})$ and $(v_j, v'_k) \in R(c_{jk})$, then either $(v_i, v'_k) \in R(c_{ik})$ or $(v_j, v_k) \in R(c_{jk})$. If neither of these two tuples exist, (v_i, v_j, v_k, v'_k) is called a *broken triangle on x_k w.r.t. x_i and x_j*.

If there exists at least one broken triangle on x_k w.r.t. x_i and x_j, (x_i, x_j, x_k) is called a *broken triple on x_k w.r.t. x_i and x_j*. Let BTP be the set of the instances for which BTP holds w.r.t. some variable ordering. The BTP property is related to the compatibility between domain values, which can be graphically visualized (Figure 1) on the microstructure graph. For example, in Figure 1 (a), there is a broken triangle on x_3 with respect to the variables x_1 and x_2 since we have $(v_1, v'_3) \notin R(c_{13})$ and $(v_2, v_3) \notin R(c_{23})$ while $(v_1, v_2) \in R(c_{12})$, $(v_1, v_3) \in R(c_{13})$ and $(v_2, v'_3) \in R(c_{23})$ hold. So (x_1, x_2, x_3) is a broken triple on x_3 w.r.t. x_1 and x_2. In contrast, in Figure 1 (b), if one of the two dashed edges (that is binary tuples) appears in the microstructure, the BTP property holds for all variable orderings.

Very recently, the property BTP has been relaxed to the Extendable-Triple Property [7] by considering four variables rather than three, and allowing some broken triangles.

Definition (ETP). A binary CSP instance P satisfies the *Extendable-Triple Property* (ETP) with respect to the variable ordering $<$ if, and only if, for all subsets of four variables (x_i, x_j, x_k, x_l) such that $i < j < k < l$, there is at most one broken triple on x_l among (x_i, x_j, x_l), (x_i, x_k, x_l) and (x_j, x_k, x_l).

In this way, a binary CSP can satisfy the ETP property while it contains two broken triples among (x_i, x_j, x_k, x_l), one on x_k, and another one on x_l, while

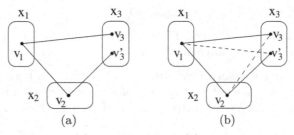

Fig. 1. A non-BTP instance (a) and a BTP one (b) w.r.t. the order $x_1 < x_2 < x_3$ if one of the dashed lines occurs.

none is possible with BTP. So, ETP strictly generalizes BTP since each instance satisfying BTP satisfies ETP while the reverse is false. So the class of instances satisfying BTP (denoted *BTP*) is strictly included in the class of instances satisfying ETP (denoted *ETP*) as indicated in Theorem 1 of [7] (*BTP* $\subsetneq$ *ETP*). As in the case of BTP, ETP allows us to define a tractable class but we need to impose an additional property related to the level of local consistency which must be verified. While the set of instances satisfying BTP define a tractable class, the set of instances satisfying ETP must also satisfy *Strong-Path-Consistency* [8], that is arc and path-consistency. Nevertheless, such instances have some of the desirable properties of instances satisfying BTP, e.g. they can be solved in polynomial time by usual algorithms such as MAC or RFL. In the next section, we introduce a new property which generalizes BTP and ETP.

3   *k*-BTP: Definition and Properties

In this section, we introduce a new property k-BTP which generalizes previous work along two axes. First, the property ETP is relaxed in the sense that we allow more broken triangles than ETP when considering subsets of four variables. But we also introduce a parameter $k \geq 2$ allowing us to consider subsets of $k + 1$ variables, with $k = 2$ corresponding to BTP and $k = 3$ corresponding to a strict generalization of ETP.

Definition (*k*-BTP). *A binary CSP instance P satisfies the property k-BTP for a given k ($2 \leq k < n$) and with respect to the variable ordering $<$ if, and only if, for all subsets of $k + 1$ variables $x_{i_1}, x_{i_2}, \ldots x_{i_{k+1}}$ such that $i_1 < i_2 < \cdots < i_{k-1} < i_k < i_{k+1}$, there is at least one triple of variables $(x_{i_j}, x_{i_{j'}}, x_{i_{k+1}})$ with $1 \leq j \neq j' \leq k$ such that there is no broken triangle on $x_{i_{k+1}}$ w.r.t. x_{i_j} and $x_{i_{j'}}$. Let k-BTP be the set of the instances for which k-BTP holds w.r.t. some variable ordering.*

One can observe that 2-BTP is exactly BTP while 3-BTP includes ETP. So, we can immediately extend Theorem 1 of [7] since $BTP \subsetneq ETP \subsetneq 3\text{-}BTP$. But above all, a more general result holds, which is an immediate consequence of the definition of k-BTP:

Theorem 1. *For all $k \geq 2$, k-BTP $\subsetneq (k+1)$-BTP*

To analyze the tractability of k-BTP, we now show that the instances of this class can be recognized in polynomial time:

Theorem 2. *Given a binary CSP instance $P = (X, D, C)$ and a constant k with $2 \leq k < n$, there is a polynomial time algorithm to find a variable ordering $<$ such that P satisfies k-BTP w.r.t. $<$, or to determine that no such ordering exists.*

Proof: As in the corresponding proofs for BTP [4] and ETP [7], we define a CSP instance P_o which is consistent if and only if a possible ordering exists. More precisely, this instance has a variable o_i with domain $\{1, \ldots, n\}$ per variable x_i of X. The value of o_i represents the position of the variable x_i in the ordering. We add a constraint involving $\{o_{i_1}, o_{i_2}, \ldots o_{i_k}, o_{i_{k+1}}\}$ and imposing the condition $o_{i_{k+1}} < max(o_{i_1}, o_{i_2}, \ldots o_{i_k})$ for each $k+1$-tuple of variables $(x_{i_1}, x_{i_2}, \ldots x_{i_k}, x_{i_{k+1}})$ such that each triple of variables $(x_{i_j}, x_{i_{j'}}, x_{i_{k+1}})$ with $1 \leq j \neq j' \leq k$ has at least one broken triangle on $x_{i_{k+1}}$ w.r.t. x_{i_j} and $x_{i_{j'}}$.

If P_o has a solution, then let $<$ be any total ordering of the variables which is a completion of the partial ordering given by the values of the variables o_i. Then for each $k+1$-tuple of variables $(x_{i_1}, x_{i_2}, \ldots x_{i_k}, x_{i_{k+1}})$, with $i_1 < \ldots < i_{k+1}$, we have at least one triple of variables $(x_{i_j}, x_{i_{j'}}, x_{i_{k+1}})$ with $1 \leq j \neq j' \leq k$ which has no broken triangle on $x_{i_{k+1}}$ w.r.t. x_{i_j} and $x_{i_{j'}}$. Indeed, if this were not the case, then the constraint $o_{i_{k+1}} < max(o_{i_1}, o_{i_2}, \ldots o_{i_k})$ would have been imposed, which is in contradiction with $i_1 < \ldots < i_{k+1}$. So, if P_o has a solution, we have an ordering satisfying the k-BTP property. Conversely, let us consider an ordering satisfying the k-BTP property and assume that P_o has no solution. It means that at least one constraint $o_{i_{k+1}} < max(o_{i_1}, o_{i_2}, \ldots o_{i_k})$ is violated. So each triple of variables $(x_{i_j}, x_{i_{j'}}, x_{i_{k+1}})$ with $1 \leq j \neq j' \leq k$ has at least one broken triangle on $x_{i_{k+1}}$, which is impossible since this ordering satisfies the k-BTP property. Hence P_o has a solution if and only if P admits an ordering satisfying the k-BTP property.

We now prove that P_o can be built and solved in polynomial time. Finding all the broken triples can be achieved in $O(n^3.d^4)$ time, while defining the constraints $o_{i_{k+1}} < max(o_{i_1}, o_{i_2}, \ldots o_{i_k})$ can be performed in $O(n^{k+1})$. So P_o can be computed in $O(n^3.d^4 + n^{k+1})$. Moreover, P_o can be solved in polynomial time by establishing generalized arc-consistency since its constraints are *max-closed* [10].   □

We analyze now the complexity of solving instances of the class k-*BTP* $(k \geq 3)$. The following theorem shows that this is NP-hard since this is true even for the smaller class $ETP \subset 3$-*BTP*.

Theorem 3. *Deciding whether an instance of the class ETP is satisfiable is NP-complete.*

Proof: It suffices to exhibit a polynomial reduction from binary CSP to its subproblem ETP. Given any binary CSP instance I, we can construct an equivalent instance I' by

1. adding a new variable x_{ij} (with domain $D_{x_{ij}} = D_{x_i}$) for each constraint c_{ij}
2. adding a new equality constraint between each x_i and x_{ij}
3. replacing each constraint $(\{x_i, x_j\}, R)$ by the constraint $(\{x_{ij}, x_j\}, R)$.

Let $<$ be any variable order in I' in which all the new variables x_{ij} occur after all the original variables x_k. Since each variable is constrained by at most two variables which precede it in this order, we can easily deduce that I' satisfies ETP. It follows from this polynomial reduction that deciding whether an instance of the class *ETP* is satisfiable is NP-complete. □

To ensure the tractability of the class k-*BTP*, we consider an additional condition which is that instances satisfy *Strong k-Consistency* [8].

Definition (Strong k-Consistency). *A binary CSP instance P satisfies i-Consistency if any consistent assignment to $i-1$ variables can be extended to a consistent assignment on any i^{th} variable. A binary CSP instance P satisfies Strong k-Consistency if it satisfies i-Consistency for all i such that $1 < i \leq k$.*

Strong k-Consistency and k-*BTP* allow us to define a new tractable class:

Theorem 4. *Let P be a binary CSP instance P such that there exists a constant k with $2 \leq k < n$ for which P satisfies both Strong k-Consistency and k-BTP w.r.t. the variable ordering $<$. Then P is consistent and a solution can be found in polynomial time.*

Proof: We consider an ordering for variable assignments corresponding to the ordering $<$. As the instance satisfies Strong k-Consistency, it satisfies arc-consistency and thus, no domain is empty and each value has a support in each other domain. Moreover, as the instance satisfies Strong k-Consistency, we have a consistent assignment on the k first variables. Now, and more generally, suppose that we have a consistent assignment $(u_1, u_2, \ldots u_{l-1}, u_l)$ for the l first variables $x_1, x_2, \ldots x_{l-1}, x_l$ in the ordering, with $k \leq l < n$. We show that this assignment can be consistently extended to the variable x_{l+1}. To show this, we must prove that $\cap_{1 \leq i \leq l} R(c_{il+1})[u_i] \neq \emptyset$, that is there is at least one value in the domain of x_{l+1} which is compatible with the assignment $(u_1, u_2, \ldots u_{l-1}, u_l)$.

We first prove this for $l = k$. Consider the consistent assignment $(u_1, u_2, \ldots u_{k-1}, u_k)$ on the k first variables. Consider a $k+1^{th}$ variable x_{k+1} appearing later in the ordering. Since P satisfies k-*BTP*, there exists at least one triple of variables $(x_j, x_{j'}, x_{k+1})$ with $1 \leq j \neq j' \leq k$ such that there is no broken triangle on x_{k+1} w.r.t. x_j and $x_{j'}$. By Lemma 2.4 given in [4], we have:

$$(R(c_{jk+1})[u_j] \subseteq R(c_{j'k+1})[u_{j'}])$$

or

$$(R(c_{j'k+1})[u_{j'}] \subseteq R(c_{jk+1})[u_j])$$

Without loss of generality, assume that we have $R(c_{jk+1})[u_j] \subseteq R(c_{j'k+1})[u_{j'}]$ and $j < j'$. Since P satisfies Strong k-Consistency, we know that the sub-assignment of $(u_1, u_2, \ldots, u_j, \ldots u_{k-1}, u_k)$ on $k-1$ variables excluding the assignment $u_{j'}$ for $x_{j'}$ can be consistently extended to x_{k+1}. Moreover, we know that

$R(c_{jk+1})[u_j] \subseteq R(c_{j'k+1})[u_{j'}]$ and by arc-consistency, $R(c_{i_j i_{k+1}})[u_j] \neq \emptyset$. Thus, $(u_1, u_2, \ldots, u_j, \ldots, u_{j'}, \ldots, u_k, u_{k+1})$ is a consistent assignment to the $k+1$ first variables.

Note that this proof holds for all subsets of $k + 1$ variables such that x_{k+1} appears later in the ordering $<$, not only for the $k + 1$ first variables $x_1, x_2, \ldots x_{k-1}, x_k$ and x_{k+1}.

Now, we prove the property for l with $k < l < n$. That is, we show that a consistent assignment $(u_1, u_2, \ldots u_{l-1}, u_l)$ can be extended to a $(l+1)^{th}$ variable. As induction hypothesis, we assume that every consistent assignment on $l - 1$ variables can be extended to a lth variable, which appears later in the considered ordering $<$.

Consider a consistent assignment $(u_1, u_2, \ldots u_{l-1}, u_l)$ on the l first variables. Let $(u_{i_1}, u_{i_2}, \ldots u_{i_k})$ be a sub-assignment on k variables of the assignment $(u_1, u_2, \ldots u_{l-1}, u_l)$. As P satisfies k-BTP, and as $k < l < n$, for all subsets of k variables $x_{i_1}, x_{i_2}, \ldots x_{i_k}$, we know that there is a triangle which is not broken in x_{l+1} w.r.t. x_{i_j} and $x_{i_{j'}}$, with x_{i_j} and $x_{i_{j'}}$ appearing in the variables $x_{i_1}, x_{i_2}, \ldots x_{i_k}$. So, without loss of generality, we can consider that $i_1 \leq i_j < i_{j'} \leq i_k \leq l$ and we have $R(c_{i_j l+1})[u_{i_j}] \subseteq R(c_{i_{j'} l+1})[u_{i_{j'}}]$. Note that x_{i_j} and $x_{i_{j'}}$ can be interchanged in the ordering if necessary.

Now, consider the consistent assignment $(u_1, u_2, \ldots u_{l-1}, u_l)$ on the l first variables. By the induction hypothesis, each partial assignment of $(u_1, u_2, \ldots u_{l-1}, u_l)$ on $l - 1$ variables can be extended to a consistent assignment on x_{l+1} with a compatible value u_{l+1}. So, consider the partial assignment on $l - 1$ variables where $u_{i_{j'}}$ does not appear. This assignment is for example $(u_1, u_2, \ldots u_{i_j}, \ldots u_{l-1}, u_l, u_{l+1})$. As we have $R(c_{i_j l+1})[u_{i_j}] \subseteq R(c_{i_{j'} l+1})[u_{i_{j'}}]$, the value $u_{i_{j'}}$ is also compatible with u_{l+1}, and thus the assignment $(u_1, u_2, \ldots u_{i_j}, \ldots u_{i_{j'}}, \ldots u_{l-1}, u_l, u_{l+1})$ on the $l+1$ first variables is a consistent assignment.

So, every consistent assignment $(u_1, u_2, \ldots u_{l-1}, u_l)$ on $(x_1, x_2, \ldots x_{l-1}, x_l)$ can be extended to a $(l+1)^{th}$ variable, for all l with $k < l < n$. And more generally, we have shown that every consistent assignment on l variables, not necessarily consecutive in the ordering (as are the l first variables), can be extended to a consistent assignment for every $(l+1)^{th}$ variable which appears after these l variables in the ordering $<$ associated with k-BTP. Thus, the induction hypothesis holds for the next step.

Note that this proof also shows that an instance which satisfies Strong k-Consistency and k-BTP (with respect to the variable ordering $<$) is consistent.

Finally, given the ordering $<$, we show that finding a solution can be performed in polynomial time. Given a consistent assignment $(u_1, u_2, \ldots u_l)$ with $l < n$, finding a compatible value u_{l+1} for the next variable x_{l+1} is feasible by searching in its domain whose size is at most d. For each value, we need to verify the constraints connecting the variable x_{l+1} which can be done in $O(e_{l+1})$ if the next variable x_{l+1} has e_{l+1} neighbors in the previous variables. Since $\Sigma_{1 \leq l < n} e_{l+1} = e$, the total cost to find a solution is $O((n + e).d)$. $\square$

In the sequel, we denote k-BTP-SkC, the class of instances satisfying k-BTP and Strong k-Consistency. One of the most interesting properties of the tractable class BTP is the fact that the instances of this class can be solved in polynomial time using classical algorithms (such as MAC or RFL) implemented in most solvers. The next property establishes that a similar result holds for k-BTP-SkC. Indeed, the proof of Theorem 4 allows us to show that algorithms such as BT (Backtracking), MAC and RFL can solve any instance of the class k-BTP-SkC in polynomial time:

Theorem 5. *Given a binary CSP instance $P = (X, D, C)$ and a variable ordering $<$ such that P satisfies k-BTP w.r.t. $<$, and is Strongly-k-Consistent, the algorithms BT, MAC and RFL find a solution of the instance P in polynomial time.*

Proof: As the instance satisfies Strong k-Consistency, BT using the ordering $<$ for the variable assignment can find a consistent assignment on $x_1, x_2, \ldots x_{k-1}$ and x_k. Moreover, given l, with $k < l < n$, it is shown in the proof of Theorem 4 that a consistent assignment $(u_1, u_2, \ldots u_{l-1}, u_l)$ on $x_1, x_2, \ldots x_{l-1}$ and x_l can be extended to a $(l+1)^{th}$ variable, that is on x_{l+1}. To find the assignment of x_{l+1}, we need to look for a compatible value in its domain. This is feasible in $O(e_{l+1}.d)$ assuming that x_{l+1} has e_{l+1} neighbors in the previous variables. So, as for the proof of Theorem 4, finding a solution of P is globally feasible in $O((n + e).d)$. If we consider now algorithms such as MAC or RFL, by the same reasoning, we show that their complexity is bounded by $O(n.(n + e).d^2)$ due to the additional cost of the arc-consistency filtering performed after each variable assignment. $\square$

In Section 5, we discuss the interest of the class k-BTP from a practical viewpoint. In the next section, we study the relationships between k-BTP and some tractable classes.

4   Relationship with Some Tractable Classes

We consider the important tractable class based on the notion of tree-decomposition of graphs [11].

Definition (Tree-Decomposition). *Given a graph $G = (X, C)$, a tree-decomposition of G is a pair (E, T) with $T = (I, F)$ a tree and $E = \{E_i : i \in I\}$ a family of subsets of X, such that each subset (called cluster or bag in Graph Theory) E_i is a node of T and satisfies:*

1. *$\cup_{i \in I} E_i = X$,*
2. *for each edge $\{x, y\} \in C$, there exists $i \in I$ with $\{x, y\} \subseteq E_i$, and*
3. *for all $i, j, k \in I$, if k is in a path from i to j in T, then $E_i \cap E_j \subseteq E_k$.*

The width of a tree-decomposition (E, T) is equal to $max_{i \in I} |E_i| - 1$. The tree-width w of G is the minimal width over all the tree-decompositions of G.

Let k-TW be the class of binary CSPs instances such that their tree-width is less than or equal to a constant k. Recently, M. Vardi asked a question about the relationships between k-TW and ETP or other generalizations of BTP. The next theorems give a first partial answer to this question.

Theorem 6. k-$TW \subsetneq (k+1)$-BTP.

Proof: We show firstly that k-$TW \subseteq (k+1)$-BTP. It is well known that if the tree-width of a binary instance of CSP is bounded by k, there is an ordering $<$ on variables, such that for $x_i \in X$, $|\{x_j \in X : j < i \text{ and } c_{ji} \in C\}| \leq k$ [2]. Now, consider a subset of $k+2$ variables $x_{i_1}, x_{i_2}, \ldots x_{i_k}, x_{i_{k+1}}, x_{i_{k+2}}$ such that $i_1 < i_2 < \cdots < i_{k-1} < i_k < i_{k+1} < i_{k+2}$. Since the tree-width is bounded by k, we know that there are at most k constraints $c_{i_j i_{k+2}} \in C$. So, there is at least one triple of variables $(x_{i_j}, x_{i_{j'}}, x_{i_{k+2}})$ with $1 \leq j \neq j' \leq k$ such that $c_{i_j i_{k+2}} \notin C$ or $c_{i_{j'} i_{k+2}} \notin C$. Without loss of generality, assume that there is no constraint $c_{i_j i_{k+2}} \in C$. Thus, there is no broken triangle on $x_{i_{k+2}}$ w.r.t. x_{i_j} and $x_{i_{j'}}$ because all the values of $D_{x_{i_j}}$ are compatible with all the values of $D_{x_{i_{k+2}}}$. So, the considered instance of CSP satisfies the property $(k+1)$-BTP. Finally, it is easy to define instances whose tree-width is strictly greater than k which satisfy the property $(k+1)$-BTP. For example, we can consider an instance of CSP with domains of size one, with the complete constraint graph, and with one solution. The tree-width of this instance is $n-1$ while it satisfies k-BTP for all possible values of k. $\qquad\square$

The cost of checking for satisfiability of instances in k-TW has a similar cost to that of achieving Strong $(k+1)$-Consistency, that is $O(n^{k+1}d^{k+1})$. Nevertheless, this does not allow us to establish a formal inclusion of k-TW in $(k+1)$-BTP-$S(k+1)C$ which is tractable while $(k+1)$-BTP is NP-complete for $k \geq 2$ by Theorem 3. But if we denote k-TW-$S(k+1)C$, the class of binary CSPs instances belonging to k-TW and which satisfy Strong $(k+1)$-Consistency, the next result holds:

Theorem 7. k-TW-$S(k+1)C \subsetneq (k+1)$-BTP-$S(k+1)C$.

The tractable class BTP has also recently been generalized in a different way to that proposed in this paper, again by noticing that not all broken triangles need to be forbidden [12]. We will show that these two generalizations are orthogonal.

Definition ($\forall\exists$-BTP). *A binary CSP instance P satisfies the property $\forall\exists$-BTP w.r.t. the variable ordering $<$ if, and only if, for each pair of variables x_i, x_k such that $i < k$, for all $v_i \in D_{x_i}$, $\exists v_k \in D_{x_k}$ such that $(v_i, v_k) \in R(c_{ik})$ and for all x_j with $j < k$ and $j \neq i$, and for all $v_j \in D_{x_j}$ and for all $v'_k \in D_{x_k}$, (v_i, v_j, v_k, v'_k) is not a broken triangle on x_k w.r.t. x_i and x_j. Let $\forall\exists$-BTP be the set of the instances for which $\forall\exists$-BTP holds w.r.t. some variable ordering.*

The class $\forall\exists$-BTP can be solved and recognized in polynomial time [12]. It represents a tractable class which strictly includes BTP since it does not forbid all broken triangles. Since k-BTP also does not forbid all broken triangles, it is natural to compare these two classes. We do this for the special case $k = 3$, but the same argument applies for any value of $k \geq 3$.

Theorem 8. *Even for sets of binary CSP instances which are strong path consistent, the properties 3-BTP and $\forall\exists$-BTP are incomparable.*

Proof: Consider an instance P^* in which each domain D_{x_k} contains a value a^* such that for all other variables x_i, for all values $v_i \in D_{x_i}$, $(v_i, a^*) \in R(c_{ik})$. Then P^* satisfies $\forall\exists$-BTP since there can be no broken triangle of the form (v_i, v_j, a^*, v_k'), the value a^* being compatible with all assignments to all other variables. It is easy to complete such an instance P^* so that it does not satisfy 3-BTP for any variable ordering by adding broken triangles on domain elements other than a^*.

Consider a 3-variable binary CSP instance P_3 with domains $\{0, 1, 2\}$ and the following three constraints: $x_1 \neq x_2$, $x_1 \neq x_3$, $x_2 \neq x_3$, i.e. a 3-colouring problem on a complete graph on three vertices. Then P_3 is strong path consistent and trivially satisfies 3-BTP (since there are only 3 variables), but P_3 does not satisfy $\forall\exists$-BTP for any ordering $i < j < k$ of the variables (due to the existence of broken triangles on assignments (x_i, a), (x_j, b), (x_k, a), (x_k, b) for all pairs of distinct colours a, b). $\square$

We now consider a very general tractable class recently discovered by Naanaa [9] and which undoubtedly deserves to be better known.

Let E be a finite set and let $\{E_i\}_{i \in I}$ be a finite family of subsets of E. The family $\{E_i\}_{i \in I}$ is said to be *independent* if and only if for all $J \subset I$,

$$\bigcap_{i \in I} E_i \subset \bigcap_{j \in J} E_j$$

(where the notation $A \subset B$ means that A is a proper subset of B). Observe that $\{E_i\}_{i \in I}$ cannot be independent if $\exists j \neq j' \in I$ such that $E_j \subseteq E_{j'}$, since in this case and with $J = I \setminus \{j'\}$ we would have

$$\bigcap_{i \in I} E_i = \bigcap_{j \in J} E_j.$$

Definition (Directional Rank). *Let P be a binary CSP instance whose variables are totally ordered by $<$. The* directional rank *of variable x_m is the size k of the largest consistent assignment $(a_1, \ldots, a_k)$ to a set of variables $x_{i_1}, \ldots, x_{i_k}$ (with $i_1 < \ldots < i_k < m$) such that the family of sets $\{R(c_{i_j m})[a_j]\}_{j=1,\ldots,k}$ is independent. The* directional rank *of P (w.r.t the ordering $<$ of its variables) is the maximum directional rank over all its variables.*

Naanaa has shown that if P is a binary CSP instance which has directional rank no greater than k and is directional strong $(k + 1)$-consistent then I is globally consistent [9]. We denote $DR\text{-}k$, the set of these instances. Naanaa points out that some known tractable classes, such as binary CSP instances with connected row convex constraints [13], have bounded directional rank.

If a binary CSP instance P is $(k + 1)$-BTP, then no variable can have a directional rank greater than k. This is because for any variable x_m and any assignments $(a_1, \ldots, a_{k+1})$ to any set of variables $x_{i_1}, \ldots, x_{i_{k+1}}$ with

$i_1 < \ldots < i_{k+1} < m$, by the definition of $(k+1)$-BTP, we must have $R(c_{i_j m})[a_j] \subseteq R(c_{i_{j'} m})[a_{j'}]$ for some $j \neq j' \in \{1, \ldots, k+1\}$; hence, as observed above, the sets $\{R(c_{i_j m})[a_j]\}_{j=1,\ldots,k+1}$ cannot be independent. It follows that the tractability of $(k+1)$-BTP-$S(k+1)C$ is also a corollary of the result of Naanaa [9]. On the other hand, the property $(k+1)$-BTP, although subsumed by DR-k, can be detected in time complexity $O(n^k d^k + n^3 d^4)$ compared to $O(n^{k+1} d^{k+1})$ for DR-k.

5   Experiments

In this section, we compare the tractable classes BTP, ETP-SPC, k-BTP-SkC and DR-k-1 (where SPC stands for Strong Path Consistency) from a practical viewpoint. We only consider the case $k = 3$, since strong k-consistency becomes too expensive in time for $k > 3$ and may add constraints of arity $k - 1$.

Tractable classes are often critized for being artificial in the sense that their underlying properties seldom occur in real instances. So, here, we first highlight the existence of instances belonging to some of these classes among the benchmark instances classically exploited for solver evaluations and comparisons. More precisely, our experiments involve 2,373 binary benchmarks from the third CSP Solver Competition[1] and cover all the benchmarks exploited in [7].

Then we will investigate the possible link between efficient solving and belonging to these tractable classes.

5.1   Instances Belonging to Tractable Classes

Since the tractable classes ETP-SPC, 3-BTP-SPC and DR-2 require strong path-consistency, we first achieve SPC on each instance before checking whether it belongs to the considered classes, in the same spirit as [14,15]. In so doing, 628 instances are detected as inconsistent and so they trivially belong to all of these tractable classes. 85 of the remaining instances belong to 3-BTP-SPC while 87 have directional rank at most two. Among these instances, we have respectively 71 and 76 instances in BTP-SPC and ETP-SPC. These differences between these tractable classes are well highlighted by some instances of the bqwh-15-106 family since we can observe all the possible configurations of the relations BTP-$SPC \subsetneq ETP$-$SPC \subsetneq 3$-BTP-$SPC \subsetneq DR$-2. For example, instance bqwh-15-106-13 belongs to all the considered tractable classes while instances bqwh-15-106-28, bqwh-15-106-16 and bqwh-15-106-76 only belong respectively to three, two or one of these tractable classes. Table 1 presents some instances belonging to classes ETP-SPC, 3-BTP-SPC or DR-2. It also provides the tree-width w of these instances and their tree-width w' once SPC is enforced. When the exact tree-width is unknown (recall that computing an optimal tree-decomposition is an NP-hard problem), we give a range. We can note the diversity of these instances (academic, random or real-world instances). Some of these instances belong to 3-BTP-SPC or DR-2 thanks to their structure. For instance,

---

[1] See http://www.cril.univ-artois.fr/CPAI08.

Table 1. Some instances belonging to *BTP-SPC*, *ETP-SPC*, 3-*BTP-SPC* or *DR*-2 after the application of SPC with their tree-width w and the tree-width w' of the instances once SPC is enforced.

Instance	n	w	w'	*BTP-SPC*	*ETP-SPC*	3-*BTP-SPC*	*DR*-2
bqwh-15-106-13	106	$[7, 48]$	104	yes	yes	yes	yes
bqwh-15-106-16	106	$[6, 45]$	99	no	no	yes	yes
bqwh-15-106-28	106	$[7, 52]$	105	no	yes	yes	yes
bqwh-15-106-76	106	$[6, 44]$	100	no	no	no	yes
bqwh-15-106-77	106	$[7, 50]$	100	no	no	yes	yes
bqwh-18-141-33	141	$[7, 64]$	134	yes	yes	yes	yes
bqwh-18-141-57	141	$[7, 66]$	137	yes	yes	yes	yes
domino-100-100	100	2	2	yes	yes	yes	yes
domino-5000-500	5000	2	2	yes	yes	yes	yes
driverlogw-04c-sat	272	$[19, 56]$	$[214, 221]$	no	no	no	yes
driverlogw-09-sat	650	$[39, 108]$	629	yes	yes	yes	yes
fapp17-0300-10	300	$[6, 153]$	$[6, 154]$	yes	yes	yes	yes
fapp18-0350-10	350	$[5, 192]$	$[12, 199]$	yes	yes	yes	yes
fapp23-1800-9	1800	$[6, 1325]$	$[41, 1341]$	yes	yes	yes	yes
graph12-w0	680	1	1	yes	yes	yes	yes
graph13-w0	916	1	1	yes	yes	yes	yes
hanoi-7	126	1	1	yes	yes	yes	yes
langford-2-4	8	7	7	yes	yes	yes	yes
lard-83-83	83	82	82	no	no	yes	yes
lard-91-91	91	90	90	no	no	yes	yes
os-taillard-4-100-0	16	$[3, 9]$	15	yes	yes	yes	yes
os-taillard-4-100-9	16	$[3, 9]$	15	yes	yes	yes	yes
scen5	400	$[11, 32]$	$[167, 188]$	no	no	yes	yes

graph12-w0 and hanoi-7 have an acyclic constraint graph while the tree-width of domino-100-100 and crossword-m1-uk-puzzle01 is two. However, most instances have a tree-width greater than two. Moreover, in most cases, the application of SPC may significantly increase the original tree-width of these instances. For example, the tree-width of instance driverlogw-09-sat is initially bounded by 108 and is equal to 629 after the application of SPC. This increase is explained by the pairs of values which are forbidden by SPC. When SPC forbids a pair of values (v_i, v_j) for a given pair of variables (x_i, x_j), it removes (v_i, v_j) from the relation $R(c_{ij})$ if the constraint c_{ij} exists. However, if the constraint c_{ij} does not exist yet, SPC must first add it to the problem. In such a case, depending on the added constraints and their number, the tree-width may significantly increase. Note that the considered instances whose tree-width is initially at most two have a tree-width unchanged by the application of SPC.

5.2 Link Between Efficient Solving and Belonging to Tractable Classes

In this subsection, our aim is not to provide a new module based on tractable classes in order to improve the efficiency of solvers but to see whether we can exploit some tractable classes to explain the efficiency of solvers on some instances. Indeed, we think that tractable classes are more useful from a practical viewpoint if they are implicitly handled by classical solvers than by ad-hoc methods (as is generally the case). For instance, it is well kwown that MAC can solve in backtrack-free manner any binary CSP whose constraint network is acyclic without knowing that the instance has this particular feature [16].

Most state-of-the-art solvers rely on variants of MAC or RFL algorithms. In the following, we focus our study on MAC but we have observed similar results for RFL.

As far as solving is concerned, all the instances belonging to 3-BTP-SPC or DR-2 are solved in a backtrack-free manner by MAC except the instance driverlogw-04c-sat which needs one backtrack. Note that MAC has no knowledge about the variable ordering used to satisfy 3-BTP or to obtain a directional rank of at most two. In most cases, we have observed that the ordering CSP instance built in the proof of Theorem 2 in order to compute a suitable variable ordering has no constraints. So any variable ordering is suitable. In contrast, for about a dozen instances, this CSP has several constraints but remains clearly under-constrained and the constraint network has several connected components. This ensues that the ordering CSP in general a huge number of solutions. So it is very likely that MAC exploits implicitly one of these suitable variable orderings. For example, the ordering CSP for checking whether the bqwh-15-106-76 instance (which has 106 variables) has a directional rank at most two has 65 connected components and admits more than 33 million solutions.

Some of the instances are solved efficiently by MAC in a backtrack-free manner even though they do not belong to one of the studied tractable classes. Hence, we now consider the notion of backdoor [17] with the aim in view to provide some explanation about this efficiency in the same spirit as [7]. A *backdoor* is a set of variables defined with respect to a class such that once the backdoor variables are assigned, the problem falls in the class. Here, we are interested in back-doors which are discovered implicitly by MAC when it assigns some variables. Indeed, after some assignments and the associated filtering, the remaining part of the problem may become tractable. So we assess the number of variables which must be assigned before MAC finds implicitly a backdoor w.r.t. one of the studied classes. In practice, over the 50 considered instances, we observe that MAC finds a backdoor w.r.t. BTP after having assigned more variables than for the other considered classes. The numbers of assigned variables required to find a backdoor respectively for ETP and 3-BTP are very close, and even equal in most cases. By considering DR-2, we save a few variables compared to ETP and 3-BTP. For example, MAC needs to assign at most five variables before finding a backdoor w.r.t. to 3-BTP or DR-2 for 14 instances compared to 12 and 4 instances, respectively, for ETP and BTP^2. Of course, the resulting instances do not necessarily satisfy strong path-consistency and so we cannot exploit directly Theorem 5 to explain the efficiency of MAC. Nevertheless, when the instance is 3-BTP and

---

[2] Note that these instances do not include all the instances mentioned in [7] since some of them belong to 3-BTP-SPC and/or DR-2.

strong path-consistent after having assigned some variables, MAC may exploit implicitly a suitable variable ordering since, as evoked above, the corresponding ordering CSP often admits a large number of solutions. Furthermore Theorem 5 provides sufficient conditions so that MAC solves some instances in polynomial time, but these conditions are not always necessary. For instance, MAC solves the instances which belong to BTP in polynomial time without requiring a suitable variable ordering or the satisfaction of strong path-consistency. Hence, one part of the explanation of the practical efficiency of MAC may lie in its ability to exploit implicitly different tractable classes.

6 Conclusion

This paper introduces a novel family of tractable classes for binary CSPs, denoted k-BTP whose tractability is associated with a given level of strong k-consistency. It is based on a hierarchy of classes of instances with the BTP class as the base case. While BTP is defined on subsets of 3 variables, the k-BTP class is defined on sets of $k+1$ variables, while relaxing the restrictive conditions imposed by BTP which is the class 2-BTP. We showed that k-BTP inherits some of the desirable properties of BTP, such as polynomial solvability using standard algorithms such as MAC. We also showed that k-BTP strictly generalizes the class of instances whose tree-width is bounded by a constant and we analyzed the relationships with the class based on the notion of *directional rank* recently introduced by Naanaa. To assess the practical interest of the k-BTP class, an experimental analysis is presented focusing on the particular case of 3-BTP. This analysis shows a significant advantage of 3-BTP compared to BTP and to CSPs of bounded tree-width.

Further research is required to determine if the condition corresponding to strong k-consistency is actually necessary or whether a weaker condition would suffice. Indeed, experiments showed that MAC can solve without backtracking certain instances belonging to 3-BTP even when they do not verify the corresponding level of consistency. From a practical point of view, an interesting challenge is to find the minimum (generally) required level of consistency among different kinds of local consistencies such as PIC [18], maxRPC [19] or SAC [20]. Note that, from a theoretical point of view, we can easily deduce from Theorem 3 that any local consistency that only performs domain filtering (e.g. PIC, maxRPC, SAC) cannot be sufficient (assuming P$\neq$NP) since ETP is invariant under domain filtering operations.

Moreover, studying a relaxation of the k-BTP condition needs to be addressed so as to further expand the class of instances that can be solved in polynomial time, but along different avenues to the one proposed in [9], even if further theoretical and experimental research are clearly required to fully appreciate all the consequences of Naanaa's result. Finally, it could be interesting to investigate a similar approach to the one introduced in [21] which provides a novel polynomial-time reduction operation based on the merging of domain values.

References

1. Rossi, F., van Beek, P., Walsh, T.: Handbook of Constraint Programming. Elsevier (2006)
2. Dechter, R., Pearl, J.: Tree-Clustering for Constraint Networks. Artificial Intelligence **38**, 353–366 (1989)
3. Gottlob, G., Leone, N., Scarcello, F.: A Comparison of Structural CSP Decomposition Methods. Artificial Intelligence **124**, 243–282 (2000)
4. Cooper, M.C., Jeavons, P., Salamon, A.: Generalizing constraint satisfaction on trees: hybrid tractability and variable elimination. Artificial Intelligence **174**, 570–584 (2010)
5. Sabin, D., Freuder, E.: Contradicting conventional wisdom in constraint satisfaction. In: Proceedings of ECAI, pp. 125–129 (1994)
6. Nadel, B.: Tree Search and Arc Consistency in Constraint-Satisfaction Algorithms. Search in Artificial Intelligence, pp. 287–342. Springer-Verlag (1988)
7. Jégou, P., Terrioux, C.: The extendable-triple property: a new CSP tractable class beyond BTP. In: Proceedings of AAAI, pp. 3746–3754 (2015)
8. Freuder, E.: A Sufficient Condition for Backtrack-Free Search. Journal of the ACM **29**(1), 24–32 (1982)
9. Naanaa, W.: Unifying and extending hybrid tractable classes of csps. Journal of Experimental and Theoretical Artificial Intelligence **25**(4), 407–424 (2013)
10. Jeavons, P., Cooper, M.: Tractable constraints on ordered domains. Artificial Intelligence **79**(2), 327–339 (1995)
11. Robertson, N., Seymour, P.D.: Graph minors II: Algorithmic aspects of treewidth. Algorithms **7**, 309–322 (1986)
12. Cooper, M.C.: Beyond consistency and substitutability. In: O'Sullivan, B. (ed.) CP 2014. LNCS, vol. 8656, pp. 256–271. Springer, Heidelberg (2014)
13. Deville, Y., Barette, O., van Hentenryck, P.: Constraint satisfaction over connected row convex constraints. Artificial Intelligence **109**(1–2), 243–271 (1999)
14. El Mouelhi, A., Jégou, P., Terrioux, C.: Hidden tractable classes: from theory to practice. In: Proceedings of ICTAI, pp. 437–445 (2014)
15. El Mouelhi, A., Jégou, P., Terrioux, C.: Hidden Tractable Classes: from Theory to Practice. Constraints (2015)
16. Sabin, D., Freuder, E.: Understanding and Improving the MAC Algorithm. In: Smolka, G. (ed.) CP 1997. LNCS, vol. 1330, pp. 167–181. Springer, Heidelberg (1997)
17. Williams, R., Gomes, C.P., Selman, B.: Backdoors to typical case complexity. In: Proceedings of IJCAI, pp. 1173–1178 (2003)
18. Freuder, E., Elfe, C.D.: Neighborhood inverse consistency preprocessing. In: Proceedings of AAAI, pp. 202–208 (1996)
19. Debruyne, R., Bessière, C.: From restricted path consistency to max-restricted path consistency. In: Smolka, G. (ed.) CP 1997. LNCS, vol. 1330, pp. 312–326. Springer, Heidelberg (1997)
20. Debruyne, R., Bessière, C.: Domain Filtering Consistencies. Journal of Artificial Intelligence Research **14**, 205–230 (2001)
21. Cooper, M.C., El Mouelhi, A., Terrioux, C., Zanuttini, B.: On broken triangles. In: O'Sullivan, B. (ed.) CP 2014. LNCS, vol. 8656, pp. 9–24. Springer, Heidelberg (2014)

The Unary Resource with Transition Times

Cyrille Dejemeppe, Sascha Van Cauwelaert$^{(\boxtimes)}$, and Pierre Schaus

UCLouvain, ICTEAM, Place Sainte Barbe 2, 1348 Louvain-la-Neuve, Belgium
{cyrille.dejemeppe,sascha.cauwelaert,pierre.schaus}@uclouvain.be

Abstract. Transition time constraints are ubiquitous in scheduling problems. They are said to be sequence-dependent if their durations depend on both activities between which they take place. In this context, we propose to extend the Θ-tree and Θ-Λ-tree data structures introduced by Vilím in order to strengthen the bound computation of the earliest completion time of a set of activities, by taking into account the sequence dependent transition time constraints. These extended structures can be substituted seamlessly in the state-of-the-art Vilím's filtering algorithms for unary resource constraints (Overload Checking, Detectable Precedences, Not-First/Not-Last and Edge-Finding algorithms) without changing their $\mathcal{O}(n \log(n))$ time complexities. Furthermore, this new propagation procedure is totally independent from additional constraints or the objective function to optimize. The proposed approach is able to reduce the number of nodes by several order of magnitudes on some instances of the job-shop with transition times problem, without introducing too much overhead on other instances for which it is less effective.

Keywords: Scheduling · Transition times · Global constraints · Constraint Programming

1 Introduction

This work extends the classic unary/disjunctive resource propagation algorithms to include propagation over *sequence-dependent* transition times between activities. A wide range of real-world scheduling problems from the industry involves transition times between activities. An example is the quay crane scheduling problem in container terminals [21] where the crane is modeled as a unary resource and transition times represent the moves of the crane on the rail to move from one position to another along the vessel to load/unload containers.

We introduce filtering algorithms to tighten the bounds of (non-preemptive) activities while taking into account the transition times between them. These filtering algorithms are extensions of the unary resource propagation algorithms (Overload Checking, Detectable Precedences, Not-First/Not-Last, Edge-Finding) introduced in [18]. All these algorithms rely on an efficient computation of the earliest completion time (ect) of a group of activities using the so-called Theta tree and Theta-Lambda tree data structures. We demonstrate the efficiency of the filtering on job-shop with transition times problem instances.

© Springer International Publishing Switzerland 2015
G. Pesant (Ed.): CP 2015, LNCS 9255, pp. 89–104, 2015.
DOI: 10.1007/978-3-319-23219-5_7

In Section 2, we give an overview of the tackled problems and of current state-of-the-art techniques to solve them. In Section 3, we explain the requirements needed to integrate transition times propagation. Section 4 explains how to obtain lower bounds for the time spent by transitions between activities from a set. Then, Section 5 describes how to integrate this bound to efficiently compute the *ect* of a set of activities with extended Θ-tree structures. Section 6 then explains how classic unary algorithms can consider transition times by using the extended Θ-tree structures. Finally, we report results obtained by the new propagation procedure in Section 7.

2   Background

In Constraint Programming (CP), a scheduling problem is modeled by associating three variables to each activity A_i: start_i, end_i, and duration_i representing respectively the starting time, ending time and processing time of A_i. These variables are linked together by the following relation:

$$\text{start}_i + \text{duration}_i = \text{end}_i$$

Depending on the considered problem, global constraints linking the activity variables are added to the model. In this work, we are interested in the unary resource constraint. A *unary resource*, sometimes referred to as a *machine*, is a resource allowing only a single activity to use it at any point in time. As such, all activities demanding the same unary resource cannot overlap in time:

$$\forall i, j \ \ i \neq j : (\text{end}_i \leq \text{start}_j) \vee (\text{end}_j \leq \text{start}_i)$$

The unary resource can be generalized by requiring transition times between activities. A transition time $tt_{i,j}$ is a minimal amount of time that must occur between two activities A_i and A_j if $A_i \prec A_j$ (precedes). These transition times are described in a matrix $\mathcal{M}$ in which the entry at line i and column j represents the minimum transition time between A_i and A_j, $tt_{i,j}$. We assume that transition times respect the triangular inequality. That is, inserting an activity between two activities always increases the time between these activities:

$$\forall i, j, k \ \ i \neq j \neq k : tt_{i,j} \leq tt_{i,k} + tt_{k,j}$$

The unary resource with transition times imposes the following relation:

$$\forall i, j : (\text{end}_i + tt_{i,j} \leq \text{start}_j) \vee (\text{end}_j + tt_{j,i} \leq \text{start}_i) \tag{1}$$

2.1   Related Work

As described in a recent survey [2], scheduling problems with transition times can be classified in different categories. First the activities can be in *batch* (i.e. a machine allows several activities of the same batch to be processed simultaneously) or not. Transition times may exist between successive batches.

A CP approach for batch problems with transition times is described in [18]. Secondly the transition times may be *sequence-dependent* or *sequence-independent*. Transition times are said to be sequence-dependent if their durations depend on both activities between which they occur. On the other hand, transition times are sequence-independent if their duration only depend on the activity after which it takes place. The problem category we study in this article is non-batch sequence-dependent transition time problems.

Several methods have been proposed to solve such problems. Ant Colony Optimization (ACO) approaches were used in [9] and [15] while [6], [4], [13] and [10] propose Local Search and Genetic Algorithm based methods. [13] proposes a propagation procedure with the Iterative Flattening Constraint-Based Local Search technique. The existing CP approaches for solving sequence-dependent problems are [8], [3], [20] and [11].

Focacci et al [8] introduce a propagator for job-shop problems involving alternative resources with non-batch sequence-dependent transition times. In this approach a successor model is used to compute lower-bounds on the total transition time. The filtering procedures are based on a minimum assignment algorithm (a well known lower bound for the Travelling Salesman Problem). In this approach the total transition time is a constrained variable involved in the objective function (the makespan).

In [3], a Travelling Salesman Problem with Time Window (TSPTW) relaxation is associated to each resource. The activities used by a resource are represented as vertices in a graph and edges between vertices are weighted with corresponding transition times. The TSPTW obtained by adding time windows to vertices from bounds of corresponding activities is then resolved. If one of the TSPTW is found un-satisfiable, then the corresponding node of the search tree is pruned. A similar technique is used in [5] with additional propagation.

In [20], an equivalent model of multi-resource scheduling problem is proposed to solve sequence-dependent transition times problems. Finally, in [11], a model with a reified constraint for transition times is associated to a specific search to solve job-shop with sequence-dependent transition times problems.

To the best of our knowledge, the CP filtering introduced in this article is the first one proposing to extend all the classic filtering algorithms for unary resources (Overload Checking [7], Detectable Precedences [17], Not-First/Not-Last [19] and Edge Finding [19]) by integrating transition times, independently of the objective function of the problem. This filtering can be used in any problem involving a unary resource with sequence-dependent transition times.

2.2 Unary Resource Propagators in CP

The earliest starting time of an activity A_i denoted est_i, is the time before which A_i cannot start. The latest starting time of A_i, lst_i, is the time after which A_i cannot start. The domain of $start_i$ is thus the interval $[est_i; lst_i]$. Similarly the earliest completion time of A_i, ect_i, is the time before which A_i cannot end and the latest completion time of A_i, lct_i, is the time after which A_i cannot end.

The domain of end_i is thus the interval $[ect_i; lct_i]$. These definitions can be extended to a set of activity Ω. For example, est_Ω is defined as follows:

$$est_\Omega = \min\{est_j | j \in \Omega\}$$

The propagation procedure for the unary resource constraint introduced in [18] contains four different propagation algorithms all running with time complexity in $\mathcal{O}(n\log(n))$: Overload Checking (OC), Detectable Precedences (DP), Not-First/Not-Last (NF/NL) and Edge Finding (EF). These propagation algorithms all rely on an efficient computation of the earliest completion time of a set of activities Ω using data structures called *Theta Tree* and *Theta-Lambda Tree* introduced in [18]. Our contribution is a tighter computation of the lower bound ect_Ω taking into account the transition times between activities.

3   Transition Times Extension Requirements

The propagation procedure we introduce in this article relies on the computation of ect_Ω, the earliest completion time of a set of activities. This value depends on the transition times between activities inside Ω. Let Π_Ω be the set of all possible permutations of activities in Ω. For a given permutation $\pi \in \Pi_\Omega$ (where $\pi(i)$ is the activity taking place at position i), we can define the total time spent by transition times, tt_π, as follows:

$$tt_\pi = \sum_{i=1}^{|\Omega|-1} tt_{\pi(i),\pi(i+1)}$$

A lower bound for the earliest completion time of Ω can then defined as:

$$ect_\Omega^{NP} = \max_{\Omega' \subseteq \Omega}\left\{ est_{\Omega'} + p_{\Omega'} + \min_{\pi \in \Pi_{\Omega'}} tt_\pi \right\} \tag{2}$$

Unfortunately, computing this value is NP-hard. Indeed, computing the optimal permutation $\pi \in \Pi$ minimizing tt_π is equivalent to solving a TSP. Since embedding an exponential algorithm in a propagator is generally impractical, a looser lower bound can be used instead:

$$ect_\Omega = \max_{\Omega' \subseteq \Omega}\{est_{\Omega'} + p_{\Omega'} + \underline{tt}(\Omega')\}$$

where $\underline{tt}(\Omega')$ is a lower bound of the total time consumed by transition times between activities in Ω':

$$\underline{tt}(\Omega') \leq \min_{\pi \in \Pi_{\Omega'}} tt_\pi$$

Our goal is to keep the overall $\mathcal{O}(n\log(n))$ time complexity of Vilím's algorithms. The lower bound $\underline{tt}(\Omega')$ must therefore be available in constant time for a given set Ω'. Our approach to obtain constant time lower-bounds for a given set Ω' during search is to base its computation solely on the cardinality $|\Omega'|$.

More precisely, for each possible subset of cardinality $k \in \{1, \ldots, n\}$, we pre-compute the smallest transition time permutation of size k on Ω:

$$\underline{tt}(k) = \min_{\{\Omega' \subseteq \Omega: \, |\Omega'| = k\}} \left\{ \min_{\pi \in \Pi_{\Omega'}} tt_\pi \right\}$$

For each k, the lower bound computation thus requires to solve a resource constrained shortest path problem (also NP-hard) with a fixed number of edges k and with a free origin and destination. The next section proposes several ways of pre-computing efficient lower bounds $\underline{tt}(k)$ for $k \in \{1, \ldots, n\}$. Our formula to compute a lower bound for the earliest completion time of a set of activities (making use of pre-computed lower-bounds of transition times) becomes:

$$ect_\Omega^\diamond = \max_{\Omega' \subseteq \Omega} \left\{ est_{\Omega'} + p_{\Omega'} + \underline{tt}(|\Omega'|) \right\} \tag{3}$$

4   Lower Bound of Transitions Times

The computation of $\underline{tt}(k)$ for all $k \in \{1, \ldots, n\}$ is NP-hard. This is a constrained shortest path problem (for $k = n$ it amounts to solving a TSP) in a graph where each node corresponds to an activity and directed edges between nodes represent the transition time between corresponding activities. Although these computations are achieved at the initialization of the constraint, we propose to use polynomial lower bounding procedures instead. Several approaches are used and since none of them is dominated any other one, we simply take the maximum of the computed lower bounds.

Minimum Weight Forest. A lower bound for $\underline{tt}(k)$ is a minimal subset of edges of size k taken from this graph. We propose to strengthen this bound by using Kruskal's algorithm [12] to avoid selecting edges forming a cycle. We stop this algorithm as soon as we have collected k edges. The result is a set of edges forming a minimum weight forest (i.e. a set of trees) with exactly k edges.

Dynamic Programming. We can build the layered graph with exactly k layers and each layer containing all the activities. Arcs are only defined between two successive layers with the weights corresponding to the transition times. A shortest path on this graph between the first and last layer can be obtain with Dijkstra. By construction this shortest path will use exactly k transitions, the relaxation being that a same activity or transition can be used several times.

Minimum Cost Flow. Another relaxation is to keep the degree constraint but relax the fact that selected edges must form a contiguous connected path. This relaxation reduces to solving a minimum cost flow problem of exactly k units on the complete bipartite graph formed by the transitions.

Lagrangian Relaxation. As explained in Chapter 16 of [1], the Lagrangian relaxation (a single Lagrangian multiplier is necessary for the exactly k transitions constraint) combined with a sub-gradient optimization technique can easily be applied to compute a lower bound on the constrained shortest path problem.

5   Extending the Θ-tree with Transition Times

As introduced in [17], the $\mathcal{O}(n \log n)$ propagation algorithms for unary resource use the so-called Θ-tree data structure. We propose to extend it in order to integrate transition times while keeping the same time complexities for all its operations.

A Θ-tree is a balanced binary tree in which each leaf represents an activity from a set Θ and internal nodes gather information about the set of (leaf) activities under this node. For an internal node v, we denote by Leaves(v), the leaf activities under v. Leaves are ordered in non-decreasing order of the est of the activities. That is, for two activities i and j, if $est_i < est_j$, then i is represented by a leaf node that is at the left of the leaf node representing j. This ensures the property:

$$\forall i \in Left(v), \forall j \in Right(v) : est_i \leq est_j$$

where left(v) and right(v) are respectively the left and right children of v, and Left(v) and Right(v) denote Leaves(left(v)) and Leaves(right(v)).

A node v contains precomputed values about Leaves(v): ΣP_v represents the sum of the durations of activities in Leaves(v) and ect_v is the ect of Leaves(v). More formally, the values maintained in an internal node v are defined as follows:

$$\Sigma P_v = \sum_{j \in \text{Leaves}(v)} p_j$$

$$ect_v = ect_{\text{Leaves}(v)} = \max_{\Theta' \subseteq \text{Leaves}(v)} \{est_{\Theta'} + p_{\Theta'}\}$$

For a given leaf l representing an activity i, the values of ΣP_l and ect_l are p_i and ect_i, respectively. In [18] Vilím has shown that for a node v these values only depends on the values defined in both its *left(v)* and *right(v)* child. The incremental update rules introduced in [18] are:

$$\Sigma P_v = \Sigma P_{left(v)} + \Sigma P_{right(v)}$$

$$ect_v = \max \left\{ ect_{right(v)}, ect_{left(v)} + \Sigma P_{right(v)} \right\}$$

An example of a classic Θ-tree is given in Figure 1.

When transition times are considered, the ect_v value computed in the internal nodes of the Θ-tree may only be a loose lower-bound since it is only based on the earliest start times and the processing times. We strengthen the estimation of the earliest computation times (denoted ect^*) by also considering transition times. We add another value inside the nodes: n_v is the cardinality of Leaves(v) ($n_v = |\text{Leaves}(v)|$). The new update rules for the internal nodes of a Θ-tree are:

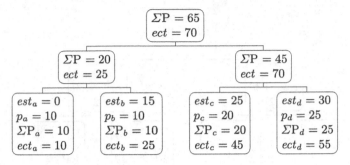

Fig. 1. Classic Θ-tree as described in [18].

$$\Sigma P_v = \Sigma P_{left(v)} + \Sigma P_{right(v)}$$
$$n_v = n_{left(v)} + n_{right(v)}$$
$$ect_v^* = \begin{cases} \max\{ect_{right(v)}^*, ect_{left(v)}^* + \Sigma P_{right(v)} + \underline{tt}(n_{right(v)}+1)\} & : v \text{ internal} \\ ect_v & : v \text{ leaf} \end{cases}$$

As an example, let us consider the set of four activities used in the Θ-tree example of Figure 1. Let us assume that the associated transition times are as defined in the matrix $\mathcal{M}$ of Figure 2. The lower bounds for set of activities of different cardinality are reported next to the matrix. With the new update rules defined above, we obtain the extended Θ-tree presented in Figure 3. Note that the values of ect^* in the internal nodes are larger than the values of ect reported in the classic Θ-tree (Figure 1).

$$\mathcal{M} = \begin{pmatrix} 0 & 10 & 13 & 18 \\ 12 & 0 & 15 & 15 \\ 10 & 18 & 0 & 20 \\ 19 & 11 & 16 & 0 \end{pmatrix}$$

Lower Bound	$k=1$	$k=2$	$k=3$
Min Weight Forest	10	20	31
Dynamic Programming	10	20	32
Min Cost Flow	10	20	33
Lagrangian Relaxation	10	20	32
$\underline{tt}(k)$	10	20	33

Fig. 2. Example of transition time matrix and associated lower bounds of transition times permutations.

Lemma 1. $ect_v \leq ect_v^* \leq ect_{Leaves(v)}^\diamond = \max_{\Theta' \subseteq Leaves(v)}\{est_{\Theta'} + p_{\Theta'} + \underline{tt}(|\Theta'|)\}$

Proof. The proof is similar to the proof of Proposition 7 in [18], by also integrating the inequality $\underline{tt}(|\Theta'|) \geq \underline{tt}(|\Theta' \cap Left(v)|) + \underline{tt}(|\Theta' \cap Right(v)|)$, which is itself a direct consequence of the fact that $\underline{tt}(k)$ is monotonic in k.

Since the new update rules are also executed in constant time for one node, we keep the time complexities of the initial Θ-tree structure from [18] which are at worst $\mathcal{O}(n \log(n))$ for the insertion of all activities inside the tree.

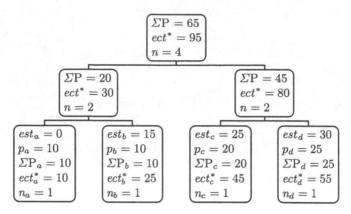

Fig. 3. Extended Θ-tree for transition times. The ect^* values reported in the internal nodes have been computed using the update rule of the extended Θ-tree.

Extending the Θ-Λ-tree with Transition Times

The Edge-Finding (EF) algorithm requires an extension of the original Θ-tree, called Θ-Λ-tree [18]. This extension is used to obtain an efficient EF algorithm. In this extension, in addition to the activities included in a Θ-tree, activities can be marked as *gray nodes*. Gray nodes represent activities that are not really in the set Θ. However, they allow to easily compute ect_Θ if *one* of the gray activities were included in Θ. If we consider the set of gray activities Λ such that $\Lambda \cap \Theta = \emptyset$, we are interested in computing the largest ect obtained by including *one* activity from Λ into Θ:

$$\overline{ect}_{(\Theta,\Lambda)} = \max_{i \in \Lambda} \; ect_{\Theta \cup \{i\}}$$

In addition to ΣP_v, ect_v, the Θ-Λ-tree structure also maintains $\overline{\Sigma P}_v$ and $\overline{ect}_v$, respectively corresponding to ΣP_v and ect_v, *if* the single gray activity in the sub-tree rooted by v maximizing ect_v *were* included:

$$\overline{ect}^*_{(\Theta,\Lambda)} = \max \left\{ ect^*_\Theta, \; \max_{i \in \Lambda} \left\{ ect^*_{\Theta \cup \{i\}} \right\} \right\}$$

The update rule for $\overline{\Sigma P}_v$ remains the same as the one described in [18]. However, following a similar reasoning as the one used for the extended Θ-tree, we add the n_v value, and update rules are modified for $\overline{ect}_v$ and $\overline{n}_v$. The rules become:

$$\overline{\Sigma P}_v = \max \left\{ \overline{\Sigma P}_{left(v)} + \Sigma P_{right(v)}, \Sigma P_{left(v)} + \overline{\Sigma P}_{right(v)} \right\}$$

$$\overline{ect}^*_v = \max \left\{ \begin{array}{l} \overline{ect}^*_{right(v)}, \\ \overline{ect}^*_{left(v)} + \Sigma P_{right(v)} + \underline{tt}(n_{right(v)} + 1), \\ ect^*_{left(v)} + \overline{\Sigma P}_{right(v)} + \underline{tt}(\overline{n}_{right(v)} + 1) \end{array} \right\}$$

$$\overline{n}_v = \left\{ \begin{array}{ll} n_v + 1 & \text{if the subtree rooted in } v \text{ contains a gray node} \\ n_v & \text{otherwise} \end{array} \right.$$

This extended Θ-Λ-tree allows us to efficiently observe how the ect^* of a set of activities is impacted if a single activity is added to this set. This information allows the EF algorithm to perform propagation efficiently[1]. An example of Θ-Λ-tree based on the example from Figure 3 and Figure 2 is displayed in Figure 4.

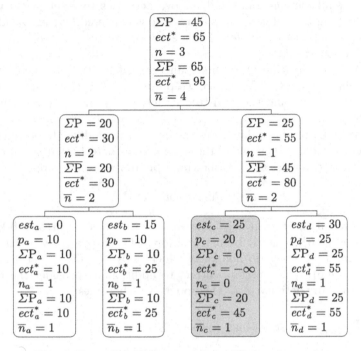

Fig. 4. Extended Θ-Λ-tree with modified update rules.

Similarly to the reasoning applied for the Θ-tree, the time complexities remain the same as the ones for the original Θ-Λ-tree structure from [18], which are at worst $\mathcal{O}(n \log(n))$.

6 Disjunctive Propagation Algorithms with Transition Times

In [18], a propagation procedure for the unary resource constraint is defined. This propagation procedure consists of a propagation loop including Overload Checking (OC), Detectable Precedences (DP), Not-First/Not-Last (NF/NL) and Edge Finding (EF) propagation algorithms. The first three rely on the Θ-tree while the latter employs the Θ-Λ-tree. Some small modifications can be done to these algorithms to obtain an efficient propagation procedure making use of knowledge about transition times.

---

[1] Finding the "responsible" activity $\arg\max_i ect_{\Theta \cup \{i\}}$ (required by EF) is done similarly to [18].

6.1 Extension of Classic Unary Resource Propagation Algorithms

The four mentioned propagation algorithms use a Θ-tree or a Θ-Λ-tree to compute ect_Θ on a set of activities Θ. OC checks if $ect_\Theta > lct_\Theta$. DP, NF/NL and EF rely on a set of rules that potentially allow to update the est or lct of an activity. They all incrementally add/remove activities to a set of activities Θ while maintaining the value ect_Θ. When a rule is triggered by the consideration of a given activity, the est or lct of this activity can be updated according to the current value of ect_Θ.

These four propagation algorithms can be used for the propagation of the transition time constraints. To do so, we propose to substitute in the filtering algorithms the Θ-tree and the Θ-Λ-tree structures by their extended versions. In the presence of transition times, $ect^*/\overline{ect}^*$ is indeed a stronger bound than $ect/\overline{ect}$. Furthermore, the update rules can be slightly modified to obtain an even stronger propagation. When one of these algorithms detects that an activity i is after all activities in a set Θ, the following update rule can be applied:

$$est_i \leftarrow \max\{est_i, ect_\Theta^*\}$$

In addition to all the transitions between activities of Θ - already taken into account in ect_Θ^* - there must be a transition between one activity and i (not necessarily from Θ, as we do not know which activity will be just before i in the final schedule). It is therefore correct to additionally consider the minimal transition from any activity to i. The update rule becomes:

$$est_i \leftarrow \max\left\{est_i, ect_\Theta^* + \min_{j \neq i} tt_{j,i}\right\}$$

An analogous reasoning can be applied to the update rule of the lct of an activity.

Similarly to the fix point propagation loop proposed in [18] for the unary resource constraint, the four extended propagation algorithms are combined to achieve an efficient propagation on transition time constraints. This allows to obtain a global propagation procedure instead of the conjunction of pairwise transition constraints described by Equation 1. The approach has however the disadvantage that the computation of ect_Θ^* integrates a lower bound. This prevents having the guarantee that sufficient propagation is achieved. The loop must thus also integrate the conjunction of pairwise transition constraints given in Equation 1. However, experimental results provided in Section 7 exhibits that the supplementary global constraint reasoning can provide a substantial filtering gain.

6.2 Detectable Precedences Propagation Example

Let us consider a small example (inspired from an example of [18]) with 3 activities, A, B and C whose domains are illustrated in Figure 5. The corresponding transition matrix and lower bounds are given in Figure 6.

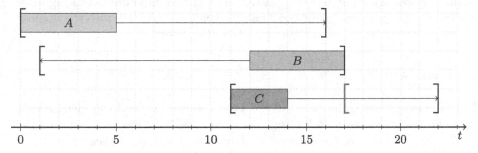

Fig. 5. Example of extended *Detectable Precedences* with transition times. The extended version updates est_C from 11 to 17, while the joint use of transition time binary constraints with the original unary constraint is not able to make this deduction.

$$\mathcal{M} = \begin{pmatrix} 0\ 4\ 6 \\ 2\ 0\ 5 \\ 4\ 3\ 0 \end{pmatrix}$$

Lower Bound	$k=1$	$k=2$
Min Weight Forest	2	5
Dynamic Programming	2	5
Min Cost Flow	2	5
Lagrangian Relaxation	2	5
$\underline{tt}(k)$	2	5

Fig. 6. Transition times for activities from Figure 5

(a) Classic Θ-tree

(b) Extended Θ-tree

Fig. 7. Comparison of classic and extended Θ-tree on the example described in Figures 5 and 6.

From this example, the *Detectable Precedences* algorithm will eventually build a Θ-tree containing activities A and B. Figures 7a and 7b respectively show the classic and the extended Θ-trees.

As one can see, ect^* is larger than ect as it is not agnostic about the transition time constraints. Furthermore, the update rule of est_C also includes the minimal transition time from any activity to C. This leads to the following update of est_C:

$$est_C = \max\left\{ est_C, ect_\Theta^* + \min_{i \neq C} tt_{i,C} \right\}$$
$$= \max\{11, 12 + 5\} = 17$$

We finally obtain an updated est_C, as shown by the red bold bracket in Figure 5. Notice that the joint use of the constraints given in Equation 1 with the original unary constraint of [18] would not make this deduction.

7   Evaluation

To evaluate our constraint, we used the OscaR solver [14] and ran instances on AMD Opteron processors (2.7 GHz). For each considered instance, we used the 3 following filterings for the unary constraint with transition times:

1. Binary constraints[2] (ϕ_b) given in Equation 1.
2. Binary constraints given in Equation 1 with the Unary global constraint of [18] (ϕ_{b+u}).
3. The constraint introduced in this article (ϕ_{uTT}). Based on our experience, we slightly changed the propagation loop order: Edge-finding is put in first position.

Considered Benchmarks. We constructed instances considering transition times from famous JobShop benchmarks. For a given benchmark $\mathcal{B}$, in each instance, we added generated transition times between activities, while ensuring that triangular inequality always hold. From $\mathcal{B}$, we generated new benchmarks $\mathcal{B}_{(a,b)}$ inside which the instances are expanded by transition times uniformly picked between $a\%$ and $b\%$ of $\overline{D}$, where $\overline{D}$ is the average duration of all activities in the original instance.

We generated instances from the well-known Taillard's instances[3]. From each instance, we generated 2 instances for a given pair (a, b), where the following pairs were used: $(50, 100)$, $(50, 150)$, $(50, 200)$, $(100, 150)$, $(100, 200)$ and $(150, 200)$. This allowed us to create 960 new instances[4].

Comparison of the 3 models

In order to present fair results regarding the benefits that are only provided by our constraint, we first followed the methodology introduced in [16]. Afterwards, we made measurements using a static search strategy, as it cannot be influenced by the additional pruning provided by our constraint.

Potential of the Constraint. In brief, the approach presented in [16] proposes to pre-compute a search tree using the filtering that prunes the less - the *baseline*

---

[2] For efficiency reason, dedicated propagators have been implemented instead of posting reified constraint.

[3] Available at http://mistic.heig-vd.ch/taillard/problemes.dir/ordonnancement.dir/ordonnancement.html.

[4] Available at http://becool.info.ucl.ac.be/resources/benchmarks-unary-resource-transition-times

propagator - and then to *replay* this search tree using the different studied filtering procedures. The point is to only measure the time gain provided by the propagation, by decoupling the gain provided by the search strategy (while still being able to use dynamic ones) from the one provided by the propagation. We used ϕ_b as the baseline filtering, and the *SetTimes* (st) search strategy to construct the search tree, as this strategy is recognized to provide good performances in Scheduling. The search tree construction time was limited to 600 seconds. We then constructed *performance profiles* as described in [16]. Basically, those are cumulative distribution functions of a performance metric τ. Here, τ is the ratio between the solution time (or number of backtracks) of a target approach (i.e. ϕ_{b+u} or ϕ_{uTT}) and that of the baseline (i.e. ϕ_b). For time (similar for number of backtracks), the function is defined as:

$$F_{\phi_i}(\tau) = \frac{1}{|\mathcal{M}|} \left| \left\{ M \in \mathcal{M} : \frac{t(replay(\text{st}), M \cup \phi_i)}{t(replay(\text{st}), M)} \leq \tau \right\} \right| \qquad (4)$$

where $\mathcal{M}$ is the set of considered instances while $t(replay(\text{st}), M \cup \phi_i)$ and $t(replay(\text{st}), M)$ are respectively the time required to replay the generated search tree with the studied model (model using ϕ_i, i.e. ϕ_{b+u} or ϕ_{uTT}) and with the baseline model.

Figures 8a and 8b respectively provide the profiles for time and backtrack for all the 960 instances[5]. Figure 8c provides a "long-term" view of Figure 8a.

From Figure 8a, we can first conclude that ϕ_{b+u} is clearly worse than ϕ_{uTT} and ϕ_b from a time perspective. Moreover, Figure 8b shows that ϕ_{b+u} rarely offers more pruning than ϕ_b.

In comparison, we can see from Figure 8a that for ~ 20% of the instances, ϕ_{uTT} is about 10 times faster than ϕ_b, and that we solve ~ 35% of the instances faster (see $F_{\phi_{uTT}}(1)$). Moreover, it offers more pruning for ~ 75% of the instances (see Figure 8b).

From Figure 8c, we can see that the constraint does not have too much overhead, as ϕ_{uTT} is at worst about 7.5 times slower than ϕ_b for ~ 45% percent of the instances ($F_{\phi_{uTT}}(7.5) - F_{\phi_{uTT}}(1)$). It is a bit slower for the remaining

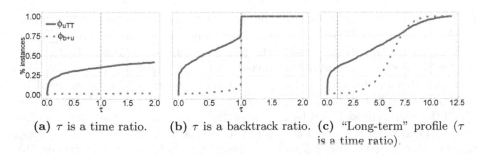

(a) τ is a time ratio.   (b) τ is a backtrack ratio.   (c) "Long-term" profile (τ is a time ratio).

Fig. 8. Performance profiles for the 960 generated instances.

---

[5] When instances were separated by number of jobs, the profiles had similar shapes.

Table 1. Best time results for ϕ_{uTT} compared to ϕ_b. The problem is to find the given makespan m using a binary static search strategy. Time is in seconds.

Instance	m	ϕ_{uTT}		ϕ_b		ϕ_{b+u}	
		Time	#Fails	Time	#Fails	Time	#Fails
15_15-3_225_50_100-1	2,344	**1.12**	**2,442**	117.92	980,330	432.07	911,894
50_15-8_750_50_100-2	6,682	**2.11**	**744**	182.27	1,127,272	999.79	1,127,272
20_15-7_300_150_200-2	4,784	**0.24**	**449**	17.63	168,466	62.27	168,466
15_15-6_225_50_100-1	2,398	**3.90**	**5,593**	187.93	889,079	534.20	602,591
50_20-3_1000_50_150-2	7,387	**2.96**	**1,709**	126.61	584,407	829.25	584,407
100_20-4_2000_150_200-1	18,595	**11.59**	**885**	340.32	332,412	1225.44	206,470
30_15-3_450_50_200-1	4,643	**1.97**	**1,178**	39.23	226,700	314.34	226,700
15_15-5_225_100_150-2	3,320	**0.91**	**2,048**	16.40	119,657	63.38	119,657
50_20-2_1000_50_100-1	6,979	**3.79**	**1,680**	63.16	878,162	4.63	1,695
30_15-10_450_100_200-1	5,586	**0.74**	**687**	9.24	106,683	41.25	106,683

Table 2. Worst time results for ϕ_{uTT} compared to ϕ_b. The problem is to find the given makespan m using a binary static search strategy. Time is in seconds.

Instance	m	ϕ_{uTT}		ϕ_b		ϕ_{b+u}	
		Time	#Fails	Time	#Fails	Time	#Fails
15_15-10_225_50_200-2	2,804	645.26	**546,803**	**127.38**	**546,803**	572.81	**546,803**
50_15-9_750_50_200-1	6,699	954.77	**164,404**	**174.63**	164,437	1,108.43	164,437
20_20-5_400_100_150-2	4,542	213.54	**78,782**	**38.26**	78,968	180.20	78,968
20_20-8_400_100_150-2	4,598	147.55	**164,546**	**26.42**	164,576	175.69	164,576
15_15-2_225_50_100-2	2,195	178.37	**96,821**	**31.23**	**96,821**	139.84	**96,821**
20_20-6_400_100_200-1	4,962	11.15	**8,708**	**1.94**	8,745	11.87	8,745
30_20-8_600_50_200-1	5,312	18.63	**6,665**	**3.15**	6,687	19.93	6,687
20_15-10_300_50_200-2	3,571	85.84	**61,185**	**14.24**	**61,185**	65.12	**61,185**
50_20-8_1000_100_200-1	9,186	286.61	**88,340**	**46.17**	**88,340**	180.23	**88,340**
20_15-1_300_100_150-1	3,557	189.37	**208,003**	**29.55**	209,885	157.33	209,885

$\sim 20\%$, which roughly corresponds to the percentage of instances for which ϕ_{uTT} provides no extra pruning (see $F_{\phi_{uTT}}(1)$ in Figure 8b).

Evaluation Over a Static Search Strategy. We here present results in a more "traditional" fashion. We compute the best makespan m that can be obtained with ϕ_b within 600 seconds, using the following binary static search strategy: fixed variable order, left branch assigns $start_i$ to est_i, right branch removes est_i from the domain of $start_i$. Then, the time and number of failures required by each model to find this solution are computed. We filtered out instances for which the solution was found by ϕ_b in less than 1 seconds and we computed the time ratio between ϕ_{uTT} and ϕ_b. From this perspective, the 10 best and worst results are reported in tables 1 and 2, respectively. On the 10 best instances, the gains (the number of failures and time) are significant (sometimes two orders of magnitude). On the 10 worst instances, the times obtained with ϕ_{uTT} are similar to the results using the classical unary resource (i.e. ϕ_{b+u}), while they are at worst around 6.4 times slower than the simple binary decomposition (i.e. ϕ_b).

8 Conclusion

In this paper, we proposed to extend classic unary resource propagation algorithms such that they consider transition times. We first stated that a lower bound of the time taken by transitions between activities from a set Ω is required to have a tighter bound of ect_Ω. We described several possible methods to compute these lower bounds. We then proposed to extend the Θ-tree and Θ-Λ-tree structures to integrate these lower bounds. These extended structures can then be used in unary propagation algorithms: OC, DP, NF/NL and EF. The new obtained propagation procedure has the advantage that it can be used conjointly with any other constraint and that it is completely independent from the objective to optimize. We have demonstrated that the additional pruning achieved by this propagation can dramatically reduce the number of nodes (and thus the time taken to solve the problem) on a wide range of instances.

Future work would analyze the possibility to integrate tighter incremental lower bounds in Θ-tree and Θ-Λ-tree structures. The order and real usefulness of the propagators (OC, DP, NF/NL, EF) should also be studied in order to acquire the most efficient fixpoint propagation loop. Finally, we would like to experiment on a new update rule in Θ-tree and Θ-Λ-tree to be able to obtain tighter lower bounds for ect_Ω.

References

1. Ahuja, R.K., Magnanti, T.L., Orlin, J.B.: Network Flows: Theory, Algorithms, and Applications. Prentice-Hall Inc, Upper Saddle River (1993)
2. Allahverdi, A., Ng, C., Cheng, T.E., Kovalyov, M.Y.: A survey of scheduling problems with setup times or costs. European Journal of Operational Research **187**(3), 985–1032 (2008)
3. Artigues, C., Belmokhtar, S., Feillet, D.: A new exact solution algorithm for the job shop problem with sequence-dependent setup times. In: Régin, J.-C., Rueher, M. (eds.) CPAIOR 2004. LNCS, vol. 3011, pp. 37–49. Springer, Heidelberg (2004)
4. Artigues, C., Buscaylet, F., Feillet, D.: Lower and upper bound for the job shop scheduling problem with sequence-dependent setup times. In: Proceedings of the Second Multidisciplinary International Conference on Scheduling: Theory and Applications, MISTA 2005 (2005)
5. Artigues, C., Feillet, D.: A branch and bound method for the job-shop problem with sequence-dependent setup times. Annals of Operations Research **159**(1), 135–159 (2008)
6. Balas, E., Simonetti, N., Vazacopoulos, A.: Job shop scheduling with setup times, deadlines and precedence constraints. Journal of Scheduling **11**(4), 253–262 (2008)
7. Baptiste, P., Le Pape, C., Nuijten, W.: Constraint-based scheduling: applying constraint programming to scheduling problems, vol. 39. Springer Science & Business Media (2001)

8. Focacci, F., Laborie, P., Nuijten, W.: Solving scheduling problems with setup times and alternative resources. In: AIPS, pp. 92–101 (2000)
9. Gagné, C., Price, W.L., Gravel, M.: Scheduling a single machine with sequence dependent setup time using ant colony optimization. Faculté des sciences de l'administration de l'Université Laval, Direction de la recherche (2001)
10. González, M.A., Vela, C.R., Varela, R.: A new hybrid genetic algorithm for the job shop scheduling problem with setup times. In: ICAPS, pp. 116–123 (2008)
11. Grimes, D., Hebrard, E.: Job shop scheduling with setup times and maximal time-lags: a simple constraint programming approach. In: Lodi, A., Milano, M., Toth, P. (eds.) CPAIOR 2010. LNCS, vol. 6140, pp. 147–161. Springer, Heidelberg (2010)
12. Kruskal, J.B.: On the shortest spanning subtree of a graph and the traveling salesman problem. Proceedings of the American Mathematical society **7**(1), 48–50 (1956)
13. Oddi, A., Rasconi, R., Cesta, A., Smith, S.F.: Exploiting iterative flattening search to solve job shop scheduling problems with setup times. PlanSIG2010, p. 133 (2010)
14. OscaR Team: OscaR: Scala in OR (2012). https://bitbucket.org/oscarlib/oscar
15. Tahar, D.N., Yalaoui, F., Amodeo, L., Chu, C.: An ant colony system minimizing total tardiness for hybrid job shop scheduling problem with sequence dependent setup times and release dates. In: Proceedings of the International Conference on Industrial Engineering and Systems Management, pp. 469–478 (2005)
16. Van Cauwelaert, S., Lombardi, M., Schaus, P.: Understanding the potential of propagators. In: Proceedings of the Twelfth International Conference on Integration of Artificial Intelligence and Operations Research techniques in Constraint Programming (2015)
17. Vilím, P.: $O(n\log n)$ filtering algorithms for unary resource constraint. In: Régin, J.-C., Rueher, M. (eds.) CPAIOR 2004. LNCS, vol. 3011, pp. 335–347. Springer, Heidelberg (2004)
18. Vilım, P.: Global constraints in scheduling. Ph.D. thesis, Charles University in Prague, Faculty of Mathematics and Physics, Department of Theoretical Computer Science and Mathematical Logic, KTIML MFF, Universita Karlova, Praha (2007)
19. Vilím, P., Barták, R., Čepek, O.: Extension of o (n log n) filtering algorithms for the unary resource constraint to optional activities. Constraints **10**(4), 403–425 (2005)
20. Wolf, A.: Constraint-based task scheduling with sequence dependent setup times, time windows and breaks. GI Jahrestagung **154**, 3205–3219 (2009)
21. Zampelli, S., Vergados, Y., Van Schaeren, R., Dullaert, W., Raa, B.: The berth allocation and quay crane assignment problem using a CP approach. In: Schulte, C. (ed.) CP 2013. LNCS, vol. 8124, pp. 880–896. Springer, Heidelberg (2013)

A Global Constraint for a Tractable Class of Temporal Optimization Problems

Alban Derrien[1], Jean-Guillaume Fages[2], Thierry Petit[1,3],
and Charles Prud'homme[1 (✉)]

[1] TASC (CNRS/INRIA), Mines Nantes, 44307 Nantes, France
{alban.derrien,charles.prudhomme}@mines-nantes.fr
[2] COSLING S.A.S., 44307 Nantes, France
jg.fages@cosling.com
[3] Foisie School of Business, WPI, Worcester, MA 01609, USA
TPetit@wpi.edu

Abstract. This paper is originally motivated by an application where the objective is to generate a video summary, built using intervals extracted from a video source. In this application, the constraints used to select the relevant pieces of intervals are based on Allen's algebra. The best state-of-the-art results are obtained with a small set of ad hoc solution techniques, each specific to one combination of the 13 Allen's relations. Such techniques require some expertise in Constraint Programming. This is a critical issue for video specialists. In this paper, we design a generic constraint, dedicated to a class of temporal problems that covers this case study, among others. *ExistAllen* takes as arguments a vector of tasks, a set of disjoint intervals and any of the 2^{13} combinations of Allen's relations. *ExistAllen* holds if and only if the tasks are ordered according to their indexes and for any task at least one relation is satisfied, between the task and at least one interval. We design a propagator that achieves bound-consistency in $O(n + m)$, where n is the number of tasks and m the number of intervals. This propagator is suited to any combination of Allen's relations, without any specific tuning. Therefore, using our framework does not require a strong expertise in Constraint Programming. The experiments, performed on real data, confirm the relevance of our approach.

1 Introduction

The study of temporal relations between elements of a process is a very active topic of research, with a wide range of applications: biomedical informatics [5,20], law [17], media [2,4,8] etc. Temporal reasoning enables to analyze the content of a document in order to infer high-level information. In particular, a summary of a tennis match may be generated from the match recording with Artificial Intelligence techniques [2,4]. The summarization requires to extract some noticeable time intervals and annotate them with qualitative attributes using signal recognition techniques. Then, the summary generation may be formulated as a Constraint Satisfaction Problem (CSP) where variables are the video segments to

© Springer International Publishing Switzerland 2015
G. Pesant (Ed.): CP 2015, LNCS 9255, pp. 105–120, 2015.
DOI: 10.1007/978-3-319-23219-5_8

be displayed whereas constraints stem from different considerations: displaying relevant information, balancing the selected content, having nice transitions etc. Then, the CSP may be solved quite efficiently with a Constraint-Programming (CP) solver. At first sight, this seems an easy task.

Unfortunately, things get harder when it comes to practice. Designing a CP model requires some expert knowledge, in order to achieve good performances on hard problems. This is particularly true when one has to design a global constraint that would be missing in the solver. For instance, the summarization model of [2] relied on many disjunctions that are poorly propagated by constraint engines and thus lead to unsatisfiable performances. Therefore, the authors collaborated with CP experts to design ad hoc global constraints, which lead to significant speedups [4]. Alternatively, one may have used temporal logic models, to benefit from solution techniques and solvers dedicated the temporal CSP [6,7,9,18]. As the video summarization gets more sophisticated, all these approaches suffer from the need of specific and often intricate propagators/models. This is a critical issue for video specialists, who are rarely CP experts. Furthermore, one may need to include in her model other features available in state-of-the-art constraint solvers, such as standard global constraints and predefined search strategies. What is missing is a both expressive and efficient global constraint for modeling a relevant class of problems on time intervals.

We introduce the *ExistAllen* constraint, defined on a vector of tasks $\mathcal{T}$ and a set of disjoint intervals $\mathcal{I}$, respectively of size n and m. Given a subset $\mathcal{R}$ of Allen's relations [1], *ExistAllen* is satisfied if and only if the two following properties are satisfied:

1. For any task in $\mathcal{T}$ at least one relation in $\mathcal{R}$ is satisfied between this task and at least one interval in $\mathcal{I}$.
2. Tasks in $\mathcal{T}$ are ordered according to their indexes in the vector given as argument, *i.e.*, for any integer i, $1 \leq i < n$, the task T_i should end before or at the starting time of T_{i+1}.

In the context of video-summarization, tasks in $\mathcal{T}$ are the video segments that compose the summary. Fixed video sequences in $\mathcal{I}$ are extracted from the source according to some precise features. In this way, it is possible to constrain the content of the summary with qualitative information.

Considering the invariability of task processing times, we introduce a bound-consistency propagator for this constraint, suited to any of the 2^{13} subsets of Allen's relations. The time complexity of the most natural algorithm for this propagator is $O(n \times m)$. We propose an improved algorithm, running in $O(n+m)$ time. While *ExistAllen* may be used in different contexts, e.g., online scheduling, this paper is motivated by video-summarization. Our experiments on the Boukadida et al.'s application [4] demonstrate that using our generic constraint and its linear propagator is significantly better than the models built with standard constraints of the solver, and competitive with the ad hoc global constraint approach.

Table 1. Allen's temporal algebra relations.

Symbol	Relation	
$T^{(1)} \, p \, T^{(2)} \, (T^{(2)} \, pi \, T^{(1)})$	$T^{(1)} \, precedes \, T^{(2)}$	$T^{(1)}$ $\longmapsto$ $\quad$ $T^{(2)}$ $\longmapsto$
$T^{(1)} \, m \, T^{(2)} \, (T^{(2)} \, mi \, T^{(1)})$	$T^{(1)} \, meets \, T^{(2)}$	$T^{(1)}$ $\longmapsto$ $T^{(2)}$ $\longmapsto$
$T^{(1)} \, o \, T^{(2)} \, (T^{(2)} \, oi \, T^{(1)})$	$T^{(1)} \, overlaps \, T^{(2)}$	$T^{(1)}$ $\longmapsto$ $T^{(2)}$ $\longmapsto$
$T^{(1)} \, s \, T^{(2)} \, (T^{(2)} \, si \, T^{(1)})$	$T^{(1)} \, starts \, T^{(2)}$	$T^{(1)}$ $\longmapsto$ $T^{(2)}$ $\longmapsto$
$T^{(1)} \, d \, T^{(2)} \, (T^{(2)} \, di \, T^{(1)})$	$T^{(1)} \, during \, T^{(2)}$	$T^{(1)}$ $\longmapsto$ $T^{(2)}$ $\longmapsto$
$T^{(1)} \, f \, T^{(2)} \, (T^{(2)} \, fi \, T^{(1)})$	$T^{(1)} \, finishes \, T^{(2)}$	$T^{(1)}$ $\longmapsto$ $T^{(2)}$ $\longmapsto$
$T^{(1)} \, eq \, T^{(2)}$	$T^{(1)} \, equal \, to \, T^{(2)}$	$T^{(1)}$ $\longmapsto$ $T^{(2)}$ $\longmapsto$

2   Background

In this section we give some background and fix the notations used in this paper.

2.1   Temporal Constraint Networks

Temporal reasoning has been an important research topic for the last thirty years. One may distinguish qualitative temporal reasoning, based on relations between intervals and/or time points, from quantitative reasoning where duration of a given event is represented in a numerical fashion. Allen's algebra [1] represents qualitative temporal knowledge by interval constraint networks. An interval constraint network is a directed graph where nodes represent intervals and edges are labelled with disjunctions of Allen's relations. Table 1 details those relations. Many state-of-the-art papers deal with generic solving techniques and tractability of temporal networks [6,7,9,18], including temporal problems with quantified formulas [10,12]. Most of these methods make no strong restriction on the constraint network to be solved.

A few techniques, more specialized, focus on optimization problems. The two most related to this paper are the following. Kumar et al. [11] consider temporal problems with "taboo" regions, minimizing the number of intersections between tasks and a set of fixed intervals. These problems occur in scheduling applications, among others. In the context of video summarization, a recent paper [4] proposes the idea of using global constraints involving a set of ordered intervals. Each constraint is restricted to one specific relation in the 2^{13} combinations of Allen's relations. The propagators are not described.

2.2   Constraint Programming (CP)

CP is a problem solving framework where relations between variables are stated in the form of constraints, which together form a constraint network. Each variable x has a domain $D(x)$, whose minimum value is $\underline{x}$ and maximum value is $\overline{x}$. A *task* T in a set $\mathcal{T}$ is an object represented by three integer variables: s_T, its starting time, e_T, its ending time, and p_T, its processing time. The task should satisfy the constraint $s_T + p_T = e_T$. An interval I in a set $\mathcal{I}$ is a fixed task, defined by integer values instead of variables. A *propagator* is an algorithm associated with a constraint, stated on a set of variables. This propagator removes from domains values that cannot be part of a solution to that constraint. The notion of consistency characterizes propagator effectiveness. In this paper, we consider *bound*($\mathbb{Z}$)-*consistency* [3]. When domains are exclusively represented by their bounds (i.e., have no holes), *bound*($\mathbb{Z}$)-*consistency* ensures that for each variable x, $\underline{x}$ and $\overline{x}$ can be part of a solution of the constraint.

3   The *ExistAllen* Constraint

This section introduces the *ExistAllen* constraint and its propagator. Let $\mathcal{T} = \{T_1, T_2, \ldots, T_n\}$ be a set of tasks, such that any task T_{i+1} must be scheduled at or after the end of task T_i. Similarly, we define a set of ordered Intervals $\mathcal{I} = \{I_1, I_2, \ldots, I_m\}$. From a subset $\mathcal{R}$ of Allen's relations, *ExistAllen* ensures that any task in $\mathcal{T}$ is related to at least one interval in $\mathcal{I}$.

Definition 1 (*ExistAllen*). $ExistAllen(\mathcal{T}, \mathcal{R}, \mathcal{I}) \Leftrightarrow$

$$\forall T \in \mathcal{T}, \bigvee_{R \in \mathcal{R}} \bigvee_{I \in \mathcal{I}} T\,R\,I$$

$$\wedge \quad (\forall i, 1 \leq i < n, s_{T_{i+1}} \geq e_{T_i})$$

Example 1. When summarizing the video of a tennis match, it may be required that each segment (task) selected to be part the summary contains an applause. Applause intervals can be preprocessed [4]. Such requirement can then be formulated with a *ExistAllen* constraint, where $\mathcal{T}$ are the selected segments, $\mathcal{I}$ are the applause segments and $\mathcal{R}$ is set to $\{fi, di, eq, si\}$.

A symmetry can be used, where the problem is seen in a mirror: starting variables become ending variables and the lower bounds filtering of the mirror relation is applied to the upper bounds (e.g. *starts* is propagated onto upper bounds using *finishes*, see Table 1). Therefore, we put the focus on the algorithms for the lower bounds of starting/ending task variables in $\mathcal{T}$. We consider here that processing times are exclusively updated by the constraints $s_T + p_T = e_T$.

3.1   Basic Filtering: One Relation, One Task, One Interval

The basic filtering rule specific to each Allen's relation can be derived from time-point logic relations [19], in order to state lower and upper bounds of starting and ending times of tasks. For instance, consider a task $T \in \mathcal{T}$, an interval $I \in \mathcal{I}$, and the relation *starts*. The relation is satisfied if and only if two conditions are met:

$$T \; s \; I \;\Leftrightarrow\; s_T = s_I \;\wedge\; e_T < e_I$$

The only filtering on the lower bounds of task T induced by the two conditions of relation *starts* is $\underline{s_T} \geq s_I$. On the same basis, we define in Table 2 the conditions and filtering rules for the 13 Allen's algebra relations between a task T and an interval I.

Table 2. Allen's algebra lower bound filtering on each variable of a task T.

Relation	Conditions	Filtering
$T \, p \, I$	$e_T < s_I$	
$T \, pi \, I$	$e_I < s_T$	$\underline{s_T} > e_I$
$T \, m \, I$	$e_T = s_I$	$\underline{e_T} \geq s_I$
$T \, mi \, I$	$s_T = e_I$	$\underline{s_T} \geq e_I$
$T \, o \, I$	$s_T < s_I$	
	$e_T < e_I$	
	$e_T > s_I$	$\underline{e_T} > s_I$
$T \, oi \, I$	$s_T < e_I$	
	$s_T > s_I$	$\underline{s_T} > s_I$
	$e_T > e_I$	$\underline{e_T} > e_I$
$T \, s \, I$	$s_T = s_I$	$\underline{s_T} \geq s_I$
	$e_T < e_I$	

Relation	Conditions	Filtering
$T \, si \, I$	$s_T = s_I$	$\underline{s_T} \geq s_I$
	$e_T > e_I$	$\overline{e_T} > e_I$
$T \, d \, I$	$s_T > s_I$	$\underline{s_T} > s_I$
	$e_T < e_I$	
$T \, di \, I$	$s_T < s_I$	
	$e_T > e_I$	$\underline{e_T} > e_I$
$T \, f \, I$	$s_T > s_I$	$\underline{s_T} > s_I$
	$e_T = e_I$	$\underline{e_T} \geq e_I$
$T \, fi \, I$	$s_T < s_I$	
	$e_T = e_I$	$\underline{e_T} \geq e_I$
$T \, eq \, I$	$s_T = s_I$	$\underline{s_T} \geq s_I$
	$e_T = e_I$	$\underline{e_T} \geq e_I$

3.2   A First Propagator

We propose a propagator based on two procedures. Again, we only present the algorithms adjusting lower bounds of starting times and ending times.

The main procedure, EXISTALLENQUADRATIC (Algorithm 1), takes as arguments the sets $\mathcal{T}$ and $\mathcal{I}$ and a subset $\mathcal{R}$ of Allen's relations. Algorithm 1 considers all tasks in $\mathcal{T}$ and checks the relations according to intervals in $\mathcal{I}$. At the end of the procedure, the bounds of variables in $\mathcal{T}$ are updated according to the earliest support. If no support has been found, a domain is emptied and the solver raises a failure exception. The order between the tasks is maintained by the procedure PROPAGATE at line 10, whose propagation is obvious.

Algorithm 1 calls the procedure ALLENALLRELATION(Algorithm 2, line 6), which performs the check for one task and one interval, with all the relations in $\mathcal{R}$. In order to define a procedure instead of a function returning a pair of bounds, we use two global variables s^* and e^*, storing permanently the two lowest adjustments, respectively for the lower bounds of the starting and ending

```
Require: Global Variable : s*, e*
 1: procedure EXISTALLENQUADRATIC(T, R, I)
 2:     for i = 1 to n do                              ▷ Loop over tasks
 3:         s* ← s̄_{T_i} + 1                ▷ Intialize to a value out of the domain
 4:         e* ← ē_{T_i} + 1
 5:         for j = 1 to m do                           ▷ Loop over Intervals
 6:             ALLENALLRELATION(T_i, R, I_j)
 7:         end for
 8:         s_{T_i} ← s*
 9:         e_{T_i} ← e*
10:         if i < n − 1 then PROPAGATE(e_{T_i}, ≤, s_{T_{i+1}}) end if
11:     end for
12: end procedure
```

Algorithm 1: Main procedure.

time of the current task. An adjustment of the bound of one such variable is made in Algorithm 2 if and only if the current relation gives a support which is less than the lowest previously computed support.

```
Require: Global Variable : s*, e*
 1: procedure ALLENALLRELATION(T, R, I)
 2:     for all r ∈ R do
 3:         if CHECKCONDITION(T, r, I) then
 4:             s* ← min(s*, SEEKSUPPORTSTART(T, r, I))
 5:             e* ← min(e*, SEEKSUPPORTEND(T, r, I))
 6:         end if
 7:     end for
 8: end procedure
```

Algorithm 2: Update of s^* and e^*.

The function CHECKCONDITION(T, r, I) (line 3) returns true if and only if a support can be found. Consider again the example of relation *starts*. From the condition induced by Table 2, col. 2, we have:

$$\text{CHECKCONDITION}(T, s, I) \Leftrightarrow \underline{s_T} \leq s_I \wedge \overline{s_T} \geq s_I \wedge \underline{e_T} < e_I.$$

If the conditions are met then a support exists. SEEKSUPPORTSTART(T, s, I) returns the lowest support for the starting time variable, that is, s_I. As no filtering is directly induced for the ending time variable, the minimal support returned, for the relation s by SEEKSUPPORTEND(T, s, I) is $\max(\underline{e_T}, s_I + \underline{p_T})$. For each Allen's algebra relation, the three functions are similarly derived from Table 2.

Lemma 1. *The time complexity of Algorithm 1 is in $O(n \times m)$.*

Proof. As the number of Allen's relations is constant (at most 13), and the call of the three functions (lines 3, 4 and 5 in Algorithm 2) are in $O(1)$, the whole filtering of lower bounds is performed in $O(n \times m)$ time complexity.

Theorem 1. *The propagator based on Algorithm 1 and its symmetric calls for upper bounds of starting/ending variables in $\mathcal{T}$, ensure* bounds($\mathbb{Z}$)-*consistency if processing times are fixed integers.*

Proof. After the execution of Algorithm 1, the optimal update has been done for each lower bound, according to the current upper bounds. The filtering of upper bounds is symmetric. Therefore, providing that durations of tasks are fixed, the propagator ensures *bound($\mathbb{Z}$)-consistency* when a fixpoint is reached.

A fixpoint is not necessarily reached after running Algorithm 1 twice, once to filter lower bounds and once to filter upper bounds. Indeed, the pass on upper bounds can filter values which were previously supports for lower bounds. Let's consider ExistAllen($\mathcal{T}, \mathcal{R}, \mathcal{I}$) depicted in Figure 1 wherein: $\mathcal{T} = \{T_1 = \langle s_{T_1} = [0,2], p_{T_1} = [2,4], e_{T_1} = [4,6] \rangle, T_2 = \langle s_{T_2} = [5,6], p_{T_2} = [4,5], e_{T_2} = [10,10] \rangle\}$, $\mathcal{R} = \{d, di\}$ and $\mathcal{I} = \{I_1 = [1,5], I_2 = [6,8]\}$.

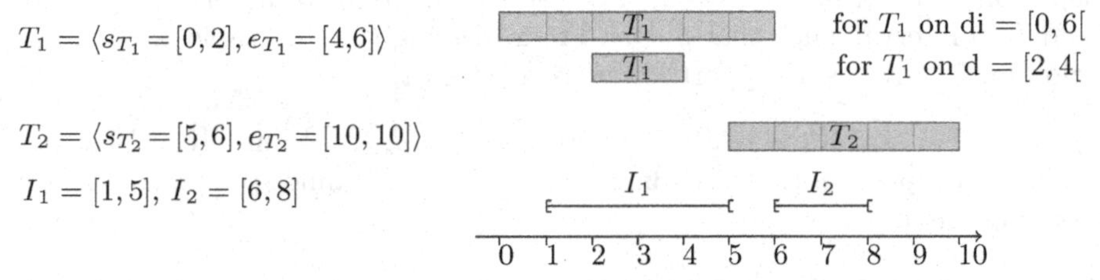

$$T_1 = \langle s_{T_1} = [0,2], e_{T_1} = [4,6] \rangle$$

for T_1 on di = $[0,6[$

for T_1 on d = $[2,4[$

$$T_2 = \langle s_{T_2} = [5,6], e_{T_2} = [10,10] \rangle$$

$$I_1 = [1,5], I_2 = [6,8]$$

Fig. 1. Several phases may be required to get a fixpoint when the processing times of tasks are not fixed.

We now simulate the calls of Algorithm 1 required to reach a fixpoint. No values are filtered during the run of Algorithm 1 on lower bounds, since di provides the minimal support for s_{T_1}, d provides the minimal support for e_{T_1} and di provides the minimal supports for T_2. On the run of Algorithm 1 on upper bounds, since no relation provides support for 6 from s_{T_2}, the value is removed. The ordering constraint (Algorithm 1, line 10) removes 6 from e_{T_1}. Thus, di is no longer valid to provide a support for T_1. Consequently, a second call to Algorithm 1 on lower bounds has to be done and the minimal value for s_{T_1} will then be 2, given by relation d.

However, when task processing times are constants, the minimum support of the ending date of an activity is the minimum support of its starting date plus the constant. Therefore, the issue mentioned in the previous example cannot occur. The fixpoint can be reached in two passes. One may note that, in this case, we could improve the algorithm by ordering Allen's relations. As our target application involves variable processing times, we do not detail this simplification.

3.3   A Linear Propagator

This section introduces an improved propagator, running in $O(n + m)$ time complexity.

First, one may observe that the satisfaction of the relation *precedes* can be done in constant time. Indeed, if a task T can precede the last interval I_m, the lower bounds are a support. And if not, task T cannot precede any interval. The same way relation *precedes inverse* can be symmetrically checked with the first intervals. Therefore, to simplify the presentation and without loss of generality, we now consider that relations p and pi can be isolated and treated separately. For each task, they can be checked in constant time. In this section we exclude them from $\mathcal{R}$.

Second, as the sets $\mathcal{T}$ and $\mathcal{I}$ are chronologically ordered, we can exploit dominance properties that lead to a linear propagator. We now provide those properties and their proof, as well as the EXISTALLENLINEAR procedure. As in Section 3.2, we focus on lower bounds of starting/ending variables of tasks in $\mathcal{T}$.

Property 1. An interval I which starts after a support y for the ending time of a task T cannot provide a support for e_T lower than y.

Proof. Consider a task T, $\mathcal{R} \subseteq \{m, mi, o, oi, s, si, d, di, f, fi, eq\}$, y a support for e_T and I an interval, such that $y < s_I$. From filtering rules in Table 2, col. 3, no relation in $\mathcal{R}$ provides a support lower than y for e_T.   □

Property 2. For any of the relations in $\{mi, oi, s, si, d, f, eq\}$, an interval I which starts after a support y for the ending time of a task T cannot provide a support for s_T lower than y.

Proof. Similar to proof of Property 1.   □

Property 3. For any of the relations in $\{m, o, di, fi\}$, an interval I which starts after a support for the ending time of a task T can provide a support for s_T. The lowest support for T is then given by the interval I in $\mathcal{I}$ having the lowest s_I.

Proof. Let T be a task and r be a relation, $r \in R = \{m, o, di, fi\}$, x be a support for s_T, y be a support for e_T and I be the interval in $\mathcal{I}$ with the lowest s_I, such that $s_I > y$. For each $r \in R$, we distinguish two cases.

Case 1 (T is related to I with relation r, T r I). As the support for e_T is at least s_I and no rule on starting time is explicitly defined by r (Table 2, col. 3), then the support for s_T is at least $s_I - \overline{p_T}$. Given that all intervals from $\mathcal{I}$ are chronologically ordered, no interval greater than I can provide a lower x.

Case 2 (T is not related to I with relation r: ¬(T r I)). Consider the m relation. Task T can meet interval I only if $s_I \in [\underline{e_T}, \overline{e_T}]$ (Table 2 col. 2). As it exists a support y with a value $e^* < s_I$ and the relation m is not satisfied, we have $\overline{e_T} < s_I$. Given that all intervals are chronologically ordered, no interval with a greater index can meet the task T. A similar reasoning can be applied to the other relations in R.   □

Thanks to Properties 1, 2 and 3, we can improve the Algorithm 1 by stopping the iteration over intervals in $\mathcal{I}$ for a given task T (Line 5) if $e^* < s_I$. We now provide two properties from which, for a given task T, the iteration over intervals in $\mathcal{I}$ does not have to always start at the first interval of $\mathcal{I}$.

Property 4. An interval I which ends before the ending time of a task T_i cannot provide a support for the next task T_{i+1}.

Proof. Let T_i be a task and I an interval such that $e_I < \underline{e_{T_i}}$. Then $e_I < \underline{s_{T_{i+1}}}$. T_{i+1} cannot be in relation with I. $\qquad\square$

Property 5. Given a task T, there exists at most one interval between the interval I_i with the highest ending time such that $e_{I_i} < \underline{e_T}$ and the interval I_j with the lowest starting time such that $s_{I_j} > \underline{e_T}$.

Proof. Let T be a task, let I_i be the interval with the highest e_I such that $e_I < \underline{e_T}$, we have then $e_{I_{i+1}} > \underline{e_T}$, and let I_j be the interval with the lowest s_{I_j} such that $s_{I_j} > \underline{e_T}$, we have then $s_{I_{j-1}} < \underline{e_T}$. Given that all intervals are chronologically ordered, and that $s_{I_{j-1}} < e_{I_{i+1}}$, we have that $j - 1 \leq i + 1$, that is $j - i \leq 2$. As the difference between indices i and j is at most 2, there is at most one interval between i and j. $\qquad\square$

Thanks to Properties 4 and 5, we know that the next task cannot be in relation with any interval whose index is lower than or equal to $j - 2$. We can improve the Algorithm 1 by starting the iteration over intervals for a given task at $j - 1$.

Require: Global Variable : s^*, e^*
1: **procedure** EXISTALLENLINEAR($\mathcal{T}$, R, $\mathcal{I}$)
2:     $j \leftarrow 0$
3:     **for** i = 1 to n **do**   ▷ Loop over Tasks
4:         $s^* \leftarrow \overline{s_{T_i}} + 1$ ▷ Intialize to a value out of the domain
5:         $e^* \leftarrow \overline{e_{T_i}} + 1$
6:         **repeat**   ▷ Loop over Intervals
7:             $j \leftarrow j + 1$
8:             ALLENALLRELATION(T_i, $\mathcal{R}$, I_j)
9:         **until** $j < m$ and $e^* < s_{I_j}$ ▷ Stop iteration, see Properties 1, 2 and 3
10:        $s_{T_i} \leftarrow s^*$
11:        $\overline{e_{T_i}} \leftarrow e^*$
12:        **if** $i < n - 1$ **then** PROPAGATE(e_{T_i}, $\leq$, $s_{T_{i+1}}$) **end if**
13:            $j \leftarrow max(0, j - 2)$ ▷ Set next interval index, see Properties 4 and 5
14:    **end for**
15: **end procedure**

Algorithm 3: Linear Algorithm for Main Procedure.

By construction, Algorithms 3 and 1 do exactly the same filtering.

Theorem 2. *The time complexity of Algorithm 3 is in $O(n + m)$.*

Proof. The number of evaluated intervals is equal to $m+2\times(n-1)$: every time a new task is evaluated, the algorithm goes two intervals back. The new algorithm is then in $O(n+m)$.   □

3.4   Improvements

In this section, we describe three improvements brought to the *ExistAllen* constraint in order to improve its efficiency in practice.

First, the Algorithm 2 can be adapted to store, for a given task, the index of the first interval which satisfied the conditions of a relation. Indeed, intervals located before that interval do not provide a support for the task (and they will never do in the current search sub-tree). By doing so, useless calls to CHECK-CONDITION can be avoided since they will always return false. In practice, an operation is added after the line 7 in Algorithm 3 to put in j the maximum between j and the first satisfying interval for the task evaluated. These indices are automatically updated upon backtrack.

Similarly, the tasks whose variables have been modified since the last call of the procedure have to be memorized. Thus, the for-loop (line 3-14 in Algorithm 3) can start from the index of the first modified task. Moreover, following tasks that have not been modified can be skipped safely.

Finally, our generic framework enables to replace some remarkable combinations of the Allen's algebra relations with *meta-relations*. By doing so, even if the complexity of the Algorithm 3 remains the same, the number of operations made to check conditions and seek supports for the combined relations may be reduced. For instance, a "*contains*" meta-relation, as described in Example 1, which expresses $\{fi, di, eq, si\}$, can save up to three calls of the methods in the for-loop in Algorithm 2, lines 2-7. Note that since $\{p, pi\}$ are handled in a particular way by the constraint, it is even more efficient to limit the combinations to relations in $\{m, mi, o, oi, s, si, d, di, f, fi, eq\}$. Adding meta-relations is easy in our implementation since we use a facade design pattern to define the 13 relations. Some meta-relations may require to define their inverse, in order to filter on upper bounds. This is not the case for "*contains*", though. Indeed, the mirror relation of fi is si, the mirror relation of si is fi, while the mirror relation of di is di and the mirror relation of eq is eq.

4   Evaluation

The main contribution of this work is an "expressivity gain", which leads to reducing the investment necessary to build and maintain a model. Nevertheless, it is important to check if this gain does not come at the price of efficiency. In this section, we empirically evaluate the impact of the proposed filtering algorithm. First, we recall the video summarization problem. Second, we show that the expressive *ExistAllen* constraint we introduced is very competitive with the state-of-the-art dedicated approach.

4.1   Problem Description

The video summarization problem of [4] consists in extracting audio-visual features and computing segments from an input video. The goal is to provide a summary of the video. More precisely, they consider tennis match records.

Several features (games, applause, speech, dominant color, etc.) are extracted as a preprocessing step, in order to compute time intervals that describe the video. Then, the problem is to compute segments of the video that will constitute the summary. The number of segments to compute and their minimal and maximal duration are given as parameter, as well as the summary duration. In this case-study, the purpose is to build a tennis match summary with a duration between four and five minutes, composed of ten segments, whose duration varies between 10 and 120 seconds.

In order to obtain a good summary (from the end-user point of view), this process is subject to constraints, such as:

- (1a) a segment should not cut a speech interval,
- (1b) a segment should not cut a game interval,
- (2) each selected segment must contain an applause interval,
- (3) the cardinality of the intersection between the segments and the dominant color intervals must be at least one third of the summary,

On the assumption that an applause indicates an interesting action, the presence of applause in the summary must be maximized, *i.e.*, the cardinality of the intersection between the computed segments and the applause time intervals must be as large as possible.

Table 3. Match features.

Name	Total duration	# Speech	# Applause	# Dominant color	# Games
M_1	2h08	571	271	1323	156
M_2	1h22	332	116	101	66
M_3	3h03	726	383	223	194

4.2   Benchmark

We consider the model implementation as well as a 3-instance dataset (see Table 3) kindly provided by Boukadida et. al. [4].

The model is encoded using integer variables. A segment is represented by three variables to indicate its start, duration and end. If constraints (1a) and (1b) are easily ensured by forbidding values for the segment start and end variables, most constraints have been encoded using ad hoc propagators. This holds on constraint (2), whereas it could be handled with a single *ExistAllen* constraint wherein $\mathcal{T}$ is the set of selected segments, $\mathcal{R}$ is equal to $\{fi, di, eq, si\}$ and $\mathcal{I}$ is the set of applause time intervals. Therefore, to evaluate the practical impact of the linear-time *ExistAllen* propagator, four models are considered.

1. `decomp`: constraint (2) is explicitly represented by the disjunction depicted in Section 3, Definition 1, using primitive constraints of the solver.
2. `allen(n.m)`: constraint (2) is represented by an *ExistAllen* constraint, using the quadratic propagator presented in section 3.2,
3. `allen(n+m)`: constraint (2) is represented by an *ExistAllen* constraint, using the linear propagator described in sections 3.3 and 3.4.
4. `ad hoc`: constraint (2) is represented with the dedicated constraints of [4]. Such model is given for reference only as neither its complexity nor its consistency level are known.

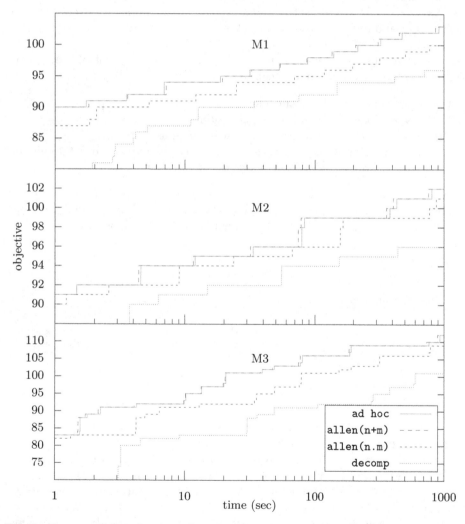

Fig. 2. Comparative evaluation of `decomp`, `ad hoc`, `allen(n.m)` and `allen(n+m)` on the three matches $M1$, $M2$ and $M3$ with a static search heuristic. The plots report the evolution of the objective value to be maximized with respect to the cpu time in seconds. The x-axis are in logscale.

Each of the four models has been implemented in Choco-3.3.0 [16]. Each of the instances was executed with a 15 minutes timeout, on a Macbook Pro with 8-core Intel Xeon E5 at 3Ghz running a MacOS 10.10, and Java 1.8.0_25. Each instance was run on its own core, each with up to 4096MB of memory.

In order to compare the efficiency on the four models, we first consider a static search heuristic: the variables representing the segment bounds are selected in a lexicographic order and assigned to their lower bounds. In this way, we ensure that the same search tree is explored, finding the same solutions in the same

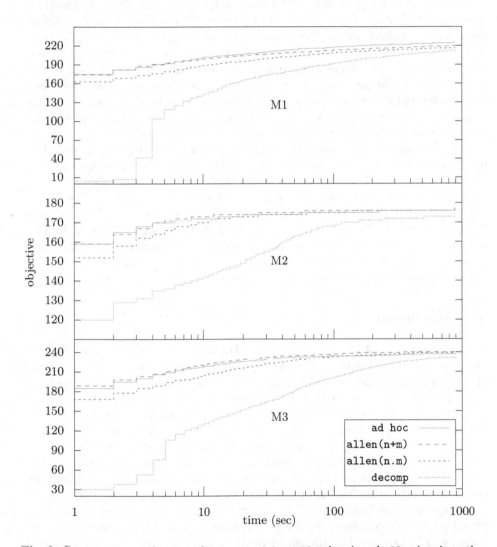

Fig. 3. Comparative evaluation of `decomp`, `ad hoc`, `allen(n.m)` and `allen(n+m)` on the three matches $M1$, $M2$ and $M3$ with ABS and PGLNS. The plots report the evolution of the objective value to be maximized with respect to the cpu time in seconds. The x-axis are in logscale.

order, and that only the running time is evaluated. The comparative evaluations of the four models are in reported in Figure 2. Each plot is associated with an instance and indicates the improvement of the objective function over time. Recall that the x-axis are in logscale. The three plots are similar.

First of all, a strict reformulation (decomp) is clearly not competitive with the models with specific constraints: decomp is always the slowest. This has to do with the huge number of constraints and additional variables it requires to express the *ExistAllen* constraint. As an example in the match $M1$ where there are 10 tasks, 271 intervals and 4 relations, each triplet $\langle T, R, I \rangle$ is expressed by three new binary constraints, each of them reified by a new boolean variable. Second, the quadratic propagator improves the filtering of a strict reformulation. Third, as expected, the performances are even better with the linear propagator. For information purpose, the results of the ad hoc model (as used in [4]) are reported. allen(n+m), our generic constraint is shown to be the most effective model, mildly faster than ad hoc. This confirms that our generic linear propagator, besides being expressive and flexible and offering guarantees on its complexity and consistency level, is very efficient in practice.

The four models have also been evaluated with the configuration described in [4], that is, using Activity-based search [13] combined with Propagation-Guided Large Neighborhood Search [14].[1] Due to the intrinsic randomness of ABS and PGLNS, each resolution was run 30 times. Thus, the average objective values are reported in Figure 3. Recall that the x-axis are in logscale. The dynamic strategy offered by the combination of ABS and PGLNS enables to reduce the differences between the various models. Although to a slightly lesser extent, the order between efficiency of the four models is preserved when applying a more aggressive search strategy heuristic.

5   Conclusion

We introduced *ExistAllen*, a generic constraint defined on a vector of tasks and a set of disjoint intervals, which applies on any of the 2^{13} combinations of Allen's algebra relations. This constraint is useful to tackle many problems related to time intervals, such as the video-summarization problem [4], used as a case study. From a technical viewpoint, we proposed a generic propagator that achieves bound-consistency in $O(n + m)$ worst-case time, where n is the number of tasks and m the number of intervals, whereas the most natural implementation requires $O(n \times m)$ worst-case time. Our experiments demonstrate that using our technique is very competitive with the best ad hoc approach, specific to one particular combination of relations, while being much more expressive.

Future work includes the extension of this approach to several task sets, in order to tackle problems beyond the context of video-summarization. In the Satellite Data Download Management Problem [15], Earth observation satellites acquire data that need to be downloaded to the ground. The download of data is subject to temporal constraints, such as fitting in visibility windows. Using

---

[1] The configuration for ABS and PGLNS are the default one described in [13] and [14].

ExistAllen constraint, the visibility windows are the intervals, the amount of acquired data fixes the duration of the download tasks, and the relation required is *during*. The tasks can be ordered with respect to the type of acquired data and their need to be scheduled.

Acknowledgments. The authors thank Haykel Boukadida, Sid-Ahmed Berrani and Patrick Gros for helping us modeling the tennis match summarization problem and for having providing us their dataset.

References

1. Allen, J.F.: Maintaining knowledge about temporal intervals. Commun. ACM **26**(11), 832–843 (1983)
2. Berrani, S.-A., Boukadida, H., Gros, P.: Constraint satisfaction programming for video summarization. In: Proceedings of the 2013 IEEE International Symposium on Multimedia, ISM 2013, Washington, DC, USA, pp. 195–202. IEEE Computer Society (2013)
3. Bessière, C.: Constraint propagation. Research report 06020. In: Rossi, F., van Beek, P., Walsh, T. (eds.) Handbook of Constraint Programming, LIRMM, chapter 3. Elsevier (2006)
4. Boukadida, H., Berrani, S.-A., Gros, P.: A novel modeling for video summarization using constraint satisfaction programming. In: Bebis, G., et al. (eds.) ISVC 2014, Part II. LNCS, vol. 8888, pp. 208–219. Springer, Heidelberg (2014)
5. Bramsen, P., Deshp, P., Lee, Y.K., Barzilay, R.: Finding temporal order in discharge summaries, pp. 81–85 (2006)
6. Choueiry, B.Y., Lin, X.: An efficient consistency algorithm for the temporal constraint satisfaction problem. AI Commun. **17**(4), 213–221 (2004)
7. Dechter, R., Meiri, I., Pearl, J.: Temporal constraint networks. Artif. Intell. **49**(1–3), 61–95 (1991)
8. Ibrahim, Z.A.A., Ferrane, I., Joly, P.: Temporal relation analysis in audiovisual documents for complementary descriptive information. In: Detyniecki, M., Jose, J.M., Nürnberger, A., van Rijsbergen, C.J.K. (eds.) AMR 2005. LNCS, vol. 3877, pp. 141–154. Springer, Heidelberg (2006)
9. Koubarakis, M.: From local to global consistency in temporal constraint networks. Theor. Comput. Sci. **173**(1), 89–112 (1997)
10. Koubarakis, M., Skiadopoulos, S.: Querying temporal and spatial constraint networks in PTIME. Artif. Intell. **123**(1–2), 223–263 (2000)
11. Satish Kumar, T.K., Cirillo, M., Koenig, S.: Simple temporal problems with taboo regions. In: des Jardins, M., Littman, M.L. (eds.) Proceedings of the Twenty-Seventh AAAI Conference on Artificial Intelligence, July 14–18, 2013, Bellevue, Washington, USA. AAAI Press (2013)
12. Ladkin, P.B.: Satisfying first-order constraints about time intervals. In: Shrobe, H.E., Mitchell, T.M., Smith, R.G. (eds.) Proceedings of the 7th National Conference on Artificial Intelligence. St. Paul, MN, August 21–26, 1988, pp. 512–517. AAAI Press/The MIT Press (1988)
13. Michel, L., Van Hentenryck, P.: Activity-based search for black-box constraint programming solvers. In: Beldiceanu, N., Jussien, N., Pinson, É. (eds.) CPAIOR 2012. LNCS, vol. 7298, pp. 228–243. Springer, Heidelberg (2012)

14. Perron, L., Shaw, P., Furnon, V.: Propagation guided large neighborhood search. In: Proceedings of the Principles and Practice of Constraint Programming - CP 2004, 10th International Conference, CP 2004, Toronto, Canada, pp. 468–481, September 27–October 1, 2004

15. Pralet, C., Verfaillie, G., Maillard, A., Hebrard, E., Jozefowiez, N., Huguet, M.-J., Desmousceaux, T., Blanc-Paques, P., Jaubert, J.: Satellite data download management with uncertainty about the generated volumes. In: Chien, S., Do, M.B., Fern, A., Ruml, W. (eds.) Proceedings of the Twenty-Fourth International Conference on Automated Planning and Scheduling, ICAPS 2014, Portsmouth, New Hampshire, USA, June 21–26, 2014. AAAI (2014)

16. Prud'homme, C., Fages, J.-G., Lorca, X.: Choco3 Documentation. http://www.choco-solver.org. TASC, INRIA Rennes, LINA CNRS UMR 6241, COSLING S.A.S. (2014)

17. Schilder, F.: Event extraction and temporal reasoning in legal documents. In: Schilder, F., Katz, G., Pustejovsky, J. (eds.) Annotating, Extracting and Reasoning about Time and Events. LNCS (LNAI), vol. 4795, pp. 59–71. Springer, Heidelberg (2007)

18. van Beek, P., Cohen, R.: Exact and approximate reasoning about temporal relations. Computational Intelligence 6, 132–144 (1990)

19. Vilain, M., Kautz, H., van Beek, P.: Readings in qualitative reasoning about physical systems. In: Constraint Propagation Algorithms for Temporal Reasoning: A Revised Report, pp. 373–381. Morgan Kaufmann Publishers Inc., San Francisco (1990)

20. Zhou, L., Hripcsak, G.: Methodological review: Temporal reasoning with medical data-a review with emphasis on medical natural language processing. J. of Biomedical Informatics 40(2), 183–202 (2007)

Exploiting GPUs in Solving (Distributed) Constraint Optimization Problems with Dynamic Programming

Ferdinando Fioretto[1,2]([✉]), Tiep Le[1], Enrico Pontelli[1], William Yeoh[1], and Tran Cao Son[1]

[1] Department of Computer Science, New Mexico State University,
Las Cruces, NM, USA
{ffiorett,tile,epontell,wyeoh,tson}@cs.nmsu.edu
[2] Department of Mathematics and Computer Science,
University of Udine, Udine, Italy

Abstract. This paper proposes the design and implementation of a dynamic programming based algorithm for (distributed) constraint optimization, which exploits modern massively parallel architectures, such as those found in modern Graphical Processing Units (GPUs). The paper studies the proposed algorithm in both centralized and distributed optimization contexts. The experimental analysis, performed on unstructured and structured graphs, shows the advantages of employing GPUs, resulting in enhanced performances and scalability.

1 Introduction

The importance of constraint optimization is outlined by the impact of its application in a range of *Constraint Optimization Problems* (COPs), such as supply chain management (e.g., [15,27]) and roster scheduling (e.g., [1,8]). When resources are distributed among a set of autonomous agents and communication among the agents are restricted, COPs take the form of *Distributed Constraint Optimization Problems* (DCOPs) [21,33]. In this context, agents coordinate their value assignments to maximize the overall sum of resulting constraint utilities. DCOPs are suitable to model problems that are distributed in nature, and where a collection of agents attempts to optimize a global objective within the confines of localized communication. They have been employed to model various distributed optimization problems, such as meeting scheduling [20,32,35], resources allocation [13,36], and power network management problems [17].

Dynamic Programming (DP) based approaches have been adopted to solve COPs and DCOPs. The *Bucket Elimination* (BE) procedure [10] iterates over the variables of the COP, reducing the problem at each step by replacing a

---

This research is partially supported by the National Science Foundation under grant number HRD-1345232. The views and conclusions contained in this document are those of the authors and should not be interpreted as representing the official policies, . either expressed or implied, of the sponsoring organizations, agencies, or the U.S. government.

G. Pesant (Ed.): CP 2015, LNCS 9255, pp. 121–139, 2015.
DOI: 10.1007/978-3-319-23219-5_9

variable and its related utility functions with a single new function, derived by optimizing over the possible values of the replaced variable. The *Dynamic Programming Optimization Protocol* (DPOP) [25] is one of the most efficient DCOP solvers, and it can be seen as a distributed version of BE, where agents exchange newly introduced utility functions via messages.

The importance of DP-based approaches arises in several optimization fields including constraint programming [2,28]. For example, several *propagators* adopt DP-based techniques to establish constraint consistency; for instance, (1) the *knapsack* constraint propagator proposed by Trick applies DP techniques to establish arc consistency on the constraint [31]; (2) the propagator for the *regular* constraint establishes arc consistency using a specific digraph representation of the DFA, which has similarities to dynamic programming [24]; (3) the *context free grammar* constraint makes use of a propagator based on the CYK parser that uses DP to enforce generalized arc consistency [26].

While DP approaches may not always be appropriate to solve (D)COPs, as their time and space requirements may be prohibitive, they may be very effective in problems with particular structures, such as problems where their underlying constraint graphs have small induced widths or distributed problems where the number of messages is crucial for performance, despite the size of the messages. The structure exploited by DP-based approaches in constructing solutions makes it suitable to exploit a novel class of massively parallel platforms that are based on the *Single Instruction Multiple Thread* (SIMT) paradigm—where multiple threads may concurrently operate on different data, but are all executing the same instruction at the same time. The SIMT-based paradigm is widely used in modern *Graphical Processing Units* (GPUs) for general purpose parallel computing. Several libraries and programming environments (e.g., *Compute Unified Device Architecture* (CUDA)) have been made available to allow programmers to exploit the parallel computing power of GPUs.

In this paper, we propose a design and implementation of a DP-based algorithm that exploits parallel computation using GPUs to solve (D)COPs. Our proposal aims at employing GPU hardware to speed up the inference process of DP-based methods, representing an alternative way to enhance the performance of DP-based constraint optimization approaches. This paper makes the following contributions: (1) We propose a novel design and implementation of a centralized and a distributed DP-based algorithm to solve (D)COPs, which harnesses the computational power offered by parallel platforms based on GPUs; (2) We enable the use of concurrent computations between CPU(s) and GPU(s), during (D)COP resolution; and (3) We report empirical results that show significant improvements in performance and scalability.

2   Background

2.1   Centralized Constraint Optimization Problems (COPs)

A (centralized) *Constraint Optimization Problem* (COP) is defined as $(\mathbf{X}, \mathbf{D}, \mathbf{C})$ where: $\mathbf{X} = \{x_1, \ldots, x_n\}$ is a set of variables; $\mathbf{D} = \{D_1, \ldots, D_n\}$ is a set of

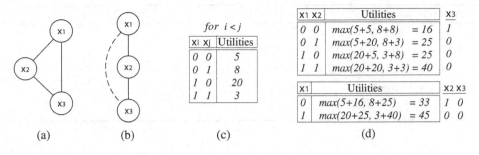

Fig. 1. Example (D)COP (a-c) and *UTIL* phase computations in DPOP (d).

domains for the variables in $\mathbf{X}$, where D_i is the set of possible values for the variable x_i; $\mathbf{C}$ is a finite set of utility functions on variables in $\mathbf{X}$, with f_i : $\times_{x_j \in \mathbf{x}^i} D_j \rightarrow \mathbb{R}^+ \cup \{-\infty\}$, where $\mathbf{x}^i \subseteq \mathbf{X}$ is the set of variables relevant to f_i, referred to as the *scope* of f_i, and $-\infty$ is used to denote that a given combination of values for the variables in $\mathbf{x}^i$ is not allowed.[1] A *solution* is a value assignment for a subset of variables from $\mathbf{X}$ that is consistent with their respective domains; i.e., it is a partial function $\theta : \mathbf{X} \rightarrow \bigcup_{i=1}^{n} D_i$ such that, for each $x_j \in \mathbf{X}$, if $\theta(x_j)$ is defined, then $\theta(x_j) \in D_j$. A solution is *complete* if it assigns a value to each variable in $\mathbf{X}$. We will use the notation σ to denote a complete solution, and, for a set of variables $\mathbf{V} = \{x_{i_1}, \ldots, x_{i_h}\} \subseteq \mathbf{X}$, $\sigma_{\mathbf{V}} = \langle \sigma(x_{i_1}), \ldots, \sigma(x_{i_h}) \rangle$, where $i_1 < \cdots < i_h$. The goal for a COP is to find a complete solution σ^* that maximizes the total problem utility expressed by its utility functions, i.e., $\sigma^* = \mathrm{argmax}_{\sigma \in \Sigma} \sum_{f_i \in \mathbf{C}} f_i(\sigma_{\mathbf{x}^i})$, where Σ is the *state space*, defined as the set of all possible complete solutions.

Given a COP P, $G_P = (\mathbf{X}, E_{\mathbf{C}})$ is the *constraint graph* of P, where $\{x, y\} \in E_{\mathbf{C}}$ iff $\exists f_i \in \mathbf{C}$ such that $\{x, y\} \subseteq \mathbf{x}^i$. Fig. 1(a) shows the constraint graph of a simple COP with three variables, x_1, x_2, and x_3. The domain of each variable is the set $\{0, 1\}$. Fig. 1(c) describes the utility functions of the COP.

Definition 1 (Projection). *The* projection *of a utility function f_i on a set of variables $\mathbf{V} \subseteq \mathbf{x}^i$ is a new utility function $f_{i|\mathbf{V}} : \mathbf{V} \rightarrow \mathbb{R}^+ \cup \{-\infty\}$, such that for each possible assignment $\theta \in \times_{x_j \in \mathbf{V}} D_j$, $f_{i|\mathbf{V}}(\theta) = \max\limits_{\sigma \in \Sigma, \sigma_{\mathbf{V}} = \theta} f_i(\sigma_{\mathbf{x}^i})$.*

In other words, $f_{i|\mathbf{V}}$ is constructed from the tuples of f_i, removing the values of the variable that do not appear in $\mathbf{V}$ and removing duplicate values by keeping the maximum utility of the original tuples in f_i.

Definition 2 (Concatenation). *Let us consider two assignments θ', defined for variables V, and θ'', defined for variables W, such that for each $x \in V \cap W$ we have that $\theta'(x) = \theta''(x)$. Their* concatenation *is an assignment $\theta' \cdot \theta''$ defined for $V \cup W$, such as for each $x \in V$ (resp. $x \in W$) we have that $\theta' \cdot \theta''(x) = \theta'(x)$ (resp. $\theta' \cdot \theta''(x) = \theta''(x)$).*

---

[1] For simplicity, we assume that tuples of variables are built according to a predefined ordering.

Algorithm 1. BE

---

1 **for** $i \leftarrow n$ **downto** 1 **do**
2     $B_i \leftarrow \{f_j \in \mathbf{C} \mid x_i \in \mathbf{x}^j \wedge i = \max\{k \mid x_k \in \mathbf{x}^j\}\}$
3     $\hat{f}_i \leftarrow \pi_{-x_i}\left(\sum_{f_j \in B_i} f_j\right)$
4     $\mathbf{X} \leftarrow \mathbf{X} \setminus \{x_i\}$
5     $\mathbf{C} \leftarrow (\mathbf{C} \cup \{\hat{f}_i\}) \setminus B_i$

We define two operations on utility functions:

- The *aggregation* of two functions f_i and f_j, is a function $f_i + f_j : \mathbf{x}^i \cup \mathbf{x}^j \rightarrow \mathbb{R}^+ \cup \{-\infty\}$, such that $\forall \theta' \in \times_{x_k \in \mathbf{x}^i} D_k$ and $\forall \theta'' \in \times_{x_k \in \mathbf{x}^j} D_k$, if $\theta' \cdot \theta''$ is defined, then we have that $(f_i + f_j)(\theta' \cdot \theta'') = f_i(\theta') + f_j(\theta'')$.
- *Projecting out* a variable $x_j \in \mathbf{x}^i$ from a function f_i, denoted as $\pi_{-x_j}(f_i)$, produces a new function with scope $\mathbf{x}^i \setminus \{x_j\}$, and defined as the projection of f_i on $\mathbf{x}^i \setminus \{x_j\}$, i.e., $\pi_{-x_j}(f_i) = f_{i|\mathbf{x}^i \setminus \{x_j\}}$.

Bucket Elimination (BE): BE [10,11] is a dynamic programming based procedure that can be used to solve COPs. Algorithm 1 illustrates its pseudocode. Given a COP $(\mathbf{X}, \mathbf{D}, \mathbf{C})$ and an ordering $o = \langle x_1, \ldots, x_n \rangle$ on the variables in $\mathbf{X}$, we say that a variable x_i has a higher *priority* with respect to variable x_j if x_i appears after x_j in o. BE operates from the highest to lowest priority variable. When operating on variable x_i, it creates a bucket B_i, which is the set of all utility functions that involve x_i as the highest priority variable in their scope (line 2). The algorithm then computes a new utility function $\hat{f}_i$ by aggregating the functions in B_i and projecting out x_i (line 3). Thus, x_i can be removed from the set of variables $\mathbf{X}$ to be processed (line 4) and the new function $\hat{f}_i$ replaces in $\mathbf{C}$ all the utility functions that appear in B_i (line 5). In our example, BE operates, in order, on the variables x_3, x_2, and x_1. When x_3 is processed, the bucket B_3 is $\{f_{13}, f_{23}\}$, and the $\hat{f}_3$ utility function is shown in Fig. 1(d) top. The rightmost column shows the values for x_3 after its projection. BE updates the sets $\mathbf{X} = \{x_1, x_2\}$ and $\mathbf{C} = \{f_{12}, \hat{f}_3\}$. When x_2 is processed, $B_2 = \{f_{12}, \hat{f}_3\}$ and $\hat{f}_2$ is shown in Fig. 1(d) bottom. Thus, $\mathbf{X} = \{x_1\}$ and $\mathbf{C} = \{\hat{f}_2\}$. Lastly, the algorithm processes x_1, sets $B_1 = \{\hat{f}_2\}$, and $\hat{f}_1$ contains one value combination $\sigma^* = \langle 1, 0, 0 \rangle$, which corresponds to an optimal solution to the problem.

The complexity of the algorithm is bounded by the time needed to process a bucket (line 3), which is exponential in number of variables in the bucket.

2.2 Distributed Constraint Optimization Problems (DCOPs)

In a *Distributed Constraint Optimization Problem* (DCOP) [21,25,33], the variables, domains, and utility functions of a COP are distributed among a collection of *agents*. A DCOP is defined as $(\mathbf{X}, \mathbf{D}, \mathbf{C}, \mathbf{A}, \alpha)$, where $\mathbf{X}, \mathbf{D}$, and $\mathbf{C}$ are defined as in a COP, $\mathbf{A} = \{a_1, \ldots, a_p\}$ is a set of *agents*, and $\alpha : \mathbf{X} \rightarrow \mathbf{A}$ maps each variable to one agent. Following common conventions, we restrict our attention to binary utility functions and assume that α is a bijection: Each agent controls exactly one variable. Thus, we will use the terms "variable" and "agent" interchangeably and assume that $\alpha(x_i) = a_i$. This is a common assumption in the DCOP literature as there exist pre-processing techniques that transform a general DCOP into this more restrictive DCOP [7,34]. In DCOPs, solutions are defined as for COPs, and many solution approaches emulate those proposed in the COP literature. For example, ADOPT [21] is a distributed version of *Iterative Deepening Depth First Search*, and DPOP [25] is a distributed version of BE. The main difference is in the way the information is shared among agents. Typically, a DCOP agent knows exclusively its domain and the functions involving its variable. It can communicate exclusively with its neighbors (i.e., agents directly connected to it in the constraint graph[2]), and the exchange of information takes the form of messages. Given a DCOP P, a *DFS pseudo-tree* arrangement for G_P is a spanning tree $T = \langle \mathbf{X}, E_T \rangle$ of G_P such that if $f_i \in \mathbf{C}$ and $\{x, y\} = \mathbf{x}^i$, then x and y appear in the same branch of T. Edges of G_P that are *in* (resp. *out* of) E_T are called *tree edges* (resp. *backedges*). The tree edges connect parent-child nodes, while backedges connect a node with its *pseudo-parents* and its *pseudo-children*. We use $N(a_i) = \{a_j \in \mathbf{A} \mid \{x_i, x_j\} \in E_T\}$ to denote the neighbors of agent a_i; C_i, PC_i, P_i, and PP_i to denote the set of children, pseudo-children, parent, and pseudo-parents of agent a_i; and $sep(a_i)$ to denote the *separator* of agent a_i, which is the set of ancestor agents that are constrained (i.e., they are linked in G_P) with agent a_i or with one of its descendant agents in the pseudo-tree. Fig. 1(b) shows one possible pseudo-tree for the problem, where the agent a_1 has one pseudo-child a_3 (the dotted line is a backedge).

Dynamic Programming Optimization Protocol (DPOP): DPOP [25] is a dynamic programming based DCOP algorithm that is composed of three phases. **(1)** *Pseudo-tree generation*: Agents coordinate to build a pseudo-tree, realized through existing distributed pseudo-tree construction algorithms [16]. **(2)** *UTIL propagation*: Each agent, starting from the leaves of the pseudo-tree, computes the optimal sum of utilities in its subtree for each value combination of variables in its separator. The agent does so by aggregating the utilities of its functions with the variables in its separator and the utilities in the *UTIL* messages received from its child agents, and then projecting out its own variable. In our example problem, agent a_3 computes the optimal utility for each value combination of variables x_1 and x_2 (Fig. 1(d) top), and sends the utilities to its parent agent a_2 in a *UTIL* message. When the root agent a_1 receives the *UTIL* message from each of its children, it computes the maximum utility of the entire problem. **(3)** *VALUE propagation*: Each agent, starting from the root of the pseudo-tree, determines the optimal value for its variable. The root agent does so by choosing the value of its variable from its *UTIL* computations—selecting the value with the maximal utility. It sends the selected value to its children in a *VALUE* message. Each agent, upon receiving a *VALUE* message, determines the value for its variable that results in the maximum utility given the variable

---

[2] The *constraint graph* of a DCOP is equivalent to that of the corresponding COP.

assignments (of the agents in its separator) indicated in the *VALUE* message. Such assignment is further propagated to the children via *VALUE* messages.

The complexity of DPOP is dominated by the *UTIL propagation* phase, which is exponential in the size of the largest separator set $sep(a_i)$ for all $a_i \in \mathbf{A}$. The other two phases require a polynomial number of linear size messages, and the complexity of the local operations is at most linear in the size of the domain.

Observe that the *UTIL propagation* phase of DPOP emulates the BE process in a distributed context [6]. Given a pseudo-tree and its *preorder listing o*, the *UTIL* message generated by each DPOP agent a_i is equivalent to the aggregated and projected function $\hat{f}_i$ in BE when x_i is processed according to the ordering o.

2.3　Graphical Processing Units (GPUs)

Modern *GPUs* are multiprocessor devices, offering hundreds of computing cores and a rich memory hierarchy to support graphical processing. We consider the NVIDIA *CUDA* programming model [29], which enables the use of the multiple cores of a graphics card to accelerate general (non-graphical) applications. The underlying model of parallelism is *Single-Instruction Multiple-Thread* (SIMT), where the same instruction is executed by different threads that run on identical cores, grouped in *Streaming Multiprocessors* (SMs), while data and operands may differ from thread to thread.

A typical CUDA program is a C/C++ program. The functions in the program are distinguished based on whether they are meant for execution on the CPU (referred to as the *host*) or in parallel on the GPU (referred as the *device*). The functions executed on the device are called *kernels*, and are executed by several *threads*. To facilitate the mapping of the threads to the data structures being processed, threads are grouped in *blocks*, and have access to several memory levels, each with different properties in terms of speed, organization, and capacity. CUDA maps blocks (coarse-grain parallelism) to the SMs for execution. Each SM schedules the threads in a block (fine-grain parallelism) on its computing cores in chunks of 32 threads (*warps*) at a time. Threads in a block can communicate by reading and writing a common area of memory (*shared memory*). Communication between blocks and communication between the blocks and the host is realized through a large slow *global memory*. The development of CUDA programs that efficiently exploit SIMT parallelism is a challenging task. Several factors are critical in gaining performance. Memory levels have significantly different sizes (e.g., registers are in the order of dozens per thread, shared memory is in the order of a few kilobytes per block) and access times, and various optimization techniques are available (e.g., *coalesced* of memory accesses to contiguous locations into a single memory transaction).

3　GPU-Based (Distributed) Bucket Elimination (GPU-(D)BE)

Our *GPU-based (Distributed) Bucket Elimination* framework, extends BE (resp. DPOP) by exploiting GPU parallelism within the *aggregation* and *projection*

Algorithm 2. GPU-(D)BE

---

(1) Generate pseudo-tree
2   GPU-INITIALIZE( )
3   if $C_i = \emptyset$ then
4   $\quad\mid\quad$ $UTIL_{x_i} \Leftarrow$ PARALLELCALCUTILS( )
(5) $\quad\mid\quad$ Send $UTIL$ message $(x_i, UTIL_{x_i})$ to P_i
6   else
7   $\quad\mid\quad$ Activate UTILMessageHandler($\cdot$)
(8) Activate VALUEMessageHandler($\cdot$)

---

operations. These operations are responsible for the creation of the functions $\hat{f}_i$ in BE (line 3 of Algorithm 1) and the $UTIL$ tables in DPOP ($UTIL$ propagation phase), and they dominate the complexity of the algorithms. Thus, we focus on the details of the design and the implementation relevant to such operations. Due to the equivalence of BE and DPOP, we will refer to the $UTIL$ *tables* and to the aggregated and projected functions $\hat{f}$ of Algorithm 1, as well as variables and agents, interchangeably. Notice that the computation of the utility for each value combination in a $UTIL$ table is independent of the computation in the other combinations. The use of a GPU architecture allows us to exploit such independence, by concurrently exploring several combinations of the $UTIL$ table, computed by the aggregation operator, as well as concurrently projecting out variables.

Algorithm 2 illustrates the pseudocode, where we use the following notations: Line numbers in parenthesis denote those instructions required exclusively in the distributed case. Starred line numbers denote those instructions executed concurrently by both the CPU and the GPU. The symbols $\leftarrow$ and $\Leftarrow$ denote sequential and parallel (multiple GPU-threads) operations, respectively. If a parallel operation requires a copy from host (device) to device (host), we write $\overset{D \leftarrow H}{\Leftarrow}$ ($\overset{H \leftarrow D}{\Leftarrow}$). Host to device (resp. device to host) memory transfers are performed immediately before (resp. after) the execution of the GPU kernel. Algorithm 2 shows the pseudocode of GPU-(D)BE for an agent a_i. Like DPOP, also GPU-(D)BE is composed of three phases; the first and third phase are executed exclusively in the distributed version. The first phase is identical to that of DPOP (line 1). In the second phase:

- Each agent a_i calls GPU-INITIALIZE() to set up the GPU kernel. For example, it determines the amount of global memory to be assigned to each $UTIL$ table and initializes the data structures on the GPU device memory (line 2).
- Each agent a_i aggregates the utilities for the functions between its variables and its separator, projects its variable out (line 4), and sends them to its parent (line 5). The *MessageHandlers* of lines 7 and 8 are activated for each new incoming message.

By the end of the second phase (line 11), the root agent knows the overall utility for each values of its variable x_i. It chooses the value that results in the maximum

Procedure UTILMessageHandler(a_k, $UTIL_{a_k}$)

---

(9) Store $UTIL_{a_k}$
10   **if** received $UTIL$ message from each child $a_c \in C_i$ **then**
11 | $UTIL_{a_i} \Leftarrow$ PARALLELCALCUTILS()
12 | **if** $P_i = NULL$ **then**
13 | | $d_i^* \leftarrow$ CHOOSEBESTVALUE($\emptyset$)
(14) | | **foreach** $a_c \in C_i$ **do**
(15) | | | $VALUE_{a_i} \leftarrow (x_i, d_i^*)$
(16) | | | Send $VALUE$ message $(a_i, VALUE_{a_i})$ to a_c
(17) | **else** Send $UTIL$ message $(a_i, UTIL_{a_i})$ to P_i

---

Procedure VALUEMessageHandler(a_k, $VALUE_{a_k}$)

---

(18) $VALUE_{a_i} \leftarrow VALUE_{a_k}$
(19) $d_i^* \leftarrow$ CHOOSEBESTVALUE($VALUE_{a_i}$)
(20) **foreach** $a_c \in C_i$ **do**
(21) | $VALUE_{a_i} \leftarrow \{(x_i, d_i^*)\} \cup \{(x_k, d_k^*) \in VALUE_{a_k} \mid x_k \in sep(a_c)\}$
(22) | Send $VALUE$ message $(a_i, VALUE_{a_i})$ to a_c

---

utility (line 13). Then, in the distributed version, it starts the third phase by sending to each child agent a_c the value of its variable x_i (lines 14-16). These operations are repeated by every agent receiving a $VALUE$ message (lines 18-22). In contrast, in the centralized version, the value assignment for each variable is set by the root agent directly.

3.1   GPU Data Structures

In order to fully capitalize on the parallel computational power of GPUs, the data structures need to be designed in such a way to limit the amount of information exchanged between the CPU host and the GPU device, and in order to minimize the accesses to the (slow) device global memory (and ensure that they are coalesced). To do so, each agent identifies the set of relevant *static entities*, i.e., information required during the GPU computation, which does not mutate during the resolution process. The static entities are communicated to the GPU once at the beginning of the computation. This allows each agent running on a GPU device to communicate with the CPU host exclusively to exchange the results of the aggregation and projection processes. The complete set of utility functions, the constraint graph, and the agents ordering, all fall in such category. Thus, each agent a_i stores:

- The set of utility functions involving exclusively x_i and a variable in a_i's separator set: $S_i = \{f_j \in \mathbf{C} \mid x_i \in \mathbf{x}^j \wedge sep(a_i) \cap \mathbf{x}^j \neq \emptyset\}$. For a given function $f_j \in S_i$, its utility values are stored in an array named $gFunc_j$.

- The domain D_i of its variable (for simplicity assumed to be all of equal cardinality).
- The set C_i of a_i's children.
- The separator sets $sep(a_i)$, and $sep(a_c)$, for each $a_c \in C_i$.

The GPU-INITIALIZE() procedure of line 2, invoked after the pseudo-tree construction, stores the data structures above for each agent on the GPU device. As a technical detail, all the data stored on the GPU global memory is organized in mono-dimensional arrays, so as to facilitate *coalesced memory accesses*. In particular, the identifier and scope of the functions in S_i as well as identifiers and separator sets of child agents in C_i are stored within a single mono-dimensional array. The utility values stored in the rows of each function are padded to ensures that a row is aligned to a memory word—thus minimizing the number of memory accesses.

GPU-INITIALIZE() is also responsible for reserving a portion of the GPU global memory to store the values for the agent's *UTIL* table, denoted by $gUtils_i$, and those of its children, denoted by $gChUtils_c$, for each $a_c \in C_i$. As a technical note, an agent's *UTIL* table is mapped onto the GPU device to store only the utility values, not the associated variables values. Its j-th entry is associated with the j-th permutation of the variable values in $sep(a_i)$, in lexicographic order. This strategy allows us to employ a simple perfect hashing to efficiently associate row numbers with variables' values and vice versa. Note that the agent's *UTIL* table size grows exponentially with the size of its separator set; more precisely, after projecting out x_i, it has $|D_i|^{sep(a_i)}$ entries. However, the GPU global memory is typically limited to a few GB (e.g., in our experiments it is 2GB). Thus, each agent, after allocating its static entities, checks if it has enough space to allocate its children's *UTIL* tables and a consistent portion (see next subsection for details) of its own *UTIL* table. In this case, it sets the *project_on_device* flag to true, which signals that both aggregate and project operations can be done on the GPU device.[3] Otherwise it sets the flag to false and bounds the device *UTIL* size table to the maximum storable space on the device. In this case, the aggregation operations are performed only partially on the GPU device.

3.2   Parallel Aggregate and Project Operations

The PARALLELCALCUTILS procedure (executed in lines 4 and 11) is responsible for performing the aggregation and projection operations, harnessing the parallelism provided by the GPU. Due to the possible large size of the *UTIL* tables, we need to separate two possible cases and devise specific solutions accordingly:

(a) When the device global memory is sufficiently large to store all a_i's children *UTIL* tables as well as a significant portion of a_i's *UTIL* table[4] (i.e., when

---

[3] If the *UTIL* table of agent a_i does not fit in the global memory, we partition such table in smaller chunks, and iteratively execute the GPU kernel until all rows of the table are processed.

[4] In our experiments, we require that at least 1/10 of the *UTIL* table can be stored in the GPU. We experimentally observed that a partitioning of the table in at most

---

Procedure ParallelCalcUtils( )

---

23   **if** *project_on_device* **then**

24   $\quad gChUTIL_{a_c} \overset{D \leftarrow H}{\Leftarrow} UTIL_{a_c}$ for all $a_c \in C_i$

25   $R \leftarrow 0 \; ; \; UTIL_{a_i} \leftarrow \emptyset$

26   **while** $R < |D_i|^{sep(a_i)}$ **do**

27   $\quad$ **if** *project_on_device* **then**

28* $\qquad UTIL'_{a_i} \overset{H \leftarrow D}{\Leftarrow}$ GPU-Aggregate-Project(R)

29   $\quad$ **else**

30* $\qquad UTIL'_{a_i} \overset{H \leftarrow D}{\Leftarrow}$ GPU-Aggregate(R)

31* $\qquad UTIL'_{a_i} \leftarrow$ AggregateCh-Project$(a_i, UTIL'_{a_i}, UTIL_{a_c})$ for all $a_c \in C_i$

32* $\quad UTIL_{a_i} \leftarrow UTIL_{a_i} \cup$ Compress$(UTIL'_{a_i})$

33   $\quad R \leftarrow R + |UTIL'_{a_i}|$

34   **return** $UTIL_{a_i}$

---

project_on_device = `true`), both aggregation and projection of the agent's *UTIL* table are performed in parallel on the GPU. The procedure first stores the *UTIL* tables received from the children of a_i into their assigned locations in the GPU global memory (lines 23–24). It then iterates through successive GPU kernel calls (line 28) until the $UTIL_{a_i}$ table is fully computed (lines 26–33). Each iterations computes a certain number of rows of the $UTIL_{a_i}$ table (R serves as counter).

(b) When the device global memory is insufficiently large to store all a_i's children *UTIL* tables as well as a significant portion of a_i's *UTIL* table (i.e., when *project_on_device* = `false`), the agent alternates the use of the GPU and the CPU to compute $UTIL_{a_i}$. The GPU is in charge of aggregating the functions in S_i (line 30), while the CPU aggregates the children *UTIL* table,[5] projecting out x_i. Note that, in this case, the $UTIL_{a_i}$ storage must include all combinations of values for the variables in $sep(x_i) \cup \{x_i\}$, thus the projection operation is performed on the CPU host. As in the previous case, the $UTIL_{a_i}$ is computed incrementally, given the amount of available GPU global memory.

To fully capitalize on the use of the GPU, we exploit an additional level of parallelism, achieved by running GPU kernels and CPU computations concurrently; this is possible when the $UTIL_{a_i}$ table is computed in multiple chunks. Fig. 2 illustrates the concurrent computations between the CPU and GPU. After transferring the children *UTIL* tables into the device memory (Init)—in case

---

10 chunks provides a good time balance between memory transfers and actual computation.

[5] The CPU aggregates only those child *UTIL* table that could not fit in the GPU memory. Those that fit in memory are integrated through the GPU computation as done in the previous point.

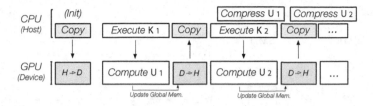

Fig. 2. Concurrent computation between host and device.

(a) only—the execution of kernel K_1 produces the update of the first chunk of $UTIL_{a_i}$, denoted by U_1 in Fig. 2, which is transferred to the CPU host. The successive parallel operations are performed asynchronously with respect to the GPU, that is, the execution of the j-th CUDA kernel K_j ($j > 1$), returns the control immediately to the CPU, which concurrently operates a compression operation on the previously computed $UTIL'_{a_i}$ chunk (line 32), referred to as U_{k-1} in Fig. 2. For case **(b)**, the CPU also executes concurrently the AGGREGATECH-PROJECT of line 31. We highlight the concurrent operations by marking with a * symbol their respective lines in the procedure PARALLELCALCUTILS.

Technical Details: We now describe in more detail how we divide the workload among parallel blocks, i.e., the mapping between the $UTIL$ table rows and the CUDA blocks. A total of $T = 64 \cdot k$ ($1 \le k \le 16$) threads (a block) are associated to the computation of T permutations of values for $sep(a_i)$. The value k depends on the architecture and it is chosen to maximize the number of concurrent threads running at the same time. In our experiments, we set $k = 3$. The number of blocks is chosen so that the corresponding aggregate number of threads does not exceed the total number of $UTIL'_{a_i}$ permutations currently stored in the device. Let h be the number of stream multiprocessors of the GPU. Then, the maximum number of $UTIL$ permutations that can be computed concurrently is $M = h \cdot T$. In our experiments $h = 14$, and thus, $M = 2688$. Fig. 3 provides an illustration of the $UTIL$ permutations computed in parallel on GPU. The blocks B_i in each row are executed in parallel on different SMs. Within each block, a total of (at most) 192 threads operate on as many entries of the $UTIL$ table.

The GPU kernel procedure is shown in lines 35-49. We surround line numbers with $|\cdot|$ to denote parts of the procedure executed by case **(b)**. The kernel takes as input the number R of the $UTIL$ table permutations computed during the previous kernel calls. Each thread identifies its entry index r_{id} within the table chunk $UTIL'_{a_i}$ (line 35). It then assigns the shared memory allocated to local arrays to store the static entities S_i, C_i, and $sep(a_c)$, for each $a_c \in C_i$. In addition it reserves the space θ to store the assignments corresponding to the $UTIL$ permutation being computed by each thread, which is retrieved using the thread entry index and the offset R (line 38). DECODE implements a *minimal perfect hash function* to convert the entry index of the $UTIL$ table to its associated variables value permutation. Each thread aggregates the functions in S_i (lines 42-44) and the $UTIL$ tables of a_i's children (lines 45-47), for each element of its domain

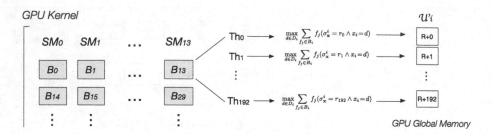

Fig. 3. GPU kernel parallel computations.

Procedure GPU-Aggregate-Project(R)

|35| $r_{id} \leftarrow$ the thread's entry index of $UTIL'_i$
|36| $d_{id} \leftarrow$ the thread's value index of D_i
|37| $\langle |\theta, S_i|, C_i, sep(x_c) \rangle \leftarrow$ AssignSharedMem()   for all $x_c \in C_i$
|38| $\theta \leftarrow$ Decode($R + r_{id}$)
|39| $util \leftarrow -\infty$
40 **foreach** $d_{id} \in D_i$ **do**
|41| $util_{d_{id}} \leftarrow 0$
|42| **foreach** $f_j \in S_i$ **do**
|43| $\rho_j \leftarrow$ Encode($\theta_{\mathbf{x}^j} \mid x_i = d_{id}$)
|44| $util_{d_{id}} \leftarrow util_{d_{id}} + gFunc_j[\rho_j]$
45   **foreach** $a_c \in C_i$ **do**
46 $\rho_c \leftarrow$ Encode($\theta_{sep(a_c)} \mid x_i = d_{id}$)
47 $util_{d_{id}} \leftarrow util_{d_{id}} + gChUtils_c[\rho_c]$
|48| $util \leftarrow \max(util, util_{d_{id}})$
|49| $gUtils_i[r_{id}] \leftarrow util$

(lines 40-48). The Encode routine converts a given assignments for the variables in the scope of a function f_j (line 43), or in the separator set of child a_c (line 46), to the corresponding array index, sorted in lexicographic order. The value for the variable x_i within each input, is updated at each iteration of the for loop. The projection operation is executed in line 48. Finally, the thread stores the best utility in the corresponding position of the array $gUtils_i$

The GPU-Aggregate procedure (called in line 30), is illustrated in lines 35-49—line numbers surrounded by $|\cdot|$. Each thread is in charge of a value combination in $sep(a_i) \cup \{x_i\}$, thus, the **foreach** loop of lines 40-48 is operated in parallel by $|D_i|$ threads. Lines 45-47 are not executed. The AggregateCh-Project procedure (line 31), which operates on the CPU, is similar to the GPU-Aggregate-Project procedure, except that lines 36-37, and 42-44, are not executed.

The proposed kernel has been the result of several investigations. We experimented with other levels of parallelism, e.g., by unrolling the for-loops among groups of threads. However, these modifications create divergent branches, which degrade the parallel performance. We experimentally observed that such degradation worsen consistently as the size of the domain increases.

3.3   General Observations

Observation 1. *GPU-DBE requires the same number of messages as those required by DPOP, and it requires messages of the same size as those required by DPOP.*

Observation 2. *The UTIL messages constructed by each GPU-DBE agent are identical to those constructed by each corresponding DPOP agent.*

The above observations follow from the pseudo-tree construction and VALUE propagation GPU-DBE phases, which are identical to those of DPOP. Thus, their corresponding messages and message sizes are identical in both algorithms. Moreover, given a pseudo-tree, each DPOP/GPU-DBE agent computes the *UTIL* table containing each combination of values for the variables in its separator set. Thus, the *UTIL* messages of GPU-DBE and DPOP are identical.

Observation 3. *The memory requirements of GPU-(D)BE is, in the worst case, exponential in the induced width of the problem (for each agent).*

This observation follows from the equivalence of the *UTIL* propagation phase of DPOP and BE [6] and from Observation 2.

Observation 4. *GPU-(D)BE is complete and correct.*

The completeness and correctness of GPU-(D)BE follow from the completeness and correctness of BE [10] and DPOP [25].

4   Related Work

The use of GPUs to solve difficult combinatorial problems has been explored by several proposals in different areas of constraint optimization. For instance, Meyer *et al.* [18] proposed a multi-GPU implementation of the *simplex tableau* algorithm which relies on a vertical problem decomposition to reduce communication between GPUs. In constraint programming, Arbelaez and Codognet [3] proposed a GPU-based version of the *Adaptive Search* that explores several *large neighborhoods* in parallel, resulting in a speedup factor of 17. Campeotto *et al.* [9] proposed a GPU-based framework that exploits both parallel propagation and parallel exploration of several large neighborhoods using local search techniques, leading to a speedup factor of up to 38. The combination of GPUs with dynamic programming has also been explored to solve different combinatorial optimization problems. For instance, Boyer *et al.* [5] proposed the use of GPUs to compute the classical DP recursion step for the knapsack problem, which led to a speedup factor of 26. Pawłowski *et al.* [23] presented a DP-based solution for the *coalition structure formation problem* on GPUs, reporting up to two orders of magnitude of speedup. Differently from other proposals, our approach aims at using GPUs to exploit SIMT-style parallelism from DP-based methods to solve general COPs and DCOPs.

5   Experimental Results

We compare our centralized and distributed versions of GPU-(D)BE with BE [10] and DPOP [25] on binary constraint networks with *random*, *scale-free*, and *regular grid* topologies. The instances for each topology are generated as follows:

Random: We create an n-node network, whose density p_1 produces $\lfloor n\,(n-1)\,p_1 \rfloor$ edges in total. We do not bound the tree-width, which is based on the underlying graph.

Scale-free: We create an n-node network based on the Barabasi-Albert model [4]: Starting from a connected 2-node network, we repeatedly add a new node, randomly connecting it to two existing nodes. In turn, these two nodes are selected with probabilities that are proportional to the numbers of their connected edges. The total number of edges is $2\,(n-2)+1$.

Regular grid: We create an n-node network arranged as a rectangular grid, where each internal node is connected to four neighboring nodes, while nodes on the grid edges (resp. corners) are connected to two (resp. three) neighboring nodes.

We generate 30 instances for each topology, ensuring that the underlying graph is connected. The utility functions are generated using random integer costs in $[0, 100]$, and the constraint tightness (i.e., ratio of entries in the utility table different from $-\infty$) p_2 is set to 0.5 for all experiments. We set as default parameters, $|\mathbf{A}|=|\mathbf{X}|=10$, $|D_i|=5$ for all variables, and $p_1=0.3$ for random networks, and $|\mathbf{A}|=|\mathbf{X}|=9$ for regular grids. Experiments for GPU-DBE are conducted using a multi-agent DCOP simulator, that simulates the concurrent activities of multiple agents, whose actions are activated upon receipt of a message. We use the publicly-available implementation of DPOP available in the FRODO framework v.2.11 [19], and we use the same framework to run the BE algorithm, in a centralized setting.

Since all algorithms are complete, our focus is on runtime. Performance of the centralized algorithms are evaluated using the algorithm's wallclock runtime, while distributed algorithms' performances are evaluated using the *simulated runtime* metric [30]. We imposed a timeout of 300s of wallclock (or simulated) time and a memory limit of 32GB. Results are averaged over all instances and are statistically significant with p-values $< 1.638\,e^{-12}$.[6] These experiment are performed on an *AMD Opteron 6276*, 2.3GHz, 128GB of RAM, which is equipped with a GPU device *GeForce GTX TITAN* with 14 multiprocessors, 2688 cores, and a clock rate of 837MHz.

---

[6] t-test performed with null hypothesis: GPU-based algorithms are faster than non-GPU ones.

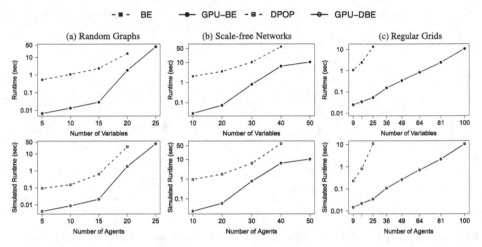

Fig. 4. Runtimes for COPs (top) and DCOPs (bottom) at varying number of variables/agents.

Fig. 4 illustrates the runtime, in seconds, for random (a), scale-free (b), and regular grid (c) topologies, varying the number of variables (resp. agents) for the centralized (resp. distributed) algorithms. The centralized algorithms (BE and GPU-BE) are shown at the top of the figure, while the distributed algorithms (DPOP and GPU-DBE) are illustrated at the bottom. All plots are in log-scale. We make the following observations:

- The GPU-based DP-algorithms (for both centralized and distributed cases) are consistently faster than the non-GPU-based ones. The speedups obtained by GPU-BE vs. BE are, on average, and minimum (showed in parenthesis) 69.3 (16.1), 34.9 (9.5), and 125.1 (42.6), for random, scale-free, and regular grid topologies, respectively. For the distributed algorithms, the speedups obtained by GPU-DBE vs. DPOP are on average (minimum) 44.7 (14.7), 22.3 (8.2), and 124.2 (38.8), for random, scale-free, and regular grid topologies, respectively.

- In terms of scalability, the GPU-based algorithms scale better than the non-GPU-based ones. In addition, their scalability increases with the level of structure exposed by each particular topology. On random graphs, which have virtually no structure, the GPU-based algorithms reach a timeout for instances with small number of variables (25 variables—compared to 20 variables for the non-GPU-based algorithms). On scale-free networks, the GPU-(D)BE algorithms can solve instances up to 50 variables,[7] while BE and DPOP reach a timeout for instances greater than 40 variables. On regular grids, the GPU-based algorithms can solve instances up to 100 variables, while the non-GPU-based ones, fail to solve any instance with 36 or more variables.

We relate these observations to the size of the separator sets and, thus, the size of the *UTIL* tables that are constructed in each problem. In our experiments,

---

[7] With 60 variables, we reported 12/30 instances solved for GPU-(D)BE.

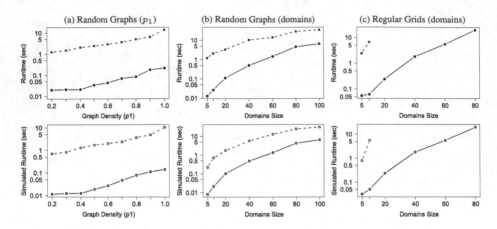

Fig. 5. Runtimes for COPs (top) and DCOPs (bottom) at varying number of variables/agents.

we observe that the average sizes of the separator sets are consistently larger in random graphs, followed by scale-free networks, followed by regular grids.

- Finally, the trends of the centralized algorithms are similar to those of the distributed algorithms: The simulated runtimes of the DCOP algorithms are consistently smaller than the wallclock runtimes of the COP ones.

Fig. 5 illustrates the behavior of the algorithms when varying the graph density p_1 for the random graphs (a), and the domains size for random graphs (b) and regular grids (c). As for the previous experiments, the centralized (resp. distributed) algorithms are shown on the top (resp. bottom) of the figure. We can observe:

- The trends for the algorithms runtime, when varying both p_1 and domains size, are similar to those observed in the previous experiments.
- GPU-(D)BE achieves better speed-up for smaller p_1 (Fig. 4 (a)). The result is explained by observing that small p_1 values correspond to smaller induced width of the underlying constraint graph. In turn, for small p_1 values, GPU-(D)BE agents construct smaller *UTIL* tables, which increases the probability of performing the complete inference process on the GPU, through the GPU-AGGREGATE-PROJECT procedure. This observation is also consistent with what observed in the previous experiments in terms of scalability.
- GPU-(D)BE achieves greater speedups in presence of large domains. This is due to the fact that large domains correspond to large *UTIL* tables, enabling the GPU-based algorithms to exploit a greater amount of parallelism, provided that the *UTIL* tables can be stored in the global memory of the GPU.

6   Conclusions and Discussions

In this paper, we presented an investigation of the use of GPUs to exploit SIMT-style parallelism from DP-based methods to solve COPs and DCOPs.

We proposed a procedure, inspired by BE (for COPs) and DPOP (for DCOPs), that makes use of multiple threads to parallelize the aggregation and projection phases. Experimental results show that the use of GPUs may provide significant advantages in terms of runtime and scalability. The proposed results are significant—the wide availability of GPUs provides access to parallel computing solutions that can be used to improve efficiency of (D)COP solvers. Furthermore, GPUs are renowned for their complex architectures (multiple memory levels with very different size and speed characteristics; relatively slow cores), which often create challenges to the effective exploitation of parallelism from irregular applications; the strong experimental results indicate that the proposed algorithms are well-suited to GPU architectures. While envisioning further research in this area, we anticipate several challenges:

- In terms of implementation, GPU programming can be more demanding when compared to a classical sequential implementation. One of the current limitations for (D)COP-based GPU approaches is the absence of solid abstractions that allow component integration, modularly, without restructuring the whole program.

- Exploiting the integration of CPU and GPU computations is a key factor to obtain competitive solvers performance. Complex and repeated calculations should be delegated to GPUs, while simpler and memory intensive operations should be assigned to CPUs. It is however unclear how to determine good tradeoffs of such integrations. For instance, repeatedly invoking many memory demanding GPU kernels could be detrimental to the overall performance, due to the high cost of allocating the device memory (e.g., shared memory). Creating lightweight communication mechanisms between CPU and GPU (for instance, by taking advantage of the asynchronism of CUDA streams) to allow active GPU kernels to be used in multiple instances could be a possible solution to investigate.

- While this paper describes the applicability of our approach to BE and DPOP, we believe that analogous techniques can be derived and applied to other DP-based approaches to solve (D)COPs—e.g., to implement the logic of DP-based propagators. We also envision that such technology could open the door to efficiently enforcing higher form of consistencies than domain consistency (e.g., *path consistency* [22], *adaptive consistency* [12], or the more recently proposed *branch consistency* for DCOPs [14]), especially when the constraints need to be represented explicitly.

References

1. Abdennadher, S., Schlenker, H.: Nurse scheduling using constraint logic programming. In: Proceedings of the Conference on Innovative Applications of Artificial Intelligence (IAAI), pp. 838–843 (1999)
2. Apt, K.: Principles of constraint programming. Cambridge University Press (2003)
3. Arbelaez, A., Codognet, P.: A GPU implementation of parallel constraint-based local search. In: Proceedings of the Euromicro International Conference on Parallel, Distributed and network-based Processing (PDP), pp. 648–655 (2014)

4. Barabási, A.-L., Albert, R.: Emergence of scaling in random networks. Science **286**(5439), 509–512 (1999)
5. Boyer, V., El Baz, D., Elkihel, M.: Solving knapsack problems on GPU. Computers & Operations Research **39**(1), 42–47 (2012)
6. Brito, I., Meseguer, P.: Improving DPOP with function filtering. In: Proceedings of the International Conference on Autonomous Agents and Multiagent Systems (AAMAS), pp. 141–158 (2010)
7. Burke, D., Brown, K.: Efficiently handling complex local problems in distributed constraint optimisation. In: Proceedings of the European Conference on Artificial Intelligence (ECAI), pp. 701–702 (2006)
8. Burke, E.K., De Causmaecker, P., Berghe, G.V., Van Landeghem, H.: The state of the art of nurse rostering. Journal of scheduling **7**(6), 441–499 (2004)
9. Campeotto, F., Dovier, A., Fioretto, F., Pontelli, E.: A GPU implementation of large neighborhood search for solving constraint optimization problems. In: Proceedings of the European Conference on Artificial Intelligence (ECAI), pp. 189–194 (2014)
10. Dechter, R.: Bucket elimination: a unifying framework for probabilistic inference. In: Learning in graphical models, pp. 75–104. Springer (1998)
11. Dechter, R.: Constraint Processing. Morgan Kaufmann Publishers Inc., San Francisco (2003)
12. Dechter, R., Pearl, J.: Network-based heuristics for constraint-satisfaction problems. Springer (1988)
13. Farinelli, A., Rogers, A., Petcu, A., Jennings, N.: Decentralised coordination of low-power embedded devices using the Max-Sum algorithm. In: Proceedings of the International Conference on Autonomous Agents and Multiagent Systems (AAMAS), pp. 639–646 (2008)
14. Fioretto, F., Le, T., Yeoh, W., Pontelli, E., Son, T.C.: Improving DPOP with branch consistency for solving distributed constraint optimization problems. In: O'Sullivan, B. (ed.) CP 2014. LNCS, vol. 8656, pp. 307–323. Springer, Heidelberg (2014)
15. Gaudreault, J., Frayret, J.-M., Pesant, G.: Distributed search for supply chain coordination. Computers in Industry **60**(6), 441–451 (2009)
16. Hamadi, Y., Bessière, C., Quinqueton, J.: Distributed intelligent backtracking. In: Proceedings of the European Conference on Artificial Intelligence (ECAI), pp. 219–223 (1998)
17. Kumar, A., Faltings, B., Petcu, A.: Distributed constraint optimization with structured resource constraints. In: Proceedings of the International Conference on Autonomous Agents and Multiagent Systems (AAMAS), pp. 923–930 (2009)
18. Lalami, M.E., El Baz, D., Boyer, V.: Multi GPU implementation of the simplex algorithm. Proceedings of the International Conference on High Performance Computing and Communication (HPCC) **11**, 179–186 (2011)
19. Léauté, T., Faltings, B.: Distributed constraint optimization under stochastic uncertainty. In: Proceedings of the AAAI Conference on Artificial Intelligence (AAAI), pp. 68–73 (2011)
20. Maheswaran, R., Tambe, M., Bowring, E., Pearce, J., Varakantham, P.: Taking DCOP to the real world: Efficient complete solutions for distributed event scheduling. In: Proceedings of the International Conference on Autonomous Agents and Multiagent Systems (AAMAS), pp. 310–317 (2004)
21. Modi, P., Shen, W.-M., Tambe, M., Yokoo, M.: ADOPT: Asynchronous distributed constraint optimization with quality guarantees. Artificial Intelligence **161**(1–2), 149–180 (2005)

22. Montanari, U.: Networks of constraints: Fundamental properties and applications to picture processing. Information sciences **7**, 95–132 (1974)
23. Pawłowski, K., Kurach, K., Michalak, T., Rahwan, T.: Coalition structure generation with the graphic processor unit. Technical Report CS-RR-13-07, Department of Computer Science, University of Oxford (2014)
24. Pesant, G.: A regular language membership constraint for finite sequences of variables. In: Proceedings of the International Conference on Principles and Practice of Constraint Programming (CP), pp. 482–495 (2004)
25. Petcu, A., Faltings, B.: A scalable method for multiagent constraint optimization. In: Proceedings of the International Joint Conference on Artificial Intelligence (IJCAI), pp. 1413–1420 (2005)
26. Quimper, C.-G., Walsh, T.: Global grammar constraints. In: Benhamou, F. (ed.) CP 2006. LNCS, vol. 4204, pp. 751–755. Springer, Heidelberg (2006)
27. Rodrigues, L.C.A., Magatão, L.: Enhancing supply chain decisions using constraint programming: a case study. In: Gelbukh, A., Kuri Morales, Á.F. (eds.) MICAI 2007. LNCS (LNAI), vol. 4827, pp. 1110–1121. Springer, Heidelberg (2007)
28. Rossi, F., van Beek, P., Walsh, T. (eds.) Handbook of Constraint Programming. Elsevier (2006)
29. Sanders, J., Kandrot, E.: CUDA by Example. An Introduction to General-Purpose GPU Programming. Addison Wesley (2010)
30. Sultanik, E., Modi, P.J., Regli, W.C.: On modeling multiagent task scheduling as a distributed constraint optimization problem. In: Proceedings of the International Joint Conference on Artificial Intelligence (IJCAI), pp. 1531–1536 (2007)
31. Trick, M.A.: A dynamic programming approach for consistency and propagation for knapsack constraints. Annals of Operations Research **118**(1–4), 73–84 (2003)
32. Yeoh, W., Felner, A., Koenig, S.: BnB-ADOPT: An asynchronous branch-and-bound DCOP algorithm. Journal of Artificial Intelligence Research **38**, 85–133 (2010)
33. Yeoh, W., Yokoo, M.: Distributed problem solving. AI Magazine **33**(3), 53–65 (2012)
34. Yokoo, M. (ed.): Distributed Constraint Satisfaction: Foundation of Cooperation in Multi-agent Systems. Springer (2001)
35. Zivan, R., Okamoto, S., Peled, H.: Explorative anytime local search for distributed constraint optimization. Artificial Intelligence **212**, 1–26 (2014)
36. Zivan, R., Yedidsion, H., Okamoto, S., Glinton, R., Sycara, K.: Distributed constraint optimization for teams of mobile sensing agents. Journal of Autonomous Agents and Multi-Agent Systems **29**(3), 495–536 (2015)

Conflict Ordering Search
for Scheduling Problems

Steven Gay[1]([⊠]), Renaud Hartert[1], Christophe Lecoutre[2], and Pierre Schaus[1]

[1] UCLouvain, ICTEAM, Place Sainte Barbe 2, 1348 Louvain-la-Neuve, Belgium
{steven.gay,renaud.hartert,pierre.schaus}@uclouvain.be
[2] CRIL-CNRS UMR 8188, Université d'Artois, 62307 Lens, France
lecoutre@cril.fr

Abstract. We introduce a new generic scheme to guide backtrack search, called Conflict Ordering Search (COS), that reorders variables on the basis of conflicts that happen during search. Similarly to generalized Last Conflict (LC), our approach remembers the last variables on which search decisions failed. Importantly, the initial ordering behind COS is given by a specified variable ordering heuristic, but contrary to LC, once consumed, this first ordering is forgotten, which makes COS conflict-driven. Our preliminary experiments show that COS – although simple to implement and parameter-free – is competitive with specialized searches on scheduling problems. We also show that our approach fits well within a restart framework, and can be enhanced with a value ordering heuristic that selects in priority the last assigned values.

1 Introduction

Backtracking search is a central complete algorithm used to solve combinatorial constrained problems. Unfortunately, it suffers from thrashing – repeatedly exploring the same fruitless subtrees – during search. Restarts, adaptive heuristics, and strong consistency algorithms are typical Constraint Programming (CP) techniques used to cope with thrashing.

Last Conflicts (LC) [9] has been shown to be highly profitable to complete search algorithms, both in constraint satisfaction and in automated artificial intelligence planning. The principle behind LC is to select in priority the last conflicting variables as long as they cannot be instantiated without leading to a failure. Interestingly enough, last conflict search can be combined with any underlying variable ordering heuristic. In *normal mode*, the underlying heuristic selects the variables to branch on, whereas in *conflict mode*, variables are directly selected in a conflict set built by last conflict.

While last conflict uses conflicts to repair the search heuristic, we show in this paper that conflicts can also be used to drive the search process by progressively replacing the initial variable heuristic. Basically, the idea behind our approach – namely, Conflict Ordering Search – is to reorder variables according to the most recent conflict they were involved in. Our experiments highlight that this simple reordering scheme, while being generic, can outperform domain specific heuristics for scheduling problems.

© Springer International Publishing Switzerland 2015
G. Pesant (Ed.): CP 2015, LNCS 9255, pp. 140–148, 2015.
DOI: 10.1007/978-3-319-23219-5_10

2   Related Works

We start by providing a quick overview of general-purpose search heuristics and schemes since our approach is definitively one of them. The simple variable ordering heuristic dom [5] – which selects variables by their domain size – has long been considered as the most robust backtrack search heuristic. However, a decade ago, modern *adaptive* heuristics were introduced. Such heuristics take into account information related to the part of the search space already explored. The two first proposed generic adaptive heuristics are impact [16] and wdeg [1]. The former relies on a measure of the effect of any assignment, and the latter associates a counter with each constraint (and indirectly, with each variable) indicating how many times any constraint led to a domain wipe-out. Counting-based heuristics [14] and activity-based search [11] are two recently introduced additional adaptive techniques to guide the search process.

Interestingly, Last Conflict (LC) [9] is a search mechanism that can be applied on top of any variable ordering heuristic. Precisely, the generalized form $LC(k)$ works by recording and assigning first the k variables involved in the k last decisions that provoked a failure after propagation. The underlying search heuristic is used when all the last k conflicting variables have been assigned. While the ability of relying on an underlying search heuristic is a strong point of LC, setting the parameter k can be problematic, as we shall see later. The related scheme introduced in this paper goes further and orders permanently all conflicting variables using the time of their last conflicts, eventually becoming independent of the helper variable ordering heuristic.

3   Guiding Search by Timestamping Conflicts

We first introduce basic concepts. Then, we introduce Conflict Ordering Search and highlight its benefits within a context of restarts. We conclude this section by discussing the differences between $LC(k)$ and COS.

3.1   Background

CSP. A Constraint Satisfaction Problem (CSP) P is a pair $(\mathcal{X}, \mathcal{C})$, where $\mathcal{X}$ is a finite set of variables and $\mathcal{C}$ is a finite set of constraints. Each variable $x \in \mathcal{X}$ has a domain $dom(x)$ that contains the allowed values for x. A valuation on a subset $X \subseteq \mathcal{X}$ of variables maps each variable $x \in X$ with a value in $dom(x)$. Each constraint $c \in \mathcal{C}$ has a scope $scp(c) \subseteq \mathcal{X}$, and is semantically defined by a set of allowed valuations on $scp(c)$; the valuations that satisfy c. A valuation on $\mathcal{X}$ is a solution of P iff it satisfies each constraint of P. A CSP is satisfiable iff it admits at least one solution.

Tree-Search. One can solve CSPs by using backtrack search, a complete depth-first exploration of the search space, with backtracking when a dead-end occurs. At each search node, a filtering process ϕ can be performed on domains by soliciting propagators associated with constraints. A CSP is in failure, denoted $\perp$, when unsatisfiability is detected by ϕ. A branching heuristic is a function that maps a non-failed CSP to an ordered sequence of constraints, called *decisions*. In this work, we only consider binary variable-based branching heuristics, i.e., heuristics that always generate sequences of decisions of the form $\langle x \in D, x \notin D \rangle$, where x is a variable of $\mathcal{X}$ and D a strict subset of $dom(x)$. A search tree is the structure explored by backtrack search through its filtering capability and its branching heuristic. A failed node in the search tree is a node where unsatisfiability has been detected by ϕ.

3.2 Conflict Ordering

Considering a variable-based branching heuristic, we can associate a failed search node with the variable involved in the decision leading to it. This allows us to timestamp variables with the number of the last conflict they caused (see Fig. 1). The basic idea behind Conflict Ordering Search is to leverage this timestamping mechanism to reorder the variables during search.

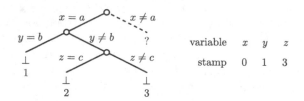

Fig. 1. Conflict numbering and timestamps associated with each variable. Variables are stamped with the number of their latest conflict (or 0 by default).

Algorithm 1 describes Conflict Ordering Search. For simplicity, we only consider classical binary branching with decisions of the form $x \leq v$ and $x > v$. We use an integer nConflicts to count the number of conflicts and a reference lastVar to the last variable involved in a decision (initially *null* at the root of the search-tree). We also consider a one-dimensional array stamps that associates with each variable $x \in \mathcal{X}$ the last time variable x was involved in a conflict. They are all initialized to 0. We suppose that ϕ corresponds to a domain filtering consistency, which is at least as strong as the partial form of arc consistency ensured by the forward checking algorithm [5].

If the resulting CSP at line 1 is trivially inconsistent ($\perp$), false is returned (line 6). If the failure is due to a previous decision (line 3), the number of conflicts is incremented and the conflicting variable timestamped with this number (lines 4 and 5). Otherwise, COS returns true if a solution has been found, i.e., the domain of each variable in $\mathcal{X}$ is a singleton (lines 7 and 8). The selection of the next variable to branch on is performed between lines 9 and 13. Here, the timestamps are used to select the unbound variable involved in the latest conflict. If no unbound variable ever conflicted, the search falls back to the bootstrapping heuristic *varHeuristic*. When a new value has been selected by the heuristic *valHeuristic*$[x]$, we recursively call *COS*. One can observe that the complexity of selecting a variable is linear in time and space, hence scaling well.[1]

Algorithm 1. $COS(P = (\mathcal{X}, \mathcal{C})$: CSP)

Output: true iff P is satisfiable

1 $P \leftarrow \phi(P)$
2 **if** $P = \bot$ **then**
3   **if** lastVar $\neq null$ **then**
4   nConflicts $\leftarrow$ nConflicts $+ 1$
5   stamps[lastVar] $\leftarrow$ nConflicts
6   **return** false
7 **if** $\forall x \in \mathcal{X}, |dom(x)| = 1$ **then**
8   **return** true
9 failed $\leftarrow \{x \in \mathcal{X} : \text{stamps}[x] > 0 \land |dom(x)| > 1\}$
10 **if** failed $= \emptyset$ **then**
11   lastVar $\leftarrow$ *varHeuristic.select*()
12 **else**
13   lastVar $\leftarrow$ argmax$_{x \in \text{failed}}\{\text{stamps}[x]\}$
14 $v \leftarrow$ *valHeuristic*[lastVar].*select*()
15 **return** $COS(P_{|\text{lastVar} \leq v}) \lor COS(P_{|\text{lastVar} > v})$

Example 1. Let us consider a toy CSP with n "white" variables, and m "black" variables. White variables have a binary domain while black variables have $\{1, 2, \ldots, m-1\}$ as domain. We also add a binary difference constraint on each pair of black variables (thus making the CSP unsatisfiable), but no constraint at all on the white variables. Let us also assume a variable ordering heuristic that selects the white variables first, then the black variables. Hence, proving unsatisfiability using this heuristic requires to prove the "black conflict" for the 2^n valuations of the white variables (see left part of Fig. 2). Using COS on top of this heuristic allows one to detect unsatisfiability quickly. Indeed, the $m - 2$ first conflicting black variables will be prioritized as a white variable cannot be involved in a conflict. The number of times the "black conflict" as to be proven thus becomes linear in the number of white variables n (see right part of Fig. 2).

COSPhase: a Variant. Because it is known that remembering last assigned values for later priority uses can be worthwhile (see for example phase saving [15] in SAT), we propose such a variant for COS. So, when a positive decision $x \leq v$ succeeds, we record its value v. Then, when branching is performed on a timestamped variable x, we exploit the associated recorded value v. If v is still in

---

[1] The time complexity could be improved if an ordered linked-list is used instead of the array stamps.

Without Conflict Ordering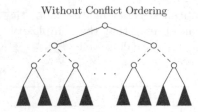

With Conflict Ordering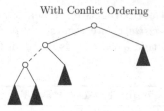

Fig. 2. Conflict Ordering Search reorders the variables to reduce the number of times the inconsistent black subtree has to be explored.

the domain of x, we use interval splitting on it, i.e., we branch with decisions $x \leq v$ and $x > v$, otherwise the value heuristic is solicited. Observe that this mechanism follows the first-fail/best-first principle.

3.3   Restarts and Timestamp Ordering

Depth-first search is far from always being the most efficient way to solve a CSP. In some cases, it may suffer from heavy-tailed distributions [4]. While restarting the search with a randomized heuristic is a typical way to avoid the worst parts of a long-tailed curve, *nogood learning* mitigates the effect of cutting a search process short by remembering parts of the search space already explored [8].[2]

In the context of COS, we observe that our approach not only remembers the sources of conflicts but also produce a variable ordering that yields a behavior similar to randomization. We thus propose to use no additional randomization when restarting, only using the natural randomizing effect of conflict ordering instead. The rationale is that while conflict ordering is good at finding a set of conflicting variables in a given context – i.e., a sequence of previous decisions – restarting with conflict ordering has the effect of trying the latest conflict set in other contexts.

Example 2. Let us consider the toy CSP described in Example 1. The time required to prove unfeasibility could be drastically reduced if a restart occurs after having explored the inconsistent black subtree at least once. Indeed, in this context, the second restart will directly explore the black search tree without even considering the white variables that have been "disculpated".

3.4   Differences with Last Conflict Search

Although similar, we show that COS and LC are rather different. Indeed, LC relies on a parameter k that corresponds to the maximum size of the conflict sets that can be captured by the search process. The value of k is of importance as setting k too low may not allow LC to capture the encountered conflict sets. For instance, $LC(k)$ cannot capture the conflict set in Example 1 if k is lower than $m - 2$. COS, however, does not require any parameter and is able to handle

---

[2] Similar frameworks are typically used by SAT solvers [12].

conflict sets of any size. While setting the parameter k to the number of decision variables may solve the problem, LC still suffers from resets of its conflict set that occur each time the conflicting variables have been successfully assigned. Conversely, COS does not forget conflicting variables and progressively reorders those variables to give priority to the recently conflicting ones. This is particularly important with restarts as LC is not designed to focus on conflict sets in such contexts (see Example 2).

4   Experiments

We have tested our approach on RCPSP (Resource-Constrained Project Scheduling Project) instances from PSPLIB [6]. We used a computer equipped with a i7-3615QM processor running at 2.30GHz. The problem has been modeled in the open-source solver OscaR [13], using precedence and cumulative constraints. Precedences are simple binary precedence constraints, and cumulative constraints use the Time-Tabling propagator presented in [2]. Both overload checking [19] or time-table edge-finding [17] were tested but energy-based reasoning does not help much on PSPLIB instances, whose optimal solutions typically waste capacity. Adding TTDR [3] helps even with learning searches, but it makes the experiments harder to reproduce, thus we chose to not use it.

4.1   Branch-and-Bound

The goal of this first experiment is to find an optimal solution using a pure branch-and-bound search. We compare five search solving methods. The first is a simple min/min scheme, which selects the variable with the smallest minimal value and chooses the smallest value (for assignment). The second one is the scheduling-specialized SetTimes heuristic with dominances [7], which is a min/min scheme that simply postpones assignments (instead of making value refutations) when branching at right, fails when a postponed task can no longer be woken up, and assigns the tasks that are not postponed and cannot be disturbed by other tasks. Finally, the last three heuristics correspond to conflict-based reasoning searches, namely, LC(k) for the best value of k, our main contribution COS, and COSPhase based on the variant presented in Section 15. All these have a min/min helper heuristic.

We have observed that the ranking of these five search methods is the same on the four RCPSP benchmarks (J30, J60, J90, J120) from PSPLIB. Results are represented on the left part of Fig. 3 where the y-axis is the cumulated number of solved instances and the x-axis is the CPU time. SetTimes is clearly an improvement on min/min, as it finishes ahead and seems to continue its course on a better slope than min/min. In turn, the well-parameterized LC (we empirically chose the best possible value for k) fares better than SetTimes. Finally, COS allows us to close a higher number of instances, and the variant COSPhase improves COS even further.

4.2   Branch-and-Bound with Restarts

As illustrated in Example 2, keeping the conflict ordering between restarts could drastically reduce search efforts. The aim of this second experiment is to compare the performance of COS if the ordering is kept between restarts or not. Experimental settings are the same as before except that we use restarts and nogood recording has explained in [10]. The first iteration is limited to 100 failures and increases by a 1.15 factor. We compared the performance of COS and COSPhase with and without reset (we add "rst-" as prefix for the resetting version). All searches rely on $\min^{rnd}/\min$ – a randomized version of $\min/\min$ that breaks ties randomly – as helper heuristic.

Results are presented in the right part of Fig. 3. First, we observe that restarts do have a positive effect as $\min^{rnd}/\min$ obtains better results than $\min/\min$ in the previous experiment. Next, we see that resetting the conflict order has a bad effect on the search process. Indeed, these variant obtain worse results than in the pure branch-and-bound framework. This highlights that using conflict-based search as a full heuristic can yield much better results than using it as a repairing patch. Finally, phase recording does not seem to help anymore.

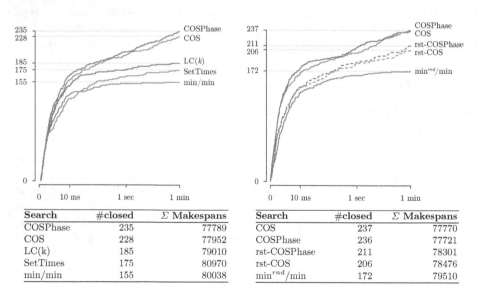

Search	#closed	Σ Makespans
COSPhase	235	77789
COS	228	77952
LC(k)	185	79010
SetTimes	175	80970
min/min	155	80038

Search	#closed	Σ Makespans
COS	237	77770
COSPhase	236	77721
rst-COSPhase	211	78301
rst-COS	206	78476
$\min^{rnd}/\min$	172	79510

Fig. 3. On the left, results obtained for pure branch-and-bound, and on the right, results obtained with branch-and-bound with restarts. Graphs at the top show the cumulated number of solved RCPSP instances from PSPLIB120. The tables at the bottom compare branching heuristics at the end of the 60s timeout, giving the number of closed instances and the sum of makespans.

4.3   Destructive Lower Bounds.

We performed similar experiments for destructive lower bounds. We added the Time-Table Edge-Finding propagator presented in [17] since it has a large impact in this case. The results are similar. We also compared COS to the recently introduced Failure Directed Search [18] by implementing it directly in CP Optimizer. Unfortunately our COS implementation in CP Optimizer was not able to obtain results competitive with FDS.

5 Conclusion

In this paper, we have proposed a general-purpose search scheme that can be combined with any variable ordering heuristic. Contrary to Last Conflict, Conflict Ordering Search is very aggressive, discarding progressively the role played by the heuristic. Besides, by means of simple timestamps, all variables recorded in the global conflict set stay permanently ordered, the priority being modified at each new conflict. We have shown that on some structured known problems our approach outperforms other generic and specific solving methods. So, COS should be considered as one new useful technique to be integrated in the outfit of constraint systems.

Acknowledgments. Steven Gay is financed by project Innoviris 13-R-50 of the Brussels-Capital region. Renaud Hartert is a Research Fellow of the Fonds de la Recherche Scientifiques - FNRS. Christophe Lecoutre benefits from the financial support of CNRS and OSEO (BPI France) within the ISI project ?Pajero?. The authors would like to thank Petr Vilím for his help with the comparison to FDS in CP Optimizer.

References

1. Boussemart, F., Hemery, F., Lecoutre, C., Sais, L.: Boosting systematic search by weighting constraints. In: Proceedings of ECAI 2004, pp. 146–150 (2004)
2. Gay, S., Hartert, R., Schaus, P.: Simple and scalable time-table filtering for the cumulative constraint. In: Pesant, G. (ed.) Proceedings of CP 2015. LNCS, vol. 9255, pp. 149–157. Springer, Heidelberg (2015)
3. Gay, S., Hartert, R., Schaus, P.: Time-table disjunctive reasoning for the cumulative constraint. In: Michel, L. (ed.) CPAIOR 2015. LNCS, vol. 9075, pp. 157–172. Springer, Heidelberg (2015)
4. Gomes, C., Selman, B., Crato, N., Kautz, H.: Heavy-tailed phenomena in satisfiability and constraint satisfaction problems. Journal of Automated Reasoning **24**, 67–100 (2000)
5. Haralick, R.M., Elliott, G.L.: Increasing tree search efficiency for constraint satisfaction problems. Artificial Intelligence **14**, 263–313 (1980)
6. Kolisch, R., Schwindt, C., Sprecher, A.: Benchmark instances for project scheduling problems. In: Project Scheduling, pp. 197–212. Springer (1999)
7. Le Pape, C., Couronné, P., Vergamini, D., Gosselin, V.: Time-versus-capacity compromises in project scheduling (1994)
8. Lecoutre, C., Sais, L., Tabary, S., Vidal, V.: Nogood recording from restarts. In: Proceedings of IJCAI 2007, pp. 131–136 (2007)

9. Lecoutre, C., Sais, L., Tabary, S., Vidal, V.: Reasonning from last conflict(s) in constraint programming. Artificial Intelligence **173**(18), 1592–1614 (2009)
10. Lecoutre, C., Sais, L., Tabary, S., Vidal, V.: Recording and minimizing nogoods from restarts. Journal on Satisfiability, Boolean Modeling and Computation **1**, 147–167 (2007)
11. Michel, L., Van Hentenryck, P.: Activity-based search for black-box constraint programming solvers. In: Beldiceanu, N., Jussien, N., Pinson, É. (eds.) CPAIOR 2012. LNCS, vol. 7298, pp. 228–243. Springer, Heidelberg (2012)
12. Moskewicz, M.W., Madigan, C.F., Zhao, Y., Zhang, L., Malik, S. Chaff: engineering an efficient SAT solver. In: Proceedings of DAC 2001, pp. 530–535 (2001)
13. OscaR Team. OscaR: Scala in OR (2012). bitbucket.org/oscarlib/oscar
14. Pesant, G., Quimper, C.-G., Zanarini, A.: Counting-based search: Branching heuristics for constraint satisfaction problems. Journal of Artificial Intelligence Research **43**, 173–210 (2012)
15. Pipatsrisawat, K., Darwiche, A.: A lightweight component caching scheme for satisfiability solvers. In: Marques-Silva, J., Sakallah, K.A. (eds.) SAT 2007. LNCS, vol. 4501, pp. 294–299. Springer, Heidelberg (2007)
16. Refalo, P.: Impact-based search strategies for constraint programming. In: Wallace, M. (ed.) CP 2004. LNCS, vol. 3258, pp. 557–571. Springer, Heidelberg (2004)
17. Vilím, P.: Timetable edge finding filtering algorithm for discrete cumulative resources. In: Achterberg, T., Beck, J.C. (eds.) CPAIOR 2011. LNCS, vol. 6697, pp. 230–245. Springer, Heidelberg (2011)
18. Vilím, P., Laborie, P., Shaw, P.: Failure-directed search for constraint-based scheduling. In: Michel, L. (ed.) CPAIOR 2015. LNCS, vol. 9075, pp. 437–453. Springer, Heidelberg (2015)
19. Wolf, A., Schrader, G.: $\mathcal{O}(n \log n)$ overload checking for the cumulative constraint and its application. In: Umeda, M., Wolf, A., Bartenstein, O., Geske, U., Seipel, D., Takata, O. (eds.) INAP 2005. LNCS (LNAI), vol. 4369, pp. 88–101. Springer, Heidelberg (2006)

Simple and Scalable Time-Table Filtering
for the Cumulative Constraint

Steven Gay$^{(\boxtimes)}$, Renaud Hartert, and Pierre Schaus

ICTEAM, UCLouvain, Place Sainte Barbe 2, 1348 Louvain-la-neuve, Belgium
{steven.gay,renaud.hartert,pierre.schaus}@uclouvain.be

Abstract. Cumulative is an essential constraint in the CP framework, and is present in scheduling and packing applications. The lightest filtering for the cumulative constraint is time-tabling. It has been improved several times over the last decade. The best known theoretical time complexity for time-table is $O(n \log n)$ introduced by Ouellet and Quimper. We show a new algorithm able to run in $O(n)$, by relying on range min query algorithms. This approach is more of theoretical rather than practical interest, because of the generally larger number of iterations needed to reach the fixed point. On the practical side, the recent synchronized sweep algorithm of Letort et al, with a time-complexity of $O(n^2)$, requires fewer iterations to reach the fix-point and is considered as the most scalable approach. Unfortunately this algorithm is not trivial to implement. In this work we present a $O(n^2)$ simple two step alternative approach: first building the mandatory profile, then updating all the bounds of the activities. Our experimental results show that our algorithm outperforms synchronized sweep and the time-tabling implementations of other open-source solvers on large scale scheduling instances, sometimes significantly.

Keywords: Constraint programming · Large-scale · Scheduling · Cumulative constraint · Time-table

1 Preliminaries

In this paper, we focus on a single cumulative resource with a discrete finite capacity $C \in \mathbb{N}$ and a set of n tasks $\Omega = \{1, \dots, n\}$. Each task i has a start time $s_i \in \mathbb{Z}$, a fixed duration $d_i \in \mathbb{N}$, and an end time $e_i \in \mathbb{Z}$ such that the equality $s_i + d_i = e_i$ holds. Moreover, each task i consumes a fixed amount of resource $c_i \in \mathbb{N}$ during its processing time. Tasks are non-preemptive, i.e., they cannot be interrupted during their processing time. In the following, we denote by $\underline{s}_i$ and $\overline{s}_i$ the earliest and the latest start time of task i and by $\underline{e}_i$ and $\overline{e}_i$ the earliest and latest end time of task i (see Fig. 1). The cumulative constraint [1] ensures that the accumulated resource consumption does not exceed the maximum capacity C at any time t (see Fig. 2): $\forall t \in \mathbb{Z} : \sum_{i \in \Omega : s_i \leq t < e_i} c_i \leq C$.

Even tasks that are not fixed convey some information that can be used by filtering rules. For instance, tasks with a tight execution window must consume some resource during a specific time interval known as *mandatory part*.

© Springer International Publishing Switzerland 2015
G. Pesant (Ed.): CP 2015, LNCS 9255, pp. 149–157, 2015.
DOI: 10.1007/978-3-319-23219-5_11

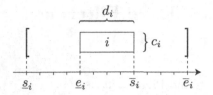
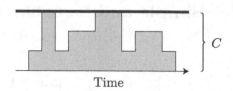

Fig. 1. Task i is characterized by its start time s_i, its duration d_i, its end time e_i, and its resource consumption c_i.

Fig. 2. Accumulated resource consumption over time. The `cumulative` constraint ensures that the maximum capacity C is not exceeded.

Definition 1 (Mandatory part). *Let us consider a task $i \in \Omega$. The mandatory part of i is the time interval $[\overline{s}_i, \underline{e}_i[$. Task i has a mandatory part only if its latest start time is smaller than its earliest end time.*

If task i has a mandatory part, we know that task i will consume c_i units of resource during all its mandatory part no matter its start time. Fig. 3 illustrates the mandatory part of an arbitrary task i.

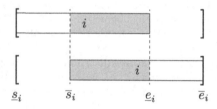

Fig. 3. Task i has a mandatory part $[\overline{s}_i, \underline{e}_i[$ if its latest start time $\overline{s}_i$ is smaller than its earliest end time $\underline{e}_i$: $\overline{s}_i < \underline{e}_i$. Task i always consumes the resource during its mandatory part no matter its start time.

By aggregation, mandatory parts allow us to have an optimistic view of the resource consumption over time. This aggregation is known as the *time-table*.

Definition 2 (Time-Table). *The time-table TT_Ω is the aggregation of the mandatory parts of all the tasks in Ω. It is defined as the following step function:*

$$\mathrm{TT}_\Omega = t \in \mathbb{Z} \longrightarrow \sum_{i \in \Omega \,|\, \overline{s}_i \leq t < \underline{e}_i} c_i. \tag{1}$$

The capacity of the resource is exceeded if $\exists t \in \mathbb{Z} : \mathrm{TT}_\Omega(t) > C$.

The time-table of a resource can be computed in $\mathcal{O}(n)$ by a sweep algorithm given the tasks sorted by latest start time and earliest end time [2,11,15].
    The *time-table filtering rule* is formalized as follows:

$$(t < \underline{e}_i) \ \wedge \ (c_i + \mathrm{TT}_{\Omega \setminus i}(t) > C) \ \Rightarrow \ t < s_i. \tag{2}$$

Observe that this filtering rule only describes how to update the start time of a task. End times are updated in a symmetrical way.

Let j be a rectangle denoted $\langle a_j, b_j, h_j \rangle$ with $a_j \in \mathbb{Z}$ (resp. $b_j \in \mathbb{Z}$) its start (resp. end) time, $h_j \in \mathbb{N}$ its height, and $b_j - a_j$ its duration (length). The time-table TT_Ω can be represented as a contiguous sequence of rectangles

$$TT_\Omega = \langle -\infty, a_1, 0 \rangle, \langle a_1, b_1, h_1 \rangle, \ldots, \langle a_m, b_m, h_m \rangle, \langle b_m, \infty, 0 \rangle \tag{3}$$

such that $b_i = a_{i+1}$ and that the following holds:

$$\forall \langle a_j, b_j, h_j \rangle \in TT_\Omega, \forall t \in [a_j, b_j[\quad : \quad TT_\Omega(t) = h_j. \tag{4}$$

We assume that the sequence is minimal, i.e., no consecutive rectangles have the same height. The maximum number of rectangles is thus limited to $2n + 1$.

Definition 3 (Conflicting Rectangle). *For a task i, a left-conflicting rectangle is a rectangle $\langle a_j, b_j, h_j \rangle \in TT_{\Omega \setminus i}$ such that $(a_j < \underline{e}) \wedge (b_j \geq \underline{s}_i) \wedge (h_j > C - c_i)$. We say that the task is in left-conflict with rectangle j. Right-conflicting rectangles are defined symmetrically.*

The time-table filtering rule can thus be rephrased as follows:

$$\forall i \in \Omega, \forall \langle a_j, b_j, h_j \rangle \in TT_{\Omega \setminus i} : j \text{ is in left-conflict with } i \Rightarrow b_j \leq s_i. \tag{5}$$

Definition 4 (Time-Table Consistency). *A cumulative constraint is left (resp. right) time-table consistent if no task has a left (resp. right) conflicting rectangle. It is time-table consistent if it is left and right time-table consistent.*

2   Existing Algorithms for Time-Tabling

Using the notion of conflicting rectangles, one can design a naive time-tabling algorithm by confronting every task to every rectangle of the profile. The following algorithms improve on this, mainly by avoiding fruitless confrontations of rectangles and tasks.

Sweep-line Algorithm. The sweep-line algorithm introduced by Beldiceanu et al. [2] introduces tasks from left to right, and builds the mandatory profile on-the-fly. This allows to confront tasks and rectangles only if they can overlap in time. It can factorize confrontations of a rectangle against several tasks, by organizing tasks in a heap. It pushes tasks to the right until they have no left-conflicting rectangle, as pictured in Figure 4(c). This algorithm runs in $O(n^2)$.

Idempotent Sweep-line Algorithm. The sweep-line algorithm by Letort et. al [11] improves on building the profile on-the-fly, by taking in consideration mandatory parts that appear dynamically as tasks are pushed. It reaches left-time-table consistency in $O(n^2)$, or $O(n^2 \log n)$ for its faster practical implementation.

Interval Tree Algorithm. The algorithm of Ouellet and Quimper [14] first builds the profile, then introduces rectangles and tasks in an interval tree. Rectangles are introduced by decreasing height, tasks by increasing height. This allows tasks and rectangles to be confronted only when their heights do conflict. For each task introduction, the tree structure decides in $\log n$ if its time domain conflicts with some rectangle. Its filtering is weaker, since it pushes a task i only after left-conflicting rectangles that overlap $[\underline{s}_i, \underline{e}_i[$, as pictured in Figure 4(b). The algorithm has time complexity $O(n \log n)$.

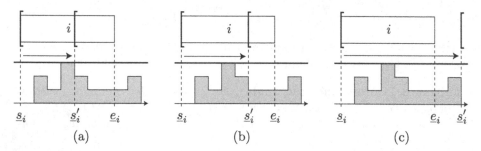

Fig. 4. Filtering obtained for (a) our linear time-tabling (b) Ouellet et al [14] and (c) Beldiceanu et al [2].

New Algorithms. In this paper, we introduce two new algorithms for time-tabling. The first one is of theoretical interest and runs in $O(n)$. It uses range-max-query algorithms to determine whether a task has a conflicting rectangle. As the algorithm of Ouellet et al [14], it confronts task i only with rectangles overlapping $[\underline{s}_i, \underline{e}_i[$, but only chooses the one with the largest height instead of the largest end. Thus, it prunes even less, as depicted in Figure 4(a).

The second algorithm is practical, and runs in $O(n^2)$. It separates profile building from task sweeping. To locate tasks on the profile, it exploits residues from previous computations, and incrementally removes fixed tasks that cannot lead to any more pruning. It uses sweeping, thus pruning as much as [2] per call, but it updates both the earliest start and latest end times of the tasks in a single execution.

3   A Linear Time-Table Filtering Algorithm

In order to obtain linear time complexity, we will confront task i to only one well-chosen rectangle, for every task i.

Proposition 1. *Suppose the mandatory profile does not overload the resource. Let i be a task, and j^* be a highest rectangle of the profile overlapping $[\underline{s}_i, \min(\underline{e}_i, \overline{s}_i)$ $[: j^* = argmax_j \{h_j \mid \langle a_j, b_j, h_j \rangle \in \mathrm{TT}_\Omega \ and \ [a_j, b_j[\cap [\underline{s}_i, \min(\underline{e}_i, \overline{s}_i)[\neq \emptyset\}$.*
*   Then j^* is in left-conflict iff $h_{j^*} + c_i > C$; otherwise i has no rectangles in left-conflict.*

Proof. If i has a mandatory part, we only need to look for conflict rectangles overlapping $[\underline{s}_i, \overline{s}_i[$, since the profile already includes the mandatory part of i. Otherwise, we need to look at $[\underline{s}_i, \underline{e}_i[$. If rectangle j^* is not in left-conflict with i, then no other rectangle can, since it would need to be higher than j^*.

To retrieve the index j^*, we must answer the question: given a vector of values and two indices on this vector, what is the index between those two indices that has the highest value? This kind of query corresponds to the range max query problem[1], it can be done in constant time, given a linear time preprocessing [7].

Example 1. Assume the vector is $values = [5, 4, 2, 1, 4, 3, 0, 8, 2, 3]$. The range max query between index 4 and index 7 is 5, denoted $rmq(values, 4, 7) = 5$. This is indeed at index 5 that there is the maximum value on the subvector $[1, 4, 3, 0]$.

In our case the vector is composed of the heights of the rectangles of the profile $heights = [h_1, h_2, \ldots, h_m]$. The two indices of the query are respectively:

- $j_1(i)$ is the index j of the rectangle $\langle a_j, b_j, h_j \rangle$ s.t. $\underline{s}_i \in [a_j, b_j[$.
- $j_2(i)$ is the index j of the rectangle $\langle a_j, b_j, h_j \rangle$ s.t. $\min(\underline{e}_i, \overline{s}_i) - 1 \in [a_j, b_j[$.

The whole algorithm is given in Algorithm 1. An example of the filtering is given in Figure 4 (a). Notice that the task is pushed after a highest conflicting rectangle, which is not as good as the filtering of [14] (Figure 4 (b)).

Algorithm 1. MinLeftTTLinearTime(Ω,C)

Input: A set of tasks Ω, capacity C.
Output: *true* iff propagation failed, i.e. if the problem is infeasible.

1 initialize TT_Ω // $\langle a_j, b_j, h_j \rangle \forall i \in \{1 \ldots m\}$
2 **if** $\max_{j \in [1;m]} h_j > C$ **then return** true
3 $heights \leftarrow [h_1, h_2, \ldots, h_m]$
4 $\forall i \in \Omega$ compute $j_1(i), j_2(i)$
5 initialize $rmq(heights)$
6 **for** $i \in \Omega$ *such that* $\underline{s} < \overline{s}$ **do**
7 | $j^* \leftarrow rmq(heights, j_1(i), j_2(i))$
8 |_ **if** $h_{j^*} + c_i > C$ **then** $\underline{s}_i \leftarrow b_j$

9 **return** false

Time Complexity. As in [6], we assume that all the time points are encoded with $w - bit$ integers and can thus be sorted in linear time. Given the sorted time points the profile TT_Ω can be computed in linear time using a sweep line algorithm, and all the indices $j_1(i), j_2(i)$ can be computed in linear time as well. The range min/max query is a well studied problem. Preprocessing in line 5 can be done in linear time, so that any subsequent query at Line 7 executes in constant time [7]. Thus, the whole algorithm executes in $O(n)$.

---
[1] a straightforward variant of the well-known range min query problem.

Discussion. Although the linear time complexity is an improvement over the $O(n \log n)$ algorithm introduced in [14], we believe that this result is more of theoretical rather than practical interest. The linear time range max query initialization hides non-negligible constants. The range max query used in Line 7 to reach this time complexity could be implemented by simply iterating on the rectangles from $j_1(i)$ to $j_2(i)$. On most problems, the interval $[\underline{s}_i, \min(\overline{s}_i, \underline{e}) - 1]$ only overlaps a few rectangles of the profile, so the $O(n)$ cost is not high in practice. Another limitation of the algorithm is that (as for the one of [14]) it may be called several times before reaching the fix-point (although it does not suffer from the slow convergence phenomenon described in [3]) either. It may be more efficient to continue pushing a task further to the right as in [2] rather than limiting ourselves to only one update per task per call to the procedure. This is precisely the objective of the algorithm introduced in the next section.

4   An Efficient $O(n^2)$ Time-Table Filtering

In this section, we introduce a practical algorithm for time-table filtering. It proceeds in two main phases: first the computation of the mandatory profile, then a per-task sweeping from left to right and from right to left. This modular design makes the algorithm simple, and its scalability comes from being able to exploit structures separately, for instance using sorting only on few tasks. We review the main phases of Algorithm 2 in execution order.

Building the Profile. Line 2 computes the mandatory profile as a sequence of rectangles. We process only those tasks in Ω that have a mandatory part. We will try to reduce that set of tasks further ahead, reducing the work in this part.

Computing Profile Indices. Line 4 computes, for all unfixed tasks i, the profile rectangle containing $\underline{s}_i$. This value is saved between consecutive calls in a residue[2]; most of the time, it is still valid and we do not have to recompute it, if not, a dichotomic search is performed, at a cost of $O(\log n)$. Note that [2] sorts tasks by $\underline{s}$ to locate tasks on the profile, at a theoretical cost of $O(n \log n)$. Here we pay $O(\log n)$ only for tasks where the residue is invalid. Similarly, line 5 computes the rectangle containing the last point of i, $\overline{e}_i - 1$.

Per-Task Sweeping. The loop in line 6 looks for left and right-conflicting rectangles for i linearly. The main difference with the global sweeping in [2] is that our method does not factorize sweeping according to height, wagering that the cost of moving tasks in height-ordered heaps is higher than that of pushing every task until no conflict remains. This main part has a worst case cost $O(n^2)$.

Fruitless Fixed Tasks Removal. After the main loop, line 24 removes fixed tasks at profile extremities that can no longer contribute to pruning. This filtering is $O(n)$. Note that Ω is kept along the search tree using a reversible sparse-set [5].

---

[2] this is similar to residual supports for AC algorithms [10].

Algorithm 2. ScalableTimeTable(Ω,C)

Input: A set of tasks Ω, capacity C.

Output: *true* iff propagation failed, i.e. if the problem is infeasible.

1   $\Omega_u \leftarrow \{i \mid \underline{s}_i < \overline{s}_i\}$ // unfixed tasks

2   initialize TT_Ω // $\langle a_j, b_j, h_j \rangle, \forall j \in \{1 \ldots m\}$

3   **if** $\max_{j \in [1;m]} h_j > C$ **then return** true

4   $\forall i \in \Omega_u$, compute $j_1(i)$ such that $\underline{s}_i \in [a_{j_1(i)}; b_{j_1(i)}[$

5   $\forall i \in \Omega_u$, compute $j_2(i)$ such that $\overline{e}_i - 1 \in [a_{j_2(i)}; b_{j_2(i)}[$

6   **for** $i \in \Omega_u$ **do**

7 | $j \leftarrow j_1(i)$

8 | $\underline{s}_i^* \leftarrow \underline{s}_i$

9 | **while** $j \leq m$ *and* $a_j < \min(\underline{s}_i^* + d_i, \overline{s}_i)$ **do**

10 | | **if** $C - c_i < h_j$ **then**

11 | | | $\underline{s}_i^* \leftarrow \min(b_j, \overline{s}_i)$ // j in left-conflict

12 | | $j \leftarrow j + 1$

13 | **if** $\underline{s}_i^* > \underline{s}_i$ **then** $\underline{s}_i \leftarrow \underline{s}_i^*$

14 |

15 | $j \leftarrow j_2(i)$

16 | $\overline{e}_i^* \leftarrow \overline{e}_i$

17 | **while** $j \geq 1$ *and* $b_j \geq \max(\overline{e}_i^* - d_i, \underline{e}_i)$ **do**

18 | | **if** $C - c_i < h_j$ **then**

19 | | | $\overline{e}_i^* \leftarrow \max(a_j, \underline{e}_i)$ // j in right-conflict

20 | | $j \leftarrow j - 1$

21 | **if** $\overline{e}_i^* < \overline{e}_i$ **then** $\overline{e}_i \leftarrow \overline{e}_i^*$

22   $s_{\min}^u \leftarrow \min_{i \in \Omega_u} \underline{s}_i$

23   $e_{\max}^u \leftarrow \max_{i \in \Omega_u} \overline{e}_i$

24   $\Omega \leftarrow \Omega \setminus \{i \in Omega \mid \overline{e}_i \leq s_{\min}^u \vee e_{\max}^u \leq \underline{s}_i\}$

25   **return** false

---

5   Experiments

We have tested our ScalableTimeTable filtering against the time-table filtering of or-tools [12], Choco3 [4] and Gecode [9] solvers. The algorithm in Choco3 and Gecode solver is the one of [2]. Similarly to our algorithm, or-tools also builds the time-table structure before filtering. To the best of our knowledge, no implementation of Letort et al [11] algorithm is publicly available. We have thus implemented the heap-based variant of the algorithm, faster in practice, with the same quality standard as our new ScalableTimeTable [13]. In the models used for this experiment, cumulative propagators enforce only the resource constraint, precedences are enforced by separate propagators.

We have generated randomly n (ranging from 100 to 12800) tasks with duration between 200 and 2000 and heights between 1 and 40. The capacity is fixed to 100. The search is a simple greedy heuristic selecting the current tasks with the smallest earliest possible start, hence there is no backtrack. The search finishes when all the tasks have been placed. This simple experimental setting guarantees that every solver has exactly the same behavior. The results are given on Figure 5. As can be seen, the time-table implementation of Choco3 and Gecode are quickly not able to scale well for more than 1600 tasks. The algorithm of or-tools, Letort et al and our new algorithm are still able to handle 12800 tasks. Surprisingly, our algorithm outperforms the one of Letort et al despite its simplicity.

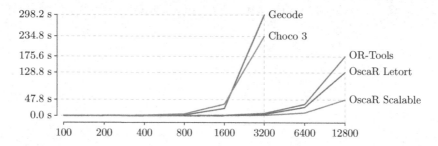

Fig. 5. Comparison of Time-Table implementations.

6   Conclusion

We have introduced an $O(n)$ time-table filtering using range min queries. We believe that the usage of range min query may be useful for subsequent research on scheduling, for instance in time-table disjunctive reasoning [8]. We introduced simple but scalable $O(n^2)$ filtering for the cumulative constraint. Our results show that despite its simplicity, it outperforms current implementations of time-table constraints in some open-source solvers and also the recent synchronized sweep algorithm. The resources related to this work are available here http:// bit.ly/cumulativett.

References

1. Aggoun, A., Beldiceanu, N.: Extending chip in order to solve complex scheduling and placement problems. Mathematical and Computer Modelling **17**(7), 57–73 (1993)
2. Beldiceanu, N., Carlsson, M.: A New multi-resource *cumulatives* constraint with negative heights. In: Van Hentenryck, P. (ed.) CP 2002. LNCS, vol. 2470, pp. 63–79. Springer, Heidelberg (2002)
3. Bordeaux, L., Hamadi, Y., Vardi, M.Y.: An analysis of slow convergence in interval propagation. In: Bessière, C. (ed.) CP 2007. LNCS, vol. 4741, pp. 790–797. Springer, Heidelberg (2007)
4. Charles Prud'homme, X.L., Fages, J.-G.: Choco3 documentation. TASC, INRIA Rennes, LINA CNRS UMR 6241, COSLING S.A.S. (2014)
5. de Saint-Marcq, V.C., Schaus, P., Solnon, C., Lecoutre, C.: Sparse-sets for domain implementation. In: CP Workshop on Techniques for Implementing Constraint Programming Systems (TRICS), pp. 1–10 (2013)

6. Fahimi, H., Quimper, C.-G.: Linear-time filtering algorithms for the disjunctive constraint. In: Twenty-Eighth AAAI Conference on Artificial Intelligence (2014)
7. Fischer, J., Heun, V.: Theoretical and practical improvements on the RMQ-problem, with applications to LCA and LCE. In: Lewenstein, M., Valiente, G. (eds.) CPM 2006. LNCS, vol. 4009, pp. 36–48. Springer, Heidelberg (2006)
8. Gay, S., Hartert, R., Schaus, P.: Time-table disjunctive reasoning for the cumulative constraint. In: Michel, L. (ed.) CPAIOR 2015. LNCS, vol. 9075, pp. 157–172. Springer, Heidelberg (2015)
9. Gecode Team. Gecode: Generic constraint development environment (2006). http://www.gecode.org
10. Lecoutre, C., Hemery, F., et al.: A study of residual supports in arc consistency. In: IJCAI, vol. 7, pp. 125–130 (2007)
11. Letort, A., Beldiceanu, N., Carlsson, M.: A Scalable Sweep Algorithm for the *cumulative* Constraint. In: Milano, M. (ed.) Principles and Practice of Constraint Programming. LNCS, pp. 439–454. Springer, Heidelberg (2012)
12. Or-tools Team. or-tools: Google optimization tools (2015). https://developers.google.com/optimization/
13. OscaR Team. OscaR: Scala in OR (2012). https://bitbucket.org/oscarlib/oscar
14. Ouellet, P., Quimper, C.-G.: Time-Table extended-edge-finding for the cumulative constraint. In: Schulte, C. (ed.) CP 2013. LNCS, vol. 8124, pp. 562–577. Springer, Heidelberg (2013)
15. Vilím, P.: Timetable edge finding filtering algorithm for discrete cumulative resources. In: Achterberg, T., Beck, J.C. (eds.) CPAIOR 2011. LNCS, vol. 6697, pp. 230–245. Springer, Heidelberg (2011)

General Bounding Mechanism for Constraint Programs

Minh Hoàng Hà[1], Claude-Guy Quimper[2]([✉]), and Louis-Martin Rousseau[1]

[1] Department of Mathematics and Industrial Engineering and CIRRELT,
École Polytechnique de Montréal, C.P. 6079, Succursale Centre-ville,
Montreal, QC H3C 3A7, Canada
{minhhoang.ha,louis-martin.rousseau}@cirrelt.net
[2] Département d'informatique et de génie logiciel and CIRRELT,
Université Laval, Quebec, Canada
claude-guy.quimper@ift.ulaval.ca

Abstract. Integer programming (IP) is one of the most successful approaches for combinatorial optimization problems. Many IP solvers make use of the linear relaxation, which removes the integrality requirement on the variables. The relaxed problem can then be solved using linear programming (LP), a very efficient optimization paradigm. Constraint programming (CP) can solve a much wider variety of problems, since it does not require the problem to be expressed in terms of linear equations. The cost of this versatility is that in CP there is no easy way to automatically derive a good bound on the objective. This paper presents an approach based on ideas from Lagrangian decomposition (LD) that establishes a general bounding scheme for any CP. We provide an implementation for optimization problems that can be formulated with knapsack and regular constraints, and we give comparisons with pure CP approaches. Our results clearly demonstrate the benefits of our approach on these problems.

Keywords: Constraint programming · Automatic bounding · Lagrangian decomposition · Knapsack constraint · Regular constraint

1 Introduction

Constraint Programming (CP) is an efficient tool for complex decision problems arising in industry, such as vehicle routing, scheduling, and resource allocation. CP solvers essentially solve satisfiability problems, that is, they determine whether or not there exists a solution to a given model. When applied to optimization problems, most CP solvers solve a sequence of satisfiability problems, requiring at each step that the solution found improves on the solution found at the previous step. The search stops when no feasible solution can be found, proving that the previous solution was indeed optimal. The ability to compute bounds on the objective function (the optimization criterion) is crucial. It allows faster termination of the final subproblem (by proving infeasibility more quickly), and it speeds up

© Springer International Publishing Switzerland 2015
G. Pesant (Ed.): CP 2015, LNCS 9255, pp. 158–172, 2015.
DOI: 10.1007/978-3-319-23219-5_12

the solution of all the subproblems (because filtering and pruning become possible when a good bound on the objective is known). In all cases where CP has successfully solved optimization problems, the modeler had to implement good bounding mechanisms. The lack of a good general-purpose bounding scheme is one of the main drawbacks of CP in comparison with integer programming (IP).

In CP, the constraints are independent, and information is communicated solely through the *domain store*, the set of possible remaining values for each variable. During the search, each constraint looks at the domain store, determines whether a solution still exists, and filters out pairs of variables and values that no longer lead to feasible solutions. This decoupling of the problem in terms of the independent constraints allows us to use a simple and efficient combinatorial algorithm for each constraint. A similar concept is used in linear programming (LP): Lagrangian decomposition (LD) relaxes the link between difficult sets of constraints and introduces in the objective a penalty vector that acts as the glue linking the remaining relatively simple subproblems. In CP, we observe that there is no such glue for the constraints (or subproblems), making any default bound computation very weak. If we look at CP through LD glasses, we see that the relaxed constraints are actually a set of implicit constraints stating that any given variable should have the same value in all the different constraints in which it appears. Therefore, to apply LD techniques to a CP model, we must penalize that different constraints assume the same variable can take different values during search. Our goal is to develop a general bounding mechanism for CP that is compatible with every CP model and transparent to the user.

This paper is organized as follows. The next section reviews related work on decomposition techniques and CP, and Section 3 presents our general bounding approach. Section 4 evaluates the method on two optimization problems: a) the knapsack problem and b) the simple shift scheduling problem that can be modeled as a combination of many regular constraints. Finally, Section 5 summarizes our conclusions.

2   Related Work

We present here some background on LD and the global CP constraints investigated in this paper.

2.1   Lagrangian Decomposition

Decomposition techniques are an efficient approach for large-scale optimization problems. One of the most popular techniques is Lagrangian relaxation (LR), a natural approach that exploits the problem structure. It relaxes the "complicated" constraints and penalizes the constraint violation in the objective function. This simplifies the solution procedure since the resulting problem is easier to solve.

Several authors have used LR coupled with CP to solve optimization problems. Sellmann and Fahle [20] propose an algorithm for the automatic recording problem. They introduce two linear substructures, and they dualize the first and propagate the second. At convergence, they use the optimal dual information to propagate the first substructure. Ouaja and Richards [15] embed LR and CP into a branch-and-bound algorithm for the traffic placement problem. At each node of the search tree, CP is used to find a solution if one exists and to prove infeasibility otherwise, and LD indicates how close the solution is to optimality. Benoist et al. [2] propose hybrid algorithms combining LR and CP for the traveling tournament problem, which includes round-robin assignment and travel optimization. They show that LR provides not only good bounds to limit the search but also information to guide new solutions or efficient branching. Sellmann [19] investigates the theoretical foundations of CP-based LR and proves that suboptimal Lagrangian multipliers can have stronger filtering abilities than optimal ones.

A drawback of LR is that the problem loses its original structure since the complicated constraints are removed from the constraint set and penalized in the objective function. Moreover, the penalized constraints must generally be linear. This seriously restricts the application of LR in the context of CP where most global constraints have a nonlinear structure. For example, it is not easy to dualize some popular global constraints such as the all-different, global cardinality, and regular constraints. To overcome this restriction, Fontaine et al. [8] generalize LR to arbitrary high-level models using the notion of satisfiability (violation) degree. In this method, a satisfiability degree (or violation degree) is defined for each hard constraint and penalized in the objective function instead of in the constraint itself. The results show that LR coupled with CP can efficiently solve some classes of graph coloring problems.

Another way to avoid the penalization of nonlinear hard constraints is to use LD. This was first introduced by Guignard and Kim [10]; it is also called variable splitting. It creates "copies" of a set of variables responsible for connecting important substructures in the model, using one copy per substructure and dualizing the equality of a variable and its copy. Since the substructures of the original problem are retained, LD provides bounds that are always at least as tight as those from LR. The method is particularly useful if there are no apparent complicating constraints or the complicating constraints have nonlinear forms.

The research on integrating LD into CP is limited. To the best of our knowledge, only Cronholm and Ajili [5] have studied this. They propose a hybrid algorithm for the multicast network design problem. Their approach enables the propagation of all the substructures at every dual iteration and also provides strong cost-based filtering. We continue their theme with our attempt to use LD to construct an automatic bounding mechanism, thus making CP tools more efficient for optimization problems. To do this, we consider each global constraint in the CP model as a substructure, and we connect them through LD to provide valid bounds. These are used to prune the nodes in the search tree. We investigate two global constraints: knapsack and regular constraints.

2.2 Knapsack and Regular Constraints

In this subsection, we discuss knapsack and regular constraints and formulate the related optimization problems.

Knapsack Constraint. Knapsack constraints are probably the most commonly used constraints for problems in mixed integer programming and CP. These linear constraints have the form $wx \leq W$ where W is scalar, $x = [x_1, x_2, ..., x_n]$ is a vector of n binary variables, and $w = [w_1, w_2, ..., w_n]$ is a vector of n coefficients.

Most CP solvers filter knapsack constraints in a straightforward manner: the domain reduction is based on interval arithmetic and simple bounding arguments. This filtering is fast and effective but requires an important branching in practice. Another knapsack filtering has been proposed by Trick [22], who uses a dynamic programming structure to represent the constraint and achieves hyper-arc consistency, thus determining infeasibility before all the variables are set. The filtering is based on a layered directed graph where the rows correspond to W values (0 through $|W|$), the columns represent variable indexes (0 through n), and the arcs represent variable-value pairs. This graph has the property that any path from (0, 0) to the nodes of the last layer corresponds to a feasible solution to the knapsack constraint.

The filtering can reduce the branching when dealing with the satisfiability version of the knapsack problem. This version, the market split problem, has no objective function (see Trick [22] for more information). An effective implementation is essential to reduce the computational time. In the next sections, we show that integrating the filtering proposed by Trick [22] with LD reduces both the branching and the computational time for optimization problems.

Fahle and Sellmann [7] introduce an optimization version of the knapsack constraint. Given a lower bound $B \in \mathbb{N}$ and an vector of n coefficients $p = [p_1, p_2, ..., p_n]$, the constraint enforces not only $wx \leq W$ but also $px \geq B$. Since achieving generalized arc consistency for this constraint is NP-hard, Fahle and Sellmann [7] introduce the notion of *relaxed consistency* for optimization constraints, and they use bounds based on LP relaxations for polynomial-time domain filtering. Using bounds with guaranteed accuracy, Sellmann [18] exploits an existing approximation algorithm to provide an *approximate consistency*. Katriel et al. [11] develop an efficient incremental version of the previous algorithm. Malitsky et al. [13] generalize the method of [11] to provide a filtering algorithm for a more general constraint, the bounded knapsack constraint, in which the variables are integer.

The knapsack constraint is the main component of the class of knapsack problems, which has many variants (see Kellerer et al. [12] for more details). The simplest is the 0/1 knapsack problem. Given a set of n items, where each item i has a weight w_i and a value p_i, we must determine which items to include in the knapsack so that the total weight is less than or equal to a given limit W and the total value is maximized. In this paper, we focus on the multidimensional knapsack problem (MKP), which is as follows:

$$\text{Maximize} \quad \sum_{i=1}^{n} p_i x_i \tag{1}$$

$$\text{subject to} \quad \sum_{i=1}^{n} w_{ij} x_i \le C_j \quad j = 1, \dots, m \tag{2}$$

$$x_i \in \{0, 1\} \tag{3}$$

Regular Constraint. The regular global constraint is generally specified by using a (nondeterministic) finite automaton π, which is defined as a 5-tuple $\pi = (Q, A, \tau, q_0, F)$ where:

- Q is the finite set of states;
- A is the alphabet;
- $\tau : Q \times A \times Q$ is the transition table;
- $q_0 \in Q$ is the initial state;
- $F \subseteq Q$ is a set of final states.

The constraint REGULAR($[X_1, \dots, X_n], \pi$) holds if the sequence of values of the variables $X_1, \dots, X_n$ is a member of the regular language recognized by a deterministic finite automaton π. The recognition is confirmed if there exists a sequence of states $q_{i_0}, \dots, q_{i_n}$ such that $q_{i_0} = q_0$, $(q_{i_k}, X_i, q_{i_{k+1}})$ is a transition in τ, and $q_{i_n} \in F$ is a final state.

Pesant [16] introduced a filtering algorithm for the regular constraint. It constructs a layered graph similar to that proposed for the knapsack constraint above, except that its rows correspond to states of the automaton. This approach was later extended to the optimization constraints SOFT-REGULAR [23], COST-REGULAR [6], and MULTICOST-REGULAR [14] to enforce bounds on the global cost of the assignment. The underlying solution methods compute the shortest and longest paths in an acyclic graph.

The regular global constraint is useful in modeling complex work regulations in the shift scheduling problem (SSP), since the deterministic finite automaton can specify the rules that regulate transitions in the sequence of activities [3,4,6,17]. The problem consists in scheduling a sequence of activities (work, break, lunch, and rest activities) for a set of employees. We will investigate a version of the SSP in which only one employee is considered and the work regulations are so complex that modeling them requires many regular constraints. Let W be a set of work activities, T a set of periods, Π a set of deterministic finite automata expressing work regulations, and p_{ij} the profit for the assignment of activity $i \in W$ at period $j \in T$. We must assign one and only one activity for each period such that the transition of the activities over the periods satisfies the work regulations defined by the automata in Π and the total profit is maximized. To state the model for the SSP in CP, we use the element constraint ELEMENT(INDEX, TABLE, VALUE), which holds if VALUE is equal to the INDEX of the TABLE, i.e., VALUE = TABLE[INDEX], to represent the assignment. Let X_j be the variable representing the activity assigned to period j. The problem is then:

$$\text{Maximize} \quad \sum_{j \in T} C_j \tag{4}$$

$$\text{subject to} \quad \text{REGULAR}(X, \pi_i) \quad \forall \pi_i \in \Pi \tag{5}$$

$$\text{ELEMENT}(X_j, p_{ij}, C_j) \quad \forall i \in W, j \in T \tag{6}$$

$$X_j \in W \quad \forall j \in T \tag{7}$$

$$C_j \in \mathbb{R}, \min_{i \in W} p_{ij} \leq C_j \leq \max_{i \in W} p_{ij} \quad \forall j \in T \tag{8}$$

where constraints (6) ensure that C_j is equal to p_{ij} if and only if $X_j = i$. This specifies the profit of each assignment.

3 Proposed Approach

3.1 Lagrangian Decomposition

We first recall the LD approach. Consider the problem of computing $\max\{z = c^\top x | C_1(x) \wedge C_2(x)\}$, where x is a set of variables, c is a set of coefficients, and $C_1(x)$ and $C_2(x)$ are two arbitrary constraints. One can obtain a relaxation, i.e., an upper bound on the solution, as follows:

$$\max\left\{z = c^\top x \,\Big|\, C_1(x) \wedge C_2(x)\right\} = \max\left\{c^\top x \,\Big|\, C_1(x) \wedge C_2(y) \wedge x = y\right\}$$

$$= \max\left\{c^\top x + u^\top(x - y) \,\Big|\, C_1(x) \wedge C_2(y) \wedge x = y\right\}$$

$$\leq \max\left\{c^\top x + u^\top(x - y) \,\Big|\, C_1(x) \wedge C_2(y)\right\}$$

$$= \max\left\{(c^\top + u^\top)x \,\Big|\, C_1(x)\right\} + \max\left\{-u^\top y \,\Big|\, C_2(y)\right\}$$

In LD, a set of variables x is duplicated by introducing an identical set of variables y, and the difference $(x - y)$ is added to the objective function with a violation penalty cost $u \geq 0$, where u is the vector of Lagrangian multipliers. The original problem is then relaxed to two subproblems: one with C_1 and the other with C_2. The two subproblems can be solved separately. Solving the resulting programs with a given parameter u provides a valid upper bound on the original objective function. The non-negative multipliers u that give the best bound can be found by solving the Lagrangian dual. The decomposition can easily be generalized to m constraints:

$$\max\left\{z = c^\top x \,\Big|\, \bigwedge_{i=1}^{m} C_i(x)\right\} = \max\left\{c^\top x^1 \,\Big|\, \bigwedge_{i=1}^{m} C_i(x^i) \bigwedge_{i=2}^{m} x^i = x^1\right\}$$

$$= \max\left\{c^\top x^1 + \sum_{i=2}^{m} u(i)^\top(x^1 - x^i) \,\Big|\, \bigwedge_{i=1}^{m} C_i(x^i) \bigwedge_{i=2}^{m} x^i = x^1\right\}$$

$$\leq \max\left\{c^\top x^1 + \sum_{i=2}^{m} u(i)^\top(x^1 - x^i) \,\Big|\, \bigwedge_{i=1}^{m} C_i(x^i)\right\}$$

$$= \max\left\{\left(c^\top + \sum_{i=2}^{m} u(i)^\top\right)x^1 \,\Big|\, C_1(x^1)\right\} + \sum_{i=2}^{m} \max\left\{-u(i)^\top x^i \,\Big|\, C_i(x^i)\right\}$$

This decomposition works well for numeric variables, i.e., variables whose domains contain scalars. In CP, we often encounter domains that contain non-numeric values. For instance, the regular constraint described in Section 2.2 applies to variables whose domains are sets of characters. In most applications, it makes no sense to multiply these characters with a Lagrangian multiplier. In this situation, we do not apply the LD method to the original variables X_i but instead to a set of binary variables $x_{i,v} \in \{0,1\}$ where $x_{i,v} = 1 \iff X_i = v$. Rather than having a Lagrangian multiplier for each variable X_i, we instead have a multiplier for each binary variable $x_{i,v}$.

Finding the optimal multipliers is the main challenge in optimizing LD since it is nondifferentiable. There are several approaches; the most common is the subgradient method (see Shor et al. [21]). Starting from an arbitrary value of the multipliers u_0, it solves subproblems iteratively for different values of u. These are updated as follows:

$$u^{k+1} = u^k + t_k(y^k - x^k) \qquad (9)$$

where the index k corresponds to the iteration number, y is a copy of variable x, and t_k is the step size. The step size is computed using the distance between the objective value of the preceding iteration, Z^{k-1}, and the estimated optimum Z^*:

$$t_k = \frac{\lambda(Z^{k-1} - Z^*)}{\|y^{k-1} - x^{k-1}\|^2} \qquad 0 \leq \lambda \leq 2. \qquad (10)$$

Here, λ is a scaling factor used to control the convergence; it is normally between 0 and 2.

Our approach integrates an automatic bounding mechanism into the branch-and-bound algorithm of CP. The idea is that, at each node of the search tree, we use LD to yield valid bounds. The approach divides the problem into many subproblems; each subproblem has one global constraint and an objective function involving Lagrangian multipliers. For each subproblem, the solver must find the assignment that satisfies the global constraint while optimizing a linear objective function. It is generally possible to adapt the filtering algorithm of the global constraints to obtain an optimal support. If a constraint takes as parameter a cost for each pair for variable/value and it constrains the cumulative cost of the assignments to be bounded, then this constraint is compatible with the bounding mechanism we propose. Soft constraints such as Cost-GCC and Cost-Regular [23] can therefore be used.

3.2   The Knapsack Constraint

For the MKP, the subproblems are 0/1 knapsack problems. They can be solved via dynamic programming, which runs in pseudo-polynomial time $\mathcal{O}(nW)$, where n is the number of items and W is the size of the knapsack. We use the algorithm by Trick [22] to filter the constraint, and we adapt it to compute the bound. Trick constructs an acyclic graph with $n + 1$ layers where L_j contains the nodes of layer j for $0 \leq j \leq n$. Each node is labeled with a pair of integers: the layer and the accumulated weight. At layer 0, we have a single node $L_0 = \{(0,0)\}$. At layer j, we have $L_j = \{(j,k) \mid 1 \in \text{dom}(X_j) \wedge (j-1, k-w_j) \in L_{j-1} \vee 0 \in \text{dom}(X_j) \wedge (j-1,k) \in L_{j-1}\}$. There is an edge labeled with value 1 between the nodes $(j-1, k-w_j)$ and (j,k) whenever these nodes exist and an edge labeled with value 0 between the nodes $(j-1,k)$ and (j,k) whenever these nodes exist. Trick's algorithm filters the graph by removing all the nodes and edges that do not lie on a path connecting the node $(0,0)$ to a node (n,k) for $0 \leq k \leq C$. After the graph has been filtered, if no edges labeled with the value 1 remain between layer $j-1$ and layer j, this value is removed from the domain of x_j. Similarly, if no edges labeled with the value 0 remain between layer $j-1$ and layer j, this value is removed from the domain of x_j.

We augment this algorithm by computing costs. We assign to each edge connecting node $(j-1, k-w_j)$ to node (j,k) a cost of $c_j + \sum_{l=2}^{m} u(l,j)$ for the graph representing the first constraint ($i = 1$), where c_j is the value of item j and m is the number of constraints; and a cost of $-u(i,j)$ where $i \geq 2$ is the index of the constraint for the graph representing the remaining constraints. We use dynamic programming to compute for each node (j,k) the cost of the longest path connecting the source node $(0,0)$ to the node (j,k):

$$M[j,k] = \begin{cases} 0 & \text{if } k = j = 0 \\ -\infty & \text{if } k > j = 0 \\ \max(M[j-1, k], M[j-1, k-w_j] + addedCost) & \text{otherwise} \end{cases}$$

where $addedCost$ is equal to $c_j + \sum_{l=2}^{m} u(l,j)$ if $i = 1$ and $-u(i,j)$ if $i > 1$.

The optimal bound for this constraint is given by $\max_{0 \leq j \leq W} M[n,j]$. Indeed, any path in the filtered graph connecting the source node $(0,0)$ to a node (n,k) for $0 \leq k \leq C$ corresponds to a valid assignment. The expression $\max_{0 \leq j \leq W} M[n,j]$ returns the largest cost of such a path and therefore the maximum cost associated with a solution of the knapsack constraint.

3.3 The Regular Constraint

We proceed similarly with the regular constraint. Pesant [16] presents an algorithm also based on a layered graph. The set L_j contains the nodes at layer j for $0 \leq j \leq n$. Each node is labeled with an integer representing the layer and a state of the automaton. The first layer contains a single node, and we have $L_0 = \{(0, q_0)\}$. The other layers, for $1 \leq j \leq n$, contain the nodes $L_j = \{(j, q_2) \mid (j-1, q_1) \in L_{j-1} \wedge a \in \text{dom}(X_j) \wedge (q_1, a, q_2) \in \tau\}$. An edge is a triplet (n_1, n_2, a) where n_1 and n_2 are two nodes and $a \in A$ is a label. Two nodes can be connected to each other with multiple edges having distinct labels. The set of edges is denoted E. There is an edge connecting node $(j-1, q_1)$ to node (j, q_2) with label a whenever these nodes exist and $(q_1, a, q_2) \in \tau \wedge a \in \text{dom}(X_j)$ holds. As with Trick's algorithm, Pesant filters the graph by removing all nodes and edges that do not lie on a path connecting the source node $(0, q_0)$ to a node (n, q) where $q \in F$ is a final state. If no edges with label a connecting a node in L_{j-1} to a node in L_j remain, the value a is pruned from the domain $\text{dom}(X_j)$.

This filtering algorithm can be augmented by associating a cost with each edge. We assume that there is a Lagrangian multiplier $u(i, j, a)$ for each binary variable x_{ja} representing the assignment of variable X_j to character $a \in A$. Here, $i \geq 1$ represents the index of the constraint. An edge (q_1, a, q_2) has a cost of $p_{aj} + \sum_{l=2}^{m} u(l, j, a)$ in the graph induced by the first constraint $(i = 1)$, and a cost of $-u(i, j, a)$ in the graph induced by the remaining constraints $(i > 1)$. Using dynamic programming, we can compute the cost of the longest path connecting node $(0, q_0)$ to a node (j, q):

$$R[j, q] = \begin{cases} 0 & \text{if } j = 0 \\ \max_{((j-1, q_1), (j, q), a) \in E}(R[j - 1, q_1] + addedCost) & \text{otherwise} \end{cases}$$

where $addedCost$ is equal to $p_{aj} + \sum_{l=2}^{m} u(l, j, a)$ if $i = 1$ and $-u(i, j, a)$ if $i > 1$.

Since every path from layer 0 to layer n in the graph corresponds to a valid assignment for the regular constraint, the optimal bound for this constraint is given by $\max_{q \in F} R[n, q]$, i.e., the greatest cost for a valid assignment.

3.4   The Subgradient Procedure

We use the subgradient procedure to solve the Lagrangian dual. The estimated optimum Z^* is set to the value of the best incumbent solution, and the scaling factor λ is set to 2. The updating strategy halves λ when the objective function does not improve in five iterations. At each node of the search tree, the number of iterations for the subgradient procedure, which gives the best trade-off between the running time and the number of nodes required in the search tree, is fixed to 60 for the MKP in which the subproblem is solved by dynamic programming and to 10 for all other cases. The subgradient procedure is terminated as soon as the resulting Lagrangian bound is inferior to the value of the best incumbent solution.

The initial multipliers u_0 at the root node are fixed to 1. At the other nodes, we use an inheritance mechanism, i.e., the value of u_0 at a node is taken from the multipliers of the parent node. This mechanism is better than always fixing the initial multipliers to a given value. We have tested two strategies: we fix u_0 either to the multipliers that give the best bound or to the last multipliers of the parent node. The results show that the latter strategy generally performs better. This is because the limited number of subgradient iterations at each node is not sufficient to provide tight bounds for the overall problem. Therefore, the more iterations performed at early nodes, the better the multipliers expected at later nodes.

4   Computational Results

This section reports some experimental results. The goal is not to present state-of-the-art results for specific problems but to show that LD could make CP tools more efficient for optimization problems. The algorithms are built around CP Optimizer 12.6 with depth-first search. All the other parameters of the solver are set to their default values. The criteria used to evaluate the solutions are the number of nodes in the search tree, the computational time, and the number of instances successfully solved to optimality.

We investigate the behavior of the algorithms in two contexts: with and without initial lower bounds. In case A, we provide the optimal solution to the model by adding a constraint on the objective function, setting it to be at least the value of the optimal solution; the goal is then to prove the optimality of this value. In case B, we start the algorithms without an initial lower bound. These two tests provide different insights, since the presence of different bounding mechanisms will impact the search and branching decisions of the solver, with unpredictable outcomes. However, once the solver has found the optimal solution, it must always perform a last search to demonstrate optimality. A better bounding mechanism will always contribute to this phase of the overall search.

4.1   Results for MKP

The experiments in this subsection analyze the performance of LD with two MKP solution methods: i) direct CP using only sum constraints, and ii) the approach of Trick [22]. For i) the Lagrangian subproblems are solved by dynamic programming, and for ii) they are solved using the filtering graph of Trick [22].

We selected two groups of small and medium MKP instances from the OR-Library [1] for the test: Weing (8 problems, each with 2 constraints) and Weish (30 problems, each with 5 constraints). We limited the computational time to one hour.

Table 1 reports the results for case B and Table 2 reports the results for case A. The instances not included in Table 1 could not be solved by any of the algorithms; a dash indicates that the algorithm could not solve the instance within the time limit.

As can be seen in Table 2, the LD approach processes much fewer nodes per seconds, but the extra effort invested in bounding significantly decreases the size of the search tree and the time to find the optimal solution, especially for the larger instances. The most efficient method is CP + LD: it can solve 30 instances. However, it generally requires more time to solve the small instances. Moreover, it is quite time-consuming compared with the original method on a few instances, e.g., Weish6, Weish7, Weish8, and Weing15. The Trick + LD approach improves on Trick for all the criteria, even for the small instances. This is because solving the Lagrangian subproblems based on the information available from the filtering algorithm is computationally less costly. Moreover, the propagation performed at each node of the search tree is limited for the direct CP formulation, and the computation of the Lagrangian bound significantly increases this. On the other hand, the Trick method relies on a filtering algorithm for which the complexity is linear in the size of the underlying graph. In this case, the additional computation for the LD bounds has less impact.

Table 1. Results for MKP without initial lower bound

Instance	# Vars	CP			CP + LD			Trick			Trick + LD		
		Nodes	Time	Nodes/Time	Nodes	Time	Nodes/Time	Nodes	Time	Nodes/Time	Nodes	Time	Nodes/Time
Weing1	28	4424	1.19	3718	860	3.95	218	4424	6.19	715	1100	6.00	183
Weing2	28	5572	1.29	4319	744	3.07	242	5572	11.94	467	744	4.30	173
Weing3	28	8650	0.55	16k	270	0.97	278	8650	15.34	564	280	1.53	183
Weing4	28	4106	1.08	3802	538	2.62	205	4106	19.72	208	538	4.14	130
Weing5	28	13k	0.58	22k	262	1.00	262	12k	21.13	615	262	1.53	171
Weing6	28	9150	1.14	8026	876	3.59	244	9150	20.50	446	1012	4.83	210
Weing7	105	-	-	-	32k	3410.04	9	-	-	-	-	-	-
Weing8	105	-	-	-	19k	147.98	128	-	-	-	19k	655.46	29
Weish1	30	35k	1.78	20k	1320	37.31	35	35k	910.82	38	1286	59.90	21
Weish2	30	40k	1.64	24k	1280	27.14	47	40k	2481.72	16	1384	57.47	24
Weish3	30	11k	0.83	11k	674	8.47	80	11k	669.01	16	760	24.35	31
Weish4	30	2342	0.80	2927	856	11.91	72	2342	186.47	13	826	29.68	28
Weish5	30	1614	0.90	1793	728	9.19	79	1614	149.78	11	644	23.74	27
Weish6	40	1.2M	12.56	96k	4286	369.13	12	-	-	-	3368	320.92	10
Weish7	40	901k	15.12	60k	3888	290.79	13	-	-	-	3482	352.67	10
Weish8	40	1.1M	19.74	56k	4004	256.00	16	-	-	-	3464	392.57	9
Weish9	40	144k	3.12	46k	2426	46.78	52	-	-	-	2212	191.83	12
Weish10	50	28M	589.20	48k	4486	264.40	17	-	-	-	3969	734.65	5
Weish11	50	6M	95.84	63k	3764	159.05	24	-	-	-	3764	595.38	6
Weish12	50	21M	355.78	59k	4738	307.37	15	-	-	-	3728	651.90	6
Weish13	50	20M	338.07	59k	4208	250.62	17	-	-	-	4374	908.53	5
Weish14	60	-	-	-	8424	645.04	13	-	-	-	8856	2472.92	4
Weish15	60	35M	720.98	49k	13k	1363.17	10	-	-	-	12k	3482.11	3
Weish16	60	-	-	-	14k	1216.45	12	-	-	-	13k	3324.81	4
Weish17	60	-	-	-	14k	1600.78	9	-	-	-	-	-	-
Weish18	70	-	-	-	-	-	-	-	-	-	-	-	-
Weish19	70	-	-	-	7750	667.02	12	-	-	-	8907	3174.07	3
Weish20	70	-	-	-	18k	2610.74	7	-	-	-	-	-	-
Weish21	70	-	-	-	15k	1642.16	9	-	-	-	-	-	-
Weish22	80	-	-	-	15k	2905.41	5	-	-	-	-	-	-
Weish23	80	-	-	-	12k	2002.74	6	-	-	-	-	-	-

Table 2 presents the results for case A. Here, the Lagrangian approaches are even more efficient. The algorithms with LD can successfully solve all 38 instances, and the search tree is relatively small. The computational time of CP + LD is still worse than that of CP on a few small instances, but the difference has decreased significantly. The main reason for this is that the optimal solutions used to compute the step size make the subgradient procedure more stable and cause it to converge more quickly. This observation suggests that adding good lower bounds as soon as possible in the solution process will improve the method.

We computed at the root of the search tree the bound that the LD provides as well as the bound returned by CPLEX using a linear relaxation (LP) and the bound that CPLEX computes when the integer variables are not relaxed and all cuts are added (ILP). Due to lack of space, we do not report the bounds for each instance. In average, the bounds for LD, LP, and ILP are 1.03%, 0,79%, and 0.22% greater than the optimal value. Even though LD does not provide the best bound, it represents a significative improvement to CP whose bound is very weak.

4.2   Results for SSP

To generate the regular constraints, we use the procedure proposed by Pesant [16] to obtain random automata. The proportion of undefined transitions in τ is set to 30% and the proportion of final states to 50%. To control the problem size, we create instances with only 2 regular constraints and 50 periods. The assignment profits are selected at random from integer values between 1 and 100. Each instance is then generated based on two parameters: the number of activities (nva) and the number of states (nbs) in the automata. For each pair (nva,nbs), we randomly generate three instances. The instances are labeled nbv-nbs-i, where

Table 2. Results for MKP: proving optimality

Instance	# Vars	CP			CP + LD			Trick			Trick + LD		
		Nodes	Time	Nodes/Time	Nodes	Time	Nodes/Time	Nodes	Time	Nodes/Time	Nodes	Time	Nodes/Time
Weing1	28	3408	0.21	16k	24	0.62	39	3408	5.25	649	24	0.78	31
Weing2	28	5070	0.21	24k	30	0.48	63	5070	16.36	310	30	0.74	41
Weing3	28	7004	0.55	13k	32	0.32	100	7004	11.78	595	34	0.42	81
Weing4	28	2344	0.14	17k	16	0.37	43	2344	10.45	224	16	0.81	20
Weing5	28	18k	0.33	55k	26	0.32	81	18k	35.30	510	26	0.56	46
Weing6	28	8038	0.22	37k	30	0.44	68	8038	27.62	291	30	0.80	38
Weing7	105	-	-	-	134	4		-	-	-	134	174.03	1
Weing8	105	-	-	-	168	3.54	47	-	-	-	520	40.95	13
Weish1	30	11k	0.27	41k	54	2.85	19	11k	447.18	25	70	5.72	12
Weish2	30	23k	0.43	53k	66	1.98	33	23k	1679.54	14	62	9.15	7
Weish3	30	8394	0.23	36k	38	2.80	14	8394	476.57	18	46	5.59	8
Weish4	30	632	0.14	4514	34	1.26	27	632	57.19	11	38	4.64	8
Weish5	30	556	0.12	4633	34	0.87	39	556	55.93	10	36	5.67	6
Weish6	40	861k	12.92	67k	56	5.54	10	-	-	-	72	16.58	4
Weish7	40	456k	7.07	64k	60	8.40	7	-	-	-	72	17.40	4
Weish8	40	707k	10.38	68k	60	9.05	7	-	-	-	54	19.52	3
Weish9	40	74k	1.26	59k	48	2.28	21	-	-	-	74	18.18	4
Weish10	50	8.6M	132.92	65k	134	29.53	5	-	-	-	192	51.11	4
Weish11	50	1.4M	22.75	62k	86	10.50	8	-	-	-	82	28.44	3
Weish12	50	9M	135.70	66k	114	24.92	5	-	-	-	122	38.23	3
Weish13	50	4.1M	60.90	67k	120	17.84	7	-	-	-	112	35.23	3
Weish14	60	-	-	-	104	27.97	4	-	-	-	330	123.89	3
Weish15	60	9.7M	173.65	56k	92	28.20	3	-	-	-	146	68.70	2
Weish16	60	-	-	-	128	37.95	3	-	-	-	450	192.18	2
Weish17	60	156M	3537.25	44k	90	34.36	3	-	-	-	176	115.41	1.5
Weish18	70	-	-	-	200	87.97	2	-	-	-	142	116.60	1.2
Weish19	70	-	-	-	200	86.98	2	-	-	-	320	228.97	1.4
Weish20	70	-	-	-	174	65.84	3	-	-	-	596	340.87	1.7
Weish21	70	-	-	-	134	48.17	3	-	-	-	354	225.53	1.6
Weish22	80	-	-	-	150	87.39	2	-	-	-	1068	1083.83	1.0
Weish23	80	-	-	-	158	83.39	1.9	-	-	-	177	350.55	0.5
Weish24	80	-	-	-	146	84.90	1.7	-	-	-	154	194.55	0.8
Weish25	80	-	-	-	238	117.47	2	-	-	-	586	539.24	1.1
Weish26	90	-	-	-	178	135.75	1.3	-	-	-	438	567.02	0.8
Weish27	90	-	-	-	152	94.48	1.6	-	-	-	258	356.33	0.7
Weish28	90	-	-	-	152	96.71	1.6	-	-	-	266	344.16	0.8
Weish29	90	-	-	-	152	104.87	1.4	-	-	-	416	601.38	0.7
Weish30	90	-	-	-	152	126.96	1.2	-	-	-	138	372.65	0.4

Table 3. Results for SSP without initial lower bound

Instance	Pesant			Pesant + LD		
	Nodes	Time	Nodes/Time	Nodes	Time	Nodes/Time
10-20-01	2.2M	1232.72	1785	156k	346.92	450
10-20-02	-	-	-	298k	741.35	402
10-20-03	1.6M	783.29	2043	158k	332.15	476
10-80-01	256k	1325.80	193	84k	1020.53	82
10-80-02	788k	3307.39	238	238k	2834.86	84
10-80-03	847k	2344.55	361	246k	2176.73	113
20-20-01	-	-	-	828k	1856.10	446
20-20-02	-	-	-	1.3M	3404.56	382
20-20-03	2.1M	2427.72	865	164k	439.97	373
20-80-01	-	-	-	373k	3944.18	95
20-80-02	-	-	-	436k	5206.81	84
20-80-03	-	-	-	228k	2561.64	89

Table 4. Results for SSP: proving optimality

Instance	Pesant			Pesant + LD		
	Nodes	Time	Nodes/Time	Nodes	Time	Nodes/Time
10-20-01	87k	42.20	2062	2564	5.94	432
10-20-02	407k	342.43	1189	44k	108.30	406
10-20-03	71k	28.60	2483	4118	7.48	551
10-80-01	20k	118.97	168	4546	71.16	64
10-80-02	25k	81.39	307	5466	63.19	87
10-80-03	26k	85.67	303	4762	58.34	82
20-20-01	343k	176.38	1945	2651	13.41	198
20-20-02	372k	297.97	1248	10k	51.63	194
20-20-03	53k	44.28	1197	1353	7.43	182
20-80-01	105k	486.89	216	10k	147.63	68
20-80-02	216k	1648.28	131	13k	211.23	62
20-80-03	16k	128.74	124	5128	72.33	71

i is the instance number. Because SSP instances are more difficult than MKP instances, the computational time is increased to 2 hours.

The experiment investigates the performance of LD when combined with the method of Pesant [16] for the SSP. As for the MKP, we test two cases: case A has initial lower bounds and case B does not. Tables 3 and 4 present the results, which demonstrate the performance of our approach. The LD method clearly improves the method of Pesant [16] for all three criteria. More precisely, for case B, it can solve 6 additional instances, reducing the size of the search tree by a factor of about 7 and the computational time by a factor of 1.5 on average. In case A, it works even better, reducing the size of the search tree by a factor of about 16 and the computational time by a factor of more than 4 on average.

This is because we can reuse the graph constructed by the filtering algorithm to solve the Lagrangian subproblems.

5   Conclusions and Future Work

We have introduced an automatic bounding mechanism based on the LD concept to improve the performance of CP for optimization problems. We tested the approach on two problems, the MKP and a synthesized version of the SSP. These rely on two well-known global constraints: knapsack and regular.

Our next step will be to apply the approach to other global constraints to solve real problems. One candidate is the global cardinality constraint, for which the subproblems can be solved in polynomial time with the aid of a graph obtained from the filtering algorithm. In addition, our approach could be improved in several ways. First, we could use the information obtained from solving the Lagrangian dual to design a strong cost-based filtering, as proposed by Cron- holm and Ajili [5]. Second, by checking the feasibility of the solutions of the Lagrangian subproblems we could find good solutions that help to limit the search and improve the convergence of the subgradient method. Moreover, we could use the solutions of the subproblems to guide the search by branching on the variable with the largest difference between itself and its copies. Finally, as reported by Guignard [9], the subgradient approach can have unpredictable convergence behavior, so a more suitable algorithm could improve the performance of our approach.

Acknowledgement. This work was financed with a Google Research Award. We would like to thank Laurent Perron for his support.

References

1. Beasley, J.E.: OR-library (2012). http://people.brunel.ac.uk/~mastjjb/jeb/info.html
2. Benoist, T., Laburthe, F., Rottembourg, B.: Lagrange relaxation and constraint programming collaborative schemes for travelling tournament problems. In: CP-AI-OR 2001, pp. 15–26. Wye College (2001)
3. Chapados, N., Joliveau, M., L'Ecuyer, P., Rousseau, L.-M.: Retail store scheduling for profit. European Journal of Operational Research **239**(3), 609–624 (2014). doi:10.1016/j.ejor.2014.05.033
4. Côté, M.-C., Gendron, B., Quimper, C.-G., Rousseau, L.-M.: Formal languages for integer programming modeling of shift scheduling problems. Constraints **16**(1), 54–76 (2011). doi:10.1007/s10601-009-9083-2
5. Cronholm, W., Ajili, F.: Strong cost-based filtering for lagrange decomposition applied to network design. In: Wallace, M. (ed.) CP 2004. LNCS, vol. 3258, pp. 726–730. Springer, Heidelberg (2004). http://www.springerlink.com/index/ur3uvyqbp0216btd.pdf
6. Demassey, S., Pesant, G., Rousseau, L.-M.: A cost-regular based hybrid column generation approach. Constraints **11**(4), 315–333 (2006). http://dblp.uni-trier.de/db/journals/constraints/constraints11.html#DemasseyPR06

7. Fahle, T., Sellmann, M.: Cost based filtering for the constrained knapsack problem. Annals of OR **115**(1–4), 73–93 (2002). doi:10.1023/A:1021193019522
8. Fontaine, D., Michel, L.D., Van Hentenryck, P.: Constraint-based lagrangian relaxation. In: O'Sullivan, B. (ed.) CP 2014. LNCS, vol. 8656, pp. 324–339. Springer, Heidelberg (2014)
9. Guignard, M.: Lagrangean relaxation. Top **11**(2), 151–200 (2003). doi:10.1007/BF02579036. ISSN: 1134–5764
10. Guignard, M., Kim, S.: Lagrangean decomposition: A model yielding stronger Lagrangean bounds. Mathematical Programming **39**(2), 215–228 (1987). doi:10.1007/BF02592954. ISSN: 00255610
11. Katriel, I., Sellmann, M., Upfal, E., Van Hentenryck, P.: Propagating knapsack constraints in sublinear time. In: Proceedings of the Twenty-Second AAAI Conference on Artificial Intelligence, Vancouver, British Columbia, Canada, July 22–26, pp. 231–236 (2007). http://www.aaai.org/Library/AAAI/2007/aaai07-035.php
12. Kellerer, H., Pferschy, U., Pisinger, D.: Knapsack Problems. Springer, Berlin (2004)
13. Malitsky, Y., Sellmann, M., Szymanek, R.: Filtering bounded knapsack constraints in expected sublinear time. In: Proceedings of the Twenty-Fourth AAAI Conference on Artificial Intelligence, AAAI 2010, Atlanta, Georgia, USA, July 11–15, 2010. http://www.aaai.org/ocs/index.php/AAAI/AAAI10/paper/view/1855
14. Menana, J., Demassey, S.: Sequencing and counting with the `multicost-regular` constraint. In: van Hoeve, W.-J., Hooker, J.N. (eds.) CPAIOR 2009. LNCS, vol. 5547, pp. 178–192. Springer, Heidelberg (2009). doi:10.1007/978-3-642-01929-6_14. ISBN: 3642019285
15. Ouaja, W., Richards, B.: A hybrid multicommodity routing algorithm for traffic engineering. Networks **43**(3), 125–140 (2004). http://dblp.uni-trier.de/db/journals/networks/networks43.html#OuajaR04
16. Pesant, G.: A regular language membership constraint for finite sequences of variables. In: Wallace, M. (ed.) CP 2004. LNCS, vol. 3258, pp. 482–495. Springer, Heidelberg (2004). http://www.springerlink.com/content/ed24kyhg561jjthj
17. Quimper, C.-G., Rousseau, L.-M.: A large neighbourhood search approach to the multi-activity shift scheduling problem. J. Heuristics **16**(3), 373–392 (2010). doi:10.1007/s10732-009-9106-6
18. Sellmann, M.: Approximated consistency for knapsack constraints. In: Rossi, F. (ed.) CP 2003. LNCS, vol. 2833, pp. 679–693. Springer, Heidelberg (2003). doi:10.1007/978-3-540-45193-8_46
19. Sellmann, M.: Theoretical foundations of CP-based lagrangian relaxation. In: Wallace, M. (ed.) CP 2004. LNCS, vol. 3258, pp. 634–647. Springer, Heidelberg (2004). http://dblp.uni-trier.de/db/conf/cp/cp/2004.html#sellmann.ISBN:3-540-23241-9
20. Sellmann, M., Fahle, T.: Constraint programming based Lagrangian relaxation for the automatic recording problem. Annals of Operations Research **118**, 17–33 (2003). doi:10.1023/A:1021845304798. ISBN: 0254-5330
21. Shor, N.Z., Kiwiel, K.C., Ruszcayński, A., Ruszcayński, A.: Minimization methods for non-differentiable functions. Springer-Verlag New York Inc., New York (1985). ISBN: 0-387-12763-1
22. Trick, M.A.: A dynamic programming approach for consistency and propagation for knapsack constraints. Annals of OR **118**(1–4), 73–84 (2003). http://dblp.uni-trier.de/db/journals/anor/anor118.html#Trick03
23. van Hoeve, W.J., Pesant, G., Rousseau, L.-M.: On global warming: Flow-based soft global constraints. J. Heuristics **12**(4–5), 347–373 (2006). doi:10.1007/s10732-006-6550-4

Smallest MUS Extraction
with Minimal Hitting Set Dualization

Alexey Ignatiev[1], Alessandro Previti[2], Mark Liffiton[3],
and Joao Marques-Silva[1,2]

[1] INESC-ID, IST, Technical University of Lisbon, Lisboa, Portugal
[2] CASL, University College Dublin, Belfield, Ireland
[3] Illinois Wesleyan University, Bloomington, IL, USA

Abstract. Minimal explanations of infeasibility are of great interest
in many domains. In propositional logic, these are referred to as Minimal Unsatisfiable Subsets (MUSes). An unsatisfiable formula can have
multiple MUSes, some of which provide more insights than others. Different criteria can be considered in order to identify a good minimal
explanation. Among these, the size of an MUS is arguably one of the
most intuitive. Moreover, computing the smallest MUS (SMUS) finds
several practical applications that include validating the quality of the
MUSes computed by MUS extractors and finding equivalent subformulae
of smallest size, among others. This paper develops a novel algorithm for
computing a smallest MUS, and we show that it outperforms all the previous alternatives pushing the state of the art in SMUS solving. Although
described in the context of propositional logic, the presented technique
can also be applied to other constraint systems.

1 Introduction

For inconsistent formulae, finding a minimal explanation of their infeasibility
is a central task in order to disclose the source of the problem. In general, an
inconsistent formula can have multiple explanations. Thus, defining a measure of
the quality of an explanation becomes necessary in order to focus on those providing more insights. For the case of a propositional formula, a natural measure
of explanation quality is the size of the computed minimal unsatisfiable subsets
(MUSes). From a query complexity perspective, extracting an MUS is in FP^{NP}.
In contrast, deciding whether there exists an MUS of size less than or equal to
k is Σ_2^{P}-complete [8,12]. As a consequence, extracting a smallest MUS (SMUS)
is in $\text{FP}^{\Sigma_2^{\text{P}}}$.

Computing an SMUS is central to a number of practical problems, including validating MUSes computed with modern MUS extractors and finding an
equivalent subformula of minimum size.

This work is partially supported by SFI PI grant BEACON (09/IN.1/ I2618),
FCT grant POLARIS (PTDC/EIA-CCO/123051/2010) and national funds through
Fundação para a Ciência e a Tecnologia (FCT) with reference UID/CEC/50021/2013.

G. Pesant (Ed.): CP 2015, LNCS 9255, pp. 173–182, 2015.
DOI: 10.1007/978-3-319-23219-5_13

This paper shows how recent work on iterative enumeration of hitting sets can be adapted to computing the smallest MUS of unsatisfiable formulae, improving upon state-of-the-art algorithms. The approach is implemented and tested on Boolean Satisfiability instances, but the technique applies equally well to computing SMUSes of any constraint system for which the satisfiability of all subsets is well-defined.

The paper is organized as follows. In Section 2 basic definitions are provided. Section 3 introduces an algorithm for the extraction of an SMUS. In Section 4 some optimizations are presented. Finally, Section 5 is dedicated to the presentation of the experimental results, and Section 6 concludes the paper.

2   Preliminaries

This section provides common definitions that will be used throughout the paper. Additional standard definitions are assumed (e.g. see [3]). In what follows, $\mathcal{F}$ denotes a propositional formula expressed in Conjunctive Normal Form (CNF). A formula in CNF is a conjunction of clauses, where each clause is a disjunction of literals. A literal is either a Boolean variable or its complement. Also, we may refer to formulae as sets of clauses and clauses as sets of literals. An assignment to variables that satisfies formula $\mathcal{F}$ is said to be a model of $\mathcal{F}$. A formula is unsatisfiable when no model exists. Unless specified explicitly, $\mathcal{F}$ is assumed to be unsatisfiable. The following definitions also apply.

Definition 1. *A subset $\mathcal{U} \subseteq \mathcal{F}$ is a minimal unsatisfiable subset (MUS) if $\mathcal{U}$ is unsatisfiable and $\forall \mathcal{U}' \subset \mathcal{U}, \mathcal{U}'$ is satisfiable. An MUS $\mathcal{U}$ of $\mathcal{F}$ of the smallest size is called a smallest MUS (SMUS).*

Definition 2. *A subset $\mathcal{C}$ of $\mathcal{F}$ is a minimal correction subset (MCS) if $\mathcal{F} \setminus \mathcal{C}$ is satisfiable and $\forall \mathcal{C}' \subseteq \mathcal{C} \wedge \mathcal{C}' \neq \emptyset, (\mathcal{F} \setminus \mathcal{C}) \cup \mathcal{C}'$ is unsatisfiable.*

Definition 3. *A satisfiable subset $\mathcal{S} \subseteq \mathcal{F}$ is a maximal satisfiable subset (MSS) if $\forall \mathcal{S}' \subseteq \mathcal{F}$ s.t. $\mathcal{S} \subset \mathcal{S}', \mathcal{S}'$ is unsatisfiable.*

An MSS can also be defined as the complement of an MCS (and vice versa). If $\mathcal{C}$ is an MCS, then $\mathcal{S} = \mathcal{F} \setminus \mathcal{C}$ represents an MSS. On the other hand, MUSes and MCSes are related by the concept of *minimal hitting set*.

Definition 4. *Given a collection Γ of sets from a universe $\mathbb{U}$, a hitting set h for Γ is a set such that $\forall S \in \Gamma, h \cap S \neq \emptyset$.*

A hitting set h is *minimal* if none of its subset is a hitting set. The minimal hitting set duality between MUSes and MCSes is well-known (e.g. see [15,17]):

Proposition 1. *Given a CNF formula $\mathcal{F}$, let MUSes($\mathcal{F}$) and MCSes($\mathcal{F}$) be the set of all MUSes and MCSes of $\mathcal{F}$, respectively. Then the following holds:*

1. *A subset $\mathcal{U}$ of $\mathcal{F}$ is an MUS iff $\mathcal{U}$ is a minimal hitting set of MCSes($\mathcal{F}$).*
2. *A subset $\mathcal{C}$ of $\mathcal{F}$ is an MCS iff $\mathcal{C}$ is a minimal hitting set of MUSes($\mathcal{F}$).*

The duality relating MUSes and MCSes is a key aspect of the algorithm presented below. In the next section, we will describe how the smallest MUS can be computed by exploiting this observation.

Algorithm 1. Basic SMUS algorithm

Input: CNF formula $\mathcal{F}$

1 **begin**
2     $\mathcal{H} \leftarrow \emptyset$
3   **while** true **do**
4         $h \leftarrow \text{MinimumHS}(\mathcal{H})$
5         $\mathcal{F}' \leftarrow \{c_i \mid e_i \in h\}$
6   **if** not $\text{SAT}(\mathcal{F}')$ **then**
7   return $\mathcal{SMUS} \leftarrow \mathcal{F}'$
8   **else**
9             $\mathcal{C} \leftarrow grow(\mathcal{F}')$
10         $\mathcal{H} \leftarrow \mathcal{H} \cup \{\mathcal{C}\}$

11 **end**

---

3   Basic Algorithm

This section describes the new SMUS algorithm. We start by providing the definition of a *minimum size hitting set*:

Definition 5. *Let Γ be a collection of sets and $MHS(\Gamma)$ the set of all minimal hitting sets on Γ. Then a hitting set $h \in MHS(\Gamma)$ is said to be a **minimum** hitting set if $\forall\, h' \in MHS(\Gamma)$ we have that $|h| \leq |h'|$.*

The algorithm is based on the following observation:

Proposition 2. *A set $\mathcal{U} \subseteq \mathcal{F}$ is an SMUS of $\mathcal{F}$ if and only if $\mathcal{U}$ is a minimum hitting set of $MCSes(\mathcal{F})$.*

By duality [17], we also have that the minimum hitting set of $MUSes(\mathcal{F})$ corresponds to a MaxSAT solution for $\mathcal{F}$. This observation has already been exploited in [6]. Algorithm 1 can be seen as the dual of the algorithm presented in [6]. $\mathcal{H}$ represents a collection of sets, where each set corresponds to an MCS on $\mathcal{F}$. Thus, elements of the sets in $\mathcal{H}$ represent clauses of $\mathcal{F}$. Let $elm(\mathcal{H})$ denote the set of elements in $\mathcal{H}$. Each element $e \in elm(\mathcal{H})$ is associated with a clause $c \in \mathcal{F}$. At the beginning of the algorithm $\mathcal{H}$ is empty (line 2). At each step, a minimum hitting set h is computed on $\mathcal{H}$ (see line 4) and the induced formula $\mathcal{F}'$ is tested for satisfiability at line 6. If $\mathcal{F}'$ is satisfiable, it is extended by the *grow* procedure into an MSS containing $\mathcal{F}'$, the complement of which is returned as the MCS $\mathcal{C}$. Then $\mathcal{C}$ is added to the collection $\mathcal{H}$. If instead $\mathcal{F}'$ is unsatisfiable then $\mathcal{F}'$ is guaranteed to be an SMUS of $\mathcal{F}$ as the following lemma states:

Lemma 1. *Let $\mathcal{K} \subseteq MCSes(\mathcal{F})$. Then a subset $\mathcal{U}$ of $\mathcal{F}$ is an SMUS if $\mathcal{U}$ is a minimum hitting set on $\mathcal{K}$ and $\mathcal{U}$ is unsatisfiable.*

Proof. Since $\mathcal{U}$ is unsatisfiable it means that it already hits every MCS in $MCSes(\mathcal{F})$ (Proposition 1). $\mathcal{U}$ is also a minimum hitting set on $MCSes(\mathcal{F})$, since

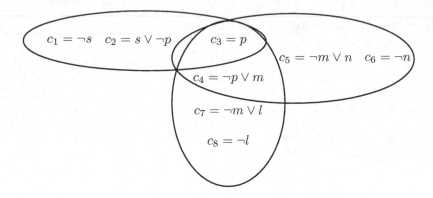

Fig. 1. Formula example

Table 1. Example SMUS computation

MinimumHS($\mathcal{H}$)	SAT($\mathcal{F}'$)	$\mathcal{F} \setminus grow(\mathcal{F}', \mathcal{F})$	$\mathcal{H} = \mathcal{H} \cup \mathcal{C}$
$\mathcal{F}' \leftarrow \{\emptyset\}$	true	$\mathcal{C} \leftarrow \{c_3\}$	$\{\{c_3\}\}$
$\mathcal{F}' \leftarrow \{c_3\}$	true	$\mathcal{C} \leftarrow \{c_2, c_4\}$	$\{\{c_3\}, \{c_2, c_4\}\}$
$\mathcal{F}' \leftarrow \{c_2, c_3\}$	true	$\mathcal{C} \leftarrow \{c_1, c_4\}$	$\{\{c_3\}, \{c_2, c_4\}, \{c_1, c_4\}\}$
$\mathcal{F}' \leftarrow \{c_3, c_4\}$	true	$\mathcal{C} \leftarrow \{c_1, c_5, c_7\}$	$\{\{c_3\}, \{c_2, c_4\}, \{c_1, c_4\}, \{c_1, c_5, c_7\}\}$
$\mathcal{F}' \leftarrow \{c_1, c_2, c_3\}$	false	$\{c_1, c_2, c_3\}$ is an SMUS of $\mathcal{F}$	

it is a minimum hitting set for $\mathcal{K} \subseteq \text{MCSes}(\mathcal{F})$ and no other added MCS can make it grow in size. Moreover, all the other hitting sets can either grow in size or remain the same. Thus, by Proposition 2 $\mathcal{U}$ must be an SMUS.   □

Lemma 1 states that it is not necessary to enumerate all MCSes of formula $\mathcal{F}$. Instead, it is enough to compute only those whose minimum hitting set is an MUS. Therefore, Algorithm 1 terminates once the first MUS is computed, which is by construction guaranteed to be of the smallest size.

It is worth noting that nothing in Algorithm 1 is specific to Boolean CNF, and in fact the algorithm can be applied to any type of constraint system for which the satisfiability of constraint subsets can be checked. The algorithm and all following additions to it are constraint agnostic.

An example of a run of Algorithm 1 is shown in Table 1 (Figure 1 illustrates the input formula). The first column contains the formula $\mathcal{F}'$ induced by the minimum hitting set on $\mathcal{H}$. Whenever $\mathcal{F}'$ is satisfiable (second column), the *grow* procedure (see line 9 of Algorithm 1) returns an MCS (third column). The last column shows the current $\mathcal{H}$, the collected set of MCSes from which a minimum hitting set will be found in the next iteration. In the last row, $\mathcal{F}' = \{c_1, c_2, c_3\}$ and since the call $\mathtt{SAT}(\mathcal{F}')$ returns false, set $\{c_1, c_2, c_3\}$ represents an SMUS. Notice that $\mathtt{MinimumHS}(\mathcal{H})$ could have returned $\{c_3, c_4, c_5\}$ instead of $\{c_1, c_2, c_3\}$. In this case, further iterations would be necessary in order to find an SMUS.

4 Additional Details

This section describes some essential details of the approach being proposed. These include computing a minimum hitting set with at most one SAT call, enumerating disjoint MCSes, and reporting an upper bound within a reasonably short time.

4.1 Reducing the Number of SAT Calls

A number of different approaches can be used for computing minimum size hitting sets [6]. This section describes a different alternative that exploits the use of SAT solvers as well as the problem structure. The proposed approach exploits incremental MaxSAT solving [5].

Let the current minimum hitting set (MHS) h have size k, and let $\mathcal{C}$ be a new set to hit. Call a SAT solver on the resulting formula, requesting an MHS of size k. If the formula is satisfiable, then we have a new MHS of size k that also hits $\mathcal{C}$. However, if the formula is unsatisfiable, this means there is no MHS of size k. Thus, an MHS of size $k + 1$ can be constructed by adding to h any element of $\mathcal{C}$. Clearly, by construction, h does not hit any element of $\mathcal{C}$. Thus, every MHS can be obtained with a *single* call to a SAT solver.

Additionally, extraction of each MCS $\mathcal{C}$ (see line 9) requires a number of SAT calls. Alternatively, one can avoid the computation of MCS $\mathcal{C}$ and use any correction subset $\mathcal{C}'$ s.t. $\mathcal{C} \subseteq \mathcal{C}'$ instead. Indeed, any MHS of the set of all correction subsets of $\mathcal{F}$ also hits every MCS of $\mathcal{F}$ [17]. Moreover, observe that a proposition similar to Lemma 1 can be proved for a partial set of correction subsets of $\mathcal{F}$. Therefore, the *grow* procedure can be skipped, and the complement to $\mathcal{F}'$ can be directly added to $\mathcal{H}$.

4.2 Disjoint MCS Enumeration

Enumeration of disjoint inconsistencies has been used in various settings in recent years. For example, disjoint unsatisfiable core enumeration in MaxSAT was proposed in [16], and nowadays it is often used in state-of-the-art MaxSAT solvers (e.g. see [7]). Enumeration of disjoint MCSes, which act as unsatisfiable cores over quantified constraints, has also proven its relevance for the SMUS problem in the context of QMaxSAT [9]. Also note that disjoint MCSes are known to have a strong impact on the performance of the branch-and-bound algorithms for SMUS by refining the lower bound on the size of the SMUS [13].

Our approach based on the minimum hitting set duality [15,17] between MCSes and MUSes of an unsatisfiable CNF formula $\mathcal{F}$ can also make use of disjoint MCSes. Indeed, since each MCS must be hit by any MUS of a CNF formula, disjoint MCSes must be hit by the smallest MUS of the formula. Therefore, given a set of disjoint MCSes $\mathcal{D}$, one can initialize set $\mathcal{H}$ with $\mathcal{D}$ instead of the empty set (see line 2 of Algorithm 1) in order to boost the performance of Algorithm 1. Note that this also simplifies the computation of the first minimum hitting set, which for a set of disjoint MCSes of size k is exactly of size k. According to the experimental results described in Section 5, this improvement has a huge impact on the overall performance of the proposed approach.

4.3   Finding Approximate Solutions

Although the approach being proposed performs better than the known alternative approaches to SMUS, this problem is computationally much harder than extracting any MUS of a CNF formula (the decision version of the SMUS problem is known to be Σ_2^P-complete, e.g. see [8,12]). One may find this complexity characterization a serious obstacle for using SMUS algorithms in practice. Indeed, a user may prefer to find an MUS close to the smallest size within a reasonable amount of time instead of waiting until the smallest MUS is found.

In order to resolve this issue and following the ideas of [10], Algorithm 1 can be extended for computing a *"good"* upper bound on the exact solution of the SMUS problem. Any MUS can be considered as an upper bound on the smallest MUS. Therefore and since extracting one MUS is a relatively simple task, enumerating MUSes within a given time limit and choosing the smallest one among them can be satisfactory for a user. MUS enumeration can be done in a way similar to the one proposed in [14]. Observe that Algorithm 1 iteratively refines a lower bound of the SMUS, and so can serve to measure the quality of approximate upper bounds.

This pragmatic policy of computing an upper bound before computing the exact solution provides a user with a temporary approximate answer within a short period of time and continues with computing the exact solution, if needed (i.e. if the user is not satisfied with the quality of the upper bound). Otherwise, the upper bound is enough and the user can stop the process.

5   Experimental Results

This section evaluates the approach to the smallest MUS problem proposed in this paper. The experiments were performed in Ubuntu Linux on an Intel Xeon E5-2630 2.60GHz processor with 64GByte of memory. The time limit was set to 800s and the memory limit to 10GByte. The approach proposed above was implemented in a prototype called FORQES (*FORmula QuintESsence extractor*). The underlying SAT solver of FORQES is Glucose 3.0[1] [1]. A weakened version

---

[1] Available from http://www.labri.fr/perso/lsimon/glucose/

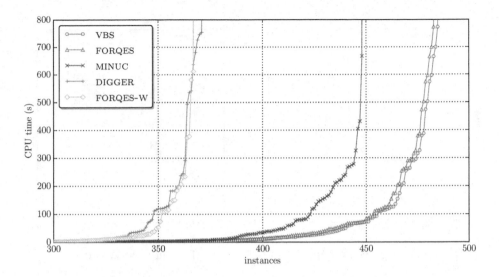

Fig. 2. Cactus plot showing the performance of FORQES, MINUC, and DIGGER

of FORQES, which does not use disjoint MCS enumeration and implements Algorithm 1, is referred to as FORQES-W.

The performance of FORQES was compared to the state-of-the-art approach to the SMUS problem that uses the quantified MaxSAT formulation of SMUS [9]. The most efficient version of the tool described in [9] performs core-guided QMaxSAT solving; in the following it is referred to as MINUC. Additionally, a well-known branch-and-bound SMUS extractor called DIGGER (see [13]) was also considered. Note that the versions of MINUC and DIGGER participating in the evaluation make use of disjoint MCS enumeration.

Several sets of benchmarks were used to assess the efficiency of the new algorithm, all used in the evaluation in [9] as well. This includes a collection of instances from automotive product configuration benchmarks [18] and two sets of circuit diagnosis instances. Additionally, we selected instances from the complete set of the MUS competitions benchmarks[2] as follows. Because extracting a smallest MUS of a CNF formula is computationally much harder than extracting any MUS of the formula, the instances that are difficult for a state-of-the-art MUS extractor were excluded. Instead, we considered only formulae solvable by the known MUS extractor MUSER-2 (e.g. see [2]) within 10 seconds. The total number of instances considered in the evaluation is 682.

Figure 2 shows a cactus plot illustrating the performance of the tested solvers on the total set of instances. FORQES exhibits the best performance, being able to solve 483 instances. MINUC comes second with 448 instances solved. Thus, FORQES solves 7.8% more instances than MINUC. DIGGER and FORQES-W have almost the same performance solving 371 and 367 instances, respectively.

---

[2] See http://www.satcompetition.org/2011/

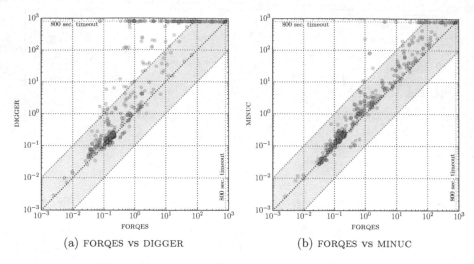

(a) FORQES vs DIGGER   (b) FORQES vs MINUC

Fig. 3. Performance of FORQES, MINUC, and DIGGER

More details on the solvers' performance can be found in Figure 3a, Figure 3b, and Figure 4a. Also note that the *virtual best solver* (VBS) among all the considered solvers is able to solve 485 instances, which is only 2 more than what FORQES can solve on its own. Interestingly, neither MINUC nor DIGGER contribute to the VBS. Although the weakened version of FORQES (FORQES-W) performs quite poorly, it is the only solver (besides the normal version of FORQES) that contributes to the VBS. This can be also seen in Figure 4a.

As it was described in Section 4.3, the approach being proposed can report an upper bound, which can be used as an approximation of the SMUS if finding the exact solution is not efficient and requires too much time. The following evaluates the quality of the upper bound reported by this pragmatic solving strategy. Given an upper bound UB computed within some time limit and the optimal value $opt \leq UB$, the closer value $\frac{UB}{opt}$ is to 1 the better the quality of UB is. Since it is often hard to find the exact optimal value, one can consider a lower bound on the exact solution instead of the optimal value in order to estimate the quality of the upper bound. Indeed, any set of disjoint MCSes found within a given timeout can be seen as a lower bound on the exact solution.

Given a CNF formula, an upper bound on its smallest MUS reported by FORQES within a time limit is computed with an external call to a known MUS enumerator called MARCO [14]. Several timeout values were tested, namely 5, 10, and 20 seconds. Figure 4b shows a cactus plot illustrating the quality of the upper bounds computed this way. It is not surprising that generally the more time is given to the solver, the better upper bound is computed. For about 400 instances, value $\frac{UB}{LB}$ is extremely close to 1 meaning that the upper bound is usually almost equal to the lower bound. As one can see, the upper bound is less than an order of magnitude larger than the lower bound for about 550, 585, and 620 instances if computed within 5, 10, and 20 seconds, respectively. Also

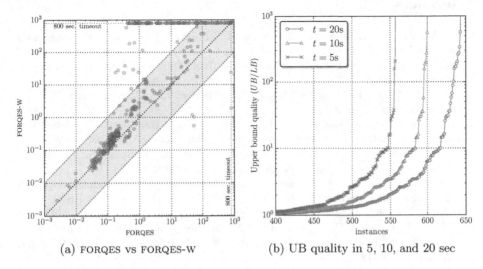

(a) FORQES vs FORQES-W   (b) UB quality in 5, 10, and 20 sec

Fig. 4. Performance of FORQES-W and UB quality

note that both lower and upper bounds are relatively easy to compute, e.g. both of them can be computed almost for 650 instances (out of 682). Moreover, they give an idea of how large the interval is between them. Given this information, a user can decide how critical it is to compute the exact solution.

In summary, the experimental results indicate that the proposed approach pushes the state of the art in SMUS solving, outperforming all the previous approaches in terms of the number of solved instances. Moreover, the scatter plots indicate that in most of the cases the proposed approach is also the fastest in comparison to others. Moreover, a pragmatic policy to report an upper bound within a short period of time can be helpful when it is hard to compute the exact solution and provide a user with an approximate solution of the SMUS problem of reasonable size.

6   Conclusion

This paper adapts recent algorithms for implicit set problems [4,6,11,19] to the case of computing the smallest MUS. A number of enhancements are developed and added to the new algorithm. Experimental results, obtained on representative problems instances, show clear gains over what currently represents the state of the art. A natural line of research is to apply the novel SMUS algorithm in concrete practical applications of SMUSes, e.g. finding smallest equivalent subformulae of smallest size.

References

1. Audemard, G., Lagniez, J.-M., Simon, L.: Improving glucose for incremental SAT solving with assumptions: application to MUS extraction. In: Järvisalo, M., Van Gelder, A. (eds.) SAT 2013. LNCS, vol. 7962, pp. 309–317. Springer, Heidelberg (2013)
2. Belov, A., Lynce, I., Marques-Silva, J.: Towards efficient MUS extraction. AI Commun. **25**(2), 97–116 (2012)
3. Biere, A., Heule, M., van Maaren, H., Walsh, T. (eds.): Handbook of Satisfiability. Frontiers in Artificial Intelligence and Applications, vol. 185. IOS Press (2009)
4. Chandrasekaran, K., Karp, R.M., Moreno-Centeno, E., Vempala, S.: Algorithms for implicit hitting set problems. In: SODA, pp. 614–629 (2011)
5. Cimatti, A., Griggio, A., Schaafsma, B.J., Sebastiani, R.: A modular approach to MaxSAT modulo theories. In: Järvisalo, M., Van Gelder, A. (eds.) SAT 2013. LNCS, vol. 7962, pp. 150–165. Springer, Heidelberg (2013)
6. Davies, J., Bacchus, F.: Solving MAXSAT by solving a sequence of simpler SAT instances. In: Lee, J. (ed.) CP 2011. LNCS, vol. 6876, pp. 225–239. Springer, Heidelberg (2011)
7. Davies, J., Bacchus, F.: Exploiting the power of MIP solvers in MAXSAT. In: Järvisalo, M., Van Gelder, A. (eds.) SAT 2013. LNCS, vol. 7962, pp. 166–181. Springer, Heidelberg (2013)
8. Gupta, A.: Learning Abstractions for Model Checking. PhD thesis, Carnegie Mellon University, June 2006
9. Ignatiev, A., Janota, M., Marques-Silva, J.: Quantified maximum satisfiability: a core-guided approach. In: Järvisalo, M., Van Gelder, A. (eds.) SAT 2013. LNCS, vol. 7962, pp. 250–266. Springer, Heidelberg (2013)
10. Ignatiev, A., Janota, M., Marques-Silva, J.: Towards efficient optimization in package management systems. In: ICSE, pp. 745–755 (2014)
11. Karp, R.M.: Implicit hitting set problems and multi-genome alignment. In: Amir, A., Parida, L. (eds.) CPM 2010. LNCS, vol. 6129, pp. 151–151. Springer, Heidelberg (2010)
12. Liberatore, P.: Redundancy in logic I: CNF propositional formulae. Artif. Intell. **163**(2), 203–232 (2005)
13. Liffiton, M.H., Mneimneh, M.N., Lynce, I., Andraus, Z.S., Marques-Silva, J., Sakallah, K.A.: A branch and bound algorithm for extracting smallest minimal unsatisfiable subformulas. Constraints **14**(4), 415–442 (2009)
14. Liffiton, M.H., Previti, A., Malik, A., Marques-Silva, J.: Fast, flexible MUS enumeration. Constraints (2015). http://dx.doi.org/10.1007/s10601-015-9183-0
15. Liffiton, M.H., Sakallah, K.A.: Algorithms for computing minimal unsatisfiable subsets of constraints. J. Autom. Reasoning **40**(1), 1–33 (2008)
16. Marques-Silva, J., Planes, J.: Algorithms for maximum satisfiability using unsatisfiable cores. In: DATE, pp. 408–413 (2008)
17. Reiter, R.: A theory of diagnosis from first principles. Artif. Intell. **32**(1), 57–95 (1987)
18. Sinz, C., Kaiser, A., Küchlin, W.: Formal methods for the validation of automotive product configuration data. AI EDAM **17**(1), 75–97 (2003)
19. Stern, R.T., Kalech, M., Feldman, A., Provan, G.M.: Exploring the duality in conflict-directed model-based diagnosis. In: AAAI, pp. 828–834 (2012)

Upper and Lower Bounds on the Time Complexity of Infinite-Domain CSPs

Peter Jonsson and Victor Lagerkvist[(✉)]

Department of Computer and Information Science, Linköping University,
Linköping, Sweden
{peter.jonsson,victor.lagerkvist}@liu.se

Abstract. The constraint satisfaction problem (CSP) is a widely studied problem with numerous applications in computer science. For infinite-domain CSPs, there are many results separating tractable and NP-hard cases while upper bounds on the time complexity of hard cases are virtually unexplored. Hence, we initiate a study of the worst-case time cmplexity of such CSPs. We analyse backtracking algorithms and show that they can be improved by exploiting sparsification. We present even faster algorithms based on enumerating finite structures. Last, we prove non-trivial lower bounds applicable to many interesting CSPs, under the assumption that the strong exponential-time hypothesis is true.

1 Introduction

The *constraint satisfaction problem* over a constraint language Γ (CSP(Γ)) is the problem of finding a variable assignment which satisfies a set of constraints, where each constraint is constructed from a relation in Γ. This problem is a widely studied computational problem and it can be used to model many classical problems such as k-colouring and the Boolean satisfiability problem. In the context of artificial intelligence, CSPs have been used for formalizing a wide range of problems, cf. Rossi et al. [30]. Efficient algorithms for CSP problems are hence of great practical interest. If the domain D is finite, then a CSP(Γ) instance I with variable set V can be solved in $O(|D|^{|V|} \cdot poly(||I||))$ time by enumerating all possible assignments. Hence, we have an obvious *upper bound* on the time complexity. This bound can, in many cases, be improved if additional information about Γ is known, cf. the survey by Woeginger [36] or the textbook by Gaspers [14]. There is also a growing body of literature concerning *lower bounds* [16,20,21,33].

When it comes to CSPs over infinite domains, there is a large number of results that identify polynomial-time solvable cases, cf. Ligozat [23] or Rossi et al. [30]. However, almost nothing is known about the time complexity of solving NP-hard CSP problems. One may conjecture that a large number of practically relevant CSP problems do not fall into the tractable cases, and this motivates a closer study of the time complexity of hard problems. Thus, we initiate such a study in this paper. Throughout the paper, we measure time

© Springer International Publishing Switzerland 2015
G. Pesant (Ed.): CP 2015, LNCS 9255, pp. 183–199, 2015.
DOI: 10.1007/978-3-319-23219-5_14

complexity in the number of variables. Historically, this has been the most common way of measuring time complexity. One reason is that an instance may be massively larger than the number of variables — a SAT instance with n variables may contain up to 2^{2n} distinct clauses if repeated literals are disallowed — and measuring in the instance size may give far too optimistic figures, especially since naturally appearing test examples tend to contain a moderate number of constraints. Another reason is that in the finite-domain case, the size of the search space is very closely related to the number of variables. We show that one can reason in a similar way when it comes to the complexity of many infinite-domain CSPs.

The relations in finite-domain CSPs are easy to represent by simply listing the allowed tuples. When considering infinite-domain CSPs, the relations need to be implicitly represented. A natural way is to consider disjunctive formulas over a finite set of basic relations. Let $\mathcal{B}$ denote some finite set of basic relations such that $\mathrm{CSP}(\mathcal{B})$ is tractable. Let $\mathcal{B}^{\vee\infty}$ denote the closure of $\mathcal{B}$ under disjunctions, and let $\mathcal{B}^{\vee k}$ be the subset of $\mathcal{B}^{\vee\infty}$ containing only disjunctions of length at most k. Consider the following example: let $D = \{true, false\}$ and let $\mathcal{B} = \{B_1, B_2\}$ where $B_1 = \{true\}$ and $B_2 = \{false\}$. It is easy to see that $\mathrm{CSP}(\mathcal{B}^{\vee\infty})$ corresponds to the Boolean SAT problem while $\mathrm{CSP}(\mathcal{B}^{\vee k})$ corresponds to the k-SAT problem.

CSPs in certain applications such as AI are often based on binary basic relations and unions of them (instead of free disjunctive formulas). Clearly, such relations are a subset of the relations in $\mathcal{B}^{\vee k}$ and we let $\mathcal{B}^{\vee=}$ denote this set of relations. We do not explicitly bound the length of disjunctions since they are bounded by $|\mathcal{B}|$. The literature on such CSPs is voluminous and we refer the reader to Renz and Nebel [29] for an introduction. The languages $\mathcal{B}^{\vee\infty}$ and $\mathcal{B}^{\vee k}$ have been studied to a smaller extent in the literature. There are both works studying disjunctive constraints from a general point of view [9,11] and application-oriented studies; examples include temporal reasoning [19,31], interactive graphics [27], rule-based reasoning [25], and set constraints (with applications in descriptive logics) [4]. We also note (see Section 2.2 for details) that there is a connection to constraint languages containing first-order definable relations. Assume Γ is a finite constraint language containing relations that are first-order definable in $\mathcal{B}$, and that the first order theory of $\mathcal{B}$ admits quantifier elimination. Then, upper bounds on $\mathrm{CSP}(\Gamma)$ can be inferred from results such as those that will be presented in Sections 3 and 4. This indicates that studying the time complexity of $\mathrm{CSP}(\mathcal{B}^{\vee\infty})$ is worthwhile, especially since our understanding of first-order definable constraint languages is rapidly increasing [3].

To solve infinite-domain CSPs, backtracking algorithms are usually employed. Unfortunately, such algorithms can be highly inefficient in the worst case. Let p denote the maximum arity of the relations in $\mathcal{B}$, let $m = |\mathcal{B}|$, and let $|V|$ denote the number of variables in a given CSP instance. We show (in Section 3.1) that the time complexity ranges from $O(2^{2^{m \cdot |V|^p} \cdot \log(m \cdot |V|^p)} \cdot poly(||I||))$ (which is doubly exponential with respect to the number of variables) for $\mathrm{CSP}(\mathcal{B}^{\vee\infty})$ to $O(2^{2^{m} \cdot |V|^p \cdot \log m} \cdot poly(||I||))$ time for $\mathcal{B}^{\vee=}$ (and the markedly better bound of $O(2^{|V|^p \log m} \cdot poly(|||I|||))$ if $\mathcal{B}$ consists of pairwise disjoint relations.) The use of heuristics can probably improve these figures in some cases, but we have not been able to find such results in the literature and it is not obvious how to analyse backtracking combined with heuristics. At this stage, we are mostly interested in obtaining a baseline: we need to know the performance of simple algorithms before we start studying more sophisticated ones. However, some of these bounds can be improved by combining backtracking search with methods for reducing the number of constraints. We demonstrate this with *sparsification* [18] in Section 3.2.

In Section 4 we switch strategy and show that disjunctive CSP problems can be solved significantly more efficiently via a method we call *structure enumeration*. This method is inspired by the enumerative method for solving finite-domain CSPs. With this algorithm, we obtain the upper bound $O(2^{|V|^p \cdot m} \cdot poly(|||I|||))$ for $\mathrm{CSP}(\mathcal{B}^{\vee \infty})$. If we additionally assume that $\mathcal{B}$ is jointly exhaustive and pairwise disjoint then the running time is improved further to $O(2^{|V|^p \cdot \log m} \cdot poly(|||I|||))$. This bound beats or equals every bound presented in Section 3. We then proceed to show even better bounds for certain choices of $\mathcal{B}$. In Section 4.2 we consider equality constraint languages over a countably infinite domain and show that such CSP problems are solvable in $O(|V|B_{|V|} \cdot poly(|||I|||))$ time, where $B_{|V|}$ is the $|V|$-th Bell number. In Section 4.3 we focus on three well-known temporal reasoning problems and obtain significantly improved running times.

We tackle the problem of determining lower bounds for $\mathrm{CSP}(\mathcal{B}^{\vee \infty})$ in Section 5, i.e. identifying functions f such that no algorithm for $\mathrm{CSP}(\mathcal{B}^{\vee \infty})$ has a better running time than $O(f(|V|))$. We accomplish this by relating CSP problems and certain complexity-theoretical conjectures, and obtain strong lower bounds for the majority of the problems considered in Section 4. As an example, we show that the temporal $\mathrm{CSP}(\{<, >, =\}^{\vee \infty})$ problem is solvable in time $O(2^{|V| \log |V|} \cdot poly(|||I|||))$ but, assuming a conjecture known as the *strong exponential time hypothesis* (SETH), not solvable in $O(c^{|V|})$ time for *any* $c > 1$. Hence, even though the algorithms we present are rather straightforward, there is, in many cases, very little room for improvement, unless the SETH fails.

2   Preliminaries

We begin by defining the constraint satisfaction problem and continue by discussing first-order definable relations.

2.1   Constraint Satisfaction

Definition 1. *Let Γ be a set of finitary relations over some set D of values. The constraint satisfaction problem over Γ (CSP(Γ)) is defined as follows:*

Instance: *A set V of variables and a set C of constraints of the form $R(v_1, \ldots, v_k)$, where k is the arity of R, $v_1, \ldots, v_k \in V$ and $R \in \Gamma$.*
Question: *Is there a function $f : V \rightarrow D$ such that $(f(v_1), \ldots, f(v_k)) \in R$ for every $R(v_1, \ldots, v_k) \in C$?*

The set Γ is referred to as the *constraint language*. Observe that we do not require Γ or D to be finite. Given an instance I of CSP(Γ) we write $||I||$ for the number of bits required to represent I. We now turn our attention to constraint languages based on disjunctions. Let D be a set of values and let $\mathcal{B} = \{B_1, \ldots, B_m\}$ denote a finite set of relations over D, i.e. $B_i \subseteq D^j$ for some $j \geq 1$. Let the set $\mathcal{B}^{\vee\infty}$ denote the set of relations defined by disjunctions over $\mathcal{B}$. That is, $\mathcal{B}^{\vee\infty}$ contains every relation $R(x_1, \ldots, x_p)$ such that $R(x_1, \ldots, x_p)$ if and only if $B_{i_1}(\mathbf{x}_1) \vee \cdots \vee B_{i_t}(\mathbf{x}_t)$ where $\mathbf{x}_1, \ldots, \mathbf{x}_t$ are sequences of variables from $\{x_1, \ldots, x_p\}$ such that the length of $\mathbf{x}_j$ equals the arity of B_{i_j}. We refer to $B_{i_1}(\mathbf{x}_1), \ldots, B_{i_t}(\mathbf{x}_t)$ as the *disjuncts* of R. We assume, without loss of generality, that a disjunct occurs at most once in a disjunction. We define $\mathcal{B}^{\vee k}$, $k \geq 1$, as the subset of $\mathcal{B}^{\vee\infty}$ where each relation is defined by a disjunction of length at most k. It is common, especially in qualitative temporal and spatial constraint reasoning, to study a restricted variant of $\mathcal{B}^{\vee k}$ when all relations in $\mathcal{B}$ has the same arity p. Define $\mathcal{B}^{\vee =}$ to contain every relation R such that $R(\mathbf{x})$ if and only if $B_{i_1}(\mathbf{x}) \vee \cdots \vee B_{i_t}(\mathbf{x})$, where $\mathbf{x} = (x_1, \ldots, x_p)$. For examples of basic relations, we refer the reader to Sections 4.2 and 4.3.

We adopt a simple representation of relations in $\mathcal{B}^{\vee\infty}$: every relation R in $\mathcal{B}^{\vee\infty}$ is represented by its defining disjunctive formula. Note that two objects $R, R' \in \mathcal{B}^{\vee\infty}$ may denote the same relation. Hence, $\mathcal{B}^{\vee\infty}$ is not a constraint language in the sense of Definition 1. We avoid tedious technicalities by ignoring this issue and view constraint languages as multisets. Given an instance $I = (V, C)$ of CSP($\mathcal{B}^{\vee\infty}$) under this representation, we let Disj(I) = $\{B_{i_1}(\mathbf{x}_1), \ldots, B_{i_t}(\mathbf{x}_t) \mid B_{i_1}(\mathbf{x}_1) \vee \cdots \vee B_{i_t}(\mathbf{x}_t) \in C\}$ denote the set of all disjuncts appearing in C.

We close this section by recapitulating some terminology. Let $\mathcal{B} = \{B_1, \ldots, B_m\}$ be a set of relations (over a domain D) such that all $B_1, \ldots, B_m$ have arity p. We say that $\mathcal{B}$ is *jointly exhaustive* (JE) if $\bigcup \mathcal{B} = D^p$ and that $\mathcal{B}$ is *pairwise disjoint* (PD) if $B_i \cap B_j = \emptyset$ whenever $i \neq j$. If $\mathcal{B}$ is both JE and PD we say that it is JEPD. Observe that if $B_1, \ldots, B_m$ have different arity then these properties are clearly not relevant since the intersection between two such relations is always empty. These assumptions are common in for example qualitative spatial and temporal reasoning, cf. [24]. Let Γ be an arbitrary set of relations with arity $p \geq 1$. We say that Γ is *closed under intersection* if $R_1 \cap R_2 \in \Gamma$ for all choices of $R_1, R_2 \in \Gamma$. Let R be an arbitrary binary relation. We define the *converse* $R^\smile$ of R such that $R^\smile = \{(y, x) \mid (x, y) \in R\}$. If Γ is a set of binary relations, then we say that Γ is *closed under converse* if $R^\smile \in \Gamma$ for all $R \in \Gamma$.

2.2 First-Order Definable Relations

Languages of the form $\mathcal{B}^{\vee\infty}$ have a close connection with languages defined over first-order structures admitting quantifier elimination, i.e. every first-order definable relation can be defined by an equivalent formula without quantifiers. We have the following lemma.

Lemma 2. *Let Γ be a finite constraint language first-order definable over a relational structure $(D, R_1, \ldots, R_m)$ admitting quantifier elimination, where*

$R_1, \ldots, R_m$ are JEPD. Then there exists a k such that (1) $CSP(\Gamma)$ is polynomial-time reducible to $CSP(\{R_1, \ldots, R_m\}^{\vee k})$ and (2) if $CSP(\{R_1, \ldots, R_m\}^{\vee k})$ is solvable in $O(f(|V|) \cdot poly(||I||))$ time then $CSP(\Gamma)$ is solvable in $O(f(|V|) \cdot poly(||I||))$ time.

Proof. Assume that every relation $R \in \Gamma$ is definable through a quantifier-free first-order formula ϕ_i over $R_1, \ldots, R_m$. Let ψ_i be ϕ_i rewritten in conjunctive normal form. We need to show that every disjunction in ψ_i can be expressed as a disjunction over $R_1, \ldots, R_m$. Clearly, if ψ_i only contains positive literals, then this is trivial. Hence, assume there is at least one negative literal. Since $R_1, \ldots, R_m$ are JEPD it is easy to see that for any negated relation in $\{R_1, \ldots, R_m\}$ there exists $\Gamma \subseteq \{R_1, \ldots, R_m\}$ such that the union of Γ equals the complemented relation. We can then reduce $CSP(\Gamma)$ to $CSP(\{R_1, \ldots, R_m\}^{\vee k})$ by replacing every constraint by its conjunctive normal formula over $R_1, \ldots, R_m$. This reduction can be done in polynomial time with respect to $||I||$ since each such definition can be stored in a table of fixed size. Moreover, since this reduction does not increase the number of variables, it follows that $CSP(\Gamma)$ is solvable in $O(f(|V|) \cdot poly(||I||))$ time whenever $CSP(\mathcal{B}^{\vee k})$ is solvable in $O(f(|V|) \cdot poly(||I||))$ time.   □

As we will see in Section 4, this result is useful since we can use upper bounds for $CSP(\mathcal{B}^{\vee k})$ to derive upper bounds for $CSP(\Gamma)$, where Γ consists of first-order definable relations over $\mathcal{B}$. There is a large number of structures admitting quantifier elimination and interesting examples are presented in every standard textbook on model theory, cf. Hodges [15]. A selection of problems that are highly relevant for computer science and AI are discussed in Bodirsky [3].

3   Fundamental Algorithms

In this section we investigate the complexity of algorithms for $CSP(\mathcal{B}^{\vee \infty})$ and $CSP(\mathcal{B}^{\vee k})$ based on branching on the disjuncts in constraints (Section 3.1) and the sparsification method (Section 3.2.) Throughout this section we assume that $\mathcal{B}$ is a set of basic relations such that $CSP(\mathcal{B})$ is in P.

3.1   Branching on Disjuncts

Let $\mathcal{B} = \{B_1, \ldots, B_m\}$ be a set of basic relations with maximum arity $p \geq 1$. It is easy to see that $CSP(\mathcal{B}^{\vee \infty})$ is in NP. Assume we have an instance I of $CSP(\mathcal{B}^{\vee \infty})$ with variable set V. Such an instance contains at most $2^{m \cdot |V|^p}$ distinct constraints. Each such constraint contains at most $m \cdot |V|^p$ disjuncts so the instance I can be solved in

$$O((m \cdot |V|^p)^{2^{m \cdot |V|^p}} \cdot poly(||I||)) = O(2^{2^{m \cdot |V|^p} \cdot \log(m \cdot |V|^p)} \cdot poly(||I||))$$

time by enumerating all possible choices of one disjunct out of every disjunctive constraint. The satisfiability of the resulting sets of constraints can be checked in polynomial time due to our initial assumptions. How does such an enumerative approach compare to a branching search algorithm? In the worst case, a branching algorithm without heuristic aid will go through all of these cases so the bound above is valid for such algorithms. Analyzing the time complexity of branching algorithms equipped with powerful heuristics is a very different (and presumably very difficult) problem.

Assume instead that we have an instance I of $\mathrm{CSP}(\mathcal{B}^{\vee k})$ with variable set V. There are at most $m \cdot |V|^p$ different disjuncts which leads to at most $\sum_{i=0}^{k}(m|V|^p)^i \leq k \cdot (m|V|^p)^k$ distinct constraints. We can thus solve instances with $|V|$ variables in $O(k^{k \cdot (m|V|^p)^k} \cdot poly(||I||)) = O(2^{k \cdot \log k \cdot (m|V|^p)^k} \cdot poly(||I||))$ time.

Finally, let I be an instance of $\mathrm{CSP}(\mathcal{B}^{\vee =})$ with variable set V. It is not hard to see that I contains at most $2^m \cdot |V|^p$ distinct constraints, where each constraint has length at most m. Non-deterministic guessing gives that instances of this kind can be solved in

$$O(m^{2^m \cdot |V|^p} \cdot poly(||I||)) = O(2^{2^m \cdot |V|^p \cdot \log m} \cdot poly(||I||))$$

time. This may appear to be surprisingly slow but this is mainly due to the fact that we have not imposed any additional restrictions on the set $\mathcal{B}$ of basic relations. Hence, assume that the relations in $\mathcal{B}$ are PD. Given two relations $R_1, R_2 \in \mathcal{B}^{\vee =}$, it is now clear that $R_1 \cap R_2$ is a relation in $\mathcal{B}^{\vee =}$, i.e. $\mathcal{B}^{\vee =}$ is closed under intersection. Let $I = (V, C)$ be an instance of $\mathrm{CSP}(\mathcal{B}^{\vee =})$. For any sequence of variables $(x_1, \ldots, x_p)$, we can assume that there is at most one constraint $R(x_1, \ldots, x_p)$ in C. This implies that we can solve $\mathrm{CSP}(\mathcal{B}^{\vee =})$ in $O(m^{|V|^p} \cdot poly(||I||)) = O(2^{|V|^p \log m} \cdot poly(||I||))$ time. Combining everything so far we obtain the following upper bounds.

Lemma 3. *Let $\mathcal{B}$ be a set of basic relations with maximum arity p and let $m = |\mathcal{B}|$. Then*

- *$\mathrm{CSP}(\mathcal{B}^{\vee \infty})$ is solvable in $O(2^{2^{m \cdot |V|^p} \cdot \log(m \cdot |V|^p)} \cdot poly(||I||))$ time,*
- *$\mathrm{CSP}(\mathcal{B}^{\vee k})$ is solvable in $O(2^{k \cdot \log k \cdot (m|V|^p)^k} \cdot poly(||I||))$ time,*
- *$\mathrm{CSP}(\mathcal{B}^{\vee =})$ is solvable in $O(2^{2^m \cdot |V|^p \cdot \log m} \cdot poly(||I||))$ time, and*
- *$\mathrm{CSP}(\mathcal{B}^{\vee =})$ is solvable in $O(2^{|V|^p \log m} \cdot poly(||I||))$ time if $\mathcal{B}$ is PD.*

A bit of fine-tuning is often needed when applying highly general results like Lemma 3 to concrete problems. For instance, Renz and Nebel [29] show that the RCC-8 problem can be solved in $O(c^{\frac{|V|^2}{2}})$ for some (unknown) $c > 1$. This problem can be viewed as $\mathrm{CSP}(\mathcal{B}^{\vee =})$ where $\mathcal{B}$ contains JEPD binary relations and $|\mathcal{B}| = 8$. Lemma 3 implies that $\mathrm{CSP}(\mathcal{B}^{\vee =})$ can be solved in $O(2^{3|V|^2})$ which is significantly slower if $c < 8^2$. However, it is well known that $\mathcal{B}$ is closed under converse. Let $I = (\{x_1, \ldots, x_n\}, C)$ be an instance of $\mathrm{CSP}(\mathcal{B}^{\vee =})$. Since $\mathcal{B}$ is closed under converse, we can always assume that if $R(x_i, x_j) \in C$, then $i \leq j$. Thus, we can solve $\mathrm{CSP}(\mathcal{B}^{\vee =})$ in $O(m^{\frac{|V|^2}{2}} \cdot poly(||I||)) = O(2^{\frac{|V|^2}{2} \log m} \cdot poly(||I||))$ time. This figure matches the bound by Renz and Nebel better when c is small.

3.2 Sparsification

The complexity of the algorithms proposed in Section 3 is dominated by the number of constraints. An idea for improving these running times is therefore to reduce the number of constraints within instances. One way of accomplishing this is by using *sparsification* [18]. Before presenting this method, we need a few additional definitions. An instance of the *k-Hitting Set problem* consists of a finite set U (the *universe*) and a collection $\mathcal{C} = \{S_1, \ldots, S_m\}$ where $S_i \subseteq U$ and $|S_i| \leq k$, $1 \leq i \leq m$. A *hitting set* for $\mathcal{C}$ is a set $C \subseteq U$ such that $C \cap S_i \neq \emptyset$ for each $S_i \in \mathcal{C}$. Let $\sigma(\mathcal{C})$ be the set of all hitting sets of $\mathcal{C}$. The k-Hitting Set problem is to find a minimal size hitting set. $\mathcal{T}$ is a *restriction* of $\mathcal{C}$ if for each $S \in \mathcal{C}$ there is a $T \in \mathcal{T}$ with $T \subseteq S$. If $\mathcal{T}$ is a restriction of $\mathcal{C}$, then $\sigma(\mathcal{T}) \subseteq \sigma(\mathcal{C})$. We then have the following result[1].

Theorem 4 (Impagliazzo et al. [18]). *For all $\varepsilon > 0$ and positive k, there is a constant K and an algorithm that, given an instance $\mathcal{C}$ of k-Hitting Set on a universe of size n, produces a list of $t \leq 2^{\varepsilon \cdot n}$ restrictions $\mathcal{T}_1, \ldots, \mathcal{T}_t$ of $\mathcal{C}$ so that $\sigma(\mathcal{C}) = \bigcup_{i=1}^{t} \sigma(\mathcal{T}_i)$ and so that for each $\mathcal{T}_i$, $|\mathcal{T}_i| \leq Kn$. Furthermore, the algorithm runs in time $poly(n) \cdot 2^{\varepsilon \cdot n}$.*

Lemma 5. *Let $\mathcal{B}$ be a set of basic relations with maximum arity p and let $m = |\mathcal{B}|$. Then $CSP(\mathcal{B}^{\vee k})$ is solvable in $O(2^{(\varepsilon + K \log k) \cdot |V|^p \cdot m} \cdot poly(||I||))$ time for every $\varepsilon > 0$, where K is a constant depending only on ε and k.*

Proof. Let $I = (V, C)$ be an instance of $CSP(\mathcal{B}^{\vee k})$. We can easily reduce $CSP(\mathcal{B}^{\vee k})$ to k-Hitting set by letting $U = \mathrm{Disj}(I)$ and $\mathcal{C}$ be the set corresponding to all disjunctions in C. Then choose some $\varepsilon > 0$ and let $\{\mathcal{T}_1, \ldots, \mathcal{T}_t\}$ be the resulting sparsification. Let $\{\mathcal{T}_1', \ldots, \mathcal{T}_t'\}$ be the corresponding instances of $CSP(\mathcal{B}^{\vee k})$. Each instance $\mathcal{T}_i'$ contains at most $K \cdot |U| \leq K \cdot |V|^p \cdot m$ distinct constraints, where K is a constant depending on ε and k, and can therefore be solved in time $O(poly(||I||) \cdot k^{K \cdot |V|^p \cdot m})$ by exhaustive search à la Section 3.1. Last, answer yes if and only if some $\mathcal{T}_i'$ is satisfiable. This gives a total running time of

$$poly(|V|^p \cdot m) \cdot 2^{\varepsilon \cdot |V|^p \cdot m} + 2^{\varepsilon \cdot |V|^p \cdot m} \cdot k^{K \cdot |V|^p \cdot m} \cdot poly(||I||) \in$$
$$O(2^{\varepsilon \cdot |V|^p \cdot m} \cdot 2^{K \cdot |V|^p \cdot m \cdot \log k} \cdot poly(||I||)) = O(2^{(\varepsilon + K \log k) \cdot |V|^p \cdot m} \cdot poly(||I||))$$

since $t \leq 2^{\varepsilon \cdot n}$. □

This procedure can be implemented using only polynomial space, just as the enumerative methods presented in Section 3.1. This follows from the fact that the restrictions $\mathcal{T}_1, \ldots, \mathcal{T}_t$ of $\mathcal{C}$ can be computed one after another with polynomial delay [10, Theorem 5.15]. Although this running time still might seem excessively slow observe that it is significantly more efficient than the $2^{k \cdot \log k \cdot (m|V|^p)^k}$ algorithm for $CSP(\mathcal{B}^{\vee k})$ in Lemma 3.

---

[1] We remark that Impagliazzo et al. [18] instead refer to the k-Hitting set problem as the k-Set cover problem.

4   Improved Upper Bounds

In this section, we show that it is possible to obtain markedly better upper bounds than the ones presented in Section 3. In Section 4.1 we first consider general algorithms for CSP($\mathcal{B}^{\vee\infty}$) based on structure enumeration, and in Sections 4.2 and 4.3, based on the same idea, we construct even better algorithms for equality constraint languages and temporal reasoning problems.

4.1   Structure Enumeration

We begin by presenting a general algorithm for CSP($\mathcal{B}^{\vee\infty}$) based on the idea of enumerating all variable assignments that are implicitly described in instances. As in the case of Section 3 we assume that $\mathcal{B}$ is a set of basic relations such that CSP($\mathcal{B}$) is solvable in $O(poly(||I||))$ time.

Theorem 6. *Let $\mathcal{B}$ be a set of basic relations with maximum arity p and let $m = |\mathcal{B}|$. Then CSP($\mathcal{B}^{\vee\infty}$) is solvable in $O(2^{m|V|^p} \cdot poly(||I||))$ time.*

Proof. Let $I = (V, C)$ be an instance of CSP($\mathcal{B}^{\vee\infty}$). Let $S = \text{Disj}(I)$ and note that $|S| \leq m|V|^p$. For each subset S_i of S first determine whether S_i is satisfiable. Due to the initial assumption this can be done in $O(poly(||I||))$ time since this set of disjuncts can be viewed as an instance of CSP($\mathcal{B}$). Next, check whether S_i satisfies I by, for each constraint in C, determine whether at least one disjunct is included in S_i. Each such step can determined in time $O(poly(||I||))$ time. The total time for this algorithm is therefore $O(2^{m|V|^p} \cdot poly(||I||))$. $\qquad\square$

The advantage of this approach compared to the branching algorithm in Section 3 is that enumeration of variable assignments is much less sensitive to instances with a large number of constraints. We can speed up this result even further by making additional assumptions on the set $\mathcal{B}$. This allows us to enumerate smaller sets of constraints than in Theorem 6.

Theorem 7. *Let $\mathcal{B}$ be a set of basic relations with maximum arity p and let $m = |\mathcal{B}|$. Then*

1. *CSP($\mathcal{B}^{\vee\infty}$) solvable in $O(2^{|V|^p \cdot \log m} \cdot poly(||I||)))$ time if $\mathcal{B}$ is JEPD, and*
2. *CSP($\mathcal{B}^{\vee\infty}$) is solvable in $O(2^{|V|^p \cdot \log(m+1)} \cdot poly(||I||)))$ time if $\mathcal{B}$ is PD.*

Proof. First assume that $\mathcal{B}$ is JEPD and let $I = (V, C)$ be an instance of CSP($\mathcal{B}^{\vee\infty}$). Observe that every basic relation has the same arity p since $\mathcal{B}$ is JEPD. Let F be the set of functions from $|V|^p$ to $\mathcal{B}$. Clearly $|F| \leq 2^{|V|^p \log m}$. For every $f_i \in F$ let $S_{f_i} = \{B_j(\mathbf{x_j}) \mid \mathbf{x_j} \in V^p, f_i(\mathbf{x_j}) = B_j\}$. For a set S_{f_i} one can then determine in $O(poly(||I||))$ time whether it satisfies I by, for every constraint in C, check if at least one disjunct in every constraint is included in S_{f_i}. Hence, the algorithm is sound. To prove completeness, assume that g is a satisfying assignment of I and let S_g be the set of disjuncts in C which are true in this assignment. For every $B_i(\mathbf{x_i}) \in S_g$ define the function f as $f(\mathbf{x_i}) = B_i$.

Since $\mathcal{B}$ is PD it cannot be the case that $f(\mathbf{x_i}) = B_i = B_j$ for some $B_j \in \mathcal{B}$ distinct from B_i. Next assume that there exists $\mathbf{x_i} \in V^p$ but no $B_i \in \mathcal{B}$ such that $B_i(\mathbf{x_i}) \in S_g$. Let $\mathcal{B} = \{B_1, \ldots, B_m\}$ and let $f_1, \ldots, f_m$ be functions agreeing with f for every value for which it is defined and such that $f_i(\mathbf{x_i}) = B_i$. Since $\mathcal{B}$ is JE it holds that f satisfies I if and only if some f_i satisfies I.

Next assume that $\mathcal{B}$ is PD but not JE. In this case we use the same construction but instead consider the set of functions F' from V^p to $\mathcal{B} \cup \{D^p\}$. There are $2^{|V|^p \cdot \log(m+1)}$ such functions, which gives the desired bound $O(2^{|V|^p \cdot \log(m+1)} \cdot poly(||I||))$. The reason for adding the additional element D^p to the domains of the functions is that if $f \in F'$, and if $f(\mathbf{x}) = D^p$ for some $\mathbf{x} \in V^p$, then this constraint does not enforce any particular values on $\mathbf{x}$.   □

4.2 Equality Constraint Languages

Let $\mathcal{E} = \{=, \neq\}$ over some countably infinite domain D. The language $\mathcal{E}^{\vee\infty}$ is a particular case of an *equality constraint language* [5], i.e. sets of relations definable through first-order formulas over the structure $(D, =)$. Such languages are of fundamental interest in complexity classifications for infinite domain CSPs, since a classification of CSP problems based on first-order definable relations over some fixed structure, always includes the classification of equality constraint language CSPs. We show that the $O(2^{|V|^2} \cdot poly(||I||))$ time algorithm in Theorem 7 can be improved upon quite easily. But first we need some additional machinery. A *partition* of a set X with n elements is a pairwise disjoint set $\{X_1, \ldots, X_m\}$, $m \leq n$ such that $\bigcup_{i=1}^m X_i = X$. A set X with n elements has B_n partitions, where B_n is the n-th *Bell number*. Let $L(n) = \frac{0.792n}{\ln(n+1)}$. It is known that $B_n < L(n)^n$ [1] and that all partitions can be enumerated in $O(nB_n)$ time [13,32].

Theorem 8. *$CSP(\mathcal{E}^{\vee\infty})$ is solvable in $O(|V|2^{|V| \cdot \log L(|V|)} \cdot poly(||I||))$ time.*

Proof. Let $I = (V, C)$ be an instance of $CSP(\mathcal{E}^{\vee\infty})$. For every partition $S_1 \cup \ldots \cup S_n$ of V we interpret the variables in S_i as being equal and having the value i, i.e. a constraint $(x = y)$ holds if and only if x and y belong to the same set and $(x \neq y)$ holds if and only if x and y belong to different sets. Then check in $poly(||I||)$ time if this partition satisfies I using the above interpretation. The complexity of this algorithm is therefore $O(|V|B_{|V|} \cdot poly(||I||)) \subseteq O(|V|L(|V|)^{|V|} \cdot poly(||I||)) = O(|V|2^{|V| \cdot \log L(|V|)} \cdot poly(||I||))$.   □

Observe that this algorithm is much more efficient than the $O(2^{|V|^2} \cdot poly(||I||))$ algorithm in Theorem 7. It is well known that equality constraint languages admit quantifier elimination [5]. Hence, we can use Lemma 2 to extend Theorem 8 to cover arbitrary equality constraint languages.

Corollary 9. *Let Γ be a finite set of relations first-order definable over $(D, -)$. Then $CSP(\Gamma)$ is solvable in $O(|V|2^{|V| \cdot \log L(|V|)} \cdot poly(||I||))$ time.*

4.3 Temporal Constraint Reasoning

Let $\mathcal{T} = \{<, >, =\}$ denote the JEPD order relations on $\mathbb{Q}$ and recall that $\mathrm{CSP}(\mathcal{T})$ is tractable [34]. Theorem 7 implies that $\mathrm{CSP}(\mathcal{T}^{\vee\infty})$ can be solved in $O(2^{|V|^2 \cdot \log 3} \cdot poly(\||I\||))$ time. We improve this as follows.

Theorem 10. $\mathrm{CSP}(\mathcal{T}^{\vee\infty})$ *is solvable in* $O(2^{|V| \log |V|} \cdot poly(\||I\||))$ *time.*

Proof. Let $I = (V, C)$ be an instance of $\mathrm{CSP}(\mathcal{T}^{\vee\infty})$. Assume $f : V \to \mathbb{Q}$ satisfies this instance. It is straightforward to see that there exists some $g : V \to \{1, \ldots, |V|\}$ which satisfies I, too. Hence, enumerate all $2^{|V| \log |V|}$ functions from V to $\{1, \ldots, |V|\}$ and answer yes if any of these satisfy the instance. $\qquad\square$

It is well known that the first-order theory of $(\mathbb{Q}, <)$ admits quantifier elimination [6, 15]. Hence, we can exploit Lemma 2 to obtain the following corollary.

Corollary 11. *Let Γ be a finite temporal constraint language over $(\mathbb{Q}, <)$. If $\mathrm{CSP}(\Gamma)$ is NP-complete, then it is solvable in* $O(2^{|V| \log |V|} \cdot poly(\||I\||))$ *time.*

We can also obtain strong bounds for Allen's interval algebra, which is a well-known formalism for temporal reasoning. Here, one considers relations between intervals of the form $[x, y]$, where $x, y \in \mathbb{R}$ is the starting and ending point, respectively. Let *Allen* be the $2^{13} = 8192$ possible unions of the set of the thirteen relations in Table 1. For convenience we write constraints such as $(\mathsf{p} \vee \mathsf{m})(x, y)$ as $x\{\mathsf{p}, \mathsf{m}\}y$, using infix notation and omitting explicit disjunction signs. The problem $\mathrm{CSP}(Allen)$ is NP-complete and all tractable fragments have been identified [22].

Given an instance $I = (V, C)$ of $\mathrm{CSP}(Allen)$ we first create two fresh variables x_i^s and x_i^e for every $x \in V$, intended to represent the startpoint and endpoint of the interval x. Then observe that a constraint $x\{r_1, \ldots, r_m\}y \in C$, where each r_i is a basic relation, can be represented as a disjunction of temporal constraints over x^s, x^e, y^s and y^e by using the definitions of each basic relation in Table 1. Applying Theorem 10 to the resulting instance gives the following result.

Table 1. The thirteen basic relations in Allen's interval algebra. The endpoint relations $x^s < x^e$ and $y^s < y^e$ that are valid for all relations have been omitted.

Basic relation		Example	Endpoints
x precedes y	p	xxx	$x^e < y^s$
y preceded by x	p^{-1}	yyy	
x meets y	m	xxxx	$x^e = y^s$
y met-by x	m^{-1}	yyyy	
x overlaps y	o	xxxx	$y^s < x^e$,
y overl.-by x	o^{-1}	yyyy	$x^e < y^e$
x during y	d	xxx	y^s,
y includes x	d^{-1}	yyyyyyy	$x^e < y^e$
x starts y	s	xxx	y^s,
y started by x	s^{-1}	yyyyyyy	$x^e < y^e$
x finishes y	f	xxx	$x^e = y^e$,
y finished by x	f^{-1}	yyyyyyy	y^s
x equals y	$\equiv$	xxxx	y^s,
		yyyy	$x^e = y^e$

Corollary 12. *CSP(Allen) is solvable in* $O(2^{2|V|(1+\log|V|)} \cdot poly(||I||))$ *time.*

Finally, we consider *branching time*. We define the following relations on the set of all points in the forest containing all oriented, finite trees where the in-degree of each node is at most one.

1. $x = y$ if and only if there is a path from x to y and a path from y to x,
2. $x < y$ if and only if and there is a path from x to y but no path from y to x,
3. $x > y$ if and only if there is a path from y to x but no path from x to y,
4. $x||y$ if and only if there is no path from x to y and no path from y to x,

These four basic relations are known as the *point algebra for branching time*. We let $\mathcal{P} = \{||, <, >, =\}$. The problem $\text{CSP}(\mathcal{P}^{\vee\infty})$ is NP-complete and many tractable fragments have been identified [8].

Theorem 13. *CSP($\mathcal{P}^{\vee\infty}$) is solvable in* $O(2^{|V|+2|V|\log|V|} \cdot poly(||I||))$ *time.*

Proof. Let $I = (V, C)$ be an instance of $\text{CSP}(\mathcal{P}^{\vee\infty})$. We use the following algorithm.

1. enumerate all directed forests over V where the in-degree of each node is at most one,
2. for every forest F, if at least one disjunct in every constraint in C is satisfied by F, answer yes,
3. answer no.

It is readily seen that this algorithm is sound and complete for $\text{CSP}(\mathcal{P}^{\vee\infty})$. As for the time complexity, recall that the number of directed labelled trees with $|V|$ vertices is equal to $|V|^{|V|-2}$ by Cayley's formula. These can be efficiently enumerated by e.g. enumerating all *Prüfer sequences* [28] of length $|V| - 2$. To enumerate all forests instead of trees, we can enumerate all labelled trees with $|V| + 1$ vertices and only consider the trees where the extra vertex is connected to all other vertices. By removing this vertex we obtain a forest with $|V|$ vertices. Hence, there are at most $2^{|V|}|V|^{|V|-1}$ directed forests over V. The factor $2^{|V|}$ stems from the observation that each forest contains at most $|V|$ edges, where each edge has two possible directions. We then filter out the directed forests containing a tree where the in degree of any vertex is more than one. Last, for each forest, we enumerate all $|V|^{|V|}$ functions from V to the forest, and check in $poly(||I||)$ time whether it satisfies I. Put together this gives a complexity of $O(2^{|V|}|V|^{|V|-1}|V|^{|V|} \cdot poly(||I||)) \subseteq O(2^{|V|+2|V|\log|V|} \cdot poly(||I||))$. $\square$

Branching time does not admit quantifier elimination [3, Section 4.2] so Lemma 2 is not applicable. However, there are closely connected constraint languages on trees that have this property. Examples include the *triple consistency problem* with important applications in bioinformatics [7].

5   Lower Bounds

The algorithms presented in Section 4 give new upper bounds for the complexity of $CSP(\mathcal{B}^{\vee\infty})$. It is natural to also ask, given reasonable complexity theoretical assumptions, how much room there is for improvement. This section is divided into Section 5.1, where we obtain lower bounds for $CSP(\mathcal{B}^{\vee\infty})$ and $CSP(\mathcal{B}^{\vee k})$ for $\mathcal{B}$ that are JEPD, and in Section 5.2, where we obtain lower bounds for Allen's interval algebra.

5.1   Lower Bounds for JEPD Languages

One of the most well-known methods for obtaining lower bounds is to exploit the *exponential-time hypothesis* (ETH). The ETH states that there exists a $\delta > 0$ such that 3-SAT is not solvable in $O(2^{\delta|V|})$ time by any deterministic algorithm, i.e. it is not solvable in subexponential time [16]. If the ETH holds, then there is an increasing sequence $s_3, s_4, \ldots$ of reals such that k-SAT cannot be solved in time $2^{s_k|V|}$ but it can be solved in $2^{(s_k+\epsilon)|V|}$ time for arbitrary $\epsilon > 0$. The *strong exponential-time hypothesis* (SETH) is the conjecture that the limit of the sequence $s_3, s_4, \ldots$ equals 1, and, as a consequence, that SAT is not solvable in time $O(2^{\delta|V|})$ for any $\delta < 1$ [16]. These conjectures have in recent years successfully been used for proving lower bounds of many NP-complete problems [26].

Theorem 14. *Let $\mathcal{B} = \{R_1, R_2, \ldots, R_m\}$ be a JEPD set of nonempty basic relations. If the SETH holds then $CSP(\mathcal{B}^{\vee\infty})$ cannot be solved in $O(2^{\delta|V|})$ time for any $\delta < 1$.*

Proof. If the SETH holds then SAT cannot be solved in $O(2^{\delta|V|})$ time for any $\delta < 1$. We provide a polynomial-time many-one reduction from SAT to $CSP(\mathcal{B}^{\vee\infty})$ which only increases the number of variables by a constant — hence, if $CSP(\mathcal{B}^{\vee\infty})$ is solvable in $O(2^{\delta|V|})$ time for some $\delta < 1$ then SAT is also solvable in $O(2^{\delta|V|})$ time, contradicting the original assumption.

Let $I = (V, C)$ be an instance of SAT, where V is a set of variables and C a set of clauses. First observe that since $m \geq 2$ and since $\mathcal{B}$ is JEPD, $\mathcal{B}$ must be defined over a domain with two or more elements. Also note that the requirement that $\mathcal{B}$ is JEPD implies that complement of $R_1(\mathbf{x})$ can be expressed as $R_2(\mathbf{x}) \vee \ldots \vee R_m(\mathbf{x})$. Now, let p denote the arity of the relations in $\mathcal{B}$. We introduce $p-1$ fresh variables $T_1, \ldots, T_{p-1}$ and then for every clause $(\ell_1 \vee \ldots \vee \ell_k) \in C$ create the constraint $(\phi_1(x_1, T_1, \ldots, T_{p-1}) \vee \ldots \vee \phi_k(x_k, T_1, \ldots, T_{p-1}))$, where $\phi_i(x_i, T_1, \ldots, T_{p-1}) = R_1(x_i, T_1, \ldots, T_{p-1})$ if $\ell_i = x_i$ and $\phi_i(x_i, T_1, \ldots, T_{p-1}) = R_2(x_i, T_1, \ldots, T_{p-1}) \vee \ldots \vee R_m(x_i, T_1, \ldots, T_{p-1})$ if $\ell_i = \neg x_i$. Hence, the resulting instance is satisfiable if and only if I is satisfiable. Since the reduction only introduces $p - 1$ fresh variables it follows that SAT is solvable in time $O(2^{\delta(|V|+p-1)}) = O(2^{\delta|V|})$. $\square$

Even though this theorem does not rule out the possibility that $CSP(\mathcal{B}^{\vee k})$ can be solved significantly faster for some k it is easy to see that $CSP(\mathcal{B}^{\vee k})$ cannot be solved in subexponential time for any $k \geq 3(|\mathcal{B}| - 1)$. First assume that the ETH holds. By following the proof of Theorem 14 we can reduce 3-SAT to $CSP(\mathcal{B}^{\vee 3(|\mathcal{B}|-1)})$, which implies that $CSP(\mathcal{B}^{\vee 3(|\mathcal{B}|-1)})$ cannot be solved in $2^{\delta n}$ time either. The bound $k = 3(|\mathcal{B}| - 1)$ might obviously feel a bit unsatisfactory and one might wonder if this can be improved. We can in fact make this much more precise by adding further restrictions to the set $\mathcal{B}$. As in the case of the equality constraint languages in Section 4.2 we let $=$ denote the equality relation on a given countably infinite domain.

Theorem 15. *Let $\mathcal{B} = \{=, R_1, \ldots, R_m\}$ be a set of binary PD, nonempty relations. If the ETH holds then $CSP(\mathcal{B}^{\vee k})$ cannot be solved in $O(2^{s_k|V|})$ time.*

Proof. We prove this result by reducing k-SAT to $CSP(\mathcal{B}^{\vee k})$ in such a way that we at most introduce one fresh variable. Let $I = (V, C)$ be an instance of k-SAT, where V is a set of variables and C a set of clauses. We know that $R_1 \subseteq \{(a, b) \mid a, b \in D \text{ and } a \neq b\}$ since $\mathcal{B}$ is PD. Introduce one fresh variable T. For every clause $(\ell_1 \vee \ldots \vee \ell_k) \in C$ create the constraint $(\phi_1 \vee \ldots \vee \phi_k)$, where $\phi_i := x_j = T$ if $\ell_i = x_j$ and $\phi_i = R_1(x_j, T)$ if $\ell_i = \neg x_j$. Let (V', C') be the resulting instance of $CSP(\mathcal{B}^{\vee k})$. We show that (V', C') has a solution if and only if (V, C) has a solution.

Assume first that (V, C) has a solution $f : V \rightarrow \{0, 1\}$. Arbitrarily choose a tuple $(a, b) \in R_1$. We construct a solution $f' : V' \rightarrow \{a, b\}$ for (V', C'). Let $f'(T) = b$, and for all $v \in V$ let $f'(v) = b$ if $f(v) = 1$ and let $f'(v) = a$ if $f(v) = 0$. Arbitrarily choose a clause $(\ell_1 \vee \ldots \vee \ell_k) \in C$ and assume for instance that ℓ_1 evaluates to 1 under the solution f. If $\ell_1 = x_i$, then $f(x_i) = 1$ and the corresponding disjunct in the corresponding disjunctive constraint in C' is $x_i = T$. By definition, $(f'(x_i), f'(T)) = (b, b)$. If $\ell_1 = \neg x_i$, then $f(x_i) = 0$ and the corresponding disjunct in the corresponding disjunctive constraint in C' is $R_1(x_i, T)$. By definition, $(f'(x_i), f'(T)) = (a, b)$ and $(a, b) \in R_1$.

Assume instead that $f' : V' \rightarrow D$ is a solution to (V', C'), and that $f'(T) = c$. We construct a solution $f : V \rightarrow \{0, 1\}$ to (V, C) as follows. Arbitrarily choose a disjunctive constraint $(d_1 \vee \cdots \vee d_k) \in C'$ and let $(\ell_1 \vee \cdots \vee \ell_k)$ be the corresponding clause in C'. Assume that $\ell_1 = x_i$. If d_1 is true under f', then let $f(x_i) = 1$ and, otherwise, $f(x_i) = 0$. If $\ell_1 = \neg x_i$, then do the opposite: $f(x_i) = 0$ if d_1 is true and $f(x_i) = 1$ otherwise. If the function f is well-defined, then f is obviously a solution to (V, C). We need to prove that there is no variable that is simultaneously assigned 0 and 1. Assume this is the case. Then there is some variable x_i such that the constraints $x_i = T$ and $R_1(x_i, T)$ are simultaneously satisfied by f'. This is of course impossible due to the fact that R_1 contains no tuple of the form (a, a). □

If we in addition require that $\mathcal{B}$ is JE we obtain substantially better lower bounds for $CSP(\mathcal{B}^{\vee \infty})$.

Theorem 16. *Let $\mathcal{B} = \{=, R_1, \ldots, R_m\}$ be a set of binary JEPD relations over a countably infinite domain. If the SETH holds then $CSP(\mathcal{B}^{\vee \infty})$ cannot be solved in $O(c^{|V|})$ time for any $c > 1$.*

Proof. First observe that the binary inequality relation $\neq$ over D can be defined as $\bigcup_{i=1}^{m} R_i$ since $\mathcal{B}$ is JEPD. In the the proof we therefore use $\neq$ as an abbreviation for $\bigcup_{i=1}^{m} R_i$. Let $I = (V, C)$ be an instance of SAT with variables $V = \{x_1, \ldots, x_n\}$ and the set of clauses C. Let K be an integer such that $K > \log c$. Assume without loss of generality that n is a multiple of K. We will construct an instance of CSP($\mathcal{B}^{\vee\infty}$) with $\frac{n}{K} + 2^K = \frac{n}{K} + O(1)$ variables. First, introduce 2^K fresh variables $v_1, \ldots, v_{2^K}$ and make them different by imposing $\neq$ constraints. Second, introduce $\frac{n}{K}$ fresh variables $y_1, \ldots, y_{\frac{n}{K}}$, and for each $i \in \{1, \ldots, \frac{n}{K}\}$ impose the constraint $(y_i = v_1 \vee y_i = v_2 \vee \cdots \vee y_i = v_{2^k})$. Let $V_1, \ldots, V_{\frac{n}{K}}$ be a partition of V such that each $|V_i| = K$. We will represent each set V_i of Boolean variables by one y_i variable over D. To do this we will interpret each auxiliary variable z_i as a K-ary Boolean tuple. Let $h : \{v_1, \ldots, v_{2^K}\} \rightarrow \{0, 1\}^K$ be an injective function which assigns a Boolean K-tuple for every variable v_i. Let g_+ be a function from $\{1, \ldots, K\}$ to subsets of $\{v_1, \ldots, v_{2^K}\}$ such that $v_i \in g(j)$ if and only if the j-th element in $h(v_i)$ is equal to 1. Define g_- in the analogous way. Observe that $|g_+(j)| = |g_-(j)| = 2^{K-1}$ for each $j \in \{1, \ldots, K\}$.

For the reduction, let $(\ell_{i_1} \vee \ldots \vee \ell_{i_{n'}})$, $\ell_{i_j} = x_{i_j}$ or $\ell_{i_j} = \neg x_{i_j}$, be a clause in C. We assume that $n' \leq n$ since the clause contains repeated literals otherwise. For each literal ℓ_{i_j} let $V_{i'} \subseteq V$ be the set of variables such that $x_{i_j} \in V_{i'}$. Each literal ℓ_{i_j} is then replaced by $\bigvee_{z \in g_+(i_j)} y_{i'} = z$ if $\ell_{i_j} = x_{i_j}$, and with $\bigvee_{z \in g_-(i_j)} y_{i'} = z$ if $\ell_{i_j} = \neg x_{i_j}$. This reduction can be done in polynomial time since a clause with n' literals is replaced by a disjunctive constraint with $n' 2^{K-1}$ disjuncts (since K is a constant depending only on c). It follows that SAT can be solved in

$$O(c^{\frac{n}{K}+O(1)} \cdot poly(\|I\|)) = O(2^{(\frac{n}{K}+O(1)) \cdot \log c} \cdot poly(\|I\|)) = O(2^{\delta \cdot n} \cdot poly(\|I\|))$$

for some $\delta < 1$, since $K > \log c$. Thus, the SETH does not hold. $\square$

As an illustrative use of the theorem we see that the temporal problem CSP($\mathcal{T}^{\vee\infty}$) is solvable in $O(2^{|V| \log |V|} \cdot poly(\|I\|))$ time but not in $O(c^{|V|})$ time for any $c > 1$ if the SETH holds. Lower bounds can also be obtained for the branching time problem CSP($\mathcal{P}^{\vee\infty}$) since there is a trivial reduction from CSP($\mathcal{T}$)$^{\vee\infty}$ which does not increase the number of variables: simply add a constraint $(x < y \vee x > y \vee x = y)$ for every pair of variables in the instance. Similarly, the equality constraint satisfaction problem CSP($\mathcal{E}^{\vee\infty}$) is not solvable in $O(c^{|V|})$ time for any $c > 1$ either, unless the SETH fails. Hence, even though the algorithms in Sections 4.2 and 4.3 might appear to be quite simple, there is very little room for improvement.

5.2   Lower Bounds for Allen's Interval Algebra

Theorems 14, 15 and 16 gives lower bounds for all the problems considered in Sections 4.2 and 4.3 except for CSP(*Allen*) since unlimited use of disjunction is not allowed in this language. It is however possible to relate the complexity of CSP(*Allen*) to the CHROMATIC NUMBER problem, i.e. the problem of computing the number of colours needed to colour a given graph.

Theorem 17. *If CSP(Allen) can be solved in $O(\sqrt{c}^{|V|})$ time for some $c < 2$, then* CHROMATIC NUMBER *can be solved in $O((c+\epsilon)^{|V|})$ time for arbitrary $\epsilon > 0$.*

Proof. We first present a polynomial-time many-one reduction from k-COLOURABILITY to CSP(*Allen*) which introduces k fresh variables. Given an undirected graph $G = (\{v_1, \dots, v_n\}, E)$, introduce the variables $z_1, \dots, z_k$ and $v_1, \dots, v_n$, and:

1. impose the constraints $z_1\{\mathsf{m}\}z_2\{\mathsf{m}\}\dots\{\mathsf{m}\}z_k$,
2. for each v_i, $1 \le i \le n$, add the constraints $v_i\{\equiv, \mathsf{s}^{-1}\}z_1$, $v_i\{\mathsf{p}, \mathsf{m}, \mathsf{f}^{-1}, \mathsf{d}^{-1}\}z_j$ $(2 \le j \le k-1)$, and $v_i\{\mathsf{p}, \mathsf{m}, \mathsf{f}^{-1}\}z_k$,
3. for each $(v_i, v_j) \in E$, add the constraint $v_i\{\mathsf{s}, \mathsf{s}^{-1}\}v_j$.

Consulting Table 1, we see that for each v_i, it holds that its right endpoint must equal the right endpoint of some z_i, and its left endpoint must equal the left endpoint of z_1. It is now obvious that the resulting instance has a solution if and only if G is k-colourable. The result then follows since there is a polynomial-time Turing reduction from CHROMATIC NUMBER to CSP(*Allen*) by combining binary search (that will evaluate $\log n$ Allen instances) with the reduction above (recall that $O(\log n \cdot c^n) \subseteq O((c + \epsilon)^n)$ for every $\epsilon > 0$) . Observe that if $k = n$ then the reduction introduces n fresh variables, which is where the constant $\sqrt{c}$ in the expression $O(\sqrt{c}^{|V|})$ stems from. CSP(*Allen*). $\qquad\square$

The exact complexity of CHROMATIC NUMBER has been analysed and discussed in the literature. Björklund et al. [2] have shown that the problem is solvable in $2^{|V|} \cdot poly(||I||)$ time. Impagliazzo and Paturi [17] poses the following question: "Assuming SETH, can we prove a $2^{n-o(n)}$ lower bound for CHROMATIC NUMBER?". Hence, an $O(\sqrt{c}^{|V|})$, $c < 2$, algorithm for CSP(*Allen*) would also be a major breakthrough for CHROMATIC NUMBER.

6   Discussion

We have investigated several novel algorithms for solving disjunctive CSP problems, which, with respect to worst-case time complexity, are much more efficient than e.g. backtracking algorithms without heuristics. These bounds can likely be improved, but, due to the lower bounds in Section 5, probably not to a great degree. Despite this, algorithms for solving infinite domain constraint satisfaction problems are in practice used in many non-trivial applications. In light of this the following research direction is particularly interesting: *how to formally analyse the time complexity of branching algorithms equipped with (powerful) heuristics?* In the case of finite-domain CSPs and, in particular, DPLL-like algorithms for the k-SAT problem there are numerous results to be found in the literature, cf. the survey by Vsemirnov et al. [35]. This is not the case for infinite-domain CSPs, even though there is a considerable amount of empirical evidence that infinite-domain CSPs can be efficiently solved by such algorithms, so one ought to be optimistic about the chances of actually obtaining non-trivial bounds. Yet, sharp formal analyses appear to be virtually nonexistent in the literature.

Another research direction is to strengthen the lower bounds in Section 5 even further. It would be interesting to prove stronger lower bounds for $CSP(\mathcal{B}^{\vee k})$ for some concrete choices of $\mathcal{B}$ and k. As an example, consider the temporal problem $CSP(\mathcal{T}^{\vee 4})$. From Theorem 15 we see that $CSP(\mathcal{T}^{\vee 4})$ is not solvable in $O(2^{s_4|V|})$ time for some $s_4 < \log 1.6$, assuming the ETH holds, since the currently best deterministic algorithm for 4-SAT runs in $O(1.6^{|V|})$ time [12]. On the other hand, if $CSP(\mathcal{T}^{\vee 4})$ is solvable in $O(\sqrt{c}^{|V|})$ time for some $c < 2$, then CHROMATIC NUMBER can be solved in $O((c + \epsilon)^{|V|})$ time for arbitrary $\epsilon > 0$. This can be proven similar to the reduction in Theorem 17 but by making use of temporal constraints instead of interval constraints. Hence, for certain choices of $\mathcal{B}$ and k it might be possible to improve upon the general bounds given in Section 5.

References

1. Berend, D., Tassa, T.: Improved bounds on Bell numbers and on moments of sums of random variables. Probability and Mathematical Statistics **30**(2), 185–205 (2010)
2. Björklund, A., Husfeldt, T., Koivisto, M.: Set partitioning via inclusion-exclusion. SIAM Journal on Computing **39**(2), 546–563 (2009)
3. Bodirsky, M.: Complexity classification in infinite-domain constraint satisfaction. Mémoire d'habilitation à diriger des recherches, Université Diderot - Paris 7. arXiv:1201.0856 (2012)
4. Bodirsky, M., Hils, M.: Tractable set constraints. Journal of Artificial Intelligence Research **45**, 731–759 (2012)
5. Bodirsky, M., Kára, J.: The complexity of equality constraint languages. Theory of Computing Systems **43**(2), 136–158 (2008)
6. Bodirsky, M., Kára, J.: The complexity of temporal constraint satisfaction problems. Journal of the ACM **57**(2), 9:1–9:41 (2010)
7. Bodirsky, M., Mueller, J.K.: The complexity of rooted phylogeny problems. Logical Methods in Computer Science **7**(4) (2011)
8. Broxvall, M., Jonsson, P.: Point algebras for temporal reasoning: Algorithms and complexity. Artificial Intelligence **149**(2), 179–220 (2003)
9. Broxvall, M., Jonsson, P., Renz, J.: Disjunctions, independence, refinements. Artificial Intelligence **140**(1–2), 153–173 (2002)
10. Calabro, C.: The Exponential Complexity of Satisfiability Problems. PhD thesis, University of California, San Diego, CA, USA (2009)
11. Cohen, D., Jeavons, P., Jonsson, P., Koubarakis, M.: Building tractable disjunctive constraints. Journal of the ACM **47**(5), 826–853 (2000)
12. Dantsin, E., Goerdt, A., Hirsch, E.A., Kannan, R., Kleinberg, J.M., Papadimitriou, C.H., Raghavan, P., Schöning, U.: A deterministic $(2 - 2/(k + 1))^n$ algorithm for k-SAT based on local search. Theoretical Computer Science **289**(1), 69–83 (2002)
13. Djokic, B., Miyakawa, M., Sekiguchi, S., Semba, I., Stojmenovic, I.: A fast iterative algorithm for generating set partitions. The Computer Journal **32**(3), 281–282 (1989)
14. Gaspers, S.: Exponential Time Algorithms - Structures, Measures, and Bounds. VDM (2010)
15. Hodges, W.: A Shorter Model Theory. Cambridge University Press, New York (1997)
16. Impagliazzo, R., Paturi, R.: On the complexity of k-SAT. Journal of Computer and System Sciences **62**(2), 367–375 (2001)

17. Impagliazzo, R., Paturi, R.: Exact complexity and satisfiability. In: Gutin, G., Szeider, S. (eds.) IPEC 2013. LNCS, vol. 8246, pp. 1–3. Springer, heidelberg (2013)
18. Impagliazzo, R., Paturi, R., Zane, F.: Which problems have strongly exponential complexity? Journal of Computer and System Sciences **63**(4), 512–530 (2001)
19. Jonsson, P., Bäckström, C.: A unifying approach to temporal constraint reasoning. Artificial Intelligence **102**(1), 143–155 (1998)
20. Jonsson, P., Lagerkvist, V., Nordh, G., Zanuttini, B.: Complexity of SAT problems, clone theory and the exponential time hypothesis. In: Proceedings of the 24th Annual ACM-SIAM Symposium on Discrete Algorithms (SODA-2013), pp. 1264–1277 (2013)
21. Kanj, I., Szeider, S.: On the subexponential time complexity of CSP. In: Proceedings of the Twenty-Seventh AAAI Conference on Artificial Intelligence (AAAI-2013) (2013)
22. Krokhin, A., Jeavons, P., Jonsson, P.: Reasoning about temporal relations: The tractable subalgebras of Allen's interval algebra. Journal of the ACM **50**(5), 591–640 (2003)
23. Ligozat, G.: Qualitative Spatial and Temporal Reasoning. Wiley-ISTE (2011)
24. Ligozat, G., Renz, J.: What Is a qualitative calculus? a general framework. In: Zhang, C., W. Guesgen, H., Yeap, W.-K. (eds.) PRICAI 2004. LNCS (LNAI), vol. 3157, pp. 53–64. Springer, Heidelberg (2004)
25. Liu, B., Jaffar, J.: Using constraints to model disjunctions in rule-based reasoning. In: Proceedings of the Thirteenth National Conference on Artificial Intelligence, AAAI 1996, Portland, Oregon, vol. 2, pp. 1248–1255 (1996)
26. Lokshtanov, D., Marx, D., Saurabh, S.: Lower bounds based on the exponential time hypothesis. Bulletin of EATCS **3**(105) (2013)
27. Marriott, K., Moulder, P., Stuckey, P.J.: Solving disjunctive constraints for interactive graphical applications. In: Walsh, T. (ed.) CP 2001. LNCS, vol. 2239, p. 361. Springer, Heidelberg (2001)
28. Prüfer, H.: Neuer beweis eines satzes über permutationen. Archiv der Mathematik und Physik **27**, 742–744 (1918)
29. Renz, J., Nebel, B.: Efficient methods for qualitative spatial reasoning. Journal of Artificial Intelligence Research **15**(1), 289–318 (2001)
30. Rossi, F., van Beek, P., Walsh, T. (eds.) Handbook of Constraint Programming. Elsevier (2006)
31. Stergiou, K., Koubarakis, M.: Backtracking algorithms for disjunctions of temporal constraints. Artificial Intelligence **120**(1), 81–117 (2000)
32. Stojmenović, I.: An optimal algorithm for generating equivalence relations on a linear array of processors. BIT Numerical Mathematics **30**(3), 424–436 (1990)
33. Traxler, P.: The time complexity of constraint satisfaction. In: Grohe, M., Niedermeier, R. (eds.) IWPEC 2008. LNCS, vol. 5018, pp. 190–201. Springer, Heidelberg (2008)
34. Vilain, M.B., Kautz, H.A.: Constraint propagation algorithms for temporal reasoning. In: Proceedings of the 5th National Conference on Artificial Intelligence (AAAI 1986), pp. 377–382 (1986)
35. Vsemirnov, M., Hirsch, E., Dantsin, E., Ivanov, S.: Algorithms for SAT and upper bounds on their complexity. Journal of Mathematical Sciences **118**(2), 4948–4962 (2003)
36. Woeginger, G.: Exact algorithms for NP-hard problems: a survey. In: Juenger, M., Reinelt, G., Rinaldi, G. (eds.) Combinatorial Optimization - Eureka! You Shrink!. LNCS, vol. 2570, pp. 185–207. Springer, Heidelberg (2000)

Generalized Totalizer Encoding
for Pseudo-Boolean Constraints

Saurabh Joshi[1]([✉]), Ruben Martins[1], and Vasco Manquinho[2]

[1] Department of Computer Science, University of Oxford, Oxford, UK
{saurabh.joshi,ruben.martins}@cs.ox.ac.uk
[2] INESC-ID/Instituto Superior Técnico, Universidade de Lisboa, Lisboa, Portugal
vasco.manquinho@inesc-id.pt

Abstract. Pseudo-Boolean constraints, also known as 0-1 Integer Linear Constraints, are used to model many real-world problems. A common approach to solve these constraints is to encode them into a SAT formula. The runtime of the SAT solver on such formula is sensitive to the manner in which the given pseudo-Boolean constraints are encoded. In this paper, we propose generalized Totalizer encoding (GTE), which is an arc-consistency preserving extension of the Totalizer encoding to pseudo-Boolean constraints. Unlike some other encodings, the number of auxiliary variables required for GTE does not depend on the magnitudes of the coefficients. Instead, it depends on the number of distinct combinations of these coefficients. We show the superiority of GTE with respect to other encodings when large pseudo-Boolean constraints have low number of distinct coefficients. Our experimental results also show that GTE remains competitive even when the pseudo-Boolean constraints do not have this characteristic.

1 Introduction

Pseudo-Boolean constraints (PBCs) or 0-1 Integer Linear constraints have been used to model a plethora of real world problems such as computational biology [13,24], upgradeability problems [3,15,16], resource allocation [27], scheduling [26] and automated test pattern generation [22]. Due to its importance and a plethora of applications, a lot of research has been done to efficiently solve PBCs. One of the popular approaches is to convert PBCs into a SAT formula [7,11,21] thus making them amenable to off-the-shelf SAT solvers. We start by formally introducing PBC, followed by a discussion on how to convert a PBC into a SAT formula.

A PBC is defined over a finite set of Boolean variables $x_1, \ldots, x_n$ which can be assigned a value 0 (*false*) or 1 (*true*). A literal l_i is either a Boolean variable x_i (positive literal) or its negation $\neg x_i$ (negative literal). A positive (resp. negative) literal l_i is said to be assigned 1 if and only if the corresponding variable x_i is assigned 1 (resp. 0). Without a loss of generality, PBC can be defined as a linear inequality of the following normal form:

© Springer International Publishing Switzerland 2015
G. Pesant (Ed.): CP 2015, LNCS 9255, pp. 200–209, 2015.
DOI: 10.1007/978-3-319-23219-5_15

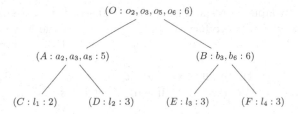

Fig. 1. Generalized Totalizer Encoding for $2l_1 + 3l_2 + 3l_3 + 3l_4 \leq 5$

$$\sum w_i l_i \leq k \qquad (1)$$

Here, $w_i \in \mathbb{N}^+$ are called coefficients or weights, l_i are input literals and $k \in \mathbb{N}^+$ is called the bound. Linear inequalities in other forms (e.g. other inequality, equalities or negative coefficients) can be converted into this normal form in linear time [8]. *Cardinality constraint* is a special case of PBC when all the weights have the value 1. Many different encodings have been proposed to encode cardinality constraints[4,5,25,28]. Linear pseudo-Boolean solving (PBS) is a generalization of the SAT formulation where constraints are not restricted to clauses and can be PBCs. A related problem to PBS is the linear pseudo-Boolean optimization (PBO) problem, where all the constraints must be satisfied and the value of a linear cost function is optimized. PBO usually requires an iterative algorithm which solves a PBS in every iteration [11,18,19,21]. Considering that the focus of the paper is on encodings rather than algorithms, we restrict ourselves to the decision problem (PBS).

This paper makes the following contributions.

- We propose an arc-consistency [12] preserving extension of Totalizer encoding [5] called Generalized Totalizer encoding (GTE) in Section 2.
- We compare various PBC encoding schemes that were implemented in a common framework, thus providing a fair comparison. After discussing related work in Section 3, we show GTE as a promising encoding through its competitive performance in Section 4.

2   Generalized Totalizer Encoding

The Totalizer encoding [5] is an encoding to convert cardinality constraints into a SAT formula. In this section, the generalized Totalizer encoding (GTE) to encode PBC into SAT is presented. GTE can be better visualized as a binary tree, as shown in Fig. 1. With the exception of the leaves, every node is represented as (*node_name* : *node_vars* : *node_sum*). The *node_sum* for every node represents the maximum possible weighted sum of the subtree rooted at that node. For any node A, a node variable a_w represents a weighted sum w of the underlying subtree. In other words, whenever the weighted sum of some of the input literals in the subtree becomes w, a_w must be set to 1. Note that for any node A, we would need one variable corresponding to every distinct weighted sum that the input literals under A can produce. Input literals are at the leaves, represented as $(node\_name : literal\_name : literal\_weight)$ with each of the terms being self explanatory.

For any node P with children Q and R, to ensure that weighted sum is propagated from Q and R to P, the following formula is built for P:

$$\left(\bigwedge_{\substack{q_{w_1} \in Q.node\_vars \\ r_{w_2} \in R.node\_vars \\ w_3 = w_1 + w_2 \\ p_{w_3} \in P.node\_vars}} (\neg q_{w_1} \vee \neg r_{w_2} \vee p_{w_3}) \right) \wedge \left(\bigwedge_{\substack{s_w \in (Q.node\_vars \cup R.node\_vars) \\ w = w' \\ p_{w'} \in P.node\_vars}} (\neg s_w \vee p_{w'}) \right) \quad (2)$$

The left part of Eqn. (2) ensures that, if node Q has witnessed a weighted sum of w_1 and R has witnessed a weighted sum of w_2, then P must be considered to have witnessed the weighted sum of $w_3 = w_1 + w_2$. The right part of Eqn. (2) just takes care of the boundary condition where weighted sums from Q and R are propagated to P without combining it with their siblings. This represents that Q (resp. R) has witnessed a weighted sum of w but R (resp. Q) may not have witnessed any positive weighted sum.

Note that node O in Fig. 1 does not have variables for the weighted sums larger than 6. Once the weighted sum goes above the threshold of k, we represent it with $k+1$. Since all the weighted sums above k would result in the constraint being not satisfied, it is sound to represent all such sums as $k+1$. This is in some sense a generalization of k-simplification described in [9,17]. For k-simplification, w_3 in Eqn. (2) would change to $w_3 = min(w_1 + w_2, k + 1)$.

Finally, to enforce that the weighted sum does not exceed the given threshold k, we add the following constraint at the root node O :

$$\neg o_{k+1} \quad (3)$$

Encoding Properties: Let A_{I_w} represent the multiset of weights of all the input literals in the subtree rooted at node A. For any given multiset S of weights, let $Weight(S) = \sum_{e \in S} e$. For a given multiset S, let $unique(S)$ denote the set with all the multiplicity removed from S. Let $|S|$ denote the cardinality of the set S. Hence, the total number of node variables required at node A is:

$$|unique\left(\{Weight(S)|S \subseteq A_{I_w} \wedge S \neq \emptyset\}\right)| \quad (4)$$

Note that unlike some other encodings [7,14] the number of auxiliary variables required for GTE does not depend on the magnitudes of the weights. Instead, it depends on how many unique weighted sums can be generated. Thus, we claim that for pseudo-Boolean constraints where the distinct weighted sum combinations are low, GTE should perform better. We corroborate our claim in Section 4 through experiments.

Nevertheless, in the worst case, GTE can generate exponentially many auxiliary variables and clauses. For example, if the weights of input literals $l_1, \ldots, l_n$

are respectively $2^0, \ldots, 2^{n-1}$, then every possible weighted sum combination would be unique. In this case, GTE would generate exponentially many auxiliary variables. Since every variable is used in at least one clause, it will also generate exponentially many clauses.

Though GTE does not depend on the magnitudes of the weights, one can use the magnitude of the largest weight to categorize a class of PBCs for which GTE is guaranteed to be of polynomial size. If there are n input literals and the largest weight is a polynomial $P(n)$, then GTE is guaranteed to produce a polynomial size formula. If the largest weight is $P(n)$, then the total number of distinct weight combinations (Eqn. (4)) is bounded by $nP(n)$, resulting in a polynomial size formula.

The best case for GTE occurs when all of the weights are equal, in which case the number of auxiliary variables and clauses is, respectively, $\mathcal{O}(n\,log_2n)$ and $\mathcal{O}(n^2)$. Notice that for this best case with k-simplification, we have $\mathcal{O}(nk)$ variables and clauses, since it will behave exactly as the Totalizer encoding [5].

Note also that the generalized arc consistency (GAC) [12] property of Totalizer encoding holds for GTE as well. GAC is a property of an encoding which allows the solver to infer maximal possible information through propagation, thus helping the solver to prune the search space earlier. The original proof [5] makes an inductive argument using the left subtree and the right subtree of a node. It makes use of the fact that, if there are q input variables set to 1 in the left child Q and r input variables are set to 1 in the right child R, then the encoding ensures that in the parent node P, the variable p_{q+r} is set to 1. Similarly, GTE ensures that if the left child Q contributes w_1 to the weighted sum (q_{w_1} is set to 1) and the right child R contributes w_2 to the weighted sum (r_{w_2} is set to 1), then the parent node P registers the weighted sum to be at least $w_3 = w_2 + w_1$ (p_{w_3} is set to 1). Hence, the GAC proof still holds for GTE.

3   Related Work

The idea of encoding a PBC into a SAT formula is not new. One of the first such encoding is described in [11,30] which uses binary adder circuit like formulation to compute the weighted sum and then compare it against the threshold k. This encoding creates $\mathcal{O}(n\,log_2k)$ auxiliary clauses, but it is not arc-consistent. Another approach to encode PBCs into SAT is to use sorting networks [11]. This encoding produces $\mathcal{O}(N\,log_2^2N)$ auxiliary clauses, where N is bounded by $\lceil log_2w_1 \rceil + \ldots + \lceil log_2w_n \rceil$. This encoding is also not arc-consistent for PBCs, but it preserves more implications than the adder encoding, and it maintains GAC for cardinality constraints.

The Watchdog encoding [7] scheme uses the Totalizer encoding, but in a completely different manner than GTE. It uses multiple Totalizers, one for each bit of the binary representation of the weights. The Watchdog encoding was the first polynomial sized encoding that maintains GAC for PBCs and it only generates $\mathcal{O}(n^3log_2n\,log_2w_{max})$ auxiliary clauses. Recently, the Watchdog encoding has been generalized to a more abstract framework with the Binary Merger encoding [20]. Using a different translation of the components of the Watchdog encoding allows the Binary Merger encoding to further reduce the number of auxiliary clauses to $\mathcal{O}(n^2 log_2^2 n \, log_2 w_{max})$. The Binary Merger is also polynomial and maintains GAC.

Other encodings that maintain GAC can be exponential in the worst case scenario, such as BDD based encodings [1,6,11]. These encodings share quite a lot of similarity to GTE, such as GAC and independence from the magnitude of the weight. One of the differences is that GTE always has a tree like structure amongst auxiliary variables and input literals. However, the crucial difference lies in the manner in which auxiliary variables are generated, and what they represent. In BDD based approaches, an auxiliary variable D_i attempts to reason about the weighted sum of the input literals either $l_i, \ldots, l_n$ or $l_1, \ldots, l_i$. On the other hand, an auxiliary variable a_w at a node A in GTE attempts to only reason about the weighted sum of the input literals that are descendants of A. Therefore, two auxiliary variables in two disjoint subtrees in GTE are guaranteed to reason about disjoint sets of input literals. We believe that such a localized reasoning could be a cause of relatively better performance of GTE as reported in Section 4. It is worth noting that the worst case scenario for GTE, when weights are of the form a^i, where $a \geq 2$, would generate a polynomial size formula for BDD based approaches [1,6,11].

As GTE generalizes the Totalizer encoding, the Sequential Weighted Counter (SWC) encoding [14] generalizes sequential encoding [28] for PBCs. Like BDD based approaches and GTE, SWC can be exponential in the worst case.

4   Implementation and Evaluation

All experiments were performed on two AMD 6276 processors (2.3 GHz) running Fedora 18 with a timeout of 1,800 seconds and a memory limit of 16 GB. Similar resource limitations were used during the last pseudo-Boolean (PB) evaluation of 2012[1]. For a fair comparison, we implemented GTE (gte) in the PBLIB [29] (version 1.2) open source library which contains a plethora of encodings, namely, Adder Networks (adder) [11,30], Sorting Networks (sorter) [11], watchdog (watchdog) [7], Binary Merger (bin-merger) [20], Sequential Weighted Counter (swc) [14], and BDDs (bdd) [1]. A new encoding in PBLIB can be added by implementing encode method of the base class Encoder. Thus, all the encodings mentioned above, including GTE, only differ in how encode is implemented while they share the rest of the whole environment. PBLIB provides parsing and normalization [11] routines for PBC and uses MINISAT 2.2.0 [10] as a backend SAT solver. When the constraint to be encoded into CNF is a cardinality constraint, we use the default setting of PBLIB that dynamically selects a cardinality encoding based on the number of auxiliary clauses. When the constraint to be encoded into CNF is a PBC, we specify one of the above encodings.

Benchmarks: Out of all 355 instances from the DEC-SMALLINT-LIN category in the last PB evaluation of 2012 (PB'12), we only considered those 214

---
[1] http://www.cril.univ-artois.fr/PB12/

Table 1. Characteristics of pseudo-Boolean benchmarks

Benchmark	#PB	#lits	k	max w_i	$\sum w_i$	#diff w_i
PB'12	164.31	32.25	27.94	12.55	167.14	6.72
pedigree	1.00	10,794.13	11,106.69	456.28	4,665,237.38	2.00

Table 2. Number of solved instances

Benchmark	Result	sorter	swc	adder	watchdog	bin-merger	bdd	gte
PB'12	SAT	72	74	73	79	79	**81**	**81**
(214)	UNSAT	74	77	83	**85**	85	84	84
pedigree	SAT	2	7	6	25	43	82	**83**
(172)	UNSAT	0	7	6	23	35	72	**75**
Total	SAT/UNSAT	146	165	172	212	242	319	**323**

instances[2] that contain at least 1 PBC. We also consider an additional set of pedigree benchmarks from computational biology [13]. These benchmarks were originally encoded in Maximum Satisfiability (MaxSAT) and were used in the last MaxSAT Evaluation of 2014[3]. Any MaxSAT problem can be converted to a corresponding equivalent pseudo-Boolean problem [2]. We generate two pseudo-Boolean decision problems (one satisfiable, another unsatisfiable) from the optimization version of each of these benchmarks. The optimization function is transformed into a PBC with the value of the bound k set to a specific value. Let the optimum value for the optimization function be k_{opt}. The satisfiable decision problem uses k_{opt} as the value for the bound k, whereas the unsatisfiable decision problem uses $k_{opt} - 1$ as the value for the bound k. Out of 200 generated instances[4], 172 had at least 1 PBC and were selected for further evaluation.

Tab. 1 shows the characteristics of the benchmarks used in this evaluation. #PB denotes the average number of PBCs per instance. #lits, k, max w_i, $\sum w_i$ and #diff w_i denote the per constraint per instance average of input literals, bound, the largest weight, maximum possible weighted sum and the number of distinct weights. PB'12 benchmarks are a mix of crafted as well as industrial benchmarks, whereas all of the pedigree benchmarks are from the same biological problem [13]. The PB'12 benchmarks have on average several PBCs, however, they are relatively small in magnitude. In contrast, the pedigree benchmarks contain one large PB constraint with very large total weighted sum. pedigree benchmarks have only two distinct values of weights, thus making them good candidates for using GTE.

Results: Tab. 2 shows the number of instances solved using different encodings. sorter, adder and swc perform worse than the remaining encodings for both sets of benchmarks. The first two are not arc-consistent therefore the SAT solver is not

---

[2] Available at http://sat.inesc-id.pt/~ruben/benchmarks/pb12-subset.zip
[3] http://www.maxsat.udl.cat/14/
[4] Available at http://sat.inesc-id.pt/~ruben/benchmarks/pedigrees.zip

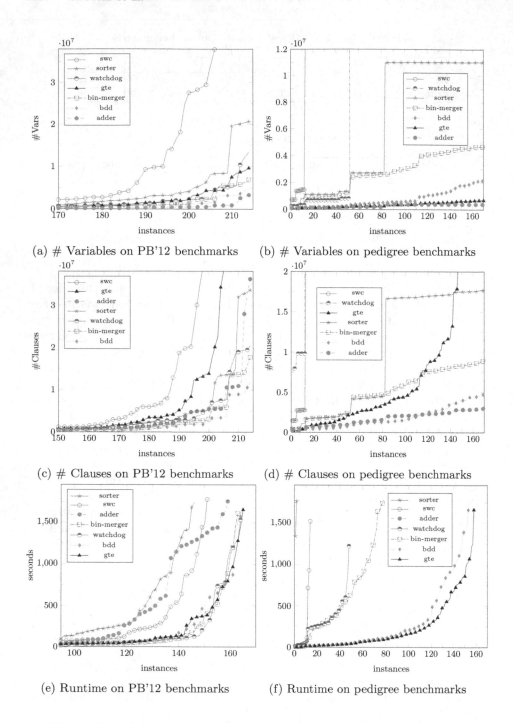

(a) # Variables on PB'12 benchmarks

(b) # Variables on pedigree benchmarks

(c) # Clauses on PB'12 benchmarks

(d) # Clauses on pedigree benchmarks

(e) Runtime on PB'12 benchmarks

(f) Runtime on pedigree benchmarks

Fig. 2. Cactus plots of number of variables, number of clauses and runtimes

able to infer as much information as with arc-consistent encodings. swc, though arc-consistent, generates a large number of auxiliary variables and clauses, which deteriorates the performance of the SAT solver.

gte provides a competitive performance to bdd, bin-merger and watchdog for PB'12. However, only the gte and bdd encodings are able to tackle pedigree benchmarks, which contain a large number of literals and only two different coefficients. Unlike other encodings, gte and bdd are able to exploit the characteristics of these benchmarks.

swc requires significantly large number of variables as the value of k increases, whereas bdd and gte keep the variable explosion in check due to reuse of variables on similar combinations (Figs. 2a and 2b). This reuse of auxiliary variables is even more evident on pedigree benchmarks (Fig. 2b) as these benchmarks have only two different coefficients resulting in low number of combinations. k-simplification also helps gte in keeping the number of variables low as all the combinations weighing more than $k + 1$ are mapped to $k + 1$.

Number of clauses required for gte is quite large as compared to some other encodings (Figs. 2c and 2d). gte requires clauses to be generated for all the combinations even though most of them produce the same value for the weighted sum, thus reusing the same variable. Though bdd has an exponential worst case, in practice it appears to generate smaller formulas (Figs. 2c and 2d).

Fig. 2e shows that gte provides a competitive performance with respect to bin-merger, watchdog and bdd. Runtime on pedigree benchmarks as shown in Fig. 2f establishes gte as the clear winner with bdd performing a close second. The properties that gte and bdd share help them perform better on pedigree benchmarks as they are not affected by large magnitude of weights in the PBCs.

5 Conclusion

Many real-world problems can be formulated using pseudo-Boolean constraints (PBC). Given the advances in SAT technology, it becomes crucial how to encode PBC into SAT, such that SAT solvers can efficiently solve the resulting formula.

In this paper, an arc-consistency preserving generalization of the Totalizer encoding is proposed for encoding PBC into SAT. Although the proposed encoding is exponential in the worst case, the new Generalized Totalizer encoding (GTE) is very competitive in relation with other PBC encodings. Moreover, experimental results show that when the number of different weights in PBC is small, it clearly outperforms all other encodings. As a result, we believe the impact of GTE can be extensive, since one can further extend it into incremental settings [23].

Acknowledgments. This work is partially supported by the ERC project 280053, FCT grants AMOS (CMUP-EPB/TIC/0049/2013), POLARIS (PTDC/EIA-CCO/123051/2010), and INESC-ID's multiannual PIDDAC UID/CEC/50021/2013.

References

1. Abío, I., Nieuwenhuis, R., Oliveras, A., Rodríguez-Carbonell, E., Mayer-Eichberger, V.: A New Look at BDDs for Pseudo-Boolean Constraints. Journal of Artificial Intelligence Research **45**, 443–480 (2012)
2. Aloul, F., Ramani, A., Markov, I., Sakallah, K.: Generic ILP versus specialized 0–1 ILP: an update. In: International Conference on Computer-Aided Design, pp. 450–457. IEEE Press (2002)
3. Argelich, J., Berre, D.L., Lynce, I., Marques-Silva, J., Rapicault, P.: Solving linux upgradeability problems using boolean optimization. In: Workshop on Logics for Component Configuration, pp. 11–22 (2010)
4. Asín, R., Nieuwenhuis, R., Oliveras, A., Rodríguez-Carbonell, E.: Cardinality Networks: a theoretical and empirical study. Constraints **16**(2), 195–221 (2011)
5. Bailleux, O., Boufkhad, Y.: Efficient CNF encoding of boolean cardinality constraints. In: Rossi, F. (ed.) CP 2003. LNCS, vol. 2833, pp. 108–122. Springer, Heidelberg (2003)
6. Bailleux, O., Boufkhad, Y., Roussel, O.: A Translation of Pseudo Boolean Constraints to SAT. Journal on Satisfiability, Boolean Modeling and Computation **2**(1–4), 191–200 (2006)
7. Bailleux, O., Boufkhad, Y., Roussel, O.: New encodings of pseudo-boolean constraints into CNF. In: Kullmann, O. (ed.) SAT 2009. LNCS, vol. 5584, pp. 181–194. Springer, Heidelberg (2009)
8. Barth, P.: Logic-based 0–1 Constraint Programming. Kluwer Academic Publishers (1996)
9. Büttner, M., Rintanen, J.: Satisfiability planning with constraints on the number of actions. In: International Conference on Automated Planning and Scheduling, pp. 292–299. AAAI Press (2005)
10. Eén, N., Sörensson, N.: An extensible SAT-solver. In: Giunchiglia, E., Tacchella, A. (eds.) SAT 2003. LNCS, vol. 2919, pp. 502–518. Springer, Heidelberg (2004)
11. Eén, N., Sörensson, N.: Translating Pseudo-Boolean Constraints into SAT. Journal on Satisfiability, Boolean Modeling and Computation **2**(1–4), 1–26 (2006)
12. Gent, I.P.: Arc consistency in SAT. In: European Conference on Artificial Intelligence, pp. 121–125. IOS Press (2002)
13. Graça, A., Lynce, I., Marques-Silva, J., Oliveira, A.L.: Efficient and accurate haplotype inference by combining parsimony and pedigree information. In: Horimoto, K., Nakatsui, M., Popov, N. (eds.) ANB 2010. LNCS, vol. 6479, pp. 38–56. Springer, Heidelberg (2012)
14. Hölldobler, S., Manthey, N., Steinke, P.: A compact encoding of pseudo-boolean constraints into SAT. In: Glimm, B., Krüger, A. (eds.) KI 2012. LNCS, vol. 7526, pp. 107–118. Springer, Heidelberg (2012)
15. Ignatiev, A., Janota, M., Marques-Silva, J.: Towards efficient optimization in package management systems. In: International Conference on Software Engineering, pp. 745–755. ACM (2014)
16. Janota, M., Lynce, I., Manquinho, V.M., Marques-Silva, J.: PackUp: Tools for Package Upgradability Solving. JSAT **8**(1/2), 89–94 (2012)
17. Koshimura, M., Zhang, T., Fujita, H., Hasegawa, R.: QMaxSAT: A Partial MaxSAT Solver. Journal on Satisfiability, Boolean Modeling and Computation **8**, 95–100 (2012)
18. Le Berre, D., Parrain, A.: The Sat4j library, release 2.2. Journal on Satisfiability, Boolean Modeling and Computation **7**(2–3), 59–64 (2010)

19. Manquinho, V., Marques-Silva, J., Planes, J.: Algorithms for weighted boolean optimization. In: Kullmann, O. (ed.) SAT 2009. LNCS, vol. 5584, pp. 495–508. Springer, Heidelberg (2009)
20. Manthey, N., Philipp, T., Steinke, P.: A more compact translation of pseudo-boolean constraints into cnf such that generalized arc consistency is maintained. In: Lutz, C., Thielscher, M. (eds.) KI 2014. LNCS, vol. 8736, pp. 123–134. Springer, Heidelberg (2014)
21. Manthey, N., Steinke, P.: npSolver - A SAT based solver for optimization problems. In: Pragmatics of SAT (2012)
22. Marques-Silva, J.: Integer programming models for optimization problems in test generation. In: Asia and South Pacific Design Automation, pp. 481–487. IEEE Press (1998)
23. Martins, R., Joshi, S., Manquinho, V., Lynce, I.: Incremental cardinality constraints for MaxSAT. In: O'Sullivan, B. (ed.) CP 2014. LNCS, vol. 8656, pp. 531–548. Springer, Heidelberg (2014)
24. Miranda, M., Lynce, I., Manquinho, V.: Inferring phylogenetic trees using pseudo-Boolean optimization. AI Communications 27(3), 229–243 (2014)
25. Ogawa, T., Liu, Y., Hasegawa, R., Koshimura, M., Fujita, H.: Modulo based CNF encoding of cardinality constraints and its application to MaxSAT solvers. In: International Conference on Tools with Artificial Intelligence, pp. 9–17. IEEE Press (2013)
26. Prestwich, S., Quirke, C.: Boolean and pseudo-boolean models for scheduling. In: International Workshop on Modelling and Reformulating Constraint Satisfaction Problems (2003)
27. Ribas, B.C., Suguimoto, R.M., Montaño, R.A.N.R., Silva, F., de Bona, L., Castilho, M.A.: On modelling virtual machine consolidation to pseudo-boolean constraints. In: Pavón, J., Duque-Méndez, N.D., Fuentes-Fernández, R. (eds.) IBERAMIA 2012. LNCS, vol. 7637, pp. 361–370. Springer, Heidelberg (2012)
28. Sinz, C.: Towards an optimal CNF encoding of boolean cardinality constraints. In: van Beek, P. (ed.) CP 2005. LNCS, vol. 3709, pp. 827–831. Springer, Heidelberg (2005)
29. Steinke, P., Manthey, N.: PBLib-A C++ Toolkit for Encoding Pseudo-Boolean Constraints into CNF. Tech. rep., Technische Universität Dresden (2014). http://tools.computational-logic.org/content/pblib.php
30. Warners, J.: A Linear-Time Transformation of Linear Inequalities into Conjunctive Normal Form. Information Processing Letters 68(2), 63–69 (1998)

Smaller Selection Networks for Cardinality Constraints Encoding

Michał Karpiński$^{(\boxtimes)}$ and Marek Piotrów

Institute of Computer Science, University of Wrocław,
Joliot-Curie 15, 50-383 WrocłAw, Poland
{karp,mpi}@cs.uni.wroc.pl

Abstract. Selection comparator networks have been studied for many years. Recently, they have been successfully applied to encode cardinality constraints for SAT-solvers. To decrease the size of generated formula there is a need for constructions of selection networks that can be efficiently generated and produce networks of small sizes for the practical range of their two parameters: n – the number of inputs (Boolean variables) and k – the number of selected items (a cardinality bound). In this paper we give and analyze a new construction of smaller selection networks that are based on the pairwise selection networks introduced by Codish and Zazon-Ivry. We prove also that standard encodings of cardinality constraints with selection networks preserve arc-consistency.

1 Introduction

Comparator networks are probably the simplest data-oblivious model for sorting-related algorithms. The most popular construction is due to Batcher [4] and it's called *odd-even* sorting network. For all practical values, this is the best known sorting network. However, in 1992 Parberry [10] introduced the serious competitor to Batcher's construction, called *pairwise* sorting network. In context of sorting, pairwise network is not better than odd-even network, in fact it has been proven that they have exactly the same size and depth. As Parberry said himself: *"It is the first sorting network to be competitive with the odd-even sort for all values of n"*. There is a more sophisticated relation between both types of network and their close resemblance. For overview of sorting networks, see Knuth [8] or Parberry [9].

In recent years new applications for sorting networks have been found, for example in encoding of *pseudo Boolean constraints* and *cardinality constraints* for SAT-solvers. Cardinality constraints take the form $x_1 + x_2 + \ldots + x_n \sim k$, where $x_1, x_2, \ldots, x_n$ are Boolean variables, k is a natural number, and $\sim$ is a relation from the set $\{=, <, \leq, >, \geq\}$. Cardinality constraints are used in many applications, the significant ones worth mentioning arise in SAT-solvers. Using cardinality constraints with cooperation of SAT-solvers we can handle many practical problems, for example, cumulative scheduling [11] and timetabling [3], that are proven to be hard. Works of Asín *et al.* [1,2] describe how to use odd-even sorting network to encode cardinality constraints into Boolean formulas. In [7] authors do the same with pseudo Boolean constraints.

© Springer International Publishing Switzerland 2015
G. Pesant (Ed.): CP 2015, LNCS 9255, pp. 210–225, 2015.
DOI: 10.1007/978-3-319-23219-5_16

It has already been observed that using selection networks instead of sorting networks is more efficient for the encoding of cardinality constraints. Codish and Zazon-Ivry [6] introduced pairwise cardinality networks, which are networks derived from pairwise sorting networks that express cardinality constraints. Two years later, same authors [5] reformulated the definition of pairwise selection networks and proved that their sizes are never worse than the sizes of corresponding odd-even selection networks. To show the difference they plotted it for selected values of n and k.

In this paper we give a new construction of smaller selection networks that are based on the pairwise selection ones and we prove that the construction is correct. We also estimate the size of our networks and compute the difference in sizes between our selection networks and the corresponding pairwise ones. The difference can be as big as $n \log n / 2$ for $k = n/2$. Finally, we analyze the standard 3(6)-clause encoding of a comparator and prove that such CNF encoding of any selection network preserves arc-consistency with respect to a corresponding cardinality constraint.

The rest of the paper is organized in the following way: in Section 2 we give definitions and notations used in this paper. In Section 3 we recall the definition of pairwise selection networks and define auxiliary bitonic selection networks that we will use to estimate the sizes of our networks. In Section 4 we present the construction of our selection networks and prove its correctness. In Section 5 we analyze the sizes of the networks and, finally, in Section 6 we examine the arc-consistency of selection networks.

2   Preliminaries

In this section we will introduce definitions and notations used in the rest of the paper.

Definition 1 (Input Sequence). *Input sequence of length n is a sequence of natural numbers $\bar{x} = \langle x_1, \ldots, x_n \rangle$, where $x_i \in \mathbb{N}$ (for all $i = 1..n$). We say that $\bar{x} \in \mathbb{N}^n$ is sorted if $x_i \geq x_{i+1}$ (for each $i = 1..n-1$). Given $\bar{x} = \langle x_1, \ldots, x_n \rangle$, $\bar{y} = \langle y_1, \ldots, y_n \rangle$ we define concatenation as $\bar{x} :: \bar{y} = \langle x_1, \ldots, x_n, y_1, \ldots, y_n \rangle$. We will use the following functions from $\mathbb{N}^n$ to $\mathbb{N}^{n/2}$:*

$$left(\bar{x}) = \langle x_1, \ldots, x_{n/2} \rangle, \qquad right(\bar{x}) = \langle x_{n/2+1}, \ldots, x_n \rangle$$

Let $n, m \in \mathbb{N}$. We define a relation $\succeq'$ on $\mathbb{N}^n \times \mathbb{N}^m$. Let $\bar{x} = \langle x_1, \ldots, x_n \rangle$ and $\bar{y} = \langle y_1, \ldots, y_m \rangle$, then:

$$\bar{x} \succeq \bar{y} \iff \forall_{i \in \{1, \ldots, n\}} \forall_{j \in \{1, \ldots, m\}} \; x_i \geq y_j$$

Definition 2 (Comparator). *Let $\bar{x} \in \mathbb{N}^n$ and let $i, j \in \mathbb{N}$, where $1 \leq i < j \leq n$. A comparator is a function $c_{i,j}$ defined as:*

$$c_{i,j}(\bar{x}) = \bar{y} \iff y_i = \max\{x_i, x_j\} \wedge y_j = \min\{x_i, x_j\} \wedge \forall_{k \neq i,j} \; x_k = y_k$$

Definition 3 (Comparator Network). *We say that $f^n : \mathbb{N}^n \to \mathbb{N}^n$ is a comparator network of order n, if it can be represented as the composition of finite number of comparators, namely, $f^n = c_{i_1,j_1} \circ \cdots \circ c_{i_k,j_k}$. The size of comparator network (number of comparators) is denoted by $|f^n|$. Comparator network of size 0 is denoted by id^n.*

Definition 4 (V-shaped and Bitonic Sequences). *A sequence $\bar{x} \in \mathbb{N}^n$ is called v-shaped if $x_1 \geq \ldots \geq x_i \leq \ldots \leq x_n$ for some i, where $1 \leq i \leq n$. A v-shaped sequence or its circular shift is traditionally called bitonic.*

Definition 5 (Sorting Network). *A comparator network f^n is a sorting network, if for each $\bar{x} \in \mathbb{N}^n$, $f^n(\bar{x})$ is sorted.*

Two types of sorting networks are of interest to us: *odd-even* and *pairwise*. Based on their ideas, Knuth [8] (for odd-even network) and Codish and Zazon-Ivry [5,6] (for pairwise network) showed how to transform them into selection networks (we name them $oe\_sel_k^n$ and $pw\_sel_k^n$ respectively).

Definition 6 (Top k Sorted Sequence). *A sequence $\bar{x} \in \mathbb{N}^n$ is top k sorted, with $k \leq n$, if $\langle x_1, \ldots, x_k \rangle$ is sorted and $\langle x_1, \ldots, x_k \rangle \succeq \langle x_{k+1}, \ldots, x_n \rangle$.*

Definition 7 (Selection Network). *A comparator network f_k^n (where $k \leq n$) is a selection network, if for each $\bar{x} \in \mathbb{N}^n$, $f_k^n(\bar{x})$ is top k sorted.*

To simplify the presentation we assume that n and k are powers of 2.

A clause is a disjunction of literals (Boolean variables x or their negation $\neg x$). A CNF formula is a conjunction of one or more clauses.

A unit propagation (UP) is a process, that for given CNF formula, clauses are sought in which all literals but one are false (say l) and l is undefined (initially only clauses of size one satisfy this condition). This literal l is set to true and the process is iterated until reaching a fix point.

Cardinality constraints are of the form $x_1 + \ldots + x_n \sim k$, where $k \in \mathbb{N}$ and $\sim$ belongs to $\{<, \leq, =, \geq, >\}$. We will focus on cardinality constraints with less-than relation, i.e. $x_1 + \ldots + x_n < k$. An encoding (a CNF formula) of such constraint preserves arc-consistency, if as soon as $k - 1$ variables among the x_i's become true, the unit propagation sets all other x_i's to false.

In [7] authors are using sorting networks for an encoding of cardinality constraints, where inputs and outputs of a comparator are Boolean variables and comparators are encoded as a CNF formula. In addition, the k-th greatest output variable y_k of the network is forced to be 0 by adding $\neg y_k$ as a clause to the formula that encodes $x_1 + \ldots + x_n < k$. They showed that the encoding preserves arc-consistency.

A single comparator can be translated to a CNF formula in the following way: let a and b be variables denoting upper and lower inputs of the comparator, and c and d be variables denoting upper and lower outputs of a comparator, then:

$$fcomp(a, b, c, d) \Leftrightarrow (c \Leftrightarrow a \vee b) \wedge (d \Leftrightarrow a \wedge b)$$

is the *full encoding* of a comparator. Notice that it consists of 6 clauses. Let f be a comparator network. Full encoding ϕ of f is a conjunction of full encoding of every comparator of f.

In [2] authors observe that in case of $\sim$ being $<$ or $\leq$, it is sufficient to use only 3 clauses for a single comparator, namely:

$$hcomp(a,b,c,d) \Leftrightarrow \underbrace{(a \Rightarrow c)}_{(c1)} \wedge \underbrace{(b \Rightarrow c)}_{(c2)} \wedge \underbrace{(a \wedge b \Rightarrow d)}_{(c3)} \tag{1}$$

We call it: *a half encoding*. In [2] it is used to translate an odd-even sorting network to an encoding that preserves arc-consistency. We show a more general result (with respect to both [7] and [2]), that the half encoding of any selection network preserves arc-consistency for the "$<$" and "$\leq$" relations. Similar results can be proved for the "$=$" relation using the full encoding of comparators and for the "$>$" or "$\geq$" relations using an encoding symmetric to $hcomp(a,b,c,d)$, namely: $(d \Rightarrow a) \wedge (d \Rightarrow b) \wedge (c \Rightarrow a \vee b)$.

3   Pairwise and Bitonic Selection Networks

Now we present two constructions for selection networks. First, we recall the definition of pairwise selection networks by Codish and Zazon-Ivry [5,6]. Secondly, we give the auxiliary construction of a *bitonic* selection network $bit\_sel_k^n$, that we will use to estimate the sizes of our improved pairwise selection network in Section 5.

Definition 8 (Domination). $\bar{x} \in \mathbb{N}^n$ *dominates* $\bar{y} \in \mathbb{N}^n$ *if* $x_i \geq y_i$ *(for* $i = 1..n$*).*

Definition 9 (Splitter). *A comparator network* f^n *is a* splitter *if for any sequence* $\bar{x} \in \mathbb{N}^n$*, if* $\bar{y} = f^n(\bar{x})$*, then* $left(\bar{y})$ *dominates* $right(\bar{y})$*.*

Observation 1. *We can construct splitter* $split^n$ *by joining inputs* $\langle i, n/2 + i \rangle$*, for* $i = 1..n/2$*, with a comparator. The size of a splitter is* $|split^n| = n/2$*.*

Lemma 1. *If* $\bar{b} \in \mathbb{N}^n$ *is bitonic and* $\bar{y} = split^n(\bar{b})$*, then* $left(\bar{y})$ *and* $right(\bar{y})$ *are bitonic and* $left(\bar{y}) \succeq right(\bar{y})$*.*

Proof. See Appendix B of [4]. □

The construction of a pairwise selection network is presented in Network 1. Notice that since a splitter is used as the third step, in the recursive calls we need to select k top elements from the first half of $\bar{y}$, but only $k/2$ top elements from the second half. The reason is that $r_{k/2+1}$ cannot be one of the first k largest elements of $l :: \bar{r}$. First, $r_{k/2+1}$ is not greater than any one of $\langle r_1, \ldots, r_{k/2} \rangle$ (by the definition of top k sorted sequence), and second, $\langle l_1, \ldots, l_{k/2} \rangle$ dominates $\langle r_1, \ldots, r_{k/2} \rangle$, so $r_{k/2+1}$ is not greater than any one of $\langle l_1, \ldots, l_{k/2} \rangle$. Based on these arguments we can make the following observation:

Network 1. $pw\_sel_k^n$; see [5,6]

Input: any $\bar{x} \in \mathbb{N}^n$
1: **if** $k = 1$ **then return** $max^n(\bar{x})$
2: **if** $k = n$ **then return** $oe\_sort^n(\bar{x})$
3: $\bar{y} \leftarrow split(\bar{x})$
4: $\bar{l} \leftarrow pw\_sel_k^{n/2}(left(\bar{y}))$ **and** $\bar{r} \leftarrow pw\_sel_{k/2}^{n/2}(right(\bar{y}))$
5: **return** $pw\_merge_k^n(\bar{l} :: \bar{r})$

Observation 2. *If $\bar{l} \in \mathbb{N}^{n/2}$ is top k sorted, $\bar{r} \in \mathbb{N}^{n/2}$ is top $k/2$ sorted and $\langle l_1, \ldots, l_{k/2} \rangle$ dominates $\langle r_1, \ldots, r_{k/2} \rangle$, then k largest elements of $\bar{l} :: \bar{r}$ are in $\langle l_1, \ldots, l_k \rangle :: \langle r_1, \ldots, r_{k/2} \rangle$.*

The last step of Network 1 merges k top elements from $\bar{l}$ and $k/2$ top elements from $\bar{r}$ with so called *pairwise merger*. We will omit the construction of this merger, because it is not relevant to our work. We would only like to note, that its size is: $|pw\_merge_k^n| = k \log k - k + 1$. Construction of the merger as well as the detailed proof of correctness of network $pw\_sel_k^n$ can be found in Section 6 of [5].

Definition 10 (Bitonic Splitter). *A comparator network f^n is a bitonic splitter if for any two sorted sequences $\bar{x}, \bar{y} \in \mathbb{N}^{n/2}$, if $\bar{z} = f^n(\bar{x} :: \bar{y})$, then (1) $left(\bar{z}) \succeq right(\bar{z})$ and (2) $left(\bar{z})$ and $right(\bar{z})$ are bitonic.*

Observation 3. *We can construct bitonic splitter bit_split^n by joining inputs $\langle i, n - i + 1 \rangle$, for $i = 1..n/2$, with a comparator. The size of a bitonic splitter is $|bit\_split^n| = n/2$.*

We now present the procedure for construction of the bitonic selection network. We use the odd-even sorting network oe_sort and the network bit_merge (also by Batcher [4]) for sorting bitonic sequences as black-boxes. As a reminder: bit_merge^n consists of two steps, first we use $\bar{y} = split^n(\bar{x})$, then recursively compute $bit\_merge^{n/2}$ for $left(\bar{y})$ and $right(\bar{y})$ (base case, $n = 2$, consists of a single comparator). The size of this network is: $|bit\_merge^n| = n \log n/2$. A bitonic selection network $bit\_sel_k^n$ is constructed by the procedure Network 2.

Theorem 1. *A comparator network $bit\_sel_k^n$ constructed by the procedure Network 2 is a selection network.*

Proof. Let $\bar{x} \in \mathbb{N}^n$ be the input to $bit\_sel_k^n$. After step one we get sorted sequences $B_1, \ldots, B_l$, where $l = n/k$. Let l_m be the value of l after m iterations. Let $B_1^m, \ldots, B_{l_m}^m$ be the blocks after m iterations. We will prove by induction that:

$P(m)$: *if $B_1, \ldots, B_l$ are sorted contain k largest elements of $\bar{x}$, then after m-th iteration of the second step: $l_m = l/2^m$, $B_1^m, \ldots, B_{l_m}^m$ are sorted and contain k largest elements of $\bar{x}$.*

If $m = 0$, then $l_0 = l$, so $P(0)$ holds. We show that $\forall_{m \geq 0} (P(m) \Rightarrow P(m + 1))$. Consider $(m+1)$-th iteration of step two. By the induction hypothesis $l_m = l/2^m$,

---

Network 2. $bit\_sel_k^n$

Input: any $\bar{x} \in \mathbb{N}^n$
1: $l \leftarrow n/k$ **and** partition input $\bar{x}$ into l consecutive blocks, each of size k, then sort each block with oe_sort^k, obtaining sorted blocks $B_1, \ldots, B_l$
2: **while** $l > 1$ **do**
3:      Collect blocks into pairs $\langle B_1, B_2 \rangle, \ldots, \langle B_{l-1}, B_l \rangle$
4:      **for all** $i \in \{1, 3, \ldots, l-1\}$ **do** $\bar{y}_i \leftarrow bit\_split^{2k}(B_i :: B_{i+1})$
5:      **for all** $i \in \{1, 3, \ldots, l-1\}$ **do** $B'_{\lceil i/2 \rceil} \leftarrow bit\_merge^k(left(\bar{y}_i))$
6:      $l \leftarrow l/2$ **and** relabel B'_i to B_i, for $1 \le i \le l$

$B_1^m, \ldots, B_{l_m}^m$ are sorted and contain k largest elements of $\bar{x}$. We will show that $(m+1)$-th iteration does not remove any element from k largest elements of $\bar{x}$. To see this, notice that if $\bar{y}_i = bit\_split^{2k}(B_i^m :: B_{i+1}^m)$ (for $i \in \{1, 3, \ldots, l_m-1\}$), then $left(\bar{y}_i) \succeq right(\bar{y}_i)$ and that $left(\bar{y}_i)$ is bitonic (by Definition 10). Because of those two facts, $right(\bar{y}_i)$ is discarded and $left(\bar{y}_i)$ is sorted using bit_merge^k. After this, $l_{m+1} = l_m/2 = l/2^{m+1}$ and blocks $B_1^{m+1}, \ldots, B_{l_{m+1}}^{m+1}$ are sorted. Thus $P(m+1)$ is true.

Since $l = n/k$, then by $P(m)$ we see that the second step will terminate after $m = \log \frac{n}{k}$ iterations and that B_1 is sorted and contains k largest elements of $\bar{x}$. □

The size of bitonic selection network is:

$$|bit\_sel_k^n| = \frac{n}{k}|oe\_sort^k| + \left(\frac{n}{k} - 1\right)\left(|bit\_split^{2k}| + |bit\_merge^k|\right)$$

$$= \frac{1}{4}n \log^2 k + \frac{1}{4}n \log k + 2n - \frac{1}{2}k \log k - k - \frac{n}{k} \quad (2)$$

4   New Smaller Selection Networks

As mentioned in the previous section, only the first $k/2$ elements from the second half of the input are relevant when we get to the merging step in $pw\_sel_k^n$. We will exploit this fact to create a new, smaller merger. We will use the concept of bitonic sequences, therefore the new merger will be called $pw\_bit\_merge_k^n$ and the new selection network: $pw\_bit\_sel_k^n$. The network $pw\_bit\_sel_k^n$ is generated by substituting the last step of $pw\_sel_k^n$ with $pw\_bit\_merge_k^n$. The new merger is constructed by the procedure Network 3.

Theorem 2. *The output of Network 3 consists of sorted k largest elements from input $\bar{l} :: \bar{r}$, assuming that $\bar{l} \in \mathbb{N}^{n/2}$ is top k sorted and $\bar{r} \in \mathbb{N}^{n/2}$ is top $k/2$ sorted and $\langle l_1, \ldots, l_{k/2} \rangle$ dominates $\langle r_1, \ldots, r_{k/2} \rangle$.*

Proof. We have to prove two things: (1) $\bar{b}$ is bitonic and (2) $\bar{b}$ consists of k largest elements from $\bar{l} :: \bar{r}$.

Network 3. $pw\_bit\_merge_k^n$

Input: $\bar{l} :: \bar{r}$, where $\bar{l} \in \mathbb{N}^{n/2}$ is top k sorted and $\bar{r} \in \mathbb{N}^{n/2}$ is top $k/2$ sorted and $\langle l_1, \ldots, l_{k/2} \rangle$ dominates $\langle r_1, \ldots, r_{k/2} \rangle$
1: $\bar{y} \leftarrow bit\_split^k(l_{k/2+1}, \ldots, l_k, r_1, \ldots, r_{k/2})$ **and** $\bar{b} \leftarrow \langle l_1, \ldots, l_{k/2} \rangle :: \langle y_1, \ldots, y_{k/2} \rangle$
2: **return** $bit\_merge^k(\bar{b})$

---

(1) Let j be the last index in the sequence $\langle k/2 + 1, \ldots, k \rangle$, for which $l_j > r_{k-j+1}$. If such j does not exist, then $\langle y_1, \ldots, y_{k/2} \rangle$ is nondecreasing, hence $\bar{b}$ is bitonic (nondecreasing). Assume that j exists, then $\langle y_{j-k/2+1}, \ldots, y_{k/2} \rangle$ is nondecreasing and $\langle y_1, \ldots, y_{k-j} \rangle$ is nonincreasing. Adding the fact that $l_{k/2} \geq l_{k/2+1} = y_1$ proves, that $\bar{b}$ is bitonic (v-shaped).

(2) By Observation 2, it is sufficient to prove that $\bar{b} \succeq \langle y_{k/2+1}, \ldots, y_k \rangle$. Since $\forall_{k/2 < j \leq k}\ l_{k/2} \geq l_j \geq \min\{l_j, r_{k-j+1}\} = y_{3k/2-j+1}$, then $\langle l_1, \ldots, l_{k/2} \rangle \succeq \langle y_{k/2+1}, \ldots, y_k \rangle$ and by Definition 10: $\langle y_1, \ldots, y_{k/2} \rangle \succeq \langle y_{k/2+1}, \ldots, y_k \rangle$. Therefore $\bar{b}$ consists of k largest elements from $\bar{l} :: \bar{r}$.

The bitonic merger in step 2 receives a bitonic sequence, so it outputs a sorted sequence, which completes the proof. □

The first step of improved pairwise merger is illustrated in Figure 1. We use $k/2$ comparators in the first step and $k \log k/2$ comparators in the second step. We get a merger of size $k \log k/2 + k/2$, which is better than the previous approach. In the following it is shown that we can do even better and eliminate $k/2$ term.

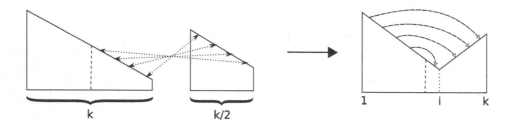

Fig. 1. Making the bitonic sequence. Arrows on the right picture show directions of inequalities. Sequence on the right is v-shaped s-dominating at point i.

The main observation is that the result of the first step of pw_bit_merge operation: $\langle b_1, b_2, \ldots, b_k \rangle$ is not only bitonic, but what we call *v-shaped s-dominating*.

Definition 11 (S-domination). *A sequence* $\bar{b} = \langle b_1, b_2, \ldots, b_k \rangle$ *is s-dominating if* $\forall_{1 \leq j \leq k/2}\ b_j \geq b_{k-j+1}$.

Lemma 2. *If* $\bar{b} = \langle b_1, b_2, \ldots, b_k \rangle$ *is v-shaped and s-dominating, then* $\bar{b}$ *is non-increasing or* $\exists_{k/2 < i < k}\ b_i < b_{i+1}$.

Proof. Assume that $\bar{b}$ is not nonincreasing. Then $\exists_{1 \leq j < k} \, b_j < b_{j+1}$. Assume that $j \leq k/2$. Since $\bar{b}$ is v-shaped, b_{j+1} must be in nondecreasing part of $\bar{b}$. If follows that $b_j < b_{j+1} \leq \ldots \leq b_{k/2} \leq \ldots \leq b_{k-j+1}$. That means that $b_j < b_{k-j+1}$. On the other hand, $\bar{b}$ is s-dominating, thus $b_j \geq b_{k-j+1}$ – a contradiction. □

We will say that a sequence $\bar{b}$ is *v-shaped s-dominating at point* i if i is the smallest index greater than $k/2$ such that $b_i < b_{i+1}$ or $i = k$ for a nonincreasing sequence.

Lemma 3. *Let* $\bar{b} = \langle b_1, b_2, \ldots, b_k \rangle$ *be v-shaped s-dominating at point* i, *then* $\langle b_1, \ldots, b_{k/4} \rangle \succeq \langle b_{k/2+1}, \ldots, b_{3k/4} \rangle$.

Proof. If $\bar{b}$ is nonincreasing, then the lemma holds. Otherwise, from Lemma 2: $k/2 < i < k$. If $i > 3k/4$, then by Definition 4: $b_1 \geq \ldots \geq b_{3k/4} \geq \ldots \geq b_i$, so the lemma holds. If $k/2 < i \leq 3k/4$, then by Definition 4: $b_1 \geq \ldots \geq b_i$, so $\langle b_1, \ldots, b_{k/4} \rangle \succeq \langle b_{k/2+1}, \ldots, b_i \rangle$. Since $b_i < b_{i+1} \leq \ldots \leq b_{3k/4}$, it suffices to prove that $b_{k/4} \geq b_{3k/4}$. By Definition 11 and 4: $b_{k/4} \geq b_{3k/4+1} \geq b_{3k/4}$. □

Definition 12 (Half Splitter). *A half splitter is a comparator network constructed by comparing inputs* $\langle k/4 + 1, 3k/4 + 1 \rangle, \ldots, \langle k/2, k \rangle$ *(normal splitter with first* $k/4$ *comparators removed). We will call it* $half\_split^k$.

Lemma 4. *If* $\bar{b}$ *is v-shaped s-dominating, then* $half\_split^k(\bar{b}) = split^k(\bar{b})$.

Proof. Directly from Lemma 3. □

Lemma 5. *Let* $\bar{b}$ *be v-shaped s-dominating. The following statements are true: (1)* $left(half\_split^k(\bar{b}))$ *is v-shaped s-dominating; (2)* $right(half\_split^k(\bar{b}))$ *is bitonic; (3)* $left(half\_split^k(\bar{b})) \succeq right(half\_split^k(\bar{b}))$.

Proof. (1) Let $\bar{y} = left(half\_split^k(\bar{b}))$. First we show that $\bar{y}$ is v-shaped. If $\bar{y}$ is nonincreasing, then it is v-shaped. Otherwise, let j be the first index from the range $\{1, \ldots, k/2\}$, where $y_{j-1} < y_j$. Since $y_j = \max\{b_j, b_{j+k/2}\}$ and $y_{j-1} \geq b_{j-1} \geq b_j$, thus $b_j < b_{j+k/2}$. Since $\bar{b}$ is v-shaped, element $b_{j+k/2}$ must be in nondecreasing part of $\bar{b}$. It follows that $b_j \geq \ldots \geq b_{k/2}$ and $b_{j+k/2} \leq \ldots \leq b_k$. From this we can see that $\forall_{j \leq j' \leq k/2} \, y_{j'} = \max\{b_{j'}, b_{j'+k/2}\} = b_{j'+k/2}$, so $y_j \leq \ldots \leq y_{k/2}$. Therefore $\bar{y}$ is v-shaped.

Next we show that $\bar{y}$ is s-dominating. Consider any j, where $1 \leq j \leq k/4$. By Definition 4 and 11: $b_j \geq b_{k/2-j+1}$ and $b_j \geq b_{k-j+1}$, therefore $y_j = b_j \geq \max\{b_{k/2-j+1}, b_{k-j+1}\} = y_{k/2-j+1}$, thus proving that $\bar{y}$ is s-dominating. Concluding: $\bar{y}$ is v-shaped s-dominating.

(2) Let $\bar{z} = right(half\_split^k(\bar{b}))$. By Lemma 4: $\bar{z} = right(split^k(\bar{b}))$. We know that $\bar{b}$ is a special case of bitonic sequence, therefore using Lemma 1 we get that $\bar{z}$ is bitonic.

(3) Let $w = half\_split^k(\bar{b})$. By Lemma 4: $\bar{w} = split^k(\bar{b})$. We know that b is a special case of bitonic sequence, therefore using Lemma 1 we get $left(\bar{w}) \succeq right(\bar{w})$. □

Network 4. $pw\_hbit\_merge_k^n$

Input: $\bar{l} :: \bar{r}$, where $\bar{l} \in \mathbb{N}^{n/2}$ is top k sorted, $\bar{r} \in \mathbb{N}^{n/2}$ is top $k/2$ sorted and $\langle l_1, \ldots, l_{k/2} \rangle$ dominates $\langle r_1, \ldots, r_{k/2} \rangle$

1: $\bar{y} \leftarrow bit\_split^k(l_{k/2+1}, \ldots, l_k, r_1, \ldots, r_{k/2})$ **and** $\bar{b} \leftarrow \langle l_1, \ldots, l_{k/2} \rangle :: \langle y_1, \ldots, y_{k/2} \rangle$
2: **return** $half\_bit\_merge^k(\bar{b})$, **where**
3: **function** $half\_bit\_merge^k(\bar{b})$
4:    **if** $k = 2$ **then return** (b_1, b_2)
5:    $\bar{b}' \leftarrow half\_split(b_1, \ldots, b_k)$
6:    **return** $half\_bit\_merge^{k/2}(left(\bar{b}')) :: bit\_merge^{k/2}(right(\bar{b}'))$

Using both $half\_split$ and Batcher's bit_merge and successively applying Lemma 5 to the resulting v-shaped s-dominating half of the output, we have all the tools needed to construct the improved pairwise merger $pw\_hbit\_merge_k^n$ (Network 4) using half splitters and then to prove that the construction is correct.

Theorem 3. *The output of Network 4 consists of sorted k largest elements from input $\bar{l} :: \bar{r}$, assuming that $\bar{l} \in \mathbb{N}^{n/2}$ is top k sorted and $\bar{r} \in \mathbb{N}^{n/2}$ is top $k/2$ sorted and $\langle l_1, \ldots, l_{k/2} \rangle$ dominates $\langle r_1, \ldots, r_{k/2} \rangle$. Moreover, $|pw\_hbit\_merge_k^n| = k \log k/2$.*

Proof. Since step 1 in Network 4 is the same as in Network 3, we can reuse the proof of Theorem 2 to deduce, that $\bar{b}$ is v-shaped and contains k largest elements from $\bar{l} :: \bar{r}$. Also, since $\forall_{1 \le j \le k/2}\ l_j \ge l_{k-j+1}$ and $l_j \ge r_j$, then $b_j = l_j \ge \max\{l_{k-j+1}, r_j\} = b_{k-j+1}$, so $\bar{b}$ is s-dominating.

We prove by the induction on k, that if $\bar{b}$ is v-shaped s-dominating, then the sequence $half\_bit\_merge^k(\bar{b})$ is sorted. For the base case, consider $k = 2$ and a v-shaped s-dominating sequence $\langle b_1, b_2 \rangle$. By Definition 11 this sequence is already sorted and we are done. For the induction step, consider $\bar{b}' = half\_split^k(\bar{b})$. By Lemma 5 we get that $left(\bar{b}')$ is v-shaped s-dominating and $right(\bar{b}')$ is bitonic. Using the induction hypothesis we sort $left(\bar{b}')$ and using bitonic merger we sort $right(\bar{b}')$. By Lemma 5: $left(\bar{b}') \succeq right(\bar{b}')$, which completes the proof of correctness.

As mentioned in Definition 12: $half\_split^k$ is just $split^k$ with the first $k/4$ comparators removed. So $half\_bit\_merge^k$ is just bit_merge^k with some comparators removed. Let's count them: in each level of recursion step we take half of comparators from $split^k$ and additional one comparator from the base case ($k = 2$). We sum them together to get:

$$1 + \sum_{i=0}^{\log k - 2} \frac{k}{2^{i+2}} = 1 + \frac{k}{4} \left(\sum_{i=0}^{\log k - 1} \left(\frac{1}{2} \right)^i - \frac{2}{k} \right) = 1 + \frac{k}{4} \left(2 - \frac{2}{k} - \frac{2}{k} \right) = \frac{k}{2}$$

Therefore we have:

$$|pw\_hbit\_merge_k^n| = k/2 + k \log k/2 - k/2 = k \log k/2$$

$\square$

The only difference between pw_sel and our pw_hbit_sel is the use of improved merger pw_hbit_merge rather than pw_merge. By Theorem 3, we conclude that $|pw\_merge_k^n| \geq |pw\_hbit\_merge_k^n|$, so it follows that:

Remark 1. $|pw\_hbit\_sel_k^n| \leq |pw\_sel_k^n|$

5   Sizes of New Selection Networks

In this section we estimate the size of $pw\_hbit\_sel_k^n$. To this end we show that the size of $pw\_hbit\_sel_k^n$ is upper-bounded by the size of $bit\_sel_k^n$ and use this fact in our estimation. We also compute the exact difference between sizes of $pw\_sel_k^n$ and $pw\_hbit\_sel_k^n$ and show that it can be as big as $n \log n/2$. Finally we show graphically how much smaller is our selection network on practical values of n and k.

We have the recursive formula for the number of comparators of $pw\_hbit\_sel_k^n$:

$$|pw\_hbit\_sel_k^n| = \begin{cases} |pw\_hbit\_sel_k^{n/2}| + |pw\_hbit\_sel_{k/2}^{n/2}| + \\ \quad + |split^n| + |pw\_hbit\_merge^k| & \text{if } k < n \\ |oe\_sort^k| & \text{if } k = n \\ |max^n| & \text{if } k = 1 \end{cases} \quad (3)$$

Lemma 6. $|pw\_hbit\_sel_k^n| \leq |bit\_sel_k^n|$.

Proof. Let $aux\_sel_k^n$ be the comparator network that is generated by substituting recursive calls in $pw\_hbit\_sel_k^n$ by calls to $bit\_sel_k^n$. Size of this network (for $1 < k < n$) is:

$$|aux\_sel_k^n| = |bit\_sel_k^{n/2}| + |bit\_sel_{k/2}^{n/2}| + |split^n| + |pw\_hbit\_merge^k| \quad (4)$$

Lemma 6 follows from Lemma 7 and Lemma 8 below, where we show that:

$$|pw\_hbit\_sel_k^n| \leq |aux\_sel_k^n| \leq |bit\_sel_k^n|$$

$\square$

Lemma 7. *For* $1 < k < n$ *(both powers of 2),* $|aux\_sel_k^n| \leq |bit\_sel_k^n|$.

Proof. We compute both values from Equation 2 and 4:

$$|aux\_sel_k^n| = \frac{1}{4}n \log^2 k + \frac{5}{2}n - \frac{1}{4}k \log k - \frac{5}{4}k - \frac{3n}{2k}$$

$$|bit\_sel_k^n| = \frac{1}{4}n \log^2 k + \frac{1}{4}n \log k + 2n - \frac{1}{2}k \log k - k - \frac{n}{k}$$

We simplify both sides to get the following inequality:

$$n - \frac{1}{2}k - \frac{n}{k} \leq \frac{1}{2}(n - k) \log k$$

which can be easily proved by induction. $\square$

Lemma 8. *For $1 \le k < n$ (both powers of 2), $|pw\_hbit\_sel_k^n| \le |aux\_sel_k^n|$.*

Proof. By induction. For the base case, consider $1 = k < n$. If follows by definitions that $|pw\_hbit\_sel_k^n| = |aux\_sel_k^n| = n - 1$. For the induction step assume that for each $(n', k') \prec (n, k)$ (in lexicographical order) the lemma holds, we get:

$$
\begin{aligned}
&|pw\_hbit\_sel_k^n| \\
&= |pw\_hbit\_sel_{k/2}^{n/2}| + |pw\_hbit\_sel_k^{n/2}| + |split^n| + |pw\_hbit\_merge^k| \\
&\qquad\qquad\qquad\qquad \textbf{(by the definition of } pw\_hbit\_sel\textbf{)} \\
&\le |aux\_sel_{k/2}^{n/2}| + |aux\_sel_k^{n/2}| + |split^n| + |pw\_hbit\_merge^k| \\
&\qquad\qquad\qquad\qquad \textbf{(by the induction hypothesis)} \\
&\le |bit\_sel_{k/2}^{n/2}| + |bit\_sel_k^{n/2}| + |split^n| + |pw\_hbit\_merge^k| \\
&\qquad\qquad\qquad\qquad \textbf{(by Lemma 7)} \\
&= |aux\_sel_k^n| \\
&\qquad\qquad\qquad\qquad \textbf{(by the definition of } aux\_sel\textbf{)}
\end{aligned}
$$

□

Let $N = 2^n$ and $K = 2^k$. We will compute upper bound for $P(n, k) = |pw\_hbit\_sel_K^N|$ using $B(n, k) = |bit\_sel_K^N|$.

Lemma 9. *Let:*

$$
P(n, k, m) = \sum_{i=0}^{m-1} \sum_{j=0}^{i} \binom{i}{j} \left((k - j)2^{k-j-1} + 2^{n-i-1} \right) + \sum_{i=0}^{m} \binom{m}{i} P(n - m, k - i).
$$

Then $\forall_{0 \le m \le \min(k, n-k)} \; P(n, k, m) = P(n, k)$.

Proof. The lemma can be easily proved by induction on m. □

Lemma 10. $P(n, k, m) \le 2^{n-2} \left(\left(k - \frac{m}{2}\right)^2 + k + \frac{7m}{4} + 8 \right) + 2^k \left(\frac{3}{2}\right)^m \left(\frac{k}{2} - \frac{m}{6}\right) - 2^k(k + 1) - 2^{n-k} \left(\frac{3}{2}\right)^m$.

Proof. Due to space restriction we only present schema of the proof. The lemma can be proven directly from the inequality below, which is a consequence of Lemma 9 and 6.

$$
P(n, k, m) \le \sum_{i=0}^{m-1} \sum_{j=0}^{i} \binom{i}{j} \left((k - j)2^{k-j-1} + 2^{n-i-1} \right) + \sum_{i=0}^{m} \binom{m}{i} B(n - m, k - i)
$$

□

Theorem 4. *For $m = \min(k, n-k)$, $P(n, k) \le 2^{n-2} \left(\left(k - \frac{m}{2} - \frac{7}{4}\right)^2 + \frac{9k}{2} + \frac{79}{16} \right) + 2^k \left(\frac{3}{2}\right)^m \left(\frac{k}{2} - \frac{m}{6}\right) - 2^k(k + 1) - 2^{n-k} \left(\frac{3}{2}\right)^m$.*

Proof. Directly from Lemma 9 and 10.   □

We will now present the *size difference* $SD(n, k)$ between pairwise selection network and our network. Merging step in $pw\_sel_K^N$ costs $2^k k - 2^k + 1$ and in $pw\_hbit\_sel_K^N$: $2^{k-1}k$, so the difference is given by the following equation:

$$SD(n, k) = \begin{cases} 0 & \text{if } n = k \\ 0 & \text{if } k = 0 \\ 2^{k-1}k - 2^k + 1 + \\ \quad + SD(n-1, k) + SD(n-1, k-1) & \text{if } 0 < k < n \end{cases} \tag{5}$$

Theorem 5. *Let* $S_{n,k} = \sum_{j=0}^{k} \binom{n-k+j}{j} 2^{k-j}$. *Then:*

$$SD(n, k) = \binom{n}{k} \frac{n+1}{2} - S_{n,k} \frac{n - 2k + 1}{2} - 2^k(k-1) - 1$$

Proof. By straightforward calculation one can verify that $S_{n,0} = 1$, $S_{n,n} = 2^{n+1} - 1$, $S_{n-1,k-1} = \frac{1}{2}(S_{n,k} - \binom{n}{k})$ and $S_{n-1,k-1} + S_{n-1,k} = S_{n,k}$. It follows that the theorem is true for $k = 0$ and $k = n$. We prove the theorem by induction on pairs (k, n). Take any (k, n), $0 < k < n$, and assume that theorem holds for every $(k', n') \prec (k, n)$ (in lexicographical order). Then we have:

$$SD(n, k) = 2^{k-1}k - 2^k + 1 + SD(n-1, k) + SD(n-1, k-1)$$

$$= 2^{k-1}k - 2^k + 1 + \binom{n-1}{k}\frac{n}{2} + \binom{n-1}{k-1}\frac{n}{2} - 2^k(k-1) - 1$$

$$- 2^{k-1}(k-2) - 1 - (S_{n-1,k}\frac{n-2k}{2} + S_{n-1,k-1}\frac{n-2k+2}{2})$$

$$= \binom{n}{k}\frac{n}{2} - S_{n,k}\frac{n-2k}{2} - S_{n-1,k-1} - 2^k(k-1) - 1$$

$$= \binom{n}{k}\frac{n+1}{2} - S_{n,k}\frac{n-2k+1}{2} - 2^k(k-1) - 1$$

□

Corollary 1. $|pw\_sel_{N/2}^N| - |pw\_hbit\_sel_{N/2}^N| = N\frac{\log N - 4}{2} + \log N + 2$, *for* $N = 2^n$.

Plots in Figure 2 show how much pw_sel and the upper bound from Theorem 4 are worse than pw_hbit_sel. Lines labeled pw_sel are plotted from $(|pw\_sel_K^N| - |pw\_hbit\_sel_K^N|)/|pw\_hbit\_sel_K^N|$ and the ones labeled *upper* are plotted from the formula $(|upper_K^N| - |pw\_hbit\_sel_K^N|)/|pw\_hbit\_sel_K^N|$, where $|upper_K^N|$ is the upper bound from Theorem 4. Both $|pw\_sel_K^N|$ and $|pw\_hbit\_sel_K^N|$ were computed directly from recursive formulas. We can see that we save the most number of comparators when k is larger than $n/2$, nevertheless for small values of n superiority of our network is apparent for any k. As for the upper bound, it gives a good approximation of $|pw\_hbit\_sel_K^N|$ when n is small, but for larger values of n it becomes less satisfactory.

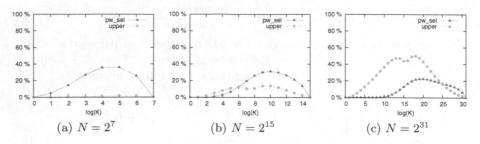

Fig. 2. The relative percentage change of the size of $pw\_sel_K^N$ and the upper bound given in Thm 4 with respect to the size of $pw\_hbit\_sel_K^N$.

6   Arc-Consistency of Selection Networks

In this section we prove that half encoding of any selection network preserves arc-consistency with respect to "less-than" cardinality constraints. The proof can be generalized to other types of cardinality constraints.

We introduce the convention, that $\langle x_1, \ldots, x_n \rangle$ will denote the input and $\langle y_1, \ldots, y_n \rangle$ will denote the output of some order n comparator network. We would also like to view them as sequences of Boolean variables, that can be set to either true (1), false (0) or undefined (X).

From now on we assume that every network f is half encoded and when we say "comparator" or "network", we view it in terms of CNF formulas. We denote $V[\phi(f)]$ to be the set of variables in encoding $\phi(f)$.

Observation 4. *A single comparator hcomp(a, b, c, d) has the following propagation properties:*

1. *If $a = 1$ or $b = 1$, then UP sets $c = 1$ (by 1.c1 or 1.c2).*
2. *If $a = b = 1$, then UP sets $c = d = 1$ (by 1.c1 and 1.c3).*
3. *If $c = 0$, then UP sets $a = b = 0$ (by 1.c1 and 1.c2).*
4. *If $b = 1$ and $d = 0$, then UP sets $a = 0$ (by 1.c3).*
5. *If $a = 1$ and $d = 0$, then UP sets $b = 0$ (by 1.c3).*

Lemma 11. *Let f_k^n be a selection network. Assume that $k - 1$ inputs are set to 1, and rest of the variables are undefined. Unit propagation will set variables $y_1, \ldots, y_{k-1}$ to 1.*

Proof. From propagation properties of $hcomp(a, b, c, d)$ we can see that if comparator receives two 1s, then it outputs two 1s, when it receives 1 on one input and X on the other, then it outputs 1 on the upper output and X on the lower output. From this we conclude that a single comparator will sort its inputs, as long as one of the inputs is set to 1. No 1 is lost, so they must all reach the outputs. Because the comparators comprise a selection network, the 1s will appear at outputs $y_1, \ldots, y_{k-1}$.   □

The process of propagating 1s we call a *forward propagation*. For the remainder of this section assume that: f_k^n is a selection network; $k-1$ inputs are set to 1, and the rest of the variables are undefined; forward propagation has been performed resulting in $y_1, \ldots, y_{k-1}$ to be set to 1.

Definition 13 (Path). *A path is a sequence of Boolean variables* $\langle z_1, \ldots, z_m \rangle$ *such that* $\forall_{1 \leq i \leq m} z_i \in V[\phi(f_k^n)]$ *and for all* $1 \leq i < m$ *there exists a comparator* $hcomp(a, b, c, d)$ *in* $\phi(f_k^n)$ *for which* $z_i \in \{a, b\}$ *and* $z_{i+1} \in \{c, d\}$.

Definition 14 (Propagation Path). *Let x be an undefined input variable. A path* $\bar{z}_x = \langle z_1, \ldots, z_m \rangle$ $(m \geq 1)$ *is a* propagation path, *if* $z_1 \equiv x$ *and* $\langle z_2, \ldots, z_m \rangle$ *is the sequence of variables that would be set to 1 by UP, if we would set $z_1 = 1$.*

Lemma 12. *If* $\bar{z}_x = \langle z_1, \ldots, z_m \rangle$ *is a propagation path for an undefined variable x, then* $z_m \equiv y_k$.

Proof. Remember that all $y_1, \ldots, y_{k-1}$ are set to 1. Setting any undefined input variable x to 1 will result in UP to set y_k to 1. Otherwise f_k^n would not be a selection network. $\qquad\square$

The following lemma shows that propagation paths are deterministic.

Lemma 13. *Let* $\bar{z}_x = \langle z_1, \ldots, z_m \rangle$ *be a propagation path. For each* $1 \leq i \leq m$ *and* $z_1' \equiv z_i$, *if* $\langle z_1', \ldots, z_{m'}' \rangle$ *is a path that would be set to 1 by UP if we would set* $z_1' = 1$, *then* $\langle z_1', \ldots, z_{m'}' \rangle = \langle z_i, \ldots, z_m \rangle$.

Proof. By induction on $l = m - i$. If $l = 0$, then $z_1' \equiv z_m \equiv y_k$ (by Lemma 12), so the lemma holds. Let $l \geq 0$ and assume that the lemma is true for z_l. Consider $z_1' \equiv z_{l-1} \equiv z_{m-i-1}$. Set $z_{m-i-1} = 1$ and use UP to set $z_{m-i} = 1$. Notice that $z_{m-i} \equiv z_2'$, otherwise there would exist a comparator $hcomp(a, b, c, d)$, for which z_{m-i-1} is equivalent to either a or b and $z_{m-i} \equiv c$ and $z_2' \equiv d$ (or vice versa). That would mean that a single 1 on the input produces two 1s on the outputs. This contradicts our reasoning in the proof of Lemma 11. By the induction hypothesis $\langle z_2', \ldots, z_{m'}' \rangle = \langle z_{m-i}, \ldots, z_m \rangle$, so $\langle z_1', \ldots, z_{m'}' \rangle = \langle z_{m-i-1}, \ldots, z_m \rangle$. $\quad\square$

For each undefined input variable x and propagation path $\bar{z}_x = \langle z_1, \ldots, z_m \rangle$ we define a directed graph $P_x = \{ \langle z_i, z_{i+1} \rangle : 1 \leq i < m \}$.

Lemma 14. *Let* $\{x_{i_1}, \ldots, x_{i_t}\}$ $(t > 0)$ *be the set of undefined input variables. Then* $T = P_{x_{i_1}} \cup \ldots \cup P_{x_{i_t}}$ *is the tree rooted at y_k.*

Proof. By induction on t. If $t = 1$, then $T = P_{x_{i_1}}$ and by Lemma 12, $P_{x_{i_1}}$ ends in y_k, so the lemma holds. Let $t > 0$ and assume that the lemma is true for t. We will show that it is true for $t + 1$. Consider $T = P_{x_{i_1}} \cup \ldots \cup P_{x_{i_t}} \cup P_{x_{i_{t+1}}}$. By the induction hypothesis $T' = P_{x_{i_1}} \cup \ldots \cup P_{x_{i_t}}$ is a tree rooted at y_k. By Lemma 12, $V(P_{x_{i_{t+1}}}) \cap V(T') \neq \emptyset$. Let $z \in V(P_{x_{i_{t+1}}})$ be the first variable, such that $z \in V(T')$. Since $z \in V(T')$, there exists j $(1 \leq j \leq t)$ such that $z \in P_{x_{i_j}}$. By Lemma 13, starting from variable z, paths $P_{x_{i_{t+1}}}$ and $P_{x_{i_j}}$ are identical. $\quad\square$

Graph T from the above lemma will be called a *propagation tree*.

Theorem 6. *If we set* $y_k = 0$, *then unit propagation will set all undefined input variables to* 0.

Proof. Let T be the propagation tree rooted at y_k. We prove by induction on the height h of a subtree T' of T, that (*) if the root of T' is set to 0, then UP sets all nodes of T' to 0. It follows that if y_k is set to 0 then UP sets all undefined input variables to 0. If $h = 0$, then $V = \{y_k\}$, so (*) is trivially true. Let $h > 0$ and assume that (*) holds. We will show that (*) holds for height $h + 1$. Let T' be the propagation tree of height $h + 1$ and let $r = 0$ be the root. Consider children of r in T' and a comparator $hcomp(a, b, c, d)$ for which $r \in \{c, d\}$:

Case 1: r has two children. The only case is when $r \equiv c = 0$. Unit propagation sets $a = b = 0$. Nodes a and b are roots of propagation trees of height h and are set to 0, therefore by the induction hypothesis all nodes in T' will be set to 0.

Case 2: r has one child. Consider two cases: (i) if $r \equiv c = 0$ and either a or b is the child of r, then UP sets $a = b = 0$ and either a or b is the root of propagation tree of height h and is set to 0, therefore by the induction hypothesis all nodes in T' will be set to 0, (ii) $r \equiv d = 0$ and either $a = c = 1$ and b is the child of r or $b = c = 1$ and a is the child of r. Both of them will be set to 0 by UP and again we get the root of propagation tree of height h that is set to 0, therefore by the induction hypothesis all nodes in T' will be set to 0. □

7   Conclusions

We have constructed a new family of selection networks, which are based on the pairwise selection ones, but require less comparators to merge subsequences. The difference in sizes grows with k and is equal to $n\frac{\log n - 4}{2} + \log n + 2$ for $k = n/2$. In addition, we have shown that any selection network encoded in a standard way to a CNF formula preserves arc-consistency with respect to a corresponding cardinality constraint. This property is important, as many SAT-solvers take advantage of the unit-propagation algorithm, making the computation significantly faster.

It's also worth noting that using encodings based on selection networks give an extra edge in solving optimization problems for which we need to solve a sequence of problems that differ only in the decreasing bound of a cardinality constraint. In this setting we only need to add one more clause $\neg y_k$ for a new value of k, and the search can be resumed keeping all previous clauses as it is. This works because if a comparator network is a k-selection network, then it is also a k'-selection network for any $k' < k$. This property is called *incremental strengthening* and most state-of-the-art SAT-solvers provide a user interface for doing this.

We are expecting that our smaller encodings of cardinality constraints should improve the performance of SAT solvers, but the statement needs an experimental evaluation. We start doing the evaluation and the results will be presented in the near future.

References

1. Asín, R., Nieuwenhuis, R., Oliveras, A., Rodríguez-Carbonell, E.: Cardinality networks and their applications. In: Kullmann, O. (ed.) SAT 2009. LNCS, vol. 5584, pp. 167–180. Springer, Heidelberg (2009)
2. Asín, R., Nieuwenhuis, R., Oliveras, A., Rodríguez-Carbonell, E.: Cardinality networks: a theoretical and empirical study. Constraints 16(2), 195–221 (2011)
3. Asín, R., Nieuwenhuis, R.: Curriculum-based course timetabling with SAT and MaxSAT. Annals of Operations Research 218(1), 71–91 (2014)
4. Batcher, K.E.: Sorting networks and their applications. In: Proc. of the April 30-May 2, 1968, Spring Joint Computer Conference, AFIPS 1968 (Spring), pp. 307–314. ACM, New York (1968)
5. Codish, M., Zazon-Ivry, M.: Pairwise networks are superior for selection. http://www.cs.bgu.ac.il/~mcodish/Papers/Sources/pairwiseSelection.pdf
6. Codish, M., Zazon-Ivry, M.: Pairwise cardinality networks. In: Clarke, E.M., Voronkov, A. (eds.) LPAR-16 2010. LNCS, vol. 6355, pp. 154–172. Springer, Heidelberg (2010)
7. Eén, N., Sörensson, N.: Translating pseudo-boolean constraints into sat. Journal on Satisfiability, Boolean Modeling and Computation 2, 1–26 (2006)
8. Knuth, D.E.: The Art of Computer Programming, Sorting and Searching, vol. 3, 2nd edn. Addison Wesley Longman Publishing Co. Inc., Redwood City (1998)
9. Parberry, I.: Parallel complexity theory. Pitman, Research notes in theoretical computer science (1987)
10. Parberry, I.: The pairwise sorting network. Parallel Processing Letters 2, 205–211 (1992)
11. Schutt, A., Feydy, T., Stuckey, P.J., Wallace, M.G.: Why Cumulative decomposition is not as bad as it sounds. In: Gent, I.P. (ed.) CP 2009. LNCS, vol. 5732, pp. 746–761. Springer, Heidelberg (2009)

PREFIX-PROJECTION Global Constraint for Sequential Pattern Mining

Amina Kemmar[1], Samir Loudni[2(✉)], Yahia Lebbah[1],
Patrice Boizumault[2], and Thierry Charnois[3]

[1] LITIO, University of Oran 1, EPSECG of Oran, Oran, Algeria
kemmami@yahoo.fr, lebbah.yahia@univ-oran.dz
[2] GREYC (CNRS UMR 6072), University of Caen, Caen, France
{samir.loudni,patrice.boizumault}@unicaen.fr
[3] LIPN (CNRS UMR 7030), University PARIS 13, Villetaneuse, France
thierry.charnois@lipn.univ-paris13.fr

Abstract. Sequential pattern mining under constraints is a challenging data mining task. Many efficient ad hoc methods have been developed for mining sequential patterns, but they are all suffering from a lack of genericity. Recent works have investigated Constraint Programming (CP) methods, but they are not still effective because of their encoding. In this paper, we propose a global constraint based on the projected databases principle which remedies to this drawback. Experiments show that our approach clearly outperforms CP approaches and competes well with ad hoc methods on large datasets.

1 Introduction

Mining useful patterns in sequential data is a challenging task. Sequential pattern mining is among the most important and popular data mining task with many real applications such as the analysis of web click-streams, medical or biological data and textual data. For effectiveness and efficiency considerations, many authors have promoted the use of constraints to focus on the most promising patterns according to the interests given by the final user. In line with [15], many efficient ad hoc methods have been developed but they suffer from a lack of genericity to handle and to push simultaneously sophisticated combination of various types of constraints. Indeed, new constraints have to be hand-coded and their combinations often require new implementations.

Recently, several proposals have investigated relationships between sequential pattern mining and constraint programming (CP) to revisit data mining tasks in a declarative and generic way [5,9,11,12]. The great advantage of these approaches is their flexibility. The user can model a problem and express his queries by specifying what constraints need to be satisfied. But, all these proposals are not effective enough because of their CP encoding. Consequently, the design of new efficient declarative models for mining useful patterns in sequential data is clearly an important challenge for CP.

© Springer International Publishing Switzerland 2015
G. Pesant (Ed.): CP 2015, LNCS 9255, pp. 226–243, 2015.
DOI: 10.1007/978-3-319-23219-5_17

To address this challenge, we investigate in this paper the other side of the cross fertilization between data-mining and constraint programming, namely how the CP framework can benefit from the power of candidate pruning mechanisms used in sequential pattern mining. First, we introduce the global constraint PREFIX-PROJECTION for sequential pattern mining. PREFIX-PROJECTION uses a concise encoding and its filtering relies on the principle of projected databases [14]. The key idea is to divide the initial database into smaller ones projected on the frequent subsequences obtained so far, then, mine locally frequent patterns in each projected database by growing a frequent prefix. This global constraint utilizes the principle of prefix-projected database to keep only locally frequent items alongside projected databases in order to remove infrequent ones from the domains of variables. Second, we show how the concise encoding allows for a straightforward implementation of the frequency constraint (PREFIX-PROJECTION constraint) and constraints on patterns such as size, item membership and regular expressions and the simultaneous combination of them. Finally, experiments show that our approach clearly outperforms CP approaches and competes well with ad hoc methods on large datasets for mining frequent sequential patterns or patterns under various constraints. It is worth noting that the experiments show that our approach achieves scalability while it is a major issue of CP approaches.

The paper is organized as follows. Section 2 recalls preliminaries. Section 3 provides a critical review of ad hoc methods and CP approaches for sequential pattern mining. Section 4 presents the global constraint PREFIX-PROJECTION. Section 5 reports experiments we performed. Finally, we conclude and draw some perspectives.

2   Preliminaries

This section presents background knowledge about sequential pattern mining and constraint satisfaction problems.

2.1   Sequential Patterns

Let $\mathcal{I}$ be a finite set of *items*. The language of sequences corresponds to $\mathcal{L}_\mathcal{I} = \mathcal{I}^n$ where $n \in \mathbb{N}^+$.

Definition 1 (sequence, sequence database). *A sequence s over $\mathcal{L}_\mathcal{I}$ is an ordered list $\langle s_1 s_2 \ldots s_n \rangle$, where s_i, $1 \leq i \leq n$, is an item. n is called the length of the sequence s. A sequence database SDB is a set of tuples (sid, s), where sid is a sequence identifier and s a sequence.*

Definition 2 (subsequence, $\preceq$ relation). *A sequence $\alpha = \langle \alpha_1 \ldots \alpha_m \rangle$ is a subsequence of $s = \langle s_1 \ldots s_n \rangle$, denoted by $(\alpha \preceq s)$, if $m \leq n$ and there exist integers $1 \leq j_1 \leq \ldots \leq j_m \leq n$, such that $\alpha_i = s_{j_i}$ for all $1 \leq i \leq m$. We also say that α is contained in s or s is a super-sequence of α. For example, the sequence $\langle BABC \rangle$ is a super-sequence of $\langle AC \rangle$: $\langle AC \rangle \preceq \langle BABC \rangle$. A tuple (sid, s) contains a sequence α, if $\alpha \preceq s$.*

Table 1. SDB_1: a sequence database example.

sid	Sequence
1	$\langle ABCBC \rangle$
2	$\langle BABC \rangle$
3	$\langle AB \rangle$
4	$\langle BCD \rangle$

The cover of a sequence p in SDB is the set of all tuples in SDB in which p is contained. The support of a sequence p in SDB is the number of tuples in SDB which contain p.

Definition 3 (coverage, support). *Let SDB be a sequence database and p a sequence. $cover_{SDB}(p) = \{(sid, s) \in SDB \mid p \preceq s\}$ and $sup_{SDB}(p) = \# cover_{SDB}(p)$.*

Definition 4 (sequential pattern). *Given a minimum support threshold minsup, every sequence p such that $sup_{SDB}(p) \geq minsup$ is called a sequential pattern [1]. p is said to be frequent in SDB.*

Example 1. Table 1 represents a sequence database of four sequences where the set of items is $\mathcal{I} = \{A, B, C, D\}$. Let the sequence $p = \langle AC \rangle$. We have $cover_{SDB1}(p) = \{(1, s_1), (2, s_2)\}$. If we consider $minsup = 2$, $p = \langle AC \rangle$ is a sequential pattern because $sup_{SDB_1}(p) \geq 2$.

Definition 5 (sequential pattern mining (SPM)). *Given a sequence database SDB and a minimum support threshold minsup. The problem of sequential pattern mining is to find all patterns p such that $sup_{SDB}(p) \geq minsup$.*

2.2   SPM under Constraints

In this section, we define the problem of mining sequential patterns in a sequence database satisfying user-defined constraints Then, we review the most usual constraints for the sequential mining problem [15].

Problem statement. Given a constraint $C(p)$ on pattern p and a sequence database SDB, the problem of constraint-based pattern mining is to find the complete set of patterns satisfying $C(p)$. In the following, we present different types of constraints that we explicit in the context of sequence mining. All these constraints will be handled by our concise encoding (see Sections 4.2 and 4.5).

- The minimum size constraint $size(p, \ell_{min})$ states that the number of items of p must be greater than or equal to ℓ_{min}.
- The item constraint $item(p, t)$ states that an item t must belong (or not) to a pattern p.
- The regular expression constraint [7] $reg(p, exp)$ states that a pattern p must be accepted by the deterministic finite automata associated to the regular expression exp.

2.3   Projected Databases

We now present the necessary definitions related to the concept of *projected databases* [14].

Definition 6 (prefix, projection, suffix). *Let* $\beta = \langle \beta_1 \ldots \beta_n \rangle$ *and* $\alpha = \langle \alpha_1 \ldots \alpha_m \rangle$ *be two sequences, where* $m \leq n$.
- *Sequence* α *is called the prefix of* β *iff* $\forall i \in [1..m], \alpha_i = \beta_i$.
- *Sequence* $\beta = \langle \beta_1 \ldots \beta_n \rangle$ *is called the projection of some sequence s w.r.t.* α, *iff (1)* $\beta \preceq s$, *(2)* α *is a prefix of* β *and (3) there exists no proper super-sequence* β' *of* β *such that* $\beta' \preceq s$ *and* β' *also has* α *as prefix.*
- *Sequence* $\gamma = \langle \beta_{m+1} \ldots \beta_n \rangle$ *is called the suffix of s w.r.t.* α. *With the standard concatenation operator "concat", we have* $\beta = concat(\alpha, \gamma)$.

Definition 7 (projected database). *Let* SDB *be a sequence database, the* α-*projected database, denoted by* $SDB|_\alpha$, *is the collection of suffixes of sequences in* SDB *w.r.t. prefix* α.

[14] have proposed an efficient algorithm, called `PrefixSpan`, for mining sequential patterns based on the concept of *projected databases*. It proceeds by dividing the initial database into smaller ones projected on the frequent subsequences obtained so far; only their corresponding suffixes are kept. Then, sequential patterns are mined in each projected database by exploring only locally frequent patterns.

Example 2. Let us consider the sequence database of Table 1 with *minsup* = 2. `PrefixSpan` starts by scanning SDB_1 to find all the frequent items, each of them is used as a prefix to get projected databases. For SDB_1, we get 3 disjoint subsets w.r.t. the prefixes $\langle A \rangle$, $\langle B \rangle$, and $\langle C \rangle$. For instance, $SDB_1|_{\langle A \rangle}$ consists of 3 suffix sequences: $\{(1, \langle BCBC \rangle), (2, \langle BC \rangle), (3, \langle B \rangle)\}$. Consider the projected database $SDB_1|_{<A>}$, its locally frequent items are B and C. Thus, $SDB_1|_{<A>}$ can be recursively partitioned into 2 subsets w.r.t. the two prefixes $\langle AB \rangle$ and $\langle AC \rangle$. The $\langle AB \rangle$- and $\langle AC \rangle$- projected databases can be constructed and recursively mined similarly. The processing of a α-projected database terminates when no frequent subsequence can be generated.

Proposition 1 establishes the support count of a sequence γ in $SDB|_\alpha$ [14]:

Proposition 1 (Support count). *For any sequence* γ *in SDB with prefix* α *and suffix* β *s.t.* $\gamma = concat(\alpha, \beta)$, $sup_{SDB}(\gamma) = sup_{SDB|_\alpha}(\beta)$.

This proposition ensures that only the sequences in SDB grown from α need to be considered for the support count of a sequence γ. Furthermore, only those suffixes with prefix α should be counted.

2.4   CSP and Global Constraints

A *Constraint Satisfaction Problem* (CSP) consists of a set X of n variables, a domain $\mathcal{D}$ mapping each variable $X_i \in X$ to a finite set of values $D(X_i)$, and a set of constraints $\mathcal{C}$. An assignment σ is a mapping from variables in X to values in their domains: $\forall X_i \in X, \sigma(X_i) \in D(X_i)$. A constraint $c \in \mathcal{C}$ is a subset of the cartesian product of the domains of the variables that are in c. The goal is to find an assignment such that all constraints are satisfied.

Domain Consistency (DC). Constraint solvers typically use backtracking search to explore the space of partial assignments. At each assignment, filtering algorithms prune the search space by enforcing local consistency properties like domain consistency. A constraint c on X is domain consistent, if and only if, for every $X_i \in X$ and for every $d_i \in D(X_i)$, there is an assignment σ satisfying c such that $\sigma(X_i) = d_i$. Such an assignment is called a support.

Global constraints provide shorthands to often-used combinatorial substructures. We present two global constraints. Let $X = \langle X_1, X_2, ..., X_n \rangle$ be a sequence of n variables.

Let V be a set of values, l and u be two integers s.t. $0 \leq l \leq u \leq n$, the constraint $\mathtt{Among}(X, V, l, u)$ states that each value $a \in V$ should occur at least l times and at most u times in X [4]. Given a deterministic finite automaton A, the constraint $\mathtt{Regular}(X, A)$ ensures that the sequence X is accepted by A [16].

3   Related Works

This section provides a critical review of ad hoc methods and CP approaches for SPM.

3.1   Ad hoc Methods for SPM

GSP [17] was the first algorithm proposed to extract sequential patterns. It uses a generate-and test approach. Later, two major classes of methods have been proposed:
- Depth-first search based on a vertical database format e.g. cSpade incorporating contraints (max-gap, max-span, length) [21], SPADE [22] or SPAM [2].
- Projected pattern growth such as PrefixSpan [14] and its extensions, e.g. CloSpan for mining closed sequential patterns [19] or Gap-BIDE [10] tackling the gap constraint.

In [7], the authors proposed SPIRIT based on GSP for SPM with regular expressions. Later, [18] introduces Sequence Mining Automata (SMA), a new approach based on a specialized kind of Petri Net. Two variants of SMA were proposed: SMA-1P (SMA one pass) and SMA-FC (SMA Full Check). SMA-1P processes by means of the SMA all sequences one by one, and enters all resulting valid patterns in a hash table for support counting, while SMA-FC allows frequency based pruning during the scan of the database. Finally, [15] provides a survey for other constraints such as regular expressions, length and aggregates. But, all these proposals, though efficient, are ad hoc methods suffering from a lack of genericity. Adding new constraints often requires to develop new implementations.

3.2 CP Methods for SPM

Following the work of [8] for itemset mining, several methods have been proposed to mine sequential patterns using CP.

Proposals. [5] have proposed a first SAT-based model for discovering a special class of patterns with wildcards[1] in a single sequence under different types of constraints (e.g. frequency, maximality, closedness). [11] have proposed a CSP model for SPM. Each sequence is encoded by an automaton capturing all subsequences that can occur in it. [9] have proposed a CSP model for SPM with wildcards. They show how some constraints dealing with local patterns (e.g. frequency, size, gap, regular expressions) and constraints defining more complex patterns such as relevant subgroups [13] and top-k patterns can be modeled using a CSP. [12] have proposed two CP encodings for the SPM. The first one uses a global constraint to encode the subsequence relation (denoted `global-p.f`), while the second one encodes explicitly this relation using additional variables and constraints (denoted `decomposed-p.f`).

All these proposals use **reified constraints** to encode the database. A reified constraint associates a boolean variable to a constraint reflecting whether the constraint is satisfied (value 1) or not (value 0). For each sequence s of SDB, a reified constraint, stating whether (or not) the unknown pattern p is a subsequence of s, is imposed: $(S_s = 1) \Leftrightarrow (p \preceq s)$. A great consequence is that the encoding of the frequency measure is straightforward: $freq(p) = \sum_{s \in SDB} S_s$. But such an encoding has a major drawback since it requires $(m = \#SDB)$ reified constraints to encode the whole database. This constitutes a strong limitation of the size of the databases that could be managed.

Most of these proposals encode **the subsequence relation** $(p \preceq s)$ using variables $Pos_{s,j}$ ($s \in SDB$ and $1 \leq j \leq \ell$) to determine a position where p occurs in s. Such an encoding requires a large number of additional variables $(m \times \ell)$ and makes the labeling computationally expensive. In order to address this drawback, [12] have proposed a global constraint `exists-embedding` to encode the subsequence relation, and used projected frequency within an ad hoc specific branching strategy to keep only frequent items before branching over the variables of the pattern. But, this encoding still relies on reified constraints and requires to impose m `exists-embedding` global constraints.

So, we propose in the next section the PREFIX-PROJECTION global constraint that fully exploits the principle of projected databases to encode both the subsequence relation and the frequency constraint. PREFIX-PROJECTION does not require any reified constraints nor any extra variables to encode the

---

[1] A wildcard is a special symbol that matches any item of $\mathcal{I}$ including itself.

subsequence relation. As a consequence, usual SPM constraints (see Section 2.2) can be encoded in a straightforward way using directly the (global) constraints of the CP solver.

4 PREFIX-PROJECTION Global Constraint

This section presents the PREFIX-PROJECTION global constraint for the SPM problem.

4.1 A Concise Encoding

Let P be the unknown pattern of size ℓ we are looking for. The symbol $\square$ stands for an empty item and denotes the end of a sequence. The unknown pattern P is encoded with a sequence of ℓ variables $\langle P_1, P_2, \ldots, P_\ell \rangle$ s.t. $\forall i \in [1 \ldots \ell], D(P_i) = \mathcal{I} \cup \{\square\}$. There are two basic rules on the domains:

1. To avoid the empty sequence, the first item of P must be non empty, so $(\square \notin D_1)$.
2. To allow patterns with less than ℓ items, we impose that $\forall i \in [1..(\ell-1)], (P_i = \square) \rightarrow (P_{i+1} = \square)$.

4.2 Definition and Consistency Checking

The global constraint PREFIX-PROJECTION ensures both subsequence relation and minimum frequency constraint.

Definition 8 (Prefix-Projection global constraint). *Let* $P = \langle P_1, P_2, \ldots, P_\ell \rangle$ *be a pattern of size* ℓ. $\langle d_1, \ldots, d_\ell \rangle \in D(P_1) \times \ldots \times D(P_\ell)$ *is a solution of* PREFIX-PROJECTION $(P, SDB, minsup)$ *iff* $sup_{SDB}(\langle d_1, \ldots, d_\ell \rangle) \geq minsup$.

Proposition 2. *A* PREFIX-PROJECTION $(P, SDB, minsup)$ *constraint has a solution if and only if there exists an assignment* $\sigma = \langle d_1, \ldots, d_\ell \rangle$ *of variables of* P *s.t.* $SDB|_\sigma$ *has at least minsup suffixes of* σ: $\#SDB|_\sigma \geq minsup$.

Proof: This is a direct consequence of proposition 1. We have straightforwardly $sup_{SDB}(\sigma) = sup_{SDB|_\sigma}(\langle\rangle) = \#SDB|_\sigma$. Thus, suffixes of $SDB|_\sigma$ are supports of σ in the constraint PREFIX-PROJECTION $(P, SDB, minsup)$, provided that $\#SDB|_\sigma \geq minsup$. $\square$

The following proposition characterizes values in the domain of unassigned (i.e. future) variable P_{i+1} that are consistent with the current assignment of variables $\langle P_1, \ldots, P_i \rangle$.

Proposition 3. *Let* $\sigma^2 = \langle d_1, \ldots, d_i \rangle$ *be a current assignment of variables* $\langle P_1, \ldots, P_i \rangle$, P_{i+1} *be a future variable. A value* $d \in D(P_{i+1})$ *appears in a solution for* PREFIX-PROJECTION $(P, SDB, minsup)$ *if and only if* d *is a frequent item in* $SDB|_\sigma$:

$$\#\{(sid, \gamma)|(sid, \gamma) \in SDB|_\sigma \wedge \langle d \rangle \preceq \gamma\} \geq minsup$$

---

[2] We indifferently denote σ by $\langle d_1, \ldots, d_i \rangle$ or by $\langle \sigma(P_1), \ldots, \sigma(P_i) \rangle$.

Proof: Suppose that value $d \in D(P_{i+1})$ occurs in $SDB|_\sigma$ more than $minsup$. From proposition 1, we have $sup_{SDB}(concat(\sigma, \langle d \rangle)) = sup_{SDB|_\sigma}(\langle d \rangle)$. Hence, the assignment $\sigma \cup \langle d \rangle$ satisfies the constraint, so $d \in D(P_{i+1})$ participates in a solution.   □

Anti-monotonicity of the Frequency Measure. If a pattern p is not frequent, then any pattern p' satisfying $p \preceq p'$ is not frequent. From proposition 3 and according to the *anti-monotonicity property*, we can derive the following pruning rule:

Proposition 4. *Let $\sigma = \langle d_1, \ldots, d_i \rangle$ be a current assignment of variables $\langle P_1, \ldots, P_i \rangle$. All values $d \in D(P_{i+1})$ that are locally not frequent in $SDB|_\sigma$ can be pruned from the domain of variable P_{i+1}. Moreover, these values d can also be pruned from the domains of variables P_j with $j \in [i+2, \ldots, \ell]$.*

Proof: Let $\sigma = \langle d_1, \ldots, d_i \rangle$ be a current assignment of variables $\langle P_1, \ldots, P_i \rangle$. Let $d \in D(P_{i+1})$ s.t. $\sigma' = concat(\sigma, \langle d \rangle)$. Suppose that d is not frequent in $SDB|_\sigma$. According to proposition 1, $sup_{SDB|_\sigma}(\langle d \rangle) = sup_{SDB}(\sigma') < minsup$, thus σ' is not frequent. So, d can be pruned from the domain of P_{i+1}.

Suppose that the assignment σ has been extended to $concat(\sigma, \alpha)$, where α corresponds to the assignment of variables P_j (with $j > i$). If $d \in D(P_{i+1})$ is not frequent, it is straightforward that $sup_{SDB|_\sigma}(concat(\alpha, \langle d \rangle)) \leq sup_{SDB|_\sigma}(\langle d \rangle) < minsup$. Thus, if d is not frequent in $SDB|_\sigma$, it will be also not frequent in $SDB|_{concat(\sigma, \alpha)}$. So, d can be pruned from the domains of P_j with $j \in [i+2, \ldots, \ell]$.   □

Example 3. Consider the sequence database of Table 1 with $minsup = 2$. Let $P = \langle P_1, P_2, P_3 \rangle$ with $D(P_1) = \mathcal{I}$ and $D(P_2) = D(P_3) = \mathcal{I} \cup \{\Box\}$. Suppose that $\sigma(P_1) = A$, PREFIX-PROJECTION$(P, SDB, minsup)$ will remove values A and D from $D(P_2)$ and $D(P_3)$, since the only locally frequent items in $SDB_1|_{<A>}$ are B and C.

Proposition 4 guarantees that any value (i.e. item) $d \in D(P_{i+1})$ present but not frequent in $SDB|_\sigma$ does not need to be considered when extending σ, thus avoiding searching over it. Clearly, our global constraint encodes the anti-monotonicity of the frequency measure in a simple and elegant way, while CP methods for SPM have difficulties to handle this property. In [12], this is achieved by using very specific propagators and branching strategies, making the integration quite complex (see [12]).

4.3 Building the Projected Databases

The key issue of our approach lies in the construction of the projected databases. When projecting a prefix, instead of storing the whole suffix as a projected subsequence, one can represent each suffix by a pair $(sid, start)$ where sid is the sequence identifier and $start$ is the starting position of the projected suffix in the sequence sid. For instance, let us consider the sequence database of Table 1.

Algorithm 1. PROJECTSDB(SDB, $ProjSDB$, α)

Data: SDB: initial database; $ProjSDB$: projected sequences; α: prefix
begin

1     $SDB|_\alpha \leftarrow \emptyset$;
2   **for** *each pair* $(sid, start) \in ProjSDB$ **do**
3        $s \leftarrow SDB[sid]$;
4        $pos_\alpha \leftarrow 1; pos_s \leftarrow start$;
5   **while** $(pos_\alpha \leq \#\alpha \wedge pos_s \leq \#s)$ **do**
6   **if** $(\alpha[pos_\alpha] = s[pos_s])$ **then**
7              $pos_\alpha \leftarrow pos_\alpha + 1$;
8           $pos_s \leftarrow pos_s + 1$;
9   **if** $(pos_\alpha = \#\alpha + 1)$ **then**
10           $SDB|_\alpha \leftarrow SDB|_\alpha \cup \{(sid, pos_s)\}$

11   **return** $SDB|_\alpha$;

As shown in example 2, $SDB|_{\langle A \rangle}$ consists of 3 suffix sequences: $\{(1, \langle BCBC \rangle),$ $(2, \langle BC \rangle), (3, \langle B \rangle)\}$. By using the *pseudo-projection*, $SDB|_{\langle A \rangle}$ can be represented by the following three pairs: $\{(1, 2), (2, 3), (3, 2)\}$. This is the principle of *pseudo-projection*, adopted in , exploited during the filtering step of our PREFIX-PROJECTION global constraint. Algorithm 1 details this principle. It takes as input a set of projected sequences $ProjSDB$ and a prefix α. The algorithm processes all the pairs $(sid, start)$ of $ProjSDB$ one by one (line 2), and searches for the lowest location of α in the sequence s corresponding to the sid of that sequence in SDB (lines 6-8).

In the worst case, PROJECTSDB processes all the items of all sequences. So, the time complexity is $O(\ell \times m)$, with $m = \#SDB$ and ℓ is the length of the longest sequence in SDB. The worst case space complexity of pseudo-projection is $O(m)$, since we need to store for each sequence only a pair $(sid, start)$, while for the standard projection the space complexity is $O(m \times \ell)$. Clearly, the pseudo-projection takes much less space than the standard projection.

4.4 Filtering

Ensuring DC on PREFIX-PROJECTION($P, SDB, minsup$) is equivalent to finding a sequential pattern of length $(\ell - 1)$ and then checking whether this pattern remains a frequent pattern when extended to any item d_ℓ in $D(P_\ell)$. Thus, finding such an assignment (i.e. support) is as much as difficult than the original problem of sequential pattern mining. [20] has proved that the problem of counting the number of maximal[3] frequent patterns in a database of sequences is #P-complete, thereby proving the NP-hardness of the problem of mining maximal frequent sequences. The difficulty is due to the exponential number of candidates that should be parsed to find the frequent patterns. Thus, finding, for every variable $P_i \in P$ and for every $d_i \in D(P_i)$, an assignment σ satisfying PREFIX-PROJECTION($P, SDB, minsup$) s.t. $\sigma(P_i) = d_i$ is of exponential nature.

---

[3] A sequential pattern p is maximal if there is no sequential pattern q such that $p \preceq q$.

Algorithm 2. FILTER-PREFIX-PROJECTION(SDB, σ, i, P, $minsup$)

Data: SDB: initial database; σ: current prefix $\langle \sigma(P_1), \ldots, \sigma(P_i) \rangle$; $minsup$: the minimum support threshold; $\mathcal{PSDB}$: internal data structure of PREFIX-PROJECTION for storing pseudo-projected databases

begin

```
1      if (i ≥ 2 ∧ σ(Pᵢ) = □) then
2          for j ← i + 1 to ℓ do
3              ⌊ Pⱼ ← □;
4          return True;

       else
5          𝒫𝒮𝒟ℬᵢ ← PROJECTSDB(SDB, 𝒫𝒮𝒟ℬᵢ₋₁, ⟨σ(Pᵢ)⟩);
6          if (#𝒫𝒮𝒟ℬᵢ < minsup) then
7              ⌊ return False ;

           else
8              ℱℐ ← GETFREQITEMS(SDB, 𝒫𝒮𝒟ℬᵢ, minsup) ;
9              for j ← i + 1 to ℓ do
10                 foreach a ∈ D(Pⱼ) s.t.(a ≠ □ ∧ a ∉ ℱℐ) do
11                     ⌊ D(Pⱼ) ← D(Pⱼ) − {a};

12             return True;
```

FUNCTION GETFREQITEMS (SDB, $ProjSDB$, $minsup$) ;

Data: SDB: the initial database; $ProjSDB$: pseudo-projected database; $minsup$: the minimum support threshold; $ExistsItem$, $SupCount$: internal data structures using a hash table for support counting over items;

begin

```
13     SupCount[] ← {0, ..., 0};  F ← ∅ ;
14     for each pair (sid, start) ∈ ProjSDB do
15         ExistsItem[] ← {false, ..., false}; s ← SDB[sid] ;
16         for i ← start to #s do
17             a ← s[i] ;
18             if (¬ExistsItem[a]) then
19                 SupCount[a] ← SupCount[a] + 1; ExistsItem[a] ← true;
20                 if (SupCount[a] ≥ minsup) then
21                     ⌊ F ← F ∪ {a};

22     return F;
```

So, the filtering of the PREFIX-PROJECTION constraint maintains a consistency lower than DC. This consistency is based on specific properties of the projected databases (see Proposition 3), and anti-monotonicity of the frequency constraint (see Proposition 4), and resembles forward-checking regarding Proposition 3. PREFIX-PROJECTION is considered as a global constraint, since all variables share the same internal data structures that awake and drive the filtering.

Algorithm 2 describes the pseudo-code of the filtering algorithm of the PREFIX-PROJECTION constraint. It is an incremental filtering algorithm that should be run when some i first variables are assigned according to the following lexicographic ordering $\langle P_1, P_2, \ldots, P_\ell \rangle$ of variables of P. It exploits internal data-structures enabling to enhance the filtering algorithm. More precisely, it uses an incremental data structure, denoted $\mathcal{PSDB}$, that stores the intermediate pseudo-projections of SDB, where $\mathcal{PSDB}_i$ ($i \in [0, \ldots, \ell]$) corresponds to the σ-projected database of the current partial assignment $\sigma = \langle \sigma(P_1), \ldots, \sigma(P_i) \rangle$ (also called prefix) of variables $\langle P_1, \ldots, P_i \rangle$, and $\mathcal{PSDB}_0 = \{(sid, 1) | (sid, s) \in SDB\}$ is the initial pseudo-projected database of SDB (case where $\sigma = \langle\rangle$). It also uses a hash table indexing the items $\mathcal{I}$ into integers $(1 \ldots \#\mathcal{I})$ for an efficient support counting over items (see function `getFreqItems`).

Algorithm 2 takes as input the current partial assignment $\sigma = \langle\sigma(P_1), \ldots, \sigma(P_i)\rangle$ of variables $\langle P_1, \ldots, P_i\rangle$, the length i of σ (i.e. position of the last assigned variable in P) and the minimum support threshold $minsup$. It starts by checking if the last assigned variable P_i is instantiated to $\square$ (line 1). In this case, the end of sequence is reached (since value $\square$ can only appear at the end) and the sequence $\langle\sigma(P_1), \ldots, \sigma(P_i)\rangle$ constitutes a frequent pattern in SDB; hence the algorithm sets the remaining $(\ell - i)$ unassigned variables to $\square$ and returns $true$ (lines 2-4). Otherwise, the algorithm computes incrementally $\mathcal{PSDB}_i$ from $\mathcal{PSDB}_{i-1}$ by calling function PROJECTSDB (see Algorithm 1). Then, it checks in line 6 whether the current assignment σ is a $legal$ prefix for the constraint (see Proposition 2). This is done by computing the size of $\mathcal{PSDB}_i$. If this size is less than $minsup$, we stop growing σ and we return $false$. Otherwise, the algorithm computes the set of locally frequent items $\mathcal{F}_\mathcal{I}$ in $\mathcal{PSDB}_i$ by calling function `getFreqItems` (line 8).

Function `getFreqItems` processes all the entries of the pseudo-projected database one by one, counts the number of first occurrences of items a (i.e. $SupCount[a]$) in each entry $(sid, start)$, and keeps only the frequent ones (lines 13-21). This is done by using $ExistsItem$ data structure. After the whole pseudo-projected database has been processed, the frequent items are returned (line 22), and Algorithm 2 updates the current domains of variables P_j with $j \geq (i+1)$ by pruning inconsistent values, thus avoiding searching over not frequent items (lines 9-11).

Proposition 5. *In the worst case, filtering with* PREFIX-PROJECTION *global constraint can be achieved in* $O(m \times \ell + m \times d + \ell \times d)$. *The worst case space complexity of* PREFIX-PROJECTION *is* $O(m \times \ell)$.

Proof: Let ℓ be the length of the longest sequence in SDB, $m = \#SDB$, and $d = \#\mathcal{I}$. Computing the pseudo-projected database $\mathcal{PSDB}_i$ can be done in $O(m \times \ell)$: for each sequence (sid, s) of SDB, checking if σ occurs in s is $O(\ell)$ and there are m sequences. The total complexity of function GETFREQITEMS is $O(m \times (\ell + d))$. Lines (9-11) can be achieved in $O(\ell \times d)$. So, the whole complexity is $O(m \times \ell + m \times (\ell + d) + \ell \times d) = O(m \times \ell + m \times d + \ell \times d)$. The space complexity of the filtering algorithm lies in the storage of the $\mathcal{PSDB}$ internal data structure. In the worst case, we have to store ℓ pseudo-projected databases. Since each pseudo-projected database requires $O(m)$, the worst case space complexity is $O(m \times \ell)$. $\square$

4.5   Encoding of SPM Constraints

Usual SPM constraints (see Section 2.2) can be reformulated in a straightforward way. Let P be the unknown pattern.
- *Minimum size constraint:* $size(P, \ell_{min}) \equiv \bigwedge_{i=1}^{i=\ell_{min}} (P_i \neq \square)$
- *Item constraint:* let V be a subset of items, l and u two integers s.t. $0 \leq l \leq u \leq \ell$. $item(P, V) \equiv \bigwedge_{t \in V} \text{Among}(P, \{t\}, l, u)$ enforces that items of V should occur

Table 2. Dataset Characteristics.

dataset	$\#SDB$	$\#\mathcal{I}$	avg $(\#s)$	$\max_{s \in SDB} (\#s)$	type of data
Leviathen	5834	9025	33.81	100	book
Kosarak	69999	21144	7.97	796	web click stream
FIFA	20450	2990	34.74	100	web click stream
BIBLE	36369	13905	21.64	100	bible
Protein	103120	24	482	600	protein sequences
data-200K	200000	20	50	86	synthetic dataset
PubMed	17527	19931	29	198	bio-medical text

at least l times and at most u times in P. To forbid items of V to occur in P, l and u must be set to 0.

- *Regular expression constraint*: let A_{reg} be the deterministic finite automaton encoding the regular expression exp. $reg(P, exp) \equiv \texttt{Regular}(P, A_{reg})$.

5   Experimental Evaluation

This section reports experiments on several real-life datasets from [3, 6, 18] of large size having varied characteristics and representing different application domains (see Table 2). Our objective is (1) to compare our approach to existing CP methods as well as to state-of-the-art methods for SPM in terms of scalability which is a major issue of existing CP methods, (2) to show the flexibility of our approach allowing to handle different constraints simultaneously.

Experimental Protocol. The implementation of our approach was carried out in the `Gecode` solver[4]. All experiments were conducted on a machine with a processor Intel X5670 and 24 GB of memory. A time limit of 1 hour has been used. For each dataset, we varied the *minsup* threshold until the methods are not able to complete the extraction of all patterns within the time limit. ℓ was set to the length of the longest sequence of SDB. The implementation and the datasets used in our experiments are available online[5]. We compare our approach (indicated by `PP`) with:

1. two CP encodings [12], the most efficient CP methods for SPM: `global-p.f` and `decomposed-p.f`;
2. state-of-the-art methods for SPM : and `cSpade`;
3. `SMA` [18] for SPM under regular expressions.

We used the author's `cSpade` implementation [6] for SPM, the publicly available implementations of by Y. Tabei [7] and the `SMA` implementation [8] for SPM under regular expressions. The implementation [9] of the two CP encodings was

---

[4] http://www.gecode.org
[5] https://sites.google.com/site/prefixprojection4cp/
[6] http://www.cs.rpi.edu/~zaki/www-new/pmwiki.php/Software/
[7] https://code.google.com/p/prefixspan/
[8] http://www-kdd.isti.cnr.it/SMA/
[9] https://dtai.cs.kuleuven.be/CP4IM/cpsm/

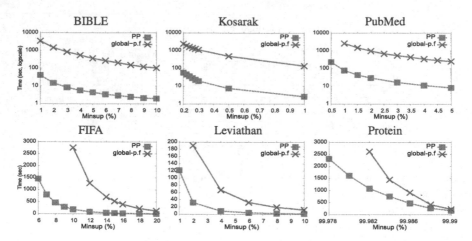

Fig. 1. Comparing PP with `global-p.f` for SPM on real-life datasets: CPU times.

carried out in the `Gecode` solver. All methods have been executed on the same machine.

(a) Comparing with CP Methods for SPM. First we compare PP with the two CP encodings `global-p.f` and `decomposed-p.f` (see Section 3.2). CPU times (in logscale for BIBLE, Kosarak and PubMed) of the three methods are shown on Fig. 1. First, `decomposed-p.f` is the least performer method. On all the datasets, it fails to complete the extraction within the time limit for all values of *minsup* we considered. Second, PP largely dominates `global-p.f` on all the datasets: PP is more than an order of magnitude faster than `global-p.f`. The gains in terms of CPU times are greatly amplified for low values of *minsup*. On BIBLE (resp. PubMed), the speed-up is 84.4 (resp. 33.5) for *minsup* equal to 1%. Another important observation that can be made is that, on most of the datasets (except BIBLE and Kosarak), `global-p.f` is not able to mine for patterns at very low frequency within the time limit. For example on FIFA, PP is able to complete the extraction for values of *minsup* up to 6% in 1, 457 seconds, while `global-p.f` fails to complete the extraction for *minsup* less than 10%.

To complement the results given by Fig. 1, Table 3 reports for different datasets and different values of *minsup*, the number of calls to the propagate routine of `Gecode` (column 5), and the number of nodes of the search tree (column 6). First, PP explores less nodes than `global-p.f`. But, the difference is not huge (gains of 45% and 33% on FIFA and BIBLE respectively). Second, our approach is very effective in terms of number of propagations. For PP, the number of propagations remains small (in thousands for small values of *minsup*) compared to `global-p.f` (in millions). This is due to the huge number of reified constraints used in `global-p.f` to encode the subsequence relation. On the contrary, our PREFIX-PROJECTION global constraint does not require any reified constraints nor any extra variables to encode the subsequence relation.

Table 3. PP vs. global-p.f.

Dataset	minsup (%)	#PATTERNS	CPU times (s)		#PROPAGATIONS		#NODES	
			PP	global-p.f	PP	global-p.f	PP	global-p.f
FIFA	20	938	**8.16**	129.54	**1884**	11649290	**1025**	1873
	18	1743	**13.39**	222.68	**3502**	19736442	**1922**	3486
	16	3578	**24.39**	396.11	**7181**	35942314	**3923**	7151
	14	7313	**44.08**	704	**14691**	65522076	**8042**	14616
	12	16323	**86.46**	1271.84	**32820**	126187396	**18108**	32604
	10	40642	**185.88**	2761.47	**81767**	266635050	**45452**	81181
BIBLE	10	174	**1.98**	105.01	**363**	4189140	**235**	348
	8	274	**2.47**	153.61	**575**	5637671	**362**	548
	6	508	**3.45**	270.49	**1065**	8592858	**669**	1016
	4	1185	**5.7**	552.62	**2482**	15379396	**1575**	2371
	2	5311	**15.05**	1470.45	**11104**	39797508	**7048**	10605
	1	23340	**41.4**	3494.27	**49057**	98676120	**31283**	46557
PubMed	5	2312	**8.26**	253.16	**4736**	15521327	**2833**	4619
	4	3625	**11.17**	340.24	**7413**	20643992	**4428**	7242
	3	6336	**16.51**	536.96	**12988**	29940327	**7757**	12643
	2	13998	**28.91**	955.54	**28680**	50353208	**17145**	27910
	1	53818	**77.01**	2581.51	**110133**	124197857	**65587**	107051
Protein	99.99	127	**165.31**	219.69	**264**	26731250	**172**	221
	99.988	216	**262.12**	411.83	**451**	44575117	**293**	390
	99.986	384	**467.96**	909.47	**805**	80859312	**514**	679
	99.984	631	**753.3**	1443.92	**1322**	132238827	**845**	1119
	99.982	964	**1078.73**	2615	**2014**	201616651	**1284**	1749
	99.98	2143	**2315.65**	−	**4485**	−	**2890**	−

(b) Comparing with ad hoc Methods for SPM. Our second experiment compares PP with state-of-the-art methods for SPM. Fig. 2 shows the CPU times of the three methods. First, cSpade obtains the best performance on all datasets (except on Protein). However, PP exhibits a similar behavior as cSpade, but it is less faster (not counting the highest values of minsup). The behavior of cSpade on Protein is due to the vertical representation format that is not appropriated in the case of databases having large sequences and small number of distinct items, thus degrading the performance of the mining process. Second, PP which also uses the concept of projected databases, clearly outperforms on all datasets. This is due to our filtering algorithm combined together with incremental data structures to manage the projected databases. On FIFA, is not able to complete the extraction for minsup less than 12%, while our approach remains feasible until 6% within the time limit. On Protein, fails to complete the extraction for all values of minsup we considered. These results clearly demonstrate that our approach competes well with state-of-the-art methods for SPM on large datasets and achieves scalability while it is a major issue of existing CP approaches.

(c) SPM under size and item constraints. Our third experiment aims at assessing the interest of pushing simultaneously different types of constraints. We impose on the PubMed dataset usual constraints such as *the minimum frequency* and the *minimum size* constraints and other useful constraints expressing some linguistic knowledge such as *the item constraint*. The goal is to retain sequential patterns which convey linguistic regularities (e.g., gene - rare disease relationships) [3]. The *size constraint* allows to remove patterns that are too small w.r.t. the number of items (number of words) to be relevant patterns. We tested this constraint with ℓ_{min} set to 3. *The item constraint* imposes that the extracted patterns must contain the item GENE and the item DISEASE. As no ad hoc

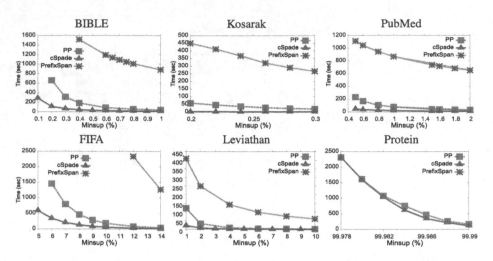

Fig. 2. Comparing PREFIX-PROJECTION with state-of-the-art algorithms for SPM.

Table 4. PP vs. `global-p.f` under minimum size and item constraints.

Dataset	minsup (%)	#PATTERNS	CPU times (s)		#PROPAGATIONS		#NODES	
			PP	global-p.f	PP	global-p.f	PP	global-p.f
PubMed	5	279	6.76	252.36	7878	12234292	2285	4619
	4	445	8.81	339.09	12091	16475953	3618	7242
	3	799	12.35	535.32	20268	24380096	6271	12643
	2	1837	20.41	953.32	43088	42055022	13888	27910
	1	7187	49.98	2574.42	157899	107978568	52508	107051

method exists for this combination of constraints, we only compare PP with `global-p.f`. Fig. 3 shows the CPU times and the number of sequential patterns extracted with and without constraints. First, pushing simultaneously the two constraints enables to reduce significantly the number of patterns. Moreover, the CPU times for PP decrease slightly whereas for `global-p.f` (with and without constraints), they are almost the same. This is probably due to the weak communication between the m `exists-embedding` reified global constraints and the two constraints. This reduces significantly the quality of the whole filtering. Second (see Table 4), when considering the two constraints, PP clearly dominates `global-p.f` (speed-up value up to 51.5). Moreover, the number of propagations performed by PP remains very small as compared to `global-p.f`. Fig. 3c compares the two methods under the minimum size constraint for different values of ℓ_{min}, with $minsup$ fixed to 1%. Once again, PP is always the best performer method (speed-up value up to 53.1). These results also confirm what we observed previously, namely the weak communication between reified global constraints and constraints imposed on patterns (i.e., size and item constraints).

(d) SPM under regular constraints. Our last experiment compares PP-REG against two variants of SMA: SMA-1P (SMA one pass) and SMA-FC (SMA Full Check). Two datasets are considered from [18]: one synthetic dataset (data-

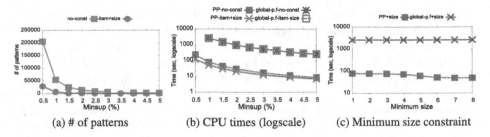

Fig. 3. Comparing PP with `global-p.f` under minimum size and item constraints on PubMed.

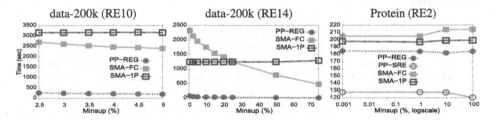

Fig. 4. Comparing PREFIX-PROJECTION with SMA for SPM under RE constraint.

200k), and one real-life dataset (Protein). For data-200k, we used two RE: RE10 $\equiv$ $A^*B(B|C)D^*EF^*(G|H)I^*$ and RE14 $\equiv A^*(Q|BS^*(B|C))D^*E(I|S)^*(F|H)G^*R$. For Protein, we used RE2 $\equiv (S|T) \cdot (R|K)$ (where . represents any symbol). Fig. 4 reports CPU-times comparison. On the synthetic dataset, our approach is very effective. For RE14, our method is more than an order of magnitude faster than SMA. On Protein, the gap between the 3 methods shrinks, but our method remains effective. For the particular case of RE2, the `Regular` constraint can be substituted by restricting the domain of the first and third variables to $\{S, T\}$ and $\{R, K\}$ respectively (denoted as PP-SRE), thus improving performances.

6 Conclusion

We have proposed the global constraint PREFIX-PROJECTION for sequential pattern mining. PREFIX-PROJECTION uses a concise encoding and provides an efficient filtering based on specific properties of the projected databases, and anti-monotonicity of the frequency constraint. When this global constraint is integrated into a CP solver, it enables to handle several constraints simultaneously. Some of them like size, item membership and regular expression are considered in this paper. Another point of strength, is that, contrary to existing CP approaches for SPM, our global constraint does not require any reified constraints nor any extra variables to encode the subsequence relation. Finally, although PREFIX-PROJECTION is well suited for constraints on sequences, it would require to be adapted to handle constraints on subsequence relations like gap.

Experiments performed on several real-life datasets show that our approach clearly outperforms existing CP approaches and competes well with ad hoc methods on large datasets and achieves scalability while it is a major issue of CP approaches. As future work, we intend to handle constraints on set of sequential patterns such as closedness, relevant subgroup and skypattern constraints.

Acknowledgments. The authors would like to thank the anonymous referees for their valuable comments. This work is partly supported by the ANR (French Research National Agency) funded projects Hybride ANR-11-BS002-002.

References

1. Agrawal, R., Srikant, R.: Mining sequential patterns. In: Yu, P.S., Chen, A.L.P. (eds.) ICDE, pp. 3–14. IEEE Computer Society (1995)
2. Ayres, J., Flannick, J., Gehrke, J., Yiu, T.: Sequential pattern mining using a bitmap representation. In: KDD 2002, pp. 429–435. ACM (2002)
3. Béchet, N., Cellier, P., Charnois, T., Crémilleux, B.: Sequential pattern mining to discover relations between genes and rare diseases. In: CBMS (2012)
4. Beldiceanu, N., Contejean, E.: Introducing global constraints in CHIP. Journal of Mathematical and Computer Modelling **20**(12), 97–123 (1994)
5. Coquery, E., Jabbour, S., Saïs, L., Salhi, Y.: A SAT-based approach for discovering frequent, closed and maximal patterns in a sequence. In: ECAI, pp. 258–263 (2012)
6. Fournier-Viger, P., Gomariz, A., Gueniche, T., Soltani, A., Wu, C., Tseng, V.: SPMF: A Java Open-Source Pattern Mining Library. J. of Machine Learning Resea. **15**, 3389–3393 (2014)
7. Garofalakis, M.N., Rastogi, R., Shim, K.: Mining sequential patterns with regular expression constraints. IEEE Trans. Knowl. Data Eng. **14**(3), 530–552 (2002)
8. Guns, T., Nijssen, S., Raedt, L.D.: Itemset mining: A constraint programming perspective. Artif. Intell. **175**(12–13), 1951–1983 (2011)
9. Kemmar, A., Ugarte, W., Loudni, S., Charnois, T., Lebbah, Y., Boizumault, P., Crémilleux, B.: Mining relevant sequence patterns with cp-based framework. In: ICTAI, pp. 552–559 (2014)
10. Li, C., Yang, Q., Wang, J., Li, M.: Efficient mining of gap-constrained subsequences and its various applications. ACM Trans. Knowl. Discov. Data 6(1), 2:1–2:39 (2012)
11. Métivier, J.P., Loudni, S., Charnois, T.: A constraint programming approach for mining sequential patterns in a sequence database. In: ECML/PKDD Workshop on Languages for Data Mining and Machine Learning (2013)
12. Negrevergne, B., Guns, T.: Constraint-based sequence mining using constraint programming. In: Michel, L. (ed.) CPAIOR 2015. LNCS, vol. 9075, pp. 288–305. Springer, Heidelberg (2015)
13. Novak, P.K., Lavrac, N., Webb, G.I.: Supervised descriptive rule discovery: A unifying survey of contrast set, emerging pattern and subgroup mining. Journal of Machine Learning Research **10** (2009)
14. Pei, J., Han, J., Mortazavi-Asl, B., Pinto, H., Chen, Q., Dayal, U., Hsu, M.: PrefixSpan: Mining sequential patterns by prefix-projected growth. In: ICDE, pp. 215–224. IEEE Computer Society (2001)
15. Pei, J., Han, J., Wang, W.: Mining sequential patterns with constraints in large databases. In: CIKM 202, pp. 18–25. ACM (2002)

16. Pesant, G.: A regular language membership constraint for finite sequences of variables. In: Wallace, M. (ed.) CP 2004. LNCS, vol. 3258, pp. 482–495. Springer, Heidelberg (2004)

17. Srikant, R., Agrawal, R.: Mining sequential patterns: Generalizations and performance improvements. In: EDBT, pp. 3–17 (1996)

18. Trasarti, R., Bonchi, F., Goethals, B.: Sequence mining automata: A new technique for mining frequent sequences under regular expressions. In: ICDM 2008, pp. 1061–1066 (2008)

19. Yan, X., Han, J., Afshar, R.: CloSpan: mining closed sequential patterns in large databases. In: Barbará, D., Kamath, C. (eds.) SDM. SIAM (2003)

20. Yang, G.: Computational aspects of mining maximal frequent patterns. Theor. Comput. Sci. **362**(1–3), 63–85 (2006)

21. Zaki, M.J.: Sequence mining in categorical domains: Incorporating constraints. In: Proceedings of the 2000 ACM CIKM International Conference on Information and Knowledge Management, McLean, VA, USA, November 6–11, pp. 422–429 (2000)

22. Zaki, M.J.: SPADE: An efficient algorithm for mining frequent sequences. Machine Learning **42**(1/2), 31–60 (2001)

On Tree-Preserving Constraints

Shufeng Kong[1], Sanjiang Li[1($\boxtimes$)], Yongming Li[2], and Zhiguo Long[1]

[1] QCIS, FEIT, University of Technology Sydney, Sydney, Australia
{Shufeng.Kong,Zhiguo.Long}@student.uts.edu.au, Sanjiang.Li@uts.edu.au
[2] College of Computer Science, Shaanxi Normal University, Xi'an, China
liyongm@snnu.edu.cn

Abstract. Tree convex constraints are extensions of the well-known row convex constraints. Just like the latter, every path-consistent tree convex constraint network is globally consistent. This paper studies and compares three subclasses of tree convex constraints which are called chain-, path- and tree-preserving constraints respectively. While the tractability of the subclass of chain-preserving constraints has been established before, this paper shows that every chain- or path-preserving constraint network is in essence the disjoint union of several independent connected row convex constraint networks, and hence (re-)establish the tractability of these two subclasses of tree convex constraints. We further prove that, when enforcing arc- and path-consistency on a tree-preserving constraint network, in each step, the network remains tree-preserving. This ensures the global consistency of the tree-preserving network if no inconsistency is detected. Moreover, it also guarantees the applicability of the partial path-consistency algorithm to tree-preserving constraint networks, which is usually more efficient than the path-consistency algorithm for large sparse networks. As an application, we show that the class of tree-preserving constraints is useful in solving the scene labelling problem.

1 Introduction

Constraint satisfaction problems (CSPs) have been widely used in many areas, such as scene labeling [10], natural language parsing [15], picture processing [16], and spatial and temporal reasoning [5,14]. Since deciding consistency of CSP instances is NP-hard in general, lots of efforts have been devoted to identify tractable subclasses. These subclasses are usually obtained by either restricting the topology of the underlying graph of the constraint network (being a tree or having treewidth bounded by a constant) or restricting the type of the allowed constraints between variables (cf. [17]).

In this paper, we are mainly interested in the second type of restriction. Montanari [16] shows that path-consistency is sufficient to guarantee that a network is globally consistent if the relations are all monotone. Van Beek and Dechter [17] generalise monotone constraints to a larger class of row convex constraints, which are further generalised to tree convex constraints by Zhang and Yap [20]. These constraints also have the nice property that every path-consistent constraint network is globally consistent.

© Springer International Publishing Switzerland 2015
G. Pesant (Ed.): CP 2015, LNCS 9255, pp. 244–261, 2015.
DOI: 10.1007/978-3-319-23219-5_18

However, neither row convex constraints nor tree convex constraints are closed under composition and intersection, the main operations of path-consistent algorithms. This means enforcing path-consistency may destroy row and tree-convexity. Deville et al. [6] propose a tractable subclass of row convex constraints, called connected row convex (CRC) constraints, which are closed under composition and intersection. Zhang and Freuder [18] also identify a tractable subclass for tree convex constraints, called *locally chain convex and strictly union closed* constraints. They also propose the important notion of consecutive constraints. Kumar [13] shows that the subclass of arc-consistent consecutive tree convex (ACCTC) constraints is tractable by providing a polynomial time randomised algorithm. But, for the ACCTC problems, "it is not known whether there are efficient deterministic algorithms, neither is it known whether arc- and path-consistency ensures global consistency on those problems." [18]

In this paper, we study and compare three subclasses of tree convex constraints which are called, respectively, chain-, path- and tree-preserving constraints. Chain-preserving constraints are exactly "locally chain convex and strictly union closed" constraints and ACCTC constraints are strictly contained in the subclass of tree-preserving constraints. We first show that every chain- or path-preserving constraint network is in essence the disjoint union of several independent CRC constraint networks and then prove that enforcing arc- and path-consistency on a tree-preserving constraint network ensures global consistency. This provides an affirmative answer to the above open problem raised in [18]. Note also that our result is more general than that of Kumar [13] as we do not require the constraint network to be arc-consistent. Moreover, when enforcing arc- and path-consistent on a tree-preserving constraint network, in each step, the network remains tree-preserving. This guarantees the applicability of the partial path-consistency algorithm [2] to tree-preserving constraint networks, which is usually more efficient than the path-consistency algorithm for large sparse networks. We further show that a large subclass of the trihedral scene labelling problem [10,12] can be modelled by tree-preserving constraints.

In the next section, we introduce basic notations and concepts that will be used throughout the paper. Chain-, path-, and tree-preserving constraints are discussed in Sections 3, 4, and 5, respectively. Application of tree-preserving constraints in the scene labelling problem is shown in Section 6. Section 7 briefly discusses the connection with majority operators [11] and concludes the paper.

2   Preliminaries

Let D be a domain of a variable x. A graph structure can often be associated to D such that there is a bijection between the vertices in the graph and the values in D. If the graph is connected and acyclic, i.e. a tree, then we say it is a *tree domain* of x. Tree domains arise naturally in e.g. scene labeling [18] and combinatorial auctions [4]. We note that, in this paper, we have a specific tree domain D_x for each variable x.

In this paper, we distinguish between tree and rooted tree. Standard notions from graph theory are assumed. In particular, the *degree* of a node a in a graph G, denoted by $\deg(a)$, is the number of neighbors of a in G.

Definition 1. *A* tree *is a connected graph without any cycle. A tree is* rooted *if it has a specified node r, called the* root *of the tree. Given a tree T, a subgraph I is called a* subtree *of T if I is connected. An empty set is a subtree of any tree.*

Let T be a (rooted) tree and I a subtree of T. I is a path *(*chain*, resp.*) *in T if each node in I has at most two neighbors (at most one child, resp.) in I. Given two nodes p, q in T, the unique path that connects p to q is denoted by $\pi_{p,q}$.*

Suppose a is a node of a tree T. A branch *of a is a connected component of $T \setminus \{a\}$.*

Throughout this paper, we always associate a subtree with its node set.

Definition 2. *A* binary constraint *has the form $(x\delta y)$, where x, y are two variables with domains D_x and D_y and δ is a binary relation from D_x to D_y, or $\delta \subseteq D_x \times D_y$. For simplicity, we often denote by δ this constraint. A value $u \in D_x$ is* supported *if there exists a value v in D_y s.t. $(u, v) \in \delta$. In this case, we say v is a* support *of u. We say a subset F of D_x is* unsupported *if every value in F is not supported. Given $A \subseteq D_x$, the* image *of A under δ is defined as $\delta(A) = \{b \in D_y : (\exists a \in A)(a, b) \in \delta\}$. For $A = \{a\}$ that contains only one value, without confusion, we also use $\delta(a)$ to represent $\delta(\{a\})$.*

A binary constraint network consists of a set of variables $V = \{x_1, x_2, ..., x_n\}$ with a finite domain D_i for each variable $x_i \in V$, and a set Δ of binary constraints over the variables of V. The usual operations on relations, e.g., intersection ($\cap$), composition ($\circ$), and inverse ($^{-1}$), are applicable to constraints. As usual, we assume that there is at most one constraint for any ordered pair of variables (x, y). Write δ_{xy} for this constraint if it exists. In this paper, we always assume δ_{xy} is the inverse of δ_{yx}, and if there is no constraint for (x, y), we assume δ_{xy} is the universal constraint.

Definition 3. *[8, 9] A constraint network Δ over n variables is k-consistent iff any consistent instantiation of any distinct $k - 1$ variables can be consistently extended to any k-th variable. We say Δ is* strongly k-consistent *iff it is j-consistent for all $j \leq k$; and say Δ is* globally consistent *if it is strongly n-consistent. 2- and 3-consistency are usually called* arc- *and* path-consistency *respectively.*

Definition 4. *Let x, y be two variables with finite tree domains $T_x = (D_x, E_x)$ and $T_y = (D_y, E_y)$ and δ a constraint from x to y. We say δ, w.r.t. T_x and T_y, is*

- tree convex *if the image of every value a in D_x (i.e. $\delta(a)$) is a (possibly empty) subtree of T_y;*
- consecutive *if the image of every edge in T_x is a subtree in T_y;*
- path-preserving *if the image of every path in T_x is a path in T_y.*
- tree-preserving *if the image of every subtree in T_x is a subtree in T_y.*

In case T_x and T_y are rooted, we say δ, w.r.t. T_x and T_y, is

- chain-preserving *if the image of every chain in T_x is a chain in T_y.*

Chain-preserving constraints are exactly those "locally chain convex and strictly union closed" constraints defined in [18].

CRC constraints are special tree convex constraints defined over chain domains. The following definition of CRC constraints is equivalent to the one given in [6].

Definition 5. *Let x, y be two variables with finite tree domains T_x and T_y, where T_x and T_y are chains. A constraint δ from x to y is* connected row convex *(CRC), w.r.t. T_x and T_y, if both δ and δ^{-1} are chain-preserving.*

The class of CRC constraints is tractable and closed under intersection, inverse, and composition [6].

Definition 6. *A binary constraint network Δ over variables in V and tree domains T_x ($x \in V$) is called tree convex, chain-, path-, or tree-preserving if every constraint $\delta \in \Delta$ is tree convex, chain-, path-, or tree-preserving, respectively. A CRC constraint network is defined similarly.*

Proposition 1. *Every chain-, path-, or tree-preserving constraint (network) is consecutive and every path-preserving constraint (network) is tree-preserving. Moreover, every arc-consistent consecutive tree convex (ACCTC) constraint (network) is tree-preserving.*

Not every consecutive tree convex constraint (or chain-preserving constraint) is tree-preserving, but such a constraint becomes tree-preserving if it is arc-consistent.

Lemma 1. *[20] Let T be a tree and suppose t_i ($i = 1, .., n$) are subtrees of T. Then $\bigcap_{i=1}^{n} t_i$ is nonempty iff $t_i \cap t_j$ is nonempty for every $1 \le i \ne j \le n$.*

Lemma 2. *Let T be a tree and t, t' subtrees of T. Suppose $\{u, v\}$ is an edge in T. If $u \in t$ and $v \in t'$, then $t \cup t'$ is a subtree of T; if, in addition, $u \notin t'$ and $v \notin t$, then $t \cap t' = \varnothing$.*

Using Lemma 1, Zhang and Yap [20] proved

Theorem 1. *A tree-convex constraint network is globally consistent if it is path-consistent.*

3   Chain-Preserving Constraints

Zhang and Freuder [18] have proved that any consistent chain-preserving network can be transformed to an equivalent globally consistent network by enforcing arc- and path-consistency. This implies that the class of chain-preserving constraints is tractable. We next show that every chain-preserving constraint network Δ can be uniquely divided into a small set of k CRC constraint networks $\Delta_1, ..., \Delta_k$ s.t. Δ is consistent iff at least one of Δ_i is consistent.

We first recall the following result used in the proof of [18, Theorem 1].

Proposition 2. *Let Δ be a chain-preserving constraint network over tree domains T_x ($x \in V$). If no inconsistency is detected, then Δ remains chain-preserving after enforcing arc-consistency.*

We note that these tree domains over variables in V may need adjustment in the process of enforcing arc-consistency. Here by *adjustment* we mean adding edges to the tree structure so that it remains connected when unsupported values are deleted.

Definition 7. *Let T be a tree with root. A chain $[a, a^*]$ in T is called an irreducible perfect chain (ip-chain) if (i) a is the root or has one or more siblings; (ii) a^* is a leaf node or has two or more children; and (iii) every node in $[a, a^*)$ has only one child.*

Note that it is possible that $a = a^*$. In fact, this happens when a is the root or has one or more siblings and has two or more children. An ip-chain as defined above is a minimum chain which satisfies (1) in the following lemma.

Lemma 3. *Suppose δ_{xy} and δ_{yx} are arc-consistent and chain-preserving w.r.t. rooted trees T_x and T_y. Assume $[a, a^*] \subseteq T_x$ is an ip-chain. Then*

$$\delta_{yx}(\delta_{xy}([a, a^*])) = [a, a^*] \tag{1}$$

and $\delta_{xy}([a, a^*])$ is also an ip-chain in T_y.

Proof. W.l.o.g., we suppose $\hat{a}$ is the parent of a, a' is a sibling of a, and $a_1, a_2, ..., a_k$ ($k \geq 2$) are the children of a^*.

Because δ_{xy} and δ_{yx} are arc-consistent and chain-preserving, $\delta_{xy}([a, a^*])$ is a non-empty chain in T_y, written $[b, b^*]$, and so is $\delta_{yx}([b, b^*])$. Suppose $\delta_{yx}([b, b^*])$ is not $[a, a^*]$. This implies that either $\hat{a}$ or one of $a_1, a_2, ..., a_k$ is in $\delta_{yx}([b, b^*])$.

Suppose $\hat{a} \in \delta_{yx}([b, b^*])$. Then there exists $\hat{b} \in [b, b^*]$ such that $(\hat{a}, \hat{b}) \in \delta_{xy}$. By $\hat{b} \in [b, b^*] = \delta_{xy}([a, a^*])$, we have $a^+ \in [a, a^*]$ s.t. $(a^+, \hat{b}) \in \delta_{xy}$. Therefore, $[\hat{a}, a^+]$ is contained in $\delta_{yx}(\delta_{xy}(\{\hat{a}\}))$. Recall that a' is a sibling of a. Because $\delta_{yx}(\delta_{xy}([\hat{a}, a']))$ contains $\hat{a}, a', a^+$, it cannot be a chain in T_x. A contradiction. Therefore, $\hat{a} \notin \delta_{yx}([b, b^*])$.

Suppose, for example, $a_1 \in \delta_{yx}([b, b^*])$. Then there exist $b' \in [b, b^*]$ s.t. $(a_1, b') \in \delta_{xy}$ and $\bar{a} \in [a, a^*]$ s.t. $(\bar{a}, b') \in \delta_{xy}$. We have $\delta_{yx}(\delta_{xy}(\{\bar{a}\}) \supseteq [\bar{a}, a_1]$ and $\delta_{yx}(\delta_{xy}([\bar{a}, a_2])$ contains $\{\bar{a}, a_1, a_2\}$, which is not a subset of a chain. Therefore, $a_i \notin \delta_{yx}([b, b^*])$.

So far, we have proved $\delta_{yx}(\delta_{xy}([a, a^*])) = [a, a^*]$. We next show $[b, b^*]$ is also an ip-chain. First, we show every node in $[b, b^*)$ has only one child. Suppose not and $b' \in [b, b^*)$ has children b_1, b_2 with $b_1 \in (b', b^*]$. Since $\delta_{xy}([\hat{a}, a^*])$ is a chain that contains $[b, b^*]$, we know $(\hat{a}, b_2)$ is not in δ_{xy}. Furthermore, as $\delta_{yx}(\{b', b_2\})$ is a chain in T_x and the image of b_2 is disjoint from $[a, a^*]$, we must have $(a_i, b_2) \in \delta_{xy}$ for some child a_i of a^*. Note that then $\delta_{xy}([a, a_i])$ contains $[b, b^*]$ and b_2 and thus is not a chain. This contradicts the chain-preserving property of δ_{xy}. Hence, every node in $[b, b^*)$ has only one child. In other words, $[b, b^*]$ is contained in an ip-chain $[u, v]$.

By the result we have proved so far, we know $\delta_{xy}(\delta_{yx}([u,v])) = [u,v]$ and $\delta_{yx}([u,v])$ is contained in an ip-chain in T_x. Because $[a,a^*] = \delta_{yx}([b,b^*]) \subseteq \delta_{yx}([u,v])$ is an ip-chain, we know $\delta_{yx}([u,v])$ is exactly $[a,a^*]$. Therefore, we have $[u,v] = \delta_{xy}(\delta_{yx}([u,v])) = \delta_{xy}([a,a^*]) = [b,b^*]$. This proves that $[b,b^*]$ is an ip-chain in T_y. □

Using the above result, we can break T_x into a set of ip-chains by deleting the edges from each node a to its children if a has two or more children. Write $\mathcal{I}_x$ for the set of ip-chains of T_x. Similar operation and notation apply to T_y. It is clear that two different ip-chains in $\mathcal{I}_x$ are disjoint and δ_{xy} naturally gives rise to a bijection from $\mathcal{I}_x$ to $\mathcal{I}_y$.

Lemma 4. *Suppose Δ is an arc-consistent and chain-preserving constraint network over tree domains T_x ($x \in V$). Fix a variable $x \in V$ and let $\mathcal{I}_x = \{I_x^1, ..., I_x^l\}$ be the set of ip-chains of T_x. Then, for every $y \neq x$ in V, the set of ip-chains in T_y is $\{\delta_{xy}(I_x^1), \delta_{xy}(I_x^2), ..., \delta_{xy}(I_x^l)\}$. Write Δ_i for the restriction of Δ to I_x^i. Then each Δ_i is a CRC constraint network and Δ is consistent iff at least one Δ_i is consistent.*

The following result asserts that the class of chain-preserving constraints is tractable.

Theorem 2. *Let Δ be a chain-preserving constraint network. If no inconsistency is detected, then enforcing arc- and path-consistency determines the consistency of Δ and transforms Δ into a globally consistent network.*

Proof. First, by Proposition 2, we transform Δ into an arc-consistent and chain-preserving constraint network if no inconsistency is detected. Second, by Lemma 4, we reduce the consistency of Δ to the consistency of the CRC constraint networks $\Delta_1, ..., \Delta_l$. By [6], we know enforcing path-consistency transforms a CRC constraint network into a globally consistent one if no inconsistency is detected. If enforcing arc- and path-consistency does not detect any inconsistency, then the result is a set of at most l globally consistent CRC networks Δ_i', the union of which is globally consistent and equivalent to Δ. □

Lemma 4 also suggests that we can use the variable elimination algorithm for CRC constraints [19] to more efficiently solve chain-preserving constraints.

4   Path-Preserving Constraints

At first glance, path-preserving constraints seem to be more general than chain-preserving constraints, but Fig. 1(a,b) show that they are in fact incomparable.

We show the class of path-preserving constraints is also tractable by establishing its connection with CRC constraints.

We have the following simple results.

Lemma 5. *Suppose δ_{xy} and δ_{yx} are path-preserving (tree-preserving) w.r.t. tree domains T_x and T_y. Let t be a subtree of T_x and δ_{xy}' and δ_{yx}' the restrictions of δ_{xy} and δ_{yx} to t. Then both δ_{xy}' and δ_{yx}' are path-preserving (tree-preserving).*

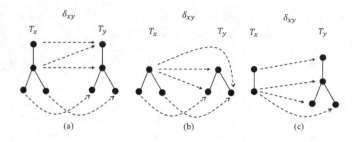

Fig. 1. (a) A chain- but not path-preserving constraint; (b) A path- but not chain-preserving constraint; (c) A tree-preserving but neither path- nor chain-preserving constraint.

Lemma 6. *Suppose δ_{xy} is nonempty and path-preserving (tree-preserving) w.r.t. tree domains T_x and T_y. If $v \in T_y$ has no support in T_x under δ_{yx}, then all supported nodes of T_y are in the same branch of v. That is, every node in any other branch of v is not supported under δ_{yx}.*

Proof. Suppose a, b are two supported nodes in T_y. There exist u_1, u_2 in T_x s.t. $u_1 \in \delta_{yx}(a)$ and $u_2 \in \delta_{yx}(b)$. By $\delta_{yx} = \delta_{xy}^{-1}$, we have $a \in \delta_{xy}(u_1)$ and $b \in \delta_{xy}(u_2)$. Hence $a, b \in \delta_{xy}(\pi_{u_1, u_2})$. Since δ_{xy} is path-preserving (tree-preserving), $\delta_{xy}(\pi_{u_1, u_2})$ is a path (tree) in T_y. If a, b are in two different branches of v, then $\pi_{a,b}$ must pass v and hence we must have $v \in \delta_{xy}(\pi_{u_1, u_2})$. This is impossible as v has no support. □

It is worth noting that this lemma does not require δ_{yx} to be path- or tree-preserving.

The following result then follows directly.

Proposition 3. *Let Δ be a path-preserving (tree-preserving) constraint network over tree domains T_x ($x \in V$). If no inconsistency is detected, then Δ remains path-preserving (tree-preserving) after enforcing arc-consistency.*

Proof. Enforcing arc-consistency on Δ only removes values which have no support under some constraints. For any $y \in V$, if v is an unsupported value in T_y, then, by Lemma 6, every supported value of T_y is located in the same branch of v. Deleting all these unsupported values from T_y, we get a subtree t of T_y. Applying Lemma 5, the restricted constraint network to t remains path-preserving (tree-preserving). □

Definition 8. *Let T be a tree. A path π in T is maximal if there exists no path π' in T that strictly contains π.*

We need three additional lemmas to prove the main result.

Lemma 7. *Suppose δ_{xy} and δ_{yx} are arc-consistent and path-preserving w.r.t. tree domains T_x and T_y. If π is a maximal path in T_x, then $\delta_{xy}(\pi)$ is a maximal path in T_y.*

Lemma 8. *Suppose δ_{xy} and δ_{yx} are arc-consistent and path-preserving w.r.t. T_x and T_y. Assume a is a node in T_x with $\deg(a) > 2$. Then there exists a unique node $b \in T_y$ s.t. $(a, b) \in \delta_{xy}$. Moreover, $\deg(a) = \deg(b)$.*

Proof. Suppose $\pi = a_0 a_1 ... a_k$ is a maximal path in T_x and $\pi^* = b_0 b_1 ... b_l$ is its image under δ_{xy} in T_y. W.l.o.g. we assume $k, l \geq 1$. Suppose a_i is a node in π s.t. $\deg(a_i) > 2$ and $a' \notin \pi$ is another node in T_x s.t. $\{a_i, a'\}$ is an edge in T_x. Suppose $\delta_{xy}(a_i) = [b_j, b_{j'}]$ and $j' > j$.

Because π is a maximal path and π^* is its image, we know $\delta_{xy}(a') \cap \pi^* = \varnothing$. Consider the edge $\{a', a_i\}$. Since $\delta_{xy}(\{a', a_i\})$ is a path in T_y, there exists a node $b' \in \delta_{xy}(a')$ s.t. either $\{b', b_j\}$ or $\{b', b_{j'}\}$ is an edge in T_y. Suppose w.l.o.g. $\{b', b_{j'}\}$ is in T_y. Note that $\pi^* = [b_0, b_l]$ is contained in the union of $\delta_{xy}([a', a_0])$ and $\delta_{xy}([a', a_k])$. In particular, b_l is in either $\delta_{xy}([a', a_0])$ or $\delta_{xy}([a', a_k])$. Let us assume $b_l \in \delta_{xy}([a', a_0])$. Then $b_l, b_j, b_{j'}, b'$ (which are not on any path) are contained in the path $\delta_{xy}([a', a_0])$, a contradiction. Therefore, our assumption that $\delta_{xy}(a_i) = [b_j, b_{j'}]$ and $j' > j$ is incorrect. That is, the image of a_i under δ_{xy} is a singleton, say, $\{b_j\}$. We next show $\deg(b_j) = \deg(a_i)$.

Because $\delta_{xy}(a_i) = \{b_j\}$, the image of each neighbor of a_i in T_x must contain a neighbor of b_j, as δ_{xy} is path-preserving. Moreover, two different neighbors a_i', a_i'' of a_i cannot map to the same neighbor b_j' of b_j. This is because the image of b_j' under δ_{yx}, which is a path in T_x, contains a_i' and a_i'', and hence also contains a_i. This contradicts the assumption $\delta_{xy}(a_i) = \{b_j\}$. This shows that $\deg(a_i) = \deg(b_j)$. $\qquad\square$

Definition 9. *Let T be a tree. A path π from a to a^* in T is called an* irreducible perfect path *(ip-path) if (i) every node on path π has degree 1 or 2; and (ii) any neighbour of a (or a^*) that is not on π has degree 3 or more.*

Let $F_x = \{a \in T_x : \deg(a) > 2\}$ and $F_y = \{b \in T_y : \deg(b) > 2\}$. Then δ_{xy}, when restricted to F_x, is a bijection from F_x to F_y. Removing all edges incident to a node in F_x, we obtain a set of pairwise disjoint paths in T_x. These paths are precisely the ip-paths of T_x. Write $\mathcal{P}_x$ for this set. Then δ_{xy} induces a bijection from $\mathcal{P}_x$ to $\mathcal{P}_y$.

Lemma 9. *Suppose Δ is an arc-consistent and path-preserving constraint network over tree domains T_x ($x \in V$). Fix a variable $x \in V$ and let $\mathcal{P}_x = \{\pi_x^1, ..., \pi_x^l\}$ be the set of ip-paths in T_x. Then, for every $y \neq x$, the set of ip-paths in T_y is $\{\delta_{xy}(\pi_x^1), ..., \delta_{xy}(\pi_x^l)\}$. Write Δ_i for the restriction of Δ to π_x^i. Then each Δ_i is a CRC constraint network and Δ is consistent iff at least one Δ_i is consistent.*

Thus the class of path-preserving constraints is tractable.

Theorem 3. *Let Δ be a path-preserving constraint network. If no inconsistency is detected, then enforcing arc- and path-consistency determines the consistency of Δ and transforms Δ into a globally consistent network.*

The proof is analogous to that of Theorem 2. Lemma 9 suggests that we can use the variable elimination algorithm for CRC constraints [19] to more efficiently solve path-preserving constraints.

5   Tree-Preserving Constraints

It is easy to see that every arc-consistent chain- or path-preserving constraint is tree-preserving, but Fig. 1(c) shows that the other direction is not always true.

In this section, we show that the class of tree-preserving constraints is tractable. Given a tree-preserving constraint network Δ, we show that, when enforcing arc- and path-consistency on Δ, in each step, the network remains tree-preserving. Hence, enforcing arc- and path-consistency on Δ will transform it to an equivalent globally consistent network if no inconsistency is detected. Moreover, we show that the partial path-consistency algorithm (PPC) of [2] is applicable to tree-preserving constraint networks. PPC is more efficient than path-consistency algorithm for large sparse constraints.

5.1   Enforcing Arc- and Path-Consistency Preserves Tree-Preserving

Unlike CRC and chain-preserving constraints, removing a value from a domain may change the tree-preserving property of a network. Instead, we need to remove a 'trunk' from the tree domain or just keep one branch.

Definition 10. *Suppose $a \neq b$ are two nodes of a tree T that are not neighbors. The* trunk *between a, b, written M_{ab}, is defined as the connected component of $T \backslash \{a, b\}$ which contains all internal nodes of $\pi_{a,b}$ (see Fig.2). The* M-contraction *of T by $M_{a,b}$, denoted by $T \ominus M_{a,b}$, is the tree obtained by removing nodes in $M_{a,b}$ and adding an edge $\{a, b\}$ to T.*

T

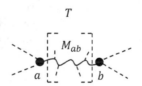

Fig. 2. M_{ab} is a trunk of tree T.

Lemma 10. *Suppose δ_{xy} and δ_{yx} are arc-consistent and tree-preserving w.r.t. tree domains T_x and T_y. Suppose a, b are two nodes in T_x s.t. $\delta_{xy}(a) \cup \delta_{xy}(b)$ is not connected in T_y. Then there exist $r, s \in T_y$ s.t. $r \in \delta_{xy}(a)$, $s \in \delta_{xy}(b)$, and $\delta_{yx}(M_{r,s}) \subseteq M_{a,b}$. Let T_y^* be the domain obtained by deleting from T_y all nodes v s.t. $\delta_{yx}(v) \subseteq M_{a,b}$. Then T_y^* becomes a tree if we add the edge $\{r, s\}$. Moreover, δ_{xy} and δ_{yx} remain arc-consistent and tree-preserving when restricted to $T_x \ominus M_{a,b}$ and T_y^*.*

Proof. Choose $r \in \delta_{xy}(a)$ and $s \in \delta_{xy}(b)$ such that the path $\pi_{r,s}$ from r to s in T_y is a shortest one among $\{\pi_{r',s'} : r' \in \delta_{xy}(a), s' \in \delta_{xy}(b)\}$. In particular, we have $\pi_{r,s} \cap (\delta_{xy}(a) \cup \delta_{xy}(b)) = \{r, s\}$. We assert that the image of every node v in $M_{r,s}$ under δ_{yx} is contained in $M_{a,b}$. Suppose otherwise and there exists u in $T_x \setminus M_{a,b}$ s.t. $(u, v) \in \delta_{xy}$. Assume that u is in the same connected component as a. Since the subtree $\delta_{yx}(\pi_{v,s})$ contains u and b, it also contains a. This implies that there is a node v' on $\pi_{v,s}$ which is in $\delta_{xy}(a)$. This is impossible as $v \in M_{r,s}$ and $\delta_{xy}(a) \cap \pi_{r,s} = \{r\}$. Therefore $\delta_{yx}(v) \subseteq M_{a,b}$ for any $v \in M_{r,s}$. Hence $\delta_{yx}(M_{r,s}) \subseteq M_{a,b}$ holds.

It is clear that, when restricted to $T_x \ominus M_{a,b}$ and $T_y \ominus M_{r,s}$, $\delta_{xy}(\{a,b\})$ is connected and so is $\delta_{yx}(\{r, s\})$. For any other edge $\{a', b'\}$ in $T_x \ominus M_{a,b}$, by $\delta_{yx}(M_{r,s}) \subseteq M_{a,b}$, $\delta_{xy}(\{a', b'\}) \cap M_{r,s} = \varnothing$ and the image of $\{a', b'\}$ is unchanged (hence connected) after the M-contraction of T_y. This shows that δ_{xy} is consecutive when restricted to $T_x \ominus M_{a,b}$. Furthermore, since every node in $T_x \ominus M_{a,b}$ is supported in $T_y \ominus M_{r,s}$, we know δ_{xy} is also tree-preserving when restricted to $T_x \ominus M_{a,b}$.

It is possible that there is a node $v \in T_y \ominus M_{r,s}$ s.t. $\delta_{yx}(v) \subseteq M_{a,b}$. We assert that any v like this has at most one branch in $T_y \setminus M_{r,s}$ s.t. there is a node v' in the branch which is supported under δ_{yx} by a node in $T_x \setminus M_{a,b}$. Because δ_{xy} is tree-preserving when restricted to $T_x \ominus M_{a,b}$, this follows immediately from Lemma 6. This implies that, if we remove all these nodes v s.t. $\delta_{yx}(v) \subseteq M_{a,b}$ from $T_y \ominus M_{r,s}$, the domain is still connected. As a consequence, the two constraints remain arc-consistent and tree preserving when restricted to $T_x \ominus M_{a,b}$ and T_y^*. □

In general, we have

Lemma 11. *Let Δ be an arc-consistent and tree-preserving constraint network over tree domains T_x ($x \in V$). Suppose $x \in V$ and $M_{a,b}$ is a trunk in T_x. When restricted to $T_x \ominus M_{a,b}$ and enforcing arc-consistency, Δ remains tree-preserving if we modify each T_y ($y \in V$) by deleting all unsupported nodes and adding some edges.*

Proof (Sketch). The result follows from Lemmas 10 and 5. One issue we need to take care of is how two trunks interact. Suppose x, y, z are three different variables and M, M' are trunks to be contracted from T_x and T_y respectively. Applying Lemma 10 to the constraints between x and z and, separately, to the constraints between y and z, we get two different trunks, say $M_{a,b}$ and $M_{c,d}$, to be contracted from the same tree domain T_z. Can we do this (i.e. applying Lemma 10) one by one? Does the order of the contractions matter? To answer these questions, we need to know what is exactly the union of two trunks. There are in essence ten configurations as shown in Fig. 3. The union of two trunks can be the whole tree, a branch, a trunk, or two disjoint trunks. If the union is the whole tree, then the network is inconsistent; if it is a branch, then we can remove it directly; if it is a trunk or two disjoint trunks, then we can use Lemma 10 to contract them one by one in either order. □

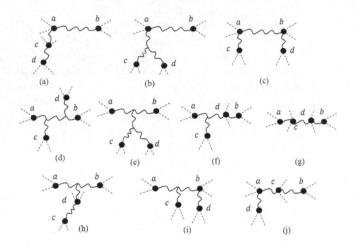

Fig. 3. Possible configurations of trunks $M_{a,b}$ and $M_{c,d}$.

Lemma 12. *Assume δ_{xy} and δ'_{xy} are two arc-consistent and tree-preserving constraints w.r.t. trees T_x and T_y. Let $\delta^*_{xy} = \delta_{xy} \cap \delta'_{xy}$. Suppose $u \in T_x$ and $\delta_{xy}(u) \cap \delta'_{xy}(u) = \varnothing$. Then the supported values of δ^*_{xy} in T_x are in at most two branches of u.*

Proof. Suppose u_1, u_2, u_3 are three supported values of δ^*_{xy} in T_x that are in three different branches of u. Take $w_i \in \delta_{xy}(u_i) \cap \delta'_{xy}(u_i)$. For each i, we have either $w_i \notin \delta_{xy}(u)$ or $w_i \notin \delta'_{xy}(u)$. Recall that π_{w_i,w_j} denotes the unique path π_{w_i,w_j} that connects w_i to w_j $(1 \leq i \neq j \leq 3)$. There are two subcases. (1) One node is on the path that connects the other two. Suppose w.l.o.g. w_3 is between w_1 and w_2. If $w_3 \notin \delta_{xy}(u)$, then there exist $v_1 \in \pi_{w_1,w_3}$ and $v_2 \in \pi_{w_3,w_2}$ s.t. $v_1, v_2 \in \delta_{xy}(u)$. This is because the image of π_{w_1,w_3} and the image of π_{w_3,w_2} (under δ_{yx}) both contain u. That is, $v_1, v_2 \in \delta_{xy}(u)$ but $w_3 \notin \delta_{xy}(u)$. Since $\delta_{xy}(u)$ is a subtree and w_3 is on the path π_{v_1,v_2}, this is impossible. The case when $w_3 \notin \delta'_{xy}(u)$ is analogous. (2) The three nodes are in three different branches of a node w. In this case, we note there exists a node between any path π_{w_i,w_j} which is a support of u under δ_{xy}. It is easy to see that w itself is a support of u under δ_{xy}. Similarly, we can show w is also a support of u under δ'_{xy}. Since both subcases lead to a contradiction, we know the supported values of δ^*_{xy} are in at most two branches of u.   □

Actually, this lemma shows

Corollary 1. *Assume δ_{xy} and δ'_{xy} are two arc-consistent and tree-preserving constraints w.r.t. trees T_x and T_y. Then those unsupported values of $\delta_{xy} \cap \delta'_{xy}$ in T_x are in a unique set of pairwise disjoint branches and trunks.*

Similar to Lemma 10, we have

Lemma 13. *Suppose δ_{xy} and δ'_{xy} are arc-consistent and tree-preserving constraints w.r.t. trees T_x and T_y and so are δ_{yx} and δ'_{yx}. Let $\delta^*_{xy} = \delta_{xy} \cap \delta'_{xy}$. Assume $\{u, v\}$ is an edge in T_x s.t. $\delta^*_{xy}(u) \cup \delta^*_{xy}(v)$ is disconnected in T_y. Then there exist $r \in \delta^*_{xy}(u)$ and $s \in \delta^*_{xy}(v)$ s.t. every node in $M_{r,s}$ is unsupported under δ^*_{yx}.*

Proof. Write $T_r = \delta^*_{xy}(u)$ and $T_s = \delta^*_{xy}(v)$. Clearly, T_r and T_s are nonempty subtrees of T_y. Since they are disconnected, there exist $r \in T_r$, $s \in T_s$ s.t. $\pi_{r,s} \cap (T_r \cup T_s) = \{r, s\}$ (see Fig. 4 for an illustration). Write $A = \delta_{xy}(u)$, $B = \delta_{xy}(v)$, $C = \delta'_{xy}(u)$ and $D = \delta'_{xy}(v)$. We show every node in $M_{r,s}$ is not supported under δ^*_{yx}.

Suppose w is an arbitrary internal node on $\pi_{r,s}$. We first show w is not supported under δ^*_{yx}. Note $w \in A \cup B$, $w \in C \cup D$, $w \notin A \cap C$, and $w \notin B \cap D$. There are two cases according to whether $w \in A$. If $w \in A$, then we have $w \notin C$, $w \in D$, and $w \notin B$. If $w \notin A$, then we have $w \in B$, $w \notin D$, and $w \in C$. Suppose w.l.o.g. $w \in A$. By $w \in A = \delta_{xy}(u)$, we have $u \in \delta_{yx}(w)$; by $w \notin B = \delta_{xy}(v)$, we have $v \notin \delta_{yx}(w)$. Similarly, we have $u \notin \delta'_{yx}(w)$ and $v \in \delta'_{yx}(w)$. Thus subtree $\delta'_{yx}(w)$ is disjoint from subtree $\delta_{yx}(w)$. This shows $\delta^*_{yx}(w) = \varnothing$ and hence w is not supported under δ^*_{yx}.

Second, suppose w_1 is an arbitrary node in $M_{r,s}$ s.t. w_1 is in a different branch of w to r and s, i.e. $\pi_{w,w_1} \cap (T_r \cup T_s) = \varnothing$. We show w_1 is not supported under δ^*_{yx} either.

Again, we assume $w \in A$. In this case, we have $u \in \delta_{yx}(w) \subseteq \delta_{yx}(\pi_{w,w_1})$ and $v \in \delta'_{yx}(w) \subseteq \delta'_{yx}(\pi_{w,w_1})$. As $\pi_{w,w_1} \cap (T_r \cup T_s) = \varnothing$, we have $\pi_{w,w_1} \cap T_r = \pi_{w,w_1} \cap A \cap C = \varnothing$. As $\pi_{w,w_1} \cap A \neq \varnothing$ and $A \cap C \neq \varnothing$, by Lemma 1, we must have $\pi_{w,w_1} \cap \delta'_{xy}(u) = \varnothing$. This shows $u \notin \delta'_{yx}(\pi_{w,w_1})$. Similarly, we can show $v \notin \delta_{yx}(\pi_{w,w_1})$. Thus subtree $\delta'_{yx}(\pi_{w,w_1})$ is disjoint from subtree $\delta_{yx}(\pi_{w,w_1})$ and, hence, $\delta^*_{yx}(\pi_{w,w_1}) = \varnothing$. This proves that w_1 is not supported under δ^*_{yx} either.

In summary, every node in $M_{r,s}$ is unsupported. □

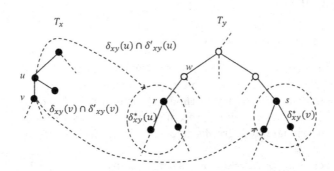

Fig. 4. Illustration of proof of Lemma 13.

Proposition 4. *[18] Assume δ_{xz} and δ_{zy} are two tree-preserving constraints w.r.t. trees T_x, T_y, and T_z. Then their composition $\delta_{xz} \circ \delta_{zy}$ is tree-preserving.*

At last, we give the main result of this section.

Theorem 4. *Let Δ be a tree-preserving constraint network. If no inconsistency is detected, then enforcing arc- and path-consistency determines the consistency of Δ and transforms Δ into a globally consistent network.*

Proof. If we can show that Δ is still tree-preserving after enforcing arc and path-consistency, then by Theorem 1 the new network is globally consistent if no inconsistency is detected.

By Proposition 3, Δ remains tree-preserving after enforcing arc-consistency. To enforce path-consistency on Δ, we need to call the following updating rule

$$\delta_{xy} \leftarrow \delta_{xy} \cap (\delta_{xz} \circ \delta_{zy}) \tag{2}$$

for $x, y, z \in V$ until the network is stable.

Suppose Δ is arc-consistent and tree-preserving w.r.t. trees T_x for $x \in V$ before applying (2). Note that if $\delta^* = \delta_{xy} \cap (\delta_{xz} \circ \delta_{zy})$ (as well as its converse) is arc-consistent, then $\delta^*(u)$ is nonempty for any node u in T_x. By Corollary 1, no branches or trunks need to be pruned in either T_x or T_y. Furthermore, by Lemma 13, $\delta^*(u) \cup \delta^*(v)$ is connected for every edge $\{u, v\}$ in T_x as there are no unsupported nodes in T_y under the converse of δ^*. Therefore δ^* is arc-consistent and consecutive, hence, tree-preserving.

If δ^* is not arc consistent, then we delete all unsupported values from T_x and T_y and enforce arc-consistency on Δ. If no inconsistency is detected then we have an updated arc-consistent and tree-preserving network by Lemma 11. Still write Δ for this network and recompute $\delta^* = \delta_{xy} \cap (\delta_{xz} \circ \delta_{zy})$ and repeat the above procedure until either inconsistency is detected or δ^* is arc-consistent. Note that, after enforcing arc-consistency, the composition $\delta_{xz} \circ \delta_{zy}$ may have changed.

Once arc-consistency of δ^* is achieved, we update δ_{xy} with δ^* and continue the process of enforcing path-consistency until Δ is path-consistent or an inconsistency is detected.   □

5.2   Partial Path-Consistency

The partial path-consistency (PPC) algorithm was first proposed by Bliek and Sam-Haroud [2]. The idea is to enforce path consistency on sparse graphs by triangulating instead of completing them. Bliek and Sam-Haroud demonstrated that, as far as CRC constraints are concerned, the pruning capacity of path consistency on triangulated graphs and their completion are identical on the common edges.

An undirected graph $G = (V, E)$ is *triangulated* or *chordal* if every cycle of length greater than 3 has a chord, i.e. an edge connecting two non-consecutive vertices of the cycle. For a constraint network $\Delta = \{v_i \delta_{ij} v_j : 1 \leq i, j \leq n\}$ over $V = \{v_1, ..., v_n\}$, the *constraint graph* of Δ is the undirected graph $G(\Delta) = (V, E(\Delta))$, for which we have $(v_i, v_j) \in E(\Delta)$ iff δ_{ij} is not a universal constraint.

Given a constraint network Δ and a graph $G = (V, E)$, we say Δ is *partial path-consistent w.r.t.* G iff for any $1 \leq i, j, k \leq n$ with $(v_i, v_j), (v_j, v_k), (v_i, v_k) \in E$ we have $\delta_{ik} \subseteq \delta_{ij} \circ \delta_{jk}$ [2].

Theorem 5. *Let Δ be a tree-preserving constraint network. Suppose $G = (V, E)$ is a chordal graph such that $E(\Delta) \subseteq E$. Then enforcing partial path-consistency on G is equivalent to enforcing path-consistency on the completion of G, in the sense that the relations computed for the constraints in G are identical.*

Proof. The proof is similar to the one given for CRC constraints [2, Theorem 3]. This is because, (i) when enforcing arc- and path-consistency on a tree-preserving constraint network, in each step, we obtain a new tree-preserving constraint network; and (ii) path-consistent tree convex constraint networks are globally consistent. □

Remark 1. Note that our definition and results of tree-preserving constraints can be straightforwardly extended to domains with acyclic graph structures (which are connected or not). We call such a structure a *forest* domain. Given a tree-preserving constraint network Δ over forest domains $F_1, ..., F_n$ of variables $v_1, ..., v_n$. Suppose F_i consists of trees $t_{i,1}, ..., t_{i,k_i}$. Note that the image of each tree, say $t_{i,1}$, of F_i under constraint R_{ij} is a subtree t of F_j. Assume t is contained in the tree $t_{j,s}$ of forest F_j. Then the image of $t_{j,s}$ under constraint R_{ji} is a subtree of $t_{i,1}$. This establishes, for any $1 \leq i \neq j \leq n$, a 1-1 correspondence between trees in F_i and trees in F_j if the image of each tree is nonempty. In this way, the consistency of Δ is reduced to the consistency of several parallel tree-preserving networks over tree domains.

6 Tree-Preserving Constraints and the Scene Labelling Problem

The scene labelling problem [10] is a classification problem where all edges in a line-drawing picture have to be assigned a label describing them. The scene labelling problem is NP-complete in general. This is true even in the case of the trihedral scenes, i.e. scenes where no four planes share a point [12].

Labels used in the scene labelling problem are listed as follows:

'+' The edge is convex which has both of its corresponding planes visible;

'−' The edge is concave which has both of its corresponding planes visible;

'→' Only one plane associated with the edge is visible, and when one moves in the direction indicated by the arrow, the pair of associated planes is *to the right*.

In the case of trihedral scenes, there are only four basic ways in which three plane surfaces can come together at a vertex [10]. A vertex projects in the picture into a 'V', 'W', 'Y' or 'T'-junction (each of these junction-types may appear with an arbitrary rotation in a given picture). A complete list of the labelled line configurations that are possible in the vicinity of a node in a picture is given in Fig. 5.

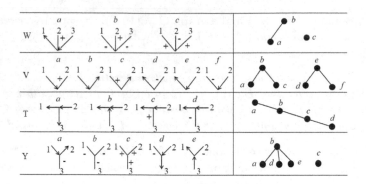

Fig. 5. Possible labelled line configurations of a junction in a picture and their corresponding forest structures.

In this section, we show that (i) every instance of the trihedral scene labelling problem can be modelled by a tree convex constraint network; (ii) a large subclass of the trihedral scene labelling problem can be modelled by tree-preserving constraints; (iii) there exists a scene labelling instance which can be modelled by tree-preserving constraints but not by chain- or CRC constraints.

A CSP for the scene labelling problem can be formulated as follows. Each junction in the line-drawing picture is a variable. The domains of the vertices are the possible configurations as shown in Fig. 5. The constraints between variables are simply that, if two variables share an edge, then the edge must be labeled the same at both ends.

Proposition 5. *Every instance of the trihedral scene labelling problem can be modelled by a tree convex constraint network. Furthermore, there are only 39 possible configurations of two neighbouring nodes in 2D projected pictures of 3D trihedral scenes, and 29 out of these can be modelled by tree-preserving constraints.*

Proof. The complete list of these configurations and their corresponding tree convex or tree-preserving constraints is attached as an appendix. Note that we do not consider T-junctions in line drawing pictures since they decompose into unary constraints. ☐

As a consequence, we know that these 29 configurations of the scene labelling problem with 2D line-drawing pictures can be solved by the path-consistency algorithm in polynomial time. Moreover, since it is NP-hard to decide if a trihedral scene labelling instance is consistent, we have the following corollary.

Corollary 2. *The consistency problem of tree convex constraint networks is NP-complete.*

We next give a scene labelling instance which can be modelled by tree-preserving constraints but not by chain-preserving or CRC constraints. Consider the line drawing in the left of the following figure and the constraints for the drawing listed in the right. One can easily verify that all constraints are tree-preserving w.r.t. the forest structures listed in Fig. 5, but, for example, δ_{21} is not chain-preserving for the forest structures illustrated in Fig. 5 and δ_{25} is not CRC.

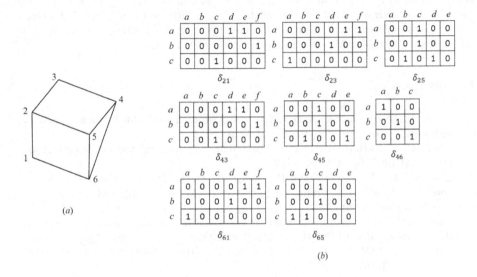

(a)

(b)

7 Further Discussion and Conclusion

In this paper, when formulating a CSP, we allow different variables have different tree domains. Feder and Vardi [7] and many other authors (see e.g. [1,11]) also considered CSPs which have a common domain D for all variables. These CSPs are called *one-sorted* in [3]. For one-sorted tree-preserving CSPs, we could also define a majority operator [11] under which the set of tree-preserving constraints is closed. This implies that the class of one-sorted tree-preserving CSPs has bounded strict width [7] and hence tractable. Indeed, such a majority operator ρ is defined as follows: for any three nodes a, b, c in a tree domain T, define $\rho(a, b, c)$ as the node d which is the intersection of paths $\pi_{a,b}$, $\pi_{a,c}$, and $\pi_{b,c}$. Following [3], it is straightforward to extend this result to multi-sorted tree-preserving CSPs.

In this paper, we identified two new tractable subclasses of tree convex constraint which are called path- and tree-preserving constraints, and proved that a chain- or path-preserving constraint network is in essence the disjoint union of several independent CRC constraint networks, and hence (re-)established the tractability of these constraints. More importantly, we proved that when enforcing arc- and path-consistency on a tree-preserving constraint network, in each step, the network remains tree-preserving. This implies that enforcing arc- and path-consistency will change a tree-preserving constraint network into a globally consistent constraint network. This also implies that the efficient partial path-consistent algorithm for large sparse networks is applicable for tree-preserving constraint network. As an application, we showed that a large class of the trihedral scene labelling problem can be modelled by tree-preserving constraints. This shows that tree-preserving constraints are useful in real world applications.

Acknowledgments. We sincerely thank the anonymous reviewers of CP-15 and IJCAI-15 for their very helpful comments. The majority operator was first pointed out to us by two reviewers of IJCAI-15. This work was partially supported by ARC (FT0990811, DP120103758, DP120104159) and NSFC (61228305).

References

1. Barto, L., Kozik, M.: Constraint satisfaction problems solvable by local consistency methods. Journal of ACM **61**(1), 3:1–3:19 (2014)
2. Bliek, C., Sam-Haroud, D.: Path consistency on triangulated constraint graphs. In: IJCAI 1999, pp. 456–461 (1999)
3. Bulatov, Andrei A., Jeavons, Peter G.: An algebraic approach to multi-sorted constraints. In: Rossi, Francesca (ed.) CP 2003. LNCS, vol. 2833, pp. 183–198. Springer, Heidelberg (2003)
4. Conitzer, V., Derryberry, J., Sandholm, T.: Combinatorial auctions with structured item graphs. In: AAAI 2004, pp. 212–218 (2004)
5. Dechter, R., Meiri, I., Pearl, J.: Temporal constraint networks. Artificial Intelligence **49**(1–3), 61–95 (1991)
6. Deville, Y., Barette, O., Hentenryck, P.V.: Constraint satisfaction over connected row convex constraints. Artificial Intelligence **109**(1–2), 243–271 (1999)
7. Feder, T., Vardi, M.Y.: The computational structure of monotone monadic snp and constraint satisfaction: A study through datalog and group theory. SIAM Journal on Computing **28**(1), 57–104 (1998)
8. Freuder, E.C.: Synthesizing constraint expressions. Communications of the ACM **21**(11), 958–966 (1978)
9. Freuder, E.C.: A sufficient condition for backtrack-free search. Journal of the ACM **29**(1), 24–32 (1982)
10. Huffman, D.A.: Impossible objects as nonsense sentences. Machine Intelligence **6**(1), 295–323 (1971)
11. Jeavons, P., Cohen, D.A., Cooper, M.C.: Constraints, consistency and closure. Artificial Intelligence **101**(1–2), 251–265 (1998)
12. Kirousis, L.M., Papadimitriou, C.H.: The complexity of recognizing polyhedral scenes. In: FOCS 1985, pp. 175–185 (1985)
13. Kumar, T.K.S.: Simple randomized algorithms for tractable row and tree convex constraints. In: AAAI 2006, pp. 74–79 (2006)
14. Li, S., Liu, W., Wang, S.: Qualitative constraint satisfaction problems: An extended framework with landmarks. Artificial Intelligence **201**, 32–58 (2013)
15. Maruyama, H.: Structural disambiguation with constraint propagation. In: ACL 1990, pp. 31–38 (1990)
16. Montanari, U.: Networks of constraints: Fundamental properties and applications to picture processing. Information Sciences **7**, 95–132 (1974)

17. Van Beek, P., Dechter, R.: On the minimality and global consistency of row-convex constraint networks. Journal of the ACM **42**(3), 543–561 (1995)
18. Zhang, Y., Freuder, E.C.: Properties of tree convex constraints. Artificial Intelligence **172**(12–13), 1605–1612 (2008)
19. Zhang, Y., Marisetti, S.: Solving connected row convex constraints by variable elimination. Artificial Intelligence **173**(12), 1204–1219 (2009)
20. Zhang, Y., Yap, R.H.C.: Consistency and set intersection. In: IJCAI 2003, pp. 263–270 (2003)

Modeling and Solving Project Scheduling with Calendars

Stefan Kreter[1], Andreas Schutt[2,3]([✉]), and Peter J. Stuckey[2,3]

[1] Operations Research Group, Institute of Management and Economics,
Clausthal University of Technology, 38678 Clausthal-Zellerfeld, Germany
stefan.kreter@tu-clausthal.de
[2] Optimisation Research Group, National ICT Australia, Melbourne, Australia
{andreas.schutt,peter.stuckey}@nicta.com.au
[3] Department of Computing and Information Systems, The University of Melbourne,
Melbourne, VIC 3010, Australia

Abstract. Resource-constrained project scheduling with the objective
of minimizing project duration (RCPSP) is one of the most studied
scheduling problems. In this paper we consider the RCPSP with general temporal constraints and calendar constraints. Calendar constraints
make some resources unavailable on certain days in the scheduling period
and force activity execution to be delayed while resources are unavailable. They arise in practice from, *e.g.*, unavailabilities of staff during
public holidays and weekends. The resulting problems are challenging
optimization problems. We develop not only four different constraint
programming (CP) models to tackle the problem, but also a specialized
propagator for the cumulative resource constraints taking the calendar
constraints into account. This propagator includes the ability to explain
its inferences so it can be used in a lazy clause generation solver. We
compare these models, and different search strategies on a challenging
set of benchmarks using a lazy clause generation solver. We close 83 of
the open problems of the benchmark set, and show that CP solutions
are highly competitive with existing MIP models of the problem.

1 Introduction

The resource-constrained project scheduling problem with general temporal and
calendar constraints (RCPSP/max-cal) is an extension of the well-known RCPSP
and RCPSP/max (see, *e.g.*, [14, Chap. 2]) through calendars. The RCPSP/max-
cal can be given as follows. For a set of activities, which require time and renewable resources for their execution, execution time intervals must be determined
in a way that minimum and maximum time lags between activities are satisfied,
the prescribed resource capacities are not exceeded, and the project duration is
minimized. The difference with RCPSP/max is that a calendar is given for each
renewable resource type that describes for each time period whether the resource
type is available or unavailable. Time periods of unavailability can occur, *e.g.*,
due to weekends or public holidays. The activities and time lags are dependent

© Springer International Publishing Switzerland 2015
G. Pesant (Ed.): CP 2015, LNCS 9255, pp. 262–278, 2015.
DOI: 10.1007/978-3-319-23219-5_19

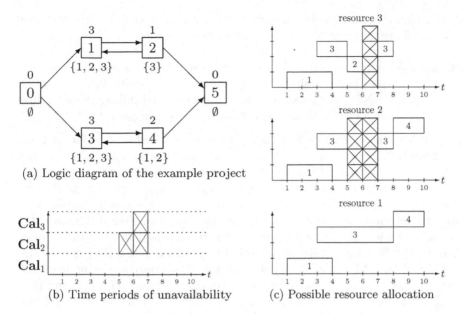

(a) Logic diagram of the example project

(b) Time periods of unavailability

(c) Possible resource allocation

Fig. 1. Illustrative Example for RCPSP/max-cal

on the resource calendars, too, and some activities can be interrupted for the duration of a break while others cannot be interrupted due to technical reasons. For the interruptible activities a start-up phase is given during which the activity is not allowed to be paused. Concerning the renewable resource types one distinguishes resource types that stay engaged or are blocked, respectively, during interruptions of activities that require it and resource types that are released and can be used to carry out other activities during interruptions.

Our motivation for developing CP models for the RCPSP/max-cal and using lazy clause generation to solve it lies in the very good results obtained by [18,19,20] solving RCPSP and RCPSP/max by lazy clause generation.

Example 1. Figure 1 shows an illustrative example with six activities and three renewable resource types. The project start (activity 0) and the project end (activity 5) are fictitious activities, *i.e.*, they do not require time or resources. A logic diagram of the project is given in Fig. 1(a) where each activity is represented by a node with the duration given above and the set of resource types used by the activity below the node. The arcs between the nodes represent time lags.

The calendars of the three renewable resource types are depicted in Fig. 1(b). If there is a box with an X for a resource type k and time t, then resource type k is not available at time t. Resource type 1 is always available and can be thought of as a machine. Resource types 2 and 3 can be thought of as different kinds of staff where resource type 2 (3) has a five-day (six-day) working week. In addition, assume that resource type 1 stays engaged or is blocked, respectively, during a break of an activity that requires resource type 1 for its execution while resource types 2 and 3 are released during interruptions of activities.

A possible resource allocation of the three renewable resource types is shown in Fig. 1(c). Activity 3 requires all renewable resource types for its execution. Since resource type 2 is not available in periods 6 and 7, activity 3 is interrupted during these periods. While resource type 1 stays engaged during the interruption, resource type 3 can be used to carry out activity 2 in period 6.   □

Few authors have dealt with calendars in project scheduling so far. A time planning method for project scheduling with the same calendar for each resource type is introduced in [23]. In [6] the RCPSP/max with different calendars for each renewable resource type is investigated for the first time but the start-up phase of the interruptible activities are not taken into account. [6] proposes methods to determine the earliest and latest start and completion times for the project activities and priority rule methods. Procedures to determine the earliest and latest start times if a start-up phase is taken into account are presented in [7] and [14, Sect. 2.11]. In addition, they sketch how priority-rule methods for the RCPSP/max can be adapted for calendars. In the approach in [7] and [14, Sect. 2.11] all resources stay engaged during interruptions of activities. Within the priority-rule methods in [6,7], and [14, Sect. 2.11] the procedures to determine the earliest and latest start times must be carried out in each iteration. Recently, a new time planning method, three binary linear model formulations, and a scatter search procedure for the RCPSP/max-cal were developed in [9]. Moreover, Kreter et al. [9] introduce a benchmark test set which is based on the UBO test set for RCPSP/max [8]. The time planning method determines all time and calendar feasible start times for the activities and absolute time lags depending on the start times of the activities once in advance and then uses this throughout the scatter search.

In CP, the works [3,4] respectively propose calendar constraints/rules for ILOG Schedule and Cosytech CHIP. The former [3] was generalized to intensity functions of activities in IBM ILOG CP Optimizer, while breaks of activities extend the length between their start and end times, only resource types that stay engaged can be modeled directly. The latter [4] introduces constraint rules in the global constraint `diffn` for parallel machine scheduling.

A practical application where calendars must be considered as well as other additional constraints can be found in batch scheduling [21]. Problems that are related to the RCPSP/max-cal are treated in [22,5]. An alternative approach to include calendars into project scheduling that makes use of calendar independent start-start, start-end, end-start, and end-end time lags is proposed in [22] and [5] studies the RCPSP with non-preemptive activity splitting, where an activity in process is allowed to pause only when resource levels are temporarily insufficient.

2   Problem Description

In this section we describe the RCPSP/max-cal formally and give an example instance. We use identifiers and definitions from [9]. In what follows, we assume that a project consists of a set $V := \{0, 1, \ldots, n, n+1\}$, $n \geq 1$, of activities, where 0 and $n+1$ represent the begin and the end of the project, respectively. Each activity i has a processing time $p_i \in \mathbb{N}_0$. Activities i with $p_i > 0$ are called real activities and the set of real activities is denoted by $V^r \subset V$. Activities 0 and $n+1$ as well as milestones, which specify significant events of the project and have a duration of $p_i = 0$, form the set $V^f = V \setminus V^r$ of fictitious activities.

A project completion deadline $\overline{d} \in \mathbb{N}$ has to be determined in order to define the time horizon of the calendars and the time axis is divided into intervals $[0, 1), [1, 2), \ldots, [\overline{d}-1, \overline{d})$ where a unit length time interval $[t-1, t)$ is also referred to as time period t. The set of renewable resource types is denoted by $\mathcal{R}$ and for each renewable resource type $k \in \mathcal{R}$ a resource capacity $R_k \in \mathbb{N}$ is given that must not be exceeded at any point in time. The amount of resource type k that is used constantly during the execution of activity $i \in V$ is given by $r_{ik} \in \mathbb{N}_0$. For fictitious activities $i \in V^f$ $r_{ik} := 0$ holds for all $k \in \mathcal{R}$. For each resource type a resource calendar is given.

Definition 1. *A calendar for resource* $k \in \mathcal{R}$ *is a step function* $\mathbf{Cal}_k(\cdot)$: $[0, \overline{d}) \rightarrow \{0, 1\}$ *continuous from the right at the jump points, where the condition*

$$\mathbf{Cal}_k(t) := \begin{cases} 1, & \text{if period } [\lfloor t \rfloor, \lfloor t+1 \rfloor) \text{ is a working period for } k \\ 0, & \text{if period } [\lfloor t \rfloor, \lfloor t+1 \rfloor) \text{ is a break period for } k \end{cases}$$

is satisfied.

With $\mathcal{R}_i := \{k \in \mathcal{R} \mid r_{ik} > 0\}$ indicating the set of resource types that is used to carry out activity $i \in V$, an activity calendar $\mathbf{C}_i(\cdot) : [0, \overline{d}) \rightarrow \{0, 1\}$ can be determined from the resource calendars as follows:

$$\mathbf{C}_i(t) := \begin{cases} \min_{k \in \mathcal{R}_i} \mathbf{Cal}_k(t), & \text{if } \mathcal{R}_i \neq \emptyset \\ 1, & \text{otherwise.} \end{cases}$$

Then, for every activity i and a point in time $t \in T := \{0, 1, \ldots, \overline{d}\}$ functions $next\_break_i(t)$ and $next\_start_i(t)$ give the start time and the end time of the next break after time t in calendar $\mathbf{C}_i$, respectively.

$$next\_break_i(t) := \min\{\tau \in T \mid \tau > t \wedge \mathbf{C}_i(\tau) = 0\}$$
$$next\_start_i(t) := \min\{\tau \in T \mid \tau > t \wedge \mathbf{C}_i(\tau) = 1 \wedge \mathbf{C}_i(\tau - 1) = 0\}$$

When calendars are present, we have to distinguish activities that can be interrupted for the duration of a break in the underlying activity calendar and activities that are not allowed to be interrupted. The set of (break-)interruptible activities is denoted by $V^{bi} \subset V$ and the set of non-interruptible activities is given by $V^{ni} = V \setminus V^{bi}$, where $V^f \subseteq V^{ni}$ holds. The execution of an activity $i \in V^{bi}$ must be interrupted at times t with $\mathbf{C}_i(t) = 0$, and the execution must be continued at the next point in time $\tau > t$ with $\mathbf{C}_i(\tau) = 1$. $S_i \in T$ indicates the start time and $E_i \in T$ represents the end of activity $i \in V$. Since the jump points in the calendars $\mathbf{Cal}_k$, $k \in \mathcal{R}$, are all integer valued, the points in time where an activity is interrupted or continued are integer valued, too. The completion time of activity $i \in V$ can be determined by $E_i(S_i) := \min\{t \mid \sum_{\tau=S_i}^{t-1} \mathbf{C}_i(\tau) = p_i\}$. For each activity $i \in V$ a start-up phase $\varepsilon_i \in \mathbb{N}_0$ is given during which activity i is not allowed to be interrupted. For all activities $i \in V^{ni}$ $\varepsilon_i := p_i$ holds. We assume that the underlying project begins at time 0, *i.e.*, $S_0 := 0$. Then, the project duration equals S_{n+1}. In addition, we assume that no activity $i \in V$ can be in execution before the project start, *i.e.*, $S_i \geq 0$, or after the project end, *i.e.*, $E_i \leq S_{n+1}$.

Between the activities a set A of minimum and maximum time lags is given. W.l.o.g. these time lags are defined between the start times of the activities (see [6,9]). For each time lag $\langle i, j \rangle \in A$, a resource set $\mathcal{R}_{ij} \subseteq \mathcal{R}$ and a length $\delta_{ij} \in \mathbb{Z}$ are given, from which we can compute a calendar $\mathbf{C}_{ij}(\cdot) : [0, \overline{d}] \to \{0, 1\}$ for each time lag by

$$\mathbf{C}_{ij}(t) := \begin{cases} \min_{k \in \mathcal{R}_{ij}} \mathbf{Cal}_k(t), & \text{if } \mathcal{R}_{ij} \neq \emptyset \\ 1, & \text{otherwise} \end{cases}$$

i.e., at least tu time units must elapse after the start of activity i before activity j can start where $tu = \min\{t \mid \sum_{\tau=S_i}^{t-1} \mathbf{C}_{ij}(\tau) = \delta_{ij}\}$.

With parameter ρ_k we indicate whether renewable resource types $k \in \mathcal{R}$ stay engaged or are blocked, respectively, during interruptions of activities that require it ($\rho_k = 1$) or are released and can be used to carry out other activities during interruptions ($\rho_k = 0$). A vector $S = (S_0, S_1, \ldots, S_{n+1})$ of all activity start times is called a schedule. Given a schedule S and point in time t the set of all real activities $i \in V^r$ that are started before but not completed at time t is called the active set and can be determined by $\mathcal{A}(S, t) := \{i \in V^r \mid S_i \leq t < E_i(S_i)\}$. Then, the resource utilization $r_k^{\mathrm{cal}}(S, t)$ of resource $k \in \mathcal{R}$ at time t according to schedule S can be computed by

$$r_k^{\mathrm{cal}}(S, t) := \sum_{i \in \mathcal{A}(S,t) \mid \mathbf{C}_i(t)=1} r_{ik} + \sum_{i \in \mathcal{A}(S,t) \mid \mathbf{C}_i(t)=0} r_{ik}\, \rho_k.$$

With the introduced notation the following mathematical formulation for the RCPSP/max-cal can be given (cf. [6]):

Minimize S_{n+1} (1)

subject to $\sum_{t=S_i}^{S_i+\varepsilon_i-1} \mathbf{C}_i(t) = \varepsilon_i$ $i \in V$ (2)

$\sum_{t=S_i}^{S_j-1} \mathbf{C}_{ij}(t) - \sum_{t=S_j}^{S_i-1} \mathbf{C}_{ij}(t) \geq \delta_{ij}$ $\langle i, j \rangle \in A$ (3)

$r_k^{\mathrm{cal}}(S, t) \leq R_k$ $k \in \mathcal{R}, t \in T \setminus \{\overline{d}\}$ (4)

$S_i \in T$ $i \in V$ (5)

The aim of the RCPSP/max-cal is to find a schedule that minimizes the project makespan (1) and satisfies the calendar constraints (2), time lags (3), and resource capacities (4).

Each project can be represented by an activity-on-node network where each activity $i \in V$ is represented by a node and each time lag $\langle i, j \rangle \in A$ is given by an arc from node i to node j with weights δ_{ij} and $\mathcal{R}_{ij}$. The activity duration as

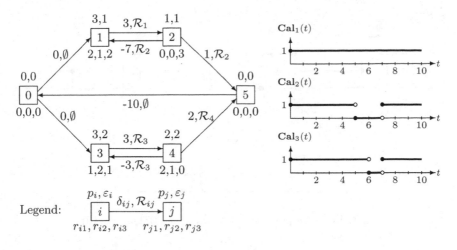

Fig. 2. Activity-on-node network and resource calendars

well as the start-up phase is given above node i in an activity-on-node network and the resource requirements of activity $i \in V$ are given below node i. For the case where time lags depend on calendars, the label-correcting and triple algorithm (see, *e.g.*, [2, Sects. 5.4 and 5.6]) can be adapted and integrated in a time planning procedure that determines a set W_i for each activity $i \in V$ containing all start times that are feasible due to the time lags and calendar constraints, *i.e.*, this procedure determines the solution space of the resource relaxation of the RCPSP/max-cal (problem (1)–(3), (5)) [9]. In addition to the sets W_i, the time planning procedure in [9] determines the "absolute" durations of each activity and time lag with respect to the activities start times. The absolute duration of an activity $i \in V$ is denoted by $p_i(S_i) := E_i(S_i) - S_i$ and the absolute time lag for $\langle i, j \rangle \in A$ by $d_{ij}(t)$ for each $t \in W_i$.

Example 2. Figure 2 shows the problem of Ex. 1 again, but now filled with information for the activites start-up phases and resource requirements as well as information for the time lags.

Activities $0, 2, 4,$ and 5 are non-interruptible while activities 1 and 3 form the set V^{bi} and therefore can be interrupted for the duration of a break in the underlying activity calendar. By applying the determination rules from above $\mathbf{Cal}_1 = \mathbf{C}_0 = \mathbf{C}_5 = \mathbf{C}_{01} = \mathbf{C}_{03} = \mathbf{C}_{50}$, $\mathbf{Cal}_2 = \mathbf{C}_1 = \mathbf{C}_3 = \mathbf{C}_4 = \mathbf{C}_{12} = \mathbf{C}_{34} = \mathbf{C}_{43} = \mathbf{C}_{45}$, and $\mathbf{Cal}_3 = \mathbf{C}_2 = \mathbf{C}_{21} = \mathbf{C}_{25}$ hold for the activity and time lag calendars. Since both time lags between activities 3 and 4 depend on the same calendar and $p_3 = \delta_{34} = -\delta_{43}$, activity 4 must be started when activity 3 ends or more precisely at the next point in time after the end of activity 3 where the calendar equals 1. The arc from the project end (node 5) to the project start (node 0) represents an upper bound on the planning horizon of $\overline{d} = 10$.

For the given example the time planning procedure from [9] determines the sets $W_0 = \{0\}$, $W_1 = \{0, 1, 2, 3, 4\}$, $W_2 = \{3, 4, 5, 7, 8, 9\}$, $W_3 = \{0, 2, 3\}$,

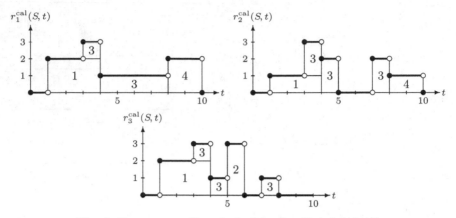

Fig. 3. Resource profiles of schedule $S = (0, 1, 5, 3, 8, 10)$

$W_4 = \{3, 7, 8\}$, and $W_5 = \{5, 6, 7, 8, 9, 10\}$. For example, activity 4 cannot start at times 5 or 6 since there is a break in calendar $\mathbf{C}_4$ from 5 to 7. Moreover, activity 4 cannot start at time 4 because it has to be executed without interruptions. Due to the time lag between activities 3 and 4, activity 3 cannot start at time 1, because if activity 3 started at time 1 activity 4 must start at time 4.

For the time- and calendar-feasible schedule $S = (0, 1, 5, 3, 8, 10)$ the resource profiles are given in Fig. 3. As already mentioned in the introduction resource type 1 stays engaged during interruptions ($\rho_1 = 1$) while resource types 2 and 3 are released during interruptions ($\rho_2 = \rho_3 = 0$). If the inequality $R_k \geq 3$ is fullfilled for each $k \in \mathcal{R}$, schedule S is resource feasible and therefore a feasible solution for the given example.   □

3   Models for RCPSP/max-cal

In this section, we present four different ways of modeling the RCPSP/max-cal. The first three approaches use only well-known constraints from finite domain propagation, while a new constraint to model the resource restrictions of the RCPSP/max-cal and a corresponding propagator are used in the fourth model.

3.1   Model `timeidx` (Time Indexed Formulation)

In preprocessing, the time planning procedure of [9] is used to determine the sets W_i of all time- and calendar-feasible start times for each activity $i \in V$ and

$$S_i \in W_i \qquad i \in V \tag{6}$$

must be satisfied. Since the absolute time lags between the activities are dependent on the start time of activity i for each $\langle i, j \rangle \in A$, element constraints are used to ensure that the correct values are taken into account.

$$\texttt{element}(S_i, \boldsymbol{d}_{ij}, d'_{ij}) \qquad \langle i, j \rangle \in A \tag{7}$$

Thereby, $\boldsymbol{d}_{ij}$ is an array that contains for all $S_i \in W_i$ the corresponding $d_{ij}(S_i)$ value. Then, the constraints modelling time lags are

$$S_j - S_i \geq d'_{ij} \qquad \langle i, j \rangle \in A \tag{8}$$

Absolute durations of the activities $i \in V$ are used and the correct assignment is ensured again by element constraints, where $\boldsymbol{p}_i$ is an array containing for all $S_i \in W_i$ the coresponding $p_i(S_i)$ value.

$$\texttt{element}(S_i, \boldsymbol{p}_i, p'_i) \qquad i \in V \tag{9}$$

We implement the resource constraints using a time-indexed decomposition with binary variables b_{it} for each real activity $i \in V^r$ and point in time $t \in T$ where b_{it} is true when i runs at t.

$$b_{it} \leftrightarrow S_i \leq t \wedge t < S_i + p'_i \qquad i \in V^r, t \in T \tag{10}$$

$$\sum_{i \in V^r} b_{it}\, r_{ik}\, (\mathbf{C}_i(t) + (1 - \mathbf{C}_i(t))\, \rho_k) \leq R_k \qquad k \in \mathcal{R}, t \in T \tag{11}$$

Model `timeidx` can now be given by: Minimize S_{n+1} subject to $(6) - (11)$.

3.2 Model 2cap (Doubling Resource Capacity)

Usually global propagators should be used to implement the resource constraints, since more information is taken into account during propagation. This model and the next make use of the global `cumulative` propagator [1] that explains its propagation [16]. If the resource $k \in \mathcal{R}$ under investigation stays engaged during interruptions of activities that require k for their execution, i.e., $\rho_k = 1$, the global cumulative propagator can be used directly with the absolute activity durations. If we regard the absolute duration of each activity $i \in V$ and assume that activity i requires r_{ik} units of resource $k \in \mathcal{R}$ with $\rho_k = 0$ at each point in time $\{S_i, \ldots, E_i(S_i) - 1\}$, there can be resource overloads at break times of an activity even if the corresponding schedule is feasible. One way to handle resources $k \in \mathcal{R}$ with $\rho_k = 0$ is to determine points in time $\mathcal{R}_k^{times}$ where there exist an activity that can be in execution and another activity that can be interrupted, double the resource capacity R_k, introduce a set V_k^d of dummy activities that require exactly R_k units of resource k at each point in time $t \in T \setminus \mathcal{R}_k^{times}$, and use the global cumulative propagator:

$$\texttt{cumulative}(S, p', r_k, R_k) \qquad k \in \mathcal{R} : (\rho_k = 1 \vee \mathcal{R}_k^{times} = \emptyset) \tag{12}$$

$$\texttt{cumulative}(S \cup S^d, p' \cup p^d, r_k \cup r_k^d, 2\,R_k) \quad k \in \mathcal{R} : (\rho_k = 0 \wedge \mathcal{R}_k^{times} \neq \emptyset) \tag{13}$$

Note that r_k is a vector containing the resource requirements on resource k of all activities $i \in V$ and that the vectors S^d, p^d, and r_k^d contain start times, absolute durations, and resource requirements on resource k, respectively, for all $j \in V_k^d$.

In addition, some decomposed constraints from (10) and (11) are required to enforce non-overload of resource k at times $\mathcal{R}_k^{times}$.

$$b_{it} \leftrightarrow S_i \leq t \wedge t < S_i + p'_i \qquad i \in V^r, t \in \bigcup_{k \in \mathcal{R}:\rho_k=0} \mathcal{R}_k^{times} \qquad (14)$$

$$\sum_{i \in V^r} b_{it} r_{ik}\, \mathbf{C}_i(t) \leq R_k \qquad k \in \mathcal{R} : \rho_k = 0, t \in \mathcal{R}_k^{times} \qquad (15)$$

For all $k \in \mathcal{R}$ with $\rho_k = 0$ the set $\mathcal{R}_k^{times}$ is defined as follows.

$$\mathcal{R}_k^{times} := \{t \in T \mid \exists i,j \in V : r_{ik} > 0 \wedge r_{jk} > 0 \wedge \min W_i \leq t < E_i(\max W_i) \wedge$$
$$\min W_j \leq t < E_j(\max W_j) \wedge \mathbf{C}_i(t) \neq \mathbf{C}_j(t)\}$$

Model 2cap can be achieved by deleting constraints (10) and (11) from model timeidx and adding constraints (12)–(15) instead.

Example 3. Regarding the example project from Fig. 2 on page 267, resource 3 is the only resource where $\mathcal{R}_k^{times} \neq \emptyset$. We can see in Fig. 3 on page 268 that in time period 6 activity 2 is in execution and activity 3 is interrupted. Hence $\mathcal{R}_3^{times} = \{5\}$. The solution presented in Fig. 3 is resource feasible for $R_3 = 3$ but cumulative does not know that activity 3 is interrupted and detects a resource overload if resource limit $R_3 = 3$ is used. By doubling the resource capacity and introducing a set V_3^d of dummy activities requiring 3 resources in all periods but 6, the cumulative of (13) does not detect a resource overload. The reason for the decomposed constraint (15) for time point 5 is clear when we imagine another activity $2'$ that requires resource type 3 for its execution and could be in execution in time period 6 just like activity 2, then for any solution where both activities 2 and $2'$ are in execution in time period 6 there is a resource overload, which the cumulative does not detect when the resource capacity is doubled. □

3.3   Model addtasks (Adding Split Tasks)

Another way to handle resources $k \in \mathcal{R}$ with $\rho_k = 0$ is to introduce for each interruptible activity $i \in V^{bi}$ a set $Add_i := \{a_1^i, a_2^i, \ldots, a_{|Add_i|}^i\}$ of additional (non-interruptible) activities that cover only those points in time $t \in \{S_i, \ldots, E_i(S_i)-1\}$ with $\mathbf{C}_i(t) = 1$, i.e., resource k is released during an interruption of activity i. For the start times and processing times of activites $a_j^i \in Add_i$ the following equalities must be guaranteed.

$$S_{a_1^i} = S_i \qquad\qquad i \in V^{bi} \qquad (16)$$

$$S_{a_j^i} = next\_start_i(S_{a_{j-1}^i}) \qquad i \in V^{bi}, j \in \{2, \ldots, |Add_i|\} \qquad (17)$$

$$p_{a_j^i} = \min(next\_break_i(S_{a_j^i}), p_i - \sum_{h=1}^{j-1} p_{a_h^i}) \quad i \in V^{bi}, j \in \{1, \ldots, |Add_i|\} \qquad (18)$$

$$r_{a_j^i,k} = r_{ik} \qquad\qquad i \in V^{bi}, j \in \{1, \ldots, |Add_i|\} \qquad (19)$$

Thereby, $next\_break_i(t)$ gives the start time of the next break after time t in calendar $\mathbf{C}_i$ and $next\_start_i(t)$ gives the end time of the next break as defined in Sect. 2. Finally, the resource requirement of each additional activity $a_j^i \in Add_i$ is set equal to r_{ik} and the global cumulative propagator can be used:

$$\texttt{cumulative}(S, p', r_k, R_k) \qquad\qquad k \in \mathcal{R} : \rho_k = 1 \qquad (20)$$

$$\texttt{cumulative}(S^a, p^a, r_k^a, R_k) \qquad\qquad k \in \mathcal{R} : \rho_k = 0 \qquad (21)$$

In constraints (21), the vectors S^a, p^a, and r_k^a contain not only the start times, durations, and resource requirements of the additional activities $a_j^i, i \in V^{bi}, j \in \{1, \ldots, |Add_i|\}$, but also the start times, durations, and resource requirements of the non-interruptible activities $i \in V^{ni}$.

Model `addtasks` can be achieved by deleting constraints (10) and (11) from model `timeidx` as well as adding constraints (16)–(21) instead.

3.4   Model `cumucal` (Global Calendar Propagator)

For our fourth model for RCPSP/max-cal, we created a global `cumulative` propagator that takes calendars into account and named it `cumulative_calendar`. The fourth model (`cumucal`) can be achieved by deleting constraints (9), (10), and (11) from model `timeidx` as well as adding constraints (22)

$$\texttt{cumulative\_calendar}(S, p, r_k, R_k, \mathbf{C}, \rho_k) \qquad\qquad k \in \mathcal{R} \qquad (22)$$

with p being the vector of all constant processing times p_i and $\mathbf{C}$ being the vector of all activity calendars $\mathbf{C}_i, i \in V$.

The `cumulative_calendar` propagator is made up of two parts, a time-table consistency check and filtering. The basic ideas of these two parts are the same as in the `cumulative` propagator of [18], but non-trivial adaptions were necessary to consider calendars. These adaptions are described in the following. The compulsory part [10] of an activity $i \in V$ is the time interval $[ub(S_i), lb(S_i) + p_i(lb(S_i)))$, where $lb(S_i)$ ($ub(S_i)$) represents the current minimum (maximum) value in the domain of S_i. If $\rho_k = 1$ for the resource $k \in \mathcal{R}$ then activity i requires r_{ik} units of resource k at each point in time of its compulsory part. Otherwise ($\rho_k = 0$), activity i requires r_{ik} units of resource k only at points in time of its compulsory part where $\mathbf{C}_i(t) = 1$. The intervals where an activity requires resource k within its compulsory part are named the *calendar compulsory parts*. At the begin of the `cumulative_calendar` propagator the calendar compulsory parts of all activities are determined and a resource profile including all these parts is built. Within the consistency check, resource overloads in this profile are detected. If an overload of the resource k occurs in the time interval $[s, e)$ involving the set of activities Ω, the following conditions hold:

$$ub(S_i) \leq s \ \wedge \ lb(S_i) + p_i(lb(S_i)) \geq e \qquad\qquad i \in \Omega$$

$$(1 - \rho_k) \cdot \mathbf{C}_i(t) + \rho_k = 1 \qquad\qquad i \in \Omega, t \in [s, e)$$

$$\sum_{i \in \Omega} r_{ik} > R_k$$

In a lazy clause generation solver integer domains are represented using Boolean variables. Each variable x with initial domain $D^0(x) = \{l, \ldots, u\}$ is represented by two sets of Boolean variables $[\![x = d]\!], l \leq d \leq u$ and $[\![x \leq d]\!], l \leq d < u$ which define which values are in $D(x)$. A lazy clause generation solver keeps the two representations of the domain in sync. In order to explain the resource overload, we use a pointwise explanation [18] at $TimeD$, which is the nearest integer to the mid-point of $[s, e)$.

$$\forall i \in \Omega : [\![back(i, TimeD + 1) \leq S_i]\!] \wedge [\![S_i \leq TimeD]\!] \rightarrow false$$

$$back(i, t) := \begin{cases} \max\{\tau \in T \mid \sum_{z=\tau}^{t-1} \mathbf{C}_i(z) = p_i\} & \text{if } \mathbf{C}_i(t-1) = 1 \\ \max\{\tau \in T \mid \sum_{z=\tau}^{t-1} \mathbf{C}_i(z) = p_i - 1\} & \text{if } \mathbf{C}_i(t-1) = 0. \end{cases}$$

The definition by cases for $back(i, t)$ is necessary to guarantee the execution of activity i at time $t-1$, if $S_i = t - back(i, t)$ holds. If for a time t with $\mathbf{C}_i(t-1) = 0$ $back(i, t)$ would be calculated with the first case, then $E_i(t - back(i, t)) < t$ and the explanation would be incorrect.

If there exists a proper subset of activities $\Omega' \subset \Omega$ with $\sum_{i \in \Omega'} r_{ik} > R_k$, the explanation of the resource overload is done on set Ω'. Sometimes more than one such subset exists. In this situation the lexicographic least set of activities is chosen as was done in [18].

Time-table filtering is also based on the resource profile of calendar compulsory parts of all activities. In a filtering without explanations the height of the calendar compulsory parts concerning one time period or a time interval is given. For an activity the profile is scanned through to detect time intervals where it cannot be executed. The lower (upper) bound of an activity's start time is updated to the first (last) possible time period with respect to those time intervals and the activity calendar. If we want to explain the new lower (upper) bound we need to know additionally which activities have the calendar compulsory parts of those time intervals.

A profile is a triple (A, B, C) where $A = [s, e)$ is a time interval, B the set of all activities that have a calendar compulsory part in the time interval A, and C the sum of the resource requirements r_{ik} of all activities in B. Here, we only consider profiles with a maximal time interval A with respect to B and C, i.e., no other profile $([s', e'), B, C)$ exists where $s' = e$ or $e' = s$.

Let us consider the case when the lower bound of the start time variable for activity i can be maximally increased from its current value $lb(S_i)$ to a new value $LB(i)$ using time-table filtering (the case of decreasing upper bounds is analogous and omitted). Then there exists a sequence of profiles $[D_1, \ldots, D_p]$ where $D_h = ([s_h, e_h), B_h, C_h)$ with $e_0 = lb(S_i)$ and $e_p = LB(i)$ such that

$$\forall h : 1 \leq h \leq p; C_h + r_{ik} > R_k \wedge s_h < e_{h-1} + p_i(e_{h-1})$$

In Sect. 2, we introduced $p_i(t)$ only for $t \in W_i$. Note that $p_i(t)$ can be calculated in the same way for $t \notin W_i$, where $p_i(t)$ takes the value $\overline{d} - t$ if less than p_i working periods are following after t in calendar $\mathbf{C}_i$. In addition, if $\rho_k = 0$ is satisfied then

$$\forall h : 1 \leq h \leq p; \exists t \in [s_h, e_h) : \mathbf{C}_i(t) = 1$$

Hence each profile D_h pushes the start time of activity i to e_h.

Again we use pointwise explanations based on single time points. Unlike the consistency case, we may need to pick a set of time points no more than the absolute duration of activity i apart to explain the increasing of the lower bound of S_i over the time interval. For a profile with length greater than the absolute processing time of activity i we may need to pick more than one time point in a profile. Let $\{t_1, \ldots, t_m\}$ be a set of time points such that $t_0 = lb(S_i)$, $t_m + 1 = LB(i)$, $\forall 1 \leq l \leq m : t_{l-1} + p_i(t_{l-1}) \geq t_l$ and there exists a mapping $P(t_l)$ of time points to profiles such that $\forall 1 \leq l \leq m : s_{P(t_l)} \leq t_l < e_{P(t_l)}$. Then we build a pointwise explanation for each time point t_l, $1 \leq l \leq m$

$$[\![back(i, t_l + 1) \leq S_i]\!] \wedge \bigwedge_{j \in B_h} ([\![back(j, t_l + 1) \leq S_j]\!] \wedge [\![S_j \leq t_l]\!]) \rightarrow [\![t_l + 1 \leq S_i]\!]$$

Example 4. We illustrate `cumulative_calendar` for the example network from Fig. 2. To explain both the time-table consistency check and the time-table filtering we are using two different cases. For the first case (consistency check), we assume that in the current search node $lb(S_1) = 3, ub(S_1) = 4, lb(S_2) = 8, ub(S_2) = 9, lb(S_3) = ub(S_3) = 3$, and $lb(S_4) = ub(S_4) = 8$ holds, *i.e.*, activities 3 and 4 are already fixed. We examine the `cumulative_calendar` for resource type 1 with a resource capacity of $R_1 = 2$. The calendar compulsory parts are $[4, 8)$ for activity 1, $[3, 8)$ for activity 3, and $[8, 10)$ for activity 4. Note that activity 2 is not taken into account since $r_{21} = 0$ and that the calendar compulsory parts equal the compulsory parts for this example because $\rho_1 = 1$. The compulsory parts of activities 1 and 3 cover the interval $[4, 8)$ and a resource overload of resource 1 occurs, since $r_{11} + r_{31} = 2 + 1 = 3 > 2 = R_1$. A pointwise explanation of the resource overload is done at $TimeD = 6$:

$$[\![3 \leq S_1]\!] \wedge [\![S_1 \leq 6]\!] \wedge [\![3 \leq S_3]\!] \wedge [\![S_3 \leq 6]\!] \rightarrow false$$

For activities $i = 1$ and $i = 3$, respectively, $C_i(TimeD - 1) = 0$ is satisfied and $back(i, TimeD)$ is calculated through the second case. Without case differentiation for $back(i, t)$ only the first case would be considered, resulting that $back(1, TimeD)$ would equal 1 and the explanation would be wrong.

For the second case (time-table filtering), we assume that in the current search node $lb(S_1) = ub(S_1) = 0, lb(S_2) = 3, ub(S_2) = 8, lb(S_3) = ub(S_3) = 3$, and $lb(S_4) = ub(S_4) = 8$ holds. We examine the `cumulative_calendar` for resource type 3 with a resource capacity of $R_3 = 3$. Activity 2 is the only task where the start time is not fixed and the consistency check detects no resource overload. The calendar compulsory parts are $[0, 3)$ for activity 1, $[3, 5), [7, 8)$ for activity 3, and $[8, 10)$ for activity 4. For the profile (A, B, C) with $A = [3, 5)$, $B = \{3\}$, and $C = 1$ the condition $C + r_{23} = 1 + 3 > 3 = R_3$ is satisfied and therefore the lower bound for variable S_2 can be increased to $LB(2) = 5$. Since the activity duration p_2 equals 1 a pointwise explanation is done for $t_0 = 3$ and $t_1 = 4$. The explanation for $t_0 = 3$ is $[\![3 \leq S_2]\!] \wedge [\![1 \leq S_3]\!] \wedge [\![S_3 \leq 3]\!] \rightarrow [\![4 \leq S_2]\!]$ and for $t_1 = 4$ it is $[\![4 \leq S_2]\!] \wedge [\![2 \leq S_3]\!] \wedge [\![S_3 \leq 4]\!] \rightarrow [\![5 \leq S_2]\!]$. $\qquad\square$

3.5   Time Granularity Considerations

All models depend on the granularity chosen for the time. If the granularity increases then the size of T increases respectively. Thus, the number of linear constraints and auxiliary Boolean variables increases for the models timeidx and 2cap, especially for the former. Moreover, filtering algorithms for the element constraints (used in all models) might be negatively affected due to a larger size of the input arrays. The implemented time-table consistency check, filtering, and explanation generation for resource overloads and start time bounds updates in cumulative_calendar depend on the granularity, too. Their respective runtime complexity are $\mathcal{O}(x \times y \log(x \times y) + x \times z)$, $\mathcal{O}(x^2 \times z \times y)$, and $\mathcal{O}(x \times z)$ where x is the number of tasks, $y - 1$ is maximal possible number of interruptions of any task and z the maximal possible absolute duration of any task.

4   Experiments and Conclusion

We conducted extensive experiments on Dell PowerEdge R415 machines running CentOS 6.5 with 2x AMD 6-Core Opteron 4184, 2.8GHz, 3M L2/6M L3 Cache and 64 GB RAM. We used MiniZinc 2.0.1 [13] and the lazy clause generation [15] solver chuffed rev 707.

A runtime limit of 10 minutes was imposed excluding runtimes needed for pre-processing, initial solution generation, and compiling the MiniZinc models to solver-dependent FlatZinc models. We used the same benchmarks and initial solutions as in [9], which are available at www.wiwi.tu-clausthal.de/en/chairs/unternehmensforschung/research/benchmark-instances/.

Since instances with 10 or 20 activities could easily be solved within a few seconds by any combination of solver, model, and search, we concentrate on instances with 50 and 100 activities. The average runtime needed for pre-processing and initial solution generation are less than a few seconds for instances with 50 activities and less than 30 seconds for instances with 100 activities, respectively.

4.1   Comparing Search Strategies

For finding the shortest project duration, we employ a branch-and-bound strategy for which we investigate following four different search combinations. Those seem likely to be most suitable based on our previous experience on solving scheduling problems using lazy clause generation (see, *e.g.*, [18,19,17]).

ff: Selects the variable with the smallest domain size and assigns the minimal value in the domain to it.

vsids: Selects the literal with the highest activity counter and sets it to true, where the literal is a part of the Boolean representation of the integer variables, *i.e.*, $[\![x = v]\!]$, $[\![x \leq v]\!]$, where x is an integer variable and $v \in \mathcal{D}(x)$. Informally, the activity counter records the recent involvement of the literal in conflicts and all activity counters are simultaneously decayed periodically.

Table 1. Comparison of search strategies on instances with 50 activities.

model	search	#opt	#feas	#inf	#un	cmp(179)		all(180)	
						avg. rt	avg. #cp	avg. rt	avg. #cp
cumucal	alt	161	0	19	0	1.13	3847	1.71	5236
cumucal	ff	160	1	19	0	7.54	16401	10.83	16310
cumucal	hs	161	0	19	0	1.33	5358	1.80	6349
cumucal	vsids	161	0	19	0	4.27	18495	4.79	19445

Table 2. Comparison of search strategies on instances with 100 activities.

model	search	#opt	#feas	#inf	#un	cmp(140)		all(180)	
						avg. rt	avg. #cp	avg. rt	avg. #cp
cumucal	alt	158	11	11	0	7.58	11498	56.18	25170
cumucal	ff	150	16	10	4	13.73	14305	82.51	16420
cumucal	hs	152	17	11	0	20.33	34900	85.20	45109
cumucal	vsids	133	36	11	0	76.87	146172	185.61	122457

The activity counter of a literal is increased during conflict analysis when the literal is related to the conflict. It is an adaption of the variable state independent decaying sum heuristic [12]. The search vsids is combined with Luby restarts [11] and a restart base of 100 conflicts.

hs: The search starts off with ff and then switches to vsids after 1000 conflicts.

alt: The search alternates between ff and vsids starting with ff. It switches from one to the other after each restart where we use the same restart policy and base as for vsids.

Tables 1 and 2 show the results of chuffed on the cumucal model using different search strategies on instances with 50 and 100 activities, respectively. The search strategies behave similar with the other models. We show the number of instances proven optimal (#opt), not proven optimal but where feasible solutions were found (#feas), proven infeasible (#inf), and where nothing was determined (#un). We compare the average runtime in seconds (avg. rt) and average number of choice points to solve (avg. #cp), on two subsets of each benchmark. The cmp subset are all the benchmarks where all solvers proved optimality or infeasibility, and all is the total set of benchmarks.

The alt search is clearly the fastest, also leading to the lowest average number of nodes explored in comparison to the rest. Interestingly, the performance of vsids significantly decays from instances with 50 activities to those ones with 100 activities in proportion to alt and ff. This decay also affects hs, but not so dramatically. The strength of the alt method is the combination of integer based search in ff which concentrates on activities that have little choice left, with the robustness of vsids which is excellent for proving optimality once a good solution is known.

Table 3. Comparison of models on instances with 50 activities.

| model | search | #opt | #feas | #inf | #un | cmp(177) | | all(180) | | |
						avg. rt	avg. #cp	avg. ft	avg. rt	avg. #cp
addtasks	alt	160	1	19	0	9.10	14232	0.84	15.39	17924
cumucal	alt	161	0	19	0	0.92	3203	0.48	1.71	5236
timeidx	alt	158	3	19	0	22.20	3484	18.65	31.82	3426
2cap	alt	161	0	19	0	1.55	5341	0.72	3.12	9619

Table 4. Comparison of models on instances with 100 activities.

| model | search | #opt | #feas | #inf | #un | cmp(138) | | all(180) | | |
						avg. rt	avg. #cp	avg. ft	avg. rt	avg. #cp
addtasks	alt	139	28	10	3	25.88	26344	4.05	139.47	35525
cumucal	alt	158	11	11	0	3.24	5037	2.45	56.18	25170
timeidx	alt	131	38	10	1	83.37	4947	78.24	196.63	4031
2cap	alt	153	16	11	0	6.13	8798	4.05	78.96	30728

Table 5. Comparison of solvers on instances with 50 activities.

| model | search | #opt | #feas | #inf | #un | cmp(170) | | all(180) | | |
						avg. rt	avg. #cp	avg. ft	avg. rt	avg. #cp
cumucal	chuffed+alt	161	0	19	0	0.59	2198	0.48	1.71	5236
timeidx	chuffed+alt	158	3	19	0	10.45	1662	18.65	31.82	3426
timeidx	ocpx+free	159	2	19	0	69.95	13383	19.97	83.92	14155
mip		153	7	18	2	222.42	—	— 750.25		—

4.2 Comparing Models

Tables 3 and 4 compare the effect of the different models using `chuffed` and the best search method `alt`. As expected, the time-indexed model, timeidx, is the worst in terms of times due to the large model size, but it propagates effectively as illustrated by the low number of explored nodes (only ever bettered by cumucal). The model `addtasks` performs worst with respect to the average number of nodes, which can be explained by the shorter activities causing weaker time-table propagation in the `cumulative` propagator. The best model is cumucal that takes the advantage of using fixed durations, since the variability is handled directly by the propagator, and because it generates the smallest model. We also show the average flattening time (avg. ft.) for all benchmarks, where clearly cumucal is advantageous.

4.3 Comparing Solvers

Table 5 compares the results obtained by `chuffed` to those obtained by Opturion CPX 1.0.2 (`ocpx`), which is available at www.opturion.com/cpx, and the best solution obtained by any mixed-integer linear programming formulation from [9] (mip), which is solved using CPLEX 12.6 on an Intel Core i7 CPU 990X with 3.47 GHz and 24GB RAM under Windows 7. For mip the runtime limit was set to 3 hours and 8 threads were used. To get an idea of the impact of the machine used, we also ran ocpx with the deterministic search ff on the same Windows machine. ocpx was more than 3 times faster on that machine. It can be seen that chuffed and ocpx clearly outperform the state-of-the-art solution approach, which is mip, and that the machine we used is even slower than the machine used in [9].

Overall the cumucal model closes all open benchmarks of size 50 and 75 of size 100, and clearly, we significantly advance the state of the art.

Acknowledgments. NICTA is funded by the Australian Government as represented by the Department of Broadband, Communications and the Digital Economy and the Australian Research Council through the ICT Centre of Excellence program. This work was partially supported by Asian Office of Aerospace Research and Development (AOARD) grant FA2386-12-1-4056.

References

1. Aggoun, A., Beldiceanu, N.: Extending CHIP in order to solve complex scheduling and placement problems. Mathematical and Computer Modelling **17**(7), 57–73 (1993)
2. Ahuja, R., Magnanti, T., Orlin, J.: Network Flows. Prentice Hall, Englewood Cliffs (1993)
3. Baptiste, P.: Constraint-Based Scheduling: Two Extensions. Master's thesis, University of Strathclyde, Glasgow, Scotland, United Kingdom (1994)
4. Beldiceanu, N.: Parallel machine scheduling with calendar rules. In: International Workshop on Project Management and Scheduling (1998)
5. Cheng, J., Fowler, J., Kempf, K., Mason, S.: Multi-mode resource-constrained project scheduling problems with non-preemptive activity splitting. Computers & Operations Research **53**, 275–287 (2015)
6. Franck, B.: Prioritätsregelverfahren für die ressourcenbeschränkte Projektplanung mit und ohne Kalender. Shaker, Aachen (1999)
7. Franck, B., Neumann, K., Schwindt, C.: Project scheduling with calendars. OR Spektrum **23**, 325–334 (2001)
8. Franck, B., Neumann, K., Schwindt, C.: Truncated branch-and-bound, schedule-construction, and schedule-improvement procedures for resource-constrained project scheduling. OR Spektrum **23**, 297–324 (2001)
9. Kreter, S., Rieck, J., Zimmermann, J.: Models and solution procedures for the resource-constrained project scheduling problem with general temporal constraints and calendars. Submitted to European Journal of Operational Research (2014)
10. Lahrichi, A.: Scheduling: The notions of hump, compulsory parts and their use in cumulative problems. Comptes Rendus de l'Académie des Sciences. Paris, Série 1, Matématique **294**(2), 209–211 (1982)
11. Luby, M., Sinclair, A., Zuckerman, D.: Optimal speedup of Las Vegas algorithms. Information Processing Letters **47**, 173–180 (1993)
12. Moskewicz, M.W., Madigan, C.F., Zhao, Y., Zhang, L., Malik, S.: Chaff: engineering an efficient SAT solver. In: Proceedings of Design Automation Conference - DAC 2001, pp. 530–535. ACM, New York (2001)

13. Nethercote, N., Stuckey, P.J., Becket, R., Brand, S., Duck, G.J., Tack, G.: MiniZinc: towards a standard CP modelling language. In: Bessière, C. (ed.) CP 2007. LNCS, vol. 4741, pp. 529–543. Springer, Heidelberg (2007)
14. Neumann, K., Schwindt, C., Zimmermann, J.: Project Scheduling with Time Windows and Scarce Resources, 2nd edn. Springer, Berlin (2003)
15. Ohrimenko, O., Stuckey, P.J., Codish, M.: Propagation via lazy clause generation. Constraints **14**(3), 357–391 (2009)
16. Schutt, A.: Improving Scheduling by Learning. Ph.D. thesis, The University of Melbourne (2011). http://repository.unimelb.edu.au/10187/11060
17. Schutt, A., Feydy, T., Stuckey, P.J.: Explaining time-table-edge-finding propagation for the cumulative resource constraint. In: Gomes, C., Sellmann, M. (eds.) CPAIOR 2013. LNCS, vol. 7874, pp. 234–250. Springer, Heidelberg (2013)
18. Schutt, A., Feydy, T., Stuckey, P.J., Wallace, M.G.: Explaining the cumulative propagator. Constraints **16**(3), 250–282 (2011)
19. Schutt, A., Feydy, T., Stuckey, P.J., Wallace, M.G.: Solving RCPSP/max by lazy clause generation. Journal of Scheduling **16**(3), 273–289 (2013)
20. Schutt, A., Feydy, T., Stuckey, P.J., Wallace, M.G.: A satisfiability solving approach. In: Schwindt, C., Zimmermann, J. (eds.) Handbook on Project Management and Scheduling, vol. 1, pp. 135–160. Springer International Publishing (2015)
21. Schwindt, C., Trautmann, N.: Batch scheduling in process industries: An application of resource-constrained project scheduling. OR Spektrum **22**, 501–524 (2000)
22. Trautmann, N.: Calendars in project scheduling. In: Fleischmann, B., Lasch, R., Derigs, U., Domschke, W., Rieder, U. (eds.) Operations Research Proceedings 2000, pp. 388–392. Springer, Berlin (2001)
23. Zhan, J.: Calendarization of timeplanning in MPM networks. ZOR - Methods and Models of Operations Research **36**, 423–438 (1992)

Deterministic Estimation of the Expected Makespan of a POS Under Duration Uncertainty

Michele Lombardi, Alessio Bonfietti[✉], and Michela Milano

DISI, University of Bologna, Bologna, Italy
{michele.lombardi2,alessio.bonfietti,michela.milano}@unibo.it

Abstract. This paper is about characterizing the expected makespan of a Partial Order Schedule (POS) under duration uncertainty. Our analysis is based on very general assumptions about the uncertainty: in particular, we assume that only the min, max, and average durations are known. This information is compatible with a whole range of values for the expected makespan. We prove that the largest of these values and the corresponding "worst-case" distribution can be obtained in polynomial time and we present an $O(n^3)$ computation algorithm. Then, using theoretical and empirical arguments, we show that such expected makespan is strongly correlated with certain global properties of the POS, and we exploit this correlation to obtain a linear-time estimator. The estimator provides accurate results under a very large variety of POS structures, scheduling problem types, and uncertainty models. The algorithm and the estimator may be used during search by an optimization approach, in particular one based on Constraint Programming: this allows to tackle a stochastic problem by solving a dramatically simpler (and yet accurate) deterministic approximation.

1 Introduction

A Partial Order Schedule (POS) is an acyclic graph $G = \langle A, E \rangle$, where A is a set of activities a_i, and E is a set of directed edges (a_i, a_j) representing end-to-start precedence relations. A POS is a flexible solution to a scheduling problem: some of the edges derive from the original Project Graph, while the remaining ones are added by an optimization approach so as to prevent potential resource conflicts.

A POS is very well suited as a solution format in the presence of duration uncertainty. Before the execution starts, each activity has a candidate start time. During execution, whenever a duration becomes known, the start times are updated (in $O(n^2)$ time) so that no precedence relation is violated: this guarantees that the schedule remains resource feasible.

In the presence of duration uncertainty, the quality of a POS should be evaluated via a stochastic metric, which in this paper is the expected value of the makespan: this is far from being a perfect choice, but still a fair one in many settings (in particular when the POS should be repeatedly executed, such as in stream processing applications).

© Springer International Publishing Switzerland 2015
G. Pesant (Ed.): CP 2015, LNCS 9255, pp. 279–294, 2015.
DOI: 10.1007/978-3-319-23219-5_20

A POS can be obtained essentially in two ways. First, in a constructive fashion, by adding edges to the graph until all potential resource conflicts are resolved: this is done by Precedence Constraint Posting (PCP) approaches [7,10,12,13]. Alternatively, one can obtain a POS from a traditional schedule with fixed start times via a post-processing step [7,8,12,13]. Despite the fact that a POS can adapt to uncertain durations, in practice the uncertainty is often largely disregarded when the graph is built. For example, all the post-processing methods operate on a schedule obtained for fixed durations. The PCP methods are capable of dealing with uncertainty, for example via scenario-based optimization or via the Sample Average Approximation [5,14]: both approaches require to sample many duration assignments and to optimize a meta-model, resulting from the combination of a set of models (one per scenario) connected by chaining constraints. Despite this, most PCP approaches in the literature either assume to have fixed durations (e.g. [7,12,13]) or rely on minimal uncertainty information to perform robust (rather than stochastic) optimization (e.g. Simple Temporal Networks with Uncertainty [11,15] or the method from [10]).

One of the main reasons for ignoring the uncertainty when searching for an optimal POS is the additional complexity. In fact, the complexity of a scenario-based approach depends on the number of samples required to have a satisfactory accuracy, which can be large for problems with many activities. The second main reason is the lack of reliable information, either because data has not been collected or because the POS must be executed in a very dynamic environment.

In this paper, we tackle the problem of estimating the expected makespan of a POS under duration uncertainty. The POS describes a solution computed with a classical CP approach with a makespan minimization objective. We use "deterministic" estimators rather than sampling-based statistics. The goal is providing an efficient alternative to scenario-based optimization in case the number of required samples is too high. Moreover, our estimators rely on very general assumptions on the uncertainty (in particular, we do not require independent durations) and are therefore well suited for cases in which extensive information is not available. Technically, we focus on computing or approximating the largest value of the expected makespan that is compatible with the available uncertainty information. Our main contributions are: first, a proof that such a expected makespan value can be obtained in polynomial time and a $O(n^3)$ computation algorithm; second, a mixed theoretical/empirical analysis that highlights interesting properties of POSs and enables the definition of a linear-time approximate estimator.

We have evaluated our estimator under an impressive variety of settings, obtaining reasonably accurate and robust results in most cases. Our algorithm could be used within a heuristic scheduling method for estimating the value of the expected makespan. Alternatively, with further research the algorithm could be turned into a propagator for a global constraint, despite this is a non-trivial task. Our estimator consists of a non-linear formula that can be embedded in a CP approach using standard building blocks (sum and min operators): the resulting CP expression is propagated in $O(n)$. Indeed, this research line stems in part from the unexpectedly good performance of a PCP approach based on Constraint Programming for robust scheduling, that we presented in [10]: our estimator in particular is designed to be used in a similar framework.

This paper is a follow-up of a previous work of ours [2], where we reached similar (although less mature) conclusions under more restrictive assumptions. The algorithm for computing the expected makespan is presented in Section 2, the analysis and the estimator are in Section 3, and Section 4 provides some concluding remarks.

2   Expected Makespan with Worst-Case Distribution

Our analysis is based on very general assumptions about the uncertainty. In particular, we assume that only the minimal, maximal, and average durations are known. Formally, the duration of each activity a_i can be modeled as a random variable D_i ranging over a known interval $[\underline{d_i}, \overline{d_i}]$ and having known expected value $\tilde{d_i}$. Note that the D_i variables are *not* assumed to be independent, and therefore they are best described in terms of their joint probability distribution.

Assuming that lower completion times are desirable, the makespan for a given instantiation of the D_i variables can be obtained by computing the longest path in G (often referred to as *critical path*). Formally, the makespan is a function $T(D)$ of the duration variables, where we refer as D to the vector of all D_i. The makespan is therefore itself stochastic, and $E\left[T(D)\right]$ denotes its expected value.

Because we rely on so little information about the uncertainty, the expected makespan is not unambiguously defined: the value of $E\left[T(D)\right]$ depends on the joint distribution of the D_i, and *there exists an infinite number of distributions that is compatible with known values of $\underline{d}$, $\overline{d}$, and $\tilde{d}$.*

The expected makespan cannot be higher than $T(\overline{d})$, i.e. than the makespan with maximal durations. Moreover, since the expected value is a linear operator, $E\left[T(D)\right]$ cannot be lower $T(\tilde{d})$, i.e. the makespan with expected durations.

Reaching stronger conclusions requires more powerful analysis tools, which will be presented in this paper. In particular, we are interested in the largest possible value of $E\left[T(D)\right]$ that is compatible with the known duration data: in absence of more information, this allows to obtain a safe estimate. The computation requires to identify a "worst-case" distribution, among the infinite set of distributions that are compatible with the given $\underline{d}$, $\overline{d}$, and $\tilde{d}$.

Properties of the Worst-case Distribution A generic (joint) probability distribution can be defined as a function:

$$P : \Omega \to [0, 1] \tag{1}$$

where Ω is a set of scenarios ω_k, each representing in our case an assignment for all D_i variables. The set Ω is called the *support* of the distribution and can have infinite size. The value $P(\omega_k)$ is the probability of scenario ω_k. The integral (or the sum, for discrete distributions) of P over Ω should be equal to 1. Our method for defining the worst-case distribution relies on a first, very important, result that is stated in the following theorem:

Theorem 1. *There exists a worst-case distribution such that, in each scenario with non-zero probability, every D_i takes either the value $\underline{d}_i$ or the value $\overline{d}_i$.*

The proof is in Appendix A. From Theorem 1 we deduce that it is always possible to define a worst-case distribution with a support Ω that is a subset of the Cartesian product $\prod_{a_i \in A} \{\underline{d}_i, \overline{d}_i\}$, and the therefore with size bounded by $2^{|A|}$.

Finding the worst-case distribution This makes it possible to model the construction of a worst-case distribution as an optimization problem. Formally, let us introduce for each scenario a decision variable $p_{\omega_k} \in [0, 1]$ representing the value of $P(\omega_k)$. The assignment of the p_{ω_k} variables must be such that the known values of the expected durations are respected:

$$\overline{d}_i \sum_{\omega_k : D_i = \overline{d}_i} p_{\omega_k} + \underline{d}_i \left(1 - \sum_{\omega_k : D_i = \overline{d}_i} p_{\omega_k}\right) = \tilde{d}_i \text{ and hence: } \sum_{\omega_k : D_i = \overline{d}_i} p_{\omega_k} = \frac{\tilde{d}_i - \underline{d}_i}{\overline{d}_i - \underline{d}_i} \quad (2)$$

Then a worst-case distribution can be found by solving:

$$\textbf{P0}: \quad \max z = \sum_{\omega_k \in \Omega} p_{\omega_k} T(D(\omega_k)) \quad (3)$$

$$\text{subject to: } \sum_{\omega_k : D_i = \overline{d}_i} p_{\omega_k} \leq \frac{\tilde{d}_i - \underline{d}_i}{\overline{d}_i - \underline{d}_i} \qquad \forall a_i \in A \quad (4)$$

$$\sum_{\omega_k \in \Omega} p_{\omega_k} \leq 1 \quad (5)$$

$$p_{\omega_k} \geq 0 \qquad \forall \omega_k \in \Omega \quad (6)$$

where $\Omega = \prod_{a_i \in A} \{\underline{d}_i, \overline{d}_i\}$. Equation (3) is the makespan definition ($D(\omega_k)$ are the durations in ω_k), Equation (4) corresponds to Equation (2), and Equation (5) ensures that the total probability mass does not exceed one. It is safe to use a $\leq$ sign in Equation (4) and (5), because increasing a p_{ω_k} can only improve the problem objective and therefore all constraints are tight in any optimal solution. P0 is linear, but has unfortunately an exponential number of variables.

Tractability of P0: Luckily, P0 has a number of nice properties that make it much easier to solve. At a careful examination, one can see that P0 is the linear relaxation of a multi-knapsack problem. Results from LP duality theory [4] imply that the optimal solution cannot contain more than $\min(n, m)$ fractional variables, where n is the number of variables ($|\Omega|$ in our case) and m is the number of knapsack constraints (i.e. $|A| + 1$).

Due to Constraint (5), in an optimal solution of P0: either 1) all variables are integer and there is a single $p_{\omega_k} = 1$; or 2) all non-zero variables are fractional. Therefore, the number of non-zero variables in an optimal solution is bounded by $|A| + 1$. This means that a worst-case distribution with a support of size at most $|A| + 1$ is guaranteed to exist.

Algorithm 1. Compute $E_{wc}[T(D)]$

Require: A POS $G = \langle A, E \rangle$, plus $\underline{d}_i$, $\overline{d}_i$, and $\tilde{d}_i$ for each activity
1: Let $T = 0$ and $s_{tot} = 1$ and let $s_i = (\tilde{d}_i - \underline{d}_i)/(\overline{d}_i - \underline{d}_i)$ for each $a_i \in A$
2: **while** $s_{tot} > 0$ **do**
3:　　　**for all** activities $a_i \in A$ **do**
4:　　　　　Let $d_i = \overline{d}_i$ if $s_i > 0$, otherwise let $d_i = \underline{d}_i$
5:　　　　Find the critical path π over G, using the d_i values as durations.
6:　　　　Let $p_\pi = \min \left\{ s_i : a_i \in \pi \text{ and } d_i = \overline{d}_i \right\}$
7:　　　　If $p_\pi = 0$, then $p_\pi = s_{tot}$
8:　　　　**for all** activities a_i on the critical path such that $d_i = \overline{d}_i$ **do**
9:　　　　　Set $s_i = s_i - p_\pi$
10:　　　Set $s_{tot} = s_{tot} - p_\pi$
11:　　　Set $T = T + p_\pi \sum_{a_i \in \pi} d_i$
12: **return** T

Furthermore, all the variables of P0 appear with identical weights in all knapsack constraints (i.e. Equation (4) and (5)). Therefore, P0 can be solved to optimality by generalizing the solution approach for the LP relaxation of the classical knapsack problem: namely, one has to repeatedly: 1) pick the variable p_{ω_k} with the highest reward $T(D(\omega_k))$ in the objective; and 2) increase the value of p_{ω_k} until one of the knapsack constraints becomes tight.

A Polynomial Time Algorithm: We present (in Algorithm 1) a practical method for the computation of $E[T(D)]$ with worst-case distribution, i.e. $E_{wc}[T(D)]$. The algorithm improves the basic process that we have just described by relying on *partial scenarios*. A partial scenario represents a group of scenarios having the same makespan, and it is identified by specifying a duration (either $\underline{d}_i$ or $\overline{d}_i$) for a subset of the activities. Algorithm 1 keeps (line 2) a global probability budget s_{tot} and a vector of probability budgets s_i to keep track of the slack in Constraint (5) and Constraints (4), respectively. Then the algorithm repeatedly identifies a partial scenario that is compatible with the remaining slack and has maximal makespan. This is done by: 1) Assigning to each activity a duration d_i. This is equal to $\overline{d}_i$ if the corresponding slack variable s_i is non-zero, otherwise, the duration is $\underline{d}_i$ (lines 4-5); and 2) finding the critical path π on the POS, using the durations d_i (line 6).

At each iteration, the partial scenario is specified by: 1) the set of activities *on the critical path* that have maximal duration $\overline{d}_i$; and 2) the set of all activities for which $d_i = \underline{d}_i$. The idea is that if both sets of activities take their prescribed duration, then the critical path is π. This process always identifies the partial scenario that is compatible with the current slack and has the largest makespan.

The partial scenario is then inserted in the joint distribution with its largest possible probability: this is determined (with an exception discussed later) by smallest slack of the activities for which $d_i = \overline{d}_i$ in the partial scenario. The probability p_π of the partial scenario is used to update the slack variables at lines 9-11. Finally, the expected makespan variable T is incremented by the probability of the partial scenario, multiplied by its associated makespan (i.e. the sum of durations of the activities on the critical path).

The mentioned exception is that the probability p_π computed at line 7 may be 0 in case all activities on π have $d_i = \underline{d}_i$. If this happens, it means that the probability of the current partial scenario is not limited by Constraints (4), but only by Constraint (5). Hence, the scenario probability can be set to the value of the remaining global slack (line 8). When the global slack becomes zero, the algorithm returns the final value of $E_{wc}[T(D)]$.

By construction, at each iteration a slack variable s_i becomes zero. If all the s_i become zero, then the next partial scenario will have a critical path with only "short" activities and the probability update at line 8 will trigger. Overall, the algorithm can perform at most $|A| + 1$ iterations, in agreement with the theoretical prediction made for problem P0. The complexity of each iteration is dominated by the $O(|A|^2)$ longest path computation. The overall complexity is therefore $O(|A|^3)$, which is a remarkable result given that in principle the worst-case distribution had to be found among a set with infinite size.

3   Estimating $E_{wc}[T(D)]$

Algorithm 1 allows an exact computation of the expected makespan with worst case distribution, but its scalability is limited by the $O(n^3)$ complexity. Moreover, the added difficulty of embedding a procedural component within an optimization approach may make the method less applicable in practice.

With the aim to address such limitations, we have devised a low-complexity, approximate, estimator based on global graph properties that are reasonably easy to measure. In detail, the estimator is given by the following formula:

$$\tau(\underline{T}, \overline{T}) = \underline{T} + (\overline{T} - \underline{T}) \frac{1}{|A|} \sum_{a_i \in A} \min\left(1, \frac{\tilde{d}_i - \underline{d}_i}{\overline{d}_i - \underline{d}_i} \frac{\sum_{a_i \in A} \overline{d}_i}{\overline{T}}\right) \tag{7}$$

Most of the terms in Equation (7) are constants that can be computed via a pre-processing step. The only input terms that change at search time are $\underline{T}$ and $\overline{T}$, which are simplified notations for $T(\underline{d}_i)$ and $T(\overline{d}_i)$: both values can be efficiently made available as decision variables in a PCP approach via (e.g.) the methods from the Constraint Programming approach in [10]. The estimator formula is non-linear, but it is not difficult to embed in a CP model by using widely available building blocks: the resulting encoding is propagated in $O(|A|)$. The estimator is obtained via a non-trivial process that is discussed in detail in the remainder of this section. Its accuracy and robustness are discussed in Section 3.3.

3.1   A Simplified POS Model

The estimator from Equation (7) is obtained by reasoning on a simplified POS model. Specifically, such model is a layered graph with constant width, and identical minimal duration $\underline{d}_i = \underline{\delta}$ and maximal duration $\overline{d}_i = \overline{\delta}$ for each activity.

A layered graph is a directed acyclic graph whose activities are organized in layers: there are edges between each activity in k-th layer and each activity in the (k+1)-th layer, and no edge between activities in the same layer. In a layered graph with constant width, all layers contain the same number of activities.

The rationale behind our simplified model is that, when a POS is constructed, new edges are added to the original graph so as to prevent possible resource conflicts. This process leads to an "iterative flattening" of the POS (see [3]), making it closer in structure to a layered graph with constant width. The assumption on the identical minimal and maximal durations is instead introduced only to increase the tractability.

Evaluating the Model: The effectiveness of using a layered graph as a model for more complex POSs is difficult to assess in a theoretical fashion, but can be checked empirically. Basically, if the model works, it should lead to reasonably accurate predictions of (e.g.) the value of $E[T(D)]$ with worst-case distribution.

This evaluation can be done by: 1) generating graphs with different number of activities, different width values, and different durations parameters; then 2) using Algorithm 1 to obtain the expected makespan with worst-case distribution. Then the results should be compared with the results obtained for a set of POSs representing solutions to real scheduling problems: if the results are similar, it will be an indication of the model quality. In the remainder of this section, we will use this approach to investigate the impact of changing the width on the value of $E_{wc}[T(D)]$.

Measuring the Width: The width w of a layered graph is (by definition) the number of activities in each layer. The width of a real POS is instead a much fuzzier concept. In our evaluation, an approximate width value is computed as:

$$w = \frac{|A|}{n_L} \qquad \text{with: } n_L = \frac{\overline{T}}{avg(\overline{d}_i)} \qquad \text{and: } avg(\overline{d}_i) = \frac{1}{|A|} \sum_{a_i \in A} \overline{d}_i \qquad (8)$$

Intuitively, we estimate the width as the ratio between the number of activities $|A|$ and the (estimated) number of layers n_L. The number of layers is obtained by dividing the makespan with maximal durations $\overline{T}$ by the average $\overline{d}_i$. By applying algebraic simplifications, we obtain the compact formula:

$$w = \frac{\sum_{a_i \in A} \overline{d}_i}{\overline{T}} \qquad (9)$$

We use the same symbol (i.e. w) for both layered graphs and real POSs, because Equation (9) returns the "true" width when applied to our simplified model.

Evaluation Setup We generated layered graphs with different numbers of activities, and for each one we varied the width from $|A|$ (single layer) to 1. In order to obtain data points for every (integer) value of w, we allow for graphs with quasi-constant width: namely, if $|A|$ is not a multiple of w, the right-most layers are permitted to contain $w - 1$ activities.

As a "real" counterpart for our empirical evaluation we employ a large bench-mark of POSs representing solutions of Resource Constrained Project Scheduling Problems (RCPSP) and Job Shop Scheduling Problems (JSSP). The POSs for the RCPSP have been obtained by solving the j30, j60, and j90 instances from the PSPlib [6], respectively having $|A| = 30, 60, 90$. The POSs for the JSSP have been obtained by solving the instances in the Taillard benchmark, from size 15×15 to 30×20: in this case the number of activities ranges from 225 to 600.

Both the RCPSP and the JSSP instances have been solved with a classical CP approach [1], the SetTimes search strategy [9], and a makespan minimization objective. The POS have been obtained via the post-processing algorithm from [13] from all the schedules found during the optimization process, so that our collection contains 5,753 POSs coming from optimal, slightly suboptimal, and quite far from optimal schedules.

The instances in the PSPLIB and the Taillard benchmarks are originally deterministic, so we had to introduce some uncertainty artificially. In particular, the problem files specify a fixed duration for each activity, that we treat as the maximal duration $\overline{d}_i$. The values of $\underline{d}_i$ and $\tilde{d}_i$ are instead generated at random.

In this particular experimentation, the minimal duration of all activities was fixed to 0 for both the real POSs and the layered graphs. The maximum duration is fixed to 1 for the layered graphs (more on this later). All the average durations $\tilde{d}_i$ were randomly generated using a large variety of different schemes (see Section 3.3 for more details).

Makespan Normalization: Intuitively, the value of $E_{wc}[T(D)]$ should depend on the graph structure, on the characteristics of the uncertainty, and on the scale of the instance. Factoring out the dependence on the scale is important in order to understand the dependence on the structure and on the uncertainty. After some attempts, we have found that is possible to normalize the scale of the expected makespan without introducing distortions by using this simple formula:

$$\text{normalized}\,(E_{wc}[T(D)]) = \frac{E_{wc}[T(D)]}{\sum_{a_i \in A} \overline{d}_i} \tag{10}$$

i.e. by dividing the original value by the sum of the maximal durations. This normalization allows to compare on a uniform scale the expected makespan of graphs with widely different durations and number of activities. For our simpli-fied model, the calculation makes the value of $\overline{d}$ completely irrelevant, which is the reason for fixing the maximal duration to 1 in the experimentation.

Some Results: We have obtained plots reporting the value of $E_{wc}[T(D)]$ over the graph width. Figure 1 shows the plots for the simplified model (with 64 activities) on the left, and for all the real POSs on the right. The values of $\tilde{d}_i$ in both cases have been generated uniformly at random between $\underline{d}_i$ (which is always 0) and $0.5\,\overline{d}_i$. The colors represent the data point density, which is benchmark dependent and unfortunately not very informative.

The results are very interesting. First, the general shape of the plots does not appear to depend on the number of activities: this can be inferred by the

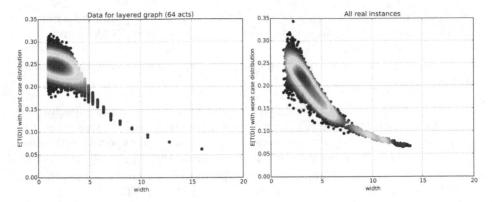

Fig. 1. Value of $E_{wc}[T(D)]$ over the graph width for the simplified model (left) and the real POSs (right)

plot for the real POSs, for which the number of activities ranges from 30 to 600. For the simplified model the plots with 16 and 32 activities (not reported) are almost indistinguishable from the plot with 64.

The most important result, however, is that the behavior of the simplified model and of the real graphs is remarkably similar, in terms of both shape and scale. This is particularly striking given the wide range of graph structures contained in our benchmark, and suggests that our simplified model is indeed well suited for studying the behavior of real POSs.

A necessary condition for the similarity to show up is that the values of the expected durations $\tilde{d}_i$ must be obtained using the same random generation scheme. This is not surprising, as it simply means that the characteristics of the uncertainty change the degree by which the value of w affects the expected makespan: this aspect will be tackled in the next section.

3.2 Deriving the Estimator

In this section, we use a mix of theoretical and empirical arguments applied to the simplified model from Section 3.1 to derive the estimator from Equation (7). In general terms, we are looking for some kind of formula that, based on global properties of the instance and the schedule can approximately predict the value of $E_{wc}[T(D)]$. Thanks to the simplicity of the layered graph model, this may be doable via Probability Theory results.

Random POS Construction: Let us assume to have a set of A of activities, with the corresponding minimal, maximal, and expected durations $\underline{d}_i$, $\overline{d}_i$, and $\tilde{d}_i$. In our simplified model all minimal durations are equal to $\underline{\delta}$ and all maximal durations are equal to δ: however, in this discussion we will use the $\underline{d}_i, \overline{d}_i$ notation whenever possible to maintain a clearer analogy with general POSs.

Since we are looking for a formula that depends on global properties, we can assume to have access to general information (e.g. the width or the number of layers), but not to the location of each activity within the graph[1]. Therefore, we will assume that the layer where each activity appears is determined at random. This is a stochastic process and we are interested in the expected value of $E_{wc}[T(D)]$ w.r.t. the random mapping: i.e., we are dealing with a double expectation in the form $E_{\mathrm{map}}[E_{wc}[T(D)]]$.

Single-layer $E_{wc}[T(D)]$: As a first step, we will focus on the inner expectation, i.e. on the computation of $E[T(D)]$ with worst case duration. At this stage, we can assume the activity positions to be known so that the expected makespan can be obtained by running Algorithm 1 (or solving P0). Since adjacent layers are fully connected, processing each layer separately is equivalent to processing the POS as whole. Now, consider the problem objective in P0:

$$\max z = \sum_{\omega_k \in \Omega} p_{\omega_k} T(\omega_k) \tag{11}$$

For a single layer of the simplified model it can be rewritten as:

$$\max z = \underline{\delta} + (\overline{\delta} - \underline{\delta}) \sum_{\omega_k : \exists D_i = \overline{\delta}} p_{\omega_k} \tag{12}$$

where we have exploited the fact that $\forall a_i \in A : \underline{d}_i = \underline{\delta}, \overline{d}_i = \overline{\delta}$, and therefore the single-layer makespan is equal to $\underline{\delta}$ unless at least one activity takes maximal duration. Therefore, the whole problem can be rewritten for a single-layer as:

$$\mathbf{P1}: \quad \max z = \underline{\delta} + (\overline{\delta} - \underline{\delta}) \sum_{\omega_k : \exists D_i = \overline{\delta}} p_{\omega_k} \tag{13}$$

$$\text{subject to: } \sum_{\omega_k : D_i = \overline{d}_i} p_{\omega_k} \le \frac{\tilde{d}_i - \underline{d}_i}{\overline{D}_i - \underline{D}_i} \qquad \forall a_i \in A' \tag{14}$$

$$\sum_{\omega_k \in \Omega'} p_{\omega_k} \le 1 \tag{15}$$

$$p_{\omega_k} \ge 0 \qquad \forall \omega_k \in \Omega' \tag{16}$$

where A' is the set of activities in the layer and $\Omega' = \prod_{a_i \in A'} \{\underline{d}_i, \overline{d}_i\}$. As already mentioned, we have not replaced $\underline{d}_i$, $\overline{d}_i$ with $\underline{\delta}$, $\overline{\delta}$ unless it was strictly necessary to obtain an important simplification.

Problem P1 is simple enough to admit a closed form solution. In particular, all scenarios where at least one activity takes the maximal duration are symmetrical. Moreover, it is always worthwhile to increase the probability of such scenarios as much as possible. Therefore the solution of P1 is always given by:

$$z^* = \underline{\delta} + (\overline{\delta} - \underline{\delta}) \min \left(1, \sum_{a_i \in A'} \frac{\tilde{d}_i - \underline{d}_i}{\overline{D}_i - \underline{D}_i} \right) \tag{17}$$

where $(\tilde{d}_i - \underline{d}_i)/(\overline{D}_i - \underline{D}_i)$ is the probability that a_i takes maximal duration.

---

[1] If we had access to this, we could use Algorithm 1.

Expectation w.r.t. the Random Mapping In our stochastic POS construction process, the choice of the activities in each layer is random. Therefore the value of $(\tilde{d}_i - \underline{d}_i)/(\overline{D_i} - \underline{D_i})$ can be seen as a random variable in Equation (17):

$$z^* = \underline{\delta} + (\overline{\delta} - \underline{\delta}) \min \left(1, \sum_{k=0}^{w-1} Q_k \right) \tag{18}$$

and selecting an activity for the current layer is equivalent to instantiating a Q_k.

For the first instantiation, the probability to pick a certain value of v_k for Q_k is given by the number of activities in A such that $(\tilde{d}_i - \underline{d}_i)/(\overline{D_i} - \underline{D_i}) = v_k$. For subsequent instantiations the probabilities will be different, because an activity cannot be inserted in two positions in the POS. In probabilistic terms, the Q_k variables are non-independent, which complicates enormously the task of defining a probabilistic model.

We address this issue by simply disregarding the constraint and allowing activity duplication to occur. We have empirically checked the accuracy of this approximation, which appear to be very good in terms of expected value. This last simplification is enough to ensure that: 1) the Q_k variables are independent; and 2) the Q_k variables have the same discrete distribution, which is defined by the frequency of occurrence of each $(\tilde{d}_i - \underline{d}_i)/(\overline{D_i} - \underline{D_i})$ among the $a_i \in A$.

Having independent and identically distributed Q_k implies that the value of Equation (18) will be identical for all the layers. Therefore a approximate formula for $E_{\text{map}}[E_{\text{wc}}[T(D)]]$ of the random POS construction process is:

$$n_L E \left[\underline{\delta} + (\overline{\delta} - \underline{\delta}) \min \left(1, \sum_{k=0}^{w-1} Q_k \right) \right] \tag{19}$$

where we recall that $n_L = |A|/w$ is the number of layers.

The Estimator Formula: Computing the expected value in Equation (19) is non-trivial, due to the presence of the "min" operator. It is however possible to address this issue by introducing some additional approximations. After exploring several alternatives, we have settled for replacing the sum of Q_k variables with a product. By exploiting the linearity of the expectation operator, we get:

$$n_L \underline{\delta} + n_L (\overline{\delta} - \underline{\delta}) E \left[\min \left(1, w Q_k \right) \right] \tag{20}$$

where it should be noted that $n_L \underline{\delta}$ and $n_L \overline{\delta}$ correspond for the simplified model to $\underline{T}$ and $\overline{T}$. Since Q_k has a discrete distribution, the expected value can be computed by: 1) multiplying by w every value in the distribution of Q_k; then 2) capping each result at 1; and finally 3) computing the average over all activities:

$$E \left[\min \left(1, w Q_k \right) \right] = \frac{1}{|A|} \sum_{a_i \in A} \min \left(1, w \frac{\tilde{d}_i - \underline{d}_i}{\overline{d}_i - \underline{d}_i} \right) \tag{21}$$

by combining Equation (20) and (21), replacing $n_L \underline{\delta}$ with $\underline{T}$, replacing $n_L \overline{\delta}$ with $\overline{T}$, and replacing w with the formula from Equation (9), we obtain the estimator original estimator formula.

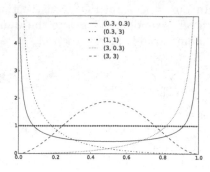

It is useful to have a visual interpretation of the α and β parameters of the beta distribution. The plot on the left depicts five different shapes of the probability density function, which correspond to the configurations used in the experiments. The α parameter is intuitively related to the probability of the lower values, while β is related to the higher values. If $\alpha = \beta = 1$, all values are equally likely and we get a uniform distribution. If $\alpha < 1$ or $\beta < 1$ the probability mass tends to cluster respectively on the lowest and highest value. Conversely, if α or β are > 1, then the probability mass tends to move to the center, and the distribution assumes a bell-like shape.

Fig. 2. Visual interpretation of the α and β parameters in the beta distribution.

3.3   Empirical Evaluation of the Estimator

We have evaluated the estimator on the same benchmark used for assessing the effectiveness of the simplified model (see Section 3.1). We have compared the accuracy of our estimator $\tau(\underline{T}, \overline{T})$ with that of two baseline predictors: 1) the makespan with worst case durations $T(\overline{d})$, which is an upper bound for $E_{wc}[T(D)]$; and 2) and the makespan with expected durations $T(\tilde{d})$, which is a lower bound. In particular $T(\tilde{d})$ proved to be a surprisingly effective estimator for the expected makespan in a previous evaluation of ours [2].

We recall that the maximal durations for each POS are specified by the instance file. The minimal and the average durations are randomly generated, in this case using a beta distribution $Beta(\alpha, \beta)$. This was chosen because it is bounded and very flexible. The parameter configurations that we have employed are presented in Figure 2. In detail, the values of $\underline{d}_i$ and $\tilde{d}_i$ are obtained as:

$$\underline{d}_i = \frac{\overline{d}_i}{2} Beta(\alpha_{\underline{d}}, \beta_{\underline{d}}) \qquad \tilde{d}_i = \underline{d}_i + (\overline{d}_i - \underline{d}_i) Beta(\alpha_{\tilde{d}}, \beta_{\tilde{d}}) \qquad (22)$$

where $\underline{d}_i$ ranges in $[0, \overline{d}_i/2]$, and $\tilde{d}_i$ ranges in $[\underline{d}_i, \overline{d}_i]$. We consider four parameter configurations for $\underline{d}_i$ and five configurations for $\tilde{d}_i$. Our goal was to generate a large variety of duration scenarios that are somehow representative of real world conditions. For instance the configuration $\alpha_{\tilde{d}} = 3, \beta_{\tilde{d}} = 3$ could model a situation where the activities are subject to many small disruptions. The configuration $\alpha_{\tilde{d}} = 0.3, \beta_{\tilde{d}} = 3$ leads to average durations very close to the minimal ones, which may be a model for unlikely, but very disruptive, machine faults.

Table 1 shows the average and standard deviation of the relative prediction error for the compared estimators. We use percentage errors computed as:

$$Err(x) = 100 \frac{E_{wc}[T(D)] - x}{E_{wc}[T(D)]} \qquad (23)$$

where x is the prediction given by an estimator. Negative errors represent over-estimations, while positive errors are under-estimations. As expected, the $T(\overline{d})$ predictor always over-estimates, while $T(\tilde{d})$ under-estimates.

As the number of activities grows, $E_{wc}[T(D)]$ gets closer to $T(\overline{d})$: intuitively, since we are reasoning with a worst-case distribution, more activities mean more things that can go wrong. This trend reflects on the performance of $T(\overline{d})$ (which gets better), and appears to have a adverse effect on the accuracy of τ and $T(\tilde{d})$.

When the minimal durations tend to be higher (i.e. when $\alpha_{\underline{d}} \geq 1$ and $\beta_{\underline{d}} < 0$), the range of the random durations is reduced and all predictions are more precise. The $T(\tilde{d})$ estimator benefits the most from this situation.

In general, the τ estimator tends to outperform the other estimators: the accuracy is much better than $T(\overline{d})$, and significantly better than $T(\tilde{d})$ in most cases. It must be said that using $T(\tilde{d})$ as an estimator for $E_{wc}[T(D)]$ lead to surprisingly good results, especially in terms of standard deviation: this is consistent with our findings form [2], and deserves further investigation. The τ estimator really shines when the $\tilde{d}_i$ values are small, i.e. for $\alpha_{\tilde{d}} < 0$: this is the situation that models the presence of machine faults, which is also the best suited for reasoning in terms of worst-case distribution (since a fault tends to affect all the activities in the system).

4   Concluding Remarks

We have tackled the problem of estimating the expected makespan of a POS under duration uncertainty, assuming access to very limited information about the probability distribution. We have provided an $O(n^3)$ algorithm for computing the largest expected makespan that is compatible with the available information, and we have proposed a linear-time estimator to approximately predict the same quantity. Both the algorithm and the estimator could be employed within an optimization approach to obtain an estimate of the expected makespan, in cases where scenario-based optimization has prohibitive complexity, or extensive duration data is not available.

Our results are very encouraging, and we believe that further improvements are possible. First, it should be possible to exploit better the striking similarity of behavior between real POSs and layered graphs, which we observed in [2] and confirmed in this paper. Second, it may be possible to obtain a more accurate and still efficient approach by exploiting additional information (e.g. a covariance matrix). Third, we have largely disregarded the strong correlation between $T(\tilde{d})$ and the expected makespan, which we had observed in [2] and was confirmed in our experimental evaluation: it should be possible to exploit this correlation in order to improve the accuracy of the estimate.

Additional possible directions for future research involve the design of an estimator for the makespan variance, or for specific quantiles in its distribution. Such an estimator would allow to state "approximate chance constraints", and allow an efficient solution of many more practical problems. Finally, it may be possible to replace our scalar estimate with ranges, similarly to what is done in statistics with confidence intervals. A the present time, we have not started to investigate any of the last two mentioned research directions.

5   Appendix A

This appendix contains the proof to Theorem 1.

Theorem 1. *There exists a worst-case distribution such that, in each scenario with non-zero probability, every D_i takes either the value $\underline{d}_i$ or the value $\overline{d}_i$.*

Proof (By contradiction and induction). Let us assume that the worst case distribution contains a scenario ω_k such that one D_i is equal to some value $v_k \in (\underline{d}_i, \overline{d}_i)$ and $P(\omega_k) > 0$. Let $\underline{\omega}_k$ and $\overline{\omega}_k$ be two scenarios that are identical to ω_k in all assignments, except that D_i takes respectively its minimal and maximal value. Let us consider shifting the probability $P(\omega_k)$ from ω_k to $\underline{\omega}_k$ and $\overline{\omega}_k$, and let $\Delta P(\underline{\omega}_k) \geq 0$ and $\Delta P(\overline{\omega}_k) \geq 0$ be the two probability variations. Then it must hold:

$$\Delta P(\overline{\omega}_k) + \Delta P(\underline{\omega}_k) = P(\omega_k) \tag{24}$$

in order to preserve the total probability mass, and:

$$\overline{d}_i \Delta P(\overline{\omega}_k) + \underline{d}_i \Delta P(\underline{\omega}_k) = v_k P(\omega_k) \tag{25}$$

to keep the expected value $\tilde{d}_i$ unchanged. We recall that in each scenario the makespan is given by the length of the critical path, and that the expected makespan is given by the integral (or sum) of $T(D(\omega_k))P(\omega_k)$. Now, let us distinguish some cases:

1) **a_i is on the critical path in ω_k.** This implies that a_i is on the critical path also in $\overline{\omega}_k$, and therefore shifting the probability causes the expected makespan to increase by $(\overline{d}_i - v_k)\Delta P(\overline{\omega}_k)$ units. Then:
   - If a_i is on the critical path in $\underline{\omega}_k$, then the expected makespan is also decreased by $(v_k - \underline{d}_i)\Delta P(\underline{\omega}_k)$ units. However, by combining Equations (24) and (25) we obtain that:

$$(\overline{d}_i - v_k)\Delta P(\overline{\omega}_k) - (v_k - \underline{d}_i)\Delta P(\underline{\omega}_k) = 0 \tag{26}$$

   i.e. the net change of the expected makespan is zero.
   - If a_i is *not* on the critical path in $\underline{\omega}_k$, then the expected makespan is decreased by less than $(v_k - \underline{d}_i)\Delta P(\underline{\omega}_k)$ units, and the net change is positive (i.e. the expected makespan is increased).
2) **a_i is *not* on the critical path in ω_k.** This implies that a_i is not on the critical path also in $\underline{\omega}_k$, and therefore that shifting probability to $\underline{\omega}_k$ leaves the expected makespan unchanged. Then:
   - If a_i is on the critical path in $\overline{\omega}_k$, the expected makespan is increased by a quantity in the range $[0, (\overline{d}_i - v_k)\Delta P(\overline{\omega}_k))$.
   - If a_i is *not* on the critical path in $\overline{\omega}_k$, the expected makespan is again unchanged.

Therefore, by reducing to zero the probability $P(\omega_k)$, the expected makespan either is increased or stays unchanged. This procedure can be repeated until there is no scenario with non-zero probability where D_i takes a value different from $\underline{d}_i$ or $\overline{d}_i$. $\square$

Table 1. Partial horizontal line

$(a_{\tilde{d}_i}, b_{\tilde{d}_i}) \longrightarrow$		(0.3,0.3)			(0.3,3)			(1,1)			(3,3)		
$(a_{\underline{d}_i}, b_{\underline{d}_i})$		τ	$\mathbf{T(\tilde{d})}$	$\mathbf{T(\overline{d})}$	τ	$\mathbf{T(\tilde{d})}$	$\mathbf{T(\overline{d})}$	τ	$\mathbf{T(\tilde{d})}$	$\mathbf{T(\overline{d})}$	τ	$\mathbf{T(\tilde{d})}$	$\mathbf{T(\overline{d})}$
J30													
(0.3,0.3)	Err	-1.08	5.06	-37.19	-3.91	9.37	-148.18	-7.75	11.32	-31.82	-12.23	15.87	-27.92
	Std	7.77	2.72	13.38	5.36	4.57	32.46	6.94	3.79	10.29	6.51	3.99	8.27
(0.3,3.0)	Err	1.55	7.54	-52.16	2.44	24.83	-352.00	-6.83	17.14	-41.85	-13.18	24.20	-35.18
	Std	10.07	3.87	19.37	6.77	8.80	113.50	8.32	4.87	13.75	8.06	4.61	10.57
(3.0,0.3)	Err	-1.62	2.66	-26.68	-3.22	3.54	-85.42	-6.38	6.22	-23.62	-9.90	9.33	-21.15
	Std	5.55	1.67	8.54	3.38	2.19	10.22	4.85	2.59	7.03	4.82	2.97	6.01
(1.0,1.0)	Err	-0.70	5.14	-37.30	-3.02	10.13	-153.98	-7.04	11.33	-32.04	-11.91	16.20	-27.98
	Std	7.71	2.66	13.60	4.79	4.52	30.28	6.42	3.64	10.17	6.28	3.88	8.10
(3.0,3.0)	Err	-0.03	5.29	-38.33	-2.17	10.25	-160.43	-6.95	11.54	-32.73	-11.72	16.46	-28.09
	Std	7.42	2.75	13.65	4.53	4.39	29.59	6.59	3.62	10.15	6.44	3.75	8.09
J60													
(0.3,0.3)	Err	-2.92	6.16	-33.85	-7.35	12.62	-133.52	-11.11	13.10	-27.88	-16.20	18.25	-23.59
	Std	6.66	2.45	10.35	5.12	4.32	23.12	5.81	3.25	7.72	5.33	3.38	6.15
(0.3,3.0)	Err	0.07	8.82	-44.82	-0.34	33.56	-268.74	-11.47	19.79	-35.42	-18.73	27.21	-29.12
	Std	8.10	3.39	14.08	6.30	7.83	66.40	7.54	4.06	9.63	6.62	3.68	7.42
(3.0,0.3)	Err	-3.19	3.32	-25.27	-6.40	4.93	-81.94	-9.31	7.54	-21.56	-13.37	10.95	-18.81
	Std	4.90	1.61	7.01	3.52	2.32	8.50	4.46	2.32	5.54	4.27	2.57	4.81
(1.0,1.0)	Err	-2.34	6.10	-33.65	-6.61	13.79	-139.56	-10.83	13.31	-28.08	-16.30	18.50	-23.87
	Std	6.57	2.47	10.04	4.82	4.53	22.94	5.77	3.21	7.57	5.50	3.32	6.21
(3.0,3.0)	Err	-1.84	6.22	-34.20	-5.66	14.53	-145.00	-10.48	13.52	-28.10	-16.10	18.69	-23.90
	Std	6.63	2.46	10.24	4.85	4.61	22.46	5.72	3.16	7.41	5.37	3.17	6.14
J90													
(0.3,0.3)	Err	-3.94	6.59	-32.30	-9.82	14.67	-125.92	-12.44	13.87	-26.23	-17.41	19.20	-22.03
	Std	6.28	2.35	9.03	5.20	4.16	19.42	5.43	2.92	6.54	4.93	3.17	5.54
(0.3,3.0)	Err	-1.16	9.62	-42.30	-1.25	38.04	-233.89	-13.54	20.81	-33.02	-20.17	28.48	-26.67
	Std	7.39	3.19	11.32	6.35	6.83	52.10	6.77	3.74	8.34	6.04	3.35	6.54
(3.0,0.3)	Err	-3.71	3.62	-24.17	-8.44	5.78	-79.91	-10.55	8.16	-20.61	-14.43	11.71	-17.88
	Std	4.59	1.55	6.08	3.70	2.42	8.03	4.26	2.25	5.03	3.88	2.51	4.24
(1.0,1.0)	Err	-3.46	6.67	-32.51	-8.80	15.97	-131.57	-12.31	14.23	-26.29	-17.24	19.42	-21.97
	Std	5.85	2.36	8.44	5.08	4.33	19.47	5.53	2.95	6.59	4.86	3.02	5.40
(3.0,3.0)	Err	-2.73	6.83	-32.88	-7.77	17.12	-136.15	-12.12	14.35	-26.59	-17.36	19.80	-22.28
	Std	5.84	2.36	8.55	5.07	4.54	19.52	5.47	2.91	6.63	5.00	2.92	5.47
Taillard													
(0.3,0.3)	Err	-7.34	11.16	-27.74	-14.69	27.53	-94.63	-12.85	20.99	-17.84	-12.29	26.59	-12.63
	Std	4.16	2.36	5.69	3.69	3.77	10.76	3.51	2.64	4.09	3.42	2.26	3.50
(0.3,3.0)	Err	-5.74	15.93	-34.45	-6.04	55.75	-134.53	-14.01	29.17	-20.96	-13.88	36.39	-14.33
	Std	4.66	3.04	6.73	4.14	3.54	17.82	4.07	2.97	4.90	3.69	2.43	3.79
(3.0,0.3)	Err	-6.97	6.92	-22.08	-13.61	13.95	-66.55	-11.20	13.85	-14.98	-10.62	17.87	-10.87
	Std	3.35	1.76	4.15	3.18	2.68	6.65	2.94	2.14	3.29	2.92	2.10	2.98
(1.0,1.0)	Err	-7.17	11.29	-28.19	-13.26	29.75	-96.61	-12.84	21.31	-17.97	-12.31	26.88	-12.65
	Std	3.95	2.44	5.27	3.73	3.46	11.00	3.44	2.54	4.00	3.36	2.34	3.46
(3.0,3.0)	Err	-6.49	11.45	-28.09	-11.98	31.37	-98.08	-12.72	21.43	-17.98	-12.28	27.13	-12.62
	Std	3.92	2.45	5.34	3.81	3.39	11.24	3.51	2.56	4.05	3.35	2.24	3.44

References

1. Baptiste, P., Le Pape, C., Nuijten, W.: Constraint-based scheduling. Kluwer Academic Publishers (2001)
2. Bonfietti, A., Lombardi, M., Milano, M.: Disregarding duration uncertainty in partial order schedules? yes, we can!. In: Simonis, H. (ed.) CPAIOR 2014. LNCS, vol. 8451, pp. 210–225. Springer, Heidelberg (2014)
3. Cesta, A., Oddi, A., Smith, S.F.: Iterative flattening: a scalable method for solving multi-capacity scheduling problems. In: AAAI/IAAI, pp. 742–747 (2000)
4. Kellerer, H., Pferschy, U., Pisinger, D.: Knapsack problems. Springer Science & Business Media (2004)
5. Kleywegt, A.J., Shapiro, A., Homem-de Mello, T.: The sample average approximation method for stochastic discrete optimization. SIAM Journal on Optimization 12(2), 479–502 (2002)
6. Kolisch, R., Sprecher, A.: Psplib-a project scheduling problem library: Or software-orsep operations research software exchange program. European Journal of Operational Research 96(1), 205–216 (1997)
7. Laborie, P.: Complete MCS-based search: application to resource constrained project scheduling. In: Proc. of IJCAI, pp. 181–186. Professional Book Center (2005)
8. Laborie, P., Godard, D.: Self-adapting large neighborhood search: application to single-mode scheduling problems. In: Proc. of MISTA (2007)
9. Le Pape, C., Couronné, P., Vergamini, D., Gosselin, V.: Time-versus-capacity compromises in project scheduling. AISB Quarterly, 19 (1995)
10. Lombardi, M., Milano, M., Benini, L.: Robust scheduling of task graphs under execution time uncertainty. IEEE Trans. Computers 62(1), 98–111 (2013)
11. Morris, P., Muscettola, N., Vidal, T.: Dynamic control of plans with temporal uncertainty. In: Proc. of IJCAI, pp. 494–499. Morgan Kaufmann Publishers Inc. (2001)
12. Policella, N., Cesta, A., Oddi, A., Smith, S.F.: From precedence constraint posting to partial order schedules: A CSP approach to Robust Scheduling. AI Communications 20(3), 163–180 (2007)
13. Policella, N., Smith, S.F., Cesta, A., Oddi, A.: Generating robust schedules through temporal flexibility. In: Proc. of ICAPS, pp. 209–218 (2004)
14. Tarim, S.A., Manandhar, S., Walsh, T.: Stochastic constraint programming: A scenario-based approach. Constraints 11(1), 53–80 (2006)
15. Vidal, T.: Handling contingency in temporal constraint networks: from consistency to controllabilities. Journal of Experimental & Theoretical Artificial Intelligence 11(1), 23–45 (1999)

A Parallel, Backjumping Subgraph Isomorphism Algorithm Using Supplemental Graphs

Ciaran McCreesh[(✉)] and Patrick Prosser

University of Glasgow, Glasgow, Scotland
c.mccreesh.1@research.gla.ac.uk, patrick.prosser@glasgow.ac.uk

Abstract. The subgraph isomorphism problem involves finding a pattern graph inside a target graph. We present a new bit- and thread-parallel constraint-based search algorithm for the problem, and experiment on a wide range of standard benchmark instances to demonstrate its effectiveness. We introduce supplemental graphs, to create implied constraints. We use a new low-overhead, lazy variation of conflict directed backjumping which interacts safely with parallel search, and a counting-based all-different propagator which is better suited for large domains.

1 Introduction

The subgraph isomorphism family of problems involve "finding a copy of" a pattern graph inside a larger target graph; applications include bioinformatics [3], chemistry [31], computer vision [12,37], law enforcement [7], model checking [33], and pattern recognition [9]. These problems have natural constraint programming models: we have a variable for each vertex in the pattern graph, with the vertices of the target graph being the domains. The exact constraints vary depending upon which variation of the problem we are studying (which we discuss in the following section), but generally there are rules about preserving adjacency, and an all-different constraint across all the variables.

This constraint-based search approach dates back to works by Ullmann [39] and McGregor [25], and was improved upon in the LV [20] and VF2 [11] algorithms. More recently, ILF [41], LAD [36] and SND [1] are algorithms which take a "deep thinking" approach, using strong inference at each stage of the search. This is powerful, but we observe LAD or SND sometimes make less than one recursive call per second with larger target graphs, and cannot always explore enough of the search space to find a solution in time. This motivates an alternative approach: on the same hardware, we will be making 10^4 to 10^6 recursive calls per core per second. The main features of our algorithm are:

1. We introduce supplemental graphs, which generalise some of the ideas in SND. The key idea is that a subgraph isomorphism $i : P \rightarrowtail T$ induces a subgraph isomorphism $F(i) : F(P) \rightarrowtail F(T)$, for certain functors F. This is

C. McCreesh—This work was supported by the Engineering and Physical Sciences Research Council [grant number EP/K503058/1].

G. Pesant (Ed.): CP 2015, LNCS 9255, pp. 295–312, 2015.
DOI: 10.1007/978-3-319-23219-5_21

used to generate implied constraints: we may now look for a mapping which is simultaneously a subgraph isomorphism between several carefully selected pairs of graphs.

2. We use weaker inference than LAD and SND: we do not achieve or maintain arc consistency. We introduce a cheaper, counting-based all-different propagator which has better scalability for large target graphs, and which has very good constant factors on modern hardware thanks to bitset encodings.

3. We describe a clone-comparing variation of conflict-directed backjumping, which does not require conflict sets. We show that an all-different propagator can produce reduced conflict explanations, which can improve backjumping.

4. We use thread-parallel preprocessing and search, to make better use of modern multi-core hardware. We explain how parallel search may interact safely with backjumping. We use explicit, non-randomised work stealing to offset poor early heuristic choices during search.

Although weaker propagation and backjumping have fallen out of fashion in general for constraint programming, here this approach usually pays off. In section 4 we show that over a large collection of instances commonly used to compare subgraph isomorphism algorithms, our solver is the single best.

2   Definitions, Notation, and a Proposition

Throughout, our graphs are finite, undirected, and do not have multiple edges between pairs of vertices, but may have loops (an edge from a vertex to itself). We write $V(G)$ for the vertex set of a graph G, and $N(G, v)$ for the neighbours of a vertex v in G (that is, the vertices adjacent to v). The *degree* of v is the cardinality of its set of neighbours. The *neighbourhood degree sequence* of v, denoted $S(G, v)$, is the sequence consisting of the degrees of every neighbour of v, from largest to smallest. A vertex is *isolated* if it has no neighbours. By $v \sim_G w$ we mean vertex v is adjacent to vertex w in graph G. We write $G[V]$ for the subgraph of G induced by a set of vertices V.

A *non-induced subgraph isomorphism* is an injective mapping $i : P \rightarrowtail T$ from a graph P to a graph T which preserves adjacency—that is, if $v \sim_P w$ then we require $i(v) \sim_T i(w)$ (and thus if v has a loop, then $i(v)$ must have a loop). The *non-induced subgraph isomorphism problem* is to find such a mapping from a given pattern graph P to a given target graph T. (The *induced subgraph isomorphism problem* additionally requires that if $v \not\sim_P w$ then $i(v) \not\sim_T i(w)$, and variants also exist for directed and labelled graphs; we discuss only the non-induced version in this paper. All these variants are NP-complete.)

If R and S are sequences of integers, we write $R \preceq S$ if there exists a subsequence of S with length equal to that of R, such that each element in R is less than or equal to the corresponding element in S. For a set U and element v, we write $U - v$ to mean $U \setminus \{v\}$, and $U + v$ to mean $U \cup \{v\}$.

A *path* in a graph is a sequence of distinct vertices, such that each successive pair of vertices are adjacent; we also allow a path from a vertex to itself, in which case the first and last vertices in the sequence are the same (and there is a *cycle*). The *distance* between two vertices is the length of a shortest path between them. We write G^d for the graph with vertex set $V(G)$, and edges between v and w if the distance between v and w in G is at most d. We introduce the notation $G^{[c,l]}$ for the graph with vertex set $V(G)$, and edges between vertices v and w (not necessarily distinct) precisely if there are at least c paths of length exactly l between v and w in G. The following proposition may easily be verified by observing that injectivity means paths are preserved:

Proposition 1. *Let $i : P \rightarrowtail T$ be a subgraph isomorphism. Then i is also*

1. *a subgraph isomorphism $i^d : P^d \rightarrowtail T^d$ for any $d \geq 1$, and*
2. *a subgraph isomorphism $i^{[c,l]} : P^{[c,l]} \rightarrowtail T^{[c,l]}$ for any $c, l \geq 1$.*

The (contrapositive of the) first of these facts is used by SND, which dynamically performs distance-based filtering during search. We will instead use the second fact, at the top of search, to generate implied constraints.

3 A New Algorithm

Algorithm 1 describes our approach. We begin (line 3) with a simple check that there are enough vertices in the pattern graph for an injective mapping to exist. We then (line 4) discard isolated vertices in the pattern graph—such vertices may be greedily assigned to any remaining target vertices after a solution is found. This reduces the number of variables which must be copied when branching. Next we construct the supplemental graphs (line 5) and initialise domains (line 6). We then (line 7) use a counting-based all-different propagator to reduce these domains further. Finally, we perform a backtracking search (line 8). Each of these steps is elaborated upon below.

3.1 Preprocessing and Initialisation

Following Proposition 1, in Algorithm 2 we construct a sequence of supplemental graph pairs from our given pattern and target graph. We will then search for a mapping which is simultaneously a mapping from each pattern graph in the

---

Algorithm 1. A non-induced subgraph isomorphism algorithm

```
1 nonInducedSubgraphIsomorphism (Graph P, Graph T) → Bool
2 begin
3     if |V(P)| > |V(T)| then return false
4     Discard isolated vertices in P
5     L ← createSupplementalGraphList(P, T)
6     D ← init(V(P), V(T), L)
7     if countingAllDifferent(D) ≠ Success then return false
8     return search(L, D) = Success
```

Algorithm 2. Supplemental graphs for Algorithm 1

---

1 `createSupplementalGraphList` (Graph $\mathcal{P}$, Graph $\mathcal{T}$) $\rightarrow$ GraphPairs
2 **begin**

3 $\quad$ **return** $\Big[(\mathcal{P}, \mathcal{T}), \quad (\mathcal{P}^{[1,2]}, \mathcal{T}^{[1,2]}), \quad (\mathcal{P}^{[2,2]}, \mathcal{T}^{[2,2]}), \quad (\mathcal{P}^{[3,2]}, \mathcal{T}^{[3,2]}),$

$\qquad\qquad (\mathcal{P}^{[1,3]}, \mathcal{T}^{[1,3]}), \quad (\mathcal{P}^{[2,3]}, \mathcal{T}^{[2,3]}), \quad (\mathcal{P}^{[3,3]}, \mathcal{T}^{[3,3]}) \Big]$

---

sequence to its paired target graph—this gives us implied constraints, leading to additional filtering during search.

Our choice of supplemental graphs is somewhat arbitrary. We observe that distances of greater than 3 rarely give additional filtering power, and constructing $G^{[c,4]}$ is computationally very expensive (for unbounded l, the construction is NP-hard). Checking $c > 3$ is also rarely beneficial. Our choices work reasonably well in general on the wide range of benchmark instances we consider, but can be expensive for trivial instances—thus there is potential room to improve the algorithm by better selection on an instance by instance basis [23].

Algorithm 3 is responsible for initialising domains. We have a variable for each vertex in the (original) pattern graph, with each domain being the vertices in the (original) target graph. It is easy to see that a vertex of degree d in the pattern graph P may only be mapped to a vertex in the target graph T of degree d or higher: this allows us to perform some initial filtering. By extension, we may use compatibility of neighbourhood degree sequences: v may only be mapped to w if $S(P, v) \preceq S(T, w)$ [41]. Because any subgraph isomorphism $P \rightarrowtail T$ is also a subgraph isomorphism $F(P) \rightarrowtail F(T)$ for any of our supplemental graph constructions F, we may further restrict initial domains by considering only the intersection of filtered domains using each supplemental graph pair individually (line 5). At this stage, we also enforce the "loops must be mapped to loops" constraint.

Following this filtering, some target vertices may no longer appear in any domain, in which case R will be reduced on line 6. If this happens, we iteratively repeat the domain construction, but do not consider any target vertex no longer

---

Algorithm 3. Variable initialisation for Algorithm 1

---

1 `init` (Vertices V, Vertices R, GraphPairs L) $\rightarrow$ Domains
2 **begin**
3 $\quad$ **repeat**
4 $\quad\quad$ **foreach** $v \in V$ **do**
5 $\quad\quad\quad$ $D_v \leftarrow \bigcap_{(P,T) \in L} \left\{ w \in R : \; v \underset{P}{\sim} v \Rightarrow w \underset{T}{\sim} w \wedge S(P, v) \preceq S(T[R], w) \right\}$
6 $\quad\quad$ $R \leftarrow \bigcup_{v \in V} D_v$
7 $\quad$ **until** R is unchanged
8 $\quad$ **return** D

Algorithm 4. Recursive search for Algorithm 1

```
1  search (GraphPairs L, Domains D) → Fail F or Success
2  begin
3  │   if D = ∅ then return Success
4  │   Dᵥ ← a domain in D with minimum size, tiebreaking on static degree in 𝒫
5  │   F ← {v}
6  │   foreach v' ∈ Dᵥ ordered by static degree in 𝒯 do
7  │   │   D' ← clone(D)
8  │   │   case assign(L, D', v, v') of
9  │   │   │   Fail F' then F ← F ∪ F'
10 │   │   │   Success then
11 │   │   │   │   case search(L, D' − Dᵥ) of
12 │   │   │   │   │   Success then return Success
13 │   │   │   │   │   Fail F' then
14 │   │   │   │   │   │   if ∄w ∈ F' such that Dᵥᵥ ≠ D'ᵥᵥ then return Fail F'
15 │   │   │   │   │   │   F ← F ∪ F'
16 │   return Fail F
```

---

in R when calculating degree sequences. (Note that for performance reasons, we do not recompute supplemental graphs when this occurs.)

3.2   Search and Inference

Algorithm 4 describes our recursive search procedure. If every variable has already been assigned, we succeed (line 3). Otherwise, we pick a variable (line 4) to branch on by selecting the variable with smallest domain, tiebreaking on descending static degree only in the original pattern graph (we tried other variations, including using supplemental graphs for calculating degree, and domain over degree, but none gave an overall improvement). For each value in its domain in turn, ordered by descending static degree in the target graph [13], we try assigning that value to the variable (line 8). If we do not detect a failure, we recurse (line 11).

The assignment and recursive search both either indicate success, or return a nogood set of variables F which cannot all be instantiated whilst respecting assignments which have already been made. This information is used to prune the search space: if a subproblem search has failed (line 13), but the current assignment did not remove any value from any of the domains involved in the discovered nogood (line 14), then we may ignore the current assignment and backtrack immediately. In fact this is simply conflict-directed backjumping [27] in disguise: rather than explicitly maintaining conflict sets to determine culprits (which can be costly when backjumping does nothing [2,14]), we lazily create the conflict sets for the variables in F' as necessary by comparing D before the current assignment with the D' created after it. Finally, as in backjumping, if

---

Algorithm 5. Variable assignment for Algorithm 5

---

1 **assign** (GraphPairs L, Domains D, Vertex v, Vertex v') $\rightarrow$ Fail F **or** Success
2 **begin**
3      $D_v \leftarrow \{v'\}$
4   **foreach** $D_w \in D - D_v$ **do**
5          $D_w \leftarrow D_w - v'$
6   **foreach** $(P, T) \in L$ **do**
7              $\lfloor$ **if** $v \sim_P w$ **then** $D_w \leftarrow D_w \cap \mathrm{N}(T, v')$
8   **if** $D_w = \emptyset$ **then return** Fail $\{w\}$
9   **return** countingAllDifferent(D)

---

none of the assignments are possible, we return with a nogood of the current variable (line 5) combined with the union of the nogoods of each failed assignment (line 9) or subsearch (line 15).

For assignment and inference, Algorithm 5 gives the value v' to the domain D_v (line 3), and then infers which values may be eliminated from the remaining domains. Firstly, no other domain may now be given the value v' (line 5). Secondly, for each supplemental graph pair, any domain for a vertex adjacent to v may only be mapped to a vertex adjacent to v' (line 7). If any domain gives a wipeout, then we fail with that variable as the nogood (line 8).

To enforce the all-different constraint, it suffices to remove the assigned value from every other domain, as we did in line 5. However, it is often possible to do better. We can sometimes detect that an assignment is impossible even if values remain in each variable's domain (if we can find a set of n variables whose domains include strictly less than n values between them, which we call a *failed Hall set*), and we can remove certain variable-value assignments that we can prove will never occur (if we can find a set of n variables whose domains include only n values between them, which we call a *Hall set*, then those values may be removed from the domains of any other variable). The canonical way of doing this is to use Régin's matching-based propagator [30].

However, matching-based filtering is expensive and may do relatively little, particularly when domains are large, and the payoff may not always be worth the cost. Various approaches to offsetting this cost while maintaining the filtering power have been considered [15]. Since we are not maintaining arc consistency in general, we instead use an intermediate level of inference which is not guaranteed to identify every Hall set: this can be thought of as a heuristic towards the matching approach. This is described in Algorithm 6.

The algorithm works by performing a linear pass over each domain in turn, from smallest cardinality to largest (line 4). The H variable contains the union of every Hall set detected so far; initially it is empty. The A set accumulates the union of domains seen so far, and n contains the number of domains contributing to A. For each new domain we encounter, we eliminate any values present in previous Hall sets (line 6). We then add that domain's values to A and increment

---

Algorithm 6. Counting-based all-different propagation

```
1  countingAllDifferent (Domains D) → Fail F or Success
2  begin
3  │   (F, H, A, n) ← (∅, ∅, ∅, 0)
4  │   foreach D_v ∈ D from smallest cardinality to largest do
5  │   │   F ← F + v
6  │   │   D_v ← D_v \ H
7  │   │   (A, n) ← (A ∪ D_v, n + 1)
8  │   │   if D_v = ∅ or |A| < n then return Fail F
9  │   └   if |A| = n then (H, A, n) ← (H ∪ A, ∅, 0)
10 └   return Success
```

---

n (line 7). If we detect a failed Hall set, we fail (line 8). If we detect a Hall set, we add those values to H, and reset A and n, and keep going (line 9).

It is important to note that this approach may fail to identify some Hall sets, if the initial ordering of domains is imperfect. However, the algorithm runs very quickly in practice: the sorting step is $\mathcal{O}(v \log v)$ (where v is the number of remaining variables), and the loop has complexity $\mathcal{O}(vd)$ (where d is the cost of a bitset operation over a target domain, which we discuss below). We validate this trade-off experimentally in the following section.

In case a failure is detected, the F set of nogoods we return need only include the variables processed so far, not every variable involved in the constraint. This is because an all-different constraint implies an all-different constraint on any subset of its variables. A smaller set of nogoods can increase the potential for backjumping (and experiments verified that this is beneficial in practice).

We have been unable to find this algorithm described elsewhere in the literature, although a sort- and counting-based approach has been used to achieve bounds consistency [28] (but our domains are not naturally ordered) and as a preprocessing step [29]. Bitsets (which we discuss below) have also been used to implement the matching algorithm [19].

3.3 Bit- and Thread-Parallelism

The use of bitset encodings for graph algorithms to exploit hardware parallelism dates back to at least Ullmann's algorithm [39], and remains an active area of research [32,40]. We use bitsets here: our graphs are stored as arrays of bit vectors, our domains are bit vectors, the neighbourhood intersection in Algorithm 5 is a bitwise-and operation, the unions in Algorithm 4 and Algorithm 6 are bitwise-or operations, and the cardinality check in Algorithm 6 is a population count (this is a single instruction in modern CPUs).

In addition to the SIMD-like parallelism from bitset encodings, we observed two opportunities for multi-core thread parallelism in the algorithm:

Graph and domain construction. We may parallelise the outer `for` loops involved in calculating neighbourhood degree sequences and in initialising the domains of variables in Algorithm 3. Similarly, constructing each supplemental graph in Algorithm 2 involves an outer **for** loop, iterating over each vertex in the input graph. These loops may also be parallelised, with one caveat: we must be able to add edges to (but not remove edges from) the output graph safely, in parallel. This may be done using an atomic "or" operation.

Search. Viewing the recursive calls made by the `search` function in Algorithm 4 as forming a tree, we may explore different subtrees in parallel. The key points are:

1. We do not know in advance whether the **foreach** loop (Algorithm 4 line 6) will exit early (either due to a solution being found, or backjumping). Thus our parallelism is speculative: we make a single thread always preserve the sequential search order, and use any additional threads to precompute subsequent entries in the loop which *might* be used. This may mean we get no speedup at all, if our speculation performs work which will not be used.

2. The **search** function, parallelised without changes, could attempt to exit early due to backjumping. We rule out this possibility by refusing to pass knowledge to the left: that is, we do not allow speculatively-found backjumping conditions to change the return value of **search**. This is for safety [38] and reproducibility: value-ordering heuristics can alter the performance of unsatisfiable instances when backjumping, and allowing parallelism to select a different backjump set could lead to an absolute slowdown [26]. To avoid this possibility, when a backjump condition is found, we *must* cancel any speculative work being done to the right of its position, and *cannot* cancel any ongoing work to the left. This means that unlike in conventional backtracking search without learning, we should *not* expect a linear speedup for unsatisfiable instances.
   (In effect we are treating the **foreach** loop as a parallel lazy fold, so that a subtree does not depend upon items to its left. Backjumping conditions are left-zero elements [21], although we do not have a unique zero.)

3. If any thread finds a solution, we *do* succeed immediately, even if this involves passing knowledge to the left. If there are multiple solutions, this can lead to a parallel search finding a different solution to the one which would be found sequentially—since the solution we find is arbitrary, this is not genuinely unsafe. However, this means we could witness a superlinear (greater than n from n threads) speedup for satisfiable instances [4].

4. For work stealing, we explicitly prioritise subproblems highest up and then furthest left in the search tree. This is because we expect our value-ordering heuristics to be weakest early on in search [18], and we use parallelism to offset poor choices early on in the search [6,24].

4 Experimental Evaluation

Our algorithm was implemented[1] in C++ using C++11 native threads, and was compiled using GCC 4.9.0. We performed our experiments on a machine with dual Intel Xeon E5-2640 v2 processors (for a total of 16 cores, and 32 hardware threads via hyper-threading), running Scientific Linux 6.6. For the comparison with SND in the following section, we used Java HotSpot 1.8.0_11. Runtimes include preprocessing and thread launch costs, but not the time taken to read in the graph files from disk (except in the case of SND, which we were unable to instrument).

For evaluation, we used the same families of benchmark instances that were used to evaluate LAD [36] and SND [1]. The "LV" family [20] contains graphs with various interesting properties from the Stanford Graph Database, and the 793 pattern/target pairs give a mix of satisfiable and unsatisfiable queries. The "SF" family contains 100 scale-free graph pairs, again mixing satisfiable and unsatisfiable queries. The remainder of these graphs come from the Vflib database [11]: the "BVG" and "BVGm" families are bounded degree graphs (540 pairs all are satisfiable), "M4D" and "M4Dr" are four-dimensional meshes (360 pairs, all satisfiable), and the "r" family is randomly generated (270 pairs, all satisfiable). We expanded this suite with 24 pairs of graphs representing image pattern queries [12] (which we label "football"), and 200 randomly selected pairs from each of a series of 2D image ("images") and 3D mesh ("meshes") graph queries [37]. The largest number of vertices is 900 for a pattern and 5,944 for a target, and the largest number of edges is 12,410 for a pattern and 34,210 for a target; some of these graphs do contain loops. All 2,487 instances are publicly available in a simple text format[2].

4.1 Comparison with Other Solvers

We compare our implementation against the Abscon 609 implementation of SND (which is written in Java) [1], Solnon's C implementation of LAD [36], and the VFLib C implementation of VF2 [11]. (The versions of each of these solvers we used could support loops in graphs correctly.)

Note that SND is not inherently multi-threaded, but the Java 8 virtual machine we used for testing makes use of multiple threads for garbage collection even for sequential code. On the one hand, this could be seen as giving SND an unfair advantage. However, nearly all modern CPUs are multi-core anyway, so one could say that it is everyone else's fault for not taking advantage of these extra resources. We therefore present both sequential (from a dedicated implementation, *not* a threaded implementation running with only a single thread) and threaded results for our algorithm.

In Fig. 1 we show the cumulative performance of each algorithm. The value of the line at a given time for an algorithm shows the total number of instances

---

[1] source code, data, experimental scripts and raw results available at
   https://github.com/ciaranm/cp2015-subgraph-isomorphism
[2] http://liris.cnrs.fr/csolnon/SIP.html

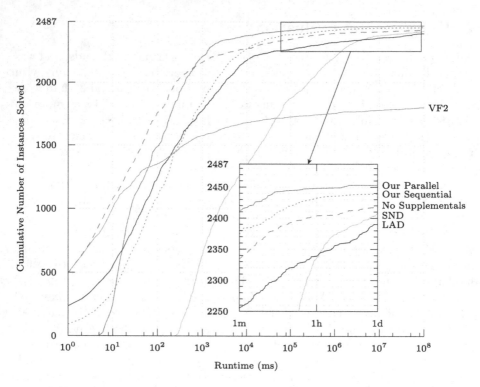

Fig. 1. Cumulative number of benchmark instances solved within a given time, for different algorithms: at time t, the value is the size of the set of instances whose runtime is at most t for that algorithm. Parallel results are using 32 threads on a 16 core hyper-threaded system.

which, individually, were solved in at most that amount of time. Our sequential implementation beats VF2 for times over 0.2s, LAD for times over 0.6s, and always beats SND. Our parallel implementation beats VF2 for times over 0.06s, LAD for times over 0.02s, and always beats SND; parallelism gives us an overall benefit from 12ms onwards. Finally, removing the supplemental graphs from our sequential algorithm gives an improvement below 10s (due to the cost of preprocessing), but is beneficial for longer runtimes.

Fig. 2 presents an alternative perspective of these results. Each point represents an instance, and the shape of a point shows its family. For the y position for an instance, we use our sequential (top graph) or parallel (bottom graph) runtime. For the x position, we use the runtime from the virtual best other solver; the colour of a point indicates which solver this is. For any point below the $x = y$ diagonal line, we are the best solver. A limit of 10^8 ms was used—points along the outer axes represent timeouts.

Although overall ours is the single best solver, VF2 is stronger on trivial instances. This is not surprising: we must spend time constructing supplemental graphs. Thus it may be worth using either VF2 or our own algorithm

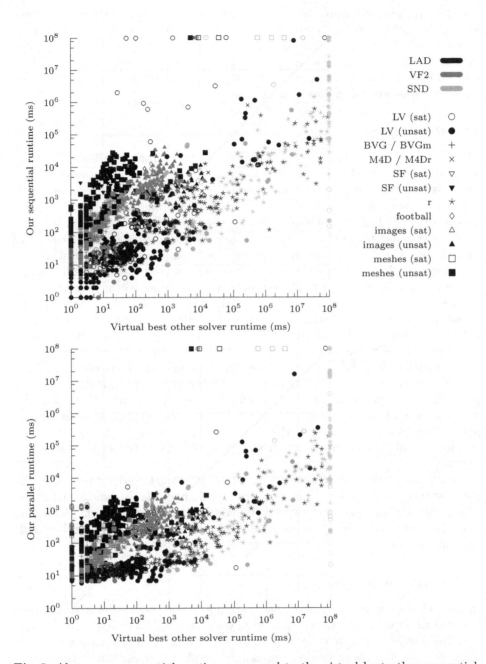

Fig. 2. Above, our sequential runtime compared to the virtual best other sequential solver, for each benchmark instance; below, the same, with our parallel runtimes and including parallel solvers. For points below the diagonal line, ours is the best solver for this instance; for points above the diagonal, the point colour indicates the best solver.

without supplemental graphs as a presolver, if short runtimes for trivial instances is desirable—this may be the case in graph database systems where many trivial queries must be run [16] (although these systems could cache the supplemental graphs for targets). These results also suggest potential scope for algorithm portfolios, or instance-specific configuration: for example, we could omit or use different supplemental graphs in some cases.

4.2   Parallelism

Fig. 3 shows, for each instance, the speedup obtained from parallelism. Except at very low sequential runtimes, we see a reasonable general improvement. For some satisfiable instances, we see strongly superlinear speedups. These instances are exceptionally hard problems [35]: we would have found a solution quickly, except for a small number of wrong turns at the top of search. Our work stealing strategy was able to avoid strong commitment to early value-ordering heuristic choices, providing an alternative to using more complicated sequential search strategies to offset this issue. (Some of these results were also visible in Fig. 2, where we timed out on satisfiable instances which another solver found trivial.)

Some non-trivial satisfiable instances exhibited a visible slowdown. This is because we were using 32 software threads, to match the advertised number of hardware hyper-threads, but our CPUs only have 16 "real" cores between them. For these instances parallelism did not reduce the amount of work required to find a solution, but did result in a lower rate of recursive calls per second on the sequential search path—this is similar to the risk of introducing a slower processing element to a cluster [38]. Even when experimenting with 16 threads, we sometimes observed a small slowdown due to worse cache and memory bus performance, and due to the overhead of modifying the code to allow for work stealing (recall that we are benchmarking against a dedicated sequential implementation).

In a small number of cases, we observe low speedups for non-trivial unsatisfiable instances. These are from cases where backjumping has a substantial effect on search, making much of our speculative parallelism wasted effort. (Additionally, if cancellation were not to be used, some of these instances would exhibit large absolute slowdowns.)

4.3   Effects of Backjumping

In Fig. 4 we show the benefits of backjumping: points below the diagonal line indicate an improvement to runtimes from backjumping. Close inspection shows that backjumping usually at least pays for itself, or gives a slight improvement. (This was not the case when we maintained conflict sets explicitly: there, the overheads lead to a small average slowdown.)

For a few instances, backjumping makes an improvement of several orders of magnitude. The effects are most visible for some of the LV benchmarks, which consist of highly structured graphs. This mirrors the conclusions of Chen and Van Beek [5], who saw that "adding CBJ to a backtracking algorithm ... can

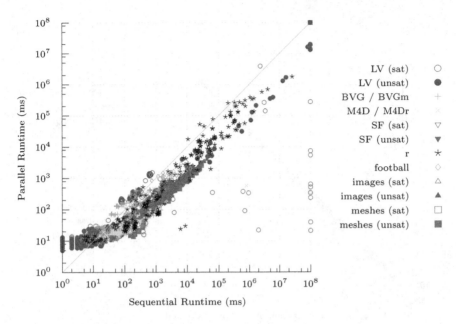

Fig. 3. The effects of parallelism, using 32 threads on a 16 core hyper-threaded system. Each point is a problem instance; points below the diagonal line indicate a speedup.

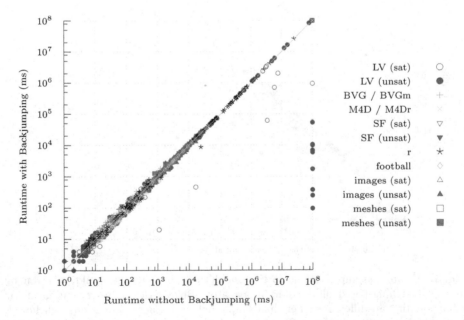

Fig. 4. The effects of backjumping. Each point is one benchmark instance; points below the diagonal line indicate a speedup.

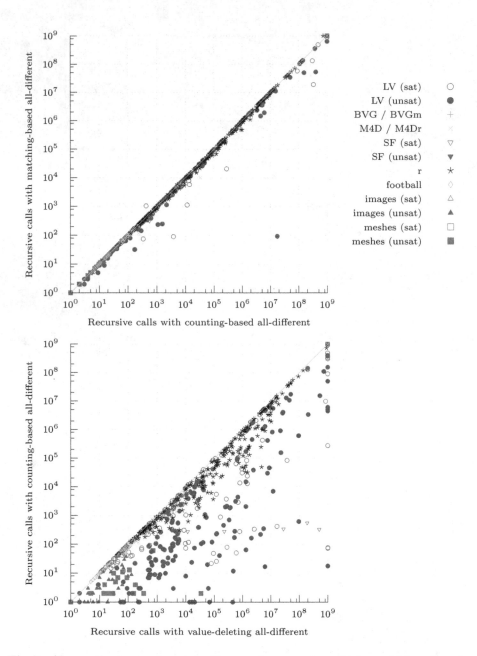

Fig. 5. Above, the improvement to the search space size which would be given by Régin's matching-based all-different propagator. Below, the improvement given by using counting all-different rather than simple deletion. Each point is one benchmark instance; the point style shows the benchmark family. Points below the diagonal line indicate a reduction in the search space size.

(still) speed up the algorithm by several orders of magnitude on hard, structured problems". Real-world graphs often have unexpected structural properties which are not present in random instances [22,34], so we consider backjumping to be worthwhile.

4.4 Comparing All-Different Propagators

We now justify our use of the counting all-different propagator. In the top half of Fig. 5 we show the benefits to the size of the search space that would be gained if we used Régin's algorithm at every step instead of our counting propagator (cutting search off after 10^9 search nodes). We see that for most graph families, there would be little to no benefit even if there was no additional performance cost. Only in a small portion of the LV graphs do we see a gain (and in one case, due to dynamic variable ordering, there is a penalty).

Thus, either our counting propagator is nearly always as good as matching, or neither propagator does very much in this domain. In the bottom half of Fig. 5 we show the benefits to the size of the search space that are gained from using counting, rather than simply deleting a value from every other domain on assignment. The large number of points below the diagonal line confirm that going beyond simple value deletion for all-different propagation is worthwhile.

5 Conclusion

Going against conventional wisdom, we saw that replacing strong inference with cheaper surrogates could pay off, and that backjumping could be implemented cheaply enough to be beneficial. We also saw parallelism give a substantial benefit. This was true even for relatively low runtimes, due to us exploiting parallelism for pre-processing as well as for search. Parallel backjumping has only been given limited attention [8,10,17]. However, a simple approach has worked reasonably well here (in contrast to stronger clause-learning systems, where successes in parallelism appear to be rare).

There is also plenty of scope for extensions of and improvement to our algorithm. We have yet to deeply investigate the possibility of constructing domain- or instance-specific supplemental graphs. Nor did we discuss directed graphs or induced isomorphisms: supplemental graphs can be taken further for these variations of the problem. In particular, composing transformations for induced isomorphisms would allow us to reason about "paths of non-edges", which may be very helpful. Finally, we have yet to consider exploiting the symmetries and dominance relations which we know are present in many graph instances.

Acknowledgments. The authors wish to thank Christine Solnon for discussions, providing the graph instances and the LAD implementation, Christophe Lecoutre for discussion and the SND implementation, and Lars Kotthoff and Alice Miller for their comments.

References

1. Audemard, G., Lecoutre, C., Samy-Modeliar, M., Goncalves, G., Porumbel, D.: Scoring-based neighborhood dominance for the subgraph isomorphism problem. In: O'Sullivan, B. (ed.) CP 2014. LNCS, vol. 8656, pp. 125–141. Springer, Heidelberg (2014). http://dx.doi.org/10.1007/978-3-319-10428-7_12
2. Bessière, C., Régin, J.: MAC and combined heuristics: two reasons to forsake FC (and CBJ?) on hard problems. In: Proceedings of the Second International Conference on Principles and Practice of Constraint Programming, Cambridge, Massachusetts, USA, August 19–22, 1996, pp. 61–75 (1996). http://dx.doi.org/10.1007/3-540-61551-2_66
3. Bonnici, V., Giugno, R., Pulvirenti, A., Shasha, D., Ferro, A.: A subgraph isomorphism algorithm and its application to biochemical data. BMC Bioinformatics 14(Suppl 7), S13 (2013). http://www.biomedcentral.com/1471-2105/14/S7/S13
4. de Bruin, A., Kindervater, G.A.P., Trienekens, H.W.J.M.: Asynchronous parallel branch and bound and anomalies. In: Ferreira, A., Rolim, J.D.P. (eds.) IRREGULAR 1995. LNCS, vol. 980, pp. 363–377. Springer, Heidelberg (1995). http://dx.doi.org/10.1007/3-540-60321-2_29
5. Chen, X., van Beek, P.: Conflict-directed backjumping revisited. J. Artif. Intell. Res. (JAIR) 14, 53–81 (2001). http://dx.doi.org/10.1613/jair.788
6. Chu, G., Schulte, C., Stuckey, P.J.: Confidence-based work stealing in parallel constraint programming. In: Gent, I.P. (ed.) CP 2009. LNCS, vol. 5732, pp. 226–241. Springer, Heidelberg (2009). http://dx.doi.org/10.1007/978-3-642-04244-7_20
7. Coffman, T., Greenblatt, S., Marcus, S.: Graph-based technologies for intelligence analysis. Commun. ACM 47(3), 45–47 (2004). http://doi.acm.org/10.1145/971617.971643
8. Conrad, J., Mathew, J.: A backjumping search algorithm for a distributed memory multicomputer. In: International Conference on Parallel Processing, ICPP 1994, vol. 3, pp. 243–246, August 1994
9. Conte, D., Foggia, P., Sansone, C., Vento, M.: Thirty years of graph matching in pattern recognition. International Journal of Pattern Recognition and Artificial Intelligence 18(03), 265–298 (2004). http://www.worldscientific.com/doi/abs/10.1142/S0218001404003228
10. Cope, M., Gent, I.P., Hammond, K.: Parallel heuristic search in Haskell. In: Selected Papers from the 2nd Scottish Functional Programming Workshop (SFP00), University of St Andrews, Scotland, July 26–28, 2000, pp. 65–76 (2000)
11. Cordella, L.P., Foggia, P., Sansone, C., Vento, M.: A (sub)graph isomorphism algorithm for matching large graphs. IEEE Trans. Pattern Anal. Mach. Intell. 26(10), 1367–1372 (2004). http://doi.ieeecomputersociety.org/10.1109/TPAMI.2004.75
12. Damiand, G., Solnon, C., de la Higuera, C., Janodet, J.C., Samuel, É.: Polynomial algorithms for subisomorphism of nD open combinatorial maps. Computer Vision and Image Understanding 115(7), 996–1010 (2011). http://www.sciencedirect.com/science/article/pii/S1077314211000816, special issue on Graph-Based Representations in Computer Vision
13. Geelen, P.A.: Dual viewpoint heuristics for binary constraint satisfaction problems. In: ECAI, pp. 31–35 (1992)
14. Gent, I.P., Miguel, I., Moore, N.C.A.: Lazy explanations for constraint propagators. In: Carro, M., Peña, R. (eds.) PADL 2010. LNCS, vol. 5937, pp. 217–233. Springer, Heidelberg (2010). http://dx.doi.org/10.1007/978-3-642-11503-5_19

15. Gent, I.P., Miguel, I., Nightingale, P.: Generalised arc consistency for the alldifferent constraint: An empirical survey. Artificial Intelligence **172**(18), 1973–2000 (2008). http://www.sciencedirect.com/science/article/pii/S0004370208001410, special Review Issue
16. Giugno, R., Bonnici, V., Bombieri, N., Pulvirenti, A., Ferro, A., Shasha, D.: Grapes: A software for parallel searching on biological graphs targeting multi-core architectures. PLoS ONE **8**(10), e76911 (2013) http://dx.doi.org/10.1371%2Fjournal.pone.0076911
17. Habbas, Z., Herrmann, F., Merel, P.P., Singer, D.: Load balancing strategies for parallel forward search algorithm with conflict based backjumping. In: Proceedings of the 1997 International Conference on Parallel and Distributed Systems, pp. 376–381, December 1997
18. Harvey, W.D., Ginsberg, M.L.: Limited discrepancy search. In: IJCAI (1), pp. 607–615. Morgan Kaufmann, San Francisco (1995)
19. Kessel, P.V., Quimper, C.: Filtering algorithms based on the word-RAM model. In: Proceedings of the Twenty-Sixth AAAI Conference on Artificial Intelligence, July 22–26, 2012, Toronto, Ontario, Canada. (2012). http://www.aaai.org/ocs/index.php/AAAI/AAAI12/paper/view/5135
20. Larrosa, J., Valiente, G.: Constraint satisfaction algorithms for graph pattern matching. Mathematical Structures in Computer Science **12**(4), 403–422 (2002). http://dx.doi.org/10.1017/S0960129501003577
21. Lobachev, O.: Parallel computation skeletons with premature termination property. In: Schrijvers, T., Thiemann, P. (eds.) FLOPS 2012. LNCS, vol. 7294, pp. 197–212. Springer, Heidelberg (2012). http://dx.doi.org/10.1007/978-3-642-29822-6_17
22. MacIntyre, E., Prosser, P., Smith, B.M., Walsh, T.: Random constraint satisfaction: theory meets practice. In: Maher, M.J., Puget, J.-F. (eds.) CP 1998. LNCS, vol. 1520, p. 325. Springer, Heidelberg (1998). http://dx.doi.org/10.1007/3-540-49481-2_24
23. Malitsky, Y.: Instance-Specific Algorithm Configuration. Springer (2014). http://dx.doi.org/10.1007/978-3-319-11230-5
24. McCreesh, C., Prosser, P.: The shape of the search tree for the maximum clique problem and the implications for parallel branch and bound. ACM Trans. Parallel Comput. **2**(1), 8:1–8:27 (2015). http://doi.acm.org/10.1145/2742359
25. McGregor, J.J.: Relational consistency algorithms and their application in finding subgraph and graph isomorphisms. Inf. Sci. **19**(3), 229–250 (1979). http://dx.doi.org/10.1016/0020-0255(79)90023-9
26. Prosser, P.: Domain filtering can degrade intelligent backtracking search. In: Proceedings of the 13th International Joint Conference on Artifical Intelligence, IJCAI 1993 ,vol. 1, pp. 262–267. Morgan Kaufmann Publishers Inc., San Francisco (1993). http://dl.acm.org/citation.cfm?id=1624025.1624062
27. Prosser, P.: Hybrid algorithms for the constraint satisfaction problem. Computational Intelligence **9**, 268–299 (1993). http://dx.doi.org/10.1111/j.1467-8640.1993.tb00310.x
28. Puget, J.: A fast algorithm for the bound consistency of alldiff constraints. In: Proceedings of the Fifteenth National Conference on Artificial Intelligence and Tenth Innovative Applications of Artificial Intelligence Conference, AAAI 1998, IAAI 1998, July 26–30, 1998, Madison, Wisconsin, USA, pp. 359–366 (1998). http://www.aaai.org/Library/AAAI/1998/aaai98-051.php

29. Quimper, C.-G., Walsh, T.: The all different and global cardinality constraints on set, multiset and tuple variables. In: Hnich, B., Carlsson, M., Fages, F., Rossi, F. (eds.) CSCLP 2005. LNCS (LNAI), vol. 3978, pp. 1–13. Springer, Heidelberg (2006). http://dx.doi.org/10.1007/11754602_1

30. Régin, J.: A filtering algorithm for constraints of difference in CSPs. In: Proceedings of the 12th National Conference on Artificial Intelligence, Seattle, WA, USA, July 31– August 4, 1994, vol. 1, pp. 362–367 (1994). http://www.aaai.org/Library/AAAI/1994/aaai94-055.php

31. Régin, J.C.: Développement d'outils algorithmiques pour l'Intelligence Artificielle. Application à la chimie organique. Ph.D. thesis, Université Montpellier 2 (1995)

32. San Segundo, P., Rodriguez-Losada, D., Galan, R., Matia, F., Jimenez, A.: Exploiting CPU bit parallel operations to improve efficiency in search. In: 19th IEEE International Conference on Tools with Artificial Intelligence, ICTAI 2007, vol. 1, pp. 53–59, October 2007

33. Sevegnani, M., Calder, M.: Bigraphs with sharing. Theoretical Computer Science **577**, 43–73 (2015). http://www.sciencedirect.com/science/article/pii/S0304397515001085

34. Slater, N., Itzchack, R., Louzoun, Y.: Mid size cliques are more common in real world networks than triangles. Network Science **2**, 387–402 (2014). http://journals.cambridge.org/article_S2050124214000228

35. Smith, B.M., Grant, S.A.: Modelling exceptionally hard constraint satisfaction problems. In: Smolka, G. (ed.) CP 1997. LNCS, vol. 1330, pp. 182–195. Springer, Heidelberg (1997). http://dx.doi.org/10.1007/BFb0017439

36. Solnon, C.: Alldifferent-based filtering for subgraph isomorphism. Artif. Intell. **174**(12–13), 850–864 (2010). http://dx.doi.org/10.1016/j.artint.2010.05.002

37. Solnon, C., Damiand, G., de la Higuera, C., Janodet, J.C.: On the complexity of submap isomorphism and maximum common submap problems. Pattern Recognition **48**(2), 302–316 (2015). http://www.sciencedirect.com/science/article/pii/S0031320314002192

38. Trienekens, H.W.: Parallel Branch and Bound Algorithms. Ph.D. thesis, Erasmus University Rotterdam (1990)

39. Ullmann, J.R.: An algorithm for subgraph isomorphism. Journal of the ACM (JACM) **23**(1), 31–42 (1976)

40. Ullmann, J.R.: Bit-vector algorithms for binary constraint satisfaction and subgraph isomorphism. J. Exp. Algorithmics **15**, 1.6:1.1–1.6:1.64 (2011). http://doi.acm.org/10.1145/1671970.1921702

41. Zampelli, S., Deville, Y., Solnon, C.: Solving subgraph isomorphism problems with constraint programming. Constraints **15**(3), 327–353 (2010). http://dx.doi.org/10.1007/s10601-009-9074-3

Automated Auxiliary Variable Elimination Through On-the-Fly Propagator Generation

Jean-Noël Monette[(✉)], Pierre Flener, and Justin Pearson

Department of Information Technology, Uppsala University, Uppsala, Sweden
{jean-noel.monette,pierre.flener,justin.pearson}@it.uu.se

Abstract. Model flattening often introduces many auxiliary variables. We provide a way to eliminate some of the auxiliary variables occurring in exactly two constraints by replacing those two constraints by a new equivalent constraint for which a propagator is automatically generated on the fly. Experiments show that, despite the overhead of the preprocessing and of using machine-generated propagators, eliminating auxiliary variables often reduces the solving time.

1 Introduction

Constraint-based modelling languages such as Essence [6] and MiniZinc [12] enable the modelling of problems at a higher level of abstraction than is supported by most constraint solvers. The transformation from a high-level model to a low-level model supported by a solver is often called *flattening* and has been the subject of intense research in order to produce good low-level, or *flattened*, models (see, e.g., [14]). By decomposing complex expressions into a form accepted by a solver, flattening often introduces many auxiliary variables into the flattened model. Those auxiliary variables and the propagators of the constraints in which they appear may have a large negative impact on the efficiency of solving (as we will show in Section 5).

In this paper, we propose a fully automated way to address this problem by removing from the flattened model some auxiliary variables that appear in exactly two constraints and replacing those two constraints by a new equivalent constraint for which a propagator is generated. Given a flattened model, our approach is fully automated and online. It can be summarised as follows:

1. Identify frequent patterns in the flattened model consisting of two constraints sharing an auxiliary variable.
2. For a pattern, define a new constraint predicate that involves all variables appearing in the pattern except for the shared auxiliary variable.
3. For a pattern, replace all its occurrences in the flattened model by instantiating the new constraint predicate.
4. For each new predicate, generate a propagator description in the indexical language of [11] and compile it for the targeted constraint solver.
5. Solve the modified flattened model using the constraint solver extended with the new indexical-based propagators.

© Springer International Publishing Switzerland 2015
G. Pesant (Ed.): CP 2015, LNCS 9255, pp. 313–329, 2015.
DOI: 10.1007/978-3-319-23219-5_22

Our experiments in Section 5 show that our approach is useful for instances that are hard to solve, reducing the average time by 9% for those taking more than one minute and sometimes more than doubling the search speed.

The rest of the paper assumes that the low-level language is FlatZinc [1] and that the constraint solver is Gecode [7], as our current implementation is based on FlatZinc and Gecode. However, the ideas presented here are applicable to other low-level modelling languages and other constraint solvers.

The paper is organised as follows. We start with some preliminaries in Section 2. Then we describe our approach in Section 3 and present a complete example in Section 4. In Section 5, we show that this approach is effective in practice. In Section 6, we discuss the merits and limitations of our approach as well as some alternatives. Finally, we conclude in Section 7.

2   Preliminaries

A *constraint-based model* is a formal way to describe a combinatorial problem. It is composed of a set of variables, each with a finite domain in which it must take its value, a set of constraints, and an objective. A *solution* is an assignment to all variables so that all constraints are satisfied. The *objective* is either to find a solution, or to find a solution in which a given variable is minimised, or to find a solution in which a given variable is maximised.

A *constraint predicate*, or simply *predicate*, is defined by a name and a *signature* that lists the *formal arguments* of the predicate. The arguments can be constants, variables, or arrays thereof. For simplicity, we identify a predicate with its name and we often do not give the types of its arguments. We specify the *semantics* of a predicate P by a logic formula involving the arguments of P and constants. For example, one could write $\text{PLUS}(X, Y, Z) \triangleq X = Y + Z$, for the predicate PLUS with formal arguments X, Y, and Z. A predicate whose signature has arrays of variables is said to be of unfixed arity, or n-ary. Unfixed-arity predicates are usually referred to as global constraints [2] in constraint programming. Each constraint in a constraint-based model is the *instantiation* of a constraint predicate to some *actual arguments*. We denote formal arguments in upper case and actual arguments inside a model in lower case.

2.1   MiniZinc and FlatZinc

MiniZinc is a solver-independent constraint-based modelling language for combinatorial problems. Before being presented to a solver, a MiniZinc model is transformed into a FlatZinc model by a process called *flattening*. The MiniZinc flattener inlines function and predicate calls, decomposes expressions, and unrolls loops to provide the solver with a model that is a conjunction of primitive constraints (i.e., constraints that are recognised by the targeted solver) over simple arguments (i.e., only variables, constants, or arrays thereof). To do so, the flattener may introduce auxiliary variables, which are annotated in the FlatZinc model with `is_introduced_var`.

```
1 var int: w;                      1 var int: w;
2 var int: y;                      2 var int: y;
3 var int: z;                      3 var int: z;
4                                  4 var int: x₁ ::var_is_introduced;
5                                  5 var bool: x₂ ::var_is_introduced;
6                                  6 var bool: x₃ ::var_is_introduced;
7 constraint w≠max(y,z) ∨ 1≤z;     7 constraint x₁ = max(y,z);
8                                  8 constraint x₂ ≡ w ≠ x₁;
9                                  9 constraint x₃ ≡ 1 ≤ z;
10                                 10 constraint x₂ ∨ x₃;
11 solve satisfy;                  11 solve satisfy;
```

$$\text{(a) MiniZinc model} \qquad\qquad \text{(b) FlatZinc model}$$

Fig. 1. MiniZinc model (1a) and FlatZinc model resulting from its flattening (1b)

For example, the MiniZinc model of Figure 1a is flattened into the FlatZinc model of Figure 1b, modulo editorial changes. The single constraint expression of the MiniZinc model (line 7 in Figure 1a) is flattened by introducing three auxiliary variables (lines 4–6 in Figure 1b) and posting four primitive constraints: the constraints in lines 7–9 functionally define the auxiliary variables representing parts of the original constraint expression, and the disjunction of line 10 corresponds to the top-level expression.

For 75 of the 488 benchmark instances we used (see Section 5 for details), no auxiliary variables are introduced by flattening. For 264 instances, flattening multiplies the number of variables in the model by more than 5. For 7 instances, flattening even multiplies the number of variables by more than 100.

2.2 Patterns, Occurrences, and Extensions

A *pattern* is here a new constraint predicate with signature $Y_1, ..., Y_n, K_1, ..., K_m$, where the Y_i are $n \geq 2$ variable identifiers and the K_i are $m \geq 0$ constant identifiers. Its semantics is specified by the conjunction of two existing predicates P_A and P_B applied to arguments $A_1, ..., A_p$ and $B_1, ..., B_q$, for some $p, q \geq 2$, such that each A_i and each B_i is either one of $Y_1, ..., Y_n, K_1, ..., K_m$, or a unique local and existentially quantified variable X, or a constant, and such that X appears in both $A_1, ..., A_p$ and $B_1, ..., B_q$. We reserve the identifier X for the local and existentially quantified variable. Hence, for simplicity, we will often omit writing the existential quantification of X in predicate semantics.

An *occurrence* of a pattern in a model is made of two constraints C_1 and C_2 sharing a variable x that appears only in C_1 and C_2 such that P_A can be instantiated to C_1 and P_B to C_2 with X being instantiated to x.

For example, the pattern $P(Y_1, Y_2, Y_3, Y_4) \triangleq \exists X : X = \max(Y_1, Y_2) \wedge Y_3 \equiv Y_4 \neq X$ occurs in the model of Figure 1b, with C_1 being $x_1 = \max(y, z)$ in line 7, C_2 being $x_2 \equiv w \neq x_1$ in line 8, and X being x_1. Hence this occurrence of P is equivalent to the constraint $P(y, z, x_2, w)$.

```
1  def PLUS(vint X, vint Y, vint Z){
2    propagator{
3      X in (min(Y)+min(Z)) .. (max(Y)+max(Z)) ;
4      Y in (min(X)-max(Z)) .. (max(X)-min(Z)) ;
5      Z in (min(X)-max(Y)) .. (max(X)-min(Y)) ;
6    }
7    checker{ X == Y + Z }
8  }
```

Fig. 2. Indexical propagator description for $X = Y + Z$

A pattern P_1 is said to be a *specialisation* of another pattern P_2 if one can define P_1 in terms of P_2 by properly instantiating some of the arguments of P_2.

For example, consider the patterns $P_1(Y_1, Y_2) \triangleq (X \vee Y_1) \wedge X \equiv 1 \leq Y_2$ and $P_2(Y_1, Y_2, K_1) \triangleq (X \vee Y_1) \wedge X \equiv K_1 \leq Y_2$, both of which occur on lines 9 and 10 of Figure 1b with X being x_3. Then P_1 is a specialisation of P_2 because $P_1(Y_1, Y_2) \Leftrightarrow P_2(Y_1, Y_2, 1)$.

2.3 Indexicals

Indexicals [20] are used to describe concisely propagation in propagator-based constraint solvers. The core indexical expression takes the form 'X in σ' and restricts the domain of variable X to be a subset of the set-valued expression σ, which depends on the domains of other variables. Figure 2 presents an indexical propagator description in the language of [11] for the constraint predicate $\text{PLUS}(X, Y, Z) \triangleq X = Y + Z$; it involves three indexical expressions, one for each variable (lines 3–5), the min and max operators referring to the domain of the argument variable. Figure 2 also contains a *checker* (line 7), which is used to test whether the constraint holds on ground instances and can be seen as a specification of the predicate semantics.

Our indexical compiler [11] takes an indexical propagator description like the one in Figure 2 and compiles it into an executable propagator for a number of constraint solvers, including Gecode. The compiled propagator is said to be *indexical-based*. The experimental results in [11] show that an indexical-based propagator uses between 1.2 and 2.7 times the time spent by a hand-crafted propagator on a selection of n-ary constraint predicates.

3 Our Approach

Given a FlatZinc model, our approach, whose implementation is referred to as `var-elim-idxs`, adds a preprocessing step before solving the model. This preprocessing is summarised in Algorithm 1. We do not describe lines 1, 5, and 6, as they involve purely mechanical and well-understood aspects. The core of the algorithm iteratively identifies frequent patterns (line 2), replaces them in the

---

Algorithm 1. Main preprocessing algorithm of `var-elim-idxs`
  Input: a flattened model
  Output: updated flattened model and extended solver

1: Parse the flattened model
2: **while** there is a most frequent pattern P in the model (Section 3.1) **do**
3:     Replace each occurrence of P in the model by instantiating P (Section 3.2)
4:     Generate and compile an indexical propagator description for P (Section 3.3)
5: Output the updated model
6: Link the compiled indexical-based propagators with the solver

---

model (line 3), and generates propagators (line 4). Sections 3.1 to 3.3 describe this core.

The loop of lines 2 to 4 ends when no pattern occurs often enough, which is when the number of occurrences of a most frequent pattern (if any) is less than 10 or less than 5% of the number of variables in the model: under those thresholds, the effort of preprocessing is not worth it. To save further efforts, if the number of auxiliary variables is less than 10 or less than 5% of the number of variables in the model (this criterion can be checked very fast with Unix utilities such as `grep`), then the preprocessing is not performed at all. The two thresholds, 10 and 5%, have been set arbitrarily after some initial experiments.

3.1   Identification of Frequent Patterns

We now show how we identify patterns that occur frequently in a given model.

First, we collect all auxiliary variables that appear in exactly two constraints. Indeed, to be in an occurrence of a pattern, a shared variable must occur only in those two constraints. Most auxiliary variables appear in exactly two constraints: this is the case of 84% of the auxiliary variables appearing in the 488 benchmark instances we used. For example, this is also the case for all auxiliary variables in Figure 1b. However, due to common subexpression elimination [13,14], some auxiliary variables appear in more than two constraints. We do not consider such variables for elimination as common subexpression elimination usually increases the amount of filtering and eliminating those variables might cancel this benefit.

Then, for each collected variable x, we create a pattern as follows. Let $P_A(a_1, \ldots, a_p)$ and $P_B(b_1, \ldots, b_q)$ be the two constraints in which x appears. The pattern is such that the two predicates are $P_A(A_1, \ldots, A_p)$ and $P_B(B_1, \ldots, B_q)$, where each A_i is defined by the following rules, and similarly for each B_i:

  – If a_i is x, then A_i is X.
  – If a_i is a variable other than x, then A_i is Y_k for the next unused k.
  – If $a_i \in \{-1, 0, 1, \textbf{true}, \textbf{false}\}$, then A_i is a_i.
  – If a_i is a constant not in $\{-1, 0, 1, \textbf{true}, \textbf{false}\}$, then A_i is K_k for the next unused k.
  – If a_i is an array, then A_i is an array of the same length where each element is defined by applying the previous rules to the element of a_i at the same position.

The purpose of the third rule is to allow a better handling of some special constants, which may simplify the generated propagators. For example, linear (in)equalities can be propagated more efficiently with unit coefficients.

In general, there might be other shared variables between $P_A(a_1, \ldots, a_p)$ and $P_B(b_1, \ldots, b_q)$ besides x but, to keep things simple, we consider them separately, i.e., a new argument Y_k with an unused k is created for *each* occurrence of another shared variable. Ignoring other shared variables does not affect the correctness of our approach, but the generated propagators may achieve less filtering than possible otherwise. We will revisit this issue in Section 6.

In order to avoid the creation of symmetric patterns, we sort the elements of an array when their order is not relevant: we do this currently only for the n-ary Boolean disjunction and conjunction constraints.

For example, both $(x \lor y) \land (x \equiv 1 \leq z)$ and $(w \lor u) \land (u \equiv 1 \leq v)$ are considered occurrences of $P(Y_1, Y_2) \triangleq (X \lor Y_1) \land (X \equiv 1 \leq Y_2)$, although, syntactically, the shared variables (x and u, respectively) occur in different positions in their respective disjunctions.

Finally, we count the occurrences of each created pattern. In doing so, we ignore the following patterns, expressed here in terms of criteria that are specific to FlatZinc and Gecode:

- Patterns involving an n-ary predicate with at least five variables among its arguments: currently, our approach considers only fixed-arity patterns. Hence n-ary constraints with different numbers of variables are considered as fixed-arity constraints with different arities. We only want to keep those of small arities for efficiency reasons. The threshold is set to 5 because no fixed-arity predicate in FlatZinc has more than four variable arguments.
- Patterns involving a predicate for which an indexical-based propagator is expected to perform poorly with respect to a hand-written propagator: this includes all the global constraint predicates from the MiniZinc standard library, as well as the element, absolute value, division, modulo, and multiplication predicates of FlatZinc.
- Patterns involving a `bool2int` predicate in which the Boolean variable is the shared variable, and another predicate that is not a pattern itself: this case is ignored as, in some cases, the Gecode-FlatZinc parser takes advantage of `bool2int` constraints and we do not want to lose that optimisation.

The two first criteria partially but not completely overlap: for example, the constraint `all_different`$([X, Y, Z, W])$ is not ruled out by the first criterion as it has fewer than five variables, but it is ruled out by the second one; conversely, `int_lin_eq`$(\text{coeffs}, [X, Y, Z, W, U], k)$ is ruled out by the first criterion but not by the second one.

The result of pattern identification is a pattern that occurs most in the model.

3.2   Pattern Instantiation

Having identified a most frequent pattern P, we replace each occurrence of P by an instantiation of P. More precisely, for each auxiliary variable x that appears in exactly two constraints, if the pattern created for x in Section 3.1 is P or a specialisation of P (detected by simple pattern matching), then we replace in the model the declaration of x and the two constraints in which x appears by an instantiation of P, obtained by replacing each formal argument of P by the actual arguments of the two constraints in which x appears. To achieve this, each argument A_i of P_A in the semantics of P is considered together with the argument a_i of the instantiation of P_A in which x appears, in order to apply the following rules, and similarly for each B_i:

- If A_i is X, then variable a_i, which is x, is not in the instantiation of P.
- If A_i is Y_k, then Y_k is instantiated to the variable a_i.
- If $A_i \in \{-1, 0, 1, \textbf{true}, \textbf{false}\}$, then a_i is not in the instantiation of P.
- If A_i is K_k, then K_k is instantiated to the constant a_i.
- If A_i is an array, then the previous rules are applied to each element of A_i.

For example, consider the pattern $P_1(Z_1, Z_2) \triangleq (X \equiv 1 \leq Z_1) \land (Z_2 \lor X)$. Variable x_3 of Figure 1b appears in $x_3 \equiv 1 \leq z$ (line 9) and $x_2 \lor x_3$ (line 10). Then P_1 can be instantiated to $P_1(z, x_2)$ and this constraint replaces lines 6, 9, and 10 in Figure 1b.

Due to the sorting of the elements for the n-ary Boolean conjunction and disjunction predicates, some occurrences of a pattern may disappear before their instantiation. Consider the MiniZinc-level expression $z_1 > 0 \lor z_2 > 0$. Flattening introduces two auxiliary Boolean variables, say b_1 and b_2, together with the three constraints $b_1 \equiv z_1 > 0$, $b_2 \equiv z_2 > 0$, and $b_1 \lor b_2$. Hence there are two occurrences of the pattern $P(Y_1, Y_2) \triangleq (X \equiv Y_1 > 0) \land (X \lor Y_2)$ but only one of them will actually be replaced, say the one in which X is b_1. After replacing it, the model contains the two constraints $P(z_1, b_2)$ and $b_2 \equiv z_2 > 0$, changing the pattern to which b_2 belongs. This new pattern might be replaced in a later iteration of the loop in Algorithm 1.

3.3   Indexical Propagator Description Generation and Compilation

The generation of a propagator for a pattern uses our indexical compiler [11] and performs the following steps:

1. Translation of the pattern into a checker in the indexical syntax.
2. Elimination of the shared variable from the checker.
3. Generation of an indexical propagator description, based on the checker.
4. Compilation [11] of the indexical propagator description into an actual propagator, written in C++ in the case of Gecode.

Step 1 only involves a change of syntax, and Step 4 has already been described in [11]. Hence, we will focus here on Steps 2 and 3.

Let X be the variable to eliminate in Step 2. Here, variable elimination can take two forms. First, if X is constrained to be equal to some expression ϕ in one of the two conjuncts, i.e., if the checker can be written as $P(\ldots, X, \ldots) \land X = \phi$ for some predicate P, then all occurrences of X in the checker are replaced by

ϕ, i.e., the checker becomes $P(\ldots, \phi, \ldots)$. As Boolean variables are considered a special case of integer variables in the indexical compiler, this rule also covers the case of X being a Boolean variable constrained to be equivalent to a Boolean expression. Second, if both conjuncts are disjunctions, one involving a Boolean variable X and the other $\neg X$, i.e., if the checker can be written as $(\delta_1 \vee X) \wedge (\delta_2 \vee \neg X)$ for some Boolean expressions δ_1 and δ_2, then applying the resolution rule yields a single disjunction without X, i.e., the checker becomes $\delta_1 \vee \delta_2$.

The generation in Step 3 of an indexical propagator description from a checker works by syntactic transformation: rewriting rules are recursively applied to the checker expression and its subexpressions in order to create progressively a collection of indexical expressions. The whole transformation has more than 250 rules. We limit our presentation to the most representative ones.

The rule for a conjunction $\gamma_1 \wedge \gamma_2$ concatenates the rewriting of γ_1 and γ_2. The rule for a disjunction $\delta_1 \vee \delta_2$ is such that δ_2 is propagated only once δ_1 can be shown to be unsatisfiable, and conversely. The rule for an equality $\phi_1 = \phi_2$ creates expressions that force the value of ϕ_1 to belong to the possible values for ϕ_2, and conversely. If ϕ_1 is a variable, say Y, then the rule creates the indexical Y in $\mathtt{UB}(\phi_2)$, where $\mathtt{UB}(\phi_2)$ is possibly an over-approximation of the set containing all possible values of the expression ϕ_2, based on the domains of the variables appearing in ϕ_2. If ϕ_1 is not a variable but a compound expression, then it must be recursively decomposed. Consider the case of ϕ_1 being $Y_1 + Y_2$: two indexical expressions are created, namely Y_1 in $\mathtt{UB}(\phi_2 - Y_2)$ and Y_2 in $\mathtt{UB}(\phi_2 - Y_1)$. The other rules cover all the other expressions that can appear in a checker. The function $\mathtt{UB}(.)$ is also defined by rules. As an example, $\mathtt{UB}(\phi_1 - \phi_2)$ is rewritten as $\mathtt{UB}(\phi_1) \ominus \mathtt{UB}(\phi_2)$, where $\ominus$ is pointwise integer set substraction: $S \ominus T = \{s - t \mid s \in S \wedge t \in T\}$.

This generation mechanism packaged with our indexical compiler has been used in [9,17] to prototype propagators rapidly for newly identified constraints, but it has never been described before. It is very similar to the compilation of projection constraints in Nicolog [18], but generalised to n-ary constraints.

4 Example: Ship Schedule

To illustrate our approach, we now consider the `ship-schedule.cp.mzn` model from the MiniZinc Challenge 2012. The objective of the problem is to find which boats are sailing and when, in order to satisfy several port constraints, e.g., tides, tugboat availability, and berth availability, as well as to maximise the total weight that can be transported. The FlatZinc model produced by flattening the MiniZinc model with the `7ShipsMixed.dzn` data file, which represents an instance with 7 boats and 74 time slots, contains 7,187 constraints and 5,978 variables, among which 4,848 are auxiliary, i.e., 81% of the variables.

When given this flattened model, `var-elim-idxs` iteratively identifies the patterns reported in Table 1. Note that pattern P_0 is used in pattern P_1. The loop of Algorithm 1 ends upon identifying pattern P_4, which occurs less often than 5% of the number of variables and is not instantiated. In total, 4 new constraint

Table 1. Patterns found in `ship-schedule.cp.mzn` with `7ShipsMixed.dzn`

Predicate	Definition			Frequency
$P_0(Y_1, Y_2)$	$\triangleq (X \vee Y_1)$	$\wedge$	$X \equiv (Y_2 = 0)$	892
$P_1(Y_1, Y_2, Y_3, Y_4)$	$\triangleq P_0(X, Y_4)$	$\wedge$	$X \equiv (Y_1 \wedge Y_2 \wedge Y_3)$	612
$P_2(Y_1, Y_2, Y_3, Y_4, K_1)$	$\triangleq (X \vee \neg(Y_1 \wedge Y_2 \wedge Y_3))$	$\wedge$	$X \equiv (Y_4 = K_1)$	612
$P_3(Y_1, Y_2, Y_3, K_1)$	$\triangleq (X \vee \neg(Y_1 \wedge Y_2))$	$\wedge$	$X \equiv (Y_3 = K_1)$	276
$P_4(Y_1, Y_2, Y_3, Y_4)$	$\triangleq (X \vee Y_1)$	$\wedge$	$X \equiv (Y_2 \wedge Y_3 \wedge Y_4)$	146

```
1   def P_1(vint Y_1::Bool, vint Y_2::Bool, vint Y_3, vint Y_4){
2       checker{
3           0==Y_4 or (1 <= Y_1 and 1 <= Y_2 and 1 <= Y_3)
4       }
5       propagator(gen)::DR{
6           once(not 0 memberof dom(Y_4)){
7               Y_1 in 1 .. sup;
8               Y_2 in 1 .. sup;
9               Y_3 in 1 .. sup;
10          }
11          (max(Y_1) < 1 or max(Y_2) < 1 or max(Y_3) < 1) -> Y_4 in {0};
12      }
13  }
```

Fig. 3. Indexical propagator description generated for pattern P_1 of Table 1 from `ship-schedule.cp.mzn` with `7ShipsMixed.dzn`. Note that the indexical compiler treats Boolean variables as a special case of integer variables with domain $\{0,1\}$, where 1 represents **true**. For example, 1 $<=$ `Y_1` means `Y_1` is **true**.

predicates (P_0 to P_3) are introduced in the model, eliminating 2,392 variables, i.e., 40% of the variables. Among those, 222 variables are eliminated thanks to pattern specialisation, i.e., because the pattern created for such a variable is not exactly one of the patterns shown in Table 1 but a specialisation of one of those, for example a specialisation of P_2 with $K_1 = 1$.

The final model contains 3,586 variables, of which 2,456 are auxiliary, and 4,795 constraints, of which 1,780 use the generated propagators. Many auxiliary variables are not eliminated because either they appear in more than two constraints, or they appear in too infrequent patterns, such as P_4 in Table 1, or they appear in constraints that are not handled by `var-elim-idxs`, such as n-ary constraints with $n \geq 5$, the multiplication constraint, and the `bool2int` constraint in the case explained in Section 3.1.

The new constraint predicates are handled by the indexical compiler to specify their checkers, generate indexical propagator descriptions, and compile the latter into Gecode propagators. Figure 3 shows the generated indexical propagator description for pattern P_1 of Table 1. It is interesting to note that the auxiliary variable X eliminated by instantiating P_1 represents the truth of $Y_1 \wedge Y_2 \wedge Y_3$.

Hence the propagator for $X \equiv (Y_1 \wedge Y_2 \wedge Y_3)$ in the original flattened model needs to watch both when all conjuncts become **true** to set X to **true** and when some conjunct becomes **false** to set X to **false**. In contrast, the propagator of Figure 3 is only interested in the falsity of the conjuncts to restrict Y_4 to 0. The generated C++ code of the indexical-based propagators for Gecode is 579 lines long.

While the unmodified Gecode-FlatZinc interpreter solves the flattened model in 111.5 seconds, `var-elim-idxs` solves it in 89.1 seconds, divided into 4.9 seconds of preprocessing and 84.2 seconds of actual solving.

5   Experimental Evaluation

We implemented the preprocessing step of `var-elim-idxs`[1] in Scala, using our indexical compiler[2] and the FlatZinc parser of the OscaR project[3] as that parser is written in Scala as well. For the experiments, we used Gecode 4.4.0[4] and the MiniZinc flattener 2.0.1.[5] The experiments were carried out inside a VirtualBox virtual machine running Ubuntu 14.04 LTS 32-bit, with access to one core of a 64-bit Intel Core i7 at 3 GHz and 1 GB of RAM.

We tested `var-elim-idxs` on 489 of the 500 FlatZinc instances from the MiniZinc Challenges 2010 to 2014. We excluded 11 instances that could not be flattened, for lack of memory or because of a syntax error.[6] We ran both Gecode and `var-elim-idxs` once on each instance with a time-out of 10 minutes per instance. Unless otherwise noted, 'Gecode' refers here to running an unmodified version of Gecode and '`var-elim-idxs`' refers to running our preprocessing step followed by running the extended version of Gecode.

For 94 instances, the preprocessing is not run at all because the `grep` command detects that there are too few auxiliary variables: the behaviour of Gecode and `var-elim-idxs` is identical as the time spent by `grep` is negligible. For 172 instances, the preprocessing is run but does not identify any frequent enough pattern: the behaviour of Gecode and `var-elim-idxs` is identical except for the extra time spent on preprocessing, discussed in the last paragraph of this section.

Table 2 reports the results, aggregated per MiniZinc model, on the 223 instances in which the preprocessing identifies frequent patterns. We refer to those instances as *modified* instances. The bottom of the table presents the aggregated results over all modified instances for which the total time of Gecode is respectively more than 1 and 60 seconds. The node rate ratio for an instance is computed as r_v/r_g, where $r_x = n_x/t_x$ with n_x being the number of nodes of the search tree visited before time-out, and t_x being the total time in the column *ratio total* and the search time in the column *ratio search*; the subscript $x = $ 'g'

---

[1] https://bitbucket.org/jmonette/var-elim-idxs
[2] https://bitbucket.org/jmonette/indexicals
[3] http://www.oscarlib.org
[4] http://www.gecode.org/
[5] http://www.minizinc.org/2.0/
[6] The *sugiyama* model could not be parsed by MiniZinc 2.0.1. It could be parsed by MiniZinc 1.6, and the problem is corrected in the development version of MiniZinc.

Table 2. Results for the 223 modified instances, aggregated per MiniZinc model, with the following columns: name of the model; number of instances; mean and standard deviation of the percentage of auxiliary variables; mean and standard deviation of the percentage of variables eliminated by `var-elim-idxs` (over *all* variables of the model); geometric mean and geometric standard deviation of the node rate ratio *including* preprocessing; as well as geometric mean and geometric standard deviation of the node rate ratio *excluding* preprocessing. The models are ordered by decreasing ratio excluding preprocessing (column 'ratio search').

name	inst.	% aux.	% elim.	ratio total	ratio search
l2p	5	93 (1)	73 (4)	1.27 (1.92)	2.30 (1.12)
amaze3	5	92 (1)	10 (1)	0.71 (1.98)	1.71 (2.83)
league	11	94 (4)	30 (13)	1.37 (1.71)	1.45 (1.64)
openshop	5	96 (1)	96 (1)	0.92 (2.15)	1.35 (1.13)
ship-schedule	15	82 (1)	37 (4)	0.52 (3.07)	1.32 (1.07)
wwtpp-real	10	75 (1)	70 (2)	0.27 (14.75)	1.32 (1.39)
radiation	10	64 (1)	32 (1)	1.15 (1.15)	1.22 (1.07)
wwtpp-random	5	75 (0)	62 (0)	0.46 (8.31)	1.21 (1.29)
javarouting	5	88 (1)	82 (1)	1.16 (1.03)	1.18 (1.02)
solbat	30	98 (0)	16 (0)	0.65 (2.56)	1.17 (1.34)
amaze	6	55 (1)	36 (1)	1.12 (1.05)	1.16 (1.01)
project-planning	6	66 (0)	31 (2)	1.12 (1.04)	1.13 (1.04)
open-stacks	5	82 (0)	43 (2)	0.63 (2.69)	1.08 (1.04)
traveling-tppv	5	83 (0)	28 (0)	1.04 (1.01)	1.05 (1.02)
fjsp	3	75 (8)	21 (3)	0.92 (1.23)	1.04 (1.08)
tpp	6	88 (1)	24 (2)	0.97 (1.05)	1.03 (1.01)
smelt	4	72 (3)	9 (2)	1.03 (1.08)	1.03 (1.08)
train	9	53 (1)	23 (2)	1.03 (1.06)	1.03 (1.06)
pattern-set-mining	1	66 (0)	29 (0)	1.02 (1.00)	1.03 (1.00)
mspsp	2	72 (2)	54 (5)	1.02 (1.02)	1.02 (1.02)
on-call-rostering	4	66 (6)	24 (4)	1.01 (1.01)	1.02 (1.02)
carpet-cutting	2	57 (0)	37 (0)	1.01 (1.02)	1.02 (1.02)
jp-encoding	5	92 (0)	20 (0)	1.00 (1.01)	1.01 (1.01)
cyclic-rcpsp	10	91 (3)	75 (5)	1.00 (1.02)	1.00 (1.02)
rcpsp-max	6	97 (1)	96 (1)	0.99 (1.02)	1.00 (1.02)
rcpsp	4	96 (3)	94 (4)	0.99 (1.02)	0.99 (1.02)
elitserien	5	71 (0)	24 (0)	0.94 (1.07)	0.99 (1.02)
liner-sf-repositioning	4	85 (0)	10 (0)	0.97 (1.02)	0.99 (1.02)
rectangle-packing	5	88 (0)	49 (1)	0.97 (1.04)	0.98 (1.05)
stochastic-fjsp	2	81 (0)	5 (0)	0.94 (1.00)	0.98 (1.03)
still-life-wastage	5	87 (1)	5 (0)	0.88 (1.14)	0.97 (1.01)
amaze2	6	93 (0)	28 (10)	0.86 (1.17)	0.87 (1.17)
fillomino	2	94 (0)	6 (0)	0.40 (2.27)	0.86 (1.07)
roster	5	82 (0)	63 (0)	0.78 (1.42)	0.79 (1.41)
mario	10	92 (0)	22 (1)	0.56 (1.87)	0.78 (1.33)
Total	223	83 (13)	38 (26)	0.81 (2.54)	1.12 (1.38)
Total (> 1 s.)	207	83 (14)	38 (26)	1.00 (1.45)	1.12 (1.31)
Total (> 60 s.)	174	83 (14)	40 (26)	1.09 (1.27)	1.11 (1.27)

Table 3. Results for the 172 unmodified instances, with the columns of Table 2

	inst.	% aux.	% elim.	ratio total	ratio search
Total	172	80 (20)	0 (0)	0.82 (1.80)	1.00 (1.08)
Total (> 1 s.)	159	79 (20)	0 (0)	0.95 (1.13)	1.00 (1.01)
Total (> 60 s.)	110	76 (20)	0 (0)	1.00 (1.02)	1.00 (1.01)

refers to Gecode and 'v' to `var-elim-idxs`. We use node rates in order to have a meaningful measure for both the instances that are solved before time-out and those that are not. This assumes that the explored search trees are the same, which is discussed in Section 6. A ratio larger than 1 means that `var-elim-idxs` is faster than Gecode. We do not consider ratios between 0.97 and 1.03 to represent a significant change.

The percentage of auxiliary variables is generally very high, with an average of 83%, but on average only 70% of all the variables are auxiliary and appear in exactly two constraints. The percentage of eliminated variables varies a lot, from as little as 5% to 96%, effectively eliminating all auxiliary variables in the case of the *openshop* model. On average, `var-elim-idxs` extends Gecode with 2.3 new propagators, with a maximum of 9 propagators for an instance of the *cyclic-rcpsp* model.

The column *ratio search* shows that preprocessing generally either improves the node rate during search (ratio larger than 1.03) or leaves it almost unchanged (ratio between 0.97 and 1.03). The node rate can be more than doubled: see the *l2p* model. For the four models at the bottom of the table, however, the performance is worse after preprocessing. On average, the node rate during search is 1.12 times higher. The geometric standard deviation is generally low, i.e., close to 1.0, for instances of the same MiniZinc model, except when some of the instances are solved very fast, partly due to measurement errors.

The column *ratio total* shows that, when also counting the time for preprocessing, the results are still promising. On average, the node rate is 0.81 times lower using preprocessing. This number is strongly affected by instances that are solved very fast. If we take into account only the 207 instances that originally take more than one second to solve, then the node rate of `var-elim-idxs` is on average identical to the one of Gecode. If we take into account only the 174 instances that originally take more than one minute to solve, then the node rate of `var-elim-idxs` is on average 1.09 times higher.

Interestingly, Table 2 also shows that there is no strong correlation between the number of eliminated variables and the node rate ratio. For instance, nearly all auxiliary variables of the *rcpsp* model are eliminated but the node rate ratio is close to 1, despite the fact that the number of propagator calls is divided by two. This probably indicates that the generated indexical-based propagator suffers from some inefficiencies.

The median preprocessing time for the 223 modified instances is 4.4 seconds, roughly equally divided between the time spent by our code in Scala and the time spent by the `g++` compiler. The minimum time is 2.9 seconds, of which a closer analysis shows that more than 2.5 seconds are actually spent in the set-up, such as loading classes or parsing header files, independently of the size of the instance or the number of identified patterns. It is important to note that neither the g++ compiler nor our indexical compiler were developed for such a use-case, as compilation is usually performed offline. The median preprocessing time for the 172 instances unmodified by preprocessing is 0.9 seconds. The minimum time is 0.7 seconds, again mostly spent in loading classes. The largest preprocessing time observed is 30 seconds, in the case of a very large *nmseq* instance. Table 3 reports aggregated results for the 172 unmodified instances: the cost of uselessly running the preprocessing is largely unnoticeable for unmodified instances that take more than 60 seconds to be solved.

6 Discussion

In the light of the experimental results, this section discusses more thoroughly the merits and limitations of our approach.

6.1 Related Work

Dealing with the introduction of auxiliary variables is an important challenge for developers of both solvers and modelling languages.

Variable Views. The initial purpose of variable views [3,16] and domain views [19] is to reduce the number of implemented propagators in a solver, but they can also be used to eliminate some auxiliary variables. A view allows one to post a constraint on an argument that is a function of a variable instead of a variable. If a constraint is of the form $x = f(y)$, where x and y are variables, then one can introduce a view $v = f(y)$ and replace the variable x by the view v. Compared with our approach, views have the benefits that they are not limited to variables appearing in two constraints and that they do not require generating new propagators, hence that they can eliminate variables that appear, for example, in global constraints. Views are however in general limited to *unary* functions, except in [3]. More importantly, to the best of our knowledge, no solver automatically transforms a flattened constraint into a view.

Flattening and Modelling Techniques. Common subexpression elimination [13,14] reduces the number of auxiliary variables by merging into a single variable all variables that represent the same expression. This also has the effect of increasing the amount of filtering. Hence, as explained in Section 3.1, we do not eliminate such variables.

Half-reification [5] is a modelling technique that replaces constraints of the form $B \equiv \phi$ by $B \implies \phi$, where B is a Boolean variable and ϕ a Boolean expression. Although this does not reduce the number of variables, it can reduce solving time by having simpler constraints. However, there are no half-reified constraint predicates in FlatZinc. Our approach enables some optimisation in the spirit of half-reification, as shown in the example of Section 4.

Model globalisation [10] aims at replacing a collection of constraints at the MiniZinc level by an equivalent global constraint. Such a replacement usually reduces the number of auxiliary variables and increases the amount of filtering, provided the global constraint is not decomposed during flattening. Globalisation may improve solving time much more than our approach but it is an offline and interactive process, hence orthogonal to our online and automated approach.

Propagator Generation. Our approach uses our indexical compiler to generate propagators. The generation of stateless propagators [8] is an alternative that can yield much faster propagators. It is however limited by the size of the domains, as the constraint is essentially represented extensionally. It is meant to be used offline, as are other approaches to propagator generation, such as [4].

6.2 Properties and Extensions

Unlike most of the approaches in Section 6.1, our approach is entirely online and automated. We review here some of its properties and discuss possible extensions.

Search Tree Shape. Given a search strategy and enough time, the search trees explored by Gecode on the original model and by `var-elim-idxs` on the modified model are the same if all the following conditions are respected:

– The search strategy does not depend on the propagation order.
– The search strategy does not need to branch on the auxiliary variables.
– The generated propagators do the same filtering as the replaced propagators.

Except in the case of the *roster* model, where the search strategy is incompletely specified, the two first conditions are respected for all the instances we used in Section 5. The third condition is more difficult to check, but seems generally respected: out of the 84 modified instances that did not time out, only 7 instances had a different and always larger search tree, namely one *fjsp* instance, two *league* instances, one *roster* instance, and three *still-life-wastage* instances.

Domains. Our approach assumes that the domains of the eliminated variables are non-constraining because the shared variable X is existentially quantified without specifying a domain. This is why we restricted ourselves to variables introduced by the MiniZinc flattener, annotated with `var_is_introduced`, as the domains proved to be non-constraining for those variables. However, auxiliary variables may also be introduced manually to simplify a model. A sound way to extend our approach to such variables while retaining correctness is to verify that the domain is non-constraining before considering a variable x for replacement, by only considering how the propagators of the two constraints in which x appears reduce the domain of x given the domains of their other variables. This would also let us apply our approach to other modelling languages that do not have the `var_is_introduced` annotation, such as, e.g., XCSP [15].

Instances and Problems. We made a deliberate choice to work at the Flat-Zinc, rather than MiniZinc, level for two reasons. First, it is much simpler to work with a flat format than with a rich modelling language. Second, it might not be clear before or during flattening what the frequent patterns are. This choice led us to work with individual instances. However, instances from the same Mini-Zinc model share the same frequent patterns. Hence, when several instances of the same MiniZinc model must be solved successively, most of the results of the preprocessing of the first instance can actually be reused to reduce the preprocessing time of the following ones. In particular, when the preprocessing does not modify the FlatZinc model, detecting this on small instances saves the potentially high cost of unnecessarily parsing large instances.

Improved Propagator Generation. As seen in Table 2, `var-elim-idxs` does not remove all the auxiliary variables. Partly, this is not a limitation of our approach but of its implementation. Increasing the reach of our approach amounts to improving the generation of the propagators in order to handle efficiently more constraints, including n-ary ones. This can be done by improving our indexical compiler [11] or by using other techniques such as [8] or [3], but such improvements are orthogonal to this paper. Our experiments show that our approach is already practical as it is.

Increased Filtering. When identifying patterns, if more than one shared variable is identified, then it is possible to generate propagators achieving more filtering. Our approach can be extended to multiple shared variables. However, for it to be worthwhile, it is necessary to ensure that the propagator generation takes advantage of multiple occurrences of a variable other than X in the checker. This is currently not the case but an interesting line of future work.

7 Conclusion

We presented a new approach to eliminate many of the auxiliary variables introduced into a flattened constraint-based model. Our approach adds a preprocessing step that modifies the flattened model and extends the solver with propagators generated on the fly for new constraint predicates. This is made possible through the generation of indexical-based propagators from logical formulas. Experiments with our prototype implementation show that our approach makes a solver about 9% faster on average, and sometimes more than 2 times faster, for instances that take more than one minute to solve. This indicates that our preprocessing should be activated for instances that are difficult to solve, which are the ones for which it is important to decrease solving time.

Acknowledgements. This work is supported by grants 2011-6133 and 2012-4908 of the Swedish Research Council (VR). We thank the anonymous reviewers for their constructive and insightful comments.

References

1. Becket, R.: Specification of FlatZinc. http://www.minizinc.org/downloads/doc-1.6/flatzinc-spec.pdf
2. Beldiceanu, N., Carlsson, M., Demassey, S., Petit, T.: Global constraint catalogue: Past, present, and future. Constraints **12**(1), 21–62 (2007). The catalogue is at http://sofdem.github.io/gccat
3. Correia, M., Barahona, P.: View-based propagation of decomposable constraints. Constraints **18**(4), 579–608 (2013)
4. Dao, T.B.H., Lallouet, A., Legtchenko, A., Martin, L.: Indexical-based solver learning. In: Van Hentenryck, P. (ed.) CP 2002. LNCS, vol. 2470, pp. 541–555. Springer, Heidelberg (2002)
5. Feydy, T., Somogyi, Z., Stuckey, P.J.: Half reification and flattening. In: Lee, J. (ed.) CP 2011. LNCS, vol. 6876, pp. 286–301. Springer, Heidelberg (2011)
6. Frisch, A.M., Grum, M., Jefferson, C., Martinez Hernandez, B., Miguel, I.: The design of ESSENCE: a constraint language for specifying combinatorial problems. In: IJCAI 2007, pp. 80–87. Morgan Kaufmann (2007)
7. Gecode Team: Gecode: A generic constraint development environment (2006). http://www.gecode.org
8. Gent, I.P., Jefferson, C., Linton, S., Miguel, I., Nightingale, P.: Generating custom propagators for arbitrary constraints. Artificial Intelligence **211**, 1–33 (2014)
9. Hassani Bijarbooneh, F.: Constraint Programming for Wireless Sensor Networks. Ph.D. thesis, Department of Information Technology, Uppsala University, Sweden (2015). http://urn.kb.se/resolve?urn=urn:nbn:se:uu:diva-241378
10. Leo, K., Mears, C., Tack, G., Garcia de la Banda, M.: Globalizing constraint models. In: Schulte, C. (ed.) CP 2013. LNCS, vol. 8124, pp. 432–447. Springer, Heidelberg (2013)
11. Monette, J.-N., Flener, P., Pearson, J.: Towards solver-independent propagators. In: Milano, M. (ed.) CP 2012. LNCS, vol. 7514, pp. 544–560. Springer, Heidelberg (2012)
12. Nethercote, N., Stuckey, P.J., Becket, R., Brand, S., Duck, G.J., Tack, G.: MiniZinc: towards a standard CP modelling language. In: Bessière, C. (ed.) CP 2007. LNCS, vol. 4741, pp. 529–543. Springer, Heidelberg (2007). The MiniZinc toolchain is available at http://www.minizinc.org
13. Nightingale, P., Akgün, Ö., Gent, I.P., Jefferson, C., Miguel, I.: Automatically improving constraint models in savile row through associative-commutative common subexpression elimination. In: O'Sullivan, B. (ed.) CP 2014. LNCS, vol. 8656, pp. 590–605. Springer, Heidelberg (2014)
14. Rendl, A., Miguel, I., Gent, I.P., Jefferson, C.: Automatically enhancing constraint model instances during tailoring. In: Bulitko, V., Beck, J.C. (eds.) SARA 2009. AAAI Press (2009)
15. Roussel, O., Lecoutre, C.: XML representation of constraint networks: Format XCSP 2.1. CoRR abs/0902.2362 (2009). http://arxiv.org/abs/0902.2362
16. Schulte, C., Tack, G.: View-based propagator derivation. Constraints **18**(1), 75–107 (2013)
17. Scott, J.D.: Rapid prototyping of a structured domain through indexical compilation. In: Schaus, P., Monette, J.N. (eds.) Domain Specific Languages in Combinatorial Optimisation (CoSpeL workshop at CP 2013) (2013). http://cp2013.a4cp.org/workshops/cospel

18. Sidebottom, G., Havens, W.S.: Nicolog: A simple yet powerful cc(FD) language. Journal of Automated Reasoning **17**, 371–403 (1996)
19. Van Hentenryck, P., Michel, L.: Domain views for constraint programming. In: O'Sullivan, B. (ed.) CP 2014. LNCS, vol. 8656, pp. 705–720. Springer, Heidelberg (2014)
20. Van Hentenryck, P., Saraswat, V., Deville, Y.: Design, implementation, and evaluation of the constraint language cc(FD). Tech. Rep. CS-93-02, Brown University, Providence, USA (January 1993), revised version in Journal of Logic Programming **37**(1–3), 293–316 (1998). Based on the unpublished manuscript Constraint Processing in cc(FD) (1991)

Automatically Improving SAT Encoding of Constraint Problems Through Common Subexpression Elimination in Savile Row

Peter Nightingale[✉], Patrick Spracklen, and Ian Miguel

School of Computer Science, University of St Andrews, St Andrews, UK
{pwn1,jlps,ijm}@st-andrews.ac.uk

Abstract. The formulation of a Propositional Satisfiability (SAT) problem instance is vital to efficient solving. This has motivated research on preprocessing, and inprocessing techniques where reformulation of a SAT instance is interleaved with solving. Preprocessing and inprocessing are highly effective in extending the reach of SAT solvers, however they necessarily operate on the lowest level representation of the problem, the raw SAT clauses, where higher-level patterns are difficult and/or costly to identify. Our approach is different: rather than reformulate the SAT representation directly, we apply automated reformulations to a higher level representation (a constraint model) of the original problem. Common Subexpression Elimination (CSE) is a family of techniques to improve automatically the formulation of constraint satisfaction problems, which are often highly beneficial when using a conventional constraint solver. In this work we demonstrate that CSE has similar benefits when the reformulated constraint model is encoded to SAT and solved using a state-of-the-art SAT solver. In some cases we observe speed improvements of over 100 times.

1 Introduction

The Propositional Satisfiability Problem (SAT) is to find an assignment to a set of Boolean variables so as to satisfy a given Boolean formula, typically expressed in conjunctive normal form [4]. SAT has many important applications, such as hardware design and verification, planning, and combinatorial design [14]. Powerful, robust solvers have been developed for SAT employing techniques such as conflict-driven learning, watched literals, restarts and dynamic heuristics for backtracking solvers [15], and sophisticated incomplete techniques such as stochastic local search [22].

The formulation of a SAT problem instance is vital to efficient solving. This has motivated research on preprocessing [7,27], and inprocessing [12] where reformulation of the SAT instance is interleaved with solving. Both techniques are highly effective in extending the reach of SAT solvers, however they necessarily operate on the lowest level representation of the problem, the raw SAT clauses, where higher-level patterns are difficult and/or costly to identify.

© Springer International Publishing Switzerland 2015
G. Pesant (Ed.): CP 2015, LNCS 9255, pp. 330–340, 2015.
DOI: 10.1007/978-3-319-23219-5_23

Our approach is different: rather than reformulate the SAT representation directly, we apply automated reformulations to a higher level representation of the original problem. An increasingly popular means of deriving SAT formulations is by taking a constraint model and employing a set of automated encoding steps to produce an equivalent SAT formulation [28]. Constraint satisfaction is a formalism closely related to SAT in which we seek an assignment of values to decision variables so as to satisfy a set of constraints [21]. Constraint modelling languages typically support decision variables with richer domains and a richer set of constraints than the CNF used with SAT. Hence, an input problem can be expressed conveniently in a higher level constraint language, while employing efficient SAT solvers to find solutions.

Common Subexpression Elimination (CSE) is a very well established technique in compiler construction [5]. In that context the value of a previously-computed expression is used to avoid computing the same expression again. Shlyakhter et al [23] exploited identical subformulae during grounding out of quantified Boolean formulae. Similarly, it is a useful technique in the automatic improvement of constraint models, where it acts to reduce the size of a constraint model by removing redundant variables and constraints [1,10,19,20]. This in turn can create a stronger connection between different parts of the model, resulting in stronger inference and reduced search during constraint solving.

Earlier work applied CSE directly to SAT formulations, with limited success [29]. Herein we establish the success of an alternative approach in which CSE is applied to a constraint model prior to SAT encoding. We apply CSE to a constraint problem instance expressed in the constraint modelling language ESSENCE', which includes integer (as well as Boolean) decision variables, a set of infix operators on integer and Boolean expressions, and various global constraints and functions [20]. The reformulated constraint model is automatically encoded to SAT using the SAVILE ROW system, yielding substantial improvements in SAT solver runtime over encoding without CSE.

Our method has the advantage of allowing us to exploit patterns present in the high level description of the problem that are obscured in the SAT formulation, and so very difficult to detect using SAT pre/inprocessing approaches. As a simple example, a decision variable in a constraint model typically requires a collection of SAT variables and clauses to encode. If, via CSE, we are able to reduce two such variables to one then the set of Boolean variables and clauses to encode the second variable will never be added to the SAT formulation. Performing the equivalent step directly on the SAT formulation would require the potentially very costly step of identifying the structure representing the second variable then proving its equivalence to the structure encoding the first.

In performing CSE on a constraint model preparatory to SAT encoding, we have modified and enhanced existing constraint model CSE approaches to take into account that the eventual target is a SAT formulation. Firstly the set of candidate expressions for CSE differs when the target is SAT. Secondly, implied constraints that are added to elicit common subexpressions are removed following CSE if they are unchanged. In addition we describe for the first time

---

Algorithm 1. Identical-CSE(AST, ST)

Require: AST: Abstract syntax tree representing the model
Require: ST: Symbol table containing CSP decision variables
1: $newcons \leftarrow$ empty list {Collect new constraints}
2: $map \leftarrow$ empty hash table mapping expressions to lists
3: populateMap(AST, map)
4: **for all** key in map in decreasing size order **do**
5:     $ls \leftarrow map(key)$ {ls is a list of identical AST nodes}
6:     $ls \leftarrow$ filter(isAttached, ls) {Remove AST nodes no longer contained in AST or $newcons$}
7:     **if** length(ls) > 1 **then**
8:         $e \leftarrow$ head(ls)
9:         $bnds \leftarrow$ bounds(e)
10:         $aux \leftarrow$ ST.newAuxVar($bnds$)
11:         $newc \leftarrow$ ($e = aux$) {New constraint defining aux}
12:         $newcons$.append($newc$)
13:         **for all** $a \in ls$ **do**
14:             Replace a with copy(aux) within AST or $newcons$
15: AST $\leftarrow$ AST $\wedge$ fold($\wedge$, $newcons$)

Algorithm 2. populateMap(A, map)

Require: A: Reference to an abstract syntax tree
Require: map: Hash table mapping expressions to lists
1: **if** A is a candidate for CSE **then**
2:     Add A to list $map[A]$
3: **for all** $child \in A$.Children() **do**
4:     populateMap($child$, map)

---

an identical CSE algorithm that is independent of general flattening, allowing flexibility to extract common subexpressions would not ordinarily be flattened and to control the order of CSE.

2   CSE for SAT Encoding

The simplest form of CSE that we consider is Identical CSE, which extracts sets of identical expressions. Suppose $x \times y$ occurs three times in a model. Identical CSE would introduce a new decision variable a and new constraint $x \times y = a$. The three original occurrences of $x \times y$ would be replaced by a. In SAVILE ROW, Identical CSE is implemented with Algorithm 1. Andrea Rendl's Tailor [10,20] and MiniZinc [13,26] also implement Identical CSE, however (in contrast to Tailor and MiniZinc) our algorithm is not tied to the process of flattening nested expressions into primitive expressions supported directly by the constraint solver. This is advantageous because it allows us to identify and exploit common subexpressions in expressions that do not need to be flattened. The SMT solver CVC4 merges identical subtrees in its abstract syntax tree [3]. It is not clear whether this affects the search or is simply a memory saving feature.

The first step is to recursively traverse the model (by calling Algorithm 2) to collect sets of identical expressions. Algorithm 2 collects only expressions that are candidates for CSE. Atomic variables and constants are not candidates. Compound expressions are CSE candidates by default, however when the target is a SAT encoding we exclude all compound expressions that can be encoded as a single SAT literal. This avoids creating a redundant SAT variable that is equal to (or the negation of) another SAT variable, thus improving the encoding. The following expressions are not candidates: $x = c$, $x \neq c$, $x \leq c$, $x < c$, $x \geq c$, $x > c$, $\neg x$ (where x is a decision variable and c is a constant).

The second step of Identical CSE is to iterate through sets of expressions in decreasing size order (line 4). When an expression e is eliminated by CSE, the number of occurrences of any expressions contained in e is reduced. Therefore eliminating long expressions first may obviate the need to eliminate short expressions. For each set (of size greater than one) of identical expressions a new decision variable aux is created, and each of the expressions is replaced with aux. One of the expressions e in the set is used to create a new constraint $e = aux$. Crucially the new constraint contains the original object e so it is possible to extract further CSEs from within e.

Prior to running Identical CSE the model is simplified by evaluating all constant expressions and placing it into negation normal form. In addition some type-specific simplifications are performed (eg $x \leftrightarrow$ true rewrites to x). Commutative expressions (such as sums) are sorted to make some equivalent expressions syntactically identical.

In our previous work we investigated Associative-Commutative CSE (AC-CSE) for constraint solvers [19] and in that context Identical CSE was always enabled. Identical CSE is complementary to AC-CSE.

Active CSE. Active CSE extends Identical CSE by allowing non-identical expressions to be extracted using a single auxiliary variable. For example, suppose we have $x = y$ and $x \neq y$ in the model. We can introduce a single Boolean variable a and a new constraint $a \leftrightarrow (x = y)$, then replace $x = y$ with a and $x \neq y$ with $\neg a$. For solvers that support negation (such as SAT solvers) $\neg a$ can be expressed in the solver input language with no further rewriting, so we have avoided encoding both $x = y$ and $x \neq y$.

The Active CSE algorithm implemented in SAVILE ROW is an extension of Algorithm 1. The algorithm works as follows: for each candidate expression e a simple transformation is applied to it (for example producing $\neg e$). The transformed expression is placed into the normal form and commutative subexpressions are sorted. The algorithm then queries map to discover expressions matching the transformed expression.

Active CSE as implemented in SAVILE ROW 1.6.3 applies four transformations: Boolean negation, arithmetic negation, multiply by 2, and multiply by -2. Rendl implemented Boolean negation active CSE in her Tailor system, along with active reformulations based upon De Morgan's laws and Horn clauses [20]. In SAVILE ROW, the use of negation normal form obviates the use of the latter two. To our knowledge MiniZinc [13,26] does not implement Active CSE.

Associative-Commutative CSE (AC-CSE). Nightingale et al [19] (for finite domains) and Araya et al [1] (for numerical CSP) established the use of AC-CSE for constraint models. To our knowledge neither Tailor [10,20] nor MiniZinc [13,26] implement AC-CSE. It exploits the properties of associativity and commutativity of binary operators, such as in sum constraints. For SAT encoding, our approach refines the procedure for AC-CSE given in Nightingale et al. In that procedure, implied sum constraints are added, which are deduced from global constraints in the model, such as all-different and global cardinality. These implied sums are used to trigger AC-CSE. Since large sum constraints are cumbersome to encode in SAT, and can therefore degrade performance, we add a test to check whether the implied sums are modified following AC-CSE. If not, they are deemed not to be useful and removed prior to SAT encoding.

Extended resolution [2] is gaining interest and can be viewed as AC-CSE applied directly to the disjunctive clauses of a SAT formula.

Effects of CSE on the Output Formula. We give a short example of a constraint reformulation and its effect on the SAT encoding. Suppose we have two occurrences of $x \times y$, both are contained in sums, and $x, y \in \{1 \dots 10\}$. Ordinarily we would create a new auxiliary variable ($a_1, a_2 \in \{1 \dots 100\}$) for each occurrence, and add two new constraints: $x \times y = a_1$ and $x \times y = a_2$. Both a_1 and a_2 would be encoded using just under 200 SAT variables and approximately 400 clauses each. Also, both new constraints would be encoded using 100 clauses each. In contrast, Identical CSE would create a single auxiliary variable for both occurrences of $x \times y$, and there would be one new constraint, saving hundreds of SAT variables and clauses. It is difficult to see how SAT pre/inprocessing rules could identify the structure that was exploited by Identical CSE.

3   Experimental Evaluation

Our goal is to investigate whether reformulations performed on a constraint problem instance are beneficial when the problem instance is solved by encoding to SAT and using a state-of-the-art SAT solver. To achieve this we need to ensure that the baseline encoding to SAT is sensible. Therefore we have used standard encodings from the literature such as the order encoding for sums [28] and support encoding [8] for binary constraints. Also we do not attempt to encode all constraints in the language: several constraint types are decomposed before encoding to SAT. Details are given in the SAVILE ROW tutorial 1.6.3 appendix A [18].

In our experiments we compare four configurations of SAVILE ROW: *Basic*, which includes the default options of unifying equal variables, filtering domains and aggregation; *Identical CSE*, which is Basic plus Identical CSE; *Identical & Active CSE*, which is Basic plus the two named CSE algorithms, and *Identical, Active & AC-CSE*, which is Basic plus all three CSE algorithms. Our benchmark set is the set of example problems included with SAVILE ROW 1.6.3 [18]. There are 49 problem classes including common benchmark problems such as EFPA

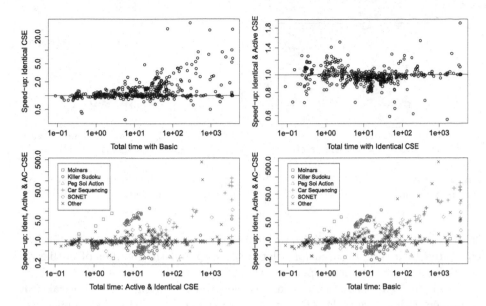

Fig. 1. Identical CSE vs Basic (upper left), Active & Identical CSE vs Identical CSE (upper right), Identical, Active & AC-CSE vs Identical & Active CSE (lower left), same vs Basic (lower right).

[11] and car sequencing [6] as well as less common problems such as Black Hole solitaire [9]. In total there are 492 problem instances.

Experiments were run with 32 processes in parallel on a machine with two 16-core AMD Opteron 6272 CPUs at 2.1 GHz and 256 GB RAM. We used the SAT solver Lingeling [12] which was winner of the Sequential, Application SAT+UNSAT track of the SAT competition 2014. We downloaded lingeling-ayv-86bf266-140429.zip from http://fmv.jku.at/lingeling/. We used default options for Lingeling so inprocessing was switched on. All times reported include SAVILE ROW time and Lingeling's reported time, and are a median of 10 runs with 10 different random seeds given to Lingeling. A time limit of one hour was applied. We used a clause limit of 100 million, and for instances that exceeded the clause limit we treated them as if they timed out at 3600s (to allow comparison with others). Of 492 instances, 7 reached the clause limit with *Basic* and 6 with the other configurations.

Summary Plots for Full Set of Benchmarks. In Figures 1–2 we present a summary view of the performance of our CSE methods over our full set of 49 benchmark problem classes. Figure 1 (upper left) compares the basic encoding with identical CSE. On easier problem instances CSE has a limited effect, but as problem difficulty increases so does the potential of identical CSE to reduce search effort very significantly - in some cases by over 20 times. There are a small number of outliers among the harder instances where identical CSE degrades overall performance. We conjecture that this is due to the change in problem

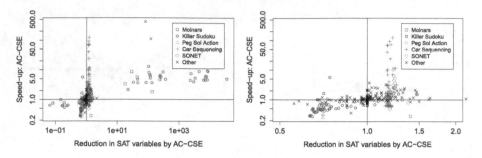

Fig. 2. Identical, Active & AC-CSE vs Identical & Active CSE plotted against reduction in SAT variables. The plot on the right is a subset of the left.

structure affecting the heuristics of Lingeling. The degradation effect is limited compared with the potential for a large speedup and the number of outliers is small. The geometric mean speed-up is 1.24.

In Figure 1 (upper right) we compare identical CSE alone with identical CSE combined with active CSE. The results show that this additional step is largely neutral or incurs a very small overhead of performing the active CSE checks, but that there are a number of occasions where active CSE significantly enhances identical CSE. Again, there are a small number of outliers where performance is significantly degraded, which we again believe to be due to a bad interaction with the SAT solver search strategy. The geometric mean speed-up is 0.98, indicating a very small average slow-down.

Figure 1 (lower left and right) plots the utility of AC-CSE. In some cases we see a very considerable improvement in performance, however there are also cases where performance is degraded. Five notable problem classes have been separated in the plots. Of these, Killer Sudoku is the most ambiguous, with clusters of instances both above and below the break-even line. For Car Sequencing and SONET, some of the easier instances are below the break-even line, but the more difficult instances exhibit a speed-up. Peg Solitaire Action is degraded on all seven instances. Molnars exhibits a speed up with one exception. Over all instances the geometric mean speed-up is 1.24.

To partly explain these results, we measured the size of the formula produced with and without AC-CSE. Figure 2 has the same y-axis as Figure 1 (lower left) but with a different x-axis: the ratio of the number of variables in the SAT formula. Values of x above 1 indicate that applying AC-CSE has reduced the number of SAT variables. For Killer Sudoku, there is a clear link between the number of SAT variables in the formula and the speed up quotient. It is also worth noting that for all instances of Peg Solitaire Action applying AC-CSE both increases the number of SAT variables and degrades performance. On the other hand Car Sequencing and SONET show no correlation between speed up

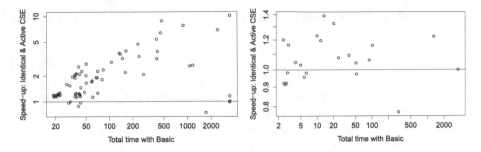

Fig. 3. Identical & Active CSE vs Basic: Car Sequencing (left) and SONET (right).

quotient and reduction of SAT variables, indicating that the number of SAT variables alone is a very coarse measure of difficulty.

Case Study 1: Car Sequencing Our ESSENCE' model of Car Sequencing [24] uses a sequence of integer variables $x[1 \ldots n]$ to represent the sequence of cars on a production line. For each option (to be fitted at a station on the production line) we have a limit on the proportion of cars: at most p of any q adjacent cars may have the option installed so as not to overload the station. To model this we employ overlapping sums of length q containing $x[i] \in S$, where S is the set of car classes that have the option installed, and i is an index within the subsequence of length q.

The number of each car class to build is enforced with a global cardinality constraint [17] on x. Also, for each option we know how many cars require that option in total (t) thus we add a further *implied* constraint: $\sum_{i=1}^{n}(x[i] \in S) = t$. We experiment with the 80 instances used in Nightingale [17]. Identical and Active CSE are both able to extract the expressions $x[i] \in S$ that appear in several sum constraints, avoiding multiple SAT encodings of the same set-inclusion constraint. Figure 3 (left) plots the time improvement of Active CSE compared with the Basic encoding. The improvement is substantial and increases with the difficulty of the problem instances.

AC-CSE is able to extract common subexpressions among the sum constraints for a given option. The p of q constraints overlap with each other and also with the implied constraint for the option. Figure 1 (lower left) plots the time improvement of adding AC-CSE to identical and active CSE. The additional improvement is substantial, with many instances becoming solvable within one hour and a peak speed up of over 100 times. With the Basic encoding 13 of the 80 instances time out at one hour. In contrast, when combining Identical, Active and AC-CSE we found that only two instances timed out. The other 11 are solved within one hour, most with very substantial speed-ups.

Case Study 2: SONET The SONET problem [16,25] is a network design problem where each node is installed on a set of *rings* (fibre-optic connections). If two nodes are required to be connected, there must exist a ring on which they are both installed. We use the simplified SONET problem where each ring has unlimited data capacity (Section 3 of [25]). Rings are indistinguishable so we use lexicographic ordering constraints to order the rings in non-decreasing order. This is an optimisation problem: the number of node-ring connections is minimised. The problem formulation and set of 24 instances are exactly as described in Nightingale et al [19].

Figure 3 (right) compares Identical and Active CSE to the Basic encoding on this problem class. The initial formulation of SONET contains no identical or active common subexpressions, however each decomposition of a lexicographic ordering constraint has identical subexpressions that are exploited by Identical CSE, causing the modest gains seen in the plot. There are four groups of constraints in SONET: the objective function, the constraints ensuring nodes are connected when required, a constraint for each ring limiting the number of nodes, and the symmetry breaking constraints. Apart from symmetry breaking all constraints are sums and all three groups overlap, therefore AC-CSE is successful on this problem as shown in Figure 1 (lower left).

4   Conclusion

Common Subexpression Elimination has proven to be a valuable tool in the armoury of reformulations applied to constraint models, however hitherto there has only been limited success in applying CSE to SAT formulations [29]. We have shown how CSE can be used to improve SAT formulations derived through an automated encoding process from constraint models. Our approach has the advantage that it can identify and exploit structure present in a constraint model that is subsequently obscured by the encoding process, while still taking advantage of powerful SAT solvers. The result is a method that, when applicable, can produce a very significant reduction in search effort.

We have evaluated our approach on a wide range of benchmark problems. On some instances we observed improvements of SAT solver speed of over 50 times. On the car sequencing problem, for example, the peak speed increase is over 100 times. With the basic approach, 13 of 80 car sequencing instances could not be solved in one hour, whereas with the full CSE approach only two instances could not be solved.

Acknowledgments. We wish to thank the EPSRC for funding this work through grants EP/H004092/1 and EP/M003728/1, and Christopher Jefferson for helpful discussions.

References

1. Araya, I., Neveu, B., Trombettoni, G.: Exploiting common subexpressions in numerical CSPs. In: Stuckey, P.J. (ed.) CP 2008. LNCS, vol. 5202, pp. 342–357. Springer, Heidelberg (2008)
2. Audemard, G., Katsirelos, G., Simon, L.: A restriction of extended resolution for clause learning sat solvers. In: Proceedings of the Twenty-Fourth AAAI Conference on Artificial Intelligence (2010)

3. Barrett, C., Conway, C.L., Deters, M., Hadarean, L., Jovanović, D., King, T., Reynolds, A., Tinelli, C.: CVC4. In: Gopalakrishnan, G., Qadeer, S. (eds.) CAV 2011. LNCS, vol. 6806, pp. 171–177. Springer, Heidelberg (2011)
4. Biere, A., Heule, M., van Maaren, H.: Handbook of Satisfiability, vol. 185. IOS Press (2009)
5. Cocke, J.: Global common subexpression elimination. ACM Sigplan Notices 5(7), 20–24 (1970)
6. Dincbas, M., Simonis, H., Van Hentenryck, P.: Solving the car-sequencing problem in constraint logic programming. In: Proceedings of the 8th European Conference on Artificial Intelligence (ECAI 1988), pp. 290–295 (1988)
7. Eén, N., Biere, A.: Effective preprocessing in SAT through variable and clause elimination. In: Bacchus, F., Walsh, T. (eds.) SAT 2005. LNCS, vol. 3569, pp. 61–75. Springer, Heidelberg (2005)
8. Gent, I.P.: Arc consistency in SAT. In: Proceedings of the 15th European Conference on Artificial Intelligence (ECAI 2002), pp. 121–125 (2002)
9. Gent, I.P., Jefferson, C., Kelsey, T., Lynce, I., Miguel, I., Nightingale, P., Smith, B.M., Tarim, S.A.: Search in the patience game 'black hole'. AI Communications 20(3), 211–226 (2007)
10. Gent, I.P., Miguel, I., Rendl, A.: Tailoring solver-independent constraint models: a case study with ESSENCE' and MINION. In: Miguel, I., Ruml, W. (eds.) SARA 2007. LNCS (LNAI), vol. 4612, pp. 184–199. Springer, Heidelberg (2007)
11. Huczynska, S., McKay, P., Miguel, I., Nightingale, P.: Modelling equidistant frequency permutation arrays: an application of constraints to mathematics. In: Gent, I.P. (ed.) CP 2009. LNCS, vol. 5732, pp. 50 64. Springer, Heidelberg (2009)
12. Järvisalo, M., Heule, M.J.H., Biere, A.: Inprocessing rules. In: Gramlich, B., Miller, D., Sattler, U. (eds.) IJCAR 2012. LNCS, vol. 7364, pp. 355–370. Springer, Heidelberg (2012)
13. Leo, K., Tack, G.: Multi-pass high-level presolving. In: Proceedings of the 24th International Joint Conference on Artificial Intelligence (IJCAI) (to appear, 2015)
14. Marques-Silva, J.: Practical applications of boolean satisfiability. In: 9th International Workshop on Discrete Event Systems (WODES 2008), pp. 74–80 (2008)
15. Moskewicz, M.W., Madigan, C.F., Zhao, Y., Zhang, L., Malik, S.: Chaff: engineering an efficient SAT solver. In: Proceedings of the 38th Annual Design Automation Conference, pp. 530–535. ACM (2001)
16. Nightingale, P.: CSPLib problem 056: Synchronous optical networking (SONET) problem. http://www.csplib.org/Problems/prob056
17. Nightingale, P.: The extended global cardinality constraint: An empirical survey. Artificial Intelligence 175(2), 586–614 (2011)
18. Nightingale, P.: Savile Row, a constraint modelling assistant (2015). http://savilerow.cs.st-andrews.ac.uk/
19. Nightingale, P., Akgün, Ö., Gent, I.P., Jefferson, C., Miguel, I.: Automatically improving constraint models in Savile Row through associative-commutative common subexpression elimination. In: O'Sullivan, B. (ed.) CP 2014. LNCS, vol. 8656, pp. 590–605. Springer, Heidelberg (2014)
20. Rendl, A.: Effective Compilation of Constraint Models. Ph.D. thesis, University of St Andrews (2010)
21. Rossi, F., van Beek, P., Walsh, T. (eds.) Handbook of Constraint Programming. Elsevier (2006)
22. Shang, Y., Wah, B.W.: A discrete lagrangian-based global-search method for solving satisfiability problems. Journal of Global Optimization 12(1), 61–99 (1998)

23. Shlyakhter, I., Sridharan, M., Seater, R., Jackson, D.: Exploiting subformula sharing in automatic analysis of quantified formulas. In: Sixth International Conference on Theory and Applications of Satisfiability Testing (SAT 2003) (2003), poster
24. Smith, B.: CSPLib problem 001: Car sequencing. http://www.csplib.org/Problems/prob001
25. Smith, B.M.: Symmetry and search in a network design problem. In: Barták, R., Milano, M. (eds.) CPAIOR 2005. LNCS, vol. 3524, pp. 336–350. Springer, Heidelberg (2005)
26. Stuckey, P.J., Tack, G.: MiniZinc with functions. In: Gomes, C., Sellmann, M. (eds.) CPAIOR 2013. LNCS, vol. 7874, pp. 268–283. Springer, Heidelberg (2013)
27. Subbarayan, S., Pradhan, D.K.: NiVER: non-increasing variable elimination resolution for preprocessing SAT instances. In: Hoos, H.H., Mitchell, D.G. (eds.) SAT 2004. LNCS, vol. 3542, pp. 276–291. Springer, Heidelberg (2005)
28. Tamura, N., Taga, A., Kitagawa, S., Banbara, M.: Compiling finite linear CSP into SAT. Constraints **14**(2), 254–272 (2009)
29. Yan, Y., Gutierrez, C., Jeriah, J.C., Bao, F.S., Zhang, Y.: Accelerating SAT solving by common subclause elimination. In: Proceedings of the Twenty-Ninth AAAI Conference on Artificial Intelligence (AAAI 2015), pp. 4224–4225 (2015)

Exact Sampling for Regular and Markov Constraints with Belief Propagation

Alexandre Papadopoulos[1,2]([✉]), François Pachet[1,2],
Pierre Roy[1], and Jason Sakellariou[1,2]

[1] Sony CSL, 6 Rue Amyot, 75005 Paris, France
roy@csl.sony.fr
[2] Sorbonne Universités, UPMC University Paris 06, UMR 7606, LIP6,
75005 Paris, France
{alexandre.papadopoulos,jason.sakellariou}@lip6.fr, pachet@csl.sony.fr

Abstract. Sampling random sequences from a statistical model, subject to hard constraints, is generally a difficult task. In this paper, we show that for Markov models and a set of REGULAR global constraints and unary constraints, we can perform perfect sampling. This is achieved by defining a factor graph, composed of binary factors that combine a Markov chain and an automaton. We apply a simplified version of belief propagation to sample random sequences satisfying the global constraints, with their correct probability. Since the factor graph is linear, this procedure is efficient and exact. We illustrate this approach to the generation of sequences of text or music, imitating the style of a corpus, and verifying validity constraints, such as syntax or meter.

Keywords: Global constraints · Unary constraints · Markov constraints · Belief propagation · Sampling

1 Introduction

Generating novel sequences, such as text or music, that imitate a given style is usually achieved by replicating statistical properties of a corpus. This inherently stochastic process can be typically performed by sampling a probability distribution. In practice, we often need to impose additional properties on sequences, such as syntactic patterns for text, or meter for music, that are conveniently stated using constraint satisfaction approaches. However, typical constraint satisfaction procedures are not concerned with the distribution of their solutions. On the other hand, traditional sampling algorithms are generally not suited to satisfy hard constraints, since they can suffer from high rejection rates or lack coverage of the solution space. Both issues can be avoided, in some cases, by taking advantage of constraint programming techniques.

In this paper, we show how to sample Markov sequences subject to a conjunction of REGULAR constraints [22], i.e., constraints stated with an automaton, as well as additional unary constraints. Regular grammars can express parts-of-speech patterns on text. In music, METER [26] constrains Markov temporal

© Springer International Publishing Switzerland 2015
G. Pesant (Ed.): CP 2015, LNCS 9255, pp. 341–350, 2015.
DOI: 10.1007/978-3-319-23219-5_24

sequences to be metrically correct. METER can also be expressed as an automaton, as we explain further in this paper. We achieve this result by defining a tree-structured factor graph composed of unary and binary factors. The variables of this graphical model represent the elements of the sequence, and binary factors encode a type of conjunction between the Markov model and the automaton. We apply belief propagation to sample sequences with their right probability.

1.1 Related Work

The combination of statistical and logical methods has been an active research direction in artificial intelligence in the last few years. In constraint programming, stochastic techniques are often used for guiding search, but less for characterising solutions. Some work studies the impact of search heuristics on solution diversity [27], but such endeavours tend to use optimisation techniques [11,13]. Conversely, introducing constraints to probabilistic graphical models is problematic since hard constraints introduce many zero probabilities, and this causes typical sampling algorithms to suffer from high rejection rates. To overcome such issues, Gogate and Dechter proposed SampleSearch [10], with a guaranteed uniform sampling of the solutions of a CSP, using a complete solver to reduce rejection rates. Likewise, Ermon et. al [8] use a constraint solver in a blackbox scheme, and sample the solution space uniformly, often with better performance. In SAT, Markov logic networks is a well established formalism that unifies probabilistic and deterministic properties [7,25]. MC-SAT [24] samples from a non-uniform distribution of the satisfying assignments of a SAT formula. Such methods, with applications in verification, model checking, or counting problems, are general but expensive. The solution we propose, which focuses on a specific setting, is not derivable from such general methods, and is both tractable and exact.

2 Sequence Generation with Markov constraints

A Markov chain is a stochastic process, where the probability for state X_i, a random variable, depends only on the last state X_{i-1}. Each random variable X_i takes values amongst an *alphabet*, denoted $\mathcal{X}$. Seen as a generative process, a Markov chain produces sequence $X_1, \ldots, X_n$ with a probability $P(X_1) \cdot P(X_2|X_1) \cdots P(X_n|X_{n-1})$. Order k Markov chains have a longer memory: the Markov property states that $P(X_i|X_1, \ldots, X_{i-1}) = P(X_i|X_{i-k}, \ldots, X_{i-1})$. They are equivalent to order 1 Markov chains on an alphabet composed of k-grams, and therefore we assume only order 1 Markov chains.

Markov chains have been classically used for generating sequences that imitate a given style [5,14,23]. A Markov chain is trained by learning the transition probabilities on a corpus. For example, a musical piece can be represented as a sequence of complex objects, constituted of pitch, duration, metrical position, and more [17]. A Markov chain trained on this corpus will produce musical sequences in the style of the composer. With text, we can use a Markov chain whose alphabet is the set of words of the corpus, to generate new sentences in the style of its author.

Markov generation can be controlled using Markov constraints. This allows us to specify additional properties that a sequence should verify. For example, METER [26] imposes that sequences of notes are metrically correct. Often, such constraints can be conveniently stated using a REGULAR constraint [22], defined with an automaton $\mathcal{A} = \langle Q, \Sigma, \delta, q_0, F \rangle$, where Q is a set of states, Σ an alphabet defining labels on transitions, δ the transition function linking a state $q \in Q$ and a label $a \in \Sigma$ to the successor state $q' = \delta(q, a)$, $q_0 \in Q$ the initial state, and $F \subseteq Q$ the set of accepting states. In this case, we have $\Sigma = \mathcal{X}$, i.e. transitions are labelled using states of the Markov chain, so that the automaton recognises admissible Markov sequences. Combining Markov constraints with other constraints, we can restrict the solution space in any desirable way [15], but without any guarantee that the generated sequences will reflect the original distribution in any way. In the specific case of unary constraints, we can have this guarantee [2]. The result presented here can be seen as a further generalisation of this result to a set of REGULAR constraints. A specific implementation of this idea was used to generate non-plagiaristic sequences [18].

3   Background on Belief Propagation

Let $X_1, \ldots, X_n$ be n discrete random variables, and let $p(X_1, \ldots, X_n)$ be a distribution of the random sequence $X_1, \ldots, X_n$. A graphical model [21] is a compact representation of p as the product of m factors holding on a subset of the variables, i.e. $p(X_1, \ldots, X_n) = \prod_{j=1}^{m} f_j(S_j)$, where the factor f_j is a function holding on a subset $S_j \subseteq \{X_1, \ldots, X_n\}$ of the variables. CSPs can be seen as graphical models, where solutions are uniformly distributed.

Belief propagation, specifically the sum-product algorithm [20] is an algorithm for performing statistical inference, based on a *factor graph* representation. A factor graph is a bipartite undirected graph $G = (X, F, E)$, representing the factorisation of a probability function. Nodes represent either variables or factors, and edges connect factors to the variables to which that factor applies: $X = \{X_1, \ldots, X_n\}$, $F = \{f_1, \ldots, f_m\}$, and an edge (X_i, f_j) is in E iff $X_i \in S_j$.

Example 1. Consider a probability function holding on three variables X_1, X_2, X_3, defined as the product of four factors $p(X_1, X_2, X_3) = f_1(X_1, X_2) \cdot f_2(X_2, X_3) \cdot f_3(X_1, X_3) \cdot f_4(X_3)$. The corresponding factor graph is shown on Figure 1.

The main use of factor graph in statistical inference is to compute marginals. Marginals are defined for each variable: $p_i(X_i) = \sum_{\{X_j | j \neq i\}} p(X_1, \ldots, X_n)$. Once marginals have been computed, sampling can be performed easily. When the factor graph is a *tree*, computing marginals is polynomial. Tree factor graphs correspond to Berge-acyclic constraint networks, and such results generalise the well-known results in constraints [3, 6, 9].

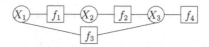

Fig. 1. The factor graph for the function $p(X_1, X_2, X_3) = f_1(X_1, X_2) \cdot f_2(X_2, X_3) \cdot f_3(X_1, X_3) \cdot f_4(X_3)$.

4   Belief Propagation for Markov and Regular

We apply those techniques to the problem of sampling constrained Markov sequences, and describe belief propagation in the case where we impose sequences to be recognised by an automaton $\mathcal{A}$, i.e. to belong to the language $\mathcal{L}(\mathcal{A})$ of words recognised by $\mathcal{A}$. This is equivalent to sampling the target distribution p_{target} defined as:

$$p_{target}(X_1, \ldots, X_n) \propto \begin{cases} P(X_2|X_1) \cdots P(X_n|X_{n-1}) \cdot & \text{if } X_1 \cdots X_n \in \mathcal{L}(\mathcal{A}) \\ P_1(X_1) \cdots P_n(X_n) \\ 0 & \text{otherwise} \end{cases}$$

We use the symbol $\propto$ to indicate that the equality holds after normalisation, so that p_{target} defines a probability function. $P(X_2|X_1) \cdots P(X_n|X_{n-1})$ gives the typical order 1 Markov probability of the sequences $X_1, \ldots, X_n$, provided it is accepted by the automaton. Additionally, we add unary constraints P_i, i.e. factors biasing each variable X_i individually. Implicitly, there is a big factor holding on the full sequence $X_1, \ldots, X_n$ taking value 1 when $X_1 \cdots X_n \in \mathcal{L}(\mathcal{A})$, and value 0 otherwise, corresponding to a hard global constraint. Consequently, the factor graph of p_{target} is not a tree.

We propose a reformulation of $p_{target}(X_1, \ldots, X_n)$ into a new function p_{reg} of $Y_1, \ldots, Y_n$, where the new Y_i variables take values $(a, q) \in \mathcal{X} \times Q$, where $a \in \mathcal{X}$ is a state of the Markov chain, and $q \in Q$ is a state of the automaton. Recall that transitions of the automaton are also labelled with elements of $\mathcal{X}$. This function p_{reg} is composed of simple binary factors, and its factor graph, which is tree structured, is shown on Figure 2.

Fig. 2. The factor graph of the distribution on Markov sequences accepted by an automaton $\mathcal{A}$, defined by $p_{reg}(Y_1, \ldots, Y_n)$

We define a binary factor combining the Markov transition probabilities with the valid transitions from the automaton, as follows:

$$f((a,q),(a',q')) \propto \begin{cases} P(a'|a), & \text{if } q' = \delta(q,a'), \\ 0 & \text{otherwise} \end{cases}$$

This factor gives the probability for choosing, from state q, the transition labelled with a', which reaches q' (denoted by $q' = \delta(q,a')$). This probability depends on the label a of the transition that was used to reach q, and is given by the Markov transition probability from a to a'. This factor is applied along the sequence, i.e. $f_i = f, \forall 1 \leq i < n,$. The binary factors imply that non-zero probability sequences correspond to a walk in the automaton. Unary factors g_i additionally impose that such walks start from the initial state (enforced by g_1) and end at an accepting state (enforced by g_n), while taking into account the unary constraints of p_{target} (enforced by all g_i):

$$g_1((a,q)) \propto \begin{cases} P_1(a), & \text{if } q = \delta(q_0,a) \\ 0, & \text{otherwise.} \end{cases} \qquad g_n((a,q)) \propto \begin{cases} P_n(a), & \text{if } q \in F \\ 0, & \text{otherwise.} \end{cases}$$

Other unary factors are simply defined as $g_i((a,q)) \propto P_i(a)$.

Theorem 1. *Sampling p_{target} is equivalent to sampling p_{reg}, and projecting each resulting sequence $(a_1,q_1), \ldots, (a_n,q_n)$ to $a_1, \ldots, a_n$.*

Proof. We prove there is a one-to-one correspondence between non-zero probability sequences of p_{reg} and p_{target}, and that corresponding sequences have the same probability.

Let $(a_1,q_1), \ldots, (a_n,q_n)$ be a sequence such that $p_{reg}((a_1,q_1), \ldots, (a_n,q_n)) \geq 0$. This means that q_1 is the successor of the initial state q_0 for a_1 (from the definition of g_1), q_i is the successor of state q_{i-1} for a_i, for each $i > 1$ (from the definition of f), and q_n is an accepting state (from the definition of g_n). In other words, $a_1, \ldots, a_n$ is accepted by the automaton, and, according to the definitions of the factors, with probability exactly equal to p_{target}.

Conversely, suppose that $a_1, \ldots, a_n$ is a sequence with a non-zero p_{target} probability. Since $\mathcal{A}$ is deterministic, there exists a unique sequence of states $q_0, q_1, \ldots, q_n$, with $q_n \in F$, that recognises $a_1, \ldots, a_n$, and therefore a unique sequence $(a_1,q_1), \ldots, (a_n,q_n)$ with a p_{reg} probability equal to $p_{target}(a_1, \ldots, a_n)$. $\square$

In order to sample sequences from this factor graph, we adapt the general sum-product algorithm [21], and simplify it for the following reasons: the factor graph has no cycle (removing any issue for converging to a fixed point), the factor graph is almost a linear graph (induced by the sequence), factors are only unary and binary, and the procedure is used only for sampling individual sequences. This algorithm is shown on Algorithm 1 for self-containedness. It computes the backward messages $m_{i\leftarrow}$, the forward messages $m_{i\rightarrow}$, and the actual sequence $y_1, \ldots, y_n$, all highlighted in blue in the algorithm. The exact justification of the algorithm is a well-established result [12,20], and we only give an intuitive explanation. During the backward phase, $m_{i\leftarrow}$ contains the marginal of Y_i of the product of all factors of p_{reg} holding on $Y_i, \ldots, Y_n$. This represents the impact on Y_i of the sub-factor graph "to the right" of Y_i, in the same way that arc-consistency guarantees that a value can be extended to a full instantiation. Eventually, $m_{1\leftarrow}$ is the marginal of Y_1 of all p_{reg}, and a value is drawn randomly according to this distribution. The full sequence is generated during the forward phase. At each iteration, $p_i(Y_i)$ is the marginal over Y_i of p_{reg} given the partial instantiation. In order to sample several sequences, the backward phase needs to be performed only once, and the forward phase will sample a new random sequence every time, with its correct probability. From a constraint programming point of view, computing the marginals at each step is a generalisation to random variables of computing arc-consistent domains. The time for sampling one sequence is bounded by $O(n \cdot (|\mathcal{X}||Q|)^2)$.

Algorithm 1. Sum-product algorithm for sampling Markov with Regular

---

Data: Function $p_{reg}(Y_1, \ldots, Y_n)$ and its factor graph
Result: A sequence $y_1, \ldots, y_n$, with probability $p_{reg}(y_1, \ldots, y_n)$

// Backward phase
$m_{n\leftarrow} \leftarrow g_n$
for $i \leftarrow n-1$ to 1 do
    foreach $y \in \mathcal{X} \times Q$ do
        $m_{i\leftarrow}(y) \leftarrow \sum_{y' \in \mathcal{X} \times Q} g_i(y) \cdot f_i(y, y') \cdot m_{i+1\leftarrow}(y')$
    Normalise $m_{i\leftarrow}$

// Forward phase
$p_1 \leftarrow m_{1\leftarrow}$
$y_1 \leftarrow$ Draw with probability $p_1(y_1)$
for $i \leftarrow 2$ to n do
    foreach $y \in Q$ do
        $m_{i\rightarrow}(y) \leftarrow f_{i-1}(y_{i-1}, y)$
    Normalise $m_{i\rightarrow}$
    foreach $y \in \mathcal{X} \times Q$ do $p_i(y) \leftarrow m_{i\rightarrow}(y) \cdot m_{i\leftarrow}(y)$
    $y_i \leftarrow$ Draw with probability $p_i(y_i)$
return $(y_1, \ldots, y_n)$

---

5   Examples

If no automaton is imposed, our model, which imposes only unary constraints, is equivalent to the model in [16]. We compared the new model with our old model, and observed it behaves equivalently, with the benefit of an improved efficiency. We generated sequences of 16 notes with a Markov order 1 in the style of Bill Evans, with two unary constraints constraining the first and last note. Our old model could sample an average of 450 sequences per second, while our new model produces an average of 1200 sequences per second, almost three times more.

Automata can be handy for expressing patterns on text or music. For example, in music, a semiotic structure is a symbolic description of higher level patterns, from manual annotation or using pattern discovery techniques [4]. Semiotic structures can be easily stated using automata. In text, automata can be used to state syntactic rules over sequences of words.

METER [26], in its most basic form, imposes that a sequence of variables have a fixed total duration D, assuming each value has a specific duration, and assuming the existence of a special padding value with a null duration, which is used only at the end of the sequence. METER can also be encoded using REGULAR. We build an automaton $\langle Q, \mathcal{X}, \delta, q_0, F \rangle$ where each state represents a partial duration between 0 and D, i.e. $Q = \{q_0, \ldots, q_D\}$. For every element $e \in \mathcal{X}$ of the Markov chain, we add a transition from q_{o_1} to q_{o_2} labelled by e iff $o_2 = o_1 + d(e)$, where $d(e)$ is the duration of e. Finally, we set $F = \{q_D\}$. This ensures that any accepting sequence will have a total duration of D exactly. By imposing this automaton to the Markov model, we can sample metrically correct Markov sequences with their correct probabilities. We tested this with a toy problem: produce sequences of a variable number of words, but with fixed number of syllables equal to 36, i.e. the duration of a word is its number of syllables. We are able to sample around in average 1100 sequences per second, against 230 sequences per second produced by a CP model with a single METER constraint, almost five times more.

In previous work, we introduced MAXORDER, which limits the maximum order of generated sequences, i.e. the length of exact copies made from the input corpus [19]. This constraint was filtered by computing a particular automaton and propagating it using REGULAR. We can use this automaton with the model of this paper, in order to sample Markov sequences with a maximum order guarantee. Furthermore, by computing the intersection between the METER and the MAXORDER automaton, we can also impose meter on such sequences.

6 Evaluation

We compare our fixed-length belief propagation model with a random walk in the automaton. The purpose of this experiment is to show that a random walk in the automaton does not sample sequences correctly, and confirm empirically that our belief propagation-based model is correct, with a limited time penalty.

We used each method to sample sequences of words based on Pushkin's *Eugene Onegin*, of length 8, of Markov order 1 and with a max order less than 4, imposed using a max order automaton [19]. We implemented our experiments in Oracle Java 7, and ran them on an iMac with a 3.4GHz Intel Core i7 CPU, and 16GB RAM. The automaton was computed in about 200ms. For the random walk method, we imposed the length by rejecting shorter sequences. In total, we sampled over 20 million sequences. Of those, 5 million were unique sequences. The baseline random walk procedure generated an average of 5500 sequences per second (counting only non-rejected sequences), while the belief propagation-based method generated an average of 3500 sequences per second. For comparison, our REGULAR-based CP model produced only about 50 sequences per second. We filtered those that were generated over 50 times, of which there were about 47000 with random walk, and about 35000 with belief propagation. We estimated the probability of a sequence by computing the sum of the probability of all unique sequences found by either method, and use this for normalising.

We plot our results on Figure 3. Each point on either graph corresponds to a sequence. Its value on the x-axis is its probability, estimated as described previously, while the values on the y-axis is the empirical probability, i.e. the frequency at which the specific sequence has been sampled compared to the total number of sequences. Figure 3(a) shows that the baseline sampling approach performs poorly: many sequences, even of similar probability, are over or under-represented. On the other hand, Figure 3(b) provides a striking empirical confirmation of the correctness of the belief propagation model.

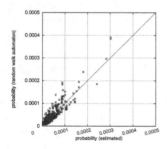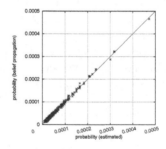

(a) Random walk in the automaton   (b) Belief propagation

Fig. 3. Sampling with random walk in the automaton compared to belief propagation.

7   Conclusion

We defined a belief propagation model for sampling Markov sequences that are accepted by a given automaton. To this aim, we introduced a tree-structured factor graph, on which belief propagation is polynomial and exact. This factor graph uses binary factors, which encode a type of conjunction between the underlying Markov model and the given automaton. We showed that this procedure allows us to sample sequences faster than equivalent CP models, and demonstrated that such sequences are sampled with their exact probabilities.

This result can be used for sequence generation problems in which users want a set of solutions that are both probable in a given statistical model, and satisfy hard regular constraints. More generally, we believe that this approach offers an interesting bridge between statistical inference and constraint satisfaction.

Acknowledgments. This research is conducted within the Flow Machines project funded by the ERC under the European Unions 7th Framework Programme (FP/2007-2013) / ERC Grant Agreement n. 291156. We thank Ricardo de Aldama, funded by the Lrn2Cre8 project (FET grant agreement no. 610859) for discussions about sampling and CSP.

References

1. Proceedings, The Twenty-First National Conference on Artificial Intelligence and the Eighteenth Innovative Applications of Artificial Intelligence Conference, July 16–20, 2006, Boston, Massachusetts, USA. AAAI Press (2006)
2. Barbieri, G., Pachet, F., Roy, P., Esposti, M.D.: Markov constraints for generating lyrics with style. In: Raedt, L.D., Bessière, C., Dubois, D., Doherty, P., Frasconi, P., Heintz, F., Lucas, P.J.F. (eds.) Frontiers in Artificial Intelligence and Applications, ECAI, vol. 242, pp. 115–120. IOS Press (2012)
3. Beeri, C., Fagin, R., Maier, D., Yannakakis, M.: On the Desirability of Acyclic Database Schemes. J. ACM **30**(3), 479–513 (1983)
4. Bimbot, F., Deruty, E., Sargent, G., Vincent, E.: Semiotic structure labeling of music pieces: Concepts, methods and annotation conventions. In: Gouyon, F., Herrera, P., Martins, L.G., Müller, M. (eds.) Proceedings of the 13th International Society for Music Information Retrieval Conference, ISMIR 2012, Mosteiro S.Bento Da Vitória, Porto, Portugal, October 8–12, 2012, pp. 235–240. FEUP Edições (2012). http://ismir2012.ismir.net/event/papers/235-ismir-2012.pdf
5. Brooks, F.P., Hopkins, A., Neumann, P.G., Wright, W.: An experiment in musical composition. IRE Transactions on Electronic Computers **6**(3), 175–182 (1957)
6. Dechter, R., Pearl, J.: Tree Clustering for Constraint Networks. Artif. Intell. **38**(3), 353–366 (1989)
7. Domingos, P.M., Kok, S., Poon, H., Richardson, M., Singla, P.: Unifying logical and statistical AI. In: Proceedings, The Twenty-First National Conference on Artificial Intelligence and the Eighteenth Innovative Applications of Artificial Intelligence Conference, July 16–20, 2006, Boston, Massachusetts, USA [1], pp. 2–9. http://www.aaai.org/Library/AAAI/2006/aaai06-001.php
8. Ermon, S., Gomes, C.P., Selman, B.: Uniform solution sampling using a constraint solver as an oracle. In: de Freitas, N., Murphy, K.P. (eds.) Proceedings of the Twenty-Eighth Conference on Uncertainty in Artificial Intelligence, Catalina Island, CA, USA, August 14–18, 2012, pp. 255–264. AUAI Press (2012). https://dslpitt.org/uai/displayArticleDetails.jsp?mmnu=1&smnu=2&article_id=2288&proceeding_id=28
9. Freuder, E.C.: A Sufficient Condition for Backtrack-Free Search. J. ACM **29**(1), 24–32 (1982)
10. Gogate, V., Dechter, R.: Studies in solution sampling. In: Fox, D., Gomes, C.P. (eds.) Proceedings of the Twenty-Third AAAI Conference on Artificial Intelligence, AAAI 2008, Chicago, Illinois, USA, July 13–17, 2008, pp. 271–276. AAAI Press (2008). http://www.aaai.org/Library/AAAI/2008/aaai08-043.php
11. Hebrard, E., Hnich, B., O'Sullivan, B., Walsh, T.: Finding diverse and similar solutions in constraint programming. In: Veloso, M.M., Kambhampati, S. (eds.) Proceedings, The Twentieth National Conference on Artificial Intelligence and the Seventeenth Innovative Applications of Artificial Intelligence Conference, July 9–13, 2005, Pittsburgh, Pennsylvania, USA, pp. 372–377. AAAI Press / The MIT Press (2005). http://www.aaai.org/Library/AAAI/2005/aaai05-059.php

12. Mezard, M., Montanari, A.: Information, physics, and computation. Oxford University Press (2009)
13. Nadel, A.: Generating diverse solutions in SAT. In: Sakallah, K.A., Simon, L. (eds.) SAT 2011. LNCS, vol. 6695, pp. 287–301. Springer, Heidelberg (2011)
14. Nierhaus, G.: Algorithmic composition: paradigms of automated music generation. Springer (2009)
15. Pachet, F., Roy, P.: Markov constraints: steerable generation of markov sequences. Constraints 16(2), 148–172 (2011)
16. Pachet, F., Roy, P., Barbieri, G.: Finite-length markov processes with constraints. In: Walsh, T. (ed.) IJCAI, pp. 635–642. IJCAI/AAAI (2011)
17. Pachet, F., Roy, P.: Imitative leadsheet generation with user constraints. In: Schaub, T., Friedrich, G., O'Sullivan, B. (eds.) ECAI 2014–21st European Conference on Artificial Intelligence, 18–22 August 2014, Prague, Czech Republic - Including Prestigious Applications of Intelligent Systems (PAIS 2014). Frontiers in Artificial Intelligence and Applications, vol. 263, pp. 1077–1078. IOS Press (2014). http://dx.doi.org/10.3233/978-1-61499-419-0-1077
18. Papadopoulos, A., Pachet, F., Roy, P.: Generating non-plagiaristic markov sequences with max order sampling. In: Degli Esposti, M., Altmann, E., Pachet, F. (eds.) Universality and Creativity in Language (forthcoming). Lecture Notes in Morphogenesis. Springer (2015)
19. Papadopoulos, A., Roy, P., Pachet, F.: Avoiding plagiarism in markov sequence generation. In: Brodley, C.E., Stone, P. (eds.) Proceedings of the Twenty-Eighth AAAI Conference on Artificial Intelligence, July 27–31, 2014, Québec City, Québec, Canada, pp. 2731–2737. AAAI Press (2014). http://www.aaai.org/ocs/index.php/AAAI/AAAI14/paper/view/8574
20. Pearl, J.: Reverend bayes on inference engines: A distributed hierarchical approach. In: Waltz, D.L. (ed.) Proceedings of the National Conference on Artificial Intelligence. Pittsburgh, PA, August 18–20, 1982, pp. 133–136. AAAI Press (1982). http://www.aaai.org/Library/AAAI/1982/aaai82-032.php
21. Pearl, J.: Probabilistic reasoning in intelligent systems - networks of plausible inference. Morgan Kaufmann series in representation and reasoning. Morgan Kaufmann (1989)
22. Pesant, G.: A regular language membership constraint for finite sequences of variables. In: Wallace, M. (ed.) CP 2004. LNCS, vol. 3258, pp. 482–495. Springer, Heidelberg (2004)
23. Pinkerton, R.C.: Information theory and melody. Scientific American (1956)
24. Poon, H., Domingos, P.M.: Sound and efficient inference with probabilistic and deterministic dependencies. In: Proceedings, The Twenty-First National Conference on Artificial Intelligence and the Eighteenth Innovative Applications of Artificial Intelligence Conference, July 16–20, 2006, Boston, Massachusetts, USA [1], pp. 458–463. http://www.aaai.org/Library/AAAI/2006/aaai06-073.php
25. Richardson, M., Domingos, P.: Markov logic networks. Machine Learning 62(1–2), 107–136 (2006). http://dx.doi.org/10.1007/s10994-006-5833-1
26. Roy, P., Pachet, F.: Enforcing meter in finite-length markov sequences. In: desJardins, M., Littman, M.L. (eds.) Proceedings of the Twenty-Seventh AAAI Conference on Artificial Intelligence, July 14–18, 2013, Bellevue, Washington, USA. AAAI Press (2013). http://www.aaai.org/ocs/index.php/AAAI/AAAI13/paper/view/6422
27. Schreiber, Y.: Value-ordering heuristics: search performance vs. solution diversity. In: Cohen, D. (ed.) CP 2010. LNCS, vol. 6308, pp. 429–444. Springer, Heidelberg (2010)

Randomness as a Constraint

Steven D. Prestwich[1]([✉]), Roberto Rossi[2], and S. Armagan Tarim[3]

[1] Insight Centre for Data Analytics, University College Cork, Cork, Ireland
s.prestwitch@cs.ucc.ie
[2] University of Edinburgh Business School, Edinburgh, UK
[3] Department of Management, Cankaya University, Ankara, Turkey

Abstract. Some optimisation problems require a random-looking solution with no apparent patterns, for reasons of fairness, anonymity, undetectability or unpredictability. Randomised search is not a good general approach because problem constraints and objective functions may lead to solutions that are far from random. We propose a constraint-based approach to finding pseudo-random solutions, inspired by the Kolmogorov complexity definition of randomness and by data compression methods. Our "entropy constraints" can be implemented in constraint programming systems using well-known global constraints. We apply them to a problem from experimental psychology and to a factory inspection problem.

1 Introduction

For some applications we require a list of numbers, or some other data structure, that is (or appears to be) *random*, while also satisfying certain constraints. Examples include the design of randomised experiments to avoid statistical bias [13], the generation of random phylogenetic trees [14], quasirandom (low discrepancy) sequences for efficient numerical integration and global optimisation [24], randomised lists without repetition for use in experimental psychology [10], random programs for compiler verification [6], and the random scheduling of inspections for the sake of unpredictability [30].

An obvious approach to obtaining a random-looking solution is simply to use a randomised search strategy, such as stochastic local search or backtrack search with a randomised value ordering. In some cases this works, for example it can generate a random permutation of a list, but in general there are several drawbacks with the randomised search approach:

- If only random-looking solutions are acceptable then the constraint model is not correct, as it permits solutions that are unacceptable. The correctness of a constraint model should be independent of the search strategy used to solve it.
- Randomised search can not prove that a random-looking solution does not exist, or prove that a solution is as random-looking as possible.
- Unless the randomised search is carefully designed (see [9] for example) it is likely to make a biased choice of solution.

G. Pesant (Ed.): CP 2015, LNCS 9255, pp. 351–366, 2015.
DOI: 10.1007/978-3-319-23219-5_25

– Even if we sample solutions in an unbiased way, there is no guarantee that such a solution will look random. Although randomly sampled unconstrained sequences are almost certain to appear random (almost all long sequences have high *algorithmic entropy* [2]) a constrained problem might have mostly regular-looking solutions. Similarly, optimising an objective function might lead to regular-looking solutions. In Section 3.2 we give examples of both phenomena.

Instead it would be useful to have available a constraint `israndom(`v`)` that forces a vector v of variables to be random, which could simply be added to a constraint model. First, however, we must define what we mean by *random*.

In information theory, randomness is a property of the data source used to generate a data sequence, not of a single sequence. The *Shannon entropy* of the source can be computed from its symbol probabilities using Shannon's well-known formula [32]. But in algorithmic information theory, randomness can be viewed as a property of a specific data sequence, and its *Kolmogorov complexity*, or *algorithmic entropy*, is defined as the length of the smallest algorithm that can describe it. For example the sequence 1111111111 may have the same probability of occurring as 1010110001 but it has lower algorithmic entropy because it can be described more simply (write 1 ten times). Algorithmic entropy formally captures the intuitive notion of whether a list of numbers "looks random", making it useful for our purposes. We shall refer to algorithmic entropy simply as *entropy* by a slight abuse of language.

Having chosen (algorithmic) entropy as our measure of randomness, we would like to have available a constraint of the form `entropy(`v, e`)` to ensure that the entropy of v is at least e. Unfortunately, defining such a constraint is impossible because algorithmic entropy is uncomputable [5]. Instead we take a pragmatic approach by defining constraints that eliminate patterns exploited by well-known data compression algorithms, which can be combined as needed for specific applications. There is a close relationship between algorithmic entropy and compressibility: applying a compression algorithm to a sequence of numbers, and measuring the length of the compressed sequence, gives an upper bound on the algorithmic entropy of the original sequence. Thus by excluding readily-compressible solutions we hope to exclude low-entropy (non-random) solutions.

The paper is organised as follows. Section 2 presents constraint-based approaches to limiting the search to high-entropy solutions. Section 3 applies these ideas to two problems. Section 4 discusses related work. Section 5 concludes the paper and discusses future work.

2   Entropy Constraints

We require constraints to exclude low-entropy solutions, which we shall call *entropy constraints*. This raises several practical questions: what types of pattern can be excluded by constraints, how the constraints can be implemented, and what filtering algorithms are available? We address these problems below. In our experiments we use the Eclipse constraint logic programming system [1].

2.1   Non-uniform Distributions

Data in which some symbols occur more often than others have non-uniform probability distributions. Huffman coding [15] and arithmetic coding [29] are compression methods that exploit this feature by encoding symbols with a variable number of bits (a *prefix code* with many bits for rare symbols and few bits for common symbols).

To eliminate this type of pattern from our solutions we define some simple entropy constraints. We assume that all variables have the same domain $\{0, \ldots, m-1\}$ but it is easy to generalise to different domains. Given lower bounds $\boldsymbol{\ell} = \langle \ell_0, \ldots, \ell_{m-1} \rangle$ and upper bounds $\boldsymbol{u} = \langle u_0, \ldots, u_{m-1} \rangle$ on the frequencies of symbols $0, \ldots, m-1$ we define a *frequency entropy constraint* $\texttt{freq}(\boldsymbol{v}, m, \boldsymbol{\ell}, \boldsymbol{u})$. It can be directly implemented by a global cardinality constraint [27] $\texttt{GCC}(\boldsymbol{v}, \langle 0, \ldots, m-1 \rangle, \boldsymbol{\ell}, \boldsymbol{u})$.

To see the effect of these constraints, consider a CSP with 100 finite domain variables $v_0, \ldots, v_{99} \in \{0, 1, \ldots, 9\}$ and no problem constraints. We use the default backtrack search heuristic to find the *lex-least solution* which we take as a proxy for the lowest-entropy solution. The lex-least solution is the solution that is least under the obvious lexicographical ordering (order by increasing value of the first variable, breaking ties using the second variable and so on, under the static variable ordering $v_1, v_2, \ldots$). If even the lex-least solution has high entropy then we believe that other solutions will too. To test this we applied randomised search to a problem with entropy constraints, and never obtained a solution that could be compressed further than the lex-least solution.

If we add $\texttt{freq}(\boldsymbol{v}, 10, \boldsymbol{0}, \boldsymbol{14})$[1] to this problem we obtain the lex-least solution

0000000000000001111111111111112222222222222233333333
333333444444444444455555555555555666666666666677

We shall estimate the entropy of this lex-least solution by compressing it using the well-known gzip[2] compression algorithm. Though gzip does not necessarily give an accurate entropy estimate it has been successfully used for this purpose (see for example [4]). Denoting the entropy of solutions by ϵ and measuring it in bytes, the lex-least solution has $\epsilon = 43$. For a random sequence of digits in the range 0–9 we typically obtain ϵ values of 80–83 so the lex-least solution is far from random, as can be observed. With no entropy constraints the lex-least solution is simply 0 repeated 100 times, which gzip compresses from 100 bytes to 26 bytes. There is no paradox in compressing these *random* numbers: integers in the range 0–9 can be represented by fewer than 8 bits (1 byte). Note that gzip compresses an empty text file to 22 bytes, so most of the 26 bytes is decompression metadata.

2.2   Repeated Strings

For the compression of discrete data probably the best-known methods are based on *adaptive dictionaries*. These underlie compression algorithms such as Linux

---

[1] We use $\boldsymbol{i}$ with integer i to denote a vector $\langle i, \ldots, i \rangle$ of appropriate length.
[2] http://www.gzip.org/

`compress`, V.24 bis, GIF, PKZip, Zip, LHarc, PNG, gzip and ARJ [31] and use algorithms described in [34–36]. These methods detect repeated k-grams (blocks of k adjacent symbols) and replace them by pointers to dictionary entries. For example the string 011101011100 contains two occurrences of the 5-gram 01110, which can be stored in a dictionary and both its occurrences replaced by a pointer to that entry.

We could generalise the `freq` constraint to limit the number of occurrences of every possible k-gram, but as there are m^k of them this is impractical unless k is small. A more scalable approach is as follows. Given an integer $k \geq 2$ and an upper bound t on the number of occurrences of all k-grams over symbols $\{0, \ldots, m-1\}$ in a vector v of variables, we define a constraint $\mathtt{dict}(v, m, k, t)$. We shall call this a *dictionary entropy constraint*. It can be implemented by the $\mathtt{Multi\text{-}Inter\text{-}Distance}(x, t, p)$ global constraint [20] with $p = n - k + 1$ (the number of x variables), $x = \langle x_0, \ldots, x_{n-k} \rangle$, and the $x_i = \sum_{j=0}^{k-1} m^j v_{i+j}$ are auxiliary integer variables representing k-grams, where n is the sequence length. This global constraint enforces an upper bound t on the number of occurrences of each value within any p consecutive x variables. In the special case $t = 1$ we can instead use $\mathtt{alldifferent}(x)$ [28].

To test this idea on the artificial problem of Section 2.1 we add $\mathtt{dict}(v, 10, k, 1)$ dictionary constraints for various k-values (but not the frequency constraints of Section 2.1). The results in Table 1 show that as we reduce k the solution contains fewer obvious patterns, and $k = 2$ gives a solution that (to gzip) is indistinguishable from a purely random sequence, as 80 bytes is within the range of ϵ for random sequences of this form.

Table 1. Lex-least solutions with dictionary entropy constraints

k	lex-least solution	ϵ
50	00000000000000000000000000000000000000000000000001	
	00000000000000000000000000000000000000000000000000	29
25	00000000000000000000000001000000000000000000000000	
	20000000000000000000000003000000000000000000000000	35
12	00000000000010000000000002000000000000300000000000040	
	00000000000500000000000006000000000000700000000000008000	48
6	00000010000020000030000040000050000060000070000080	
	00009000011000012000013000014000015000016000017000	60
3	000100200300400500600700800901101201301401501601701	
	180190210220230240250260270280290310320330340350360	71
2	00102030405060708091121314151617181922324252627282	
	93343536373839445464748495565758596676869778798890	80

In this example there are sufficient digrams (2-grams) to avoid any repetition, but for smaller m this will not be true, and in the general case we might need to use larger t or larger k. Note that we must choose $t_k \geq \lceil n/m^k \rceil$ otherwise the problem will be unsatisfiable.

2.3 Correlated Sources

Though gzip and related compression algorithms often do a very good job, they are not designed to detect all patterns. A solution with no repeated k-grams might nevertheless have low entropy and noticeable patterns. For example the following sequence of 100 integers in the range 0–9 compresses to 80 bytes, and is therefore indistinguishable from a random sequence to gzip:

0123456789024681357903691472580481592637051627384994837261507362951840852741963097531864209876543210

Yet it was written by hand following a simple pattern and is certainly not a random sequence, as becomes apparent if we examine the differences between adjacent symbols:

1 1 1 1 1 1 1 1 1 -9 2 2 2 2 -7 2 2 2 2 -9 3 3 3 -8 3 3 -5 3 3 -8 4 4 -7 4 4 -7 4
-3 4 -7 5 -4 5 -4 5 -4 5 -4 5 0 -5 4 -5 4 -5 4 -5 4 -5 7 -4 3 -4 7 -4 -4 7 -4 -4 8
-3 -3 5 -3 -3 8 -3 -3 -3 9 -2 -2 -2 -2 7 -2 -2 -2 -2 9 -1 -1 -1 -1 -1 -1 -1 -1 -1

The same differences often occur together but gzip is not designed to detect this type of pattern. As another example, the high-entropy ($k = 2$) solution found in Section 2.2 has differences

jjkilhmgnfoepdqcrbsbjjkilhmgnfoepdqcrcjjkilhmgnfoe
pdqdjjkilhmgnfoepejjkilhmgnfofjjkilhmgngjjkilhmhj

where differences $-9 \ldots +9$ are represented by symbols a–s. The differences also look quite random: gzip compresses this list of symbols to 92 bytes which is typical of a random sequence of 99 symbols from a–s. Yet they have a non-uniform distribution: a does not occur at all while j occurs 15 times.

In data compression an example of a *correlated source* of data is one in which each symbol depends probabilistically on its predecessor. This pattern is exploited in speech compression methods such as DPCM [7] and its variants. Another application is in lossless image compression, where it is likely that some regions of an image contain similar pixel values. This is exploited in the JPEG lossless compression standard [33], which predicts the value of a pixel by considering the values of its neighbours. Greater compression can sometimes be achieved by compressing the differences between adjacent samples instead of the samples themselves. This is called *differential encoding*.

We can confound differential compressors by defining new variables $v_i^{(1)} = v_i - v_{i+1} + m - 1$ with domains $\{0, \ldots, 2(m-1)\}$ to represent the differences between adjacent solution variables (shifted to obtain non-negative values), and applying entropy constraints from Sections 2.1 and 2.2 to the $v_i^{(1)}$. We shall call these *differential [frequency, dictionary] entropy constraints*. We use the notation $v^{(1)}$ because later on we shall consider differences of differences $v^{(2)}$ and so on.

Adding a differential frequency constraint $\mathtt{freq}(v^{(1)}, 18, 0, 10)$ and a differential dictionary constraint $\mathtt{dict}(v^{(1)}, 18, 3, 1)$ to the earlier constraints we get differences

jkilhmgnfoepdqcrbsbjkjilikhmhlgngmfofnepeodqdpcrcq
djkkhliimgoenfpejlhnfmhjklglgofmiiingjmhkhkjljijk

which has $\epsilon = 90$, and lex-least solution

> 00102030405060708091122132314241525162617271828192
> 93345353463837394464847556857495876596697867799889

which has $\epsilon = 80$: both ϵ values indicate totally random sequences of the respective symbol sets. However, in some ways this solution still does not look very random: its initial values are 0, and roughly the first third of the sequence has a rather regular pattern. This is caused by our taking the lex-least solution, and by there being no problem constraints to complicate matters. In such cases we could use a SPREAD-style constraint [22] to prevent too many small values from occurring at the start of the sequence, or perhaps resort to randomised search. Note that on this artificial example randomised search usually finds high-entropy solutions even without entropy constraints, but it is not *guaranteed* to do so.

2.4   Using Entropy Constraints

By expressing the randomness condition as constraints we ensure that *all* solutions are incompressible by construction. Therefore the search method used to find the sequences does not matter and we can use any convenient and efficient search algorithm, such as backtrack search (pruned by constraint programming or mathematical programming methods) or metaheuristics (such as tabu search, simulated annealing or a genetic algorithm). As long as the method is able to find a solution it does not matter if the search is biased, unless we require several evenly distributed solutions. But in the latter case we could define a new problem $\mathcal{P}'$ whose solution is a set of solutions to the original problem $\mathcal{P}$, with constraints ensuring that the $\mathcal{P}$-solutions are sufficiently distinct.

All our entropy constraints are of only two types: freq and dict. Both can be implemented via well-known Constraint Programming global constraints, or in integer linear programs via reified binary variables, or in SAT via suitable clauses. (In our experiments of Section 3 we use the second method because our constraint solver does not provide the necessary global constraints.) We can relate them by a few properties (proofs omitted):

$$\mathtt{dict}(\boldsymbol{v}, m, k, t) \Rightarrow \mathtt{dict}(\boldsymbol{v}, m, k+1, t)$$
$$\mathtt{dict}(\boldsymbol{v}^{(i)}, m, k, t) \Rightarrow \mathtt{dict}(\boldsymbol{v}^{(i-1)}, m, k+1, t)$$
$$\mathtt{freq}(\boldsymbol{v}^{(i)}, m, 0, t) \Rightarrow \mathtt{dict}(\boldsymbol{v}^{(i-1)}, m, 2, t)$$

From these we can deduce

$$\mathtt{freq}(\boldsymbol{v}^{(i)}, m, \boldsymbol{0}, t) \Rightarrow \mathtt{dict}(\boldsymbol{v}, m, i+1, t)$$

which provides an alternative way of limiting k-gram occurrences. But higher-order differential constraints should be used with caution. For example we could use $\mathtt{freq}(\boldsymbol{v}^{(2)}, m, \boldsymbol{0}, \boldsymbol{1})$ instead of $\mathtt{dict}(\boldsymbol{v}, m, 3, 1)$ as both prevent the trigram 125 from occurring more than once. But the former is stronger as it also prevents trigrams 125 and 668 from both occurring: the trigrams have the order-1 differences (1,3) and (0,2) respectively, hence the same order-2 difference (2). If we do not consider 125 and 668 to be similar in any relevant way for our application then using the freq constraint is unnecessarily restrictive.

3   Applications

We consider two applications. Section 3.1 describes a known problem from experimental psychology, and Section 3.2 describes an artificial factory inspection problem where the inspection schedule must be unpredictable.

3.1   Experimental Psychology

Experimental psychologists often need to generate randomised lists under constraints [10]. An example of such an application is word segmentation studies with a continuous speech stream. The problem discussed in [10] has a multiset W of 45 As, 45 Bs, 90 Cs and 90 Ds, and no two adjacent symbols can be identical. There is an additional constraint: that the number of CD digrams must be equal to the number of As. From this multiset must be generated a randomised list.

Generating such a list was long thought to be a simple task and a standard list randomisation algorithm was used: randomly draw an item from W and add it to the list, unless it is identical to the previous list item in which case replace it and randomly draw another item; halt when W is empty. But it was shown in [10] that this algorithm can lead to a large bias when the frequencies of the items are different, as in the problem considered. The bias is that the less-frequent items A and B appear too often early in the list, and not often enough later in the list. The bias effect is particularly bad for short lists generated by such randomisation-without-replacement methods. The bias can ruin the results of an experiment by confounding frequency effects with primacy or recency effects.

The solution proposed by [10] is to create *transition frequency* and *transition probability* tables, and to use these tables to guide sequence generation (we do not give details here). This is therefore a solved problem, but the authors state that *the correct ... table corresponding to the constraints of a given problem can be notoriously hard to construct*, and it would be harder to extend their method to problems with more complex constraints.

Generating such lists is quite easy using our method. For the above example we create a CSP with 270 variables v_i each with domain $\{0, 1, 2, 3\}$ representing A, B, C and D respectively. To ensure the correct ratio of items we use frequency constraints $\texttt{freq}(v, 4, \langle 45, 45, 90, 90 \rangle, \langle 45, 45, 90, 90 \rangle)$. We add a constraint to ensure that there are 45 CD digrams. To ensure a reasonably even spread of values we add constraints $\texttt{freq}(v_i, 4, \mathbf{0}, \langle 11, 11, 22, 22 \rangle)$ to each fifth $v_1, \ldots, v_5$ of v. Finally we add constraints $\texttt{dict}(v_i, 4, 5, 1)$ to each fifth.

Backtrack search turned out to be very slow so we use a local search algorithm, implemented in the constraint solver to take advantage of its filtering algorithms. The algorithm is not intended to be particularly competitive or interesting, but we describe it here for the sake of reproducibility. In an initial iteration it attempts to assign a value to each variable in turn using the ordering $v_1, v_2, \ldots$, and halts when domain wipeout occurs. We refer to this initial choice of value for each variable the *default*. In the next iteration it makes a few random assignment choices, but in most cases it choose the default. If any choice leads to domain wipeout then all alternatives are tried for that variable (but backtracking to other variables does not occur). If the new set of choices assign no fewer variables than the defaults then they become the new defaults. If all variables are assigned in an iteration then the algorithm has found a feasible solution and it halts.

The solution we found after a few tens of seconds is shown in Figure 1, with $\epsilon = 118$. In further tests local search achieved ϵ in the range 114–124, whereas without entropy constraints it achieved ϵ of 100–115.

```
CBDCDCADCBDBCBCABCDCBDCDBCABADCBCDCABDCBCADCDCDCACADAD
CDADCBDABDCDCBCDCDCDABCABDCDBCDBADCABDBDCACDCACDBCACDC
ADCDCACDCBDACDCADCDCBDCABCDBDBDACACDCDCDBDCACDBCDCADBD
CDCDADCACDACDCDCDCBDCDBDBDBDCDCACDBDBCDBCADCADADCDCADC
BCDBCDCDACACBCDCBCDBDCDCDCBCADCDCADCBDADCADACABDADBDAD
```

Fig. 1. A solution to the experimental psychology problem

[10] noted a systematic bias in the A:C ratio between fifths of the sequence: approximately 0.59, 0.58, 0.53, 0.49, 0.32. In ours we have 0.47, 0.50, 0.47, 0.44 and 0.59. Our approach has much in common with the Mix system used by psychologists but also important differences (see the discussion in Section 4).

3.2   Factory Inspection

Next we consider an artificial factory inspection problem. Suppose an inspector must schedule 20 factory inspections over 200 days. We represent this by finite domain variables $v_1, \ldots, v_{20} \in \{1, \ldots, 200\}$. The inspection plan should appear as random as possible so that the owners can not predict when they will be inspected.

To make the problem more realistic we could add constraints preventing different factories from being inspected on certain days, or introduce an objective function. We do this below, but first we shall simply restrict all inspections to certain available days which are unevenly distributed through the year: 1–40, 70–100, 130–150 and 180–190. We order the v_i in strictly ascending order using a global constraint $\mathtt{ordered}(v)$.

The lex-least solution is

  1 2 3 4 5 6 7 8 9 10 11 12 13 14 15 16 17 18 19 20

with differences

  1 1 1 1 1 1 1 1 1 1 1 1 1 1 1 1 1 1 1

All the inspections take place in the first 20 days, as is clearer if we visualise the schedule using two rows of 100 days each, the dark colour indicating scheduled inspections, the light colour indicating days in which an inspection could be scheduled, and white indicating days in which no inspection could be scheduled:

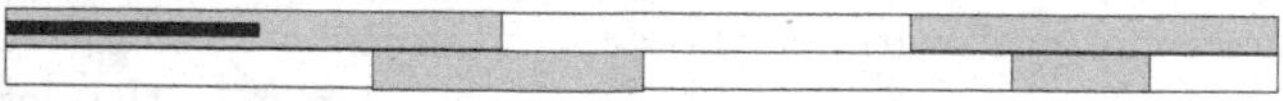

We could simply randomise the value ordering in each variable assignment during backtrack search, but this is unlikely to find the highest-entropy solutions. A typical solution found in this way is

    73 139 142 144 146 147 148 149 150 180
    181 182 183 184 185 186 187 188 189 190

with differences

    66 3 2 2 1 1 1 1 30 1 1 1 1 1 1 1 1 1 1

and the schedule:

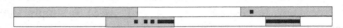

This solution also looks non-random. Because the variables are ordered, choosing each value randomly in turn causes clustering in the high values. We could randomise the search in a more clever way, for example by biasing earlier assignments to lower values, or branching under a different variable ordering. For such a simple problem this would not be hard to do, but the more complicated the constraint network is the harder this task becomes. Stochastic local search might typically find a high-entropy solution but, as pointed out in Section 1, randomised search alone is not enough so we shall apply entropy constraints.

As the v_i are all different we use difference variables. Applying entropy constraints $\texttt{freq}(v^{(1)}, 198, \mathbf{0}, \mathbf{2})$ and $\texttt{freq}(v^{(2)}, 395, \mathbf{0}, \mathbf{1})$ gives the lex-least solution

    1 2 3 5 9 11 16 20 28 31 39 70 73 82 87 99 130 136 150 180

and the schedule

This is much more random-looking, indicating that an appropriate use of entropy constraints can yield higher-entropy solutions than random search, as mentioned in Section 1.

To confirm our intuition that this solution is sufficiently random we apply gzip. However, we do not compress the solutions in integer form as it contains spaces between numbers and uses multiple characters to represent integers: this helps to hide patterns from gzip so that its entropy estimate is poor. Instead we compress the binary schedule representations (as used above) of:

  (i) the lex-least solution without entropy constraints
  (ii) a randomised backtrack search solution
  (iii) the lex-least solution with entropy constraints
  (iv) mean results for random binary sequences of length 200 containing exactly 20 ones, without the restriction to certain ranges of days

We also estimate their *Approximate Entropy* (ApEn), a measure of the regularity and predictability of a sequence of numbers. ApEn was originally developed to analyse medical data [23] but has since found many applications. We use the definition given in [2] for a sequence n symbols from an alphabet of size m:

$$\text{ApEn}(k) = \begin{cases} H(k) - H(k-1) & \text{if } k > 1 \\ H(1) & \text{if } k = 1 \end{cases}$$

where

$$H(k) = -\sum_{i=1}^{m^k} p_i \log_2 p_i$$

and p_i is the probability of k-gram i occurring in the sequence, estimated as its observed frequency f_i divided by $n - k + 1$. $H(1)$ is simply the Shannon entropy measured in bits, and $\text{ApEn}(k)$ is a measure of the entropy of a block of k symbols conditional on knowing the preceding block of $k-1$ symbols. ApEn is useful for estimating the regularity of sequences, it can be applied to quite short sequences, and it often suffices to check up to $\text{ApEn}(3)$ to detect regularity [2].

Results are shown in Table 2. The compression and ApEn results for (iii) are almost as good as those of (iv). This indicates not only that the inspections are unpredictable by the factory owners, but that the owners can not even detect in hindsight (by gzip and ApEn) the fact that the inspector had time constraints. Note that 0.47 is the theoretical $\text{ApEn}(k)$ for all $k \geq 1$, for a binary source with probabilities 0.1 and 0.9 as in this example.

Table 2. Inspection schedule entropies

solution	ϵ	ApEn(k)		
		$k = 1$	$k = 2$	$k = 3$
(i)	29	0.47	0.03	0.03
(ii)	39	0.47	0.28	0.24
(iii)	54	0.47	0.46	0.42
(iv)	56	0.47	0.46	0.45

We apply a further statistical test of randomness to the binary representation: the *Wald-Wolfowitz runs test*. A *run* is an unbroken substring of 0s (or 1s) with preceding and subsequent 1s (or 0s) except at the start and end of the string. For example the string 01100010 contains five runs: 0, 11, 000, 1 and 0. A randomly chosen string is unlikely to have a very low or very high number of runs, and this can be used as a test of randomness. For a random binary sequence with 180 zeroes and 20 ones, we can be 95% confident that the sequence is random if there are between 33 and 41 runs. The lex-least solution has 36 runs so it passes the test. Thus the lex-least solution passes several tests of randomness. (One might ask why we did not also generate constraints to enforce ApEn and the runs test: we discuss this point in Section 5.)

This example is perhaps too simple as it contains only one constraint and no objective function. In fact if we replace the **ordered** constraint by an `alldifferent` constraint then local search without entropy constraints finds a high-entropy solution with $\epsilon = 55$, ApEn(1)=0.47, ApEn(2)=0.45 and ApEn(3)=0.45 — though backtrack search with randomised value ordering only achieves $\epsilon = 48$, ApEn(1)=0.47, ApEn(2)=0.46 and ApEn(3)=0.35. We should not have to manipulate the constraint model or choose the right solver to find high-entropy solutions, and there might be a good reason for using `ordered` instead of `alldifferent` (for example it breaks permutation symmetry), but to make the problem more interesting we now consider two alternative complications.

Firstly, suppose that we have some additional constraints:

- no two inspections may occur on consecutive days: strengthen the `ordered` constraint to $v_{i+1} > v_i + 1$ $(i = 1 \ldots 19)$
- there must be at least 13 inspections in the first 100 days: $v_{13} \leq 100$
- there must be at least 2 inspections in the fourth time period: $v_{19} \geq 180$

Without entropy constraints local search now finds a lower-entropy solution with $\epsilon = 44$, ApEn(1)=0.47, ApEn(2)=0.46 and ApEn(3)=0.31, but with entropy constraints it finds this schedule:

with high entropy: $\epsilon = 57$, ApEn(1)=0.49, ApEn(2)=0.47 and ApEn(3)=0.47.

Secondly, instead of additional constraints suppose we have an objective function. Consider a scenario in which the factories are all in country A, the inspector lives in a distant country B, and flights between A and B are expensive. We might aim to minimise the number of flights while still preserving the illusion of randomness. To do this we could maximise the number of inspections on consecutive days. If we require an objective value of at least 10 then we are unable to find a solution under the above entropy constraints. We are forced to use less restrictive entropy constraints such as $\texttt{freq}(v^{(1)}, 198, 0, 10)$ and $\texttt{dict}(v^{(1)}, m, 2, 1)$ yielding the following schedule by local search:

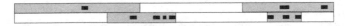

with 10 nights in a hotel, $\epsilon = 50$, ApEn(1)=0.47, ApEn(2)=0.38 and ApEn(3)=0.32. This illustrates the fact that imposing additional criteria might force us to accept lower-entropy solutions.

4   Related Work

[30] proposed *statistical constraints* to enforce certain types of randomness on a solution: solutions should pass statistical tests such as the t-test or Kolmogorov-Smirnov. An inspection plan found in using the latter test is shown in Figure 2. If this is the only test that might be applied by an observer then we are done. However, the schedule exhibits a visible pattern that could be used to predict

the next inspection with reasonable certainty. The pattern is caused by the deterministic search strategy used. It might be important to find an inspection schedule that is not predictable by visual inspection, or by a machine learning algorithm. In this case statistical tests are not enough and we must also enforce randomness.

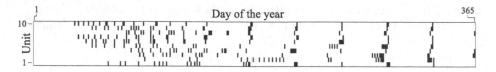

Fig. 2. A partially-random inspection plan that passes a statistical test

In the field of hardware verification SAT solvers have been induced to generate random stimuli: see for example [16] for a survey of methods such as adding randomly-chosen XOR constraints. Constraint solvers have also been used for the same purpose, a recent example being [19]. These approaches aim to generate an unbiased *set* of solutions, as do the methods of [6,8,9,11,12], whereas we aim to maximise the algorithmic entropy of a *single* solution. But (as pointed out in Section 2.4) we could obtain a similar result by defining a new problem $\mathcal{P}'$ whose solution is a set of solutions to the original problem $\mathcal{P}$, and add entropy constraints to $\mathcal{P}'$.

[3] describes a software system called Mix for generating constrained randomised number sequences. It implements a hand-coded local search algorithm with several types of constraint that are useful for psychologists, including constraints that are very similar to our `freq` and `dict` constraints (*maximum repetition* and *pattern* constraints respectively). However, no connection is made to Kolmogorov complexity or data compression, Mix does not use a generic constraint solver or metaheuristic, it does not use differential constraints (though it has other constraints we do not have), and it is designed for a special class of problem.

The **SPREAD** constraint [22] has something in common with our frequency constraints but with a different motivation. It balances distributions of values, for example spreading the load between periods in a timetable. It has efficient filtering algorithms but it does not aim to pass compression-based randomness tests.

Markov Constraints [21] express the Markov condition as constraints, so that constraint solvers can generate Markovian sequences. They have been applied to the generation of musical chord sequences.

5   Discussion

We proposed several types of entropy constraint to eliminate different types of pattern in a solution, leading to high-entropy solutions as estimated by compression algorithms and the Approximate Entropy function. These are complementary to statistical tests of the kind explored in [30]. All our constraints are based on well-known global constraints and can also be implemented in MIP or SAT. Note that instead of specifying bounds on the occurrences of symbols and k-grams we could allow the user to specify bounds on the Approximate Entropy $\mathrm{ApEn}(k)$ for various k. However, we believe that the former approach is more intuitive.

Using constraints to represent randomness makes it easy to generate random-looking solutions with special properties: we simply post constraints for randomness and for the desired properties, then any solution is guaranteed to satisfy both. However, applying entropy constraints is something of an art involving a compromise between achieving high entropy, satisfying the problems constraints and possibly optimising an objective function. Even with few or no problem constraints we must take care not to exclude so many patterns that no solutions remain, as Ramsey theory [25] shows that any sufficiently large object must contain some structure. In fact adding entropy constraints does not necessarily preserve satisfiability. If a problem has no sufficiently random-looking solutions then entropy constraints might eliminate all solutions. However, an advantage of this is that (as mentioned in Section 1) we can prove that no such solutions exist: this cannot be done with the randomised search approach. Alternatively we could take an optimisation approach by treating entropy constraints as soft constraints, and searching for the most random-looking solution.

Of course our solutions are only pseudorandom, not truly random. They were generated by restricting repeated symbols and k-grams in order to be incompressible to a certain class of data compression algorithms. It could be objected that they might fail other tests of randomness that are important to applications. Our response to this argument is: we can turn these other tests into additional constraints. For example if our solution in Section 3.2 had failed the Wald-Wolfowitz runs test, we could have added a constraint to ensure that it passed the test, as follows. Suppose we have a sequence of n binary numbers, with n_0 zeroes and n_1 ones ($n_0 + n_1 = n$). Under a normal approximation (valid for $n_0, n_1 \geq 10$) the expected number of runs is

$$\mu = 1 + \frac{2n_0 n_1}{n}$$

and the variance of this number is

$$\sigma^2 = \frac{2n_0 n_1 (2n_0 n_1 - n)}{n^2(n-1)} = \frac{(\mu - 1)(\mu - 2)}{n - 1}$$

To test for randomness with 95% confidence we require that the observed number of runs R is within $\mu \pm 1.96\sigma$. To implement this test as a constraint on binary variables $v_1, \ldots, v_n$ we define new binary variables $b_i = \mathrm{reify}(v_i = v_{i+1})$ and post a constraint

$$\mu - 1.96\sigma \leq \left(1 + \sum_{i=0}^{n-2} b_i\right) \leq \mu + 1.96\sigma$$

If we do not know the values of n_0 and n_1 in advance, the constraint implementation can create auxiliary integer variables n_0, n_1 and real variables μ, σ, and post additional constraints:

$$\sum v_i = n_1 \qquad n_0 + n_1 = n$$
$$\mu n = n + n_0 n_1 \qquad \sigma^2(n-1) = (\mu - 1)(\mu - 2)$$

There is another possible objection to our approach — in fact to the whole idea of eliminating patterns in solutions. It can be argued that a solution with a visible pattern is statistically no less likely to occur than a solution with no such pattern, and that patterns are merely psychological artefacts. For example if we generate random binary sequences of length 6 then 111111 is no less random than 010110 because both have the same probability of occurring. Considering the latter to be "more random" than the former is a form of Gambler's Fallacy, in which (for example) gamblers assume that numbers with obvious patterns are less likely to occur than random-looking numbers. But if we wish to convince humans (or automated software agents designed by humans) that a solution was randomly generated then we must reject patterns that appear non-random to humans. This intuition is made concrete by the ideas of algorithmic information theory [5]. We do not expect all readers to agree with our view: randomness is a notoriously slippery concept [2] whose nature is beyond the scope of this paper.

There are several interesting directions for future work. We hope to find new applications, possibly with other patterns to be eliminated. We could try to devise randomness constraints using ideas from other literatures. One approach is to take randomness tests applied to number sequences and turn them into constraints. For example we might implement *spectral tests* [17] which are considered to be powerful. But they are complex and we conjecture that they are unlikely to lead to efficient filtering (though this is far from certain and would be an interesting research direction). Moreover, they seem better suited to very long sequences of numbers: far longer than the size of typical solutions to optimisation problems. For evaluating pseudo-random number generators there is a well-known set of simpler tests: the *Die-Hard Tests*,[3] later extended to the *Die Harder Tests*[4] and distilled down to three tests in [18]. However, these tests are also aimed at very long sequences of numbers, and again it is not obvious how to derive constraints from them.

Acknowledgement. This publication has emanated from research supported in part by a research grant from Science Foundation Ireland (SFI) under Grant Number SFI/12/RC/2289. S. Armagan Tarim is supported by the Scientific and Technological Research Council of Turkey (TUBITAK).

---

[3] http://www.stat.fsu.edu/pub/diehard/
[4] http://www.phy.duke.edu/~rgb/General/dieharder.php

References

1. Apt, K.R., Wallace, M.: Constraint Logic Programming Using Eclipse. Cambridge University Press (2007)
2. Beltrami, E.: What Is Random? Chance and Order in Mathematics and Life. Copernicus (1999)
3. van Casteren, M., Davis, M.H.: Mix, a Program for Pseudorandomization. Behaviour Research Methods **38**(4), 584–589 (2006)
4. Cilibrasi, R., Vitányi, P.M.B.: Clustering by Compression. IEEE Transactions on Information Theory **51**(4), 1523–1545 (2005)
5. Chaitin, G.J.: Algorithmic Information Theory. Cambridge University Press (1987)
6. Claessen, K., Duregård, J., Pałka, M.H.: Generating constrained random data with uniform distribution. In: Codish, M., Sumii, E. (eds.) FLOPS 2014. LNCS, vol. 8475, pp. 18–34. Springer, Heidelberg (2014)
7. Cutler, C.C.: Differential Quantization for Television Signals. U. S. Patent 2, 605, 361, July 1952
8. Dechter, R., Kask, K., Bin, E., Emek, R.: Generating random solutions for constraint satisfaction problems. In: Proceedings of the 18th National Conference on Artificial Intelligence, pp. 15–21 (2002)
9. Ermon, S., Gomes, C.P., Selman, B.: Uniform solution sampling using a constraint solver as an oracle. In: Proceedings of the Twenty-Eighth Conference on Uncertainty in Artificial Intelligence, pp. 255–264. AUAI Press (2012)
10. French, R.M., Perruchet, P.: Generating Constrained Randomized Sequences: Item Frequency Matters. Behaviour Research Methods **41**(4), 1233–1241 (2009)
11. Hebrard, E., Hnich, B., O'Sullivan, B., Walsh, T.: Finding diverse and similar solutions in constraint programming. In: Proceedings of the 20th National Conference on Artificial Intelligence (2005)
12. Van Hentenryck, P., Coffrin, C., Gutkovich, B.: Constraint-based local search for the automatic generation of architectural tests. In: Gent, I.P. (ed.) CP 2009. LNCS, vol. 5732, pp. 787–801. Springer, Heidelberg (2009)
13. Hinkelmann, K., Kempthorne, O.: Design and Analysis of Experiments I and II. Wiley (2008)
14. Housworth, E.A., Martins, E.P.: Random Sampling of Constrained Phylogenies: Conducting Phylogenetic Analyses When the Philogeny is Partially Known. Syst. Biol. **50**(5), 628–639 (2001)
15. Huffman, D.A.: A Method for the Construction of Minimum Redundancy Codes. Proceedings of the IRE **40**, 1098–1101 (1951)
16. Kitchen, N., Kuehlmann, A.: Stimulus generation for constrained random simulation. In: Proceedings of the 2007 IEEE/ACM International Conference on Computer-Aided Design, pp. 258–265. IEEE Press (2007)
17. Knuth, D.E.: The Art of Computer Programming. Seminumerical Algorithms, 2nd edn., vol. 2, p. 89. Addison-Wesley (1981)
18. Marsaglia, G., Tsang, W.W.: Some Difficult-to-pass Tests of Randomness. Journal of Statistical Software **7**(3) (2002)
19. Naveh, R., Metodi, A.: Beyond feasibility: CP usage in constrained-random functional hardware verification. In: Schulte, C. (ed.) CP 2013. LNCS, vol. 8124, pp. 823–831. Springer, Heidelberg (2013)
20. Ouellet, P., Quimper, C.-G.: The multi-inter-distance constraint. In: Proceedings of the 22nd International Joint Conference on Artificial Intelligence, pp. 629–634 (2011)

21. Pachet, F., Roy, P.: Markov Constraints: Steerable Generation of Markov Sequences. Constraints **16**(2), 148–172 (2011)
22. Pesant, G., Régin, J.-C.: SPREAD: a balancing constraint based on statistics. In: van Beek, P. (ed.) CP 2005. LNCS, vol. 3709, pp. 460–474. Springer, Heidelberg (2005)
23. Pincus, S.M., Gladstone, I.M., Ehrenkranz, R.A.: A Regularity Statistic for Medical Data Analysis. Journal of Clinical Monitoring and Computing **7**(4), 335–345 (1991)
24. Press, W.H., Flannery, B.P., Teukolsky, S.A., Vetterling, W.T.: Numerical Recipes in C, 2nd edn. Cambridge University Press, UK (1992)
25. Ramsey, F.P.: On a Problem of Formal Logic. Proceedings London Mathematical Society s2 **30**(1), 264–286 (1930)
26. Refalo, P.: Impact-based search strategies for constraint programming. In: Wallace, M. (ed.) CP 2004. LNCS, vol. 3258, pp. 557–571. Springer, Heidelberg (2004)
27. Régin, J.-C.: Generalized Arc Consistency for Global Cardinality Constraint. In: 14th National Conference on Artificial Intelligence, pp. 209–215 (1996)
28. Régin, J.-C.: A filtering algorithm for constraints of difference in CSPs. In: Proceedings of the 12th National Conference on Artificial Intelligence, Vol. 1, pp. 362–367 (1994)
29. Rissanen, J.J., Langdon, G.G.: Arithmetic Coding. IBM Journal of Research and Development **23**(2), 149–162 (1979)
30. Rossi, R., Prestwich, S., Tarim, S.A.: Statistical constraints. In: 21st European Conference on Artificial Intelligence (2014)
31. Sayood, K.: Introduction to Data Compression. Morgan Kaufmann (2012)
32. Shannon, C.E.: A Mathematical Theory of Communication. Bell System Technical Journal **27**(3), 379–423 (1948)
33. Wallace, G.K.: The JPEG Still Picture Compression Standard. Communications of the ACM **34**, 31–44 (1991)
34. Welch, T.A.: A Technique for High-Performance Data Compression. IEEE Computer, 8–19, June 1984
35. Ziv, J., Lempel, A.: A Universal Algorithm for Data Compression. IEEE Transactions on Information Theory IT **23**(3), 337–343 (1977)
36. Ziv, J., Lempel, A.: Compression of Individual Sequences via Variable-Rate Coding. IEEE Transactions on Information Theory IT **24**(5), 530–536 (1978)

Quasipolynomial Simulation of DNNF by a Non-determinstic Read-Once Branching Program

Igor Razgon[✉]

Department of Computer Science and Information Systems,
Birkbeck, University of London, London, U.K.
igor@dcs.bbk.ac.uk

Abstract. We prove that DNNFs can be simulated by Non-deterministic Read-Once Branching Programs (NROBPs) of quasi-polynomial size. As a result, all the exponential lower bounds for NROBPs immediately apply for DNNFs.

1 Introduction

Decomposable Negation Normal Forms (DNNFs) [3] is a well known formalism in the area of propositional knowledge compilation notable for its efficient representation of CNFs with bounded structural parameters. The DNNFs lower bounds are much less understood. For example, it has been only recently shown that DNNFs can be exponentially large on (monotone 2-) CNFs [2]. Prior to that, it was known that on monotone functions DNNFs are not better than monotone DNNFs [6]. Hence all the lower bounds for monotone circuits apply for DNNFs. However, using monotone circuits to obtain new DNNF lower bounds is hardly an appropriate methodology because, in light of [2], on monotone functions, DNNFs are much weaker than monotone circuits.

In this paper we show that DNNFs are strongly related to Non-deterministic Read-Once Branching Programs (NROBPs) that can be thought as Free Binary Decision Diagrams (FBDDs) with OR-nodes. In particular, we show that a DNNF can be transformed into a NROBP with a quasi-polynomial increase of size. That is, all the exponential lower bounds known for NROBPs (see e.g. [5,8]) apply for DNNFs. As NROBPs can be linearly simulated by DNNFs (using a modification of the simulation of FBDDs by DNNFs proposed in [4]), we believe that the proposed result makes a significant progress in our understanding of complexity of DNNFs. Indeed, instead of trying to establish exponential lower bounds directly for DNNFs, we can now do this for NROBPs , which are much better understood from the lower bound perspective.

In the proposed simulation, we adapt to unrestricted DNNFs the approach that was used in [1] for quasi-polynomial simulation of *decision* DNNFs by FBDDs. For the adaptation, we find it convenient to represent NROBPs in a form where variables carry no labels and edges are labelled with literals. In particular, each input node u of the DNNF is represented in the resulting NROBP as an edge

© Springer International Publishing Switzerland 2015
G. Pesant (Ed.): CP 2015, LNCS 9255, pp. 367–375, 2015.
DOI: 10.1007/978-3-319-23219-5_26

labelled with the literal of u and these are *the only edges* that are labelled (compare with [1] where the labelling is 'pertained' to OR nodes, which is impossible for unrestricted DNNFs where the OR nodes can have an arbitrary structure).

The most non-trivial aspect of the simulation is the need to transform an AND of two NROBPs Z_1 and Z_2 into a single NROBP. Following [1], this is done by putting Z_1 'on top' of Z_2. However, this creates the problem that Z_1 becomes unusable 'outside' this construction (see Section 4.1. and, in particular, Figure 2 of [1] for illustration of this phenomenon). Similarly to [1], we address this problem by introducing multiple copies of Z_1.

Formal Statement of the Result. A DNNF Z^* is a directed acyclic graph (DAG) with many roots (nodes of in-degree 0) called *input* nodes and one leaf (node of out-degree 0) called the *output* node. The input nodes are labelled with literals, the rest are AND, and OR nodes such that each AND node has the *decomposability* property defined as follows. Let us define $Var(u)$ for a node u of Z^* as the set of variables x such that Z^* has a path from a node labelled by x to u. Then, if u is an AND node of Z^* and v and w are two different in-neighbours of u then $Var(v) \cap Var(w) = \emptyset$. Let Z_u^* be the subgraph of Z^* induced by a node u and all the nodes from which u can be reached. Then the function $F[Z_u^*]$ computed by Z_u^* is defined as follows. If u is an input node then $F[Z_u^*] = x$, where x is the literal labelling u. If u is an OR or an AND node with in-neighbours $v_1, \ldots v_q$ then $F[Z_u^*] = F[Z_{v_1}^*] \vee \cdots \vee F[Z_{v_q}^*]$, or $F[Z_u^*] = F[Z_{v_1}^*] \wedge \cdots \wedge F[Z_{v_q}^*]$, respectively. The function $F[Z^*]$ computed by Z^* is $F[Z_{out}^*]$, where out is the output node of Z^*. In the rest of the paper we assume that the AND nodes of Z^* are binary. This assumption does not restrict generality of the result since an arbitrary DNNF can be transformed into one with binary AND nodes with a quadratic increase of size.

A Non-deterministic Read-once Branching Program (NROBP) is a DAG Z with one root (and possibly many leaves). Some edges of Z are labelled with literals of variables in the way that each variable occurs at most once on each path P of Z. We denote by $A(P)$ the set of literals labelling the edges of a path P of Z. To define a function $F[Z]$ computed by Z, let us make a few notational agreements. First, we define a truth assignment to a set of variables as the set of literals of these variables that become true as result of the assignment. For example, the assignment $\{x_1 \leftarrow true, x_2 \leftarrow false, x_3 \leftarrow true\}$ is represented as $\{x_1, \neg x_2, x_3\}$. For a function F on a set Var of variables, we say that an assignment S of Var *satisfies* F is $F(S) = true$. Now, let S be an assignment of variables labelling the edges of Z. Then S satisfies $F[Z]$ if and only if there is a root-leaf path P of Z with $A(P) \subseteq S$. A DNNF and a NROBP for the same function are illustrated on Figure 1.

Remark. The above definition of NROBP is equivalent to FBDD with OR-nodes in the sense that each of them can simulate the other with a linear increase of size.

Our main result proved in the next section is the following.

Theorem 1. *Let Z^* be a DNNF with m nodes computing a function F of n variables. Then F can be computed by a NROBP of size $O(m^{\log n + 2})$.*

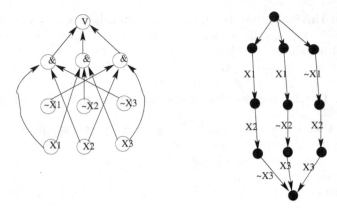

Fig. 1. DNNF and NROBP for function $(x_1 \wedge x_2 \wedge \neg x_3) \vee (x_1 \wedge \neg x_2 \wedge x_3) \vee (\neg x_1 \wedge x_2 \wedge x_3)$

2 Proof of Theorem 1

This section is organised as follows. We first present a transformation of a DNNF Z^* into a graph Z, then state two auxiliary lemmas about properties of special subgraphs of Z, their proofs postponed to Section 2.1, and then prove Theorem 1.

The first step of the transformation is to fix one in-coming edge of each AND-node u of Z^* as the *light* edge of u. This is done as follows. Let u_1 and u_2 be two in-neighbours of u such that $|Var(u_1)| \leq |Var(u_2)|$. Then (u_1, u) is the *light edge* of u and (u_2, u) is the *heavy edge* of u. (Of course if both in-neighbours of u depend on the same number of variables then u_1 and u_2 can be chosen arbitrarily.) We say that an edge (v, w) of Z^* is *a* light edge if w is an AND-node and (v, w) is its light edge. Let P be a path from u to the output node *out* of Z^*. Denote the set of light edges of P by $le(P)$. Denote by $LE(u)$ the set of all such $le(P)$ for a $u - out$ path P.

Now we define a graph Z consisting of the following nodes.

- (u, le, in) for all $u \in V(Z^*)$ (recall that if G is a graph, $V(G)$ denotes the set of nodes of G) and $le \in LE(u)$. The 'in' in the third coordinate stands for 'internal' to distinguish from the 'leaf' nodes defined in the next item.
- For each input node u of Z^* and for each $le \in LE(u)$, Z has a node (u, le, lf) where 'lf' stands for 'leaf'. We say that (u, le, lf) is the *leaf corresponding to* (u, le, in).

When we refer to a node of Z with a single letter, we use bold letters like $\mathbf{u}, \mathbf{v}$ to distinguish from nodes u, v of Z^*. We denote by $mnode(\mathbf{u}), coord(\mathbf{u}), type(\mathbf{u})$, the respective components of $\mathbf{u}$, that is $\mathbf{u} = (mnode(\mathbf{u}), coord(\mathbf{u}), type(\mathbf{u}))$. We also call the components *the main node* of $\mathbf{u}$, *the coordinate* of $\mathbf{u}$ and the *type* of $\mathbf{u}$. The nodes of Z whose type is *in* are *internal* nodes and the nodes whose type is lf are *leaf* nodes. The leaf nodes are not necessarily leaves of Z but rather leaves of special subgraphs of Z that are important for the proof.

Setting the Environment for Definition of Edges of Z. We explore the nodes u of Z^* topologically sorted from the input to the output and *process* each internal node $\mathbf{u}$ of Z with $u = mnode(\mathbf{u})$. In particular, we introduce out-neighbours of $\mathbf{u}$, possibly, together with labelling of respective edges, the set of nodes $Leaves(\mathbf{u})$, and a subgraph $Graph(\mathbf{u})$ of Z which will play a special role in the proof. The detailed description of *processing* of $\mathbf{u}$ is provided below.

- Suppose that u is an input node. Let y be the literal labelling u in Z^* and let $\mathbf{u}'$ be the leaf corresponding to $\mathbf{u}$.
  1. Introduce an edge $(\mathbf{u}, \mathbf{u}')$ and label this edge with y.
  2. Set $Leaves(\mathbf{u}) = \{\mathbf{u}'\}$.
  3. Define $Graph(\mathbf{u})$ as having node set $\{\mathbf{u}, \mathbf{u}'\}$ and the edge $(\mathbf{u}, \mathbf{u}')$.
- Suppose that u is an OR node. Let $v_1, \dots v_q$ be the in-neighbours of u in Z^*. Let $\mathbf{v_1}, \dots, \mathbf{v_q}$ be the internal nodes of Z with $v_1, \dots, v_q$ being the respective main nodes and with $coord(\mathbf{v_i}) = coord(\mathbf{u})$ for all $1 \leq i \leq q$.
  1. Introduce edges $(\mathbf{u}, \mathbf{v_1}), \dots, (\mathbf{u}, \mathbf{v_q})$.
  2. Set $Leaves(\mathbf{u}) = Leaves(\mathbf{v_1}) \cup \cdots \cup Leaves(\mathbf{v_q})$.
  3. $Graph(\mathbf{u})$ is obtained from $Graph(\mathbf{v_1}) \cup \cdots \cup Graph(\mathbf{v_q})$ by adding node $\mathbf{u}$ plus the edges $(\mathbf{u}, \mathbf{v_1}), \dots, (\mathbf{u}, \mathbf{v_q})$.
- Suppose u is an AND node. Let u_1, u_2 be two in-neighbours of u in Z^* and assume that the edge (u_1, u) is the light one. Let $\mathbf{u_1}, \mathbf{u_2}$ be two internal nodes of Z whose respective main nodes are u_1 and u_2 and $coord(\mathbf{u_1}) = coord(\mathbf{u}) \cup \{(u_1, u)\}$ and $coord(\mathbf{u_2}) = coord(\mathbf{u})$.
  1. Introduce edges $(\mathbf{u}, \mathbf{u_1})$ and $(\mathbf{w}, \mathbf{u_2})$ for each $\mathbf{w} \in Leaves(\mathbf{u_1})$.
  2. Set $Leaves(\mathbf{u}) = Leaves(\mathbf{u_2})$.
  3. $Graph(\mathbf{u})$ is obtained from $Graph(\mathbf{u_1}) \cup Graph(\mathbf{u_2})$ by adding node $\mathbf{u}$ and the edges described in the first item.

Remark. Let us convince ourselves that the nodes $\mathbf{v_1}, \dots, \mathbf{v_q}$, and $\mathbf{u_1}, \mathbf{u_2}$ with the specified coordinates indeed exist. Indeed, suppose that u is an OR-node of Z^* and let v be an in-neighbour of u. Let P be a path from u to the output node of Z^*. Then $le((v, u) + P) = le(P)$ confirming possibility of choice of nodes $\mathbf{v_1}, \dots, \mathbf{v_q}$. Suppose that u is an AND-node and let (u_1, u) and (u_2, u) be the light and heavy edges of u respectively. For a P as before, $le((u_1, u) + P) = \{(u_1, u)\} \cup le(P)$ and $le((u_2, u) + P)) = le(P)$ confirming that the nodes $\mathbf{u_1}$ and $\mathbf{u_2}$ exist. Thus the proposed processing is well-defined.

Lemma 1. *Let $\mathbf{u} \in V(Z)$ with $type(\mathbf{u}) = in$ and let $u = mnode(\mathbf{u})$. Then the following statements hold.*

1. *$Graph(\mathbf{u})$ is a DAG.*
2. *$\mathbf{u}$ is the (only) root of $Graph(\mathbf{u})$ and $Leaves(\mathbf{u})$ is the set of leaves of $Graph(\mathbf{u})$.*
3. *If u is an OR-node and $\mathbf{v_1}, \dots, \mathbf{v_q}$ are as in the description of processing of $\mathbf{u}$ then each root-leaf path P of $Graph(\mathbf{u})$ is of the form $(\mathbf{u}, \mathbf{v_i}) + P'$ where P' is a root-leaf path of $Graph(\mathbf{v_i})$.*

4. *Suppose u is an AND node and let $\mathbf{u_1}, \mathbf{u_2}$ be as in the description of processing of $\mathbf{u}$. Then each root-leaf path P of $Graph(\mathbf{u})$ is of the form $(\mathbf{u}, \mathbf{u_1}) + P_1 + (\mathbf{w}, \mathbf{u_2}) + P_2$, where P_1, P_2 are root-leaf paths of $Graph(\mathbf{u_1})$ and $Graph(\mathbf{u_2})$, respectively and $\mathbf{w}$ is the last node of P_1.*
5. $Var(\mathbf{u}) = Var(u)$ *where $Var(\mathbf{u})$ is the set of all variables labelling the edges of $Graph(\mathbf{u})$.*
6. $Graph(\mathbf{u})$ *is read-once (each variable occurs at most once on each path).*

It follows from the first, second, and the last statements of Lemma 1 that $Graph(\mathbf{u})$ is a NROBP. Therefore, we can consider the function $F[Graph(\mathbf{u})]$ computed by $Graph(\mathbf{u})$.

Lemma 2. *For each $\mathbf{u} \in V(Z)$ with $type(\mathbf{u}) = in$, $F[Graph(\mathbf{u})] = F[Z_u^*]$ where $u = mnode(\mathbf{u})$ and Z_u^* is as defined in the first paragraph of formal statement part of the introduction.*

Proof of Theorem 1. Let out be the output node of Z^* and $\mathbf{out} = (out, \emptyset, in)$ be a node of Z. $Graph(\mathbf{out})$ is a NROBP by Lemma 1 and, by Lemma 2, it computes function $F(Z_{out}^*)$. By definition, $Z_{out}^* = Z^*$ and hence $Graph(\mathbf{out})$ computes the same function as Z^*.

To upper-bound the size of Z^*, observe that for each $\mathbf{u} \in V(Z)$, $|coord(\mathbf{u})| \leq \log n$. Indeed, let us represent $coord(\mathbf{u})$ as $(u_1, u_1'), \ldots, (u_q, u_q')$, a sequence of edges occurring in this order on a path of Z^*. Then each (u_i, u_i') is the light edge of an AND-node u_i'. By the decomposability property of DNNF, $|Var(u_i)| \leq |Var(u_i')|/2$. Also, since Z^* has a path from u_i' to u_{i+1}, $|Var(u_i')| \leq |Var(u_{i+1})|$. Applying this reasoning inductively, we conclude that $|Var(u_1)| \leq |Var(u_q')|/2^q$. Since $|Var(u_1)| \geq 1$ and $|Var(u_q')| \leq n$, it follows that $|coord(\mathbf{u})| = q \leq \log n$. Thus $coord(\mathbf{u})$ is a set of light edges of Z^* of size at most $\log n$. Since there is at most one light edge per element of Z^*, there are at most m light edges in total. Thus the number of possible second coordinates for a node of Z is $\sum_{i=1}^{\log n} \binom{m}{i} \leq m^{\log n + 1}$. As the number of distinct first and third coordinates is at most m and 2, respectively, the result follows. ∎

2.1   Proofs of Auxiliary Lemmas for Theorem 1

Proof of Lemmas 1 and 2 requires two more auxiliary lemmas.

Lemma 3. *Let $\mathbf{u} \in V(Z)$ with $type(\mathbf{u}) = in$ and let $u = mnode(\mathbf{u})$. Then for each $\mathbf{v} \in V(Graph(\mathbf{u}))$, $coord(\mathbf{u}) \subseteq coord(\mathbf{v})$. Moreover, if $type(\mathbf{v}) = lf$ then $coord(\mathbf{u}) = coord(\mathbf{v})$ if and only if $\mathbf{v} \in Leaves(\mathbf{u})$.*

Proof. By induction on nodes $\mathbf{u}$ of Z according to the topological sorting of the nodes $u = mnode(\mathbf{u})$ of Z^* from input to output nodes. That is if v is an neighbour of u then for any node $\mathbf{v}$ with $v - mnode(\mathbf{v})$ the lemma holds by the induction assumption.

If u is an input node then $V(Graph(\mathbf{u}))$ consists of $\mathbf{u}$ and the leaf corresponding to $\mathbf{u}$, hence the first statement holds by construction. Otherwise,

$V(Graph(\mathbf{u}))$ consists of $\mathbf{u}$ itself and the union of all $V(Graph(\mathbf{v}))$, where, following the description of processing of $\mathbf{u}$, $\mathbf{v}$ is one of $\mathbf{v_1}, \dots, \mathbf{v_q}$ if u is an OR-node and $\mathbf{v}$ is either $\mathbf{u_1}$ or $\mathbf{u_2}$ if u is an AND-node. For each such $\mathbf{v}$ it holds by definition that $coord(\mathbf{u}) \subseteq coord(\mathbf{v})$. That is, each node $\mathbf{w} \neq \mathbf{u}$ of $Graph(\mathbf{u})$ is in fact a node of such a $Graph(\mathbf{v})$. By the induction assumption, $coord(\mathbf{v}) \subseteq coord(\mathbf{w})$ and hence $coord(\mathbf{u}) \subseteq coord(\mathbf{w})$ as required.

Using the same inductive reasoning, we show that for each $\mathbf{w} \in Leaves(\mathbf{u})$, $coord(\mathbf{w}) = coord(\mathbf{u})$. This is true by construction if u is an input node. Otherwise, $Leaves(\mathbf{u})$ is defined as the union of one or more $Leaves(\mathbf{v})$ such that $coord(\mathbf{v}) = coord(\mathbf{u})$ by construction. Then, letting $\mathbf{v}$ be such that $\mathbf{w} \in Leaves(\mathbf{v})$ we deduce, by the induction assumption that $coord(\mathbf{w}) = coord(\mathbf{v}) = coord(\mathbf{u})$.

It remains to prove that for each $\mathbf{w} \in V(Graph(\mathbf{u})) \setminus Leaves(\mathbf{u})$ such that $type(\mathbf{w}) = lf$, $coord(\mathbf{u}) \subset coord(\mathbf{w})$. This is vacuously true if u is an input node. Otherwise, the induction assumption can be straightforwardly applied as above if $\mathbf{w} \in Graph(\mathbf{v})$ for some $\mathbf{v}$ as above and $\mathbf{w} \notin Leaves(\mathbf{v})$. The only situation where it is not the case is when u is an AND node and $\mathbf{v} = \mathbf{u_1}$ where $\mathbf{u_1}$ is as in the description of processing of an AND node. In this case $coord(\mathbf{u}) \subset coord(\mathbf{u_1})$ by construction and, since $coord(\mathbf{u_1}) \subseteq coord(\mathbf{w})$, by the first statement of this lemma, $coord(\mathbf{u}) \subset coord(\mathbf{w})$. ■

Lemma 4.   1. *For each internal $\mathbf{u} \in V(Z)$, the out-going edges of $\mathbf{u}$ in Z are exactly those that have been introduced during processing of $\mathbf{u}$.*
  2. *For each $\mathbf{v} \in V(Z)$ with $type(\mathbf{v}) = lf$, the out-degree of $\mathbf{v}$ in Z is at most 1. Moreover the out-degree of $\mathbf{v}$ is 0 in each $Graph(\mathbf{u'})$ such that $\mathbf{v} \in Leaves(\mathbf{u'})$.*
  3. *Let $\mathbf{u}$ be an internal node and let $(\mathbf{w_1}, \mathbf{w_2})$ be an edge where $\mathbf{w_1} \in V(Graph(\mathbf{u})) \setminus Leaves(\mathbf{u})$. Then $(\mathbf{w_1}, \mathbf{w_2})$ is an edge of $Graph(\mathbf{u})$.*

Proof. The first statement follows by a direct inspection of the processing algorithm. Indeed, the only case where an edge $(\mathbf{u}, \mathbf{v})$ might be introduced during processing of a node $\mathbf{u'} \neq \mathbf{u}$ is where $mnode(\mathbf{u'})$ is an AND node. However, in this case $type(\mathbf{u})$ must be lf in contradiction to our assumption.

Consider the second statement. Consider an edge $(\mathbf{v}, \mathbf{w})$ such that $type(\mathbf{v}) = lf$. Suppose this edge has been created during processing of a node $\mathbf{u}$. Then $u = mnode(\mathbf{u})$ is an AND-node. Further, let $\mathbf{u_1}, \mathbf{u_2}$ be as in the description of processing of $\mathbf{u}$. Then $\mathbf{v} \in Leaves(\mathbf{u_1})$ and $\mathbf{w} = \mathbf{u_2}$. By construction, $coord(\mathbf{w}) = coord(\mathbf{u})$ and by Lemma 3, $coord(\mathbf{v}) = coord(\mathbf{u_1})$. Hence, by definition of $\mathbf{u_1}$, $coord(\mathbf{w}) \subset coord(\mathbf{v})$. Suppose that $\mathbf{v} \in Leaves(\mathbf{u'})$ for some internal $\mathbf{u'} \in V(Z)$. Then, by Lemma 3, $coord(\mathbf{v}) = coord(\mathbf{u'})$ and hence $coord(\mathbf{w}) \subset coord(\mathbf{u'})$. It follows from Lemma 3 that $\mathbf{w}$ is not a node of $Graph(\mathbf{u'})$ and hence $(\mathbf{v}, \mathbf{w})$ is not an edge of $Graph(\mathbf{u'})$. Thus the out-degree of $\mathbf{v}$ in $Graph(\mathbf{u'})$ is 0.

Continuing the reasoning about edge $(\mathbf{v}, \mathbf{w})$, we observe that $coord(\mathbf{w}) = coord(\mathbf{v}) \setminus \{(u_1, u)\}$ where $u_1 = mnode(\mathbf{u_1})$. Notice that all the edges of $coord(\mathbf{v}) = coord(\mathbf{u_1})$ lie on a path from u_1 to the output node of Z^* and (u_1, u) occurs first of them. Due to the acyclicity of Z^*, (u_1, u) is uniquely defined (in terms of $\mathbf{v}$) and hence so is $coord(\mathbf{w})$. Furthermore, as specified above, $mnode(\mathbf{w}) = mnode(\mathbf{u_2})$ which is the neighbour u other than u_1. Since (u_1, u) is uniquely defined, u and u_1 are uniquely defined as its head and tail and hence so is $mnode(\mathbf{u_2})$. Finally, by construction, we know that $\mathbf{w}$ is an internal node. Thus all three components of $\mathbf{w}$ are uniquely defined and hence so is $\mathbf{w}$ itself. That is, $\mathbf{v}$ can have at most one neighbour.

The third statement is proved by induction analogous to Lemma 3. The statement is clearly true if $u = mnode(\mathbf{u})$ is an input node of Z^* because then $Graph(\mathbf{u})$ has only one edge. Assume this is not so. If $\mathbf{w_1} = \mathbf{u}$ then the statement immediately follows from the first statement of this lemma and the definition of $Graph(\mathbf{u})$. Otherwise, $\mathbf{w_1} \in V(Graph(\mathbf{v}))$ where $\mathbf{v}$ is as defined in the proof of Lemma 3. If $\mathbf{w_1} \notin Leaves(\mathbf{v})$ then, by the induction assumption, $(\mathbf{w_1}, \mathbf{w_2})$ is an edge of $Graph(\mathbf{v})$ and hence of $Graph(\mathbf{u})$. This may be not the case only if $\mathbf{u}$ is an AND node and $\mathbf{w_1} \in Leaves(\mathbf{u_1})$ ($\mathbf{u_1}$ and $\mathbf{u_2}$ are as in the description of processing of $\mathbf{u}$). By definition of $Graph(\mathbf{u})$, $(\mathbf{w_1}, \mathbf{u_2})$ is an edge of $Graph(\mathbf{u})$ and, according to the previous paragraph, $\mathbf{w_1}$ does not have other outgoing edges. Thus it remains to assume that $\mathbf{w_2} = \mathbf{u_2}$ and the statement holds by construction. ■

Proof Sketch of Lemma 1 All the statements except 3 and 4 are proved by induction like in Lemma 3. If $u = mnode(\mathbf{u})$ is an input node then $Graph(\mathbf{u})$ is a labelled edge for which all the statements of this lemma are clearly true. So, we assume below that u is either an OR-node or an AND-node.

Statement 1. Assume that $Graph(\mathbf{u})$ does have a directed cycle C. Since $Graph(\mathbf{u})$ is loopless by construction, C contains at least one node $\mathbf{w} \neq \mathbf{u}$. By construction, $\mathbf{w}$ is a node of some $Graph(\mathbf{v})$ where $\mathbf{v}$ is as defined in the proof of Lemma 3. Then C intersects with $Leaves(\mathbf{v})$. Indeed, if we assume the opposite then, applying the last statement of Lemma 4 inductively starting from $\mathbf{w}$, we conclude that all the edges of C belong to $Graph(\mathbf{v})$ in contradiction to its acyclicity by the induction assumption. Now, C does not intersect with $Leaves(\mathbf{u})$ because they have out-degree 0 by Lemma 4. Thus if C intersects with $Graph(\mathbf{v})$ then $Leaves(\mathbf{v})$ cannot be a subset of $Leaves(\mathbf{u})$. This is only possible if $\mathbf{u}$ is an AND-node and $\mathbf{v} = \mathbf{u_1}$ ($\mathbf{u_1}$ and $\mathbf{u_2}$ are as in the description of the processing of $\mathbf{u}$). Let $\mathbf{w'} \in Leaves(\mathbf{u_1})$ be a node of C. By construction, $\mathbf{u_2}$ is an out-neighbour of $\mathbf{w'}$ and by Lemma 4, $\mathbf{w}$ does not have other out-neighbours in Z. Thus $\mathbf{u_2}$ is the successor of $\mathbf{w'}$ in C and hence C intersects with $Graph(\mathbf{u_2})$ while $Leaves(\mathbf{u_2}) \subseteq Leaves(\mathbf{u})$ in contradiction to what we have just proved. Thus C does not exists.

Statement 2. It is easy to verify by induction that $Graph(\mathbf{u})$ has a path from $\mathbf{u}$ to the rest of vertices, hence besides $\mathbf{u}$, $Graph(\mathbf{u})$ does not have any other roots. Now $\mathbf{u}$ itself is a root by the acyclicity proved above. Since vertices of $Leaves(\mathbf{u})$ have out-degree 0 in $Graph(\mathbf{u})$, clearly, they are all leaves. Suppose that some $\mathbf{w} \in Graph(\mathbf{u}) \setminus Leaves(\mathbf{u})$ is a leaf of $\mathbf{u}$. By construction, $\mathbf{w} \neq \mathbf{u}$ and hence $\mathbf{w}$ is a node of some $Graph(\mathbf{v})$ as above. Then $\mathbf{w}$ is a leaf of $Graph(\mathbf{v})$ because the latter is a subgraph of $Graph(\mathbf{u})$ and hence, by the induction assumption,

$w \in Leaves(\mathbf{v})$. Hence, as in the previous paragraph, we need $\mathbf{v}$ such that $Leaves(\mathbf{v}) \nsubseteq Leaves(\mathbf{u})$ and we conclude as above that $\mathbf{v} = \mathbf{u_1}$. But then $\mathbf{u_2}$ is an out-neighbour of w and hence w cannot be a leaf of $\mathbf{u}$, a contradiction showing that the leaves of $Graph(\mathbf{u})$ are precisely $Leaves(\mathbf{u})$.

Important Remark. In the rest of the section we use $\mathbf{u}$ and the root of $Graph(\mathbf{u})$ as well as $Leaves(\mathbf{u})$ and the leaves of $Graph(\mathbf{u})$ interchangeably without explicit reference to statement 2 of this lemma.

Statements 3 and 4. Suppose that u is an OR node and let P' be the suffix of P starting at the second node $\mathbf{v}$ of P. By statement 1 of Lemma 4, $\mathbf{v}$ is some $\mathbf{v_i}$ as in the description of processing of $\mathbf{u}$. Hence vertices of $Leaves(\mathbf{v}) \subseteq Leaves(\mathbf{u})$ have out-degree 0 in $Graph(\mathbf{u})$ (statement 2 of Lemma 4) and do not occur in the middle of P'. Applying statement 3 of Lemma 4 inductively to P' starting from $\mathbf{v}$, we observe that P' is a path of $Graph(\mathbf{v})$. The last node of P' is a leaf of $Graph(\mathbf{v})$ because it is a leaf of $Graph(\mathbf{u})$. Thus statement 3 holds.

Suppose that u is an AND-node. Then by statement 1 of Lemma 4, the second node of P is $\mathbf{u_1}$. Hence, one of $Leaves(\mathbf{u_1})$ must be an intermediate node of P. Indeed, otherwise, by inductive application of statement 3 of Lemma 4, the suffix of P starting at $\mathbf{u_1}$ is a path of $Graph(\mathbf{u_1})$. In particular, the last node w' of P is a node of $Graph(\mathbf{u_1})$. However, by Lemma 3, $coord(w') = coord(u) \subset coord(\mathbf{u_1})$. Hence, by Lemma 3, $coord(w')$ cannot belong to $Graph(\mathbf{u_1})$, a contradiction. Let $w \in Leaves(\mathbf{u_1})$ be the first such node of P. By inductive application of the last statement fo Lemma 4, the subpath P_1 of P between $\mathbf{u_1}$ and w is a root-leaf path of $Graph(\mathbf{u_1})$. By construction and the second statement of Lemma 4, $\mathbf{u_2}$ is the successor of w in P. Let P_2 be the suffix of P starting at $\mathbf{u_2}$. Arguing as for the OR case we conclude that P_2 is a root-leaf path of $\mathbf{u_2}$. Thus statement 4 holds.

Statement 5. We apply induction, taking into account that $Var(u)$ is the union of all $Var(v)$ where v is an in-neighbour of and $Var(\mathbf{u})$ is the union of all $Var(\mathbf{v})$ where $\mathbf{v}$ is as defined in the proof of Lemma 3. The details are omitted due to space constraints.

Statement 6. Let P be a root-leaf path of $Graph(\mathbf{u})$. Suppose $u = mnode(\mathbf{u})$ is an OR-node. Then $P = (\mathbf{u}, \mathbf{v_i}) + P'$, the notation as in statement 3. P' is read-once by the induction assumption and the edge $(\mathbf{u}, \mathbf{v_i})$ is unlabelled by construction. Hence P is read-once. If u is an AND-node then $P = (\mathbf{u}, \mathbf{u_1}) + P_1 + (\mathbf{w}, \mathbf{u_2}) + P_2$, all the notation as in statement 4. P_1 and P_2 are read-once by the induction assumption, edges $(\mathbf{u}, \mathbf{u_1})$ and $(\mathbf{w}, \mathbf{u_2})$ are unlabelled. The variables of P_1 and of P_2 are respective subsets of $Var(\mathbf{u_1})$ and $Var(\mathbf{u_2})$ equal to $Var(u_1)$ and $Var(u_2)$, respectively, by statement 5 which, in turn, do not intersect due to the decomposability property of AND nodes of Z^*. Hence the variables of P_1 and P_2 do not intersect and P is read-once. ∎

Proof of Lemma 2. By induction as in Lemma 3. If $u = mnode(\mathbf{u})$ is an input node then $F[Z_u^*] = x$ where x is the literal labelling u and $Graph(\mathbf{u})$ is a single edge labelled with x, hence $F[Graph(\mathbf{u})] = x$.

In the rest of the proof, we assume that the set of variables of all the considered functions is Var, the set of variables of Z^*. Introduction of redundant variables will simplify the reasoning because we can now make the induction step without modifying the considered set of variables.

Assume that u is an OR node. If S satisfies $F[Z_u^*]$ then there is an in-neighbour v of u such that S satisfies $F[Z_v^*]$. By construction there is an out-neighbour $\mathbf{v}$ of $\mathbf{u}$ such that $v = mnode(\mathbf{v})$. By the induction assumption, S satisfies $F[Graph(\mathbf{v})]$. Let P' be a $\mathbf{v} - Leaves(\mathbf{v})$ path such that $A(P') \subseteq S$. Then, by construction $(\mathbf{u}, \mathbf{v}) + P'$ is a $\mathbf{u} - Leaves(\mathbf{u})$ path with the edge $(\mathbf{u}, \mathbf{v})$ unlabelled. Hence $A(P) = A(P') \subseteq S$ and hence S satisfies $F[Graph(\mathbf{u})]$. Conversely, if S satisfies $F[Graph(\mathbf{u})]$ then there is a $\mathbf{u} - Leaves(\mathbf{u})$ path P with $A(P) \subseteq S$. By statement 3 of Lemma 1, $P = (\mathbf{u}, \mathbf{v_i}) + P'$, the notation as in the statement. $A(P') \subseteq A(P)$ and hence S satisfies $F[Graph(\mathbf{v_i})]$ and, by the induction assumption, S satisfies $F[Z_{v_i}^*]$ where $v_i = mnode(\mathbf{v_i})$. By definition of $\mathbf{v_i}$, v_i is an in-neighbour of u, hence S satisfies $F[Z_u^*]$.

Assume that u is an AND node. Let $\mathbf{u_1}, \mathbf{u_2}$ be as in the description of processing of $\mathbf{u}$ with $u_1 = mnode(\mathbf{u_1})$ and $u_2 = mnode(\mathbf{u_2})$. Suppose that S satisfies $F[Z_u^*]$. Then S satisfies both $F[Z_{u_1}^*]$ and $F[Z_{u_2}^*]$. Hence, by the induction assumption S satisfies $F[Graph(\mathbf{u_1})]$ and $F[Graph(\mathbf{u_2})]$. For each $i \in \{1,2\}$, let P_i be a $\mathbf{u_i} - Leaves(\mathbf{u_i})$ path of $Graph(\mathbf{u_i})$ with $A(P_i) \subseteq S$. Let $\mathbf{w}$ be the last node of P_1. Then $P = (\mathbf{u}, \mathbf{u_1}) + P_1 + (\mathbf{w}, \mathbf{u_2}) + P_2$ is a $\mathbf{u} = Leaves(\mathbf{u})$ path with the edges $(\mathbf{u}, \mathbf{u_1})$ and $(\mathbf{w}, \mathbf{u_2})$ unlabelled. Hence $A(P) = A(P_1) \cup A(P_2) \subseteq S$ and thus S satisfies $F[Graph(\mathbf{u})]$. Conversely, suppose that S satisfies $F[Graph(\mathbf{u})]$ and let P be a $\mathbf{u} - Leaves(\mathbf{u})$ path with $A(P) \subseteq S$. Then by statement 4 of Lemma 1, $P = (\mathbf{u}, \mathbf{u_1}) + P_1 + (\mathbf{w}, \mathbf{u_2}) + P_2$, the notation as in the statement. Clearly, $A(P_1) \subseteq S$ and $A(P_2) \subseteq S$, hence S satisfies both $F[Z_{u_1}^*]$ and $F[Z_{u_2}^*]$ by the induction assumption and thus S satisfies $F[Z_u^*]$. ∎

References

1. Beame, P., Li, J., Roy, S., Suciu, D.: Lower bounds for exact model counting and applications in probabilistic databases. In: Proceedings of the Twenty-Ninth Conference on Uncertainty in Artificial Intelligence, Bellevue, WA, USA, August 11–15, 2013 (2013)
2. Bova, S., Capelli, F., Mengel, S., Slivovsky, F.: Expander cnfs have exponential DNNF size. CoRR, abs/1411.1995 (2014)
3. Darwiche, A.: Decomposable negation normal form. J. ACM **48**(4), 608–647 (2001)
4. Darwiche, A., Marquis, P.: A knowledge compilation map. J. Artif. Intell. Res. (JAIR) **17**, 229–264 (2002)
5. Jukna, S.: Boolean Function Complexity: Advances and Frontiers. Springer-Verlag (2012)
6. Krieger, M.P.: On the incompressibility of monotone DNFs. Theory Comput. Syst. **41**(2), 211–231 (2007)
7. Oztok, U., Darwiche, A.: On compiling CNF into decision-DNNF. In: O'Sullivan, B. (ed.) CP 2014. LNCS, vol. 8656, pp. 42–57. Springer, Heidelberg (2014)
8. Wegener, I.: Branching Programs and Binary Decision Diagrams. SIAM (2000)

MiniSearch: A Solver-Independent Meta-Search Language for MiniZinc

Andrea Rendl[1]([✉]), Tias Guns[2], Peter J. Stuckey[3], and Guido Tack[1]

[1] National ICT Australia (NICTA) and Faculty of IT, Monash University,
Melbourne, Australia
andrea.rendl@nicta.com.au, guido.tack@monash.edu
[2] KU Leuven, Leuven, Belgium
tias.guns@cs.kuleuven.be
[3] National ICT Australia (NICTA) and University of Melbourne, Victoria, Australia
pstuckey@unimelb.edu.au

Abstract. Much of the power of CP comes from the ability to create complex hybrid search algorithms specific to an application. Unfortunately there is no widely accepted standard for specifying search, and each solver typically requires detailed knowledge in order to build complex searches. This makes the barrier to entry for exploring different search methods quite high. Furthermore, search is a core part of the solver and usually highly optimised. Any imposition on the solver writer to change this part of their system is significant.

In this paper we investigate how powerful we can make a uniform language for meta-search *without placing any burden on the solver writer*. The key to this is to only interact with the solver when a solution is found. We present MINISEARCH, a meta-search language that can directly use any FLATZINC solver. Optionally, it can interact with solvers through an efficient C++ API. We illustrate the expressiveness of the language and performance using different solvers on a number of examples.

1 Introduction

When using constraint programming (CP) technology, one often needs to exert some control over the meta-search mechanism. Meta-search, such as Branch-and-Bound search (BaB) or Large Neighbourhood Search (LNS), happens on top of CP tree search, and aids finding good solutions, often by encoding meta-information a modeller has about the problem.

Unfortunately, there is no widely accepted standard for controlling search or meta-search. A wide range of high-level languages have been proposed that are quite similar in how constraints are specified. However, they differ significantly in the way search can be specified. This ranges from built-in minimisation and maximisation only to fully programmable search. The main trade-off in developing search specification languages is the expressivity of the language versus the required integration with the underlying solver. Fully programmable search, including meta-search, is most expressive but requires deep knowledge of and

© Springer International Publishing Switzerland 2015
G. Pesant (Ed.): CP 2015, LNCS 9255, pp. 376–392, 2015.
DOI: 10.1007/978-3-319-23219-5_27

tight integration with a specific solver. Languages like OPL [24] and COMET [13] provide convenient abstractions for programmable search for the solvers bundled with these languages. On the other hand, some solver-independent languages such as Esra [3], Essence [5] and Essence' [6] do not support search specifications. Zinc [12] and MiniZinc [16] have support for specifying variable and value ordering heuristics to CP solvers, but also no real control over the search.

Search combinators [19] was recently proposed as a generic meta-search language for CP solvers. It interacts with the solver *at every node* in the search tree. While very expressive, it requires significant engineering effort to implement for existing solvers, since typically the solvers' search engines are highly optimised and tightly integrated with other components, such as the propagation engine and state space maintenance.

In this paper, we introduce MiniSearch, a new combinator-like meta-search language that has three design objectives: a *minimal solver interface* to facilitate solver support, *expressiveness*, and, most importantly, *solver-independence*.

The objective to obtain a minimal solver interface stems from lessons learnt in the design of search combinators that interact with the solver at every *node*. In contrast, MiniSearch interacts with the underlying solving system only *at every solution*, which is a minimal interface. At every solution, constraints can be added or constraints in a well-defined *scope* can be removed, before asking for the next solution. If the underlying solver does not support dynamic adding and removing of constraints, MiniSearch can emulate this behaviour, for little overhead.

Despite the lightweight solver interface, MiniSearch is surprisingly expressive and supports many meta-search strategies such as BaB search, lexicographic BaB, Large Neighbourhood Search variants, AND/OR search, diverse solution search, and more. MiniSearch can also be used to create interactive optimisation applications. Moreover, since MiniSearch builds upon MiniZinc, all MiniZinc language features and built-ins can be used, for instance to formulate custom neighbourhoods.

Solver-independence is the most important contribution of MiniSearch. All solvers that can read and solve FlatZinc, which the majority of CP solvers do [15,22], can be used with MiniSearch. Moreover, solvers that provide native meta-search variants, such as branch-and-bound, can declare so and avoid executing the MiniSearch decomposition instead. At the language level, this is similar to the handling of global constraints in MiniZinc. Thus, solvers can apply their strengths during meta-search, despite the minimal interface.

2   The MiniSearch language

MiniSearch is a high level meta-search language based on MiniZinc 2.0 [23]. MiniZinc is used for formulating the model and the constraints posted during search, with language extensions for specifying the search. A MiniSearch specification is provided, together with the **search** keyword, as an argument to MiniZinc's **solve** item. It is executed by the MiniSearch kernel (see Sec. 4).

With the built-in language extensions summarised in Tab. 1, users can define functions such as the following branch-and-bound (BaB) minimisation:

Table 1. MINISEARCH built-ins

MINISEARCH built-ins	Description
next()	find the next solution
post(c)	post the MINIZINC constraint c in the current scope
scope(s)	open a local scope containing search s
s_1 /\ s_2	run s_1 and iff successful, run s_2
s_1 \/ s_2	run s_1 and iff it fails, run s_2
if s then s_1 else s_2	if s is successful, run s_1, otherwise s_2
repeat(s)	repeat search s until **break** is executed
repeat (i in 1..N)(s)	repeat search s N times or until **break** is executed
break()	break within a **repeat**
fail()	return 'failure'
skip()	return 'success'
time_limit(ms,s)	run s until timelimit ms is reached
print(S)	print MINIZINC output string S
print()	print solution according to model output specification
$c := v$	assign parameter c the value v
commit()	commit to last found solution in function scope
sol(v)	return solution value of variable v
hasSol()	returns true if a solution has been found

```
include "minisearch.mzn";
var int: obj; % other variables and constraints not shown
solve search minimize_bab(obj);
output ["Objective: "++show(obj)];

function ann: minimize_bab(var int: obj) =
  repeat( if next() then commit() /\ print() /\ post(obj < sol(obj))
          else break endif );
```

The **include** item on line 1 includes the built-in MINISEARCH function declarations. This is necessary for any MINIZINC model that uses MINISEARCH. Line 3 contains the MINIZINC **solve** item followed by the new **search** keyword and a user-defined MINISEARCH function that takes a variable representing the objective as argument. Line 4 is the MINIZINC **output** item, specifying how solutions should be printed. Lines 7–8 contain the actual MINISEARCH specification. We will explain the different built-ins in more detail below, but the specification can be read as follows: repeatedly try to find the next solution; and if that is successful, commit to the solution, print it and add the constraint that the objective must have a lower value than the current solution. If unsuccessful, break out of the repeat.

All MINISEARCH built-ins are typed as functions returning *annotations*. Semantically, however, every MINISEARCH built-in returns a value that represents either 'success' or 'failure', with respect to finding a solution. The handling of these implicit return values is done by the MINISEARCH interpreter (Sec. 4.1).

2.1 MiniSearch Built-Ins Involving the Solver

Communication with the solver is restricted to three forms: invoking the solver, adding constraints/variables to the model, and scopes for temporary variables and constraints.

Invoking the solver. The MiniSearch instruction for finding the next solution is `next()`. It is successful if a solution has been found, and fails otherwise. The variable/value labelling strategy (such as first-fail on the smallest domain value) can be set in two ways: either by annotating the `solve` item (as in standard MiniZinc), which sets the labelling globally, for every call to `next()`. Otherwise, by annotating any MiniSearch function call, such as `minimize_bab`, with a labelling strategy. Note, however, that as in MiniZinc, solvers may ignore these annotations, for example if the labelling is not supported.

Solvers may declare support for native meta-search strategies, as with global constraints, in which case these MiniSearch functions are treated as built-ins.

Adding constraints and variables. A constraint is added by calling the `post()` built-in with a constraint as argument. Constraints can be formulated using the same MiniZinc constructs as in the model, including global constraints, user-defined functions and predicates. Variables can be dynamically added during search too, using the MiniZinc `let` construct (see AND/OR search in Sec. 3.3).

Search Scopes. Search scopes define the lifespan of constraints and variables in the model. MiniSearch has an implicit *global* search scope that contains all variables and constraints of the model. A new search scope can be created by using the `scope(s)` built-in that takes a MiniSearch specification `s` as an argument. When entering a scope, search is assumed to start from the root again. Whenever the execution leaves a scope, all constraints and variables that were added in the scope are removed from the model and the solver. Execution in the enclosing scope resumes from the point where it left off.

2.2 MiniSearch Control Built-Ins

All MiniSearch built-ins have an implicit return value that represents either 'success' (*true*) or 'failure' (*false*). Using this concept, we introduce MiniSearch control built-ins. All built-ins execute their arguments in order.

And, Or, Repeat. The /\-built-in runs its arguments in order and stops to return *false* as soon as one of its arguments fails. Similarly, the \/-built-in stops and returns *success* as soon as one of its arguments succeeds. Existing control mechanisms of MiniZinc such as `if then else endif` expressions can be used as well. The `repeat(s)` built-in takes a MiniSearch specification `s` and repeats it until a `break` built-in is executed; returns *false* if a break happened, otherwise returns what `s` returned. The delimited variant `repeat(i in 1..N)(s)` will execute `s` for N iterations (or until `break` is executed).

Time-Limits. The built-in `time_limit(ms,s)` imposes a time limit `ms` (in milliseconds) on any MiniSearch specification `s`. This way, `s` stops whenever the time limit is reached, returning its current status. Time-limits are handled transparently by the MiniSearch kernel as an exception.

Assigning values to constants. In standard MiniZinc constant parameters such as `int: N=10;` cannot change their value. However, in MiniSearch we often want to change constants across different iterations. For this purpose, we added the assignment operator `:=` which may only be used inside a MiniSearch specification. It overwrites that constant's current value by the value supplied.

2.3 Solution Management

The strength of any meta-search language lies in using intermediate solutions to guide the remaining search. For instance, branch-and-bound needs to access the objective to post further constraints, and a Large Neighbourhood Search thaws some of the variables in a solution to continue in that neighbourhood.

To facilitate working with solutions, the most recently found solution is always accessible in MiniSearch using the `sol` built-in, where `sol(x)` returns the value of `x` in the last solution. MiniSearch also provides a `hasSol()` built-in to test whether a solution exists.

User-defined functions. When a MiniSearch strategy is defined as a MiniZinc function, a local solution scope is created. This means that any solution found by a call to `next()` inside the function is visible for the code in the function body, but not for the caller of the function when the function returns. This architecture allows for calls to `next()` to be *encapsulated*, i.e., a function can make "tentative" calls to next in a nested search scope and only commit if these succeed. Sect. 3.3 shows how AND/OR search can be implemented based on this principle. In order to make the current solution accessible to the caller, the function must call the `commit` built-in. A function returns 'success' if it called `commit()` at least once, and 'failure' otherwise, and the last solution committed by the function will then become the current solution of the caller.

Printing Solutions & Debugging. The `print()` function without any arguments prints the last found solution in the format specified in the model's `output` item. Alternatively, `print(s)` provides more fine-grained control over the output. It prints the string `s`, which can be constructed dynamically from values in the solution using calls to `sol`. MiniSearch can be debugged using `print()` and MiniZinc's `trace()` function to display values of parameters, variables, and arbitrary expressions during search. Furthermore, the MiniSearch interpreter uses the C++ stack, so C++ debuggers can be used to follow the meta-search.

2.4 A Library of Search Strategies

Using the MiniSearch built-ins, we have defined and collected the most common meta-search approaches in the standard library `minisearch.mzn`.[1]

---

[1] The MiniSearch library is part of the open-source MiniSearch implementation [14].

These meta-search approaches can be used within any MiniZinc model that includes the library. In the next section we present some of these meta-searches in detail.

3 MiniSearch Examples

Despite MiniSearch's limited communication with the solver, it provides enough power to implement many useful complex searches that we illustrate here.

3.1 Lexicographic BaB

In multi-objective optimisation, lexicographic optimisation can be used if the objectives can be ranked according to their importance. The idea is to minimise (or maximise) an array of objectives lexicographically. Lexicographic optimisation can be more efficient than the commonly used approach of obtaining a single objective term by multiplying the components of the lexicographic objective with different constants, as the latter approach leads to large objective values, and potentially overflow.

Analogous to the implementation of branch-and-bound we post the global constraint `lex_less` so that the next solution is lexicographically smaller than the previous one. Below we show the respective MiniSearch specification.

```
function ann: minimize_lex(array[int] of var int: objs) =
   next() /\ commit() /\ print() /\
   repeat( scope(
              post(lex_less(objs, [sol(objs[i]) | i in index_set(objs)])) /\
              if next() then commit() /\ print() else break endif ) );
```

In line 2 we search for an initial solution and, if successful, repeatedly open a new scope (line 3). Then, we post the lexicographic (lex) constraint (line 4) and search for another solution in line 5. This way, in each iteration of `repeat`, we add one lex constraint, and all previously added lex constraints are removed due to the scope. This is not required but beneficial, since posting several lex-constraints can cause overhead if many intermediate solutions are found.

3.2 Large Neighbourhood Search (LNS)

Large area neighbourhood search (LNS) [20] is an essential method in the toolkit of CP practitioners. It allows CP solvers to find very good solutions to very large problems by iteratively searching large neighbourhoods (close) to optimality.

Randomised LNS explores a random neighbourhood of a given size, which can be surprisingly effective in practice as long as the neighbourhood size is chosen correctly. Cumulative scheduling is an example of a successful application [7].

The following MiniSearch specification of randomised LNS takes the objective variable, an array of decision variables that will be searched on, the number of iterations, the destruction rate (the size of the neighbourhood) and a time limit for exploring each neighbourhood. We have two scopes: in the global scope, we post BaB style constraints (line 10); in the sub-scope (line 5), we search the neighbourhoods. The predicate uniformNeighbourhood defines the neighbourhood: for each search variable we decide randomly whether to set it to its solution value of the previous solution (line 15).

```
function ann: lns(var int: obj, array[int] of var int: vars,
                  int: iterations, float: destrRate, int: exploreTime) =
  repeat (i in 1..iterations) (
    print("Iteration "++show(i)++"\n") /\
    scope(
      post(uniformNeighbourhood(vars,destrRate)) /\
      time_limit(exploreTime, minimize_bab(obj)) /\
      commit() /\ print()
    ) /\
    post(obj < sol(obj))
  );
predicate uniformNeighbourhood(array[int] of var int: x, float: destrRate) =
  if hasSol() then
    forall(i in index_set(x)) (
      if uniform(0.0,1.0) > destrRate then x[i] = sol(x[i]) else true endif )
  else true endif;
```

Adaptive LNS modifies the neighbourhood size over the course of the iterations, depending on the success of previous iterations. Below is a simple variant, where the neighbourhood size parameter nSize (line 3) is step-wise enlarged each time no solution is found (line 9). The **fail** command fails the current conjunction and will hence avoid that the post command on line 10 is executed.

```
function ann: adaptive_lns(var int: obj, array[int] of var int: vars,
                  int: iterations, int: initRate, int: exploreTime) =
  let { int: nSize = initRate, int: step = 1; } in
  repeat (i in 1..iterations) (
    print("Iteration "++show(i)++", rate="++show(nSize)++"\n") /\
    scope( ( post(uniformNeighbourhood(vars,nSize/100.0)) /\
             time_limit(exploreTime, minimize_bab(obj)) /\
             commit() /\ print()
           ) \/ (nSize := nSize + step /\ fail) )
    /\ post(obj < sol(obj))   );
```

Custom Neighbourhoods can sometimes be effective if they capture some insight into the problem structure. For instance, in a Vehicle Routing Problem (VRP), we might want to keep certain vehicle tours or vehicle-customer assignments. Below we show such a custom neighbourhood that is easily specified in MINISEARCH. The predicate keepTour (line 1) posts the tour constraints of a given vehicle number vNum. If a solution exists, the neighbourhood predicate (line 4) determines the number of customers of each vehicle (line 7), and then randomly chooses to keep the vehicle's tour (line 8) where a high customer usage results in a higher chance of keeping the vehicle. This predicate can be used instead of the uniform neighbourhood in the LNS specifications above.

```
predicate keepTour(int: vNum) =
  forall (i in 1..nbCustomers where sol(vehicle[i]) == vNum)
       ( successor[i] = sol(successor[i]) );
predicate vehicleNeighbourhood() =
  if hasSol() then
    forall (v in 1..nbVehicles) (
      let {int: usage = sum(c in 1..nbCustomers) (sol(vehicle[c]) == v) }
      in
      if usage > uniform(0,nbCustomers) then
        keepTour(v) % higher usage -> higher chance of keeping the
      vehicle
        else true endif            )
    else true endif;
```

3.3 AND/OR Search

Search in CP instantiates variables according to a systematic variable labelling, corresponding to an OR tree. AND/OR search decomposes problems into a master and several conjunctive slave sub-problems. An AND/OR search tree consists of an OR tree (master problem), an AND node with one branch for each sub-problem, and OR trees underneath for the sub-problems. A prominent example is stochastic two-stage optimisation [18], where the objective is to find optimal first-stage variable assignments (master problem) such that all second-stage variable assignments (sub-problems) are optimal for each scenario. AND/OR search for stochastic optimisation is called *policy based search* [27]. AND/OR search is also applied in other applications, such as graphical models [11].

Below is a MiniSearch example of AND/OR search for stochastic two-stage optimisation. The variables and constraints of each scenario (sub-problem) are added incrementally during search.

```
function ann: policy_based_search_min(int:sc) =
  let {
    array[1..sc] of int: sc_obj = [ 0 | i in 1..sc ];
    int: expectedCosts = infinity;
  } in (
    repeat (
      if next() then  % solution for master
        repeat (s in 1..sc) (
          scope( % a local scope for each subproblem
            let {
              array[int] of var int: recourse; % subproblem variables
              var 0..maxCosts: scenarioCosts;
            } in (
              post(setFirstStageVariables() /\ % assign master variables

                   secondStageCts(s,recourse)) /\ % subproblem
      constraints
              if minimize_bab(scenarioCosts) then
                sc_obj[s] := sol(scenarioCosts)
              else print("No solution for scenario "++show(s)++"\n") /\
                break endif
            ) )
        ) /\ % end repeat
        if expectedCosts > expectedValue(sc_obj) then
          expectedCosts := expectedValue(sc_obj) /\
          commit()  % we found a better AND/OR solution
        else skip() endif
      else break endif % no master solution
    ) );
```

```
28 % the following predicates are defined in the model according to the
      problem class
29 predicate setFirstStageVariables();
30 predicate secondStageCts(int: scenario, array[int] of var: y);
31 function int: expectedValue(var int: sc_obj);
```

Lines 3-4 initialise parameters that represent the costs for each scenario/sub-problem and the combined, expected cost of the master and subproblems. Line 7 searches for a solution to the master problem (OR tree), and if this succeeds, we continue with the AND search by finding the optimal solution for each subproblem, based on the master solution (line 9): we create each subproblem in a local scope, and add the respective variables (line 11- 12) and constraints (line 15). Furthermore, we set the master variables to the values in the master solution (line 14). Then we search for the minimal solution for the scenario (line 16), and, if successful, store it in `sc_obj` (line 17). If we find a solution for each scenario, then we compute the combined objective (expectedValue) and compare it to the incumbent solution (line 22). If we found a better solution, we store its value (line 23) and commit to it (line 24) and continue in the master scope (line 7). Otherwise, if we find no solution for one of the scenarios (line 18), the master solution is invalid. We therefore break (line 19) and continue in the master problem scope, searching for the next master solution (line 7).

3.4   Diverse Solutions

Sometimes we don't just require a satisfying or optimal solution, but a diverse set of solutions [9]. The MiniSearch specification below implements the greedy approximation method that iteratively constructs a set of K diverse solutions:

```
1  function ann: greedy_maxDiverseKset(array[int] of var int: Vars, int: K) =
2    let { array[int,int] of int: Store = array2d(1..K, index_set(Vars),
3                                    [0 | x in 1..K*length(Vars)]),
4          int: L = 0 % current length
5    } in
6    next() /\ commit() /\ print() /\
7    repeat(
8      L := L+1 /\ repeat(i in 1..length(Vars)) (Store[L,i] := sol(Vars[i])) /\
9      if L <= K then
10       scope(
11         let {var int: obj;} in
12         post(obj = sum(j in 1..L,i in index_set(Vars))(Store[j,i] != Vars[i])) /\
13         maximize_bab(obj) /\ commit() /\ print() )
14       else print(show(Store)++"\n") /\ break endif
15   );
```

The first few lines initialise the `Store`, which will contain the K diverse solutions, as well as the current *length* of the store up to which it already contains solutions. On line 8 the length is increased and the previously found solution is saved. If the length does not exceed K, we construct a new objective on line 12. This objective expresses how different a solution is from the previously found solutions using the Hamming distance. This objective is then maximised, resulting in the next most diverse solution, and the process repeats.

3.5   Interactive Optimisation

Interactive optimisation lets users participate in the solving process by inspecting solutions, adding constraints, and then re-solving. This has been shown to improve the trust of end-users into decision support systems, and a number of successful application have been implemented [2].

MINISEARCH supports interactive optimisation by calling MINIZINC built-ins that ask for user input. The MINIZINC library contains functions such as `read_int()` which accept keyboard input. More advanced input facilities can be added through user-defined MINIZINC built-ins that execute arbitrary C++ code, such as consulting a user through a graphical user interface or other means.

The following is an example of an interactive Vehicle Routing Problem solver implemented using MINISEARCH. We use a Large Neighbourhood search where in every N-th iteration, the user can chose a vehicle route that should be kept. This code can be used with the LNS implementations described earlier.

```
predicate interactiveNeighbourhood(int: iteration, int: N) =
  if iteration mod N = 0 then
    let { string: msg = "Enter a vehicle tour to keep (0 for none):\n";
          int: n = read_int(msg)
    } in  if n > 0 then keepTour(n) else true endif
  else true endif;
```

4   The MINISEARCH Kernel

This section describes the architecture of the MINISEARCH *kernel*, the engine that executes MINISEARCH specifications, interacting with both the MINIZINC compiler and the backend solver.

First, let us briefly review how MINIZINC models are solved. The MINIZINC compiler (usually invoked as `mzn2fzn`) takes text files containing the model and instance data and compiles them into FLATZINC, a low-level language supported by a wide range of solvers. The compiler generates FLATZINC that is specialised for the capabilities of the particular target solver using a solver-specific *library* of predicate declarations. Solvers read FLATZINC text files and produce text-based output for the solutions. Version 2.0 of MINIZINC is based on the `libminizinc` C++ library, which provides programmatic APIs, eliminating the need to communicate through text files. The library also defines an API for invoking solvers directly in C++. MINISEARCH is built on top of these APIs.

4.1   The MINISEARCH Interpreter

The MINISEARCH kernel implements an *interpreter* that processes MINISEARCH specifications and handles the communication between the MINIZINC compiler and the solver. The interpreter assumes that the solver interface is *incremental*, i.e. variables and constraints can be added dynamically during search. We provide an *emulation layer* (see Sec. 4.2) for solvers that do not support incremental operations, including solvers that cannot be accessed through the C++ interface.

The MINISEARCH interpreter is a simple stack-based interpreter that maintains the following state: A stack of *solutions*, one for each function scope; a stack of *time-outs* and *breaks*; and a stack of *search scopes*, containing the solver state of each scope. The interpreter starts by compiling the MINIZINC model into FLATZINC, and then interprets each MINISEARCH built-in as follows:

- `next()` invokes the solver for a new solution; if successful, it replaces the top-most solution on solution stack. If a time-out has been set, the call to the solver is only allowed to run up to the time-out. Returns `true` iff a new solution was found.
- `commit()` replaces the parent solution (2nd on the stack) with the current solution. This commits the current solution into the parent function scope. Returns `true`. Aborts if not in a function call.
- *function calls* duplicate the top of the solution stack before executing the function body. Return `true` if the function committed a new solution.
- `time_limit(l,s)` adds a new time-out `now+l` to the stack, executes `s`, and then pops the time-out. During the execution of `s`, calls to `next` and `repeat` check whether any time-outs `t` have expired (`t > now`), and if so they immediately break. Returns whatever `s` returned.
- `repeat(s)` pushes a break scope with a Boolean flag on the stack, then repeats the execution of `s` as long as the break flag is false and pops the break scope when `s` is finished. The `break` construct sets the break flag in the current break scope (similar to a time limit).
- `post(c)` compiles the MINIZINC expression `c` into FLATZINC. The compilation is incremental, i.e., the result of compiling `c` is added to the existing FLATZINC. The interpreter then adds the newly generated variables and constraints to the current solver instance.
- `scope(s)` creates a new local scope. The current implementation copies the flat model and creates a new solver instance based on this flat model.
- Other operations (`/\`,`\/`,`print`) are interpreted with respect to their semantics.

4.2   Emulating Advanced Solver Behaviour

A key goal of this work is to make MINISEARCH available for any solver that supports FLATZINC. Current FLATZINC solvers, however, neither support incrementally adding constraints and variables, nor do they implement the `libminizinc` C++ API. We therefore need to *emulate* the incremental API.

In order to emulate incrementality, we can implement the dynamic addition of variables and constraints using a restart-based approach, re-running the solver on the entire updated FLATZINC. To avoid re-visiting solutions, after each call to `next` the emulation layer adds a no-good to the model that excludes the solution that was just found. This emulation reduces the requirements on the solver to simply being able to solve a given FLATZINC model.

In order to emulate the C++ API, we generate a textual representation of the FLATZINC and call an external solver process. The emulator then converts the textual output of the process back into `libminizinc` data structures. Using this emulation of the incremental C++ solver API, any current FLATZINC solver can be used with the MINISEARCH kernel.

4.3 Built-in Primitives

Solvers can declare native support for a MINISEARCH search specification. For every MINISEARCH call such as `f(x,y,z)`, the kernel will check whether the declaration of `f` has a function body. If it does not have a function body, the function is considered to be a solver built-in, and it is executed by passing the call to a generic `solve` function in the C++ solver API. This is similar to the handling of global constraints in MINIZINC, where solvers can either use their own primitives for a given global constraint, or use the respective MINIZINC decomposition of the global constraint. This way solvers can easily apply their own primitives, but can use alternatives if they do not support a certain feature.

5 Experiments

In our experimental evaluation, we analyse MINISEARCH on practical examples to study the efficiency and compare the overhead of MINISEARCH for different interfaces. Furthermore, we study the benefits of specialised, heuristic MINISEARCH approaches to standard branch-and-bound optimisation; the only available option in solver-independent CP modelling languages.

5.1 Experimental Setup

The problems and instances are taken from the MINIZINC benchmarks repository[2]. The source code of MINISEARCH and the respective solver interfaces will be released in September 2015 [14].

Experiments are run on Ubuntu 14.04 machines with eight i7 cores and 16GB of RAM. The MINISEARCH kernel is based on the MINIZINC 2.0.1 toolchain. The native Gecode interface and incremental C++ interface were implemented using the latest Gecode source code (version 4.3.3+, 20 April 2015). The FZN solvers used are: Gecode (20 April 2015), Choco 3.3.0, Or-tools source (17 February 2015), and Opturion CPX 1.0.2. All solvers use the same variable/value ordering heuristic that is part of the MINIZINC model.

5.2 Overhead of Different MINISEARCH Interfaces

First, we study the comparative overhead of MINISEARCH with respect to the different interface levels. We do this by analysing the performance of a meta-search approach on the same solver, Gecode. We compare the following four approaches: the solver's *native* approach (without MINISEARCH) through

---

[2] https://github.com/MiniZinc/minizinc-benchmarks

Table 2. Rectangle Packing. Times (sec) averaged over 10 runs. Comparing FLATZINC-solvers with MINISEARCH BaB through FLATZINC (MS-F), MINISEARCH BaB through incremental API (MS-Inc), and native BaB with FLATZINC (Nat-F).

Rectangle Size (N)	Gecode			or-tools		choco		Opturion CPX	
	MS-F	MS-Inc	Nat-F	MS-F	Nat-F	MS-F	Nat-F	MS-F	Nat-F
14	0.7	0.1	0.2	0.5	0.2	6.4	1.0	0.3	0.1
15	0.8	0.3	0.4	0.6	0.3	6.0	1.3	0.4	0.2
16	1.3	0.5	0.5	1.1	0.5	9.1	1.6	0.6	0.2
17	4.5	2.9	2.9	5.2	3.8	16.6	5.7	0.9	0.4
18	37.4	37.4	36.8	41.2	41.5	38.0	29.8	6.7	6.2
19	77.5	43.2	42.0	59.0	32.8	59.2	28.8	12.7	7.7
20	96.4	97.9	94.1	99.2	98.9	93.9	83.0	6.4	5.6
21	472.4	469.7	462.5	416.3	410.0	250.1	239.2	83.8	82.1

FLATZINC (Nat-F), the *incremental API* interface to MINISEARCH (MS-Inc) and the *non-incremental text-based* (MS-F) interface to MINISEARCH. We use Rectangle Packing [21] as problem benchmark, and standard BaB as meta-search, since it is available natively in FLATZINC for Gecode.

The results are summarised in Table 2, columns 2–4. It shows the runtimes (in seconds) taken to find the optimal solution for the different approaches using Gecode. The native approach through the FlatZinc interface is fastest, though MINISEARCH through the incremental API interface performs quite similarly. Moreover, MINISEARCH through the FZN interface is, as expected, slower, but only by a small factor. These results are very promising, since they show that the overhead of the MINISEARCH interfaces is low.

In addition, we analyse the overhead of MINISEARCH for other FLATZINC solvers in Table 2, columns 5–10. We ran the Rectangle Packing problem for the FLATZINC solvers or-tools, Choco, and Opturion CPX on a MINIZINC model using both native branch-and-bound (Nat-F), and MINISEARCH branch-and-bound (MS-F), both through the FLATZINC interface. Overall the MINISEARCH FLATZINC interface, while slower, still gives decent performance. Choco seems to incur most overhead of the repeated restarts.

5.3 Heuristic Search

MINISEARCH strategies are not restricted to complete search. For example, different variants of Large Neighbourhood Search can be tried. To demonstrate the benefits of having heuristic search at hand, we compare the Randomised LNS approach from Section 3.2 with the standard out-of-the-box branch-and-bound approach for a selection of FLATZINC solvers on the Capacitated Vehicle Routing problem (CVRP).

We run each solver on (1) the standard MINIZINC model that uses branch-and-bound, and (2) the MINIZINC model with the MINISEARCH specification of Randomised LNS. We use the Augerat et al [1] CVRP instance sets A,B and P, where the 27 Set-A instances contain random customer locations and demands,

Table 3. Average relative improvement of MINISEARCH LNS over standard BaB for the achieved objective within a time limit for different solvers. The figures are averages over the Augerat et al [1] CVRP instance sets A,B and P.

CVRP Instance-Set	Gecode	or-tools	choco	Opt. CPX
Set-A	11.60%	11.76%	11.17%	12.11%
Set-B	13.38%	11.82%	12.62%	14.92%
Set-P	9.78%	10.53%	7.98%	11.35%

the 22 Set-B instances have clustered locations and the 23 Set-P instances are modified versions of instances from the literature. The LNS approach uses a destroy rate of 0.3 (the neighbourhood size is 30%) and an exploration timeout (for each neighbourhood) of 5 seconds. Both the LNS and BaB approach have an overall 2 minute timeout, and we report the quality of the best solution found on time-out.

We can see that LNS provides better solutions than BaB, and Table 3 reports the *average* improvement of the objective of LNS over BaB on the three instance sets. We observe an improvement between 8-15%, which is considerable within the given time limit. Furthermore, we see that the improvements are similar for all solvers across the instances sets, in particular for the Set-B instances. This demonstrates the added value of MINISEARCH search strategies to the existing out-of-the-box optimisation approaches of standard modelling languages.

6   Related Work

The starting point for MINISEARCH is search combinators [19], and MINISEARCH is indeed a combinator language. The MINISEARCH language is more restricted than search combinators. In particular, search combinators can interact with the search at every node of the tree, essentially *replacing* the solver's built-in search, while MINISEARCH only interacts with the solver's search at the granularity of solutions. Examples not possible in MINISEARCH are the use of average depth of failure or custom limits based on how many search nodes with a particular property have been encountered, and heuristics such as limited discrepancy search [8] and iterative deepening. Other search combinator expressions are however expressible with MINISEARCH, as witnessed by the library of strategies presented in this paper.

Neither search combinators nor MINISEARCH support fully programmable search. Especially not variable and value ordering heuristics, which interact with the internal data structures of the solver. These kinds of search strategies are important, but cannot be implemented without deep hooks into the solver.

Objective CP [26] can express complex meta-search at the model level, which can then be executed by any of the Objective CP solvers. For example a CP-style search can be implemented using a MIP solver. So it also provides solver-independent search, but through a much more fine grained interface, supporting interaction at each search node. The fine grained interface however means that adding a new solver to Objective CP is complex, in particular supporting the reverse mapping of information from solver objects to model objects.

Possibly the closest system to MINISEARCH is OPL Script [25], a scripting language for combining models. It essentially also communicates to solvers via models and solutions. Distinct from MINISEARCH is that it is object-oriented and provides models as first-class objects, and hence supports the construction and modification of multiple models, as opposed to our simpler mechanisms. It can communicate more information than just variables and constraints, including dual values, bases, and various statistics. However, the language is tightly linked to the CP/MIP solver ILOG CPLEX CP Optimizer. MINISEARCH on the other hand can use any FlatZinc solver.

AMPL Script [4] is similar in many respects to OPL script, though a more basic scripting language, not supporting function definitions for example. It does not have a concept of scopes, but instead supports adding, dropping and restoring variables and constraints from the model dynamically. It interacts with the solver through requests to solve, which continue until some termination condition arises, like optimality proven or a timeout. Hence its natural interaction level is more like `minimize_bab()` than `next()`.

Any meta-search that can be implemented in MINISEARCH could also be implemented directly in a solver such as Gecode, Choco, Comet, or CP Optimizer, although the implementation burden for doing so could be significant. Similarly, APIs for CP embedded in scripting languages such as Numberjack [10] (through Python) could offer a MINISEARCH-like interface. The benefits of MINISEARCH are that it is programming language- and solver-independent, that it abstracts away solution and timeout handling. Furthermore, meta-search languages are easier to use when they extend the language for expressing constraints, in this case MiniZinc, since this is often a vital component of the search.

7   Conclusion and Future Work

In this paper we present MINISEARCH, a solver-independent lightweight language to specify meta-search in MINIZINC models. The first contribution of MINISEARCH is its *expressiveness*: we showed how to formulate different complex search approaches in MINISEARCH, such as Large Neighbourhood Search or AND/OR search. It reuses the MINIZINC modelling language for specifying constraints and variables to add during search. It even enables the implementation of interactive optimisation approaches where the user guides the search process.

The second contribution is the *minimal interface* of MINISEARCH. Since the kernel interprets all statements, the communication between the solver and MINISEARCH is only necessary at every *solution* (instead of at every search *node* as in search combinators). This makes it more realistic to implement for solver developers. Furthermore, in our experiments we observed that both the incremental and the text-based interface provide satisfactory performance.

The third and most important contribution is *solver-independence*. In contrast to existing search languages, MINISEARCH can already use any solver that can process FLATZINC, which is the majority of CP solvers. This allows users to test different solvers with complex meta-searches without having to commit to one single solver.

For future work, we plan to extend MINISEARCH with parallelism, so that independent searches can be executed in parallel and advanced parallel meta-search approaches such as Embarrassingly Parallel Search [17] can be specified. Furthermore, we plan to integrate portfolio search into MINISEARCH, so that different search functions can be executed by different solvers.

Acknowledgments. NICTA is funded by the Australian Government through the Department of Communications and the Australian Research Council through the ICT Centre of Excellence Program. T.G. is supported by a Postdoc grant from the Research Foundation – Flanders.

References

1. Augerat, P., Belenguer, J., Benavent, E., Corberan, A., Naddef, D., Rinaldi, G.: Computational results with a branch and cut code for the capacitated vehicle routing problem. Technical Report 949-M. Universite Joseph Fourier, Grenoble (1995)
2. Belin, B., Christie, M., Truchet, C.: Interactive design of sustainable cities with a distributed local search solver. In: Simonis, H. (ed.) CPAIOR 2014. LNCS, vol. 8451, pp. 104–119. Springer, Heidelberg (2014)
3. Flener, P., Pearson, J., Ågren, M.: Introducing ESRA, a relational language for modelling combinatorial problems (Abstract). In: Rossi, F. (ed.) CP 2003. LNCS, vol. 2833, pp. 971–971. Springer, Heidelberg (2003)
4. Fourer, R., Gay, D.M., Kernighan, B.W.: AMPL: A Modeling Language for Mathematical Programming. Cengage Learning (2002)
5. Frisch, A.M., Harvey, W., Jefferson, C., Hernández, B.M., Miguel, I.: Essence : A constraint language for specifying combinatorial problems. Constraints **13**(3), 268–306 (2008)
6. Gent, I.P., Miguel, I., Rendl, A.: Tailoring solver-independent constraint models: a case study with ESSENCE′ and MINION. In: Miguel, I., Ruml, W. (eds.) SARA 2007. LNCS (LNAI), vol. 4612, pp. 184–199. Springer, Heidelberg (2007)
7. Godard, D., Laborie, P., Nuijten, W.: Randomized large neighborhood search for cumulative scheduling. In: Proceedings of the Fifteenth International Conference on Automated Planning and Scheduling (ICAPS 2005), Monterey, California, USA, June 5–10 2005, pp. 81–89 (2005)
8. Harvey, W.D., Ginsberg, M.L.: Limited discrepancy search. In: Proceedings of the 14th IJCAI, pp. 607–613 (1995)
9. Hebrard, E., Hnich, B., O'Sullivan, B., Walsh, T.: Finding diverse and similar solutions in constraint programming. In: Veloso, M.M., Kambhampati, S. (eds.) AAAI, pp. 372–377. AAAI Press / The MIT Press (2005)
10. Hebrard, E., O'Mahony, E., O'Sullivan, B.: Constraint programming and combinatorial optimisation in numberjack. In: Lodi, A., Milano, M., Toth, P. (eds.) CPAIOR 2010. LNCS, vol. 6140, pp. 181–185. Springer, Heidelberg (2010)
11. Marinescu, R., Dechter, R.: AND/OR branch-and-bound search for combinatorial optimization in graphical models. Artif. Intell. **173**(16–17), 1457–1491 (2009)

12. Marriott, K., Nethercote, N., Rafeh, R., Stuckey, P.J., de la Banda, M.G., Wallace, M.: The design of the Zinc modelling language. Constraints **13**(3), 229–267 (2008)
13. Michel, L., Van Hentenryck, P.: The comet programming language and system. In: van Beek, P. (ed.) CP 2005. LNCS, vol. 3709, p. 881. Springer, Heidelberg (2005)
14. MiniSearch release. http://www.minizinc.org/minisearch
15. MiniZinc challenge. http://www.minizinc.org/challenge.html
16. Nethercote, N., Stuckey, P.J., Becket, R., Brand, S., Duck, G.J., Tack, G.R.: MiniZinc: towards a standard CP modelling language. In: Bessière, C. (ed.) CP 2007. LNCS, vol. 4741, pp. 529–543. Springer, Heidelberg (2007)
17. Régin, J.-C., Rezgui, M., Malapert, A.: Embarrassingly parallel search. In: Schulte, C. (ed.) CP 2013. LNCS, vol. 8124, pp. 596–610. Springer, Heidelberg (2013)
18. Ruszczyński, A., Shapiro, A.: Stochastic Programming. Handbooks in operations research and management science. Elsevier (2003)
19. Schrijvers, T., Tack, G., Wuille, P., Samulowitz, H., Stuckey, P.J.: Search combinators. Constraints **18**(2), 269–305 (2013)
20. Shaw, P.: Using constraint programming and local search methods to solve vehicle routing problems. In: Maher, M.J., Puget, J.-F. (eds.) CP 1998. LNCS, vol. 1520, pp. 417–431. Springer, Heidelberg (1998)
21. Simonis, H., O'Sullivan, B.: Search strategies for rectangle packing. In: Stuckey, P.J. (ed.) CP 2008. LNCS, vol. 5202, pp. 52–66. Springer, Heidelberg (2008)
22. Stuckey, P.J., Feydy, T., Schutt, A., Tack, G., Fischer, J.: The MiniZinc challenge 2008–2013. AI Magazine **35**(2), 55–60 (2014)
23. Stuckey, P.J., Tack, G.: Minizinc with functions. In: Gomes, C., Sellmann, M. (eds.) CPAIOR 2013. LNCS, vol. 7874, pp. 268–283. Springer, Heidelberg (2013)
24. Van Hentenryck, P.: The OPL Optimization Programming Language. MIT Press, Cambridge (1999)
25. Van Hentenryck, P., Michel, L.: OPL script: composing and controlling models. In: Apt, K.R., Kakas, A.C., Monfroy, E., Rossi, F. (eds.) Compulog Net WS 1999. LNCS (LNAI), vol. 1865, pp. 75–90. Springer, Heidelberg (2000)
26. Van Hentenryck, P., Michel, L.: The objective-CP optimization system. In: Schulte, C. (ed.) CP 2013. LNCS, vol. 8124, pp. 8–29. Springer, Heidelberg (2013)
27. Walsh, T.: Stochastic Constraint Programming. In: van Harmelen, F. (ed.) ECAI, pp. 111–115. IOS Press (2002)

Two Clause Learning Approaches
for Disjunctive Scheduling

Mohamed Siala[1,2,3]($\boxtimes$), Christian Artigues[2,4], and Emmanuel Hebrard[2,4]

[1] Insight Centre for Data Analytics, University College Cork, Cork, Ireland
mohamed.siala@insight-centre.org, siala@laas.fr
[2] CNRS, LAAS, 7 Avenue du Colonel Roche, F-31400 Toulouse, France
[3] Univ. de Toulouse, INSA, LAAS, F-31400 Toulouse, France
[4] Univ. de Toulouse, LAAS, F-31400 Toulouse, France
{artigues,hebrard}@laas.fr

Abstract. We revisit the standard hybrid CP/SAT approach for solving disjunctive scheduling problems. Previous methods entail the creation of redundant clauses when lazily generating atoms standing for bounds modifications. We first describe an alternative method for handling lazily generated atoms without computational overhead. Next, we propose a novel conflict analysis scheme tailored for disjunctive scheduling. Our experiments on well known Job Shop Scheduling instances show compelling evidence of the efficiency of the learning mechanism that we propose. In particular this approach is very efficient for proving unfeasibility.

1 Introduction

Disjunctive scheduling refers to a large family of scheduling problems having in common the *Unary Resource Constraint*. This constraint ensures that a set of tasks run in sequence, that is, without any time overlap. The traditional constraint programming (CP) approaches for this problem rely on tailored propagation algorithms (such as *Edge-Finding* [6,17,24]) and search strategies (such as *Texture* [20]). The technique of *Large Neighborhood Search* [22] (LNS) was also extensively used in this context [7,25].

A different type of approaches emerged recently, based on the so-called *Conflict-Driven Clause Learning* (CDCL) algorithm for SAT [16]. This procedure uses *resolution* to learn a new clause for each conflict during search. Recent constraint programming approaches similarly trade off strong propagation-based inference for a way to learn during search. For instance, a hybrid CP/SAT method, *Lazy Clause Generation* (LCG) [8,9,18] was shown to be extremely efficient on the more general *Resource Constrained Project Scheduling Problem* [21] (*RCPSP*). Even simple heuristic weight-based learning was shown to be very efficient on disjunctive scheduling [10–13]. The so called *light model* combines minimalist propagation with a slight variant of the *weighted degree* [5] heuristic.

In this paper, we propose a hybrid CP/SAT method based on this light model. Similarly to LCG, our approach mimics CDCL. However, it differs from

© Springer International Publishing Switzerland 2015
G. Pesant (Ed.): CP 2015, LNCS 9255, pp. 393–402, 2015.
DOI: 10.1007/978-3-319-23219-5_28

LCG in two main respects: First, as the time horizon can be large, literals representing changes in the bounds of the tasks domains should be generated "lazily" during conflict analysis as it was proposed in [8]. However, handling domain consistency through clauses entails redundancies and hence suboptimal propagation. We use a dedicated propagation algorithm, running in constant amortized time, to perform this task. The second contribution is a novel conflict analysis scheme tailored for disjunctive scheduling. This technique could be applied to any problem where search can be restricted to a predefined set of Boolean variables. This is the case of disjunctive scheduling since once every pair of tasks sharing the same resource is sequenced, we are guaranteed to find a complete solution in polynomial time. Most methods therefore only branch on the relative order of tasks sharing a common resource. We propose to use this property to design a different conflict analysis method where we continue *resolution* until having a nogood with only Boolean variables standing for task ordering. As a result, we do not need to generate domain atoms.

We compare the two methods experimentally and show that the benefit of not having to generate new atoms during search outweigh in many cases the more expressive language of literals available in traditional hybrid solvers. The novel approach is very efficient, especially for proving unfeasibility. We were able to improve the lower bounds on several well known *Job Shop Scheduling Problem* (*JSSP*) benchmarks. However, a method implemented within IBM CP-Optimizer studio recently found, in general, better bounds [25].

The rest of the paper is organized as follows: We give a short background on hybrid CP/SAT solving in Section 2. Next, we briefly describe the baseline CP model and strategies that we use in Section 3. We introduce in Section 4 our new lazy generation approach. In Section 5, we present our novel conflict analysis scheme. Last, we give and discuss the experimental results in Section 6.

2   Hybrid CP/SAT Solving

In this section, we briefly describe the basic mechanisms and notations of hybrid CP/SAT used in modern Lazy Clause Generation solvers [8,9,18].

Let $\mathcal{X} = [x_1, .., x_n]$ be a sequence of variables. A *domain* $\mathcal{D}$ maps every variable $x \in \mathcal{X}$ to a finite set $\mathcal{D}(x) \subset \mathbb{Z}$. $\mathcal{D}(x)$ is a *range* if $\max(\mathcal{D}(x)) - \min(\mathcal{D}(x)) + 1 = |\mathcal{D}(x)|$. A *constraint* C is defined by a relation over the domains of a sequence of variables. In this context, each constraint is associated to a *propagator* to reduce the domains with respect to its relation. Propagators are allowed to use only *domain operations* of a given type (often $[\![x = v]\!]$, $[\![x \neq v]\!]$, $[\![x \leq v]\!]$, and $[\![x \geq v]\!]$). Every domain operation is associated to a *literal* p that can have an *atom* (i.e. Boolean variable) generated "eagerly" before starting the search, or "lazily" when learning a conflict involving this change. Let p be a literal corresponding to a domain operation. $level(p)$ is the number of nodes in the current branch of the search tree when p becomes true (i.e., when this domain operation takes place). Moreover, $rank(p)$ is the number of such domain operations that took place at the same node of the search tree before p. Last, a literal p is associated to a logical implication of the form $\wp \Rightarrow p$ where $\wp$ is a conjunction of literals sufficient for a given propagator to deduce p. We call $\wp$ an *explanation* of p. We assume that the function $explain(p)$ returns the explanation $\wp$ for p. The explanations can be computed at the moment of propagation (i.e., *clausal* and *forward* explanations in [8]), or on demand during conflict analysis (*backward*). The notion of explanation is extended to a failure $\perp$ in the natural way.

Whenever a failure occurs during search, a conflict analysis procedure is used to compute a nogood, that is, an explanation of the failure which is not yet logically implied by the model. Algorithm 1 depicts the conflict analysis procedure based on the 1-UIP scheme [27]. The nogood Ψ is initialized in Line 1 to the explanation for the failure. It is returned once it contains a single literal that has been propagated at the current level (Line 2). When this is not the case, the last propagated literal is selected and its explanation is resolved with the current nogood in Line 3. The final nogood Ψ can be seen as $\bigwedge_{i \in [1,n]} p_i \Rightarrow p$. The solver then backtracks to the level $l = \max(level(p_i))$.

Algorithm 1. 1-UIP-with-Propagators

1   $\Psi \leftarrow explain(\perp)$;
2   **while** $|\{q \in \Psi \mid level(q) = current\ level\}| > 1$ **do**
    $p \leftarrow \arg\max_q(\{rank(q) \mid level(q) = current\ level \ \wedge \ q \in \Psi\})$;
3     $\Psi \leftarrow \Psi \cup \{q \mid q \in explain(p) \wedge level(q) > 0\} \setminus \{p\}$;
   return Ψ ;

Consider now the lazy generation of atoms (i.e. after computing Ψ). When $[\![x \leq u]\!]$ has to be generated, the clauses $\neg[\![x \leq a]\!] \vee [\![x \leq u]\!]$; $\neg[\![x \leq u]\!] \vee [\![x \leq b]\!]$ where a and b are the nearest generated bounds to u with $a < u < b$ must be added to maintain the consistency of $[\![x \leq u]\!]$ with respect to previously generated atoms. In this case, the clause $\neg[\![x \leq l]\!] \vee [\![x \leq u]\!]$ becomes redundant.

3   A Simple CP Model for Job Shop Scheduling

A JSSP consists in sequencing the tasks from n jobs on m machines. The jobs $\mathcal{J} = \{J_i \mid 1 \leq i \leq n\}$ define sequences of tasks to be scheduled in a given order. Moreover, each of the m tasks of a job must be processed by a different machine. We denote T_{ik} the task of job J_i requiring machine M_k. Each task T_{ik} is associated to a processing duration p_{ik}. Let t_{ik} be the variable representing the starting time of task T_{ik}. The processing interval of a task T_{ik} is $[t_{ik}, t_{ik} + p_{ik})$. For all $k \in [1, m]$, the Unary Resource Constraint for machine M_k ensures that for any $i \neq j \in [1, n]$, there is no overlap between the processing interval of T_{ik} and T_{jk}. We use a simple decomposition with $O(n^2)$ Boolean variables δ_{kij} $(i < j \in [1, n])$ per machine M_k using the following DISJUNCTIVE constraints:

$$\delta_{kij} = \begin{cases} 0 \Leftrightarrow t_{ik} + p_{ik} \leq t_{jk} \\ 1 \Leftrightarrow t_{jk} + p_{jk} \leq t_{ik} \end{cases} \tag{1}$$

A JSSP requires also a total order on the tasks of each job. We enforce this order with simple precedence constraints. Last, the objective is to minimize the total scheduling duration, i.e., the makespan. To this purpose, we have an objective variable C_{max} subject to precedence constraints with the last task of each job.

We use the same search strategy as that proposed in [10,13]. First, we compute a greedy upper bound on the makespan. Then we use a dichotomic search to reduce the gap between lower and upper bounds. Each step of this dichotomic phase corresponds to the decision problem with the makespan variable C_{max} fixed to the mid-point value. Each step is given a fixed time cutoff, and exceeding it is interpreted as the instance being unsatisfiable. Therefore, the gap might not be closed at the end of this phase, so a classic branch and bound procedure is used until either closing the gap or reaching a global cutoff.

We branch on the Boolean variables of the DISJUNCTIVE constraints following [13] using the solution guided approach [3] for value ordering. We use the weighed degree heuristic $taskDom/tw$ in a similar way to [10,13] in addition to VSIDS [16]. The former branches first on variables occurring most in constraints triggering failures. The latter favors variables involved in conflict analysis.

4   Lazy Generation of Atoms

In this section we describe an efficient propagator to maintain the consistency of lazily generated atoms. Recall that domain clauses are likely to be redundant in this case (Section 2). Instead of generating clauses to encode the different relationships between the newly generated atoms, we propose to encode such relations through a dedicated propagator in order to avoid this redundancy and hence the associated overhead. We introduce the DOMAINFAITHFULNESS constraint. Its role is twofold: firstly, it simulates *Unit-Propagation* (UP) as if the atoms were generated eagerly; secondly it performs a complete channeling between the range variable and all its generated domain atoms.

We consider only the lazy generation of atoms of the type $[\![x \leq u]\!]$ since all propagators in our models perform only bound tightening operations. Nevertheless, the generalization with $[\![x = v]\!]$ is quite simple and straightforward. Let x be a range variable, $[v_1, \ldots, v_n]$ be a sequence of integer values, and $[b_1 \ldots b_n]$ be a sequence of lazily generated Boolean variables s.t. b_i is the atom $[\![x \leq v_i]\!]$. We define DOMAINFAITHFULNESS$(x, [b_1 \ldots b_n], [v_1, \ldots, v_n])$ as follows: $\forall i, b_i \leftrightarrow x \leq v_i$.

We describe now the data structures and the procedures that allow to maintain Arc Consistency [4] (AC) with a constant amortized time complexity down a branch of the search tree. We assume without loss of generality that $n \geq 1$ and that $v_i < v_{i+1}$. The filtering is then organized in two parts:

1. Simulating UP as if the atoms were eagerly generated with all domain clauses. That is, whenever an atom b_i becomes assigned to 1 (respectively 0), the atom b_{i+1} (respectively b_{i-1}) is assigned to 1 (respectively 0).
2. Performing a complete channeling between x and $b_1, \ldots, b_n$: We sketch the process related to the upper bound of x. A similar process in applied with the lower bound. There are two cases to distinguish:

(a) Changing the upper bound of x w.r.t. newly assigned atoms: When an atom $b_i \leftrightarrow [\![x \leq v_i]\!]$ becomes assigned to 1, we check if v_i can be the new upper bound of x. Note that a failure can be triggered if $v_i < \min(x)$.

(b) In the case of an upper bound propagation event, one has to assign some atoms to be consistent with the new upper bound u of x. Every atom b_i such that $v_i \geq u$ has to be assigned to 1. Let $i_{ub} = \arg\max_k(v_k \mid \mathcal{D}(b_k) = \{1\})$. Observe first that the part simulating UP (1) guarantees that all atoms b_i where $i > i_{ub}$ are already assigned to 1. Consider now the set $\varphi = \{b_{i_{ub}}, b_{i_{ub}-1}, b_{i_{ub}-2} \ldots, b_{last_{ub}}\}$ where $last_{ub} = \arg\min_k(v_k \mid v_k \geq u)$. It is now sufficient to assign every atom in φ to 1 in order to complete the filtering.

Finally, since down a branch of the search tree the upper bound can only decrease, we can compute the current value of i_{ub}, that is, $\arg\max_k(v_k \mid \mathcal{D}(b_k) = \{1\})$ by exploring the sequence of successive upper bounds from where we previously stopped. Therefore, each atom is visited at most once down a branch. This filtering can thus be performed in constant amortized time, that is, in $O(n)$ time down a branch, however, we must store i_{ub} and i_{lb} as "reversible" integers.

5 Learning Restricted to Task Ordering

Here we introduce a learning scheme as an alternative to lazy generation for disjunctive scheduling problems. Recall that we branch only on Boolean variables coming from the DISJUNCTIVE constraints. It follows that every bound literal p s.t. $level(p) > 0$ has a non-empty explanation. We can exploit precisely this property in order to avoid the generation of any bound atom.

The first step in this framework is to perform conflict analysis as usual by computing the 1-UIP nogood Ψ. Next, we make sure that the latest literal in Ψ is not a bound literal. Otherwise, we keep explaining the latest literal in Ψ until having such UIP. We know that this procedure terminates because the worst case would reach the last decision which corresponds to a UIP that is not a bound literal. Let Ψ^* be the resulting nogood.

Consider now $\Im$ to be the set of bound literals in Ψ^*. Instead of generating the corresponding atoms, we start a second phase of conflict analysis via the procedure Substitute($\Im, \Psi$) with ($\Im, \Psi^*$) as input. Algorithm 2 details this procedure. In each step of the main loop, a bound literal p from $\Im$ is chosen (Line 1) and replaced in Ψ with its explanation (Line 2). $\Im$ is updated later at each step in Line 3. The final nogood Ψ involves only Boolean variables. Note that the backjump level remains the same as in Ψ^* since for every p there must exists a literal in $explain(p)$ with the same level of p.

The advantage of this approach is that since no atom need to be generated during search, we do not need to maintain the consistency between tasks' domains and bounds atoms. In particular, it greatly reduces the impact that the size of the scheduling horizon has on the space and time complexities. Note, however, that there may be a slight overhead in the conflict analysis step, compared to the lazy generation mode, since there are more literals to explain. Moreover,

Algorithm 2. Substitute($\mathfrak{I}, \Psi$)

$visited \leftarrow \emptyset$;
while $|\mathfrak{I}| > 0$ **do**
1      $p \leftarrow$ choose $p \in \mathfrak{I}$;
     $visited \leftarrow visited \cup \{p\}$;
2      $\Psi \leftarrow \Psi \cup \{q \mid q \in explain(p) \wedge level(q) > 0\} \setminus visited$;
3      $\mathfrak{I} \leftarrow \{q \mid q \in \Psi \wedge q$ is a bound litteral$\}$;
return Ψ ;

---

since the language of literal is not as rich in this method, shorter proofs may be possible with the standard approach.

6 Experimental Results

We first compare the two approaches described in this paper, that is, the implementation of lazy generation using DOMAINFAITHFULNESS (*lazy*) as well as the new learning scheme (*disj*) against the the CP model described in [11] on two well known benchmarks for the JSSP: Lawrence [14] and Taillard [23]. Then, we compare the lower bounds found by our approaches with the best known lower bounds. All models are implemented within Mistral-2.0 and are tested on Intel Xeon E5-2640 processors. The source code, the detailed results, and the experimental protocol are available at **http://siala.github.io/jssp/details.pdf**.

We denote CP(*task*) the CP model using the *taskDom/tw* heuristic. For the hybrid models, we use the notation H(θ, σ) where $\theta \in \{vsids, task\}$ is the variable ordering and $\sigma \in \{lazy, disj\}$ indicates the type of learning used.

6.1 Empirical Evaluation on the Job Shop Scheduling Problem

We ran every method 10 times using randomized geometric restarts [26]. Each dichotomy step is limited to 300s and 4×10^6 propagation calls. The total runtime for each configuration is limited to 3600s. The results are summarized in Table 1.

Table 1 is divided in two parts. The first part concerns instances that are mostly proven optimal by our experiments. We report for these instances the average percentage of calls where optimality was proven %O, the average CPU time T, and the number of nodes explored by second (nds/s). The second part concerns the rest of the instances (i.e. hard instances). We report for each data set the speed of exploration (nds/s) along with the average *percentage relative deviation* (*PRD*) of each model. The PRD of a model m for an instance C is computed with the formula: $100 * \frac{C_m - C_{best}}{C_{best}}$, where C_m is the minimum makespan found by model m for this instance ; and C_{best} is the minimum makespan known in the literature [1,2,25]. The bigger the PRD, the less efficient a model is.

Consider first small and medium sized instances, i.e., la-01-40, tai-01-10, and tai-11-20. Table 1 shows clearly that the learning scheme that we introduced (i.e. H(*vsids/task*, *disj*)) dominates the other models on these instances. For example H(*vsids, disj*) proves 91.5% of Lawrence instances to optimality whereas

Table 1. Summary of the results

Instances	CP($task$)			H($vsids, disj$)			H($vsids, lazy$)			H($task, disj$)			H($task, lazy$)		
	Mostly proven optimal														
	%O	T	nds/s	%O	T	nds/s	%O	T	nds/s	%O	T	nds/s	%O	T	nds/s
la-01-40	87	522	8750	**91.5**	437	6814	88	632	2332	90.50	434	5218	88.75	509	2694
tai-01-10	89	768	5875	**90**	517	4975	88	1060	1033	90	634	3572	84	1227	1013
	Hard instances														
	PRD		nds/s	PRD		nds/s	PRD		nds/s	PRD		nds/s	PRD		nds/s
tai-11-20	1.8432		4908	**1.1564**		3583	1.3725		531	1.2741		2544	1.2824		489
tai-21-30	1.6131		3244	0.9150		2361	1.0841		438	0.9660		1694	**0.8745**		409
tai-31-40	5.4149		3501	4.0210		2623	**3.7350**		580	4.0536		1497	3.8844		510
tai-41-50	7.0439		2234	4.8362		1615	**4.6800**		436	4.9305		1003	5.0136		390
tai-51-60	3.0346		1688	3.2449		2726	3.7809		593	**1.1156**		1099	1.1675		575
tai-61-70	6.8598		1432	6.5890		2414	5.4264		578	3.9637		866	**3.6617**		533

CP($task$) and H($vsids, lazy$) achieve a ratio of 87% and 88%, respectively. Moreover, VSIDS generally performs better than weighted degree on these instances, although this factor does not seem as important as the type of learning.

Consider now the rest of instances (tai-21-30 to tai-61-70). The impact of clause learning is more visible on these instances. For example, with tai-31-40, CP($task$) has a PRD of 5.4149 while the worst hybrid model has a PRD of 4.0536. The choice of the branching strategy seems more important on the largest instances. For instance, on tai-51-60, the PRD of H($task, disj$) is 1.1156 while H($vsids, disj$) has a PRD of 3.2449.

Table 1 shows also that the CP model is often faster than any hybrid model (w.r.t. nds/s). This is expected because of the overhead of propagating the learnt clauses in the hybrid models. Lazy generation ($lazy$) slows the search down considerably compared to the mechanism we introduced ($disj$).

Regarding the clauses size, we observed that the new method constantly learns shorter clauses. For example when both methods use the heuristic VSIDS, the average size on tai-11..20 is 31 for the disjunctive learning whereas the standard approach has an average size of 43. This may seem surprising, but several bounds literals may share part of their explanations.

Overall, our hybrid models outperform the CP model in most cases. The novel conflict analysis shows promising results especially for small and medium sized instances. It should be noted that we did not find new upper bounds for hard instances. However, our experiments show that our best model deviates only of a few percents from the best known bounds in the literature. Note that the state-of-the-art CP method [25] uses a timeout of 30000s (similarly to [19]) with a maximum of 600s per iteration in the bottom-up phase for improving the lower bound. Moreover, they use the best known bounds as an extra information along with a double threading phase. In our experiments, we use a single thread and a fixed time limit of 3600s per instance.

6.2   Lower Bound Computation

We ran another set of experiments with a slightly modified dichotomy phase. Here we assume that the outcome of step i is satisfiable instead of unsatisfiable when the cutoff is reached. Each dichotomy step is limited to 1400s and the overall time limit is 3600s. We used the set of all 22 open Taillard instances before the results of [25] and we found 7 new lower bounds (cf. Table 2). However, most

Table 2. New Lower Bounds for Taillard Instances

tai13		tai21		tai23		tai25		tai26		tai29		tai30	
new	old	new	old	new	old	new	old	new	old	new	old	new	old
1305	1282	**1613**	1573	**1514**	1474	**1544**	1518	**1561**	1558	**1576**	1525	**1515**	1485

of these bounds were already improved by Vilím et al. [25]. The main difference is that their algorithm uses the best known bounds as an additional information before starting search. Our models, however, are completely self-contained in the sense where search is started from scratch (see Section 3).

We computed the average PRD w.r.t. the best known lower bound including those reported in [25]. Excluding the set tai40-50, the best combination, H($vsids, disj$) using Luby restarts [15], finds lower bounds that are 2.78% lower, in average, than the best known bound in the literature.

Finally, in order to find new lower bounds, we launched again the model H($vsids, disj$) with 2500s as a timeout for each dichotomy step and 7200s as a global cutoff. We found a new lower bound of 1583 for tai-29 (previously 1573) and one of 1528 for tai-30 (previously 1519).

7   Conclusion

We introduced two hybrid CP/SAT approaches for solving disjunctive scheduling problems. The first one avoids a known redundancy issue regarding lazy generation without computational overhead, whereas the second constitutes a 'hand-crafted' conflict analysis scheme for disjunctive scheduling.

These methods fit naturally the CDCL framework and, in fact, similar techniques appear to have been implemented in existing hybrid solvers, such as Chuffed, although they have never been published nor documented. Experimental evaluation shows the promise of these techniques. In particular, the novel learning scheme is extremely efficient for proving unfeasibility. Indeed, we improved the best known lower bounds of two Taillard instances.

Acknowledgments. This work is funded by CNRS and Midi-Pyrénées region (grant number 11050449). The authors would like to thank the Insight Centre for Data Analytics for kindly granting us access to its computational resources.

References

1. Best known lower/upper bounds for Taillard Job Shop instances. http://optimizizer.com/TA.php (accessed April 15, 2015)
2. Taillard, É.: http://mistic.heig-vd.ch/taillard/problemes.dir/ordonnancement.dir/jobshop.dir/best_lb_up.txt (accessed April 15, 2015)
3. Christopher, J.: Beck. Solution-guided multi-point constructive search for job shop scheduling. Journal of Artificial Intelligence Research **29**(1), 49–77 (2007)
4. Bessiere, C.: Constraint propagation. In: van Beek, P., Rossi, F., Walsh, T. (eds.) Handbook of Constraint Programming, volume 2 of Foundations of Artificial Intelligence, pp. 29–83. Elsevier (2006)
5. Boussemart, F., Hemery, F., Lecoutre, C., Sais, L.: Boosting systematic search by weighting constraints. In: Proceedings of the 16th European Conference on Artificial Intelligence, ECAI 2004, Valencia, Spain, pp. 146–150 (2004)
6. Carlier, J., Pinson, É.: An algorithm for solving the job-shop problem. Management Science **35**(2), 164–176 (1989)
7. Danna, E.: Structured vs. unstructured large neighborhood search: a case study on job-shop scheduling problems with earliness and tardiness costs. In: Rossi, F. (ed.) CP 2003. LNCS, vol. 2833, pp. 817–821. Springer, Heidelberg (2003)
8. Feydy, T., Schutt, A., Stuckey, P.J.: Semantic learning for lazy clause generation. In: Proceedings of TRICS Workshop: Techniques foR Implementing Constraint programming Systems, TRICS 2013, Uppsala, Sweden (2013)
9. Feydy, T., Stuckey, P.J.: Lazy clause generation reengineered. In: Gent, I.P. (ed.) CP 2009. LNCS, vol. 5732, pp. 352–366. Springer, Heidelberg (2009)
10. Grimes, D., Hebrard, E.: Job shop scheduling with setup times and maximal time-lags: a simple constraint programming approach. In: Lodi, A., Milano, M., Toth, P. (eds.) CPAIOR 2010. LNCS, vol. 6140, pp. 147–161. Springer, Heidelberg (2010)
11. Grimes, D., Hebrard, E.: Models and strategies for variants of the job shop scheduling problem. In: Lee, J. (ed.) CP 2011. LNCS, vol. 6876, pp. 356–372. Springer, Heidelberg (2011)
12. Grimes, D., Hebrard, E.: Solving variants of the job shop scheduling problem through conflict-directed search. INFORMS Journal on Computing **27**(2), 268–284 (2015)
13. Grimes, D., Hebrard, E., Malapert, A.: Closing the open shop: contradicting conventional wisdom. In: Gent, I.P. (ed.) CP 2009. LNCS, vol. 5732, pp. 400–408. Springer, Heidelberg (2009)
14. Lawrence, S.R.: Supplement to resource constrained project scheduling: an experimental investigation of heuristic scheduling techniques. Technical report, Graduate School of Industrial Administration, Carnegie Mellon University, Pittsburgh, PA (1984)
15. Luby, M., Sinclair, A., Zuckerman, D.: Optimal speedup of Las Vegas algorithms. Information Processing Letters **47**(4), 173–180 (1993)
16. Matthew, W., Moskewicz, C.F., Madigan, Y.Z., Zhang, L., Malik, S.: Chaff: engineering an efficient SAT solver. In: Proceedings of the 38th Annual Design Automation Conference, DAC 2001, Las Vegas, Nevada, USA, pp. 530–535 (2001)
17. Nuijten, W.: Time and resource constrained scheduling: a constraint satisfaction approach. Ph.D thesis, Eindhoven University of Technology (1994)
18. Ohrimenko, O., Stuckey, P.J., Codish, M.: Propagation via Lazy Clause Generation. Constraints **14**(3), 357–391 (2009)

19. Pardalos, P.M., Shylo, O.V.: An algorithm for the job shop scheduling problem based on global equilibrium search techniques. Computational Management Science **3**(4), 331–348 (2006)

20. Sadeh, N.M.: Lookahead techniques for micro-opportunistic job-shop scheduling. Ph.D thesis, Carnegie Mellon University, Pittsburgh, PA, USA (1991)

21. Schutt, A., Feydy, T., Stuckey, P.J., Wallace, M.G.: Solving rcpsp/max by lazy clause generation. Journal of Scheduling **16**(3), 273–289 (2013)

22. Shaw, P.: Using constraint programming and local search methods to solve vehicle routing problems. In: Maher, M.J., Puget, J.-F. (eds.) CP 1998. LNCS, vol. 1520, pp. 417–431. Springer, Heidelberg (1998)

23. Taillard, É.: Benchmarks for basic scheduling problems. European Journal of Operational Research **64**(2), 278–285 (1993). Project Management anf Scheduling

24. Vilím, P.: Edge finding filtering algorithm for discrete cumulative resources in $\mathcal{O}(kn \log (n))$. In: Gent, I.P. (ed.) CP 2009. LNCS, vol. 5732, pp. 802–816. Springer, Heidelberg (2009)

25. Vilím, P., Laborie, P., Shaw, P.: Failure-directed search for constraint-based scheduling. In: Michel, L. (ed.) CPAIOR 2015. LNCS, vol. 9075, pp. 437–453. Springer, Heidelberg (2015)

26. Walsh, T.: Search in a small world. In: Proceedings of the 16th International Joint Conference on Artificial Intelligence, IJCAI 1999, Stockholm, Sweden, pp. 1172–1177 (1999)

27. Zhang, L., Madigan, C.F., Moskewicz, M.H., Malik, S.: Efficient conflict driven learning in a boolean satisfiability solver. In: Proceedings of the 2001 IEEE/ACM International Conference on Computer-aided Design, ICCAD 2001, San Jose, California, pp. 279–285 (2001)

Bounding an Optimal Search Path with a Game of Cop and Robber on Graphs

Frédéric Simard, Michael Morin$^{(\boxtimes)}$, Claude-Guy Quimper, François Laviolette, and Josée Desharnais

Department of Computer Science and Software Engineering, Université Laval, Québec, QC, Canada
{Frederic.Simard.10,Michael.Morin.3}@ulaval.ca,
{Claude-Guy.Quimper,Francois.Laviolette,Josee.Desharnais}@ift.ulaval.ca

Abstract. In search theory, the goal of the Optimal Search Path (OSP) problem is to find a finite length path maximizing the probability that a searcher detects a lost wanderer on a graph. We propose to bound the probability of finding the wanderer in the remaining search time by relaxing the problem into a stochastic game of cop and robber from graph theory. We discuss the validity of this bound and demonstrate its effectiveness on a constraint programming model of the problem. Experimental results show how our novel bound compares favorably to the DMEAN bound from the literature, a state-of-the-art bound based on a relaxation of the OSP into a longest path problem.

Keywords: Optimal search path · Cop and robber · Constraint relaxation · Pursuit games

1 Introduction

The Optimal Search Path (OSP) problem [1,2] is a classical problem from search theory [3] of which the SAROPS [4] and the SARPlan [5] are two examples of successful systems. The goal of an OSP is to find the best finite length path a searcher has to take in order to maximize the probability of finding a lost wanderer moving randomly on a graph. It is a common assumption, in search theory, to consider unconstrained searcher's motion. The assumption holds, for instance, in contexts where the searcher is fast in comparison to the target of interest. There are, however, many cases, such as in land search and rescue [3], where the searcher is as slow as the moving wanderer. Washburn [6] also notes how bounded speeds (for searcher and wanderer) is used in modeling searches for hostile submarines or friendly vessels.

Most of the early work on the OSP problem consisted in computing bounds for a depth-first branch and bound algorithm initially developed by Stewart [2].[1] At the opposite of stochastic constraint programming, the OSP requires to compute a single scenario given by the decision variables and the probability that

---

[1] A review of the bounds developed prior to 1993 is found in the work of Washburn [6].

© Springer International Publishing Switzerland 2015
G. Pesant (Ed.): CP 2015, LNCS 9255, pp. 403–418, 2015.
DOI: 10.1007/978-3-319-23219-5_29

this scenario works is automatically computed through constraint propagation. Recent works on the OSP problem include generalization of the problem to consider, for instance, non-uniform travel time [7], searching from a distance [8], and path dependent detections [9]. Along with a novel OSP generalization, Lau et al. [7] propose the DMEAN bound. The bound is proved to effectively prune the branch-and-bound search tree.

In this paper, we present bounding techniques for a constraint programming (CP) model of the OSP problem introduced in [10]. We discuss, in Section 2, the DMEAN bound which has never been implemented in CP. Then, as our main contribution, we develop a bound on the OSP objective function (see Section 3). This bound, we call the Copwin bound, is obtained by relaxing the OSP into a game of cop and robber [11–14] which pertains to a well-studied game from graph theory (see Section 3). In Section 4, we show the benefits of using our novel Copwin bound by providing experimental results obtained on an existing CP model of the problem [10]. We conclude in Section 5.

2   The Optimal Search Path Problem

The OSP models a search operation where one searcher searches for a lost, and possibly moving, search object. The problem is known to be $\mathcal{NP}$-hard [1]. It is used, in search theory, as a model for path-constrained search operations [3] and can be formulated as follows. A searcher and a wanderer are moving on the vertices of a graph $G = (V(G), E(G))$. Their respective initial positions correspond to some vertex of $V(G)$ or of some subset of it. Each of their moves, that we can assume simultaneous, is performed along one of the edges $e \in E(G)$ of the graph and consumes one unit of time. There is a maximal number T of steps taken for the search operation. The wanderer $\mathcal{W}$ is invisible to the searcher $\mathcal{S}$, but each time they share a vertex $v \in V(G)$, the latter has a given probability of detecting and removing him[2] from the graph:

$\text{pod}(v) :=$ probability of detecting $\mathcal{W}$ when both $\mathcal{W}$ and $\mathcal{S}$ are on vertex v.

Let $\pi_t = v_1, v_2, \ldots, v_t$ be a path of length $t \leq T$ representing the t first moves of the searcher ($v_i \in V(G)$, $i = 1, \ldots, t$). We define $\pi_{t[..k]} := v_1, v_2, \ldots, v_k$ for $k \leq t$. We define the following probabilities for a searcher $\mathcal{S}$ and a wanderer $\mathcal{W}$:

$\text{poc}_{\pi_t}(v) :=$ probability[3] of $\mathcal{W}$ being on vertex v at time t and not having previously been detected, given that $\mathcal{S}$'s first t steps were along π_t;

$\text{cos}_{\pi_t} \quad :=$ Cumulative probability Of Success of detecting $\mathcal{W}$ up to time t when $\mathcal{S}$ follows π_t, i.e., it is given as $\sum_{i=1}^{t} \text{poc}_{\pi_{t[..t-1]}}(v_i)\text{pod}(v_i)$, where $\text{poc}_{\pi_{t[..0]}}$ is the initial distribution of the wanderer;

---

[2] Following the convention of cop and robber games on graphs, we suppose that the searcher (or cop) is a woman and that the wanderer (or robber) is a man.

[3] $\text{poc}_t(v)$ also stands for the probability of containment [10].

$\cos^*_{\pi_t} :=$ maximal cumulative probability of success up to T if $\mathcal{S}$'s t first steps are along π_t, that is, $\max\{\cos_{\pi_T} \mid \pi_T \text{ has prefix } \pi_t\}$.

The goal of the searcher is to find the path π_T with maximal cumulative probability of success $\cos_{\pi_T}$. This probability is actually $\cos^*_{\pi_0}$, where π_0 is the empty path.

The wanderer is passive in that he does not react to the searcher's moves. It is the usual assumption to model the wanderer's motion by using a Markovian transition matrix M with source vertices on rows and destination vertices on columns. Thus, $M(r, r')$ is the probability of a wanderer's move from vertex r to vertex r' within one time step. Each time the searcher searches a vertex r' along path π_t and the search operation does not stop, then either the wanderer is not there, or he is undetected; the latter happens with probability $1 - \text{pod}(r')$. That is, as long as the search goes on, the wanderer is not found and the knowledge about his location evolves according to:

$$\text{poc}_{\pi_t}(r') = \text{poc}_{\pi_{t[..t-1]}}(v_t)(1 - \text{pod}(v_t))M(v_t, r') \tag{1}$$

$$+ \sum_{r \in V(G) \setminus \{v_t\}} \text{poc}_{\pi_{t[..t-1]}}(r)M(r, r').$$

The next section summarizes the CP model of the OSP used in [10].

2.1   Modeling the OSP in Constraint Programming

A CP model for the OSP problem follows directly from the problem's definition. We define, for each step t ($1 \leq t \leq T$), a *decision variable* $PATH_t$ that represents the searcher's position. The $PATH_t$ domain is an enumerated domain such that $\text{dom}(PATH_t) = V(G)$. The searcher's path is constrained by a possible limited choice for its initial position $PATH_1$ and by the fact that

$$(PATH_{t-1}, PATH_t) \in E(G) \tag{2}$$

for any steps $t \in \{2, \ldots, T\}$. The encoding in the CP-framework is done via a sequence of TABLE constraints [15].

We define two sets of *implicit variables* with bounded domains for probabilities: $POC_t(v)$ with $\text{dom}(POC_t(v)) = [0, 1]$, and $POS_t(v)$ with $\text{dom}(POS_t(v)) = [0, 1]$ for $1 \leq t \leq T$ and for $v \in V(G)$. If the searcher is located in a vertex v, then she obtains a success probability as modeled by the following constraint set defined over all time steps $t \in \{1, \ldots, T\}$ and vertices $v \in V(G)$, where $\text{pod}(v)$ is the known probability of detection in vertex v:

$$PATH_t = v \Rightarrow POS_t(v) = POC_t(v)\text{pod}(v). \tag{3}$$

If the searcher is not located in vertex v at step t, then the probability of detecting the object in v at step t is null. This is modeled by the following constraint set defined over all time steps $t \in \{1, \ldots, T\}$ and vertices $v \in V(G)$:

$$PATH_t \neq v \Rightarrow POS_t(v) = 0. \tag{4}$$

The value of the $POC_1(r)$ variables is fixed given the initial distribution on the wanderer's location, a value known for all $r \in V(G)$. For all subsequent time steps $t \in \{2, \ldots, T\}$ and vertices $v \in V(G)$, the evolution of the searcher's knowledge on the wanderer's location is modeled by the following constraint set:

$$POC_t(v) = \sum_{r \in V(G)} M(r, r') \left(POC_{t-1}(r) - POS_{t-1}(r)\right). \tag{5}$$

A bounded domain variable Z with $\mathrm{dom}(Z) = [0, 1]$ is added to represent the objective value to maximize which corresponds to the cumulative overall probability of success as defined above by $\cos_{\pi_T}$ where π_T is an assignment to the $PATH_t$ variables. Following [10], we observe that the searcher is in one vertex at a time leading to

$$Z = \sum_{t=1}^{T} \max_{v \in V(G)} POS_t(v) \tag{6}$$

and improving filtering over the usual double sum definition of the objective. The objective is finally

$$\max_{PATH_1, \ldots, PATH_T} Z, \tag{7}$$

subject to Constraints (2) to (6) which leads to $\cos_{\pi_0}^*$ value.

2.2   The *DMEAN* Bound from the Literature

Martins [16] proposes to bound the objective function using the MEAN bound. Lau et al. [7] improve on this bound by presenting the discounted MEAN bound (DMEAN) which tightens the MEAN at a reasonable computational cost. In the literature, these bounds are usually applied in an OSP-specific depth-first branch-and-bound algorithm. In this section, we show how to use the DMEAN bound in our OSP CP model. We chose to implement DMEAN among other bounds from the literature since both DMEAN and our novel bound are based on graph theory problems. While DMEAN aims at solving a longest path problem, the Copwin bound, as will be explained, is based on pursuit games. Just as DMEAN, MEAN is based on the maximization of the expected number of detections in the remaining time. The core idea of the MEAN bound is to construct a directed acyclic graph (DAG) at time $t \in \{1, \ldots, T\}$ on vertices consisting of pairs (r, k) where $r \in V(G)$ and $t < k \leq T$. A directed edge is added from vertex (r, k) to vertex $(r', k+1)$ if and only if $(r, r') \in E(G)$. The maximal expected number of detections Z_{DAG} from step t to T is obtained by following the longest path on the DAG, which can be computed in time linear on $|V(G)| + |E(G)|$. Given the beginning π_t of a path, it is sufficient to know $\mathrm{poc}_{\pi_t}(v)$ for all $v \in V(G)$ to be able to compute the expected detection at step k with $t \leq k \leq T$ in vertex $r \in V(G)$. An incident edge into a node $(r', k+1)$ is weighted as follows:

$$R_{\mathrm{MEAN}}(r', k+1) = \sum_{v \in V(G)} \mathrm{poc}_{\pi_t}(v) M^{k+1-t}(v, r') \mathrm{pod}(r') \tag{8}$$

where M^{k+1-t} is the transition matrix M to the power of $k + 1 - t$.

The DMEAN bound is based on the additional observation that a wanderer moving from vertex r to r' has survived a detection on r to possibly being captured on vertex r'. On each outgoing edge from a node (r, k) into a node $(r', k+1)$, if $k = t$ we let the weight become $R_{\text{DMEAN}}(r', k+1) = R_{\text{MEAN}}(r', k+1)$. Otherwise, if $k > t$, we let:

$$R_{\text{DMEAN}}(r', k + 1) = R_{\text{MEAN}}(r', k + 1) - R_{\text{MEAN}}(r, k)M(r, r')\text{pod}(r'). \qquad (9)$$

That is, the wanderer has a probability $R_{\text{MEAN}}(r', k + 1)$ of reaching vertex r' on time step $k + 1$ and being detected there, from which we substract his probability $R_{\text{MEAN}}(r, k)M(r, r')\text{pod}(r')$. This last term is the probability the wanderer had of being on vertex r at the preceding time step k, transiting on r' and being captured there. The $R_{\text{DMEAN}}(r', k+1)$ value thus corresponds to the probability that the wanderer reaches r, survives an additional search, reaches r' and gets caught. Given the beginning of a path π_t, an admissible[4] bound for the OSP is obtained by first solving a longest path problem of which Z_{DAG} is the optimal value and then by summing Z_{DAG} and $\cos_{\pi_t}$ [7].

Remark 1. Given a graph with n vertices and m edges, DMEAN asks for the construction of a DAG with nT new vertices and mT new edges. Then, if the current time is k, the longest path can be solved in $\mathcal{O}\left((T - k + 1)(n + m)\right)$ steps.

It is convenient, to apply the DMEAN bound in our CP model of the OSP, to order the path variables in a static order of the searcher's positions in time. This is the natural order to solve an OSP since the $PATH_t$ variables are the only decision variables and since the implicit variables representing the probabilities, including the objective variable Z that represents $\cos^*_{\pi_0}$, are entirely determined by the searcher's path. When opening a node for a searcher's position at a time t, we are able to compute, using the chosen bound, a tighter upper bound on the domain of the objective variable. The DMEAN bound is proved admissible [7], that is, it never underestimates the real objective value. Whenever the bound computed on the opened node of the searcher's position at a time t is lower than the best known lower bound on the objective value (e.g., the objective value of the current incumbent solution) that searcher's move is proven unpromising.

3   Bounding the Optimal Search Path Using Search Games on Graphs

The cop and robber game on a graph consists in finding a winning strategy for a cop to catch a robber, considering perfect information for both players (i.e., each player knows her/his position and her/his adversary's position). We focus on a recent variant where the robber is random (or drunk) [17–19]. We show how a valid upper bound on the objective function of the OSP can be derived

---

[4]   A bound is *admissible* if and only if it never underestimate (resp. overestimate) the value of the objective to maximize (resp. minimize).

by considering, after t steps, the best possible scenario for the searcher that corresponds to allowing her the ability of seeing the wanderer for the remainder of the game.

3.1   A Game of Cop and Drunk Robber

In this game, a cop and a robber move in turn on a graph G, the cop moving first. In contrast with the OSP, the cop has probability 1 of catching the robber when sharing the same vertex, and she *sees* the robber, who, as does the wanderer in the previous game, walks randomly according to a stochastic transition matrix M. The value $M(r, r')$ is the probability that the robber moves to r' provided he is in r at the beginning of the turn. It is positive only if $(r, r') \in E(G)$. The game ends whenever the cop and the robber share the same vertex. The following definition generalizes Nowakowski and Winkler's [11] relational characterization for the classic cop and robber game to the stochastic one. We want to define the relation $w_n^M(r, c)$ as standing for the probability that the cop catches the robber within n moves given their positions (r, c) and the robber's random movement model M. Its definition is based upon the observation that in order to maximize her capture probability, the cop only needs to average her probability of capture at the next turn on the robber's possible transitions.

Definition 1. *Given r, c the respective positions of the robber and the cop in G, M the robber's random walk matrix, we define:*

$$w_0^M(r, c) := 1 \text{ if } r = c; \quad \text{otherwise, it is } 0;$$

$$w_n^M(r, c) := \begin{cases} 1 & \text{if } c \in N[r], \ n \geq 1; \\ \max\limits_{c' \in N[c]} \sum\limits_{r' \in N[r]} M(r, r') w_{n-1}^M(r', c) & \text{if } c \notin N[r], \ n \geq 1. \end{cases} \quad (10)$$

where $N[z] := \{z\} \cup \{v \in V(G) \mid (v, z) \in E(G)\}$ is the closed neighbourhood of z.

The following proposition gives sense to the previous definition.

Proposition 1. *If the cop plays first on G and M governs the robber's random walk, then $w_n^M(r, c)$ is the probability a cop starting on vertex c captures the drunk robber on his start vertex r in n steps or less.*

Proof. By induction. The base case, with $n = 0$, is clear. Suppose the proposition holds for $n - 1 \geq 0$ with $n > 0$ and let us now prove it for n. If $c \in N[r]$, then $w_n^M(r, c) = 1$ and the result follows because the cop indeed catches the robber. If $c \notin N[r]$, then let the cop move to some vertex c'. The position of the robber at the end of the round is r' with probability $M(r, r')$. The probability that the cop catches the robber depends on this last one's next move and on w_{n-1} following the expression $\sum_{r' \in N[r]} M(r, r') w_{n-1}^M(r', c')$. Hence, the best possible move for the cop is $\operatorname{argmax}_{c' \in N[c]} \sum_{r' \in N[r]} M(r, r') w_{n-1}^M(r', c')$. The wanted probability is thus $w_n^M(r, c)$.   □

Proposition 1 could provide a trivial upper bound on the probability of finding the wanderer, but a tighter one is presented in the following section. The bound we will derive is inspired from Proposition 1 but is closer to the OSPs objective function.

3.2   Markov Decision Process (MDP) and the OSP Bound

In Section 3.1, we presented a game where the robber is visible, a setting that we will use as a bound on the OSP's objective function. We formulate this game as a MDP from which we derive the Copwin bound and the proof that it is admissible. As a recall, a MDP [20–22] is a stochastic decision process generalizing Markov chains. In a MDP, an agent evolves in a state space S. Every time the agent finds itself in a state $s \in S$, it takes an action a in a set A of possible actions. The system then randomly transits to another valid state s' according to a given transition probability $\mathbb{P}[s' \mid s, a]$ (where $\mid$ denotes a conditional). In order to guide the agent's choice of actions, a reward function $R : S \times A \times S \to \mathbb{R}$ is defined that assigns to every triple (s, a, s') a real number. The goal of the agent is then to maximize its expected reward within a time bound T. We first formulate the game of cop and robber defined above as an MDP and then deduce a valid OSP bound. The following definition presents the MDP; its solution encodes the optimal strategy for the cop.

Definition 2 (MDP Cop and Drunk Defending Robber). *Let G be a graph, M a stochastic transition matrix on $V(G)$ and T the maximal number of time steps. We define a MDP $\mathcal{M} = (S, A, P, R)$ as follows. A state is a triplet (r, c, t) consisting in both positions of the robber and the cop $r, c \in V(G)$ in addition to the current time $t \in \{1, \ldots, T+1\}$.*

$$S := (V(G) \cup \{\text{jail}\}) \times V(G) \times \{1, 2, \ldots, T+1\}.$$

The set of actions is $V(G)$, the vertices on which the cop moves.

$$A := V(G).$$

From a pair of positions of the cop and the robber at a certain time, once the cop chooses an action another state is chosen randomly with probability P.

$$P\Big[(r', c', t') \mid (r, c, t), a\Big] := 0 \text{ whenever } a \neq c' \text{ or } c' \notin N[c] \text{ or } t' \neq t+1$$

otherwise, P is defined as:

$$P\Big[(r', c', t+1) \mid (r, c, t), c'\Big] := \begin{cases} 1 & \text{if } r = r' = \text{jail}; \\ \text{pod}(r) & \text{if } r = c', \ r' = \text{jail}; \\ (1 - \text{pod}(r))M(r, r') & \text{if } r = c', \ r' \neq \text{jail}; \\ M(r, r') & \text{if } r \notin \{c, c', \text{jail}\}. \end{cases} \quad (11)$$

$$R((r', c', t') \mid (r, c, t), a) := \begin{cases} 1 & \text{if } r' = \text{jail} \neq r, t \leq T; \\ 0 & \text{otherwise.} \end{cases} \quad (12)$$

The game is initialized as follows: the cop chooses her initial position c on a subset X of the graph vertices, and then the initial position r of the robber is picked at random according to the probability distribution poc_{π_0}, which results in an initial state $(r, c, 1)$ for the MDP. A turn starts with a cop move. If the cop transits to the robber state ($r = c'$), then there is probability $pod(r)$ that she catches the robber, which results in the robber going to jail ($r' = $ jail) and staying there for the rest of the game. The cop then receives a reward of 1. If the catch fails ($r' \neq$ jail, with probability $1 - pod(r)$) or if the cop did not transit to the robber state ($r \neq c'$), the robber is still free to roam, following M. Note that the state transition probabilities (11) are non-null only when valid moves are considered (time goes up by one and $a = c' \in N[c]$). Note also that, when the robber is in jail, no more reward is given to the cop. In the MDP $\mathcal{M}$, the cop's goal is to find a strategy, also called policy, to maximize her expected reward, that is, the probability to capture the robber before a total of T steps is reached. A strategy in $\mathcal{M}$ consists in first choosing an initial position (for the cop) and then in following a function $f : S \to A$ that, given the current state, tells the cop which action to take, that is, which state to transit to. Note that, since the current position of the robber is included in the current state of the MDP, the cop therefore has full information on the system when she is elaborating her strategy.

Because of $\mathcal{M}$'s Markov property, whenever a strategy f is fixed, one can compute the *value* $u_f(r, c, t)$ of each state (r, c, t) of the MDP, that is, the expected reward the cop can obtain from that state when following the strategy f. The optimal strategy, noted u^*, is therefore the one that gives the highest value on all states (r, c, t). The cop's optimal strategy consists in moving to the robber's position if possible, and then trying to capture him. If the robber is not positioned on one of the cop's neighbours, the cop moves to the position that maximizes her probability of capture in the remaining time allowed. Similarly to Proposition 1, the value of this optimal strategy is:

$$
u^*(r, c, t) = \begin{cases} \max_{c' \in N[c]} \sum_{r' \in N[r]} M(r, r') u^*(r', c', t+1) & \text{if } r \notin N[c], t < T; \\ 1 - (1 - pod(r))^{T+1-t} & \text{if } r \in N[c], t \leq T; \\ 0 & \text{if } r \notin N[c], t = T. \end{cases} \tag{13}
$$

If $r \notin N[c]$, the cop, who moves first, must choose a next state that will result in the best probability of eventual capture, given the robber's present position, and knowing that the robber's next move is governed by M. If $r \in N[c]$, the cop tries to catch the robber with probability of success $pod(r)$; if she fails, the robber will transit to one of his neighbours, and hence the cop can keep on trying to catch the robber until success or until the maximal time has been reached. It is important to note here that the robber is completely visible and the game is not played simultaneously, hence why the cop can follow the robber. Equation (13) is analogous to (10) with time steps reversed and $pod(r) = 1$ for all vertices. The formula follows from the fact that at the beginning of time step t, the cop has $T + 1 - t$ remaining attempts.

Since the optimal probability of capture in the OSP problem is always lower than the optimal probability of capture in the cop and robber game, we have the following proposition:

Proposition 2. *The probability $\cos^*_{\pi_0}$ of finding the wanderer is always at most that of catching the robber:*

$$\cos^*_{\pi_0} \leq \max_{c \in X} \sum_{r \in V(G)} \mathrm{poc}_{\pi_0}(r) u^*(r, c, 1),$$

where X is the subset of $V(G)$ of possible initial positions of the cop.

Proof. Clearly, $\cos^*_{\pi_0}$ is bounded by the optimal probability of capture of the cop and robber game. In the MDP, the optimal probability of capture is obtained if the cop's first choice maximizes his probability of capture considering that at that moment the robber is not yet positioned on the graph but will be according to the probability distribution poc_{π_0}.   □

Unfortunately, Proposition 2's bound is of no use in a branch-and-bound attempt for solving the OSP problem, because a bound for each π_t and each $t = 0, \ldots, T$ is needed. The next proposition generalizes it appropriately.

Proposition 3. *Let $\pi_t = v_1, v_2, \ldots, v_t$. Then*

$$\cos^*_{\pi_t} \leq \cos_{\pi_t} + \max_{c' \in N[v_t]} \sum_{r' \in V(G)} \mathrm{poc}_{\pi_t}(r') u^*(r', c', t+1), \tag{14}$$

where $N[v_0]$ is interpreted as the possible initial positions of the cop.

Proof. As in the preceding proof, $\cos^*_{\pi_t}$ is bounded by the optimal reward obtained when first playing the OSP game along π_t and then (if the wanderer is not yet detected) switching to the cop and robber game: this is done by making the wanderer (robber) visible to the searcher (cop). When starting this second phase (at the $t+1$ step), the cop must choose his next position in order to maximize the probability of capture; at this moment, the robber is not yet visible but his next position is governed by poc_{π_t}. If the cop chooses c', his probability of capture will be $\sum_{r' \in V(G)} \mathrm{poc}_{\pi_t}(r') u^*(r', c', t+1)$ and the result follows.   □

Remark 2. An important aspect of Copwin is its ability to be mostly precomputed. For any vertices $r, c \in V(G)$ and time $t \in \{1, \ldots, T+1\}$, the values $u^*(r, c, t)$ are computed recursively (starting with $t = T$) and then stored. Then, when the bound is called the next value of Equation (14) depends on the neighbours of the searcher's position and the vertices of the graph, requiring $\mathcal{O}((\Delta + 1)n)$ extra operations on an n vertex graph of maximal degree Δ. Since DMEAN's complexity is $\mathcal{O}((T - k + 1)(n + m))$, Copwin is faster on most time steps and on many instances.

We note that Pralat and Kehagias [18] also formulated the game of cop and visible drunk robber as an MDP, but instead of using this framework to compute the probability of capture of the robber it was formulated to obtain the expected number of time steps before capture.

Applying the Copwin bound in CP requires filtering the upper bound of the objective function variable Z. The solver can be configured to branch in a static order of the $PATH_t$ variable, which is, as discussed in Section 2.2, the natural order of the decision variables to solve an OSP. We proved that our bound is admissible in that it never underestimates the real objective value. In the next section, we use the Copwin bound jointly with a CP model of the OSP and compare its performance to that of the DMEAN bound from literature.

4   Experiments

All experiments were run on the supercomputer Colosse from Université Laval using the Choco 2.1.5 solver [23] along with the Java Universal Network/Graph (JUNG) 2.0.1 framework [24]. The total allowed CPU time is 5 minutes (or 5 million backtracks) per run. We used a total of six graphs representing different search environments. In the grid G^+, the nodes are connected to their North, South, East, and West neighbour. The grid G^* has the same connections than G^+ but also adds connectivity to the North-East, North-West, South-East, and South-West neighbours. The last grid is a typical hexagonal grid we call G^H with the searcher moving on its edges. Grid-like accessibility constraints are common in search theory problems in that they model aerial or marine search operations where the searcher has few accessibility constraints but her own physical constraints. This justifies the presence of grids in our OSP benchmarks. We also added a fourth grid with supplementary constraints on accessibility. This last grid G^V is a hexagonal grid with a supplementary node in the center of each hexagon. We then had removed edges randomly to simulate supplementary accessibility constraints. The last two graphs, G^L and G^T, are real underground environments. The G^L graph is the graph of the Université Laval's tunnels. The G^T graph is the graph of the London Underground subway (also known as the Tube). Both neighbours of an intermediate station, i.e., a node of degree 2, were connected together while the node was removed. This practice results in a more interesting graph environment for searches.

For each instance, we selected a $p \in \{0.3, 0.6, 0.9\}$ and put $\text{pod}(r) = p$ for all vertex r. For the wanderer's Markovian transition matrix, we pick $\rho \in \{0.3, 0.6, 0.9\}$ and considered that at any position v, the wanderer has probability ρ of staying in v and probability $1 - \frac{\rho}{\deg(v)}$ to move to one of his $\deg(v)$ neighbours. We uniformly distributed the remaining probability mass $1 - \rho$ on the neighbouring vertices. Finally, we chose $T \in \{5, 7, 9, 11, 13, 15, 17, 19, 21, 23, 25\}$. This leads to a total of 594 instances.

We conducted the experiment in three parts. In the first two parts, we compare our novel bound to the DMEAN bound we reviewed in Section 2.2. First, we implemented the two bounds as separate constraints on the upper bound of the objective value. Both bounds are evaluated on the same problem instances. This enables us to assess the bounds performance with respect to the objective value of the best incumbent solution found by the solver and to the time required to attain it.

In the second part, we implemented both bounds as a single constraint on the upper bound of the objective value. Both bounds were computed simultaneously by the constraint. The minimal value of both bounds was used to update the upper bound of the domain of the Z variable representing the objective if required. That is, whenever that minimal value is lower than the upper bound on the domain of the Z variable. If the domain of Z is emptied, then one of the bounds produced a value that is lower than the best known lower bound on the objective thus leading to a backtrack. For each bounding technique (DMEAN and Copwin), we count how many times, during the search, the bounding of the objective value causes a backtrack. We gave a single point to a bound that leads to a backtrack that way. We note that on occasions both bounds could receive a point. In all cases, the solver branches on the $PATH_t$ decision variables in their natural order, i.e., in ascending order of the time steps. We thus were able to record the exact time step at which each backtrack occurred. Such an experiment enables one to empirically assess which bound is more efficient at pruning the search tree no matter without considering any time constraints.

We show, in the third part, how the bounds perform when paired with the total detection (TD) heuristic used as a value-selection heuristic [10]. The TD heuristic is on many account similar to the bound derived in this paper which gives its theoretical justification. However, rather than using the cop and robber games theory to bound the objective value, TD is a heuristic that assigns values to the variables $PATH_t$. Even in this context, the Copwin bound performs and stays competitive.

4.1   Results and Discussion

Figure 1 shows the best incumbent objective value (COS) for all OSP problem instances. Hence, COS is the estimation of $\cos^*_{\pi_0}$ at the end of the solving process. Each dot represents the solution to a single OSP instance. The figure compares the objective value obtained when the solver uses the DMEAN bound against the objective value obtained when the solver uses the Copwin bound for all instances. The closer a dot is to the diagonal line, the closer it is to being the same value for both bounds which is viewed as both bounds having had the same effectiveness on this instance. It does appear the Copwin bound tends to lead the solver to better objective values in the total allowed time. Thus, Copwin is not only faster than DMEAN, it is sharp enough to help the solver to exploit more promising branches.

Figure 2 shows the number of times each bound empties the domain of the objective value variable in the search tree for each type of graph. We chose to plot the score dependant on the time step (or level in the search tree) to better analyse where branches were being pruned. It appears that on average

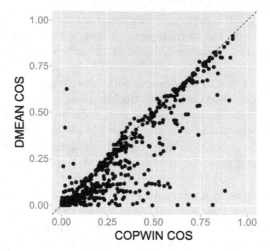

Fig. 1. Two by two comparison of DMEAN best incumbent COS value against Copwin best incumbent COS value for all 594 OSP instances; the incumbent solution of each instance is represented by a dot on the graph.

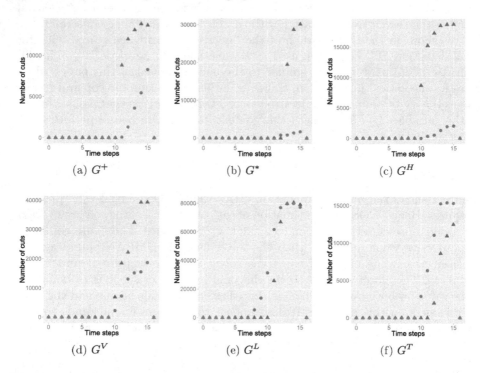

Fig. 2. Total number of cuts in the solver's search tree for both bounds against searcher's path length with $T = 17$; results for each instance are aggregated on motion models ρ and on detection models pod(r). ▲ refers to the use of Copwin whereas ● refers to DMEAN.

Copwin empties the objective value variable's domain more often than DMEAN independently of the graph, even though on some instances such as G^L both lead to a great number of prunings. This is a strong indication that Copwin performs well on all accounts on most instances: for a fixed resolution time, it leads to better objective values while pruning more branches than the DMEAN bound.

We present on Figure 3, the objective value of the best so far solution against computational time for both bounding strategies. We present these comparisons on three representative instances: one in which Copwin is better than DMEAN,

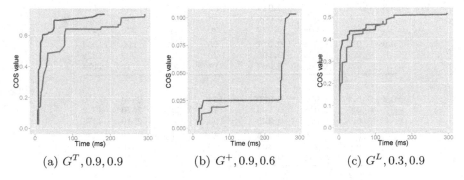

(a) $G^T, 0.9, 0.9$ (b) $G^+, 0.9, 0.6$ (c) $G^L, 0.3, 0.9$

Fig. 3. The COS value achieved and the time it was found for each bound. Red lines are associated with Copwin and blue ones with DMEAN. Each title refers to the graph type, the instance pod(r), the instance ρ and maximal time steps of 17.

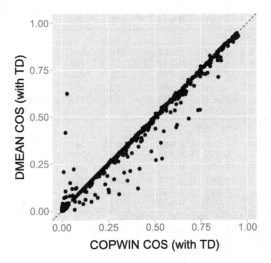

Fig 4. Comparison of the best incumbent COS value obtained with the TD heuristic for the Copwin bound against the best COS value obtained with the TD heuristic for the DMEAN bound for all OSP problem instances; the incumbent solution of each instance is represented by a dot on the graph.

Table 1. Results on OSP problem instances; bold values are better.

pod(r) ρ		Objective (COS)		Time[†] (s)		Backtracks	
		DMEAN	Copwin	DMEAN	Copwin	DMEAN	Copwin
G^T with $T = 17$							
0.3	0.3	0.111	**0.113**	298	197	3578	2791
	0.6	**0.139**	0.129	200	144	2045	2157
	0.9	**0.362**	0.354	275	208	2973	3304
0.9	0.3	0.252	**0.258**	131	287	1400	4613
	0.6	0.318	**0.325**	259	236	3082	3544
	0.9	0.736	0.736	292	185	3543	2706
G^+ with $T = 17$							
0.3	0.3	0.011	**0.014**	96	37	283	118
	0.6	0.007	**0.036**	88	281	243	1307
	0.9	0.001	**0.022**	90	111	276	497
0.9	0.3	0.030	**0.077**	95	296	290	1394
	0.6	0.020	**0.103**	95	272	283	1229
	0.9	0.001	**0.047**	26	262	19	1174
G^V with $T = 17$							
0.3	0.3	0.035	**0.079**	154	296	2510	7240
	0.6	0.038	**0.105**	146	206	2510	4922
	0.9	0.036	**0.229**	158	196	2510	4898
0.9	0.3	0.099	**0.215**	159	240	2510	5539
	0.6	0.112	**0.304**	155	264	2510	6040
	0.9	0.109	**0.531**	161	219	2510	4996
G^L with $T = 17$							
0.3	0.3	**0.344**	0.318	273	145	9224	7749
	0.6	**0.417**	0.359	287	139	9666	7476
	0.9	**0.516**	0.480	297	122	9298	6462
0.9	0.3	0.638	**0.651**	278	119	11304	6249
	0.6	0.713	0.713	289	146	11603	7890
	0.9	0.811	**0.833**	249	162	9489	8727
G^* with $T = 17$							
0.3	0.3	0.069	**0.072**	270	286	1264	2264
	0.6	0.107	**0.117**	273	183	1261	1482
	0.9	0.235	**0.324**	276	281	1254	2461
0.9	0.3	0.192	**0.205**	274	252	1264	2077
	0.6	0.304	**0.333**	264	219	1261	1842
	0.9	0.671	**0.711**	259	231	1253	1925
G^H with $T = 17$							
0.3	0.3	0.023	**0.087**	258	241	522	618
	0.6	0.015	**0.122**	255	277	519	742
	0.9	0.001	**0.318**	250	233	514	623
0.9	0.3	0.064	**0.227**	274	258	522	686
	0.6	0.043	**0.342**	260	286	520	719
	0.9	0.003	**0.816**	248	280	516	680

† The time to last incumbent solution.

one in which it is worse and another where both bounds lead to a similar performance. These graphs confirm the intuition that, most of the time, Copwin finds good solutions more quickly than DMEAN.

Figure 4 summarizes further experiments where Copwin and DMEAN were paired with the TD heuristic. Even though both bounds lead the solver to greater objective values, we observe that Copwin's objective values are on average slightly better. Thus, the behavior observed on Figure 1 is preserved when a good heuristic is added.

As an addition to the graphs of Figures 1 to 4, we include Table 1 which contains the results of the experiments of the Copwin versus the DMEAN bound for all graphs with $T = 17$.

5   Conclusion

We tackled the OSP problem from search theory using constraint programming. As a first contribution, we provided the first implementation of the DMEAN bound from the search theory literature in CP. As our main contribution, we developed the Copwin bound, a novel and competitive bound based on MDPs. This bound is derived from a simple and elegant relaxation of a search problem into a graph theory pursuit game. Involving a polynomial computational cost, the Copwin bound leads to an improved solver performance on the vast majority of OSP problem instances in our benchmark. Altough we used the bound on a CP model of the OSP, it remains a general technique applicable to other OSP algorithms.

Acknowledgments. This research is funded by the Natural Sciences and Engineering Research Council of Canada (NSERC) and the Fonds de recherche du Québec - Nature et technologies (FRQ-NT).

Computations were made on the supercomputer Colosse from Université Laval, managed by Calcul Québec and Compute Canada. The operation of this supercomputer is funded by the Canada Foundation for Innovation (CFI), NanoQubec, RMGA and the Fonds de recherche du Québec - Nature et technologies (FRQ-NT).

References

1. Trummel, K., Weisinger, J.: The complexity of the optimal searcher path problem. Operations Research **34**(2), 324–327 (1986)
2. Stewart, T.: Search for a moving target when the searcher motion is restricted. Computers and Operations Research **6**(3), 129–140 (1979)
3. Stone, L.: Theory of Optimal Search. Academic Press, New York (2004)
4. Netsch, R.: The USCG search and rescue optimal planning system (SAROPS) via the commercial/joint mapping tool kit (c/jmtk). In: Proceedings of the 24th Annual ESRI User Conference, vol. 9, August 2004
5. Abi-Zeid, I., Frost, J.: A decision support system for canadian search and rescue operations. European Journal of Operational Research **162**(3), 636–653 (2005)
6. Washburn, A.R.: Branch and bound methods for a search problem. Naval Research Logistics **45**(3), 243–257 (1998)
7. Lau, H., Huang, S., Dissanayake, G.: Discounted MEAN bound for the optimal searcher path problem with non-uniform travel times. European Journal of Operational Research **190**(2), 383–397 (2008)

8. Morin, M., Lamontagne, L., Abi-Zeid, I., Lang, P., Maupin, P.: The optimal searcher path problem with a visibility criterion in discrete time and space. In: Proceedings of the 12th International Conference on Information Fusion, pp. 2217–2224 (2009)

9. Sato, H., Royset, J.O.: Path optimization for the resource-constrained searcher. Naval Research Logistics, 422–440 (2010)

10. Morin, M., Papillon, A.-P., Abi-Zeid, I., Laviolette, F., Quimper, C.-G.: Constraint programming for path planning with uncertainty. In: Milano, M. (ed.) CP 2012. LNCS, vol. 7514, pp. 988–1003. Springer, Heidelberg (2012)

11. Nowakowski, R., Winkler, P.: Vertex-to-vertex pursuit in a graph. Discrete Mathematics **43**(2), 235–239 (1983)

12. Quilliot, A.: Problème de jeux, de point fixe, de connectivité et de représentation sur des graphes, des ensembles ordonnés et des hypergraphes. Ph.D thesis, Université de Paris VI (1983)

13. Bonato, A., Nowakowski, R.: The game of cops and robbers on graphs. American Mathematical Soc. (2011)

14. Fomin, F.V., Thilikos, D.M.: An annotated bibliography on guaranteed graph searching. Theoretical Computer Science **399**(3), 236–245 (2008)

15. Beldiceanu, N., Demassey, S.: Global constraint catalog (2014). http://sofdem. github.io/gccat/ (accessed 2015–04)

16. Martins, G.H.: A new branch-and-bound procedure for computing optimal search paths. Master's thesis, Naval Postgraduate School (1993)

17. Kehagias, A., Mitsche, D., Prałat, P.: Cops and invisible robbers: The cost of drunkenness. Theoretical Computer Science **481**, 100–120 (2013)

18. Kehagias, A., Prałat, P.: Some remarks on cops and drunk robbers. Theoretical Computer Science **463**, 133–147 (2012)

19. Komarov, N., Winkler, P.: Capturing the Drunk Robber on a Graph. arXiv preprint arXiv:1305.4559 (2013)

20. Bäuerle, N., Rieder, U.: Markov Decision Processes with Applications to Finance. Springer (2011)

21. Puterman, M.L.: Markov decision processes: discrete stochastic dynamic programming, vol. 414. John Wiley & Sons (2009)

22. Barto, A.G., Sutton, R.S.: Reinforcement learning: An introduction. MIT press (1998)

23. Laburthe, F., Jussien, N.: Choco solver documentation. École de Mines de Nantes (2012)

24. O'Madadhain, J., Fisher, D., Nelson, T., White, S., Boey, Y.B.: The JUNG (Java universal network/graph) framework (2010)

Restricted Path Consistency Revisited

Kostas Stergiou[✉]

Department of Informatics and Telecommunications Engineering,
University of Western Macedonia, Kozani, Greece
kstergiou@uowm.gr

Abstract. Restricted path consistency (RPC) is a strong local consistency for binary constraints that was proposed 20 years ago and was
identified as a promising alternative to arc consistency (AC) in an early
experimental study of local consistencies for binary constraints. However, and in contrast to other strong local consistencies such as SAC
and maxRPC, it has been neglected since then. In this paper we revisit
RPC. First, we propose RPC3, a new lightweight RPC algorithm that is
very easy to implement and can be efficiently applied throughout search.
Then we perform a wide experimental study of RPC3 and a light version
that achieves an approximation of RPC, comparing them to state-of-the-
art AC and maxRPC algorithms. Experimental results clearly show that
restricted RPC is by far more efficient than both AC and maxRPC when
applied throughout search. These results strongly suggest that it is time
to reconsider the established perception that MAC is the best general
purpose method for solving binary CSPs.

1 Introduction

Restricted path consistency (RPC) is a local consistency for binary constraints
that is stronger than arc consistency (AC). RPC was introduced by Berlandier
[4] and was further studied by Debruyne and Bessiere [7,8]. An RPC algorithm
removes all arc inconsistent values from a domain $D(x)$, and in addition, for any
pair of values (a, b), with $a \in D(x)$ and $b \in D(y)$ s.t. b is the only support for a
in a $D(y)$, it checks if (a, b) is path consistent. If it is not then a is removed from
$D(x)$. In this way some of the benefits of path consistency are retained while
avoiding its high cost.

Although RPC was identified as a promising alternative to AC as far back
as 2001 [8], it has been neglected by the CP community since then. In contrast,
stronger local consistencies such as max restricted path consistency (maxRPC)
[7] and singleton arc consistency (SAC) [8] have received considerable attention
in the past decade or so [1–3,5,9,11,13,14]. However, despite the algorithmic
developments on maxRPC and SAC, none of the two outperforms AC when
maintained during search, except for specific classes of problems. Therefore,
MAC remains the predominant generic algorithm for solving binary CSP's.

In this paper we revisit RPC and make two contributions compared to previous works that bring the state-of-the-art regarding RPC up to date. The first
is algorithmic and the second experimental.

© Springer International Publishing Switzerland 2015
G. Pesant (Ed.): CP 2015, LNCS 9255, pp. 419–428, 2015.
DOI: 10.1007/978-3-319-23219-5_30

The two algorithms that have been proposed for RPC, called RPC1 [4] and RPC2 [7], are based on the AC algorithms AC4 and AC6 respectively. As a result they suffer from the same drawbacks as their AC counterparts. Namely, they use heavy data structures that are too expensive to maintain during search. In recent years it has been shown that in the case of AC lighter algorithms which sacrifice optimality display a better performance when used inside MAC compared to optimal but heavier algorithms such as AC4, AC6, AC7, and AC2001/3.1. Hence, the development of the residue-based version of AC3 known as AC3^r [10,12]. A similar observation has been made with respect to maxRPC [1]. Also, it has been noted that cheap approximations of local consistencies such as maxRPC and SAC are more cost-effective than the full versions. In the case of maxRPC, the residue-based algorithm lmaxRPC3^r, which achieves an approximation of maxRPC, is the best choice when applying maxRPC [1].

Following these trends, we propose RPC3, an RPC algorithm that makes use of residues in the spirit of AC^r and lmaxRPC^r and is very easy to implement. As we will explain, for each constraint (x, y) and each value $a \in D(x)$, RPC3 stores two residues that correspond to the two most recently discovered supports for a in $D(y)$. This enables the algorithm to avoid many redundant constraint checks. We also consider a restricted version of the algorithm (simply called rRPC3) that achieves a local consistency property weaker than RPC, but still stronger than AC, and is considerably faster in practice.

Our second and most important contribution concerns experiments. Given that the few works on RPC date from the 90s, the experimental evaluations of the proposed algorithms were carried out on limited sets of, mainly random, problems. Equally importantly, there was no evaluation of the algorithms when used during search to maintain RPC. We carry out a wide evaluation on benchmark problems from numerous classes that have been used in CSP solver competitions. Surprisingly, results demonstrate that an algorithm that applies rRPC3 throughout search is not only competitive with MAC, but it clearly outperforms it on the overwhelming majority of tested instances, especially on structured problems. Also, it clearly outperforms lmaxRPC3^r. This is because RPC, and especially its restricted version, achieves a very good balance between the pruning power of maxRPC and the low cost of AC.

Our experimental results provide strong evidence of a local consistency that is clearly preferable to AC when maintained during search. Hence, perhaps it is time to reconsider the common perception that MAC is the best general purpose solver for binary problems.

2   Background

A *Constraint Satisfaction Problem* (CSP) is defined as a triplet $(\mathcal{X}, \mathcal{D}, \mathcal{C})$ where: $\mathcal{X} = \{x_1, \ldots, x_n\}$ is a set of n variables, $\mathcal{D} = \{D(x_1), \ldots, D(x_n)\}$ is a set of domains, one for each variable, with maximum cardinality d, and $\mathcal{C} = \{c_1, \ldots, c_e\}$ is a set of e constraints. In this paper we are concerned with binary CSPs. A binary constraint c_{ij} involves variables x_i and x_j.

At any time during the solving process if a value a_i has not been removed from the domain $D(x_i)$, we say that the value is *valid*. A value $a_i \in D(x_i)$ is *arc consistent* (AC) iff for every constraint c_{ij} there exists a value $a_j \in D(x_j)$ s.t. the pair of values (a_i, a_j) satisfies c_{ij}. In this case a_j is called an *support* of a_i. A variable is AC iff all its values are AC. A problem is AC iff there is no empty domain in D and all the variables in X are AC.

A pair of values (a_i, a_j), with $a_i \in D(x_i)$ and $a_j \in D(x_j)$, is *path consistent* PC iff for any third variable x_k constrained with x_i and x_j there exists a value $a_k \in D(x_k)$ s.t. a_k is a support of both a_i and a_j. In this case a_j is a *PC-support* of a_i in $D(x_j)$ and a_k is a *PC-witness* for the pair (a_i, a_j) in $D(x_k)$.

A value $a_i \in D(x_i)$ is *restricted path consistent* (RPC) iff it is AC and for each constraint c_{ij} s.t. a_i has a single support $a_j \in D(x_j)$, the pair of values (a_i, a_j) is *path consistent* (PC) [4]. A value $a_i \in D(x_i)$ is *max restricted path consistent* (maxRPC) iff it is AC and for each constraint c_{ij} there exists a support a_j for a_i in $D(x_j)$ s.t. the pair of values (a_i, a_j) is *path consistent* (PC) [7]. A variable is RPC (resp. maxRPC) iff all its values are RPC (resp. maxRPC). A problem is RPC (resp. maxRPC) iff there is no empty domain and all variables are RPC (resp. maxRPC).

3   The RPC3 Algorithm

The RPC3 algorithm is based on the idea of seeking two supports for a value, which was first introduced in RPC2 [7]. But in contrast to RPC2 which is based on AC6, it follows an AC3-like structure, resulting in lighter use of data structures, albeit with a loss of optimality. As explained below, we can easily obtain a restricted but more efficient version of the algorithm that only approximates the RPC property. Crucially, the lack of heavy data structures allows for the use of the new algorithms during search without having to perform expensive restorations of data structures after failures.

In the spirit of AC^r, RPC3 utilizes two data structures, R^1 and R^2, which hold residual data used to avoid redundant operations. Specifically, for each constraint c_{ij} and each value $a_i \in D(x_i)$, $R^1_{x_i, a_i, x_j}$ and $R^2_{x_i, a_i, x_j}$ hold the two most recently discovered supports of a_i in $D(x_j)$. Initially, all residues are set to a special value NIL, considered to precede all values in any domain.

The pseudocode of RPC3 is given in Algorithm 1 and Function 2. Being coarse-grained like AC3, Algorithm 1 uses a propagation list Q, typically implemented as a fifo queue. We use a constraint-oriented description, meaning that Q handles pairs of variables involved in constraints. A variable-based one requires minor modifications.

Once a pair of variables (x_i, x_j) is removed from Q, the algorithm iterates over $D(x_i)$ and for each value a_i first checks the residues $R^1_{x_i, a_i, x_j}$ and $R^2_{x_i, a_i, x_j}$ (line 5). If both are valid then a_i has at least two supports in $D(x_j)$. Hence, the algorithm moves to process the next value in $D(x_i)$. Otherwise, function *findTwoSupports* is called. This function will try to find two supports for a_i in $D(x_j)$. In case it finds none then a_i is not AC and will thus be deleted (line 13).

Algorithm 1. RPC3:boolean

1: **while** $Q \neq \emptyset$ **do**
2:     $Q \leftarrow Q - \{(x_i, x_j)\}$;
3:     Deletion $\leftarrow$ FALSE;
4:     **for each** $a_i \in D(x_i)$ **do**
5:         **if** both $R^1_{x_i, a_i, x_j}$ and $R^2_{x_i, a_i, x_j}$ are valid **then**
6:             **continue**;
7:         **else**
8:             **if** only one of $R^1_{x_i, a_i, x_j}$ and $R^2_{x_i, a_i, x_j}$ is valid **then**
9:                 $R \leftarrow$ the valid residue;
10:            **else**
11:                $R \leftarrow$ NIL;
12:            **if** $findTwoSupports(x_i, a_i, x_j, R) =$ FALSE **then**
13:                remove a_i from $D(x_i)$;
14:                Deletion $\leftarrow$ TRUE;
15:    **if** $D(x_i) = \emptyset$ **then**
16:        **return** FALSE;
17:    **if** Deletion = TRUE **then**
18:        **for each** $(x_k, x_i) \in C$ s.t. $(x_k, x_i) \notin Q$ **do**
19:            $Q \leftarrow Q \cup \{(x_k, x_i)\}$;
20:            **for each** $(x_l, x_k) \in C$ s.t. $x_l \neq x_i$ **and** $(x_l, x_i) \in C$ **and** $(x_l, x_k) \notin Q$ **do**
21:                $Q \leftarrow Q \cup \{(x_l, x_k)\}$;
22: **return** TRUE;

In case it finds only one then it will check if a_i is RPC. If it is not then it will be deleted. Function *findTwoSupports* takes as arguments the variables x_i and x_j, the value a_i, and a parameter R, which is set to the single valid residue of a_i in $D(x_j)$ (line 9) or to NIL if none of the two residues is valid.

Function *findTwoSupports* iterates over the values in $D(x_j)$ (line 3). For each value $a_j \in D(x_j)$ it checks if the pair (a_i, a_j) satisfies constraint c_{ij} (this is what function *isConsistent* does). If both residues of a_i in $D(x_j)$ are not valid then after a support is found, the algorithm continues to search for another one. Otherwise, as soon as a support is found that is different than R, the function returns having located two supports (lines 9-11).

If only one support a_j is located for a_i then the algorithm checks if the pair (a_i, a_j) is path consistent. During this process it exploits the residues to save redundant work, if possible. Specifically, for any third variable x_k that is constrained with both x_i and x_j, we first check if one of the two residues of a_i is valid and if a_j is consistent with that residue (line 16). If this is the case then we know that there is a PC-witness for the pair (a_i, a_j) in $D(x_k)$ without having to iterate over $D(x_k)$. If it is not the case then the check is repeated for the residues of a_j in $D(x_k)$. If we fail to verify the existense of a PC-witness in this way then we iterate over $D(x_k)$ checking if any value a_k is consistent with both a_i and a_j. If a PC-witness is found, we proceed with the next variable that is constrained with both x_i and x_j. Otherwise, the function returns false, signaling that a_i is not RPC.

Function 2. $findTwoSupports(x_i, a_i, x_j, R)$:**Boolean**

1: **if** R = NIL **then** oneSupport $\leftarrow$ FALSE;
2: **else** oneSupport $\leftarrow$ TRUE;
3: **for each** $a_j \in D(x_j)$ **do**
4:    **if** isConsistent(a_i, a_j) **then**
5:       **if** oneSupport = FALSE **then**
6:          oneSupport $\leftarrow$ TRUE;
7:          $R^1_{x_i,a_i,x_j} \leftarrow a_j$;
8:       **else**
9:          **if** $a_j \neq R$ **then**
10:             $R^2_{x_i,a_i,x_j} \leftarrow a_j$;
11:             **return** TRUE;
12: **if** oneSupport = FALSE **then**
13:    **return** FALSE
14: **else**
15:    **for each** $x_k \in X$, $x_k \neq x_i$ and $x_k \neq x_j$, s.t. $(x_k, x_i) \in C$ and $(x_k, x_j) \in C$ **do**
16:       **if** there is a valid residue $R^*_{x_i,a_i,x_k}$ **and** isConsistent$(R^*_{x_i,a_i,x_k}, a_j)$ **or if** there is a valid residue $R^*_{x_j,a_j,x_k}$ **and** isConsistent$(R^*_{x_j,a_j,x_k}, a_i)$
17:       **then continue**;
18:       PCwitness $\leftarrow$ FALSE;
19:       **for each** $a_k \in D(x_k)$ **do**
20:          **if** isConsistent(a_i, a_k) **and** isConsistent(a_j, a_k) **then**
21:             PCwitness $\leftarrow$ TRUE;
22:             **break**;
23:       **if** PCwitness = FALSE **then**
24:          **return** FALSE;
25: **return** TRUE;

Moving back to Algorithm 1, if at least one value is deleted from a domain $D(x_i)$, some pairs of variables must be enqueued so that the deletions are propagated. Lines 18-19 enqueue all pairs of the form (x_k, x_i). This ensures that if a value in a domain $D(x_k)$ has lost its last support in $D(x_i)$, it will be processed by the algorithm when the pair (x_k, x_i) is dequeued, and it will be removed. In addition, it ensures that if a value in $D(x_k)$ has been left with only one support in $D(x_i)$, that is not a PC-support, it will be processed and deleted once (x_k, x_i) is dequeued. This means that if we only enqueue pairs of the form (x_k, x_i), we can achieve stronger pruning than AC. However, this is not enough to achieve RPC. We call the version of RPC3 that only enqueues such pairs *restricted RPC3* (rRPC3).

To achieve RPC, for each pair (x_k, x_i) that is enqueued, we also enqueue all pairs of the form (x_l, x_k) s.t. x_l is constrained with x_i. This is because after the deletions from $D(x_i)$ the last PC-witness in $D(x_i)$ for some pair of values for variables x_k and x_l may have been deleted. This may cause further deletions from $D(x_l)$.

The worst-case time complexity of RPC3, and rRPC3, is $O(ned^3)$[1]. The space complexity is determined by the space required to store the residues, which is $O(ed)$. The time complexities of algorithms RPC1 and RPC2 are $O(ned^3)$ and $O(ned^2)$ respectively, while their space complexities, for stand-alone use, are $O(ed^2)$ and $O(end)$. RPC3 has a higher time complexity than RPC2, and a lower space complexity than both RPC1 and RPC2. But most importantly, RPC3 does not require the typically quite expensive restoration of data structures after failures when used inside search. In addition, this means that its space complexity remains $O(ed)$ when used inside search, while the space complexities of RPC1 and RPC2 will be even higher than $O(ed^2)$ and $O(end)$.

4   Experiments

We have experimented with 17 classes of binary CSPs taken from C.Lecoutre's XCSP repository: *rlfap, graph coloring, qcp, qwh, bqwh, driver, job shop, haystacks, hanoi, pigeons, black hole, ehi, queens, geometric, composed, forced random, model B random*. A total of 1142 instances were tested. Details about these classes of problems can be found in C.Lecoutre's homepage. All algorithms used the dom/wdeg heuristic for variable ordering [6] and lexicographic value ordering. The experiments were performed on a FUJITSU Server PRIMERGY RX200 S7 R2 with Intel(R) Xeon(R) CPU E5-2667 clocked at 2.90GHz, with 48 GB of ECC RAM and 16MB cache.

We have compared search algorithms that apply rRPC3 (resp. RPC3) during search to a baseline algorithm that applies AC (i.e. MAC) and also to an algorithm that applies lmaxRPC. AC and lmaxRPC were enforced using the AC^r and lmaxRPC3 algorithms respectively. For simplicity, the four search algorithms will be denoted by AC, rRPC, RPC, and maxRPC hereafter. Note that a MAC algorithm with AC^r and dom/wdeg for variable ordering is considered as the best general purpose solver for binary CSPs.

A timetout of 3600 seconds was imposed on all four algorithms for all the tested instances. Importantly, **rRPC only timed out on instances where AC and maxRPC also timed out**. On the other hand, there were several cases where rRPC finished search within the time limit but one (or both) of AC and maxRPC timed out. There were a few instances where RPC timed out while rRPC did not, but the opposite never occured.

Table 1 summarizes the results of the experimental evaluation for specific classes of problems. For each class we give the following information:

- The mean node visits and run times from non-trivial instances that were solved by all algorithms within the time limit. We consider as trivial any instance that was solved by all algorithms in less than a second.
- Node visits and run time from the single instance where AC displayed its best performance compared to rRPC.

---

[1] The proof is quite simple but it is omitted for space reasons.

- Node visits and run time from the single instance where maxRPC displayed its best performance compared to rRPC.
- Node visits and run time from a representative instance where rRPC displayed good performance compared to AC, excluding instances where AC timed out.
- The number of instances where AC, RPC, maxRPC timed out while rRPC did not. This information is given only for classes where at least one such instance occured.
- The number of instances where AC, rRPC, RPC, or maxRPC was the winning algorithm, excluding trivial instances.

Comparing AC to rRPC we can note the following. rRPC is more efficient in terms of mean run time performance on all classes of structured problems with the exception of *queens*. The difference in favor of rRPC can be quite stunning, as in the case of *qwh* and *qcp*. The numbers of node visits in these classes suggest that rRPC is able to achieve considerable extra pruning, and this is then reflected on cpu times.

Importantly, in all instances of 16 classes (i.e. all classes apart from *queens*) AC was at most 1.7 times faster than rRPC. In contrast, there were 7 instances from *rlfap* and 12 from *graph coloring* where AC timed out while rRPC finished within the time limit. The mean cpu time of rRPC on these *rlfap* instances was 110 seconds while on the 12 *graph coloring* instances the cpu time of rRPC ranged from 1.8 to 1798 seconds. In addition, there were numerous instances where rRPC was orders of magnitude faster than AC. This is quite common in *qcp* and *qwh*, as the mean cpu times demonstrate, but such instances also occur in *graph coloring*, *bqwh*, *haystacks* and *ehi*.

Regarding random problems, rRPC achieves similar performance to AC on *geometric* (which is a class with some structure) and is slower on *forced random* and *model B random*. However, the differences in these two classes are not significant. The only class where there are significant differences in favour of AC is *queens*. Specifically, AC can be up to 5 times faster than rRPC on some instances, and orders of magnitude faster than both RPC and maxRPC. This is because all algorithms spend a lot of time on propagation but evidently the strong local consistencies achieve little extra pruning. Importantly, the low cost of rRPC makes its performance reasonable compared to the huge run times of RPC and maxRPC.

The comparison between RPC and AC follows the same trend as that of rRPC and AC, but importantly the differences in favour of RPC are not as large on structured problems where AC is inefficient, while AC is clearly faster on random problems, and by far superior on dense structured problems like *queens* and *pigeons*.

Comparing the mean performance of rRPC to RPC and maxRPC we can note that rRPC is more efficient on all classes. There are some instances where RPC or/and maxRPC outperform rRPC due to their stronger pruning, but the differences in favour of RPC and maxRPC are rarely significant. In contrast, rRPC can often be orders of magnitude faster. An interesting observation that

Table 1. Node visits (n), run times in secs (t), number of timeouts (#TO) (if applicable), and number of wins (winner) in summary. The number in brackets after the name of each class gives the number of instances tested.

class	AC		rRPC		RPC		maxRPC	
	(n)	(t)	(n)	(t)	(n)	(t)	(n)	(t)
rlfap (24)								
mean	29045	64.1	11234	32.3	10878	39.0	8757	134.0
best AC	12688	9.3	11813	14.5	10048	18.2	5548	33.2
best maxRPC	8405	10.1	3846	4.4	3218	6.86	1668	4.8
good rRPC	19590	28.8	5903	8.2	5197	10.4	8808	23.7
#TO	7				3		6	
winner	1		18		1		0	
qcp (60)								
mean	307416	345.4	37725	44.8	43068	167.1	49005	101.3
best AC	36498	63.5	36286	73.7	57405	354.3	63634	173.5
best maxRPC	20988	16.8	7743	7.6	4787	11.9	1723	1.8
good rRPC	1058477	761	65475	53.5	67935	162.9	54622	63.1
winner	2		8		0		4	
qwh (40)								
mean	205232	1348.2	20663	46.2	28694	177.9	24205	64.7
best AC	6987	4.5	3734	3.0	5387	11.9	3174	3.2
best maxRPC	231087	461.4	30926	72.3	30434	187.6	13497	35.8
good rRPC	445771	859.6	35923	79.5	56965	375.4	37582	103.9
winner	0		9		0		6	
bqwh (200)								
mean	28573	7.6	7041	2.4	5466	2.9	6136	2.7
best AC	5085	1.2	4573	1.4	4232	2.0	3375	1.3
best maxRPC	324349	85.3	122845	46.1	56020	33.8	64596	29.3
good rRPC	83996	22.5	7922	2.6	10858	5.6	9325	4.2
winner	2		36		13		20	
graph coloring (177)								
mean	322220	88.6	261882	60.4	192538	73.7	263227	138.79
best AC	1589650	442.5	1589650	743.8	1266416	930.8	1589650	1010.0
best maxRPC	1977536	647.9	1977536	762.4	1265930	613.7	1977536	759.8
good rRPC	31507	189.1	3911	15.6	2851	18.2	10477	62.7
#TO	12				1		5	
winner	8		35		17		0	
geometric (100)								
mean	111611	58.4	54721	58.6	52416	97.3	38227	190.9
best AC	331764	203.1	169871	218.1	160428	358.1	113878	696.2
best maxRPC	67526	28.1	31230	28.1	30229	53.5	20071	73.6
good rRPC	254304	123.3	119248	117.4	117665	203.0	84410	363.3
winner	12		11		1		0	
forced random (20)								
mean	348473	143.5	197994	177.2	191114	309.8	154903	455.4
best AC	1207920	491.5	677896	596.7	654862	1040.4	538317	1551.4
best maxRPC	26729	7.6	12986	9.0	12722	13.5	9372	22.4
good rRPC	455489	201.6	270345	258.5	262733	462.0	207267	651.2
winner	20		0		0		0	
model B random (40)								
mean	741124	194.3	383927	224.5	361926	346.5	28871	1044.9
best AC	2212444	669.8	1197369	805.7	1136257	1283.8	-	TO
best maxRPC	345980	81.2	181127	94.4	171527	142.7	130440	405.8
good rRPC	127567	32.4	43769	24.4	41328	39.1	51455	171.5
#TO	0				0		14	
winner	39		1		0		0	
queens (14)								
mean	2092	14.2	797	59.2	2476	1032.2	953	2025.2
best AC	150	9.2	149	43.0	149	499.1	149	3189.2
best maxRPC	7886	10.8	2719	11.5	9425	910.5	3341	367.9
good rRPC	7886	10.8	2719	11.5	9425	910.5	3341	367.9
#TO	0				0		1	
winner	4		0		0		0	

requires further investigation is that in some cases the node visits of rPRC are fewer than RPC and and/or maxRPC despite the weaker pruning. This usually occurs on soluble instances and suggests that the interaction with the dom/wdeg heuristic can guide search to solutions faster.

Finally, the classes not shown in Table 1 mostly include instances that are either very easy or very hard (i.e. all algorithms time out). Specifically, instances in *composed* and *hanoi* are all trivial, and the ones in *black hole* and *job shop* are either trivial or very hard. Instances in *ehi* typically take a few seconds for AC and under a second for the other three algorithms. Instances in *haystacks* are very hard except for a few where AC is clearly outperformed by the other three algorithms. For example, in *haystacks-04* AC takes 8 seconds and the other three take 0.2 seconds. Instances in *pigeons* are either trivial or very hard except for a few instances where rRPC is the best algorithm followed by AC. For example on *pigeons-12* AC and rRPC take 709 and 550 seconds respectively, while RPC and maxRPC time out. Finally, *driver* includes only 7 instances. Among them, 3 are trivial, rRPC is the best algorithm on 3, and AC on 1.

5   Conclusion

RPC was recognized as a promising alternative to AC but has been neglected for the past 15 years or so. In this paper we have revisited RPC by proposing RPC3, a new algorithm that utilizes ideas, such as residues, that have become standard in recent years when implementing AC or maxRPC algorithms. Using RPC3 and a restricted variant we performed the first wide experimental study of RPC when used inside search. Perhaps surprisingly, results clearly demostrate that rRPC3 is by far more efficient than state-of-the-art AC and maxRPC algorithms when applied during search. This challenges the common perception that MAC is the best general purpose solver for binary CSPs.

References

1. Balafoutis, T., Paparrizou, A., Stergiou, K., Walsh, T.: New algorithms for max restricted path consistency. Constraints **16**(4), 372–406 (2011)
2. Balafrej, A., Bessiere, C., Bouyakh, E., Trombettoni, G.: Adaptive singleton-based consistencies. In: Proceedings of the Twenty-Eighth AAAI Conference on Artificial Intelligence, pp. 2601–2607 (2014)
3. Barták, R., Erben, R.: A new algorithm for singleton arc consistency. In: Proceedings of the Seventeenth International Florida Artificial Intelligence, pp. 257–262 (2004)
4. Berlandier, P.: Improving domain filtering using restricted path consistency. In: Proceedings of IEEE CAIA 1995, pp. 32–37 (1995)
5. Bessiere, C., Cardon, S., Debruyne, R., Lecoutre, C.: Efficient Algorithms for Singleton Arc Consistency. Constraints **16**, 25–53 (2011)
6. Boussemart, F., Hemery, F., Lecoutre, C., Sais, L.: Boosting systematic search by weighting constraints. In: Proceedings of ECAI 2004, Valencia, Spain (2004)

7. Debruyne, R., Bessière, C.: From restricted path consistency to max-restricted path consistency. In: Smolka, G. (ed.) CP 1997. LNCS, vol. 1330, pp. 312–326. Springer, Heidelberg (1997)
8. Debruyne, R., Bessière, C.: Domain Filtering Consistencies. JAIR **14**, 205–230 (2001)
9. Grandoni, F., Italiano, G.F.: Improved algorithms for max-restricted path consistency. In: Rossi, F. (ed.) CP 2003. LNCS, vol. 2833, pp. 858–862. Springer, Heidelberg (2003)
10. Lecoutre, C., Hemery, F.: A study of residual supports in arc cosistency. In: Proceedings of IJCAI 2007, pp. 125–130 (2007)
11. Lecoutre, C., Prosser, P.: Maintaining singleton arc consistency. In: 3rd International Workshop on Constraint Propagation and Implementation (CPAI 2006), pp. 47–61 (2006)
12. Likitvivatanavong, C., Zhang, Y., Bowen, J., Shannon, S., Freuder, E.: Arc consistency during search. In: Proceedings of IJCAI 2007, pp. 137–142 (2007)
13. Prosser, P., Stergiou, K., Walsh, T.: Singleton consistencies. In: Dechter, R. (ed.) CP 2000. LNCS, vol. 1894, pp. 353–368. Springer, Heidelberg (2000)
14. Vion, J., Debruyne, R.: Light algorithms for maintaining max-RPC during search. In: Proceedings of SARA 2009 (2009)

Machine Learning of Bayesian Networks Using Constraint Programming

Peter van Beek[✉] and Hella-Franziska Hoffmann

Cheriton School of Computer Science, University of Waterloo,
Waterloo, ON N2L 3G1, Canada
vanbeek@cs.uwaterloo.ca

Abstract. Bayesian networks are a widely used graphical model with diverse applications in knowledge discovery, classification, prediction, and control. Learning a Bayesian network from discrete data can be cast as a combinatorial optimization problem and there has been much previous work on applying optimization techniques including proposals based on ILP, A* search, depth-first branch-and-bound (BnB) search, and breadth-first BnB search. In this paper, we present a *constraint-based* depth-first BnB approach for solving the Bayesian network learning problem. We propose an improved constraint model that includes powerful dominance constraints, symmetry-breaking constraints, cost-based pruning rules, and an acyclicity constraint for effectively pruning the search for a minimum cost solution to the model. We experimentally evaluated our approach on a representative suite of benchmark data. Our empirical results compare favorably to the best previous approaches, both in terms of number of instances solved within specified resource bounds and in terms of solution time.

1 Introduction

Bayesian networks are a popular probabilistic graphical model with diverse applications including knowledge discovery, classification, prediction, and control (see, e.g., [1]). A Bayesian network (BN) can either be constructed by a domain expert or learned automatically from data. Our interest here is in the learning of a BN from discrete data, a major challenge in machine learning. Learning a BN from discrete data can be cast as a combinatorial optimization problem—the well-known *score-and-search* approach—where a scoring function is used to evaluate the quality of a proposed BN and the space of feasible solutions is systematically searched for a best-scoring BN. Unfortunately, learning a BN from data is NP-hard, even if the number of parents per vertex in the DAG is limited to two [2]. As well, the problem is unlikely to be efficiently approximatable with a good quality guarantee, thus motivating the use of global (exact) search algorithms over local (heuristic) search algorithms [3].

Global search algorithms for learning a BN from data have been studied extensively over the past several decades and there have been proposals based on dynamic programming [4–6], integer linear programming (ILP) [7,8], A* search

© Springer International Publishing Switzerland 2015
G. Pesant (Ed.): CP 2015, LNCS 9255, pp. 429–445, 2015.
DOI: 10.1007/978-3-319-23219-5_31

[9–11], depth-first branch-and-bound (BnB) search [12,13], and breadth-first BnB search [10,11,14,15]. In this paper, we present a *constraint-based* depth-first BnB approach for solving the Bayesian network learning problem. We propose an improved constraint model that includes powerful dominance constraints, symmetry-breaking constraints, cost-based pruning rules, and an acyclicity constraint for effectively pruning the search for a minimum cost solution to the model. We experimentally evaluated our approach on a representative suite of benchmark data. Our empirical results compare favorably to the best previous approaches, both in terms of number of instances solved within specified resource bounds and in terms of solution time.

2   Background

In this section, we briefly review the necessary background in Bayesian networks before defining the Bayesian network structure learning problem (for more background on these topics see, for example, [16,17]).

A Bayesian network (BN) is a probabilistic graphical model that consists of a labeled directed acyclic graph (DAG) in which the vertices $V = \{v_1, \ldots, v_n\}$ correspond to random variables, the edges represent direct influence of one random variable on another, and each vertex v_i is labeled with a conditional probability distribution $P(v_i \mid parents(v_i))$ that specifies the dependence of the variable v_i on its set of parents $parents(v_i)$ in the DAG. A BN can alternatively be viewed as a factorized representation of the joint probability distribution over the random variables and as an encoding of conditional independence assumptions.

The predominant method for BN structure learning from data is the *score-and-search* method[1]. Let G be a DAG over random variables V, and let $I = \{I_1, \ldots, I_N\}$ be a set of multivariate discrete data, where each instance I_i is an n-tuple that is a complete instantiation of the variables in V. A *scoring function* $\sigma(G \mid I)$ assigns a real value measuring the quality of G given the data I. Without loss of generality, we assume that a lower score represents a better quality network structure.

Definition 1. *Given a discrete data set $I = \{I_1, \ldots, I_N\}$ over random variables V and a scoring function σ, the Bayesian network structure learning problem is to find a directed acyclic graph G over V that minimizes the score $\sigma(G \mid I)$.*

Scoring functions commonly balance goodness of fit to the data with a penalty term for model complexity to avoid overfitting. Common scoring functions include BIC/MDL [18,19] and BDeu [20,21]. An important property of these (and all commonly used) scoring functions is decomposability, where the

---

[1] An alternative method, called *constraint-based* structure learning in the literature, is based on statistical hypothesis tests for conditional independence. We do not discuss it further here except to note that the method is known to scale to large instances but to have the drawback that it is sensitive to a single failure in a hypothesis test (see [17, p. 785]).

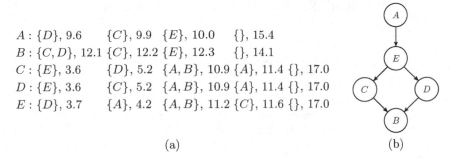

$A : \{D\}, 9.6 \quad \{C\}, 9.9 \quad \{E\}, 10.0 \quad \{\}, 15.4$
$B : \{C, D\}, 12.1 \; \{C\}, 12.2 \; \{E\}, 12.3 \quad \{\}, 14.1$
$C : \{E\}, 3.6 \quad \{D\}, 5.2 \;\; \{A, B\}, 10.9 \; \{A\}, 11.4 \; \{\}, 17.0$
$D : \{E\}, 3.6 \quad \{C\}, 5.2 \;\; \{A, B\}, 10.9 \; \{A\}, 11.4 \; \{\}, 17.0$
$E : \{D\}, 3.7 \quad \{A\}, 4.2 \;\; \{A, B\}, 11.2 \; \{C\}, 11.6 \; \{\}, 17.0$

(a)   (b)

Fig. 1. (a) Random variables and possible parent sets for Example 1; (b) minimum cost DAG structure with cost 38.9.

score of the entire network $\sigma(G \mid I)$ can be rewritten as the sum of local scores $\sum_{i=1}^{n} \sigma(v_i, parents(v_i) \mid I)$ that only depend on v_i and the parent set of v_i in G. Henceforth, we assume that the scoring function is decomposable and that, following previous work, the local score $\sigma(v_i, p \mid I)$ for each possible parent set $p \subseteq 2^{V - \{v_i\}}$ and each random variable v_i has been computed in a preprocessing step prior to the search for the best network structure. Pruning techniques can be used to reduce the number of possible parent sets that need to be considered, but in the worst-case the number of possible parent sets for each variable v_i is 2^{n-1}, where n is the number of vertices in the DAG.

Example 1. Let A, B, C, D, and E be random variables with the possible parent sets and associated scores shown in Figure 1(a). For example, if the parent set $\{C, D\}$ for random variable B is chosen there would be a directed edge from C to B and a directed edge from D to B and those would be the only incoming edges to B. The local score for this parent set is 12.1. If the parent set $\{\}$ for random variable A is chosen, there would be no incoming edges to A; i.e., A would be a source vertex. Figure 1(b) shows the minimum cost DAG with cost $15.4 + 4.2 + 3.6 + 3.6 + 12.1 = 38.9$.

3   Constraint Programming Approach

In this section, we present a constraint model and a depth-first branch-and-bound solver for the Bayesian network structure learning problem. Table 1 summarizes the notation.

Our constraint model consists of vertex variables, ordering variables, depth variables, and constraints over those variables. The ordering and depth variables, although redundant, improve the search for a solution.

Vertex (possible parent set) variables. There is a vertex variable v_i, $i \in V$, for each random variable in V and the domain of v_i, $dom(v_i)$, consists of the possible parent sets for v_i. The assignment $v_i = p$ denotes that vertex v_i has parents p in the DAG; i.e., the vertex variables represent the DAG over the set of random variables V. Associated with each domain value is a cost and the goal

Table 1. Notation for specifying constraint model.

V	set of random variables
n	number of random variables in the data set
$cost(v)$	cost (score) of variable v
$dom(v)$	domain of v
$parents(v)$	set of parents of v in the DAG
$\min(dom(v))$	the minimum value in the domain of v
$v_1, \ldots, v_n$	vertex (possible parent set) variables
$o_1, \ldots, o_n$	ordering (permutation) variables
$d_1, \ldots, d_n$	depth variables
$depth(p \mid o_1, \ldots, o_{i-1})$	depth of $p \in dom(v_j)$, where v_j occurs at position i in the ordering

is to minimize the total cost, $cost(v_1) + \cdots + cost(v_n)$, subject to the constraint that the graph is acyclic. A global constraint is introduced to enforce that the vertex variables form a DAG,

$$acyclic(v_1, \ldots, v_n), \tag{1}$$

where the constraint is satisfied if and only if the graph designated by the parent sets is acyclic. The DAG is not necessarily connected. A satisfiability checker for the acyclic constraint is given in Section 3.7, which in turn can be used to propagate the constraint.

Ordering (permutation) variables. There is an ordering variable o_i for each random variable and $dom(o_i) = V$, the set of random variables. The assignment $o_i = j$ denotes that vertex v_j is in position i in the total ordering of the variables. The ordering variables represent a permutation of the random variables. A global constraint is introduced to enforce that the order variables form a permutation of the vertex variables,

$$alldifferent(o_1, \ldots, o_n). \tag{2}$$

The alldifferent constraint is propagated by, whenever a variable becomes instantiated, simply removing that value from the domains of the other variables.

Depth variables. There is a depth variable d_i for each random variable and $dom(d_i) = \{0, \ldots, n-1\}$. The depth variables and the ordering variables are in one-to-one correspondence. The assignment $d_i = k$ denotes that the depth of the vertex variable v_j that occurs at position i in the ordering is k, where the depth is the length of the longest path from a source vertex to vertex v_j in the DAG.

Example 2. A constraint model for Example 1 would have variables $v_A, \ldots, v_E$, $o_1, \ldots, o_5$, and $d_1, \ldots, d_5$, with domains $dom(v_A) = \{\{C\}, \{D\}, \{E\}, \{\}\}, \ldots,$ $dom(v_E) = \{\{D\}, \{A\}, \{A, B\}, \{C\}, \{\}\}$, $dom(o_i) = \{A, \ldots, E\}$, and $dom(d_i) = \{0, \ldots, 4\}$.

To more formally state additional constraints, we introduce the following notation for the depth of a domain value p for a vertex variable v_j.

Definition 2. *The* depth *of a domain value p for a vertex variable v_j that occurs at position i in the ordering, denoted $depth(p \mid o_1, \ldots, o_{i-1})$, is defined as: 0 if $p = \{\}$; one plus the maximum depth of the elements of p if each element of p occurs in a parent set of a vertex variable earlier in the ordering; and ∞ otherwise.*

Example 3. In Example 2, suppose that $o_1 = A$, $v_A = \{\}$, $d_1 = 0$, $o_2 = E$, $v_E = \{A\}$, and $d_2 = 1$. The value of $depth(p \mid o_1, o_2)$ for variable C is 0 if $p = \{\}$, 1 if $p = \{A\}$, 2 if $p = \{E\}$, and ∞ if $p = \{D\}$ or $p = \{A, B\}$.

Constraints 3 & 4 establish the correspondence between the three types of variables,

$$\forall j \bullet \forall p \bullet v_j = p \iff \exists! i \bullet o_i = j \wedge d_i = depth(p \mid o_1, \ldots, o_{i-1}), \qquad (3)$$
$$\forall i \bullet \forall j \bullet o_i = j \iff \exists! p \bullet v_j = p \wedge d_i = depth(p \mid o_1, \ldots, o_{i-1}), \qquad (4)$$

where the indices i and j range over $1 \leq i, j \leq n$ and the value p ranges over $dom(v_j)$. The constraints are propagated as follows. A value $p \in dom(v_j)$ can be pruned iff $\forall i \bullet j \in dom(o_i) \Rightarrow depth(p \mid o_1, \ldots, o_{i-1}) \notin dom(d_i)$. A value $j \in dom(o_i)$ can be pruned iff $\forall p \in dom(v_j) \bullet depth(p \mid o_1, \ldots, o_{i-1}) \notin dom(d_i)$. Only bounds are maintained on the depth variables. Hence, the notation $depth(p \mid o_1, \ldots, o_{i-1}) \notin dom(d_i)$ is to be interpreted as $depth(p \mid o_1, \ldots, o_{i-1}) < \min(dom(d_i)) \vee depth(p \mid o_1, \ldots, o_{i-1}) > \max(dom(d_i))$. When propagating Constraints 3 & 4, we must determine $depth(p \mid o_1, \ldots, o_{i-1})$. In general, this is a difficult problem. We restrict ourselves to two easy special cases: (i) all of $o_1, \ldots, o_{i-1}$ have been instantiated, or (ii) some of $o_1, \ldots, o_{i-1}$ have been instantiated and all of the $p \in dom(v_j)$ are subsets of these ordering variables. We leave to future work further ways of safely approximating the depth to allow further propagation.

Example 4. In Example 2, suppose that $o_1 = A$, $v_A = \{\}$, $d_1 = 0$, $o_2 = E$, $v_E = \{A\}$, $d_2 = 1$, and that, as a result of some propagation, $\min(d_i) = 1$, $i = 3, 4, 5$. The value $\{\}$ can be pruned from each of the domains of v_B, v_C, and v_D.

One can see that the vertex variables together with the acyclic constraint are sufficient alone to model the Bayesian network structure learning problem. Such a search space over DAGs forms the basis of Barlett and Cussens' integer programming approach [8]. One can also see that the ordering (permutation) variables together with the alldifferent constraint are sufficient alone, as the minimum cost DAG for a given ordering is easily determinable. Larranaga et al. [22] were the first to propose the search space of all permutations and Teyssier and Koller [23] successfully applied it within a local search algorithm. The permutation search space also forms the basis of the approaches based on dynamic programming [4–6] and on the approaches based on searching for the shortest path in an ordering graph using A* search [9–11], depth-first branch-and-bound DFBnB search [13], and BFBnB search [10,11,15].

The unique aspects of our model lie in combining the DAG and permutation search spaces and introducing the depth variables. As is shown in the next sections, the combination of variables allows us to identify and post many additional constraints that lead to a considerable reduction in the search space.

3.1   Symmetry-Breaking Constraints (I)

Many permutations and prefixes of permutations, as represented by the ordering variables, are symmetric in that they lead to the same minimum cost DAG, or are dominated in that they lead to a DAG of equal or higher cost. The intent behind introducing the auxiliary depth variables is to rule out all but the lexicographically least of these permutations. A lexicographic ordering is defined over the depth variables—and over the ordering variables, in the case of a tie on the values of the depth variables. The following constraints are introduced to enforce the lexicographic order.

$$d_1 = 0 \tag{5}$$
$$d_i = k \iff (d_{i+1} = k \lor d_{i+1} = k+1), \qquad i = 1,\ldots,n-1 \tag{6}$$
$$d_i = d_{i+1} \implies o_i < o_{i+1}, \qquad i = 1,\ldots,n-1 \tag{7}$$

The constraints are readily propagated. Constraint 6 is a dominance constraint.

Example 5. In Example 2, consider the ordering prefix $(o_1,\ldots,o_4) = (E,C,A,D)$ with associated vertex variables $(v_E, v_C, v_A, v_D) = (\{\}, \{E\}, \{C\}, \{E\})$ and depths $(d_1,\ldots,d_4) = (0,1,2,1)$. The cost of this ordering prefix is 33.8. The ordering prefix violates Constraint 6. However, the ordering prefix $(o_1,\ldots,o_4) = (E,C,D,A)$ with depths $(d_1,\ldots,d_4) = (0,1,1,2)$ and vertex variables $(v_E,\ldots,v_D) = (\{\}, \{E\}, \{E\}, \{D\})$ satisfies the constraint and has a lower cost of 33.5.

Constraint 7 is a symmetry-breaking constraint.

Example 6. In Example 2, consider the ordering $(o_1,\ldots,o_5) = (A,E,D,C,B)$ with $(d_1,\ldots,d_5) = (0,1,2,2,3)$ and $(v_A,\ldots,v_B) = (\{\},\ldots,\{C,D\})$. The ordering violates Constraint 7. However, the symmetric ordering $(o_1,\ldots,o_5) = (A,E,C,D,B)$, which represents the same DAG, satisfies the constraint and has equal cost.

Theorem 1. *Any ordering prefix $o_1,\ldots,o_i$ can be safely pruned if the associated depth variables $d_1,\ldots,d_i$ do not satisfy Constraints 5-7.*

3.2   Symmetry-Breaking Constraints (II)

In the previous section, we identified symmetries and dominance among the ordering variables. In this section, we identify symmetries among the vertex variables. Let $[x/y]dom(v)$ be the domain that results from replacing all occurrences of y in the parent sets by x. Two vertex variables v_1 and v_2 are symmetric if $[v_1/v_2]dom(v_1) = dom(v_2)$; i.e., the domains are equal once the given substitution is applied. The symmetry is broken by enforcing that v_1 must precede v_2 in the ordering,

$$\forall i \bullet \forall j \bullet o_i = 1 \wedge o_j = 2 \implies i < j. \tag{8}$$

Example 7. In Example 2, consider vertex variables v_C and v_D. The variables are symmetric as, $[v_C/v_D]dom(v_C) = \{\{E\}, \{C\}, \{A, B\}, \{A\}, \{\}\} = dom(v_D)$.

3.3   Symmetry-Breaking Constraints (III)

A BN can be viewed as an encoding of conditional independence assumptions. Two BN structures (DAGS) are said to be I-equivalent if they encode the same set of conditional independence assumptions (see, e.g., [16, 17]). The efficiency of the search for a minimal cost BN can be greatly improved by recognizing I-equivalent partial (non-)solutions. Chickering [24, 25] provides a local transformational characterization of equivalent BN structures based on covered edges that forms the theoretical basis of these symmetry-breaking constraints.

Definition 3 (Chickering [24]). *An edge $x \to y$ in a Bayesian network is a covered edge if $parents(y) = parents(x) \cup \{x\}$.*

Theorem 2 (Chickering [24]). *Let G be any DAG containing the edge $x \to y$, and let G' be the directed graph identical to G except that the edge between x and y in G' is oriented as $y \to x$. Then G' is a DAG that is equivalent to G if and only if $x \to y$ is a covered edge in G.*

Example 8. In Figure 1(b) the edge $A \to E$ is a covered edge and the Bayesian network with the edge reversed to be $E \to A$ is an I-equivalent Bayesian network.

In what follows, we identify three cases that consist of sequences of one or more covered edge reversals and break symmetries by identifying a lexicographic ordering. Experimental evidence suggests that these three cases capture much of the symmetry due to I-equivalence. Symmetries are only broken if the costs of the two I-equivalent DAGs would be equal; otherwise there is a negative interaction with pruning based on the cost function (discussed in Section 3.8).

Case 1. Consider vertex variables v_i and v_j. If there exists domain values $p \in dom(v_i)$ and $p \cup \{v_i\} \in dom(v_j)$, this pair of assignments includes a covered edge $v_i \to v_j$; i.e., v_i and v_j would have identical parents except that v_i would also be a parent of v_j. Thus, there exists an I-equivalent DAG with the edge reversed. We keep only the lexicographically least: the pair of assignments would be permitted iff $i < j$.

Case 2. Consider vertex variables v_i, v_j, and v_k. If there exists domain values $p \in dom(v_i)$, $p \cup \{v_i\} \in dom(v_j)$, $p \cup \{v_j\} \in dom(v_k)$, where $i < j$ and $k < j$, there is a covered edge $v_i \to v_j$ and, if this covered edge is reversed, the covered edge $v_j \to v_k$ is introduced, which in turn can be reversed. Thus, there exists an

I-equivalent DAG with the edges $\{v_i \rightarrow v_j, v_j \rightarrow v_k\}$ and an I-equivalent DAG with the edges $\{v_k \rightarrow v_j, v_j \rightarrow v_i\}$. We keep only the lexicographically least: the triple of assignments would be permitted iff $i < k$.

Case 3. Consider vertex variables v_i, v_j, v_k, and v_l. If there exists domain values $p \in dom(v_i)$, $p \cup \{v_i\} \in dom(v_j)$, $p \cup \{v_i, v_j\} \in dom(v_k)$, $p \cup \{v_j, v_k\} \in dom(v_l)$, where $i < j$, $l < j$, $j < k$, there exists an I-equivalent DAG with the edges $\{v_i \rightarrow v_j, v_i \rightarrow v_k, v_j \rightarrow v_k, v_j \rightarrow v_l, v_k \rightarrow v_l\}$ and an I-equivalent DAG with the edges $\{v_l \rightarrow v_j, v_l \rightarrow v_k, v_j \rightarrow v_k, v_j \rightarrow v_i, v_k \rightarrow v_i\}$. We keep only the lexicographically least: the triple of assignments would be permitted iff $i < l$.

In our empirical evaluation, these symmetry-breaking rules were used only as a satisfiability check at each node in the search tree, as we found that propagating the I-equivalence symmetry-breaking rules did not further improve the runtime.

3.4 Dominance Constraints (I)

Given an ordering prefix $o_1, \ldots, o_{i-1}$, a domain value p for a vertex variable v_j is *consistent* with the ordering if each element of p occurs in a parent set of a vertex variable in the ordering. The domain value p assigned to vertex variable v_j that occurs at position i in an ordering should be the lowest cost p consistent with the ordering; assigning a domain value with a higher cost can be seen to be dominated as the values can be substituted with no effect on the other variables.

Theorem 3. *Given an ordering prefix $o_1, \ldots, o_{i-1}$, a vertex variable v_j, and domain elements $p, p' \in dom(v_j), p \neq p'$, if p is consistent with the ordering and $cost(p) \leq cost(p')$, p' can be safely pruned from the domain of v_j.*

Example 9. In Example 2, consider the prefix ordering $(o_1, o_2) = (C, D)$. The values $\{C\}$, $\{E\}$, and $\{\}$ can be pruned from each of the domains of v_A and v_B, and the values $\{A\}$, $\{A, B\}$, $\{C\}$, and $\{\}$ can be pruned from the domain of v_E.

3.5 Dominance Constraints (II)

Teyssier and Koller [23] present a pruning rule that is now routinely used in score-and-search approaches as a preprocessing step before search begins.

Theorem 4 (Teyssier and Koller [23]). *Given a vertex variable v_j, and domain elements $p, p' \in dom(v_j)$, if $p \subset p'$ and $cost(p) \leq cost(p')$, p' can be safely pruned from the domain of v_j.*

Example 10. In Example 2, the value $\{A, B\}$ can be pruned from the domain of v_E.

We generalize the pruning rule so that it is now applicable during the search. Suppose that some of the vertex variables have been assigned values. These assignments induce ordering constraints on the variables.

Example 11. In Example 2, suppose $v_A = \{D\}$ and $v_C = \{A, B\}$. These assignments induce the ordering constraints $D < A$, $A < C$, and $B < C$.

Definition 4. *Given a set of ordering constraints induced by assignments to the vertex variables, let* ip(p) *denote the* induced parent set *where p has been augmented with any and all variables that come before in the ordering; i.e., if $y \in p$ and $x < y$ then x is added to p.*

The generalized pruning rule is as follows.

Theorem 5. *Given a vertex variable $\bar{v}_j$, and domain elements $p, p' \in dom(v_j), p \neq p'$, if $p \subseteq$ ip(p') and $cost(p) \leq cost(p')$, p' can be safely pruned from the domain of v_j.*

Example 12. Continuing with Example 11, consider v_E with $p = \{D\}$ and $p' = \{A\}$. The induced parent set ip(p') is given by $\{A, D\}$ and $cost(p) \leq cost(p')$. Thus, p' can be pruned. Similarly, $p' = \{C\}$ can be pruned.

3.6   Dominance Constraints (III)

Consider an ordering prefix $o_1, \ldots, o_i$ with associated vertex and depth variables and let π be a permutation over $\{1, \ldots, i\}$. The cost of *completing* the partial solutions represented by the prefix ordering $o_1, \ldots, o_i$ and the permuted prefix ordering $o_{\pi(1)}, \ldots, o_{\pi(i)}$ are identical. This insight is used by Silander and Myllymäki [5] in their dynamic programming approach and by Fan et al. [9–11] in their best-first approaches based on searching for the shortest path in the ordering graph. However, all of these approaches are extremely memory intensive. Here, we use this insight to prune the search space.

Theorem 6. *Let $cost(o_1, \ldots, o_i)$ be the cost of a partial solution represented by the given ordering prefix. Any ordering prefix $o_1, \ldots, o_i$ can be safely pruned if there exists a permutation π such that $cost(o_{\pi(1)}, \ldots, o_{\pi(i)}) < cost(o_1, \ldots, o_i)$.*

Example 13. In Example 2, consider the ordering prefix $O = (o_1, o_2) = (E, A)$ with associated vertex variables $(v_E, v_A) = (\{\}, \{E\})$ and cost of 27.0. The ordering prefix O can be safely pruned as there exists a permutation $(o_{\pi(1)}, o_{\pi(2)}) = (A, E)$ with associated vertex variables $(v_A, v_E) = (\{\}, \{A\})$ that has a lower cost of 19.6.

Clearly, in its full generality, the condition of Theorem 6 is too expensive to determine exactly as it amounts to solving the original problem. However, we identify three strategies that are easy to determine and collectively were found to be very effective at pruning in our experiments while maintaining optimality.

Strategy 1. We consider permutations that differ from the original permutation only in the last l or fewer places ($l = 4$ in our experiments).

Strategy 2. We consider permutations that differ from the original permutation only in the swapping of the last variable o_i with a variable earlier in the ordering.

Strategy 3. We consider whether a permutation $o_{\pi(1)}, \ldots, o_{\pi(i)}$ of lower cost was explored earlier in the search. To be able to efficiently determine this, we use memoization for ordering prefixes and only continue with a prefix if it has a better cost than one already explored (see, e.g., [26,27]). Our implementation of memoization uses hashing with quadratic probing and the replacement policy is to keep the most recent if the table becomes too full. It is well known that there is a strong relationship between backtracking search with memoization/caching and dynamic programming using a bottom-up approach, but memoization allows trading space for time and top-down backtracking search allows pruning the search space.

3.7   Acyclic Constraint

In this section we describe a propagator for the acyclicity constraint that achieves generalized arc consistency in polynomial time. We first present and analyze an algorithm that checks satisfiability for given possible parent sets. We then explain how this algorithm can be used to achieve generalized arc consistency.

Algorithm 1 can check whether a collection of possible parent sets allows a feasible parent set assignment, i.e. an assignment that represents a DAG. Its correctness is based on the following well-known property of directed acyclic graphs that is also used in the ILP approaches [7,8].

Theorem 7. *Let G be a directed graph over vertices V and let $parents(v)$ be the parents of vertex v in the graph. G is acyclic if and only if for every non-empty subset $S \subset V$ there is at least one vertex $v \in S$ with $parents(v) \cap S = \{\}$.*

The algorithm works as follows. First, it searches possible sources for the DAG, i.e. vertices for which $\{\}$ is a possible parent set. These vertices are stored in W^0. Note that if a directed graph does not have a source, it must contain a cycle by Theorem 7. Thus, if W^0 remains empty, there is no parent set assignment satisfying the acyclicity constraint. In the next iteration, the algorithm searches for vertices that have a possible parent set consisting of possible sources only. These vertices form set W^1. Again, if there are no such vertices, then no vertex in $V \setminus W^0$ has a possible parent set completely outside $V \setminus W^0$, which violates the acyclicity characterization of Theorem 7. Thus, there is no consistent parent set assignment. We continue this process until all vertices are included in one of the W^k sets or until we find a contradicting set $V \setminus (\bigcup_{i=0}^{k} W^i)$ for some k.

Theorem 8. *We can test satisfiability of the acyclic constraint in time $O(n^2 d)$, where n is the number of vertices and d is an upper bound on the number of possible parent sets per vertex.*

Example 14. Let v_A, v_B, v_C, and v_D be vertex variables with the possible parent sets,

$$dom(v_A) = \{\{B\}, \{D\}\}, \qquad dom(v_C) = \{\{B\}, \{D\}\},$$
$$dom(v_B) = \{\{A\}, \{C\}\}, \qquad dom(v_D) = \{\{A\}, \{C\}\}.$$

Algorithm 1 returns *false* as W^0 is found to be empty.

Algorithm 1. Checking satisfiability of acyclic constraint

Input: $V = \{v_1, \ldots, v_n\}$, set $dom(v_i)$ of possible parent sets for each vertex v_i in V.

Output: *True* if there is a feasible parent set assignment and *false* otherwise. Additionally, the variables $\mathcal{S}_i$ represent a feasible assignment if one exists.

$k \leftarrow 0$;
$\mathcal{S}_i \leftarrow nil$ for all $v_i \in V$;
while $\bigcup_{j=0}^{k-1} W^j \neq V$ **do**

$\quad$ $W^k \leftarrow \{\}$;

$\quad$ **for all** *vertices v_i not in* $\bigcup_{j=0}^{k-1} W^j$ **do**

$\quad\quad$ **if** v_i *has a possible parent set $p \in dom(v_i)$ with $p \subseteq \bigcup_{j=0}^{k-1} W^j$* **then**

$\quad\quad\quad$ $\mathcal{S}_i \leftarrow p$;

$\quad\quad\quad$ $W^k \leftarrow W^k \cup \{v_i\}$;

$\quad\quad$ **end if**

$\quad$ **end**

$\quad$ **if** $W^k = \{\}$ **then return** false;

$\quad$;

$\quad$ $k \leftarrow k + 1$;

end while
return true;

The algorithm for checking satisfiability can be used to achieve generalized arc consistency by iteratively testing, for each vertex v_i, whether each $p \in dom(v_i)$ has a support. We simply substitute the set of possible parent sets $dom(v_i)$ for v_i by the set $\{p\}$. A satisfiability check on the resulting instance successfully tests whether there is a consistent parent set assignment containing $v_i = p$. If we find a parent set p that cannot appear in any feasible solution, we remove p from the corresponding domain. Note that we only prune a value p from a domain $dom(v_i)$ if $v_i = p$ cannot be part of any feasible solution. This means that $v_i = p$ can also not be part of the support of any other variable value $q \in dom(v_j)$. Therefore, the removal of p cannot cause another supported value to become unsupported. Hence, we do not have to rerun any of the tests; we can simply run the test once for every value in every domain. These considerations show that we can enforce arc consistency for the acyclicity constraint in $O(n^3 d^2)$ steps.

In our empirical evaluation, we found that achieving generalized arc consistency did not pay off in terms of reduced runtime. Hence, in the current set of experiments Algorithm 1 was used only as a satisfiability check at each node in the search tree. Instead, a limited form of constraint propagation was performed based on necessary edges between vertex variables. An edge $v_i \to v_j$ is necessary if v_i occurs in every currently valid parent set for variable v_j; i.e., $\forall p \in dom(v_j) \bullet v_i \in p$. If a directed edge $v_i \to v_j$ is a necessary edge, the directed edge $v_j \to v_i$ cannot be an edge in a valid DAG, as a cycle would be introduced. Thus, any parent set that contains v_j can be removed from the domain of v_i. Removing domain elements may introduce additional necessary edges and pruning can be based on chaining necessary edges.

Example 15. Let v_A, v_B, v_C, and v_D be vertex variables with the possible parent sets,

$$dom(v_A) = \{\{\}, \{B\}, \{C\}\} \qquad dom(v_C) = \{\{B\}\}$$
$$dom(v_B) = \{\{A\}, \{A, C\}\} \qquad dom(v_D) = \{\{A\}, \{A, C\}\}$$

Since the edge $B \rightarrow C$ is necessary, the value $\{A, C\}$ can be pruned from the domain of v_B. This causes the edge $A \rightarrow B$ to become necessary, and the values $\{B\}$ and $\{C\}$ can be pruned from the domain of v_A.

We conclude with the following observation. Let i_v be the index of the set W^i in which we include vertex v in the satisfiability algorithm. Then, i_v is a lower bound on the number that vertex v can have in any topological numbering. This lower bound can be used in propagating Constraint 4.

3.8   Solving the Constraint Model

A constraint-based depth-first branch-and-bound search is used to solve the constraint model; i.e., the nodes in the search tree are expanded in a depth-first manner and a node is expanded only if the propagation of the constraints succeeds and a lower bound estimate on completing the partial solution does not exceed the current upper bound.

The branching is over the ordering (permutation) variables and uses the static order $o_1, \ldots, o_n$. Once an ordering variable is instantiated as $o_i = j$, the associated vertex variable v_j and depth variable d_i are also instantiated.

The lower bound is based on the lower bound proposed by Fan and Yuan [11]. In brief, prior to search, the strongly connected components (SCCs) of the graph based on the top few lowest cost elements in the domains of the vertex variables are found and pattern databases are constructed based on the SCCs. The pattern databases allow a fast and often accurate lower bound estimate during the search (see [11] for details).

The initial upper bound, found before search begins, is based on the local search algorithm proposed by Teyssier and Koller [23]. The algorithm uses restarts and first-improvement moves, the search space consists of all permutations, and the neighborhood function consists of swapping the order of two variables. Of course, as better solutions are found during the search the upper bound is updated.

As a final detail, additional pruning on the vertex variables can be performed based on the (well-known) approximation of bounds consistency on a cost function that is a knapsack constraint: $z = cost(v_1) + \cdots + cost(v_n)$. Let the bounds on $cost(v_i)$ be $[l_i, u_i]$, and let lb and ub be the current lower bound and upper bound on the cost, respectively. At any point in the search we have the constraint $lb \leq z < ub$ and a value $p \in dom(v_i)$ can be pruned if $cost(p) + \sum_{j \neq i} u_j < lb$

or if $cost(p) + \sum_{j \neq i} l_j \geq ub$. Note that the expression $cost(p) + \sum_{j \neq i} l_j$ can be replaced with any lower bound on the cost of a solution that includes p and respects the current domains and instantiations, as long as the lower bound never over estimates. Fortunately, we have a fast and effective method of querying such lower bounds and we use it when performing this pruning.

4 Experimental Evaluation

In this section, we compare a bespoke C++ implementation of our constraint-based approach, called CPBayes [2], to the current state-of-the-art on benchmark instances and show that our approach compares favorably both in terms of number of instances solved within specified resource bounds and in terms of solution time.

The set of benchmark instances are derived from data sets obtained from the UCI Machine Learning Repository [3] and data generated from networks obtained from the Bayesian Network Repository [4]. Following previous work, the local score for each possible parent set and each random variable was computed in a preprocessing step (either by us or by others) prior to the search for the best network structure and we do not report the preprocessing time. Note that the computations of the possible parent sets for each variable are independent and can be determined in parallel. The BIC/MDL [18,19] and BDeu [20,21] scoring methods were used.

Table 2 shows the results of comparing CPBayes (v1.0) against Barlett and Cussens' [8] GOBNILP system (v1.4.1) based on integer linear programming, and Fan, Malone, and Yuan's [10,11,15] system (v2015) based on A* search. These two systems represent the current state-of-the-art for global (exact) approaches. Breadth-first BnB search [10,11,15] is also competitive but its effectiveness is known to be very similar to that of A*. Although for space reasons we do not report detailed results, we note that on these benchmarks CPBayes far outpaces the previous best depth-first branch-and-bound search approach [13]. GOBNILP (v1.4.1) [5] and A* (v2015) [6] are both primarily written in C/C++. A* (v2015) is the code developed by Fan et al. [10,15], but in the experiments we included our implementation of the improved lower bounds recently proposed by Fan and Yuan [11]. Thus, both CPBayes (v1.0) and A* (v2015) use exactly the same lower bounding technique (see Section 3.8). The experiments were performed on a cluster, where each node of the cluster is equipped with four AMD Opteron CPUs at 2.4 GHz and 32.0 GB memory. Resource limits of 24 hours of CPU time and 16 GB of memory were imposed both for the preprocessing step common to all methods of obtaining the local scores and again to determine the minimal cost BN using a method. The systems were run with their default values.

---

[2] CPBayes code available at: cs.uwaterloo.ca/~vanbeek/research
[3] archive.ics.uci.edu/ml/
[4] www.bnlearn.com/bnrepository/
[5] www.cs.york.ac.uk/aig/sw/gobnilp/
[6] bitbucket.org/bmmalone/

Table 2. For each benchmark, time (seconds) to determine minimal cost BN using various systems (see text), where n is the number of random variables in the data set, N is the number of instances in the data set, and d is the total number of possible parents sets for the random variables. Resource limits of 24 hours of CPU time and 16 GB of memory were imposed: OM = out of memory; OT = out of time. A blank entry indicates that the preprocessing step of obtaining the local scores for each random variable could not be completed within the resource limits.

				BDeu				BIC		
				GOBN.	A*	CPBayes		GOBN.	A*	CPBayes
Benchmark	n	N	d	v1.4.1	v2015	v1.0	d	v1.4.1	v2015	v1.0
shuttle	10	58,000	812	58.5	0.0	0.0	264	2.8	0.1	0.0
adult	15	32,561	768	1.4	0.1	0.0	547	0.7	0.1	0.0
letter	17	20,000	18,841	5,060.8	1.3	1.4	4,443	72.5	0.6	0.2
voting	17	435	1,940	16.8	0.3	0.1	1,848	11.6	0.4	0.1
zoo	17	101	2,855	177.7	0.5	0.2	554	0.9	0.4	0.1
tumour	18	339	274	1.5	0.9	0.2	219	0.4	0.9	0.2
lympho	19	148	345	1.7	2.1	0.5	143	0.5	1.0	0.2
vehicle	19	846	3,121	90.4	2.4	0.7	763	4.4	2.1	0.5
hepatitis	20	155	501	2.1	4.9	1.1	266	1.7	4.8	1.0
segment	20	2,310	6,491	2,486.5	3.3	1.3	1,053	13.2	2.4	0.5
mushroom	23	8,124	438,185	OT	255.5	561.8	13,025	82,736.2	34.4	7.7
autos	26	159	25,238	OT	918.3	464.2	2,391	108.0	316.3	50.8
insurance	27	1,000	792	2.8	583.9	107.0	506	2.1	824.3	103.7
horse colic	28	300	490	2.7	15.0	3.4	490	3.2	6.8	1.2
steel	28	1,941	113,118	OT	902.9	21,547.0	93,026	OT	550.8	4,447.6
flag	29	194	1,324	28.0	49.4	39.9	741	7.7	12.1	2.6
wdbc	31	569	13,473	2,055.6	OM	11,031.6	14,613	1,773.7	1,330.8	1,460.5
water	32	1,000					159	0.3	1.6	0.6
mildew	35	1,000	166	0.3	7.6	1.5	126	0.2	3.6	0.6
soybean	36	266					5,926	789.5	1,114.1	147.8
alarm	37	1,000					672	1.8	43.2	8.4
bands	39	277					892	15.2	4.5	2.0
spectf	45	267					610	8.4	401.7	11.2
sponge	45	76					618	4.1	793.5	13.2
barley	48	1,000					244	0.4	1.5	3.4
hailfinder	56	100					167	0.1	9.9	1.5
hailfinder	56	500					418	0.5	OM	9.3
lung cancer	57	32					292	2.0	OM	10.5
carpo	60	100					423	1.6	OM	253.6
carpo	60	500					847	6.9	OM	OT

5 Discussion and Future Work

The Bayesian Network Repository classifies networks as small (< 20 random variables), medium (20–60 random variables), large (60–100 random variables), very large (100–1000 random variables), and massive (> 1000 random variables). The benchmarks shown in Table 2 fall into the small and medium classes. We are not aware of any reports of results for exact solvers for instances beyond the medium class (Barlett and Cussens [8] report results for GOBNILP for $n > 60$, but they are solving a different problem, severely restricting the cardinality of the parent sets to ≤ 2).

Benchmarks from the small class are easy for the CPBayes and A* methods, but can be somewhat challenging for GOBNILP depending on the value of the parameter d, the total number of parent sets for the random variables. Along with the integer linear programming (ILP) solver GOBNILP, CPBayes scales fairly robustly to medium instances using a reasonable restriction on memory usage (both use only a few GB of memory, far under the 16 GB limit used in the experiments; in fairness, the scalability of the A* approach on a very large memory machine is still somewhat of an open question). CPBayes also has several other advantages, which it shares with the ILP approach, over A*, DP, and BFBnB approaches. Firstly, the constraint model is a purely declarative representation and the same model can be given to an exact solver or a solver based on local search, such as large neighborhood search. Secondly, the constraint model can be augmented with side structural constraints that can be important in real-world modeling (see [28]). Finally, the solver is an anytime algorithm since, as time progresses, the solver progressively finds better solutions.

Let us now turn to a comparison between GOBNILP and CPBayes. CPBayes scales better than GOBNILP along the dimension d which measures the size of the possible parent sets. A partial reason is that GOBNILP uses a constraint model that includes a (0,1)-variable for each possible parent set. GOBNILP scales better than CPBayes along the dimension n which measures the number of random variables. CPBayes has difficulty at the topmost range of n proving optimality. There is some evidence that $n = 60$ is near the top of the range for GOBNILP as well. Results reported by Barlett and Cussens [8] for the carpo benchmark using larger values of N and the BDeu scoring method—the scoring method which usually leads to harder optimization instances than BIC/MDL—showed that instances could only be solved by severely restricting the cardinality of the parent sets. A clear difficulty in scaling up all of these score-and-search methods is in obtaining the local scores within reasonable resource limits.

In future work on Bayesian network structure learning, we intend to focus on improving the robustness and scalability of our CPBayes approach. A direction that appears especially promising is to improve the branch-and-bound search by exploiting decomposition and lower bound caching during the search [29,30]. As well, our approach, as with all current exact approaches, assumes complete data. An important next step is to extend our approach to handle missing values and latent variables (cf. [31]).

Acknowledgments. This research was partially funded through an NSERC Discovery Grant. We thank Claude-Guy Quimper, Alejandro López-Ortiz, Mats Carlsson, and Christian Schulte for helpful discussions, and Brandon Malone and James Cussens for providing test instances and their code.

References

1. Witten, I.H., Frank, E., Hall, M.A.: Data Mining, 3rd edn. Morgan Kaufmann (2011)
2. Chickering, D., Meek, C., Heckerman, D.: Large-sample learning of Bayesian networks is NP-hard. In: Proc. of UAI, pp. 124–133 (2003)
3. Koivisto, M.: Parent assignment is hard for the MDL, AIC, and NML costs. In: Lugosi, G., Simon, H.U. (eds.) COLT 2006. LNCS (LNAI), vol. 4005, pp. 289–303. Springer, Heidelberg (2006)
4. Koivisto, M., Sood, K.: Exact Bayesian structure discovery in Bayesian networks. J. Mach. Learn. Res. **5**, 549–573 (2004)
5. Silander, T., Myllymäki, P.: A simple approach for finding the globally optimal Bayesian network structure. In: Proc. of UAI, pp. 445–452 (2006)
6. Malone, B., Yuan, C., Hansen, E.A.: Memory-efficient dynamic programming for learning optimal Bayesian networks. In: Proc. of AAAI, pp. 1057–1062 (2011)
7. Jaakkola, T., Sontag, D., Globerson, A., Meila, M.: Learning Bayesian network structure using LP relaxations. In: Proc. of AISTATS, pp. 358–365 (2010)
8. Barlett, M., Cussens, J.: Advances in Bayesian network learning using integer programming. In: Proc. of UAI, pp. 182–191 (2013)
9. Yuan, C., Malone, B.: Learning optimal Bayesian networks: A shortest path perspective. J. of Artificial Intelligence Research **48**, 23–65 (2013)
10. Fan, X., Malone, B., Yuan, C.: Finding optimal Bayesian network structures with constraints learned from data. In: Proc. of UAI, pp. 200–209 (2014)
11. Fan, X., Yuan, C.: An improved lower bound for Bayesian network structure learning. In: Proc. of AAAI (2015)
12. Tian, J.: A branch-and-bound algorithm for MDL learning Bayesian networks. In: Proc. of UAI, pp. 580–588 (2000)
13. Malone, B., Yuan, C.: A depth-first branch and bound algorithm for learning optimal bayesian networks. In: Croitoru, M., Rudolph, S., Woltran, S., Gonzales, C. (eds.) GKR 2013. LNCS, vol. 8323, pp. 111–122. Springer, Heidelberg (2014)
14. de Campos, C.P., Ji, Q.: Efficient structure learning of Bayesian networks using constraints. Journal of Machine Learning Research **12**, 663–689 (2011)
15. Fan, X., Yuan, C., Malone, B.: Tightening bounds for Bayesian network structure learning. In: Proc. of AAAI, pp. 2439–2445 (2014)
16. Darwiche, A.: Modeling and Reasoning with Bayesian Networks. Cambridge University Press (2009)
17. Koller, D., Friedman, N.: Probabilistic Graphical Models: Principles and Techniques. The MIT Press (2009)
18. Schwarz, G.: Estimating the dimension of a model. Ann. Stat. **6**, 461–464 (1978)
19. Lam, W., Bacchus, F.: Using new data to refine a Bayesian network. In: Proc. of UAI, pp. 383–390 (1994)
20. Buntine, W.L.: Theory refinement of Bayesian networks. In: Proc. of UAI, pp. 52–60 (1991)
21. Heckerman, D., Geiger, D., Chickering, D.M.: Learning Bayesian networks: The combination of knowledge and statistical data. Machine Learning **20**, 197–243 (1995)
22. Larranaga, P., Kuijpers, C., Murga, R., Yurramendi, Y.: Learning Bayesian network structures by searching for the best ordering with genetic algorithms. IEEE Trans. Syst., Man, Cybern. **26**, 487–493 (1996)

23. Teyssier, M., Koller, D.: Ordering-based search: a simple and effective algorithm for learning Bayesian networks. In: Proc. of UAI, pp. 548–549 (2005)
24. Chickering, D.M.: A transformational characterization of equivalent Bayesian network structures. In: Proc. of UAI, pp. 87–98 (1995)
25. Chickering, D.M.: Learning equivalence classes of Bayesian network structures. Journal of Machine Learning Research 2, 445–498 (2002)
26. Michie, D.: "memo" functions and machine learning. Nature 218, 19–22 (1968)
27. Smith, B.M.: Caching search states in permutation problems. In: van Beek, P. (ed.) CP 2005. LNCS, vol. 3709, pp. 637–651. Springer, Heidelberg (2005)
28. Cussens, J.: Integer programming for Bayesian network structure learning. Quality Technology & Quantitative Management 1, 99–110 (2014)
29. Kitching, M., Bacchus, F.: Symmetric component caching. In: Proc. of IJCAI, pp. 118–124 (2007)
30. Kitching, M., Bacchus, F.: Exploiting decomposition in constraint optimization problems. In: Stuckey, P.J. (ed.) CP 2008. LNCS, vol. 5202, pp. 478–492. Springer, Heidelberg (2008)
31. Friedman, N.: Learning belief networks in the presence of missing values and hidden variables. In: Proc. of ICML, pp. 125–133 (1997)

Hybridization of Interval CP and Evolutionary Algorithms for Optimizing Difficult Problems

Charlie Vanaret[1]([⊠]), Jean-Baptiste Gotteland[2],
Nicolas Durand[2], and Jean-Marc Alliot[1]

[1] Institut de Recherche en Informatique de Toulouse, 2 Rue Charles Camichel,
31000 Toulouse, France
charlie.vanaret@enseeiht.fr, jean-marc.alliot@irit.fr
[2] Ecole Nationale de l'Aviation Civile, 7 Avenue Edouard Belin,
31055 Toulouse Cedex 04, France
{gottelan,durand}@recherche.enac.fr

Abstract. The only rigorous approaches for achieving a numerical proof of optimality in global optimization are interval-based methods that interleave branching of the search-space and pruning of the subdomains that cannot contain an optimal solution. State-of-the-art solvers generally *integrate* local optimization algorithms to compute a good upper bound of the global minimum over each subspace. In this document, we propose a *cooperative* framework in which interval methods cooperate with evolutionary algorithms. The latter are stochastic algorithms in which a population of candidate solutions iteratively evolves in the search-space to reach satisfactory solutions.

Within our cooperative solver Charibde, the evolutionary algorithm and the interval-based algorithm run in parallel and exchange bounds, solutions and search-space in an advanced manner via message passing. A comparison of Charibde with state-of-the-art interval-based solvers (GlobSol, IBBA, Ibex) and NLP solvers (Couenne, BARON) on a benchmark of difficult COCONUT problems shows that Charibde is highly competitive against non-rigorous solvers and converges faster than rigorous solvers by an order of magnitude.

1 Motivation

We consider n-dimensional continuous constrained optimization problems over a hyperrectangular domain $\boldsymbol{D} = D_1 \times \ldots \times D_n$:

$$(\mathcal{P}) \quad \min_{\boldsymbol{x} \in \boldsymbol{D} \subset \mathbb{R}^n} \quad f(\boldsymbol{x})$$

$$s.t. \quad g_i(\boldsymbol{x}) \leq 0, \quad i \in \{1, \ldots, m\} \tag{1}$$

$$h_j(\boldsymbol{x}) = 0, \quad j \in \{1, \ldots, p\}$$

When f, g_i and h_j are non-convex, the problem may have multiple local minima. Such difficult problems are generally solved using generic exhaustive branch and bound (BB) methods. The objective function and the constraints are

© Springer International Publishing Switzerland 2015
G. Pesant (Ed.): CP 2015, LNCS 9255, pp. 446–462, 2015.
DOI: 10.1007/978-3-319-23219-5_32

bounded on disjoint subspaces using enclosure methods. By By keeping track of the best known upper bound $\tilde{f}$ of the global minimum f^*, subspaces that cannot contain a global minimizer are discarded (pruned).

Several authors proposed hybrid approaches in which a BB algorithm cooperates with another technique to enhance the pruning of the search-space. Hybrid algorithms may be classified into two categories [24]: *integrative* approaches, in which one of the two methods replaces a particular operator of the other method, and *cooperative* methods, in which the methods are independent and are run sequentially or in parallel. Previous works include

- integrative approaches: [34] integrates a stochastic genetic algorithm (GA) within an interval BB. The GA provides the direction along which a box is partitioned, and an individual is generated within each subbox. At each generation, the best evaluation updates the best known upper bound of the global minimum. In [9], the crossover operator is replaced by a BB that determines the best individual among the offspring.
- cooperative approaches: [28] sequentially combines an interval BB and a GA. The interval BB generates a list $\mathcal{L}$ of remaining small boxes. The GA's population is initialized by generating a single individual within each box of $\mathcal{L}$. [12] (BB and memetic algorithm) and [7] (beam search and memetic algorithm) describe similar parallel strategies: the BB identifies promising regions that are then explored by the metaheuristic. [1] hybridizes a GA and an interval BB. The two independent algorithms exchange upper bounds and solutions through shared memory. New optimal results are presented for the rotated Michalewicz ($n = 12$) and Griewank functions ($n = 6$).

In this communication, we build upon the cooperative scheme of [1]. The efficiency and reliability of their solver remain very limited; it is not competitive against state-of-the-art solvers. Their interval techniques are naive and may lose solutions, while the GA may send evaluations subject to roundoff errors. We propose to hybridize a stochastic differential evolution algorithm (close to a GA), described in Section 2, and a deterministic interval branch and contract algorithm, described in Section 3. Our hybrid solver Charibde is presented in Section 4. Experimental results (Section 5) show that Charibde is highly competitive against state-of-the-art solvers.

2   Differential Evolution

Differential evolution (DE) [29] is among the simplest and most efficient metaheuristics for continuous problems. It combines the coordinates of existing individuals (candidate solutions) with a given probability to generate new individuals. Initially devised for continuous unconstrained problems, DE was extended to mixed problems and constrained problems [23].

Let NP denote the size of the population, $W > 0$ the amplitude factor and $CR \in [0, 1]$ the crossover rate. At each generation (iteration), NP new individuals are generated: for each individual $\boldsymbol{x} = (x_1, \ldots, x_n)$, three other individuals

$u = (u_1, \ldots, u_n)$ (called *base individual*), $v = (v_1, \ldots, v_n)$ and $w = (w_1, \ldots, w_n)$, all different and different from x, are randomly picked in the population. The coordinates y_i of the new individual $y = (y_1, \ldots, y_n)$ are computed according to

$$y_i = \begin{cases} u_i + W \times (v_i - w_i) & \text{if } i = R \text{ or } r_i < CR \\ x_i & \text{otherwise} \end{cases} \tag{2}$$

where r_i is picked in $[0, 1]$ with uniform probability. The index R, picked in $\{1, \ldots, n\}$ with uniform probability for each x, ensures that at least a coordinate of y differs from that of x. y replaces x in the population if it is "better" than x (e.g. in unconstrained optimization, y is better than x if it improves the objective function).

Figure 1 depicts a two-dimensional crossover between individuals x, u (base individual), v and w. The contour lines of the objective function are shown in grey. The difference $v - w$, scaled by W, yields the direction (an approximation of the direction opposite the gradient) along which u is translated to yield y.

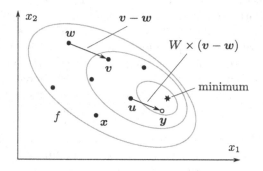

Fig. 1. Crossover of the differential evolution

3   Reliable Computations

Reliable (or rigorous) methods provide bounds on the global minimum, even in the presence of roundoff errors. The only reliable approaches for achieving a numerical proof of optimality in global optimization are interval-based methods that interleave branching of the search-space and pruning of the subdomains that cannot contain an optimal solution.

Section 3.1 introduces interval arithmetic, an extension of real arithmetic. Reliable global optimization is detailed in Section 3.2, and interval contractors are mentioned in Section 3.3.

3.1   Interval Arithmetic

An *interval* X with floating-point bounds defines the set $\{x \in \mathbb{R} \mid \underline{X} \le x \le \overline{X}\}$. $\mathbb{IR}$ denotes the set of all intervals. The *width* of X is $w(X) = \overline{X} - \underline{X}$. $m(X) = \frac{\underline{X}+\overline{X}}{2}$ is the *midpoint* of X. A *box* $\boldsymbol{X}$ is a Cartesian product of intervals. The width of a box is the maximum width of its components. The *convex hull* $\square(X,Y)$ of X and Y is the smallest interval enclosing X and Y.

Interval arithmetic [19] extends real arithmetic to intervals. Interval arithmetic implemented on a machine must be *rounded outward* (the left bound is rounded toward $-\infty$, the right bound toward $+\infty$) to guarantee conservative properties. The interval counterparts of binary operations and elementary functions produce the smallest interval containing the image. Thanks to the conservative properties of interval arithmetic, we define interval extensions (Def 1) of functions that may be expressed as a finite composition of elementary functions.

Definition 1 (Interval extension). *Let $f : \mathbb{R}^n \to \mathbb{R}$. $F : \mathbb{IR}^n \to \mathbb{IR}$ is an interval extension (or inclusion function) of f iff*

$$\forall \boldsymbol{X} \in \mathbb{IR}^n, \quad f(\boldsymbol{X}) := \{f(\boldsymbol{x}) \mid \boldsymbol{x} \in \boldsymbol{X}\} \subset F(\boldsymbol{X})$$
$$\forall \boldsymbol{X} \in \mathbb{IR}^n, \forall \boldsymbol{Y} \in \mathbb{IR}^n, \quad \boldsymbol{X} \subset \boldsymbol{Y} \Rightarrow F(\boldsymbol{X}) \subset F(\boldsymbol{Y}) \tag{3}$$

Interval extensions with various sharpnesses may be defined (Example 1). The *natural interval extension* F_N replaces the variables with their domains and the elementary functions with their interval counterparts. The *Taylor interval extension* F_T is based on the Taylor expansion at point $\boldsymbol{c} \in \boldsymbol{X}$.

Example 1 (Interval extensions). Let $f(x) = x^2 - x$, $X = [-2, 0.5]$ and $c = -1 \in X$. The exact range is $f(X) = [-0.25, 6]$ Then

- $F_N(X) = X^2 - X = [-2, 0.5]^2 - [-2, 0.5] = [0, 4] - [-2, 0.5] = [-0.5, 6]$;
- $F_T(X, c) = 2 + (2X - 1)(X + 1) = 2 + [-5, 0][-1, 1.5] = [-5.5, 7]$.

Example 1 shows that interval arithmetic often overestimates the range of a real-valued function. This is due to the *dependency problem*, an inherent behavior of interval arithmetic. Dependency decorrelates multiple occurrences of the same variable in an analytical expression (Example 2).

Example 2 (Dependency). Let $X = [-5, 5]$. Then

$$X - X = [-10, 10] = \{x_1 - x_2 \mid x_1 \in X, x_2 \in X\}$$
$$\supset \{x - x \mid x \in X\} = \{0\} \tag{4}$$

Interval extensions (F_N, F_T) have different convergence orders, that is the overestimation decreases at different speeds with the width of the interval.

3.2   Global Optimization

Interval arithmetic computes a rigorous enclosure of the range of a function over a box. The first branch and bound algorithms for continuous optimization based on interval arithmetic were devised in the 1970s [20][27], then refined during the following years [14]: the search-space is partitioned into subboxes. The objective function and the constraints are evaluated on each subbox using interval arithmetic. The subspaces that cannot contain a global minimizer are discarded and are not further explored. The algorithm terminates when $\tilde{f} - f^* < \varepsilon$.

To overcome the pessimistic enclosures of interval arithmetic, interval branch and bound algorithms have recently been endowed with filtering algorithms (Section 3.3) that narrow the bounds of the boxes without loss of solutions. Stemming from the Interval Analysis and Interval Constraint Programming communities, filtering (or contraction) algorithms discard values from the domains by enforcing local (each constraint individually) or global (all constraints simultaneously) consistencies. The resulting methods, called interval branch and contract (IBC) algorithms, interleave steps of contraction and steps of bisection.

3.3   Interval Contractors

State-of-the-art contractors (contraction algorithms) include HC4 [6], Box [32], Mohc [2], 3B [17], CID [31] and X-Newton [3]. Only HC4 and X-Newton are used in this communication.

HC4Revise is a two-phase algorithm that exploits the syntax tree of a constraint to contract each occurrence of the variables. The first phase (evaluation) evaluates each node (elementary function) using interval arithmetic. The second phase (propagation) uses projection functions to inverse each elementary function. HC4Revise is generally invoked as the revised procedure (subcontractor) of HC4, an AC3-like propagation loop.

X-Newton computes an outer linear relaxation of the objective function and the constraints, then computes a lower bound of the initial problem using LP techniques (e.g. the simplex algorithm). $2n$ additional calls may contract the domains of the variables.

4   Charibde: A Cooperative Approach

4.1   Hybridization of Stochastic and Deterministic Techniques

Our hybrid algorithm *Charibde* (Cooperative Hybrid Algorithm using Reliable Interval-Based methods and Differential Evolution), written in OCaml [16], combines a stochastic DE and a deterministic IBC for non-convex constrained optimization. Although it embeds a stochastic component, Charibde is a *fully rigorous* solver.

Previous Work. Preliminary results of a basic version of Charibde were published in 2013 [33] on classical multimodal problems (7 bound-constrained and 4 inequality-constrained problems) widely used in the Evolutionary Computation community. We showed that Charibde benefited from the start of convergence of the DE algorithm, and completed the proof of convergence faster than a standard IBC algorithm. We provided new optimal results for 3 problems (Rana, Eggholder and Michalewicz).

Contributions. In this communication, we present *two contributions*:

1. we devised a new cooperative exploration strategy MaxDist that
   – selects boxes to be explored in a novel manner;
   – periodically reduces DE's domain;
   – restarts the population within the new (smaller) domain.
   An example illustrates the evolution of the domain without loss of solutions;
2. we assess the performance of Charibde against state-of-the-art rigorous (GlobSol, IBBA, Ibex) and non-rigorous (Couenne, BARON) solvers on a benchmark of difficult problems.

Cooperative Scheme. Two independent parallel processes exchange bounds, solutions and search-space via MPI message passing (Figure 2).

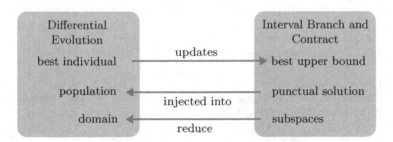

Fig. 2. Cooperative scheme of Charibde

The cooperation boils down to three main steps:

1. whenever the DE improves its best evaluation, the best individual and its evaluation are sent to the IBC to update the best known upper bound $\tilde{f}$;
2. whenever the IBC finds a better punctual solution (e.g. the center of a box), it is injected into DE's population;
3. the exploration strategy MaxDist periodically reduces the search-space of DE, then regenerates the population in the new search-space.

Sections 4.2 and 4.3 detail the particular implementations of the DE (Algorithm 1) and the IBC (Algorithm 2) within Charibde.

Algorithm 1. Charibde: Differential Evolution

---

function DIFFERENTIALEVOLUTION(f: objective function, $\mathcal{C}$: system of constraints, $\boldsymbol{D}$: search-space, NP: size of population, W: amplitude factor, CR: crossover rate)

    $P \leftarrow$ initial population, randomy generated in $\boldsymbol{D}$
    $\tilde{f} \leftarrow +\infty$
    repeat
        $(\boldsymbol{x}, f_x) \leftarrow$ MPI_ReceiveIBC()
        Insert $\boldsymbol{x}$ into P
        $\tilde{f} \leftarrow f_x$
        Generate temporary population P' by crossover
        $P \leftarrow P'$
        $(\boldsymbol{x}_{best}, f_{best}) \leftarrow$ BESTINDIVIDUAL(P)
        if $f_{best} < \tilde{f}$ **then**
            $\tilde{f} \leftarrow f_{best}$
            MPI_SendIBC($\boldsymbol{x}_{best}, f_{best}$)
        end if
    until termination criterion is met
    return best individual of P
end function

Algorithm 2. Charibde: Interval Branch and Contract

---

function INTERVALBRANCHANDCONTRACT(F: objective function, $\mathcal{C}$: system of constraints, $\boldsymbol{D}$: search-space, ε: precision)

    $\tilde{f} \leftarrow +\infty$ ▷ best known upper bound
    $\mathcal{Q} \leftarrow \{\boldsymbol{D}\}$ ▷ priority queue
    while $\mathcal{Q} \neq \varnothing$ **do**
        $(\boldsymbol{x}_{DE}, f_{DE}) \leftarrow$ MPI_ReceiveDE()
        $\tilde{f} \leftarrow \min(\tilde{f}, f_{DE})$
        Extract a box $\boldsymbol{X}$ from $\mathcal{Q}$
        Contract $\boldsymbol{X}$ w.r.t. constraints   ▷ Algorithm 3
        if $\boldsymbol{X}$ cannot be discarded **then**
            if $\overline{F(m(\boldsymbol{X}))} < \tilde{f}$ **then**   ▷ midpoint test
                $\tilde{f} \leftarrow \overline{F(m(\boldsymbol{X}))}$ ▷ update best upper bound
                MPI_SendDE($m(\boldsymbol{X}), \overline{F(m(\boldsymbol{X}))}$)
            end if
            Split $\boldsymbol{X}$ into $\{\boldsymbol{X}_1, \boldsymbol{X}_2\}$
            Insert $\{\boldsymbol{X}_1, \boldsymbol{X}_2\}$ into $\mathcal{Q}$
        end if
    end while
    return $(\tilde{f}, \tilde{\boldsymbol{x}})$
end function

4.2   Differential Evolution

Population-based metaheuristics, in particular DE, are endowed with mechanisms that help escape local minima. They are quite naturally recommended to solve difficult multimodal problems for which traditional methods struggle to converge. They are also capable of generating feasible solutions without any a priori knowledge of the topology. DE has proven greatly beneficial for improving the best known upper bound $\tilde{f}$, a task for which standard branch and bound algorithms are not intrinsically intended.

Base Individual. In the standard DE strategy, all the current individuals have the same probability to be selected as the base individual $\boldsymbol{u}$. We opted for an alternative strategy [23] that guarantees that all individuals of the population play this role once and only once at each generation: the index of the base individual is obtained by summing the index of the individual $\boldsymbol{x}$ and an offset in $\{1, \ldots, NP - 1\}$, drawn with uniform probability.

Bound Constraints. When a coordinate y_i of $\boldsymbol{y}$ (computed during the crossover) exceeds the bounds of the component D_i of the domain $\boldsymbol{D}$, the bounce-back method [23] replaces y_i with a valid coordinate y_i' that lies between the base coordinate u_i and the violated bound:

$$
y_i' = \begin{cases} u_i + \omega(\overline{D_i} - u_i) & \text{if } y_i > \overline{D_i} \\ u_i + \omega(\underline{D_i} - u_i) & \text{if } y_i < \underline{D_i} \end{cases} \tag{5}
$$

where ω is drawn in $[0, 1]$ with uniform probability.

Constraint Handling. The extension of evolutionary algorithms to constrained optimization has been addressed by numerous authors. We implemented the direct constraint handling [23] that assigns to each individual a vector of evaluations (objective function and constraints), and selects the new individual $\boldsymbol{y}$ (see Section 2) based upon simple rules:

 - $\boldsymbol{x}$ and $\boldsymbol{y}$ are feasible and $\boldsymbol{y}$ has a lower or equal objective value than $\boldsymbol{x}$;
 - $\boldsymbol{y}$ is feasible and $\boldsymbol{x}$ is not;
 - $\boldsymbol{x}$ and $\boldsymbol{y}$ are infeasible, and $\boldsymbol{y}$ does not violate any constraint more than $\boldsymbol{x}$.

Rigorous Feasibility. Numerous NLP solvers tolerate a slight violation (relaxation) of the inequality constraints (e.g. $g \leq 10^{-6}$ instead of $g \leq 0$). The evaluation of a "pseudo-feasible" solution $\boldsymbol{x}$ (that satisfies such relaxed constraints) is not a rigorous upper bound of the global minimum; roundoff errors may even lead to absurd conclusions: $f(\boldsymbol{x})$ may be lower than the global minimum, and (or) $\boldsymbol{x}$ may be very distant from actual feasible solutions in the search-space.

To ensure that an individual $\boldsymbol{x}$ is numerically feasible (i.e. that the evaluations of the constraints are nonpositive), we evaluate the constraints g_i using interval arithmetic. $\boldsymbol{x}$ is considered as feasible when the interval evaluations $G_i(\boldsymbol{x})$ are nonpositive, that is $\forall i \in \{1, \ldots, m\}, \overline{G_i(\boldsymbol{x})} \leq 0$.

Rigorous Objective Function. When x is a feasible point, the evaluation $f(x)$ may be subject to roundoff errors; the only reliable upper bound of the global minimum available is $\overline{F(x)}$ (the right bound of the interval evaluation). However, evaluating the individuals using only interval arithmetic is much costlier than cheap floating-point arithmetic.

An efficient in-between solution consists in systematically computing the floating-point evaluations $f(x)$, and computing the interval evaluation $F(x)$ only when the best known approximate evaluation is improved. $\overline{F(x)}$ is then compared to the best known reliable upper bound $\tilde{f}$: if $\tilde{f}$ is improved, $\overline{F(x)}$ is sent to the IBC. This implementation greatly reduces the cost of evaluations, while ensuring that all the values sent to the IBC are rigorous.

4.3   Interval Branch and Contract

Branching aims at refining the computation of lower bounds of the functions using interval arithmetic. Two strategies may be found in the early literature:

- the variable with the largest domain is bisected;
- the variables are bisected one after the other in a round-robin scheme.

More recently, the Smear heuristic [10] has emerged as a competitive alternative to the two standard strategies. The variable x_i for which the interval quantity $\frac{\partial F}{\partial x_i}(X)(X_i - x_i)$ is the largest is bisected.

Charibde's *main contractor* is detailed in Algorithm 3. We exploit the contracted nodes of HC4Revise to compute partial derivatives via automatic differentiation [26]. HC4Revise is a revise procedure within a quasi-fixed point algorithm with tolerance $\eta \in [0,1]$: the propagation loop stops when the box X is not sufficiently contracted, i.e. when the size of X becomes larger than a fraction ηw_0 of the initial size w_0. Most contractors include an evaluation phase that yields a lower bound of the problem on the current box. Charibde thus computes several lower bounds (natural, Taylor, LP) as long as the box is not discarded. Charibde calls ocaml-glpk [18], an OCaml binding for GLPK (GNU Linear Programming Kit). Since the solution of the linear program is computed using floating-point arithmetic, it may be subject to roundoff errors. A cheap postprocessing step [21] computes a rigorous bound on the optimal solution of the linear program, thus providing a rigorous lower bound of the initial problem.

When the problem is subject to *equality constraints* h_j ($j \in \{1, \ldots, p\}$), IBBA [22], Ibex [30] and Charibde handle a relaxed problem where each equality constraint $h_j(x) = 0$ ($j \in \{1, \ldots, p\}$) is replaced by two inequalities:

$$-\varepsilon_= \leq h_j(x) \leq \varepsilon_= \tag{6}$$

$\varepsilon_=$ may be chosen arbitrarily small.

4.4   MaxDist: A New Exploration Strategy

The boxes that cannot be discarded are stored in a priority queue $\mathcal{Q}$ to be processed at a later stage. The order in which the boxes are extracted determines

---

Algorithm 3. Charibde: contractor for constrained optimization

---

function CONTRACTION(**in-out** $\boldsymbol{X}$: box, F: objective function, **in-out** $\tilde{f}$: best upper bound, **in-out** $\mathcal{C}$: system of constraints)

    $lb \leftarrow -\infty$ ▷ lower bound

    repeat

        $w_0 \leftarrow w(\boldsymbol{X})$ ▷ initial size

        $F_{\boldsymbol{X}} \leftarrow$ HC4REVISE($F(\boldsymbol{X}) \leq \tilde{f}$)   ▷ evaluation of f/contraction

        $lb \leftarrow \underline{F_{\boldsymbol{X}}}$ ▷ lower bound by natural form

        $\boldsymbol{G} \leftarrow \nabla F(\boldsymbol{X})$ ▷ gradient by AD

        $lb \leftarrow \max(lb, \text{SECONDORDER}(\boldsymbol{X}, F, \tilde{f}, \boldsymbol{G}))$ ▷ second-order form

        $\mathcal{C} \leftarrow$ HC4($\boldsymbol{X}, \mathcal{C}, \eta$)   ▷ quasi-fixed point with tolerance η

        if use linearization **then**

            $lb \leftarrow \max(lb, \text{LINEARIZATION}(\boldsymbol{X}, F, \tilde{f}, \boldsymbol{G}, \mathcal{C}))$ ▷ simplex or X-Newton

        end if

    until $\boldsymbol{X} = \varnothing$ or $w(\boldsymbol{X}) > \eta w_0$

    return lb

end function

---

the exploration strategy of the search-space ("best-first", "largest first", "depth-first"). Numerical tests suggest that

- the "best-first" strategy is rarely relevant because of the overestimated range (due to the dependency problem);
- the "largest first" strategy does not give advantage to promising regions;
- the "depth-first" strategy tends to quickly explore the neighborhood of local minima, but struggles to escape from them.

We propose a new exploration strategy called MaxDist. It consists in extracting from $\mathcal{Q}$ the box that is *the farthest from the current solution* $\tilde{x}$. The underlying ideas are to

- explore the neighborhood of the global minimizer (a tedious task when the objective function is flat in this neighborhood) only when the best possible upper bound is available;
- explore regions of the search-space that are hardly accessible by the DE algorithm.

The distance between a point $\boldsymbol{x}$ and a box $\boldsymbol{X}$ is the sum of the distances between each coordinate x_i and the closest bound of X_i. Note that MaxDist is an adaptive heuristic: whenever the best known solution $\tilde{x}$ is updated, $\mathcal{Q}$ is reordered according to the new priorities of the boxes.

    Preliminary results (not presented here) suggest that MaxDist is competitive with standard strategies. However, the most interesting observation lies in the behavior of $\mathcal{Q}$: when using MaxDist, the maximum size of $\mathcal{Q}$ (the maximum number of boxes simultaneously stored in $\mathcal{Q}$) remains remarkably low (a few dozens compared to several thousands for standard strategies). This offers promising perspectives for the cooperation between DE and IBC: the remaining boxes of the IBC may be exploited in the DE to avoid exploring regions of the search-space that have already been proven infeasible or suboptimal.

The following numerical example illustrates how the remaining boxes are exploited to reduce DE's domain through the generations. Let

$$\min_{(x,y)\in(X,Y)} \quad -\frac{(x+y-10)^2}{30} - \frac{(x-y+10)^2}{120}$$

$$\text{s.t.} \quad \frac{20}{x^2} - y \le 0 \tag{7}$$

$$x^2 + 8y - 75 \le 0$$

be a constrained optimization problem defined on the box $X \times Y = [0, 10] \times [0, 10]$ (Figure 3a). The dotted curves represent the frontiers of the two inequality constraints, and the contour lines of the objective function are shown in solid dark. The feasible region is the banana-shaped set, and the global minimizer is located in its lower right corner.

The initial domain of DE (which corresponds to the initial box in the IBC) is first contracted with respect to the constraints of the problem. The initial population of DE is then generated within this contracted domain, thus avoiding obvious infeasible regions of the search-space. This approach is similar to that of [11]. Figure 3b depicts the contraction (the black rectangle) of the initial domain with respect to the constraints (sequentially handled by HC4Revise): $X \times Y = [1.4142, 8.5674] \times [0.2, 9.125]$.

Periodically, we compute the convex hull $\square(\mathcal{Q})$ of the remaining boxes of $\mathcal{Q}$ and replace DE's domain with $\square(\mathcal{Q})$. Note that

1. the convex hull (linear complexity) may be computed at low cost, because the size of $\mathcal{Q}$ remains small when using MaxDist;
2. by construction, MaxDist handles boxes on the rim of the remaining domain (the boxes of $\mathcal{Q}$), which boosts the reduction of the convex hull.

Figures 3c and 3d represent the *convex hull* $\square(\mathcal{Q})$ of the remaining subboxes in the IBC, respectively after 10 and 20 DE generations. The population is then randomly regenerated within the new contracted domain $\square(\mathcal{Q})$. The convex hull operation progressively eliminates local minima and infeasible regions. The global minimum eventually found by Charibde with precision $\varepsilon = 10^{-8}$ is $\tilde{f} = f(8.532424, 0.274717) = -2.825296148$; both constraints are active.

5   Experimental Results

Currently, GlobSol [15], IBBA [22] and Ibex [8] are among the most efficient solvers in rigorous constrained optimization. They share a common skeleton of interval branch and bound algorithm, but differ in the acceleration techniques. GlobSol uses the reformulation-linearization technique (RLT), that introduces new auxiliary variables for each intermediary operation. IBBA calls a contractor similar to HC4Revise, and computes a relaxation of the system of constraints

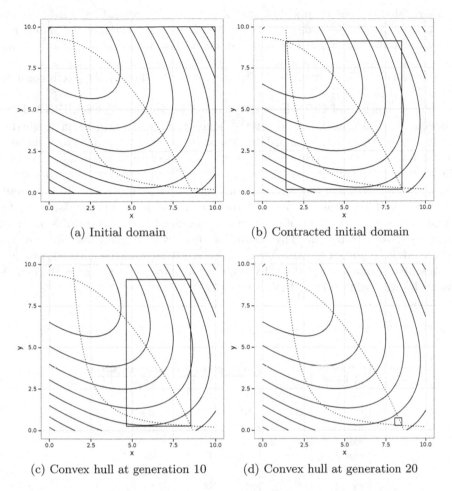

(a) Initial domain   (b) Contracted initial domain

(c) Convex hull at generation 10   (d) Convex hull at generation 20

Fig. 3. Evolution of DE's domain with the number of generations

using affine arithmetic. Ibex is dedicated to both numerical CSPs and constrained optimization; it embeds most of the aforementioned contractors (HC4, 3B, Mohc, CID, X-Newton). Couenne [5] and BARON [25] are state-of-the-art NLP solvers. They are based on a non-rigorous spatial branch and bound algorithm, in which the objective function and the constraints are over- and underestimated by convex relaxations. Although they perform an exhaustive exploration of the search-space, they cannot guarantee a given precision on the value of the optimum.

All five solvers and Charibde are compared on a subset of 11 COCONUT constrained problems (Table 1), extracted by Araya [3] for their difficulty: ex2_1_7, ex2_1_9, ex6_2_6, ex6_2_8, ex6_2_9, ex6_2_11, ex6_2_12, ex7_2_3, ex7_3_5, ex14_1_7 and ex14_2_7. Because of numerical instabilities of the ocaml-glpk LP library ("assert failure"), the results of the problems ex6_1_1, ex6_1_3 and ex_6_2_10 are not presented. The second and third columns give respectively the number of variables n and the number of constraints m. The fourth (resp. fifth) column specifies the type of the objective function (resp. the constraints): L is linear, Q is quadratic and NL is nonlinear. The logsize of the domain D (sixth column) is $\log(\prod_{i=1}^{n}(\overline{D_i} - \underline{D_i}))$.

The comparison of CPU times (in seconds) for solvers GlobSol, IBBA, Ibex, Couenne, BARON and Charibde on the benchmark of 11 problems is detailed in Table 2. Mean times and standard deviations (in brackets) are given for Charibde over 100 runs. The numerical precision on the objective function $\varepsilon = 10^{-8}$ and the tolerance for equality constraints $\varepsilon_= = 10^{-8}$ were identical for all solvers. TO (timeout) indicates that a solver could not solve a problem within one hour. The results of GlobSol (proprietary piece of software) were not available for all problems; only those mentioned in [22] are presented. The results of IBBA were also taken from [22]. The results of Ibex were taken from [3]: only the best strategy (simplex, X-NewIter or X-Newton) for each benchmark problem is presented. Couenne and BARON (only the commercial version of the code is available) were run on the NEOS server [13].

Table 1. Description of difficult COCONUT problems

Problem	n	m	Type f	g_i, h_j	Domain logsize
ex2_1_7	20	10	Q	L	$+\infty$
ex2_1_9	10	1	Q	L	$+\infty$
ex6_2_6	3	1	NL	L	$-3 \cdot 10^{-6}$
ex6_2_8	3	1	NL	L	$-3 \cdot 10^{-6}$
ex6_2_9	4	2	NL	L	-2.77
ex6_2_11	3	1	NL	L	$-3 \cdot 10^{-6}$
ex6_2_12	4	2	NL	L	-2.77
ex7_2_3	8	6	L	NL	61.90
ex7_3_5	13	15	L	NL	$+\infty$
ex14_1_7	10	17	L	NL	23.03
ex14_2_7	6	9	L	NL	$+\infty$

Charibde was run on an Intel Xeon E31270 @ 3.40GHz x 8 with 7.8 GB of RAM. BARON and Couenne were run on 2 Intel Xeon X5660 @ 2.8GHz x 12 with 64 GB of RAM. IBBA and Ibex were run on similar processors (Intel x86, 3GHz). The difference in CPU time between computers is about 10% [4], which makes the comparison quite fair.

The hyperparameters of Charibde for the benchmark problems are given in Table 3; NP is the population size, and η is the quasi-fixed point ratio. The amplitude $W = 0.7$, the crossover rate $CR = 0.9$ and the MaxDist strategy are common to all problems. Tuning the hyperparameters is generally problem-dependent, and requires structural knowledge about the problem: the population size NP may be set according to the dimension and the number of local minima,

Table 2. Comparison of convergence times (in seconds) between GlobSol, IBBA, Ibex, Charibde (mean and standard deviation over 100 runs), Couenne and BARON on difficult constrained problems

| Problem | GlobSol | Rigorous | | | Non rigorous | |
		IBBA	Ibex	Charibde	Couenne	BARON
ex2_1_7		16.7	**7.74**	34.9 (13.3)	476	16.23
ex2_1_9		154	9.07	35.9 (0.29)	**3.01**	3.58
ex6_2_6	306	1575	136	**3.3** (0.41)	TO	5.7
ex6_2_8	204	458	59.3	2.9 (0.37)	TO	TO
ex6_2_9	463	523	25.2	**2.7** (0.03)	TO	TO
ex6_2_11	273	140	7.51	**1.96** (0.06)	TO	TO
ex6_2_12	196	112	22.2	**8.8** (0.17)	TO	TO
ex7_2_3		TO	544	**1.9** (0.30)	TO	TO
ex7_3_5		TO	28.91	**4.5** (0.09)	TO	4.95
ex14_1_7		TO	406	4.2 (0.13)	13.86	**0.56**
ex14_2_7		TO	66.39	0.2 (0.04)	**0.01**	0.02
Sum	> 1442	TO	1312.32	**101.26**	TO	TO

the crossover rate CR is related to the separability of the problem, and the techniques based on linear relaxation have little influence for problems with few constraints, but are cheap when the constraints are linear.

Table 3. Hyperparameters of Charibde for the benchmark problems

Problem	NP	Bisections	Fixed-point ratio (η)	LP	X-Newton
ex2_1_7	20	RR	0.9	✓	✓
ex2_1_9	100	RR	0.8	✓	
ex6_2_6	30	Smear	0	✓	
ex6_2_8	30	Smear	0	✓	
ex6_2_9	70	Smear	0		
ex6_2_11	35	Smear	0		
ex6_2_12	35	RR	0	✓	
ex7_2_3	40	Largest	0	✓	✓
ex7_3_5	30	RR	0	✓	
ex14_1_7	40	RR	0	✓	
ex14_2_7	40	RR	0	✓	

Charibde outperforms Ibex on 9 out of 11 problems, IBBA on 10 out of 11 problems and GlobSol on all the available problems. The cumulated CPU time shows that Charibde (101.26s) improves the performances of Ibex (1312.32s) by an order of magnitude (ratio: 13) on this benchmark of 11 difficult problems. Charibde also proves highly competitive against non-rigorous solvers Couenne and BARON. The latter are faster or have similar CPU times on some of the 11 problems, however they both timeout on at least five problems (seven for Couenne, five for BARON). Overall, Charibde seems more robust and solves all the problems of the benchmark, while providing a numerical proof of optimality. Surprisingly, the convergence times do not seem directly related to the dimensions of the instances. They may be explained by the nature of the objective function and constraints (in particular, Charibde seems to struggle when the objective function is quadratic) and the dependency induced by the multiple occurrences of the variables.

Table 4 presents the best upper bounds obtained by Charibde, Couenne and BARON at the end of convergence (precision reached or timeout). Truncated digits on the upper bounds are bounded (e.g. 1.23^8_7 denotes $[1.237, 1.238]$ and -1.23^8_7 denotes $[-1.238, -1.237]$). The incorrect digits of the global minima obtained by Couenne and BARON are underlined. This demonstrates that non-rigorous solvers may be affected by roundoff errors, and may provide solutions that are infeasible or have an objective value *lower than the global minimum* (Couenne on ex2_1_9, BARON on ex2_1_7, ex2_1_9, ex6_2_8, ex6_2_12, ex7_2_3 and ex7_3_5). For the most difficult instance ex7_2_3, Couenne is not capable of finding a feasible solution with a satisfactory evaluation within one hour. It would be very informative to compute the ratio between the size of the feasible domain (the set of all feasible points) and the size of the entire domain. On the other hand, the strategy MaxDist within Charibde greatly contributes to finding an excellent upper bound of the global minimum, which significantly accelerates the interval pruning phase.

Table 4. Best upper bounds obtained by Charibde, Couenne and BARON

Problem	Charibde	Couenne	BARON
ex2_1_7	-4150.41013392^9_8	$-4150.4101\underline{2731}^8_7$	$-4150.4101\underline{6079}^8_7$
ex2_1_9	-0.3750000075	$-0.375000\underline{153}^4_3$	$-0.375001\underline{11}^1_0$
ex6_2_6	-0.00000260^3_2	0.0000071^1_0	-0.00000260^3_2
ex6_2_8	-0.02700063^{50}_{49}	-0.02700063^{50}_{49}	$-0.0270063\underline{7}^1_0$
ex6_2_9	-0.034066618^5_4	-0.034066184	$-0.0340661\underline{9}^1_0$
ex6_2_11	-0.00000267^3_2	-0.00000267^3_2	-0.00000267^3_2
ex6_2_12	0.28919475^{40}_{39}	0.28919475	$0.2891916\underline{9}^9_8$
ex7_2_3	7049.24802052^9_8	10^{50}	$7049.\underline{02029170}^7_6$
ex7_3_5	1.20671699^2_1	1.2068965	$\underline{0.23982448}^8_7$
ex14_1_7	0.0000000^{10}_{09}	0.000000000^1_0	0
ex14_2_7	0.000000000^8_7	0.000000000^1_0	0

6   Conclusion

We proposed a cooperative hybrid solver Charibde, in which a deterministic interval branch and contract cooperates with a stochastic differential evolution algorithm. The two independent algorithms run in parallel and exchange bounds, solutions and search-space in an advanced manner via message passing. The domain of the population-based metaheuristic is periodically reduced by removing local minima and infeasible regions detected by the branch and bound.

A comparison of Charibde with state-of-the-art interval-based solvers (Glob-Sol, IBBA, Ibex) and NLP solvers (Couenne, BARON) on a benchmark of difficult COCONUT problems shows that Charibde is highly competitive against non-rigorous solvers (while bounding the global minimum) and converges faster than rigorous solvers by an order of magnitude.

References

1. Alliot, J.M., Durand, N., Gianazza, D., Gotteland, J.B.: Finding and proving the optimum: cooperative stochastic and deterministic search. In: 20th European Conference on Artificial Intelligence (ECAI 2012), Montpellier, France, August 27–31, 2012 (2012)
2. Araya, I., Trombettoni, G., Neveu, B.: Exploiting monotonicity in interval constraint propagation. In: Proc. AAAI, pp. 9–14 (2010)
3. Araya, I., Trombettoni, G., Neveu, B.: A contractor based on convex interval taylor. In: Beldiceanu, N., Jussien, N., Pinson, É. (eds.) CPAIOR 2012. LNCS, vol. 7298, pp. 1–16. Springer, Heidelberg (2012)
4. Araya, I., Trombettoni, G., Neveu, B., Chabert, G.: Upper bounding in inner regions for global optimization under inequality constraints. Journal of Global Optimization **60**(2), 145–164 (2014)
5. Belotti, P., Lee, J., Liberti, L., Margot, F., Wächter, A.: Branching and bounds tighteningtechniques for non-convex minlp. Optimization Methods & Software **24**(4–5), 597–634 (2009)
6. Benhamou, F., Goualard, F., Granvilliers, L., Puget, J.F.: Revising hull and box consistency. In: International Conference on Logic Programming, pp. 230–244. MIT press (1999)
7. Blum, C., Puchinger, J., Raidl, G.R., Roli, A.: Hybrid metaheuristics in combinatorial optimization: A survey. Applied Soft Computing **11**(6), 4135–4151 (2011)
8. Chabert, G., Jaulin, L.: Contractor programming. Artificial Intelligence **173**, 1079–1100 (2009)
9. Cotta, C., Troya, J.M.: Embedding branch and bound within evolutionary algorithms. Applied Intelligence **18**(2), 137–153 (2003)
10. Csendes, T., Ratz, D.: Subdivision direction selection in interval methods for global optimization. SIAM Journal on Numerical Analysis **34**(3), 922–938 (1997)
11. Focacci, F., Laburthe, F., Lodi, A.: Local search and constraint programming. In: Handbook of metaheuristics, pp. 369–403. Springer (2003)
12. Gallardo, J.E., Cotta, C., Fernández, A.J.: On the hybridization of memetic algorithms with branch-and-bound techniques. IEEE Transactions on Systems, Man, and Cybernetics, Part B: Cybernetics **37**(1), 77–83 (2007)
13. Gropp, W., Moré, J.: Optimization environments and the NEOS server. Approximation theory and optimization, 167–182 (1997)
14. Hansen, E.: Global optimization using interval analysis. Dekker (1992)
15. Kearfott, R.B.: Rigorous global search: continuous problems. Springer (1996)
16. Leroy, X., Doligez, D., Frisch, A., Garrigue, J., Rémy, D., Vouillon, J.: The objective caml system release 3.12. Documentation and userâĂŹs manual. INRIA (2010)
17. Lhomme, O.: Consistency techniques for numeric csps. In: IJCAI, vol. 93, pp. 232–238. Citeseer (1993)
18. Mimram, S.: ocaml-glpk (2004). http://ocaml-glpk.sourceforge.net/
19. Moore, R.E.: Interval Analysis. Prentice-Hall (1966)

20. Moore, R.E.: On computing the range of a rational function of n variables over a bounded region. Computing **16**(1), 1–15 (1976)
21. Neumaier, A., Shcherbina, O.: Safe bounds in linear and mixed-integer linear programming. Mathematical Programming **99**(2), 283–296 (2004)
22. Ninin, J., Hansen, P., Messine, F.: A reliable affine relaxation method for global optimization. Groupe d'études et de recherche en analyse des décisions (2010)
23. Price, K., Storn, R., Lampinen, J.: Differential Evolution - A Practical Approach to Global Optimization. Natural Computing, Springer-Verlag (2006)
24. Puchinger, J., Raidl, G.R.: Combining metaheuristics and exact algorithms in combinatorial optimization: a survey and classification. In: Mira, J., Álvarez, J.R. (eds.) IWINAC 2005. LNCS, vol. 3562, pp. 41–53. Springer, Heidelberg (2005)
25. Sahinidis, N.V.: Baron: A general purpose global optimization software package. Journal of Global Optimization **8**(2), 201–205 (1996)
26. Schichl, H., Neumaier, A.: Interval analysis on directed acyclic graphs for global optimization. Journal of Global Optimization **33**(4), 541–562 (2005)
27. Skelboe, S.: Computation of rational interval functions. BIT Numerical Mathematics **14**(1), 87–95 (1974)
28. Sotiropoulos, D., Stavropoulos, E., Vrahatis, M.: A new hybrid genetic algorithm for global optimization. Nonlinear Analysis: Theory, Methods & Applications **30**(7), 4529–4538 (1997)
29. Storn, R., Price, K.: Differential evolution - a simple and efficient heuristic for global optimization over continuous spaces. Journal of Global Optimization, 341–359 (1997)
30. Trombettoni, G., Araya, I., Neveu, B., Chabert, G.: Inner regions and interval linearizations for global optimization. In: AAAI (2011)
31. Trombettoni, G., Chabert, G.: Constructive interval disjunction. In: Bessière, C. (ed.) CP 2007. LNCS, vol. 4741, pp. 635–650. Springer, Heidelberg (2007)
32. Van Hentenryck, P.: Numerica: a modeling language for global optimization. MIT press (1997)
33. Vanaret, C., Gotteland, J.-B., Durand, N., Alliot, J.-M.: Preventing premature convergence and proving the optimality in evolutionary algorithms. In: Legrand, P., Corsini, M.-M., Hao, J.-K., Monmarché, N., Lutton, E., Schoenauer, M. (eds.) EA 2013. LNCS, vol. 8752, pp. 29–40. Springer, Heidelberg (2014)
34. Zhang, X., Liu, S.: A new interval-genetic algorithm. In: Third International Conference on Natural Computation, ICNC 2007, vol. 4, pp. 193–197. IEEE (2007)

A General Framework for Reordering Agents Asynchronously in Distributed CSP

Mohamed Wahbi[1]([✉]), Younes Mechqrane[2], Christian Bessiere[3], and Kenneth N. Brown[1]

[1] Insight Centre for Data Analytics, University College Cork, Cork, Ireland
{mohamed.wahbi,ken.brown}@insight-centre.org
[2] Mohammed V University-Agdal Rabat, Rabat, Morocco
ymechqrane@gmail.com
[3] University of Montpellier, Montpellier, France
bessiere@lirmm.fr

Abstract. Reordering agents during search is an essential component of the efficiency of solving a distributed constraint satisfaction problem. Termination values have been recently proposed as a way to simulate the min-domain dynamic variable ordering heuristic. The use of termination values allows the greatest flexibility in reordering agents dynamically while keeping polynomial space. In this paper, we propose a general framework based on termination values for reordering agents asynchronously. The termination values are generalized to represent various heuristics other than min-domain. Our general framework is sound, complete, terminates and has a polynomial space complexity. We implemented several variable ordering heuristics that are well-known in centralized CSPs but could not until now be applied to the distributed setting. Our empirical study shows the significance of our framework compared to state-of-the-art asynchronous dynamic ordering algorithms for solving distributed CSP.

1 Introduction

Distributed artificial intelligence involves numerous combinatorial problems where multiple entities, called agents, need to cooperate in order to find a consistent combination of actions. Agents have to achieve the combination in a distributed manner and without any centralization. Examples of such problems are: traffic light synchronization [12], truck task coordination [18], target tracking in distributed sensor networks [11], distributed scheduling [16], distributed planning [6], distributed resource allocation [19], distributed vehicle routing [13], etc. These problems were successfully formalized using the distributed constraint satisfaction problem (DisCSP) paradigm.

DisCSP are composed of multiple agents, each owning its local constraint network. Variables in different agents are connected by constraints. Agents must assign values to their variables distributively so that all constraints are satisfied. To achieve this goal, agents assign values to their variables that satisfy their

© Springer International Publishing Switzerland 2015
G. Pesant (Ed.): CP 2015, LNCS 9255, pp. 463–479, 2015.
DOI: 10.1007/978-3-319-23219-5_33

own constraints, and exchange messages to satisfy constraints with variables owned by other agents. During the last two decades, many algorithms have been designed for solving DisCSP. Most of these algorithms assume a total (priority) order on agents that is static. However, it is known from centralized CSPs that reordering variables dynamically during search improves the efficiency of the search procedure. Moreover, reordering agents in DisCSP may be required in various applications (e.g., security [24]).

Agile Asynchronous Backtracking (AgileABT) [1] is a polynomial space asynchronous algorithm that is able to reorder *all* agents in the problem. AgileABT is based on the notion of *termination value*, a tuple of agents' domain sizes. Besides implementing the min-domain ([10]) dynamic variable ordering heuristic (DVO), the termination value acts as a timestamp for the orders exchanged by agents during search. Since the termination of AgileABT depends on the form of the termination value, the use of other forms of termination values (implementing other heuristics) may directly affect the termination of the algorithm.

In this paper, we generalize AgileABT to get a new framework AgileABT(α), in which α represents any measure used to implement a DVO (e.g., domain size in min-domain). AgileABT(α) is sound and complete. We define a simple condition on the measure α which guarantees that AgileABT(α) terminates. If the computation of the measure α also has polynomial space complexity, then AgileABT(α) has polynomial space complexity. This allows us to implement for the first time in Distributed CSP a wide variety of DVOs that have been studied in centralized CSP. To illustrate this, we implement a number of DVOs, including dom/deg [3] and dom/wdeg [14], and evaluate their performance on benchmark DisCSP problems.

The paper is organized as follows. Section 2 gives the necessary background on distributed CSP and dynamic reordering materials. It then discusses the Agile Asynchronous Backtracking algorithm for solving distributed CSP. Our general framework is presented and analyzed in Section 3. We show our empirical results in Section 4 and report related work in Section 5. Finally, we conclude in Section 6.

2 Background

2.1 Distributed Constraint Satisfaction Problem

The Distributed Constraint Satisfaction Problem (DisCSP) is a 5-tuple $(\mathcal{A}, \mathcal{X}, \mathcal{D}, \mathcal{C}, \varphi)$, where $\mathcal{A}$ is a set of agents $\{A_1, \ldots, A_p\}$, $\mathcal{X}$ is a set of variables $\{x_1, \ldots, x_n\}$, $\mathcal{D} = \{D_1, \ldots, D_n\}$ is a set of domains, where D_i is the initial set of possible values which may be assigned to variable x_i, $\mathcal{C}$ is a set of constraints, and $\varphi : \mathcal{X} \to \mathcal{A}$ is a function specifying an agent to control each variable. During a solution process, only the agent which controls a variable can assign it a value. A constraint $C(X) \in \mathcal{C}$, on the ordered subset of variables $X = (x_{j_1}, \ldots, x_{j_k})$, is $C(X) \subseteq D_{j_1} \times \cdots \times D_{j_k}$, and specifies the tuples of values which may be assigned simultaneously to the variables in X. For this paper, we restrict attention to binary constraints. We denote by $\mathcal{C}_i \subseteq \mathcal{C}$ all constraints that involve x_i. A *solution* is an assignment to each variable of a value from its domain, satisfying all constraints. Each agent A_i only knows constraints relevant to its variables (C_i) and the existence of other variables involved in these constraints (its *neighbors*). Without loss of generality, we assume each agent controls exactly one variable ($p = n$), so we use the terms agent and variable interchangeably and do not distinguish between A_i and x_i.

Each agent A_i stores a unique order, an ordered tuple of agents IDs, denoted by λ_i. λ_i is called the current order of A_i. Agents appearing before A_i in λ_i are the higher priority agents (predecessors) denoted by λ_i^- and conversely the lower priority agents (successors) λ_i^+ are agents appearing after A_i in λ_i. We denote by $\lambda_i[k]$ ($\forall k \in 1..n$) the ID of the agent located at position k in λ_i.

2.2   Asynchronous Backtracking - ABT

The first complete asynchronous search algorithm for solving DisCSP is *Asynchronous Backtracking* (ABT) [2,27]. In ABT, agents act concurrently and asynchronously, and do not have to wait for decisions of others. However, ABT requires a total priority order among agents. Each agent tries to find an assignment satisfying the constraints with what is currently known from higher priority agents. When an agent assigns a value to its variable, it sends out messages to lower priority agents, with whom it is constrained, informing them about its assignment. When no value is possible for a variable, the inconsistency is reported to higher agents in the form of a no-good (an unfruitful value combination). ABT computes a solution (or detects that no solution exists) in a finite time. However, the priority order of agents is static and uniform across the agents.

2.3   No-goods and Explanations

During a solution process, agents can infer inconsistent sets of assignments, called no-goods. No-goods are used to justify value removals. A no-good ruling out value v_i from the domain of a variable x_i is a clause of the form $x_j = v_j \wedge \ldots \wedge x_k = v_k \rightarrow x_i \neq v_i$. This no-good, which means that the assignment $x_i = v_i$ is inconsistent with the assignments $x_j = v_j \wedge \ldots \wedge x_k = v_k$, is used by agent A_i to justify removal of v_i from its domain D_i. The *left hand side* (*lhs*) and the *right hand side* (*rhs*) of a no-good are defined from the position of $\rightarrow$. The variables in the *lhs* of a no-good must precede the variable on its *rhs* in the current order because the assignments of these variables have been used to filter the domain of the variable in its *rhs*. These ordering constraints induced by a no-good are called safety conditions in [9]. For example, the no-good $x_j = v_j \wedge x_k = v_k \rightarrow x_i \neq v_i$ implies that $x_j \prec x_i$ and $x_k \prec x_i$ that is x_j and x_k must precede x_i in the variable ordering (i.e., $x_j, x_k \in \lambda_i^-$). We say that a no-good is *compatible* with an order λ_i if all agents in its *lhs* appear before its *rhs* in λ_i.

The current domain of a variable x_i, maintained by A_i, is composed by values not ruled out by a no-good.[1] The initial domain size (before search starts) of A_i

---

[1] To stay polynomial, A_i keeps only one no-good per removed value.

is denoted by d_i^0 while its current domain size is denoted by d_i. Let Σ_i be the conjunction of the left hand sides of all no-goods ruling out values from D_i. We explain the current domain size of D_i by the following expression $e_i : \Sigma_i \to d_i$, called explanation of x_i (e_i). Every explanation e_i induces safety conditions: $\{\forall x_m \in \Sigma_i, x_m \prec x_i\}$. When all values of a variable x_i are ruled out by some no-goods ($\Sigma_i \to 0$), these no-goods are resolved, producing a new no-good from Σ_i. There are clearly many different ways of representing Σ_i as a directed no-good (an implication). In standard backtracking search algorithms (like ABT), the variable, say x_t, that has the lowest priority in the current order (among variables in Σ_i) must change its value. x_t is called the backtracking target and the directed no-good is $ng_t : \Sigma_i \setminus x_t \to x_t \neq v_t$. In AgileABT the backtracking target is not necessarily the variable with the lowest priority within the conflicting variables in the current order.

2.4   Agile Asynchronous Backtracking

In AgileABT, an order λ is always associated with a termination value τ. A termination value is a tuple of positive integers (representing the sizes of the domains of other agents seen from A_i). When comparing two orders the strongest order is that associated with the lexicographically smallest termination value. The lexicographic order on agents IDs ($<_{lex}$) is used to break ties, the smallest being the strongest.

In AgileABT, all agents start with the same order. Then, every agent A_i is allowed to change the order asynchronously. In the following we describe AgileABT by illustrating the computation performed within agent A_i. A_i can change its current order λ_i only if it receives a stronger one from another agent or if itself proposes a new order (λ_i') stronger than its current order λ_i. A_i can only propose new orders (λ_i') when it tries to backtrack after detecting a dead-end ($\Sigma_i \to 0$).

In AgileABT, agents exchange the following types of messages to coordinate the search (where A_i is the sender):

– **ok?** message is sent by A_i to all lower agents (λ_i^+) to ask whether its assignment is acceptable. Besides the assignment, the **ok?** message contains an explanation e_i which communicates the current domain size of x_i, the current order λ_i, and the current termination value τ_i stored by A_i.
– **ngd** message is sent by A_i when all its values are ruled out by Σ_i. This message contains a directed no-good, as well as λ_i and τ_i.
– **order** message is sent to propose a new order. This message includes the order λ_i proposed by A_i accompanied by the termination value τ_i.

Each agent needs to compute the size of the domain of other variables to build its termination value. Hence, each agent A_i stores a set E_i of explanations sent by other agents. During search, A_i updates E_i to store new received explanations and to remove those that are no more relevant to the search state or not compatible with its current order λ_i. If $e_k \in E_i$, A_i uses this explanation

---

Algorithm 1. Computing termination value using heuristic α.

---

function $TV^{\alpha}(\lambda)$
1. τ is an array of length n;
2. **for** ($j \leftarrow 1$ **to** n) **do** $\tau[j] \leftarrow \alpha(\lambda[j])$;
3. **return** τ;

---

to justify the size $dom(k)$ of the current domain of x_k, i.e., d_k. Otherwise, A_i assumes that the size of the current domain of x_k is equal to d_k^0.

In AgileABT, the termination value $\tau_i = [tv_i^1, \ldots, tv_i^n]$ computed by agent A_i is such that $tv_i^k = dom(\lambda_i[k]), \forall k \in 1..n$. τ_i depends on the order λ_i and the domain sizes of agents given by the set of explanations E_i (Algorithm 1, using $TV^{\alpha}(\lambda_i)$ with $\alpha = dom$).

In standard backtracking search algorithms, the backtracking target is always the variable that has the lowest priority among the variables in the detected conflict (i.e., Σ_i). AgileABT relaxes this restriction by allowing A_i to select the target of backtracking x_t among conflicting variables Σ_i. The only restriction for selecting x_t as a backtracking target is to find an order λ_i' such that $\tau_i' = TV^{\alpha}(\lambda_i')$ with $\alpha = dom$ (Algorithm 1) is lexicographically smaller than the termination value associated with the current order λ_i and x_t is the lowest among variables in Σ_i w.r.t. λ_i'.

When a dead-end occurs, AgileABT iterates through all variables $x_t \in \Sigma_i$, considering x_t as the target of the backtracking, i.e., the directed no-good is ng_t: $\Sigma_i \setminus x_t \rightarrow x_t \neq v_t$. A_i then updates E_i to remove all explanations containing x_t (after backtracking x_t assignment will be changed). Next, it updates the explanation of x_t by considering the new generated no-good ng_t (i.e., $e_t \leftarrow [\Sigma_t \cup lhs(ng_t) \rightarrow d_t - 1]$). Finally, A_i computes a new order (λ_i') and its associated termination value (τ_i') from the updated explanations E_i. λ_i' is obtained by performing a topological sort on the directed acyclic graph (G) formed by safety conditions induced by the updated explanations E_i ($\forall x_m \in \Sigma_k | e_k \in E_i$, $(x_m, x_k) \in G$) and τ_i' is obtained from $TV^{\alpha}(\lambda_i')$ with $\alpha = dom$ (Algorithm 1). Let λ_i' be the strongest computed order over all possible targets in Σ_i. If the termination value τ_i' associated to λ_i' is lexicographically smaller than the τ_i associated to the current order λ_i, A_i reorders agents according to λ_i' and informs all agents about the new order λ_i' and its associated termination value τ_i'. The backtracking target is that used when A_i computed λ_i'. If no λ_i' stronger than λ_i exists, the backtracking target x_t is the variable that has the lowest priority among Σ_i in the current order λ_i. For more details we refer the reader to [1,25].

Example of Running AgileABT

Figure 1 presents an example of a possible execution of AgileABT on a simple problem. This problem (fig. 1a) consists of 5 agents with the following domains $\forall i \in 1..5, D_i = \{1, 2, 3, 4\}$ and 6 constraints among these agents $c_{12}: x_1 \neq x_2$, c_{13}:

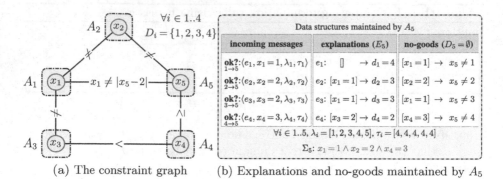

(a) The constraint graph   (b) Explanations and no-goods maintained by A_5

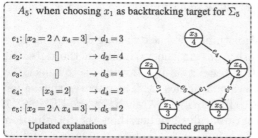

(c) A_5: updated explanations with x_1 as target   (d) A_5: selection of target (x_1)

Fig. 1. An example of a possible execution of AgileABT on a simple problem.

$x_1 \neq x_3$, c_{15}: $x_1 \neq |x_5 - 2|$, c_{25}: $x_2 \neq x_5$, c_{34}: $x_3 < x_4$, and c_{45}: $x_4 \geq x_5$. All agents start with the same initial ordering $\lambda_i = [1, 2, 3, 4, 5]$ associated with the termination value $\tau_i = [4, 4, 4, 4, 4]$ and values are chosen lexicographically. Consider the situation in A_5 after receiving **ok?** messages from other agents (fig. 1b). On receipt, explanations e_1, e_2, e_3, and e_4 are stored in E_5, and assignments $x_1 = 1$, $x_2 = 2$, $x_3 = 2$, and $x_4 = 3$ are stored in A_5 agent-view. After checking its constraints (c_{15}, c_{25}, and c_{45}), A_5 detects a dead-end ($D_5 = \emptyset$) where Σ_5: $\{x_1 = 1 \wedge x_2 = 2 \wedge x_4 = 3\}$. A_5 iterates through all variables $x_t \in \Sigma_5$, considering x_t as the target of the backtracking. Figure 1c shows the updates on the explanations stored in A_5 (E_5) when it considers x_1 as the target of the backtracking (i.e., $x_t = x_1$). A_5 updates E_5 to remove all explanations containing x_1 (i.e., e_2 and e_3) and considering the new generated no-good ng_1 in the explanation of x_1, i.e., e_1 (fig. 1c, left). Finally, A_5 computes a new order (λ_5') and its associated termination value (τ_5') from the updated explanations E_5. λ_5' is obtained by performing a topological sort on the directed acyclic graph formed by safety conditions induced by the updated explanations E_5 (fig. 1c, right). Figure 1d presents the computed orderings and their associated termination values (by topological sort) when considering each $x_t \in \Sigma_5$ as backtracking target. The strongest computed order (e.g, $\lambda_5' = [3, 4, 2, 5, 1]$, $\tau_5' = [4, 2, 4, 2, 3]$) is that computed when considering x_1 as backtracking target. Since λ_5' is stronger than λ_5, A_5 changes its current order to λ_5' and proposes this ordering to all other agents

---

Algorithm 2. AgileABT(α): Procedures for changing the order by A_i.

---

procedure processOrder(λ_j, τ_j)

1. **if** ($\tau_j \prec_{tv} \tau_i \vee (\tau_j \overset{tv}{=} \tau_i \wedge \lambda_j <_{lex} \lambda_i)$) **then** changeOrder(λ_j, τ_j) ;

procedure proposeOrder($args$)

2. $\langle \lambda_i', \tau_i' \rangle \leftarrow$ computeNewOrder($args$);
3. **if** ($\tau_i' \prec_{tv} \tau_i$) **then**
4.   changeOrder(λ_i', τ_i') ;
5.   sendMsg: $order\langle \lambda_i, \tau_i \rangle$ to all agents in $\mathcal{A}$;

procedure changeOrder(λ', τ')

6. $\lambda_i \leftarrow \lambda'$; $\tau_i \leftarrow \tau'$;
7. remove no-goods and explanations incompatible with τ_i ;

---

through **order** messages (i.e., **order**:$\langle \lambda_5', \tau_5' \rangle$). Then, A_5 sends the no-good ng_1 to agent x_1.

It has been proved that AgileABT is sound, complete and terminates [1, 25]. The termination proofs of AgileABT are based on the fact that the termination value is a tuple of positive integers (representing the expected sizes of the domains of other agents) and, as search progresses, these tuples can only decrease lexicographically. Thus, any change to the form of the termination values (i.e. implemented heuristic) may directly affect the termination of AgileABT.

3   Generalized AgileABT

In AgileABT, the termination value can be seen as an implementation of the *dom* dynamic variable ordering heuristic. In this section, we generalize AgileABT to get a new framework AgileABT(α), in which α represents any measure used to implement a DVO. The original AgileABT [1] is then equivalent to AgileABT(dom).

Due to space constraints, we only present in Algorithm 2 the pseudo-code of AgileABT(α) related to the cases where an agent (A_i) may change its current order λ_i (i.e., calling procedure changeOrder) where $\prec_{tv}$ is an ordering on the termination values and $\overset{tv}{=}$ represents the equality. A_i can change its current order λ_i and its associated termination value τ_i (procedure changeOrder, line 6) in two cases. The first case is when A_i receives a stronger order λ_j associated with the termination value τ_j from another agent A_j (processOrder, line 1 The second case occurs when A_i itself proposes a new order (λ_i') associated with a termination value τ_i' that is preferred (w.r.t. $\prec_{tv}$) to the termination value τ_i associated to its current order λ_i (procedure proposeOrder, line 3 and 4).

The soundness, completeness and polynomial space complexity[2] of AgileABT(α) are directly inherited from original AgileABT, i.e.,

---

[2] If the computation of the measure α also has polynomial space complexity.

AgileABT(*dom*). The only property that could be jeopardized is the termination of the algorithm. In the following we define a sufficient condition on termination values and the ordering $\prec_{tv}$ which guarantees that AgileABT(α) (Algorithm 2) terminates. Next, we will discuss a condition on the measure α that allows the termination values to obey the required condition.

Condition 1 *The priority ordering $\prec_{tv}$ is a well-ordering on the range of function TV^α.*

Proposition 1. *AgileABT(α) terminates if $\prec_{tv}$ obeys condition 1.*

Proof. Following the pseudo-code of AgileABT(α), an agent A_i can only change its current order in two cases (line 1 and 3). The termination values can only decrease w.r.t. the well-ordering $\prec_{tv}$, or remain the same and have a lexicographically decreasing agent order (line 1). The agent order cannot decrease lexicographically indefinitely and by condition 1 the termination values cannot decrease indefinitely w.r.t. $\prec_{tv}$. Therefore, AgileABT(α) cannot change the order indefinitely. Once the order stops changing, all agents will eventually have the same termination value and the same order to which it is attached (line 5, Algorithm 2). This order corresponds to the strongest order computed in the system so far. Since the agent order is now common and static, AgileABT(α) will behave exactly like ABT, which terminates.   □

To guarantee that AgileABT(α) terminates we need to define a well-ordering $\prec_{tv}$ on the termination values. Let α be the measure applied to the agents, and let the function TV^α be as defined in Algorithm 1. Let $\mathcal{S}$ be the range of α and let $\prec_\alpha$ be a total preference order on $\mathcal{S}$.

Definition 1. *Let λ_i and λ_j be two total agent orderings, $\tau_i = TV^\alpha(\lambda_i)$ and $\tau_j = TV^\alpha(\lambda_j)$. The termination value τ_i is preferred (w.r.t. $\prec_{tv}$) to τ_j (i.e., $\tau_i \prec_{tv} \tau_j$) if and only if τ_i is lexicographically less than τ_j (w.r.t. $\prec_\alpha$). In other words, $\tau_i \prec_{tv} \tau_j$ iff $\exists\, k \in 1..n$ such that $\alpha(\lambda_i[k]) \prec_\alpha \alpha(\lambda_j[k])$ and $\forall p \in 1..k-1$ $\alpha(\lambda_i[p]) = \alpha(\lambda_j[p])$.*

Condition 2 *The priority ordering $\prec_\alpha$ is a well-ordering on the range of measure α.*

Proposition 2. *AgileABT(α) terminates if $\prec_\alpha$ obeys condition 2.*

Proof. Suppose condition 2 is satisfied but AgileABT(α) does not terminate. Then, by proposition 1, condition 1 is not satisfied. Therefore, $\prec_{tv}$ is not a well-ordering on the range of TV^α. In other words, we can obtain an infinite decreasing sequence of termination values using a lexicographic comparison w.r.t. $\prec_\alpha$. Therefore we must have an infinite decreasing sequence of $\alpha(\lambda[k])$ values, for some $k \in 1..n$. But this contradicts condition 2 that forces $\prec_\alpha$ to be a well-ordering on $\mathcal{S}$. Therefore AgileABT(α) must terminate.   □

Algorithm 3. Compute the $wdeg$ of agent A_i ($wdeg(i)$).

// Filtering $x_j \in X$ by propagating $C(X)$
function revise($C(X), x_j$)
1. foreach ($v_j \in D_j$) do
2.   if ($\neg hasSupport(C(X), v_j, x_j)$) then $D_j \leftarrow D_j \setminus v_j$;
3. if ($D_j = \emptyset$) then $weight[C] \leftarrow weight[C] + 1$;
4. return $D_j \neq \emptyset$;

procedure computeWeight()
5. $wdeg \leftarrow 1$;
6. foreach ($C(X) \in C_i \mid nbUnassigned(X) > 1$) do $wdeg \leftarrow wdeg + weight[C]$;
7. $wdeg(i) \leftarrow \min(wdeg, W)$;

In the following, we consider a number of different heuristics that are known to be effective in reducing search in centralized CSP, but which could not before now be applied to distributed CSP. We show how the measures that inform these heuristics can obey condition 2, and thus can be applied in the general AgileABT(α) framework.

3.1 Neighborhood Based Variable Ordering Heuristics

In the first category we try to take into account the neighborhood of each agent. We implemented three DVOs (dom/deg, $dom/fdeg$ and $dom/pdeg$) to obtain respectively AgileABT(dom/deg), AgileABT($dom/fdeg$), and AgileABT($dom/pdeg$). In AgileABT(dom/deg), each agent A_i only requires to know the degree $deg(k)$ of each agent A_k in the problem. $deg(k)$ (i.e., the number of neighbors of A_k) can be obtained before the search starts as is the case for d_k^0 of each agent. Afterwards, A_i computes $\tau_i' = TV^{dom/deg}(\lambda_i')$ using $\alpha(k) = \frac{dom(k)}{deg(k)}$ (Algorithm 1, line 2). In AgileABT($dom/fdeg$) and AgileABT($dom/pdeg$) each agent A_i is required to know the set of neighbors of each agent A_k because it will need to compute the incoming degree $pdeg(k)$ and the outgoing degree $fdeg(k)$ of A_k for any proposed order. Again this information can be known in a preprocessing step before running AgileABT(α). Afterwards, A_i computes τ_i' from $TV^{dom/fdeg}(\lambda_i')$ (resp. $TV^{dom/pdeg}(\lambda_i')$) using $\alpha(k) = \frac{dom(k)}{fdeg(k)}$ (resp. $\alpha(k) = \frac{dom(k)}{pdeg(k)}$) where the incoming degree $pdeg(k)$ in λ_i' is the number of neighbors of A_k that appear before k in λ_i' and the outgoing degree $fdeg(k)$ in λ_i' is the number of neighbors of A_k that appear after k in λ_i'.

3.2 Conflict-Directed Variable Ordering Heuristic

The second category covers the conflict-directed variable ordering heuristic: $dom/wdeg$. In order to compute τ_i' using $\alpha(k) = \frac{dom(k)}{wdeg(k)}$, each agent A_i in AgileABT($dom/wdeg$) requires to know the weighted degree $wdeg(k)$ of each other agent A_k. A_i maintains its weighted degree, $wdeg(i)$, that it computes and the weighted degrees received from other agents, $wdeg(k)$. In order to compute the weighted degree, $wdeg$, (Algorithm 3) each agent A_i maintains a counter $weight[C]$ for each constraint $C(X)$ in C_i. Whenever a domain (D_j where $x_j \in X$) is wiped-out while propagating $C(X)$ (line 3, Algorithm 3), $weight[C]$ is incremented. Before assigning its variable and sending an **ok?** message to lower priority agents, A_i computes its weighted degree, $wdeg(i)$, by summing up (line 6, Algorithm 3) the weights of all constraints in C_i having at least two unassigned variables ([14]). However, to guarantee that AgileABT($dom/wdeg$) terminates we only update $wdeg(i)$ if the new computed weighted degree ($wdeg$) does not exceed a limit W on which all agents agree beforehand (line 7, Algorithm 3). In AgileABT($dom/wdeg$), whenever A_i sends an **ok?** message it attaches to this message the largest weighted degree computed so far $wdeg(i)$.

3.3   Theoretical Analysis

Proposition 3. *All measures α above are a well-ordering on a subset of $\mathbb{Q}$ w.r.t. $<$.*

Proof. We proceed by contradiction. Suppose there is an infinite decreasing sequence of values of $\alpha(k)$. In all measures above, $\alpha(k) = \frac{dom(k)}{\omega(k)}$, for some $\omega(k)$. $dom(k)$ is the expected domain size of the agent A_k. It is obvious that $dom(k)$ is a well-ordering on $\mathbb{N}$ w.r.t. $<$, and so cannot decrease indefinitely. Therefore, $\omega(k)$ must increase indefinitely. $\omega(k)$ is a positive integer whose value depends on the measure used. Two cases were explored in this paper. The first case concerns the family of degree-based heuristics (deg, $pdeg$, $fdeg$). In this case, all of the $\omega(k)$ are greater than or equal to 1 and smaller than the number of agents in the system (i.e., n) because an agent is at most constrained to $n-1$ other agents. Thus, $1 \le \omega(k) \le n-1$. The second case is related to the heuristic $wdeg$. We have outlined in section 3.2 that an agent is not allowed to increment its weight when it has reached the limit W set beforehand (Algorithm 3, line 7). Thus, $1 \le \omega(k) \le W$. In both cases $\omega(k)$ cannot increase indefinitely. Therefore for all measures presented above, of the form $\alpha(k) = \frac{dom(k)}{\omega(k)}$, cannot decrease indefinitely, and so $\alpha(k)$ is a well-ordering w.r.t. $<$.   □

4   Empirical Analysis

In this section we experimentally compare AgileABT(α)[3] using different DVO heuristics to three other algorithms: ABT, ABT_DO with nogood-triggered heuristic (ABT_DO-ng) [29] and ABT_DO with min-domain retroactive heuristic (ABT_DO_Retro(mindom)) [31]. All experiments were performed on the DisChoco 2.0 platform [26],[4] in which agents are simulated by Java threads that communicate only through message passing.

---

[3]   For AgileABT($dom/wdeg$), we fixed $W = 1,000$. But, varying W made negligible difference to the results.

[4]   http://dischoco.sourceforge.net/

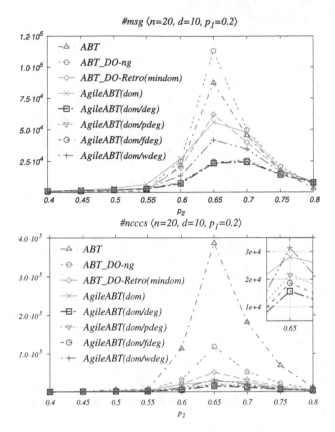

Fig. 2. Sparse uniform binary random DisCSPs

When comparing distributed algorithms, the performance is evaluated using two common metrics: the communication load and computation effort. Communication load is measured by the total number of exchanged messages among agents during algorithm execution ($\#msg$) [15]. Computation effort is measured by the number of non-concurrent constraint checks ($\#ncccs$) [28]. $\#ncccs$ is the metric used in distributed constraint solving to simulate computation time, but for dynamic reordering algorithms its variant generic $\#ncccs$ is used[30]. Algorithms are evaluated on three benchmarks: uniform binary random DisC-SPs, distributed graph coloring problems, and composed random instances. All binary table constraints in these problems are implemented using AC-2001 [4].

4.1   Uniform Binary Random DisCSPs

Uniform binary random DisCSPs are characterized by $\langle n, d, p_1, p_2 \rangle$, where n is the number of agents/variables, d is the number of values in each domain, p_1 is the network connectivity defined as the ratio of existing binary constraints to possible binary constraints, and p_2 is the constraint tightness defined as the

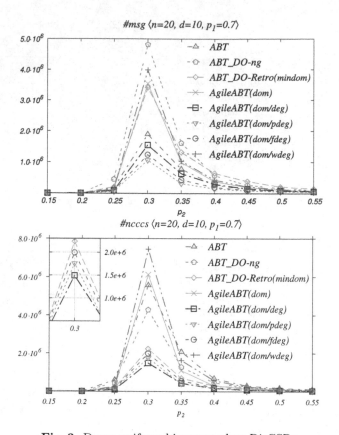

Fig. 3. Dense uniform binary random DisCSPs

ratio of forbidden value pairs to all possible pairs. We solved instances of two classes of random DisCSPs: sparse problems $\langle 20, 10, 0.2, p_2 \rangle$ and dense problems $\langle 20, 10, 0.7, p_2 \rangle$. We varied the tightness from 0.1 to 0.9 by steps of 0.05. For each pair of fixed density and tightness (p_1, p_2), we generated 20 instances, solved 5 times each. We report average over the 100 execution.

Figures 2 and 3 show the results on sparse respectively dense uniform binary random DisCSPs. In sparse problems (Figure 2), AgileABT(α) outperforms all other algorithms on both #msg and #ncccs. ABT with AC-2001 is significantly the slowest algorithm but it requires fewer messages than ABT_DO-ng. Regarding the speedup, AgileABT(α) shows almost an order of magnitude improvement compared to ABT and they all improve on ABT_DO-ng and ABT_DO_Retro(mindom). Comparing AgileABT(α) algorithms, neighborhood based heuristics (i.e., deg, $pdeg$ and $fdeg$) show an almost two-fold improvement over dom and $wdeg$ on #ncccs. This improvement is even more significant on #msg. $wdeg$ requires fewer messages than dom. In dense problems (Figure 3), neighborhood based heuristics outperform all other algorithms both on #msg and #ncccs. ABT requires almost half the #msg of ABT_DO_Retro(mindom),

AgileABT(*dom*), and AgileABT(*dom/wdeg*). ABT_DO-ng is always the algorithm that requires more messages. AgileABT(*dom/wdeg*) shows poor performance. It is slower and requires more messages than AgileABT(*dom*).

4.2   Distributed Graph Coloring Problems

Distributed graph coloring problems are characterized by $\langle n, d, p_1 \rangle$, where n, d and p_1 are as above and all constraints are binary difference constraints. We report the average on 100 instances of two classes $\langle n = 15, d = 5, p_1 = 0.65 \rangle$ and $\langle n = 25, d = 5, p_1 = 0.45 \rangle$ in Table 1. Again, AgileABT(α) using neighborhood based DVO are by far the best algorithms for solving both classes. ABT_DO-ng shows poor performance on solving those problems. ABT_DO_Retro(mindom) outperforms AgileABT(*dom*) in both classes. Comparing AgileABT(*dom*) to

Table 1. Distributed graph coloring problems

Algorithm	$\langle 15, 5, 0.65 \rangle$		$\langle 25, 5, 0.45 \rangle$	
	#*msg*	#*ncccs*	#*msg*	#*ncccs*
AgileABT(*dom/wdeg*)	90,630	188,991	2,600,016	2,783,132
AgileABT(*dom/fdeg*)	51,820	104,517	940,481	937,861
AgileABT(*dom/pdeg*)	47,949	89,514	**454,998**	**434,540**
AgileABT(*dom/deg*)	**44,083**	**78,050**	607,927	505,140
AgileABT(*dom*)	79,518	204,012	3,001,538	3,836,301
ABT_DO_Retro(mindom)	73,278	115,850	1,089,024	830,423
ABT_DO-ng	157,873	282,737	4,547,565	3,639,791
ABT	58,817	288,803	1,626,901	3,836,391

Table 2. Composed random instances

Instances	25-1-25		25-1-40	
Algorithm	#*msg*	#*ncccs*	#*msg*	#*ncccs*
AgileABT(*dom/wdeg*)	85,521	30,064	89,804	33,461
AgileABT(*dom/fdeg*)	146,668	219,980	1,337,552	2,830,906
AgileABT(*dom/pdeg*)	**57,079**	**16,043**	54,667	**17,704**
AgileABT(*dom/deg*)	122,735	309,064	740,669	2,793,670
AgileABT(*dom*)	57,451	20,944	59,859	23,405
ABT_DO_Retro(mindom)	67,022	41,401	96,783	59,980
ABT_DO-ng	1,329,257	1,614,960	$> 10^8$	$> 10^9$
ABT	2,850,137	22,042,094	9,429,088	72,524,742

AgileABT($dom/wdeg$), dom is slower than $wdeg$ but it requires fewer messages. In $\langle n = 15, d = 5, p_1 = 0.65 \rangle$, only AgileABT($\alpha$) using neighborhood based DVO outperform ABT on messages while other asynchronous dynamic ordering algorithms require more messages.

4.3 Composed Random Instances

We also evaluate all algorithms on two sets of unsatisfiable composed random instances used to evaluate the conflict-directed variable ordering heuristic in centralized CSP [7, 20].[5] Each set contains 10 different instances where each instance is composed of a main (under-constrained) fragment and some auxiliary fragments, each of which being grafted to the main one by introducing some binary constraints. Each instance contains 33 variables and 10 values per variable, and as before, each variable is controlled by a different agent. We solved each instance 5 times and present the average over 50 executions in Table 2. The results show that AgileABT($dom/pdeg$) outperforms all other algorithms in both classes. The second best algorithm for solving these instances is AgileABT(dom). ABT shows very poor performance on solving these problems followed by ABT_DO-ng that cannot solve instances in the second class (25-1-40) within the limits we fixed for all algorithms (10^8 #msg and 10^9 #$ncccs$). Regarding AgileABT(α) DVOs, $wdeg$ seems to pay off on these instances compared to dom/deg and $dom/fdeg$. In 25-1-40, AgileABT(dom/deg) outperforms ABT_DO_Retro(mindom), but the opposite happens for 25-1-25.

4.4 Discussion

Looking at all results together, we come to the straightforward conclusion that AgileABT(α) with neighbourhood-based heuristics, namely deg, $fdeg$ and $pdeg$ perform very well compared to other techniques. We think that neighbourhood-based heuristics perform well thanks to their ability to take into account the structure of the problem [3]. Distinctly, among these three heuristics $dom/pdeg$ seems to be the best one because of the limited changes on the agent at the first position. In $dom/pdeg$ $\tau[1] = dom(\lambda[1])$ because $pdeg(\lambda[1]) = 1$. Thus, the number of order changes (cost on messages) in AgileABT($dom/pdeg$) is reduced. Note that the strength of AgileABT(α) is that it enables any ordering to be identified and executed as the algorithm runs. However, each change invokes a series of coordination messages, and so too many changes of order will have a negative impact.

On the other hand the AgileABT(α) with the conflict-directed variable ordering heuristic, namely $wdeg$, shows a relatively poor performance. This fact can be explained by the limited amount of constraint propagation performed by DisCSP algorithms. Furthermore, asynchrony affects reception and treatment of *ok?* and *ngd* messages and has a direct impact on the computation of weights and new orders. For some instances of the coloring problem, the performance of

---

[5] http://www.cril.univ-artois.fr/~lecoutre/benchmarks.html

the conflict-directed heuristic varies significantly from one execution to another, indicating it is more sensitive to the asynchrony than the other heuristics.

5   Related Work

Bliek ([5]) proposed *Generalized Partial Order Dynamic Backtracking* (GPODB), an algorithm that generalizes both *Partial Order Dynamic Backtracking* (PODB) [9] and *Dynamic Backtracking* (DBT) [8]. This generalization was obtained by relaxing the safety conditions of PODB which results in additional flexibility. AgileABT has some similarities with PODB and GPODB because AgileABT also maintains a set of safety conditions. However, both PODB and GPODB are subject to the same restriction: when a dead end occurs, the backtracking target of the generated no-good must be selected such that the safety conditions induced by the new no-good satisfy all existing safety conditions. By contrast, AgileABT overcomes this restriction by allowing the violation of existing safety conditions by relaxing some explanations.

Silaghi et al. proposed Asynchronous Backtracking with Reordering (ABTR) [21–23]. In ABTR, a reordering heuristic identical to the centralized dynamic backtracking [8] was applied to ABT. The authors in [29] proposed Dynamic Ordering for Asynchronous Backtracking (ABT_DO) [29]. When an ABT_DO agent assigns a value to its variable, it can reorder lower priority agents. In other words, an agent on a position j can change order of the agents on positions $j + 1$ through n. The authors in [29] proposed three different ordering heuristics to reorder lower priority agents: random, min-domain and nogood-triggered inspired by dynamic backtracking [8]. Their experimental results show that nogood-triggered (ABT_DO-ng), where the generator of the no-good is placed just after the target of the backtrack, is best.

A new kind of ordering heuristics for ABT_DO is presented in [31] for reordering higher agents. These *retroactive* heuristics enable the generator of a no-good (backtrack) to be moved to a higher position than that of the target of the backtrack. The resulting algorithm is called ABT_DO_Retro. The degree of flexibility of these heuristics depends on the size of the no-good storage capacity K, which is predefined. Agents are limited to store no-goods with a size equal to or smaller than K. The space complexity of the agents is thus exponential in K. However, the best of those heuristics, ABT_DO_Retro(mindom) [17,31], does not need any no-good storage. In ABT_DO_Retro(mindom), the agent that generates a no-good is placed between the last and the second last agents in the no-good if its domain size is smaller than that of the agents it passes on the way up.

6   Conclusion

We proposed a general framework for reordering agents asynchronously in DisC-SPs which can implement many different dynamic variable ordering heuristics. Our general framework is sound, complete, and has a polynomial space complexity. We proved that a simple condition on the measure used in the heuristics is enough to guarantee termination. We implemented several DVOs heuristics that are well-known in centralized CSP but were not able to be used before in the distributed CSP. Our empirical study shows the significance of these DVOs on a distributed setting. In particular, it highlights the good performance of neighborhood based heuristics. Future work will focus on designing new DVOs that can be used in AgileABT(α).

Acknowledgments. This work is partly funded by Science Foundation Ireland (SFI) under Grant Number SFI/12/RC/2289.

References

1. Bessiere, C., Bouyakhf, E.H., Mechqrane, Y., Wahbi, M.: Agile asynchronous backtracking for distributed constraint satisfaction problems. In: Proceedings of ICTAI 2011, Boca Raton, Florida, USA, pp. 777–784, November 2011
2. Bessiere, C., Maestre, A., Brito, I., Meseguer, P.: Asynchronous backtracking without adding links: a new member in the ABT family. Artif. Intel. **161**, 7–24 (2005)
3. Bessiere, C., Régin, J.C.: MAC and combined heuristics: two reasons toforsake FC (and CBJ?) on hard problems. In: Freuder, E.C. (ed.) CP 1996. LNCS, vol. 1118, pp. 61–75. Springer, Heidelberg (1996)
4. Bessiere, C., Régin, J.C.: Refining the basic constraint propagation algorithm. In: Proceedings of IJCAI 2001, San Francisco, CA, USA, pp. 309–315 (2001)
5. Bliek, C.: Generalizing partial order and dynamic backtracking. In: Proceedings of AAAI 1998/IAAI 1998, Menlo Park, CA, USA, pp. 319–325 (1998)
6. Bonnet-Torrés, O., Tessier, C.: Multiply-constrained dcop for distributed planning and scheduling. In: AAAI SSS: Distributed Plan and Schedule Management, pp. 17–24 (2006)
7. Boussemart, F., Hemery, F., Lecoutre, C., Sais, L.: Boosting systematic search by weighting constraints. In: Proceedings of ECAI 2004, pp. 146–150 (2004)
8. Ginsberg, M.L.: Dynamic Backtracking. JAIR **1**, 25–46 (1993)
9. Ginsberg, M.L., McAllester, D.A.: GSAT and dynamic backtracking. In: KR, pp. 226–237 (1994)
10. Haralick, R.M., Elliott, G.L.: Increasing tree search efficiency for constraint satisfaction problems. Artif. Intel. **14**(3), 263–313 (1980)
11. Jung, H., Tambe, M., Kulkarni, S.: Argumentation as distributed constraint satisfaction: applications and results. In: Proceedings of AGENTS 2001, pp. 324–331 (2001)
12. Junges, R., Bazzan, A.L.C.: Evaluating the performance of dcop algorithms in a real world, dynamic problem. In: Proceedings of AAMAS 2008, Richland, SC, pp. 599–606 (2008)
13. Léauté, T., Faltings, B.: Coordinating logistics operations with privacy guarantees. In: Proceedings of the IJCAI 2011, pp. 2482–2487 (2011)
14. Lecoutre, C., Boussemart, F., Hemery, F.: Backjump-based techniques versus conflict-directed heuristics. In: Proceedings of IEEE ICTAI 2004, pp. 549–557 (2004)
15. Lynch, N.A.: Distributed Algorithms. Morgan Kaufmann Series (1997)
16. Maheswaran, R.T., Tambe, M., Bowring, E., Pearce, J.P., Varakantham, P.: Taking DCOP to the real world: efficient complete solutions for distributed multi-event scheduling. In: Proceedings of AAMAS 2004 (2004)

17. Mechqrane, Y., Wahbi, M., Bessiere, C., Bouyakhf, E.H., Meisels, A., Zivan, R.: Corrigendum to "Min-Domain Retroactive Ordering for Asynchronous Backtracking". Constraints **17**, 348–355 (2012)
18. Ottens, B., Faltings, B.: Coordination agent plans trough distributed constraint optimization. In: Proceedings of MASPLAN 2008, Sydney, Australia (2008)
19. Petcu, A., Faltings, B.V.: A value ordering heuristic for local search in distributed resource allocation. In: Faltings, B.V., Petcu, A., Fages, F., Rossi, F. (eds.) CSCLP 2004. LNCS (LNAI), vol. 3419, pp. 86–97. Springer, Heidelberg (2005)
20. Roussel, O., Lecoutre, C.: Xml representation of constraint networks: Format XCSP 2.1. CoRR (2009)
21. Silaghi, M.C.: Framework for modeling reordering heuristics for asynchronous backtracking. In: IEEE/WIC/ACM International Conference on Intelligent Agent Technology, IAT 2006, pp. 529–536, December 2006
22. Silaghi, M.C.: Generalized dynamic ordering for asynchronous backtracking on DisCSPs. In: Proceedings of DCR 2006 (2006)
23. Silaghi, M.C., Sam-Haroud, D., Faltings, B.: ABT with Asynchronous Reordering. In: 2nd Asia-Pacific IAT (2001)
24. Silaghi, M.C., Sam-Haroud, D., Calisti, M., Faltings, B.: Generalized english auctions by relaxation in dynamic distributed CSPs with private constraints. In: Proceedings of DCR 2001, pp. 45–54 (2001)
25. Wahbi, M.: Algorithms and Ordering Heuristics for Distributed Constraint Satisfaction Problems. John Wiley & Sons, Inc. (2013)
26. Wahbi, M., Ezzahir, R., Bessiere, C., Bouyakhf, E.H.: DisChoco 2: a platform for distributed constraint reasoning. In: Proceedings of workshop on DCR 2011, pp. 112–121 (2011). http://dischoco.sourceforge.net/
27. Yokoo, M., Durfee, E.H., Ishida, T., Kuwabara, K.: Distributed constraint satisfaction for formalizing distributed problem solving. In: Proceedings of ICDCS, pp. 614–621 (1992)
28. Zivan, R., Meisels, A.: Parallel Backtrack search on DisCSPs. In: Proceedings of DCR 2002 (2002)
29. Zivan, R., Meisels, A.: Dynamic Ordering for Asynchronous Backtracking on DisCSPs. Constraints **11**(2–3), 179–197 (2006)
30. Zivan, R., Meisels, A.: Message delay and DisCSP search algorithms. AMAI **46**(4), 415–439 (2006)
31. Zivan, R., Zazone, M., Meisels, A.: Min-Domain Retroactive Ordering for Asynchronous Backtracking. Constraints **14**(2), 177–198 (2009)

Automatically Generating Streamlined Constraint Models with ESSENCE and CONJURE

James Wetter[(✉)], Özgür Akgün, and Ian Miguel

School of Computer Science, University of St Andrews, St Andrews, UK
{jpw3,ozgur.akgun,ijm}@st-andrews.ac.uk

Abstract. Streamlined constraint reasoning is the addition of uninferred constraints to a constraint model to reduce the search space, while retaining at least one solution. Previously, effective streamlined models have been constructed by hand, requiring an expert to examine closely solutions to small instances of a problem class and identify regularities. We present a system that automatically generates many conjectured regularities for a given ESSENCE specification of a problem class by examining the domains of decision variables present in the problem specification. These conjectures are evaluated independently and in conjunction with one another on a set of instances from the specified class via an automated modelling tool-chain comprising of CONJURE, SAVILE ROW and MINION. Once the system has identified effective conjectures they are used to generate streamlined models that allow instances of much larger scale to be solved. Our results demonstrate good models can be identified for problems in combinatorial design, Ramsey theory, graph theory and group theory - often resulting in order of magnitude speed-ups.

1 Introduction

The search space defined by a constraint satisfaction problem can be vast, which can lead to impractically long search times as the size of the problem instance considered grows. An approach to mitigating this problem is to narrow the focus of the search onto promising areas of the search space using streamliners [11]. Streamliners take the form of uninferred additional constraints (i.e. constraints not proven to follow from the original problem statement) that rule out a substantial portion of the search space. If a solution lies in the remainder then it can typically be found more easily than when searching the full space. Previously, streamlined models have been produced by hand [11,13,15,16], which is both difficult and time-consuming. The principal contribution of this paper is to show how a powerful range of streamliners can be generated automatically.

Our approach is situated in the automated constraint modelling system CONJURE [2]. This system takes as input a specification in the abstract constraint specification language ESSENCE [7,8]. Figure 1 presents an example specification, which asks us to partition the integers $1 \ldots n$ into k parts subject to a set of constraints. ESSENCE supports a powerful set of type constructors, such as set, multi set, function and relation, hence ESSENCE specifications are concise and

© Springer International Publishing Switzerland 2015
G. Pesant (Ed.): CP 2015, LNCS 9255, pp. 480–496, 2015.
DOI: 10.1007/978-3-319-23219-5_34

```
language Essence 1.3

given k, l, n : int(1..)

find p: partition (numParts k) from int(1..n)
such that
 forAll s in parts(p) .
  forAll start : int(1..n-l+1) .
   forAll width : int(1..(n-start+1)/(l-1)) .
    !(forAll i : int(0..l-1) .
       (start + i*width) in s)
```

Fig. 1. ESSENCE specification of the Van Der Waerden Number Problem [24] (see Appendix A). The specification describes a certificate for the lower bound on $W(k, l)$, and will be unsatisfiable if **n** is equal to $W(k, l)$.

highly structured. Existing constraint solvers do not support these abstract decision variables directly. Therefore we use CONJURE to *refine* abstract constraint specifications into concrete constraint models, using constrained collections of primitive variables (e.g. integer, Boolean) to represent the abstract structure.

Our method exploits the structure in an ESSENCE specification to produce streamlined models automatically, for example by imposing streamlining constraints on or between the parts of the partition in the specification in Figure 1. The modified specification is refined automatically into a streamlined constraint model by CONJURE. Identifying and adding the streamlining constraints at this level of abstraction is considerably easier than working directly with the constraint model, which would involve first recognising (for example) that a certain collection of primitive variables and constraints together represent a partition — a potentially very costly step.

Our system contains a set of rules that fire when their preconditions match a given ESSENCE specification to produce candidate streamliners. Using CONJURE to refine the streamlined specifications into constraint models, solved with SAVILE ROW[1] [19] and MINION[2] [10], candidates are evaluated against instances of the specified class. Effective streamliners are combined to produce more powerful candidates. As we will show, high quality streamlined models can be produced in this way, in some cases resulting in a substantial reduction in search effort.

Herein we focus on satisfaction problems. Optimisation is an important future consideration but requires different treatment: streamliners may allow us to find good solutions quickly but exclude the optimal solution to the original model.

The rest of this paper is structured as follows. Following a summary of related work, we give some background on the ESSENCE language. Section 4 describes in detail our approach to generating streamliners automatically, then Section 5 discusses combining streamliners to produce yet more effective models. We conclude following a discussion of discovering streamliners in practice.

---

[1] http://savilerow.cs.st-andrews.ac.uk
[2] http://constraintmodelling.org/minion/

2   Related Work

Colton and Miguel [5] and Charnley et al [4] used Colton's HR system [6] to conjecture the presence of implied constraints from the solutions to small instances of several problem classes, including quasigroups and moufang loops. The Otter theorem prover[3] was used to prove the soundness of these conjectures. If proven sound, the implied constraints were added to the model to aid in the solution of larger instances of the same problem class.

Streamlined constraint reasoning differs from the approach of Charnley et al in that the conjectured constraints are not typically proven to follow from a given constraint model. Rather, they are added in the hope that they narrow the focus of the search onto an area containing solutions. When first introduced by Gomes and Sellmann [11] streamlined constraint reasoning was used to help construct diagonally ordered magic squares and spatially balanced experiment designs. For the magic squares the additional structure enforced was regularity in the distribution of numbers in the square. That is, the small numbers are not all similarly located, and likewise for large numbers. The spatially balanced experiment designs were forced to be self-symmetric: the permutations represented by the rows of the square must commute with each other. For both of these problems streamlining led to huge improvements in solve time, allowing much larger instances to be solved.

Kouril et al. refer to streamlining as "tunneling" [13]. They describe the additional constraints as tunnels that allow a SAT solver to tunnel under the great width seen early in the search tree when computing bounds on Van de Waerden numbers. They used simple constraints that force or disallow certain sequences of values to occur in the solutions. Again this led to a dramatic improvement in run time of the solver, allowing much tighter bounds to be computed.

Le Bras et al. used streamlining to help construct graceful double wheel graphs [15]. Constraints forcing certain parts of the colouring to form arithmetic sequences allowed for the construction of colourings for much larger graphs. These constraints led to the discovery of a polynomial time construction for such colourings, proving that all double wheel graphs are graceful.

Finally Le Bras et. al. made use of streamlining constraints to compute new bounds on the Erdős discrepancy problem [16]. Here constraints enforcing periodicity in the solution, the occurrence of the improved Walters sequence, and a partially multiplicative property improved solver performance, allowing the discovery of new bounds.

In all of these examples streamliners proved very valuable, but were generated by hand following significant effort by human experts. In what follows, we will show that the structure recognised and exploited by these experts is often present in abstract constraint specifications.

---

[3] http://www.cs.unm.edu/~mccune/otter/

```
language Essence 1.3

given v, b, r, k, lambda : int(1..)
where v = b, r = k
letting Obj   be new type of size v,
        Block be new type of size b

find bibd : relation (symmetric) of (Obj * Block)

such that
  forAll o  : Obj   . |bibd(o,_ )| = r,
  forAll bl : Block . |bibd(_,bl)| = k,
  forAll o1, o2 : Obj .
    o1 != o2 -> |bibd(o1,_) intersect bibd(o2,_)| = lambda
```

Fig. 2. ESSENCE specification of the square (ensured by the **where** statement) Balanced Incomplete Block Design Problem [20]. Streamliner added as an ESSENCE annotation shown underlined.

3   Background: Essence

The motivation for abstract constraint specification languages, such as Zinc [18] and ESSENCE is to address the *modelling bottleneck*: the difficulty of formulating a problem of interest as a constraint model suitable for input to a constraint solver. An abstract constraint specification describes a problem above the level at which constraint modelling decisions are made. An automated refinement system, such as CONJURE, can then be used to produce a constraint model from the specification automatically.

An ESSENCE specification, such as those given in Figures 1 and 2, identifies: the input parameters of the problem class (**given**), whose values define a problem instance; the combinatorial objects to be found (**find**); and the constraints the objects must satisfy (**such that**). In addition, an objective function may be specified (**min/maximising** — not shown in these examples) and identifiers declared (**letting**). Abstract constraint specifications must be *refined* into concrete constraint models for existing constraint solvers. Our CONJURE system employs refinement rules to convert an ESSENCE specification into the solver-independent constraint modelling language ESSENCE′. From ESSENCE′ we use SAVILE ROW to translate the model into input for a particular solver while performing solver-specific model optimisations.

A key feature of abstract constraint specification languages is the support for abstract decision variables with types such as set, multiset, relation and function, as well as *nested* types, such as set of sets and multiset of relations. This allows the problem structure to be captured very concisely. As explained below, this clarity of structure is a good basis for the conjecture of streamlining constraints.

4   From Conjectures to Streamlined Specifications

This section presents the methods used to generate streamlined models automatically. The process is driven by the decision variables in an ESSENCE specification, such as the partition in Figure 1. For each variable, the system forms conjectures of possible regularities that impose additional restrictions on the values of that variable's domain. Since the domains of ESSENCE decision variables have complex, nested types, these restrictions can have far-reaching consequences for constraint models refined from the modified specification. The intention is that the search space is reduced considerably, while retaining at least one solution. Multiple conjectures found to produce successful streamlined specifications individually can be combined in an attempt to produce a single more sophisticated streamliner. Currently conjectures are formed about each variable independently; an important future direction is to make conjectures across multiple variables.

4.1   Exploiting Essence Domain Annotations

ESSENCE allows domains to be annotated to restrict the set of acceptable values. For example, a function variable domain may be restricted to injective functions, or a partition variable domain may be restricted to regular partitions. Hence, the simplest source of streamliners is the systematic annotation of the decision variables in an input specification. This sometimes retains solutions to the original problem while improving solver performance. Consider the Balanced Incomplete Block Design problem (BIBD, Figure 2), where the decision variable is a relation. For square BIBDs we might consider a streamliner requiring a symmetric relation, achieved simply by adding the `symmetric` annotation as shown.

Figure 3 summarises an experiment with this streamliner on a set of satisfiable square BIBD instances. Original and streamlined specifications were refined with CONJURE, using the Compact heuristic [1] to select one model. For each instance SAVILE ROW was used to prepare the resulting model for MINION, which was used to find a solution. Streamlining uniformly resulted in a solvable instance, and as seen in Figure 3a in all but one instance search size is equivalent or reduced and as seen in Figure 3b the corresponding execution time is sometimes reduced.

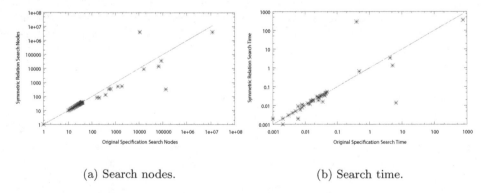

(a) Search nodes.   (b) Search time.

Fig. 3. Search effort to find the first solution to satisfiable square BIBD instances where $v = b \leq 40$: original vs streamlined specification.

4.2   Conjecture-Forming Rules

The existing ESSENCE domain annotations are, however, of limited value. They are very strong restrictions and so often remove all solutions to the original problem when added to a specification. In order to generate a larger variety of useful conjectures we employ a small set of rules, categorised into two classes:

1. First-order rules add constraints to reduce the domain of a decision variable directly.
2. Higher-order rules take other rules as arguments and use them to reduce the domain of a decision variable.

The full list of rules is shown in Table 1. A selection of the first-order rules is given in Figure 4, and a selection of the higher-order rules are given in Figure 5.

We define four first-order rules that operate on integer domains directly: 'odd', 'even', 'lowerHalf' and 'upperHalf'. Each restricts an integer domain to an appropriate subset of its values. We define six first-order rules for function domains. Two of these constrain the function to be monotonically increasing (or decreasing). The other four place the largest (or smallest) element first (or last). Functions mapping a pair of objects to a third object of the same type can be viewed as binary operators on these objects. We define three first-order rules to enforce such functions to be commutative, non-commutative, and associative respectively. Finally, we define a first-order rule for partitions called 'quasi-regular'. Partition domains in ESSENCE can have a `regular` annotation, however this can be too strong. The 'quasi-regular' streamlining rule posts a soft version

Table 1. The rules used to generate conjectures. Rows with a softness parameter specify a family of rules each member of which is defined by an integer parameter.

Class	Trigger Domain	Name	Softness
First-order	`int`	odd{even}	no
		lower{upper}Half	no
	`function int --> int`	monotonicIncreasing{Decreasing}	no
		largest{smallest}First{Last}	no
	`function (X,X) --> X`	commutative	no
		associative	no
		non-commutative	no
	`partition from X`	quasi-regular	yes
Higher-order	`set of X`	all	no
		most	yes
		half	no
		approxHalf	yes
	`function X --> Y`	range	no
		defined	no
		pre{post}fix	yes
		allBut	yes
	`function (X,X) --> Y`	diagonal	no
	`partition from X`	parts	no

Name	odd		**Name**	lowerHalf
Input	X: int		**Input**	X: int(1..u)
Output	X % 2 = 1		**Output**	X < 1 + (u - 1) / 2

Name	monotonicIncreasing		**Name**	largestFirst
Input	X: function int --> int		**Input**	X: function int(1..u) --> int
Output			**Output**	

```
        forAll i in defined(X) .          forAll i in defined(X) .
        forAll j in defined(X) .            X(min(defined(X)) <= X(i)
           i < j -> X(i) <= X(j)
```

Name	commutative		**Name**	quasi-regular
Input	X: function (D, D) --> D		**Input**	X: partition from _
Output			**Output**	

```
        forAll (i,j) in defined(X) .      minPartSize(X,
           X((i,j)) = X((j,i))               |participants(X)|/|parts(X)| - k)
                                                       /\
                                          maxPartSize(X,
                                             |participants(X)|/|parts(X)| + k)
```

Fig. 4. A selection of the first-order streamlining rules.

Name	all		**Name**	most
Parameter	R (another rule)		**Parameter**	R (another rule)
Input	X: set of _		**Parameter**	k (softness)
Output	forAll i in X . R(i)		**Input**	X: set of _
			Output	

```
                                               k >= sum i in X . toInt(!R(i))
```

Name	range		**Name**	prefix
Parameter	R (another rule)		**Parameter**	R (another rule)
Input	X: function _ --> _		**Parameter**	k (softness)
Output	R(range(X))		**Input**	X: function int(1..h) --> _
			Output	R(restrict(X, 'int(1..h-k)'))

Name	parts		**Name**	diagonal
Parameter	R (another rule)		**Parameter**	R (another rule)
Input	X: partition of _		**Input**	
Output	R(parts(X))			X: function (D1, D1) --> D2
			Output	

```
                                               { R(X')
                                               @ find X' : function D1 --> D2
                                                 such that
                                                   forAll i : D1 .
                                                     (i,i) in defined(X) -> X'(i) = X((i,i)),
                                                   forAll i : D1 .
                                                     i in defined(X') -> X'(i) = X((i,i))
                                               }
```

Fig. 5. A selection of the higher-order streamlining rules.

of the regularity constraint, which takes an integer parameter, k, to control the softness of the constraint. In our experiments we varied the value of k between 1 and 3. Larger values of k will make the constraint softer as they allow the sizes of parts in the partition to be k-apart.

Higher-order rules take another rule as an argument and lift its operation to a decision variable with a nested domain. For example, the 'all' rule for sets applies a given streamlining rule to all members of a set, if it is applicable. We define three other higher-order rules that operate on set variables: 'half', 'most' and 'approxHalf', the last two with softness parameters. For a set of integers, applying the 'half' rule with the 'even' rule as the parameter – denoted 'half(even)' – forces half of the values in the set to be even. The parameter to the higher-order rule can itself be a higher order rule, so for a set of set of integers 'all(half(even))' constrains half of the members of all inner sets to be even.

The 'defined' and 'range' rules for functions use the defined and range operators of ESSENCE to extract the corresponding sets from the function variable. Once the set is extracted the parameter rule R can be directly applied to it. The 'prefix' and 'postfix' rules work on functions that map from integers by focusing on a prefix or postfix of the integer domain. The 'parts' rule views a partition as a set of sets and opens up the possibility of applying set rules to a partition.

The 'diagonal' rule introduces an auxiliary function variable. The auxiliary variable represents the diagonal of the original variable, and it is channelled into the original variable. Once this variable is constructed the streamlining rule taken as a parameter, R, can be applied directly to the auxiliary variable. Similarly the 'allBut' rule introduces an auxiliary function variable that represents the original variable restricted to an arbitrary subset (of fixed size) of its domain. This is similar to the 'prefix' rule but allows the ignored mappings of the function to fall anywhere in the function's domain rather than only at the end.

It is important to note that allowing higher-order rules to take other higher-order rules as parameters naively can lead to infinite recursion; such as 'prefix(prefix(prefix(...)))' or 'prefix(postfix(prefix(..)))'. We considered two ways of mitigating this problem: 1) using a hard-coded recursion limit 2) only allowing one application of the same higher-order rule in a chain of rule applications and at the same level. Using a hard-coded recursion limit has two drawbacks. It still allows long lists of useless rule applications like applying 'prefix' repeatedly. It can also disallow useful but long sequences of rule applications. Instead, we implemented a mechanism where the same higher-order rule can only be applied once at the same level; that is, a rule can only be applied more than once if the domain of its input is a subcomponent of the domain of the input of the previous application. This disallows 'prefix(prefix(...))', but allows 'prefix(range(half(prefix(even),1)),1)' which works on a decision with the following type `function int --> (function int --> int)`.

In order to apply these rules to ESSENCE problem specifications, we extract a list of domains from the top level decision variables. For each domain, we find the list of all applicable rules. Each application of a streamlining rule results in a conjecture that can be used to streamline the original specification.

To see an example of an effective constraint generated by this system consider the problem of generating certificates for lower bounds on Van Der Waerden numbers shown in Figure 1. Here only one decision variable is present, a partition of integers, p. First the rule *parts* is triggered so p is treated as a set of set of integers. Next, the rule *all* is triggered such that all the members of the outer set are restricted, then another rule *approxHalf* is triggered so approximately half the members of each inner set are restricted. Finally *lowerHalf* is triggered so the domain of the integer has its upper bound halved. The complete rule that is being applied is 'parts(all(approxHalf(lowerHalf, i)))', and the resulting constraint is:

```
forAll s in parts(p) .
  |s|/2 + i >= sum x in s . toInt(x <= n/2) /\
  |s|/2 - i <= sum x in s . toInt(x <= n/2)
```

where i is the parameter given to *approxHalf*. This constraint enforces that each part of the partition consists of approximately half 'small' numbers and half 'large' numbers. Figure 8 shows that this constraint with $i = 2$ drastically reduces the number of search nodes explored when finding a single solution to the problem.

Initially these rules were applied to the variable domains declared as finds in the original specification. It was observed that the wheel like structures in the graceful graph labeling problems were not exposed in the function variables used to represent the labeling. In order to extract such structures a preprocessing step is performed that introduces restricted function variables to the rule system in order to generate a wider range of streamlining constraints.

This preprocessing step is triggered when a function variable, $f : A \rightarrow B$, is present in the find declarations. For each such variable the constraints are checked for an application of the function, $f(a)$, quantified over a strict subset of the domain, $a \in A' \subset A$. The quantification can be $\sum$, $\forall$ or $\exists$. If such an expression is present in the constraints the conjecture generation rules are also applied to a restricted version of the function, $f|_{A'}$.

This system is capable of generating novel streamliners that have not been previously reported, such as the 'quasi-regular' rule. In addition it generates some streamliners very similar to some of those previously seen in the literature. For example appliying 'restrict(range(all(odd)),C_1)' to the colouring function in the gracful double wheel labelling specification defines the same solution set as 'C_1 is odd' in La Bras et. al. [16]. Although it should be noted that the rules presented here do not generate all streamliners previously reported.

Table 2. Number of conjectures generated for a set of problem classes. The `Total` column lists all conjectures generated for each class, the `Retain` column lists the number of conjectures that retain at least one solution and the `Improve` column lists the number of conjectures that improve solver performance.

Problem Class		Total	Retain	Improve
Graceful Graph Colouring [21]	Wheel Graphs [9]	1479	1299	136
	Double Wheel Graphs [15]	2142	1942	466
	Helm Graphs [3]	1428	1296	85
	Gear Graphs [17]	1428	1214	70
Quasigroup Existence [23]	QGE3	593	570	51
	QGE4	593	572	34
	QGE5	593	582	24
	QGE6	593	569	44
	QGE7	593	555	32
Equidistant Frequency Permutation Arrays [12]		560	377	2
Van Der Waerden Numbers [24]		433	364	14
Schur Numbers [22]		437	419	0
Langfords Problem [14]		357	228	24

4.3 Experimental Analysis

To evaluate this system it was used to streamline several different problem classes, consisting of graceful graph colouring problems [21], quasigroup existence problems [23], equidistant frequency permutation arrays [12], Van Der Waerden numbers [24] with $k = 3$ and $l = 4$, Schur numbers [22] and Langford's problem [14] (see Appendix A). The instances used to evaluate the performance of the streamlined models were obtained by increasing the size of the integer parameters until the original specification was unable to compute one solution in under 100,000 search nodes. Experiments were run on a 32-core AMD Opteron 6272 at 2.1 GHz and took around 5 hours per problem class to complete.

Table 2 shows the total number of conjectures generated, the number of conjectures that retain solutions to the original problem and the number of conjectures that reduce the total number of search nodes explored after the specification is refined, tailored and solved for all instances.

Figure 6 shows the search nodes and solve time to find the first solution for the original specification vs the best streamlined model, where the quality metric compares the number of instances solved in under 100,000 search nodes and breaks ties by comparing the total number of search nodes across all instances. Both the size of the search and search time are often reduced by orders of magnitude by the streamlining constraints.

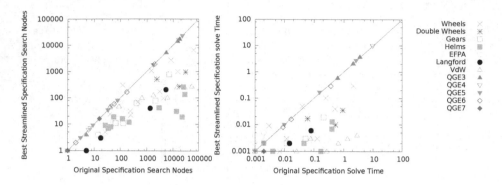

Fig. 6. The size and execution time of search required to find the first solution for a collection of problem classes for both the original specification and the streamlined specification that resulted in the smallest cumulative search size across instances. The generated conjectures often result in order of magnitude reduction in search size for harder problem instances.

5   Identifying Effective Combinations of Conjectures

It has previously been observed that applying several streamlining constraints to a model simultaneously can result in larger performance gains that any of the constraints in isolation [15]. In order to find such combinations of constraints we must consider the power set of constraints that retain solutions to the original problem. In this section we investigate finding powerful combinations of constraints automatically with the use of pruning and make a comparison between depth first search and breadth first search of the lattice of constraints.

For many of the problems considered here a large number of singleton conjectures that retain solutions are generated (see Table 2) resulting in power sets too large to be exhaustively explored in practice. Two forms of pruning were used to reduce the number of combinations to be considered:

1. if a set of conjectures fails all supersets are excluded from consideration (see Figure 7),
2. trivially conflicting conjectures are not combined, for example we avoid forcing a set to simultaneously contain only odd numbers and contain only even numbers. We associate a set of tags with each of the rules in order to implement this pruning. Rules applied to the same variable that share tags are not combined. This also removes the possibility of combining two different conjectures that differ only by a softness parameter.

These pruning rules only remove combinations that are sure to fail, or are equivalent to a smaller set of conjectures.

Even with this pruning the number of combinations to consider was found to be too large to allow exhaustive enumeration. Therefore a traversal of the lattice allowing good combinations to be identified rapidly is desired. Here we

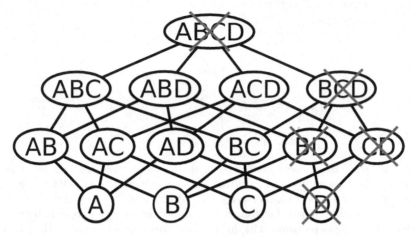

Fig. 7. The power set of singleton conjectures can be explored to identify combinations that result in powerful streamlined specifications. If small sets of conjectures that fail to retain solutions are identified all super sets can be pruned from the search vastly reducing the number of vertices to be explored.

experimented with depth first search (DFS) and breadth first search (BFS) of the lattice. In order to guide both the searches the singleton conjectures were ordered from best to worst, where the quality metric compared the number of instances solved in under 100,000 search nodes and ties were broken by comparing the total search nodes across all instances.

In order to compare the two approaches they were each allowed to evaluate 500 combinations of conjectures in addition to the singleton conjectures already evaluated for three problem classes (Van der Waerden numbers, graceful helms graphs and graceful double wheel graphs). The set of instances used for evaluation was augmented by increasing the integer parameters until the best singleton streamliner was unable to solve the instance in under 100,000 search nodes. Each combination of conjectures was evaluated by refining, tailoring and solving for the first solution using CONJURE's compact heuristic and default settings for SAVILE ROW and MINION. The experiments were performed on 32-core AMD Opteron 6272 at 2.1 GHz taking approximately 6 hours for each problem class.

Figures 8 to 10 show the best set of conjectures found by this process, where the quality metric compares the number of instances solved and ties were broken by comparing the total number of search nodes.

Figure 8 shows three singleton conjectures that were found to produce an effective streamlined specification of the Van Der Waerden numbers problem, one of which results from the chained application of four rules, whereas another results directly from a single rule. Figure 9 shows two conjectures being combined for the graceful helm graph problem, one of which results from the chained application of three rules. Figure 10 shows the combination of two conjectures for the graceful colouring of double wheel graphs. In all cases the combination of streamliners results in better performance than any of the streamliners in isolation.

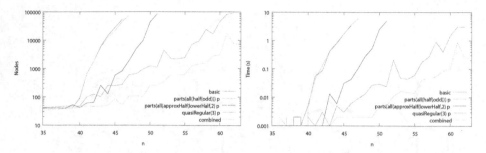

Fig. 8. Combining singleton conjectures to produce a more effective streamlined model for Van der Waerden numbers. The instances have k = 3, 1 = 4 and n varies. The first conjecture ensures odd number are evenly distributed between the parts of the partition. The second conjecture ensures the 'small' numbers are evenly distributed between the parts of the partition. The third conjecture ensures the sizes of the partition vary from each other by at most three.

On one problem, Van Der Waerden numbers, DFS performed very well, reaching a powerful set of three conjectures within the first three models evaluated. On the other problems DFS performed poorly, unable to beat the best pair found by BFS within the allotted resource budget. In all three cases DFS failed to find an improved model after the first five models it considered.

The poor performance of DFS can be attributed to two factors. First, the two best singleton conjectures do not always produce the best pair of conjectures, even when they retain solutions in combination. A better heuristic would need some notion of complementary conjectures. Second, far more combinations are pruned from the search space if failing sets are detected early. Consider the lattice of conjecture sets shown in Figure 7. If conjecture C and D fail to retain solutions when used in combination so will {A,C,D}, {B,C,D} and {A,B,C,D}. A breadth first traversal would be guaranteed to detected this failure early and would consequently never evaluate these three models. Alternatively a depth first traversal would detect this failure late, and would therefore waste time evaluating the supersets of {C,D}, all of which fail to retain solutions.

6   Discussion: Generating Streamliners in Practice

In this section, we consider the process of generating and selecting streamliners when presented with a new problem class of interest. Our methodology is a close analogue of that adopted by human experts in manual streamliner generation. Given an ESSENCE specification of the problem class, we begin by identifying suitable instances with which to evaluate candidate streamliners. These instances must be satisfiable and solvable in reasonable time so that they can be used in the evaluation of a large set of candidate streamlined specifications. This set of instances can be selected manually, or generated automatically by attempting to solve candidates using the basic specification — satisfiable instances solved within a budget are kept for streamliner evaluation.

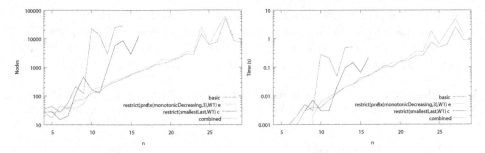

Fig. 9. Combining singleton conjectures to produce a more effective streamlined model for Graceful Helm Graphs. The parameter **n** is the size of the wheel. The first conjecture requires that the differences between the labels of the vertices on the inner loop are in decreasing order, except the last 3. The second requires that the smallest label occurring on the inner loop is the last vertex.

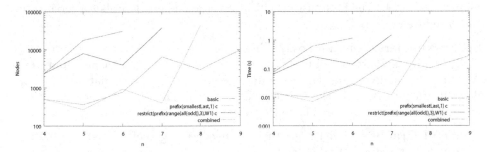

Fig. 10. Combining singleton conjectures to produce a more effective streamlined model for Graceful Double Wheel Graphs. The parameter **n** defines the size of single wheel. The first conjecture requires that the last vertex in the outer loop of the graph takes the largest value. The second requires that the difference between adjacent vertices on the inner loop are odd numbers, except the last 3.

Streamliners are then generated, combined and evaluated against the set of test instances, as described in Sections 4 and 5. This is a costly process, in the same way that a considerable effort is expended by human experts in manual streamliner generation. However, streamliners are generated for use over the entire problem class. Under the assumption that our problem class has infinitely many elements, the cost of streamliner discovery is amortised over all instances not used in the streamliner evaluation process and becomes negligible.

7 Conclusion

Streamliner generation has been the exclusive province of human experts, requiring substantial effort in examining the solutions to instances of a problem class, manually forming conjectures as to good streamliners, and then testing their efficacy in practice. In this paper we have demonstrated for the first time the automated generation of effective streamliners, achieved through the exploitation of the structure present in abstract constraint specifications written in ESSENCE. In future work we will expand our set of streamliner generation methods and explore streamliner generation in further, more complex problem classes.

Acknowledgments. This work is supported by UK EPSRC grant EP/K015745/1. We thank Ian Gent, Chris Jefferson and Peter Nightingale for helpful comments.

A   Problem Descriptions

Van Der Waerden Numbers. Van Der Waerden's theorem states that given any positive integers k and l, there exists a positive integer n such that for any partition of the integers 1 to n into k parts at least one of the parts contains an arithmetic sequence of length l. The Van Der Waerden number, $W(k, l)$, is the lowest such n [24]. The ESSENCE specification studied here describes a certificate that the given $n \neq W(k, l)$.

Schur Numbers. Given a positive integer r, there exists a positive integer s such that for any partition of the integers 1 to s at least one part is not sum free. Alternatively at least one part is a super set of $\{x, y, z\}$ such that $x + y = z$. Schur's number, $S(r)$, is the smallest such s [22]. The ESSENCE specification studied here describes a certificate that the given $s \neq S(r)$

Graceful Graphs. Given a graph with n edges a graceful labelling assigns each node in the graph a label between 0 and n such that no label is used more than once and that every pair of adjacent nodes can be uniquely identified by the absolute difference of their labels. A graceful graph is a graph that permits a graceful labelling [21]. Several classes of graph have been investigate in this context including wheels [9], double wheels [15], helms [3] and gears [17].

Quasigroup Existsence. Given a positive integer n, does there exist a quasigroup (latin square) of size n such that an additional side constraint is met. These side constraints are: QGE3 - $\forall a, b \in g$ $(a \cdot b) \cdot (b \cdot a) = a$, QGE4 - $\forall a, b \in g$ $(b \cdot a) \cdot (a \cdot b) = a$, QGE5 - $\forall a, b \in g$ $((b \cdot a) \cdot b) \cdot b = a$, QGE6 - $\forall a, b \in g$ $(a \cdot b) \cdot b = a \cdot (a \cdot b)$, QGE7 - $\forall a, b \in g$ $(b \cdot a) \cdot b = a \cdot (b \cdot a)$ [23].

Equidistant Frequency Permutation Arrays. Given v, q, λ and d, construct a set of v codewords such that each code word is of length $q \cdot \lambda$ and contains λ occurrence of each symbol in the set $\{1, 2, \ldots, q\}$. Each pair of code words must be of hamming distance d [12].

Langford's Problem. Given any positive integer n, arrange copies of the numbers between 1 and n such that for all k in $\{1 \ldots n\}$ there are k digits between occurrences of k [14].

References

1. Akgun, O., Frisch, A.M., Gent, I.P., Hussain, B.S., Jefferson, C., Kotthoff, L., Miguel, I., Nightingale, P.: Automated symmetry breaking and model selection in CONJURE. In: Schulte, C. (ed.) CP 2013. LNCS, vol. 8124, pp. 107–116. Springer, Heidelberg (2013)
2. Akgün, O., Miguel, I., Jefferson, C., Frisch, A.M., Hnich, B.: Extensible automated constraint modelling. In: AAAI 2011: Twenty-Fifth Conference on Artificial Intelligence (2011)
3. Ayel, J., Favaron, O.: Helms are graceful. Progress in Graph Theory (Waterloo, Ont., 1982), pp. 89–92. Academic Press, Toronto (1984)
4. Charnley, J., Colton, S., Miguel, I.: Automatic generation of implied constraints. In: ECAI, vol. 141, pp. 73–77 (2006)
5. Colton, S., Miguel, I.: Constraint generation via automated theory formation. In: Walsh, T. (ed.) Proceedings of the Seventh International Conference on Principles and Practice of Constraint Programming, pp. 575–579 (2001)
6. Colton, S.: Automated Theory Formation in Pure Mathematics. Ph.D. thesis, University of Edinburgh (2001)
7. Frisch, A.M., Jefferson, C., Hernandez, B.M., Miguel, I.: The rules of constraint modelling. In: Proc. of the IJCAI 2005, pp. 109–116 (2005)
8. Frisch, A.M., Harvey, W., Jefferson, C., Martínez-Hernández, B., Miguel, I.: Essence: A constraint language for specifying combinatorial problems. Constraints 13(3), 268–306 (2008). doi:10.1007/s10601-008-9047-y
9. Frucht, R.: Graceful numbering of wheels and related graphs. Annals of the New York Academy of Sciences 319(1), 219–229 (1979)
10. Gent, I.P., Jefferson, C., Miguel, I.: Minion: A fast scalable constraint solver. In: ECAI, vol. 141, pp. 98–102 (2006)
11. Gomes, C., Sellmann, M.: Streamlined constraint reasoning. In: Wallace, M. (ed.) CP 2004. LNCS, vol. 3258, pp. 274–289. Springer, Heidelberg (2004)
12. Huczynska, S., McKay, P., Miguel, I., Nightingale, P.: Modelling equidistant frequency permutation arrays: an application of constraints to mathematics. In: Gent, I.P. (ed.) CP 2009. LNCS, vol. 5732, pp. 50–64. Springer, Heidelberg (2009)
13. Kouril, M., Franco, J.: Resolution tunnels for improved SAT solver performance. In: Bacchus, F., Walsh, T. (eds.) SAT 2005. LNCS, vol. 3569, pp. 143–157. Springer, Heidelberg (2005)
14. Langford, C.D.: Problem. The Mathematical Gazette, 287–287 (1958)
15. Le Bras, R., Gomes, C.P., Selman, B.: Double-wheel graphs are graceful. In: Proceedings of the Twenty-Third international Joint Conference on Artificial Intelligence, pp. 587–593. AAAI Press (2013)
16. Le Bras, R., Gomes, C.P., Selman, B.: On the erdős discrepancy problem. In: O'Sullivan, B. (ed.) CP 2014. LNCS, vol. 8656, pp. 440–448. Springer, Heidelberg (2014)
17. Ma, K., Feng, C.: On the gracefulness of gear graphs. Math. Practice Theory 4, 72–73 (1984)
18. Marriott, K., Nethercote, N., Rafeh, R., Stuckey, P.J., de la Banda, M.G., Wallace, M.: The design of the zinc modelling language. Constraints 13(3) (2008). doi:10.1007/s10601-008-9041-4
19. Nightingale, P., Akgün, Ö., Gent, I.P., Jefferson, C., Miguel, I.: Automatically improving constraint models in Savile Row through associative-commutative common subexpression elimination. In: O'Sullivan, B. (ed.) CP 2014. LNCS, vol. 8656, pp. 590–605. Springer, Heidelberg (2014)

20. Prestwich, S.: CSPLib problem 028: Balanced incomplete block designs. http://www.csplib.org/Problems/prob028
21. Rosa, A.: On certain valuations of the vertices of a graph. In: Theory of Graphs Internat. Symposium, Rome, pp. 349–355 (1966)
22. Schur, I.: Über die kongruenz $x^m + y^m \equiv z^m$ (mod p). Jahresber. Deutsch. Math. Verein **25**, 114–117 (1916)
23. Slaney, J., Fujita, M., Stickel, M.: Automated reasoning and exhaustive search: Quasigroup existence problems. Computers & Mathematics with Applications **29**(2), 115–132 (1995)
24. Van der Waerden, B.L.: Beweis einer baudetschen vermutung. Nieuw Arch. Wisk **15**(2), 212–216 (1927)

Application Track

Constraint-Based Local Search
for Finding Node-Disjoint Bounded-Paths
in Optical Access Networks

Alejandro Arbelaez[(⊠)], Deepak Mehta, and Barry O'Sullivan

Insight Centre for Data Analytics, University College Cork, Cork, Ireland
{alejandro.arbelaez,deepak.mehta,barry.osullivan}@insight-centre.org

Abstract. Mass deployment of fibre access networks is without doubt one of the goals of many network operators around the globe. The Passive Optical Network has been proposed as a solution to help deliver mass deployment of fibre-to-the-home (FTTH), by reducing the cost per customer compared to current commercially available technologies. A major failure in the access network (e.g., fibre cut, amplifier failure, or other equipment can fail) might affect tens of thousands of customers. Therefore, protecting the network from such failures is one of the most important requirements in the deployment of FTTH solutions. In this paper we use a constraint-based local search approach to design reliable passive optical networks via node-disjoint paths whereby each customer is protected against any one node or link failure. We experiment with a set of very large size real-world networks corresponding to Ireland, Italy and the UK and demonstrate the tradeoff between cost and resiliency.

1 Introduction

Continuous growth in the amount of data transferred within national and global networks in the last decade necessitates new infrastructures and data transfer technologies [1]. In line with this, the goal of the DISCUS project [2] is to develop an end-to-end network design that can provide high-speed broadband capability of at least three orders-of-magnitude greater than today's networks to all users, reduce energy consumption by 95%, and remain economically viable.[1] The architecture is based on maximising the sharing of network infrastructure between customers by deploying a Long-Reach Passive Optical Network (LR-PON) [3] in the access part. All network points have equal bandwidth and service capability including core bandwidth (10Gbs to 100Gbs) delivered to the access network.

LR-PONs provide an economically viable solution for fibre-to-the-home network architectures [4]. In a LR-PON fibres are distributed from metro nodes (MNs) to exchange sites (ESs) through cables that form a tree distribution network. Typically due to signal loss the maximum distance from a MN to an ES is up to 90 km, and the maximum distance from an ES to the customers is up to 10 km. A major fault occurrence in a LR-PON would be a complete failure of a

---

[1] http://www.discus-fp7.eu

G. Pesant (Ed.): CP 2015, LNCS 9255, pp. 499–507, 2015.
DOI: 10.1007/978-3-319-23219-5_35

MN, which could affect tens of thousands of customers. A dual homing protection mechanism is recommended for LR-PONs whereby customers are connected to two MNs, so that whenever a single MN fails all customers are still connected to a back-up. Simply connecting two MNs to an ES is not sufficient to guarantee the connectivity because if a link or a node is used is common in the routes of fibre going from ES to its 2 MNs then both MNs would be disconnected.

Broadly speaking, there are two protection strategies for dual homing. One approach is protecting links in the tree distribution network, e.g., a fibre cut, amplifier failure, or other equipment that can fail. This strategy, known as the edge-disjoint solution, allows switching to an alternative path whenever a link in the distribution network fails. Alternatively, protection can be provided at the node level with node-disjoint paths between MNs and ESs, when an entire ES fails and all adjacent links to the node will be affected. This strategy, known as node-disjoint solution, provides a stronger protection mechanism and allows switching to an alternative path whenever a node fails. The selection of one protection mechanism over another is a matter of choice for the broadband provider, generally higher protection means higher deployment cost.

Figure 1 shows an example with two MNs F_1 and F_2 and the set of ESs $\{a, b, c, d, e, f\}$. Black and gray coloured edges are used to show the trees corresponding to the facilities F_1 and F_2 respectively, and maximum allowed distance $\lambda=12$. This example shows three scenarios satisfying the length constraint, however, each scenario shows a different level of resiliency. Dashed lines indicate edges leading to a conflict, i.e., violating node or edge disjointness, and gray nodes indicate the conflicting node when violating node-disjointness. The total solution cost for Figure 1(a) is 46. The indicated solution satisfies the length constraint (i.e., the distance from the MNs to any ES is at most $\lambda=12$) and it also satisfies node-disjoint constraints and consequently edge-disjoint constraints. Here we observe that failure of any single link or node would not affect the remaining ESs. Figure 1(b) does not satisfy node-disjoint constraints but edge-disjoint constraints. Here we observe that a single node failure could disconnect one or more ESs from the MNs F_1 and F_2, e.g., f would be disconnected from F_1 and F_2 whenever a fails. Nevertheless, the solution is resilient to any single facility or single link failure. Figure 1(c) shows a solution that violates both node and edge disjoint constraints. In this example the link $\langle e, f \rangle$ appears in both trees and the path from f is neither edge-disjoint nor node-disjoint but it is resilient to a failure of a single MN.

Network operators tend to evaluate multiple options. Their choices of providing resiliency depends on many factors, e.g., captial and operational costs, rural or sparse regions, business or residential customers etc. Ideally they would prefer a tool where one can try to impose different configurations and get feedback quickly. In particular we focus on node-disjointness as the other options can be seen as relaxation of the node-disjoint constraints. The problem is to determine the routes of cable fibres such that there are two node-disjoint paths from each ES to its two MNs, the length of each path must be below a given threshold and the total cable length required for connecting each ES to two MNs

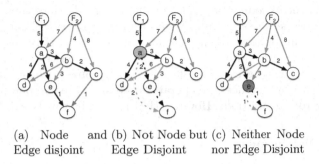

(a) Node and Edge disjoint   (b) Not Node but Edge Disjoint   (c) Neither Node nor Edge Disjoint

Fig. 1. Passive Optical Networks with and without node and edge disjoint protection.

should be minimised. The bottleneck is that the instances of the network sizes can be very huge e.g., UK has approximately 5000 ESs while Italy has more than 10,000. Solving very large instances of this problem in limited time is very challenging. We therefore apply a constraint-based local search and develop an efficient algorithm to handle node-disjoint constraints for selecting best move during LS.

2 Constraint-Based Local Search

Constraint-based local search has been recently introduced as an alternative solution to efficiently tackle many network design problems [5,6] ranging from telecommunications to transportations, and VLSI circuit design.

The local search (LS) algorithm starts with an initial solution and iteratively improves the solution by performing small changes. A move operator guides the algorithm towards better solutions. [5] defines the *node* and *subtree* operator, the algorithm moves a given node (resp. a subtree and the nodes emanating from it) from one location to another in the tree by improving the objective. These operators have been used to find distance-constrained and edge-disjoint paths. Empirical results suggest that the subtree operator outperforms the node operator. [6] defines the *arc* operator: the algorithm moves a given arc from one location to another in the tree, this operator has been used in the context of the routing and wavelength assignment problem [7] and to find edge disjoint paths in a given graph [8]. We remark that **node** and **subtree** move-operators are specifically designed for tree structures whereas the arc operator is more general and it can be applied to more general graphs. Nevertheless, the former is more efficient when we are dealing with trees, while the latter is more expensive. In this paper we limit our attention to using the subtree move-operator as it provides a better time complexity than that of the arc operator, i.e., $O(n)$ vs. $O(n^3)$ when considering the distance constraint. We refer the reader to [5] for a detailed analysis of the time complexities of the local search operators.

Other related work includes [9] where the authors proposed a dedicated algorithm to tackle the Distance Constrained Minimum Spanning Tree problem.

The authors studied two move-operators to tackle the distance constraint, informally speaking nodes are sequentially added in the tree using Prim's algorithm, and a backup route is used whenever the move cannot satisfy the length constraint. This framework is difficult to extend with side constraints, and therefore the algorithms cannot be extended to solve our problem with node and edge disjointed paths. For instance, the operators rely on a pre-computed alternative route from the root to any node. However that route might violate disjointedness.

2.1 Node-Disjointness

The LS algorithm uses the move-operators in two different phases. In the intensification phase a subtree is selected and it is moved to another location that provides the maximal improvement in the objective function. In the perturbation phase both a subtree and its new location are selected randomly in order diversify the search by adding noise to the current incumbent solution. In order to complete the intensification and diversification phases, the sub-tree operator requires four main functions: *removing* a subtree from the current solution; checking whether a given solution is *feasible* or not; *inserting* a subtree in the solution; finding the *best location* of a subtree in the current solution.

Additionally, the LS algorithm is equipped with an incremental way of maintaining information related to checking constraints and computing the cost of the assignment. For the length constraint it is necessary to maintain for each node e_j the length of the path from MN i to ES (or node in the tree) j, and the length of the path emanating from e_j down to the farthest leaf from e_j. These two lengths are updated after removing and inserting a subtree in the tree. For enforcing the edge-disjoint constraint in the two paths of a client, we must maintain the predecessors of each client in the two paths associated to the MNs, and require that the two predecessors must be different [5]. For enforcing the node-disjoint constraint between the paths of two clients it is necessary to maintain all the nodes occurring in each path except the source and target. This is done by maintaining the transitive graph.

Let M be the set of facilities or MCs. Let E be the set of clients or ESs. Let $E_i \subseteq E$ be the set of clients associated with MN $m_i \in M$. We use N to denote the set of nodes, which is equal to $M \cup E$. We use T_i to denote the tree network associated with MN i. We also use $N_i \subseteq N = E_i \cup \{m_i\}$ to denote the set of nodes in T_i. Let λ be the maximum path-length from a facility to any of its clients.

Let e_j be a node in the tree T_i associated with facility m_i. Let e_{p_j} be the predecessor of e_j, and let L_{e_j} be a list of all nodes in the subtree starting at e_j of tree T_i. We now describe the complexities of remove, insert, feasibility check and best location operations performed during search with respect to node-disjoint constraint:

Remove. To remove a subtree it is necessary to update the transitive graph. This is done by removing all the transitive links between any node in the path from the MN m_i down to e_j and all the elements in L_{e_j}. In Figure 2(a) gray lines show the links that must be removed from the transitivity graph, i.e., links that

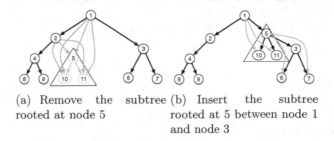

(a) Remove the subtree rooted at node 5

(b) Insert the subtree rooted at 5 between node 1 and node 3

Fig. 2. Removing/Inserting a subtree while maintaining transitive links.

reach any node in the subtree emanating from node 5. Thus, the time-complexity of removing a subtree rooted at e_j from the tree T_i is quadratic with respect to the number of clients of facility m_i.

Insert. The insertion involves adding the transitive links between each node in the path from the MN down to e_j and each element of L_{e_j}. Figure 2(b) shows the links that must be added in the transitivity graph; black links denote existing links in the tree and gray links denote the new links added by transitivity in the graph. Therefore, a subtree rooted at e_j can be inserted in T_i in quadratic time complexity with respect to the number of clients of facility m_i.

Feasibility. To verify the consistency of the node-disjoint constraint we need to check that any node occurring in the subtree starting from the node e_j is not transitively connected to any node in the path from the MN down to the potential predecessor of e_j in any other tree corresponding to other MNs. This is done by checking the occurrence of the links corresponding to the pairs of nodes occurring in the transitive graph. As the number of links can be quadratic, the time complexity is quadratic. Additionally, if the node e_j is breaking an existing arc $\langle e_p, e_q \rangle$, it is necessary to check that e_j is not transitively connected to any node in the subtree emanating from e_q.

Best Location. Selecting the best location for a given subtree involves traversing all the nodes of the tree associated with the MN and selecting the one that maximises the reduction in the cost. As the complexity of feasibility is quadratic, the time complexity of finding the best location is cubic in the number of nodes of a given tree.

The complexities for feasibility checking and best location described above do not make any assumption on the order in which different locations are explored. In the following we will show that the amortized complexities can be reduced if the locations in the tree are traversed in a particular order.

For a given subtree rooted at e_j there are two types of feasible locations. The first is a node-type location, say e_k, of the tree and the feasibility check is whether the parent of e_j can be e_k or not. This type of location is denoted by $(e_k, \emptyset)$. The second is a link-type location, say $\langle e_{p_l}, e_l \rangle$, of the tree and the feasibility check is whether we can break the link $\langle e_{p_l}, e_l \rangle$ by making e_{p_l} and e_j the parents of e_j and e_l respectively. This type of location is denoted by $(e_{p_l}, \{e_l\})$. For node-type

Algorithm 1. BestLocation(e_j, T_i, G)

1: Remove subtree rooted at e_j from T_i, update T_i and G
2: $Nlocations \leftarrow \emptyset$, $Llocations \leftarrow \emptyset$
3: Let s be a stack, let q be a queue
4: $s.push(root(T_i))$
5: **while** $s \neq \emptyset$ **do**
6:     $e_k \leftarrow s.pop()$
7:     $trans \leftarrow \{\langle e_k, e_l \rangle) \mid e_l \in L_{e_j} \wedge \langle e_k, e_l \rangle \in G\}$
8:     **if** $trans = \emptyset$ **then**
9:         $Nlocations \leftarrow Nlocations \cup \{(e_k, \emptyset)\}$
10:        **for all** $\langle e_k, e_l \rangle \in T_i$ **do**
11:           $Llocations \leftarrow Llocations \cup (e_k, \{e_l\})$
12:           $s.push(e_l)$
13:        **end for**
14:        **if** $out(e_k, T_i) = 0$ **then**
15:           $q.append((e_k, \texttt{check}))$
16: **while** $q \neq \emptyset$ **do**
17:     $(e_l, status) \leftarrow q.remove()$
18:     **if** $\langle e_{p_l}, e_l \rangle \in Llocations$ **then**
19:        **if** $\langle e_j, e_l \rangle \in G \| status = \texttt{rem}$ **then**
20:           $Llocations \leftarrow Llocations \setminus \{\langle e_{p_l}, e_l \rangle\}$
21:           $q.append((e_{p_l}, \texttt{rem}))$
22:        **else**
23:           $q.append((e_{p_l}, \texttt{check}))$
24: $Best \leftarrow \text{SelectBest}(e_j, NLocations \cup Llocations)$
25: **return** $Best$

---

locations we explore the tree in a depth-first manner, whereas for link-type locations we explore the tree in a bottom-up manner.

Algorithm 1 shows the pseudo-code to compute the feasible set of locations for a given subtree rooted at e_j and then select the best location in a tree T_i with respect to a transitive graph G. The algorithm maintains the set of feasible node-type locations with respect to the node disjoint constraint in the set $Nlocations$, and employs a depth-first search algorithm with a stack s (Lines 5-18). It explores the successors of a node e_k only when there is no existing link in the transitive graph between the node e_k and any node in L_{e_j}. Notice that the time complexity of the first while loop is quadratic in the number of the nodes of the tree. Also notice that in the first while loop, the algorithm updates the set $Llocations$ by adding all the candidate links starting from the node e_k which will be filtered in the next while loop. Moreover, for each leaf node, when the outgoing degree of node e_k is 0 (i.e. $out(e_k, T_i) = 0$), the node is appended to the queue with the status **check** as the feasible set of link locations follows a bottom-up approach.

In the second while loop, the algorithm filters the set of candidate links to determine the feasible set by employing a bottom-up approach with a queue q. It explores a node e_l only if its predecessor is a valid parent candidate for the node e_j of a given subtree. If there is a link from e_j to e_l in the transitive graph or its status is marked as rem then (e_{p_l}, e_l) is not a feasible link-location and subsequently all the links involved in the path from the facility to e_l would not be feasible link-locations. Therefore, the queue is appended with (e_{p_l}, rem) and the set *Llocations* is filtered. Otherwise, the status of the predecessor of e_l is set to check and appended to the q. Notice that the time complexity of second while loop is linear in the number of links of the tree. Thus, the amortized time complexity of feasibility checking is linear and, consequently, the amortized complexity of *BestLocation* is quadratic.

We remark that computing f_j^i for checking the length constraint can be merged with the verification of the transitive graph with the same worst-case time complexity i.e., linear. However we still incrementally maintain f_j^i and b_j^i for each node in the graph in order to amortize the complexity of the operation to depend only in the number of nodes in the subtree enumerating from e_j.

3   Empirical Evaluation

All experiments were performed on a 4-node cluster, each node features 2 Intel Xeon E5430 processors at 2.66Ghz and 12 GB of RAM memory. We report the median cost solution of 11 independent executions of the LS algorithm with a 1-hour time limit for each experiment. We would like to point out that CPLEX ran out of memory when we tried to create and solve the MIP model for nearly all instances. Hence, we limit our attention to the CBLS algorithm.

To evaluate the performance of the proposed LS we use 3 real networks from three EU countries: Ireland with 1121 ESs, the UK with 5393 ESs, and Italy with 10708 ESs. For each dataset we use [10] to identify the position of the MNs and computed four instances for each country. Ireland with 18, 20, 22, and 24 MNs; the UK with 75, 80, 85, and 90 MNs; and Italy with 140, 150, 160, and 170 MNs. We set λ=90 km. Cable fibre cost is a dominant factor in the deployment of LR-PONs, therefore we report the estimated total cost in euro to deploy edge and node disjoint distribution networks. To this end, we use the subtree operator to optimise the total cable link distance and then we extract the actual fibre cost of the solutions. [11] suggests that on average four fibres will be needed to provide broadband services to all customers for each ES, and the cost of the fibres are defined as follows: {12, €2430}, {24, €2430}, {48, €3145}, {96, €4145}, {144, €5145}, {192, €6145}, {240, €7145}, {276, €7859}, the left part indicates the number of fibres per cable and the respective cost per km.

In Table 1 we report results. As expected the cost of node-disjointness is higher than that of edge-disjointness as the former provides stronger protection. However, it is worth pointing out that the total cable link distance of deployment of a fully node disjoint solution is only up to 13.7% more expensive for Ireland, 16.3% for the UK, and 19.7% for Italy, with respect the best known edge-disjoint solutions. And with a gap between 22% and 31% with respect to the LBs,[2]

---

[2] We recall that the lower bounds are valid for both edge and node disjoint paths. Therefore we expect the node disjoint solution to be much closer to the actual optimal solution.

Table 1. Results for Ireland, UK and Italy to provide Node and Edge disjoint paths.

Country	$|M|$	LB	Cable Link distance				Fibre Cost in Mill €	
			Protection		GAP (%)			
			Node	Edge	Node	Edge	Node	Edge
	18	14809	19824	17107	25.78	13.43	55	49
Ireland	20	14845	19508	16819	24.27	11.73	54	48
$|E|=1121$	22	14990	19076	16711	22.18	10.29	53	48
	24	14570	18452	16163	22.23	9.85	51	46
	75	54720	78131	65377	30.31	16.30	223	197
UK	80	54975	77269	64565	29.20	14.85	220	194
$|E|=5393$	85	55035	75799	63517	28.00	13.35	216	191
	90	55087	74376	62163	25.93	11.38	212	188
	140	76457	111434	89418	31.43	14.49	335	293
Italy	150	76479	109954	88255	30.45	13.34	329	288
$|E|=10708$	160	76794	109708	88336	30.05	13.06	326	286
	170	77013	108288	87405	29.09	11.88	321	282

which is used as the reference baseline for comparison. Finally, we would like to highlight the cost in euro of deployment both the edge and node disjoint alternatives. We foresee that the cost of deployment node disjoint protection in LR-PON in Ireland, the UK, and Italy is respectively 9.9%, 11%, and 12% more expensive (on average) than the corresponding edge disjoint alternative.

4   Conclusions and Future Work

We have extended an existing constraint-based local search framework for finding the Distance-Constrained Node-Disjoint paths between a given pairs of nodes. The efficiency of our approach is demonstrated by experimenting with a set of problem instances provided by network operators in Ireland, the UK, and Italy. We have seen that providing node-disjointness increases the cable link distance cost up to 13% for Ireland, 16% for the UK, and 19% for Italy with respect to the best known solutions with a weaker protection mechanism. Additionally, we provided a cost estimation in euro to deploy both protection solutions. Our experiments indicate that implementing node disjointness is between 9% and 11% more expensive with respect to the best known solutions with weaker protection.

Acknowledgments. We would like to thank both *eircom* and *Telecom Italia* for providing their optical network data. We would also like to thank *atesio* for providing the Italian reference network on which the Italian experiments have been carried out. This work is supported by the EU Seventh Framework Programme (FP7/ 2007-2013) under grant agreement no. 318137 (DISCUS) and by Science Foundation Ireland (SFI) under grant 08/CE/I1423. The Insight Centre for Data Analytics is also supported by SFI under Grant Number SFI/12/RC/2289.

References

1. Benhamiche, A., Mahjoub, R., Perrot, N.: On the design of optical OFDM-based networks. In: Pahl, J., Reiners, T., Voß, S. (eds.) INOC 2011. LNCS, vol. 6701, pp. 1–6. Springer, Heidelberg (2011)
2. Ruffini, M., Wosinska, L., Achouche, M., Chen, J., Doran, N., Farjady, F., Mon-talvo, J., Ossieur, P., O'Sullivan, B., Parsons, N., Pfeiffer, T., Qiu, X.Z., Raack, C., Rohde, H., Schiano, M., Townsend, P., Wessaly, R., Yin, X., Payne, D.B.: Discus: An end-to-end solution for ubiquitous broadband optical access. IEEE Communications Magazine 52(2), 24–32 (2014)
3. Davey, R., Grossman, D., Rasztovits-Wiech, M., Payne, D., Nesset, D., Kelly, A., Rafel, A., Appathurai, S., Yang, S.H.: Long-reach passive optical networks. Journal of Lightwave Technology 27(3), 273–291 (2009)
4. Payne, D.B.: FTTP deployment options and economic challenges. In: Proceedings of the 36th European Conference and Exhibition on Optical Communication (ECOC 2009) (2009)
5. Arbelaez, A., Mehta, D., O'Sullivan, B., Quesada, L.: Constraint-based local search for edge disjoint rooted distance-constrainted minimum spanning tree problem. In: CPAIOR 2015, pp. 31–46 (2015)
6. Pham, Q.D., Deville, Y., Van Hentenryck, P.: LS(Graph): a constraint-based local search for constraint optimization on trees and paths. Constraints 17(4), 357–408 (2012)
7. Mukherjee, B.: Optical WDM Networks (Optical Networks). Springer-Verlag New York Inc., Secaucus (2006)
8. Blesa, M., Blum, C.: Finding edge-disjoint paths in networks: An ant colony optimization algorithm. Journal of Mathematical Modelling and Algorithms 6(3), 361–391 (2007)
9. Ruthmair, M., Raidl, G.R.: Variable neighborhood search and ant colony optimization for the rooted delay-constrained minimum spanning tree problem. In: Schaefer, R., Cotta, C., Kołodziej, J., Rudolph, G. (eds.) PPSN XI. LNCS, vol. 6239, pp. 391–400. Springer, Heidelberg (2010)
10. Ruffini, M., Mehta, D., O'Sullivan, B., Quesada, L., Doyle, L., Payne, D.B.: Deployment strategies for protected Long-Reach PON. Journal of Optical Communications and Networking 4, 118–129 (2012)
11. DISCUS: Project Deliverable 2.6, Architectural optimization for different geotypes

Open Packing for Facade-Layout Synthesis Under a General Purpose Solver

Andrés Felipe Barco[1]([✉]), Jean-Guillaume Fages[2], Elise Vareilles[1],
Michel Aldanondo[1], and Paul Gaborit[1]

[1] Université de Toulouse, Mines d'Albi, Route de Teillet Campus Jarlard,
81013 Albi Cedex 09, France
abarcosa@mines-albi.fr
[2] COSLING S.A.S., 2 Rue Alfred Kastler, 44307 Nantes Cedex 03, France

Abstract. Facade-layout synthesis occurs when renovating buildings to improve their thermal insulation and reduce the impact of heating on the environment. This interesting problem involves to cover a facade with a set of disjoint and configurable insulating panels. Therefore, it can be seen as a constrained rectangle packing problem, but for which the number of rectangles to be used and their size are not known *a priori*. This paper proposes an efficient way of solving this problem using constraint programming. The model is based on an open variant of the DiffN global constraint in order to deal with an unfixed number of rectangles, as well as a simple but efficient search procedure to solve this problem. An empirical evaluation shows the practical impact of every choice in the design of our model. A prototype implemented in the general purpose solver Choco is intended to assist architect decision-making in the context of building thermal retrofit.

1 Introduction

Currently buildings energetic consumption represents more than one third of the total energy consumption in developed countries [4,6,16]. One strategy for reducing such energy consumption lies on buildings thermal retrofit achieved either by an internal or an external insulation [9]. Among several options [9], an external insulation may be based on covering the entire building with an envelope made out of rectangular wood-made panels [7,23]. However, some difficulties are present when targeting such renovation in industrial proportions, e.g. in a country. These difficulties include slow conception using by hand configuration, human scheduling and craft assembly. In consequence, it is essential to assist this massive retrofit of buildings with decision support systems [10].

A crucial aspect of the retrofit automation lies in its facade layout-synthesis. Simply stated, given a rectangular facade surface and an undetermined number

The authors wish to acknowledge the TBC Générateur d'Innovation company, the Millet and SyBois companies and all partners in the CRIBA project, for their involvement in the construction of the CSP model.

G. Pesant (Ed.): CP 2015, LNCS 9255, pp. 508–523, 2015.
DOI: 10.1007/978-3-319-23219-5_36

of rectangular panels, find a solution on how to determine the number of panels, assign size to them and place them over the facade. The family of these problems is called layout synthesis [13] and are by nature combinatorial problems [5,24]. Three characteristics make this problem novel and interesting:

1. Unlike most works [12,13], the number of panels to be allocated in a non-overlapping fashion is *not known a priori*.
2. Some rectangular areas inside facades (existing windows and doors) are meant to be *completely overlapped* by panels. Each area must be covered by one and only one panel in which the corresponding hole will be manufactured once the layout definition is done.
3. Facades have *specific areas* to attach panels that are strong enough to support their weight.

Due to the rectangular geometry of panels and facades, this problem may be treated as a constrained two-dimensional packing problem [8]. By doing this, we may take advantage of the great body of literature on two-dimensional packing while exploiting technological tools, such as general purpose constraint solvers, to tackle the problem. Indeed, Constraint Programming (CP) is, arguably, the most used technology at the crossroads of Artificial Intelligence and Operations Research to address combinatorial problems. CP provides a declarative language and efficient solvers to model and solve complex decision problems where variables are subject to various constraints. It is known to solve efficiently packing problems [3] having, among other abstractions, the geometrical constraint GEOST [2]. However, as we only deal with rectangular shapes, the constraint GEOST [2] seems too complex for our need and would bring an unnecessary risk from a software maintenance point of view. Instead, we use the simpler and well known non overlapping DiffN global constraint [3]. Moreover, we exploit the possibilities, provided by general purpose CP solvers, to implement ad hoc constraints and search procedures that fit the problem structure. Thus, we consider a CP solver as an algorithm integration framework for the development of a decision support application.

Nevertheless, not having a predefined number of rectangles becomes a drawback given that the great majority of constraint programming environments implement global constraints and search engines with a fixed set of variables. In fact, performing filtering and searching using an unfixed number of variables, i.e., a dynamically changing problem, is an active research topic in the constraint programming community. In [1], the author solves the problem of unknown variables by dynamically adding variables while exploring the search tree. In essence, it introduces a setup in which constraints may be deactivated to be replaced with new activated constraints involving more or less variables. Nonetheless, even though the idea seems simple, a good implementation is intricate. Instead, our work is inspired from [22], where the authors introduce *open* global constraints. An open global constraint is an extension of an existing global constraint that includes a set variable (or an array of binary variables) to indicate the subset of decision variables the former constraint holds on. In other words, some decision variables of the problem become optional (see [11,19] and Section 4.4.16 in [18] for further information).

The aim of this paper is to propose a solution to the facade-layout synthesis problem as a two-dimensional packing problem with optional rectangles. We do so by using an open variant of the DiffN constraint [3] to handle rectangles that are *potentially* in the solution. Also, we present a simple yet efficient search heuristic which captures the problem structure. The proposed solutions are implemented using the open-source constraint environment Choco [17]. An empirical evaluation shows the practical impact of every contribution and provides a better understanding of the solving process. The paper is divided as follows. In Section 2 the facade-layout elements are introduced. In Section 3, the constraint-based definition of the problem is presented. In Section 4 we provide technical details of our implementation. In Section 5, a search heuristic that captures the problem structure is presented. Afterwards, in Section 6, we show some performance evaluation of our methods. Finally, some conclusions are discussed in Section 7.

2   Retrofit Industrialization

This work is part of project called CRIBA (for its acronym in French of Construction and Renovation in Industrialized Wood Steel) [7]. This project focuses on the industrialization of energetic renovation for residential buildings. The challenge, very ambitious, is to have a building energetic performance under $25kWh/m^2/year$ after the renovation. The complete renovation (internal and external retrofit) started at the beginning of 2015 with the internal part only.

The industrialization is based on an external new thermal envelope which wraps the whole buildings. The envelope is composed of prefabricated rectangular panels comprising insulation and cladding, and sometimes including in addition, doors, windows and solar modules. As a requirement for the renovation, facades have to be strong enough to support the weight added by the envelope.

Within CRIBA several tools, needed to industrialize the renovation process, will be developed:

a. a new method for three-dimensional building survey and modeling (building information model),
b. a configuration system for the design of the new thermal envelope (topic of this paper), and
c. a working site planning model with resource constraints.

This section introduces the problem of facade layout-synthesis from the industrial point of view.

2.1   Elements

Facades. A facade is represented by a two-dimensional coordinate plane (see Figure 1), with origin of coordinates (0,0) at the bottom-left corner of the facade, and contains rectangular zones defining:

- Perimeter of facade with it size (height and width in meters).
- Frames (existing windows and doors over the facade) play an important role as they are meant to be completely overlapped by one and only one panel. Frames are defined with:
  - Origin point (x,y) with respect to origin of facade.
  - Width and height (in meters).
- Supporting areas. As the layout problem must deal with a perpendicular space plan, gravity must be considered. It turns out that some areas over the facade have load bearing capabilities that allow us to attach panels. Supporting areas have well-defined:
  - Origin point (x,y) with respect to origin of facade.
  - Width and height (in meters).

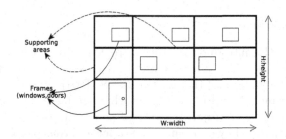

Fig. 1. Facades: Frames and supporting areas.

Rectangular panels. Panels are rectangular (see Figure 2), of varying sizes and may include different equipment (solar modules, window-holes, shutters, etc.). These panels are designed one at a time in the process of layout synthesis and manufactured in the factory prior to shipment and installation on the building site. These panels have a well-defined:

- Size (height and width in meters). Height and width are constrained by a given lower and upper bound provided that is consequence of environmental or manufacturing limitations.
- Thickness and insulation. Thermal performance of a given panel depends on several properties: Size, thickness and insulation type. Consider that the smaller the thickness of the panel the better quality should be the insulation in order to reach performance objectives.
- New frames (such as new doors and new windows). Given internal structure of rectangular panels, new frames must respect a parameterizable minimum distance (Δ) with respect to panel's borders.
- Cost depending mainly on size and attached equipment (in Euros).
- Thermal performance (in watts per meter square-kelvins, $w.m^{-2}.k^{-1}$) depending on size, chosen thickness and insulation type.

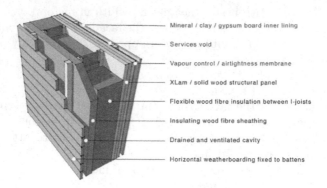

Mineral / clay / gypsum board inner lining

Services void

Vapour control / airtightness membrane

XLam / solid wood structural panel

Flexible wood fibre insulation between I-joists

Insulating wood fibre sheathing

Drained and ventilated cavity

Horizontal weatherboarding fixed to battens

Fig. 2. Rectangular parameterizable panel.

2.2   Problem Features

As mentioned in the introduction, there are three key issues reflected from
the industrial scenario. The unfixed number of panels is the most problematic.
Figure 3 depicts restrictions of the last two issues. Partially overlapping a frame,
as is the case of panel $P1$ in Figure 3.1, is forbidden due to manufacturing limitations. Also, due to the internal structure of panels, frames' border and panels'
border must be separated by a minimum distance and thus panel $P2$ is not valid.
Additionally, in oder to attach panels, the corners of each panel must match a
supporting area, thus, the panel $P3$ is not valid. Figure 3.2 presents a valid
layout plan composed of six panels where all requirements are fulfilled.

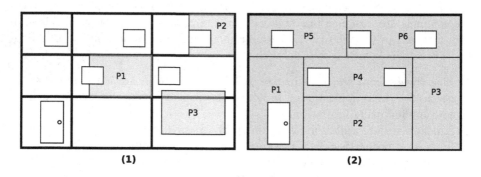

Fig. 3. Ill-configured panels and layout-plan solution.

Frames mandatory overlapping is a complex requirement for the retrofit.
Essentially, a frame (e.g. window or door) should be completely overlapped by
one and only one panel. Figure 4 presents two cases in which panels have conflicts
w.r.t. frames and possible solutions for them.

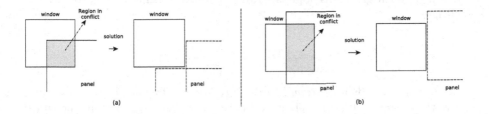

Fig. 4. Frame mandatory overlapping.

2.3   Assumptions

The following assumptions have been taken into account for the present work. First, all supporting areas are strong enough in such a way that the problem only deals with the placement of panel's corners over supporting areas. In other words, there is unlimited load-bearing capabilities in supporting areas and no capabilities in the remaining of the facade area. Second, in order to compute costs and thermal performance, we assume panel's thickness to be a constant and we consider only one insulation type. And third, we assume that costs and thermal performance depends only and are in proportion with panel's size.

2.4   Cost Function

In the industrial scenario the ranking is made w.r.t. cost and thermal performance of the layout plan. The cost of a layout plan depends on the price of isolated panels plus attached equipment. In this work however, we do not take into account attached equipment as it depends on user's preferences and not in the layout plan. Experts have provided us a way to compute the cost of an insulation envelope w.r.t. panels' size. Formula 1 expresses this knowledge.

$$c_{fac} = \sum_{i=1}^{N}(w_i \times h_i) + (\alpha - w_i - h_i) \qquad (1)$$

where w_i and h_i represent the width and height, respectively, of panel i, and α is a factor provided by the manufacturer that depends on the particular manufacturing process. As the formula express it, the term $(\alpha - w_i - h_i)$ decreases with the size of the panel. Thus manufacturing large panels is less costly, globally, than manufacturing small ones.

Now, from a thermal performance point of view having large panels is good because it minimizes the joints between panels. In fact, due to the thermal characteristics of the retrofit, the less panels, the better, because most of the thermal transfer is present in the intersection of two consecutive panels (junctions) plus facade perimeter. Therefore, the performance of a layout plan depends on the length of junctions between two consecutive panels. Computing the length of the junctions for a given envelope is straightforward, formula (2) expresses this knowledge:

$$ttc_{fac} = w_{fac} + h_{fac} + \sum_{i=1}^{N}(w_i + h_i) \tag{2}$$

where ttc_{fac} stands for *thermal transfer coefficient* for facade fac, w_{fac} and h_{fac} are the facade width and height, respectively, w_i and h_i represent the width and height of panel i, respectively, and N is the number of panels composing the envelope. According to formulas 1 and 2, using large panels is appropriated to reduce costs and improve thermal insulation. As the larger the panels the smaller the number of used panels, our optimization function lies on minimizing the number of used panels.

3   Facade-Layout Synthesis as a CSP

In this section we introduce a constraint satisfaction model for the facade-layout synthesis problem. Recall that a constraint satisfaction problem (CSP) is described in terms of a set of variables $\mathcal{V}$, a collection of potential values $\mathcal{D}$ for each variable and a set of relations $\mathcal{C}$ over those variables, referred to as constraints [14,15]. A CSP solution is an assignment of values for each variable in such a way that every constraint in $\mathcal{C}$ is satisfied.

Let F denote the set of frames and S the set of supporting areas. Let $o_{e.d}$ and $l_{e.d}$ denote the origin and length, respectively, of a given entity e in the dimension d, with $d \in \{1, 2\}$. For instance, $o_{fr.1}$ denotes the origin in the horizontal axis and $l_{fr.1}$ denotes the width of frame fr. Additionally, lb_d and ub_d respectively denote the minimum and maximum size, in dimension d, for all rectangles.

3.1   Variables

At first glance, we know that decision variables will be related to rectangles (i.e., panels). One of the major difficulties for tackling the problem using CP is that the number of rectangles to be allocated is unfixed. Therefore, we first heuristically bound this number from above. Let n denote an estimate of the maximum number of rectangles to cover the facade. Given the lower and upper bounds for rectangles' size and the facade size, we consider $n = \left\lceil \frac{fac_1 \times fac_2}{lb_1 \times lb_2} \right\rceil$. We then create a set of n *optional* rectangle variables, each one referring to a panel that may or may not belong to the solution. Each rectangles $0 \le p \le n$ is described by its presence, origin and size attributes:

- $b_p \in \{0, 1\}$ indicates whether or not rectangles p is used in the solution.
- $o_{p.d} \in [0, o_{fac.d}]$ is the origin of rectangles p in dimension d.
- $l_{p.d} \in [lb_{p.d}, ub_{p.d}]$ is the length of rectangles p in dimension d.

Note that, as a rectangle is already defined by an array of integer variables (its coordinates and size), it is more natural to extend it with a fifth binary variable representing its presence in the solution than introducing a set variable to represent all present rectangles [22]. Domains for each variable depends on the variable semantics. For instance, the origin $o_{p.1}$ is in the interval $[0, l_{fac.1}]$. Thus, for dealing with attaching points for rectangles, we use auxiliary variables to represent $o_{p.2} + l_{p.2}$ and assign as domain those points over the facade where supporting areas exists. Thus, these variables are instantiated using a domain with holes, hence avoiding posting a constraint to deal with attaching points.

Finally, we introduce an objective variable z representing the number of rectangles that are used in a solution. We then have $\left\lceil \frac{fac_1 \times fac_2}{ub_1 \times ub_2} \right\rceil \leq z \leq n$.

3.2   Business Constraints

In order to configure the layout of a given facade we use constraints to ensure relations over the variables representing entities. We shall begin the description of four constraints that are related to the problem specifications.

(C1) *Manufacturing and transportation limitations constrain panel's size with a give lower bound* lb *and upper bound* ub *in one or both dimensions:*

$$\forall p, 0 \leq p < n, d \in \{1,2\} \quad lb_d \leq l_{p.d} \leq ub_d$$

(C2) *The entire facade area must be covered with panels:*

$$\sum_{p=0}^{n-1} \left(b_p \times l_{p.1} \times l_{p.2} \right) = l_{fac.1} \times l_{fac.2}$$

It is worth noticing that this constraint will lead to a very weak filtering, because the domain of the l variables may be large and the constraint allows panel overlaps. Therefore, it will be strengthened by the search heuristic.

(C3) *Any two distinct panels that both belong to the solution do not overlap:*

$$\texttt{OpenDiffN}(b, o, l)$$

This corresponds to the *Open* [22] variant of the DiffN [3] constraint, i.e. a generalization of DiffN to handle optional rectangles.

(C4) *Each frame over the facade must be completely overlapped by one and only one panel. Additionally, frame borders and panel borders must be separated by a minimum distance denoted by* Δ:

$$\forall_f \in F, \exists p, 0 \leq p < n, d \in \{1,2\} \mid$$
$$o_{p.d} + \Delta \leq o_{f.d} \wedge o_{f.d} + l_{f.d} \leq o_{p.d} + l_{p.d} + \Delta$$

This constraint is implemented as a dedicated global constraint and will be discussed further.

3.3   Symmetry-Breaking Constraints

As we want to present to the end-user (e.g. an architect) a diverse set good if not optimal solutions, we must avoid to enumerate symmetrical ones. The following constraints break symmetries for rectangles in the solution as well as unused rectangles.

(C5) *Panels are ordered:*

$$\texttt{LexChainLessEq}(\{\{(1 - b_p), o_{p,1}, o_{p,2}\}| \; 0 \le p < n\})$$

This lexicographic constraint [21] ensures that priority is given to use the first rectangles and that rectangles that are used in the solution are ordered geometrically.

(C6) *Unused panels are arbitrarily fixed:*

$$\forall p, 0 \le p < n, \forall d \in \{1, 2\}, \quad b_p = 0 \Rightarrow o_{p,d} = 0 \wedge l_{p,d} = lb_d$$

In order to avoid wasting time on unused rectangles, we may fix their origin variables to the first possible attachment point as well as its size variables to their minimum values. In the (C6) constraint network, it is assumed that there is a valid supporting area at point (0,0).

4   Implementation

This section provides details on the model implementation. Basically, our solution follows the approach in [22] where a set variable is used to handle decision variables that are potentially in the solution. In this work however, as rectangle variables are already composed of several integer attributes, we found more natural to use an extra binary variable per rectangle instead of a set variable. Intuitively, an open constraint with boolean variables may be implemented following traditional filtering algorithms and may be enhanced by targeting the structure of the problem.

4.1   An Open Constraint for Rectangle Non-Overlapping (C3)

As can be seen in the literature, the `OpenDiffN` constraint has already been implemented (see `No-Overlap` with optional rectangles in Section 4.4.16 in [18] for instance) but we consider it is necessary to provide a brief description of its behavior. The filtering algorithm of the `OpenDiffN` checks whether two panels that are part of the solution, i.e., whose b_i is equal 1, do overlap, and proceeds to domain filtering to prevent overlaps, as traditional `DiffN` propagators do. Conversely, if two panels do overlap in space, then domain filtering on the boolean variables ensures that at least one of the two panels is not used. The overall filtering is strengthened by a constructive disjunction algorithm that computes an attaching point for the bottom left corner of each rectangle, that is valid (from the packing point of view only) with respect to already fixed panels.

4.2 A Constraint Dedicated to Frame Covering (C4)

The C4 constraint is propagated using a dedicated approach. The filtering algorithm is pretty simple and works as follows: For every frame, two *support* panels (i.e. panels the frame fits in) are computed. In case no support panel is found then the solver fails. In case only a single panel is found, then a filtering procedure is applied to enforce it to cover the frame. Finally, in case two panels have been found, then no propagation is triggered because we do not know which panel will be used to cover the facade.

4.3 Embedding Symmetry-Breaking

The lexicographic constraint has a strong influence on the model. It enables to output *different* solutions, to reduce the size of the search tree but that is not all: It is possible to speed up the other global constraints by taking that information into account while filtering. For instance, any `for` loop seeking all used rectangles ($b_i = 1$) can stop as soon as one undetermined rectangle ($b_i = \{0,1\}$) has been found because further rectangles are either undetermined or unused. Thus, it is possible to exploit the problem structure to improve the implemented constraints.

We will now see how to exploit the problem structure within search.

5 The Search Heuristic

The search heuristic is responsible of bounding rectangle's decision variables when propagation cannot infer more information. Our heuristic is described in Algorithm 1. It is a constructive approach that considers each rectangle i one by one and uses the following variable selection priority: b_i, o_{i1}, o_{i2}, l_{i1} and finally l_{i2}. We apply a traditional binary branching scheme over stated variables [20]. It means that, instead of iterating over domain values, the heuristic assigns a value to a variable and removes that value from the variable domain upon backtracking.

The originality of our method is that some decisions cannot be negated: Instruction `d.once()` in line 27 tells the solver not to try different values on failure. For instance, if $o_1 = 1$ and the node fails, it will not try to propagate $o_1 \neq 1$ and compute a new decision. Instead, it will backtrack once more (to the decision associated with the size of the previous rectangle).

The heuristic implements the following key design choices:

1. We set the b variables to 1 in order to arrive rapidly to solutions.
2. The position variables o_1 and o_2 are fixed to their lower bounds in order to leave no uncovered places between the considered rectangle and previously placed rectangles. In short, the real decision variables are b, l_1 and l_2. But o_1 and o_2 should be set in a deterministic way without backtracking. As rectangles are ordered, trying a larger value would lead to a hole on the facade, which is forbidden. Note that this is only possible because the filtering

Algorithm 1. Dedicated Search Heuristic

```
1  def Function getBranchingDecision:
2  |   int r ← −1; // compute the first unfixed rectangle;
3  |   for i ← 0 to n do
4  |   |   if |dom(bᵢ)| + |dom(oᵢ₁)| + |dom(oᵢ₂)| + |dom(lᵢ₁)| + |dom(lᵢ₂)| > 5 then
5  |   |   └  r ← i; break;
6  |   if r == −1 then
7  |   |   return null; // all rectangles are fixed (a solution has been found)
8  |   // Find the next variable-value assignment to perform
9  |   IntegerVariable var, int val;
10 |   if |dom(bᵢ)| > 1 then
11 |   |   var ← bᵢ;
12 |   |   val ← 1;
13 |   else if |dom(oᵢ₁)| > 1 then
14 |   |   var ← oᵢ₁;
15 |   |   val ← dom(oᵢ₁).lb;
16 |   else if |dom(oᵢ₂)| > 1 then
17 |   |   var ← oᵢ₂;
18 |   |   val ← dom(oᵢ₂).lb;
19 |   else if |dom(lᵢ₁)| > 1 then
20 |   |   var ← lᵢ₁;
21 |   |   val ← dom(lᵢ₁).ub;
22 |   else
23 |   |   var ← lᵢ₂;
24 |   |   val ← dom(lᵢ₂).ub;
25 |   Branch d = new Branch(var, val);
26 |   if var == oᵢ₁ ∨ var == oᵢ₂ then
27 |   |   d.once(); // prevents the solver from trying different values upon
       |   backtracking
28 |   return d;
```

is strong enough: The lower bound is indeed a valid support from the packing point of view.

3. The size variables l_1 and l_2 are set to their upper bounds in order to try rectangles as large as possible and thus cover the largest possible area. This enables to get a first solution that is very close to the optimal value.

Overall, the search mechanism combines two different principles. First, it is based on a greedy constructive approach that is efficient but limited. Second, it uses a customized (some decisions cannot be negated) backtracking algorithm to explore alternatives.

6   Evaluation

In this section we empirically evaluate our model, which has been implemented in Java, with the Choco solver [17]. We consider a dataset of 20 instances, generated from realistic specifications. The total area of the facade ranges from 10^4 to 10^6 to test the scalability of the approach. Panels are generated using a lower bound of 20 (pixels) and an upper bound of 150 (pixels), for both width and height.

6.1   A Two-Step Approach

In a first experiment we want to evaluate whether or not the maximum number of used panels is a good approximation of the optimum. Figure 5 presents the number of used panels and the number of optional panels for every instance. The maximum number of panels, which represents the worst case scenario where panels' size lower bounds are used, is never reached. We can see that the first solution is actually very far from the optimum. Further, this maximum number is an upper bound far to high: For a facade of size 2300×575, the solver handles 3220 optional panels to compute a first solution that uses only 96 panels. This means that we create too many useless variables that will slow down the solving process. Therefore, we set up a 2-step approach: First a solution is computed using our model. Second, we create a new model with the value of the previous solution as maximum number of panels. Then, we enumerate all optimal solutions.

6.2   Impact of Symmetry Breaking

In a second experiment, we measure the impact of adding symmetry-breaking constraints. More precisely we compare the time to find a first solution (Figure 6.a) and the number of computed solutions (Figure 6.b) with and without symmetry-breaking constraints. Because of the huge amount of solutions, we use a time limit

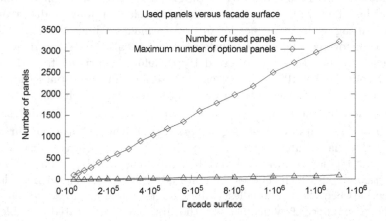

Fig. 5. Maximum number of optional panels and number of used panels in the first solution, for every instance.

of 60 seconds. As we can see it on Figure 6, symmetry-breaking constraints speed up the search. Moreover, it enables to skip thousands of solutions that are identical for the end user. This is even more important than saving computational time. Note that it seems that the number of solutions decreases when the facade area increases: this is due to the time limit. As the problem gets bigger, the solving process gets slower and enumerates less solutions in a given time.

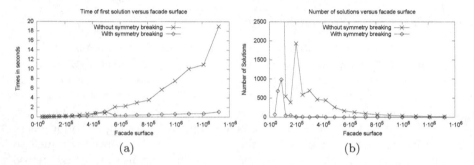

Fig. 6. Time to reach a first solution and number of computed solutions, with a 60 seconds time limit.

6.3   Search Comparison

In regard to different search heuristics, we have not found any well-suited for addressing the problem. Actually, well-known black-box search strategies such as domOverWDeg, impact-based search or activity-based search, do not perform well given the problem particularities. These heuristics are designed to solve problems in a blind way, when we have no expert knowledge of the problem. In our case, we mix very different kind of variables (booleans, positions, sizes) that we are able to group by rectangles and order. Introducing randomness on either the variable selection or the value selection may be disastrous. In particular, using arbitrary values for o_1 and o_2 makes a huge amount of possibilities for uncovered places. Nonetheless, in order to validate the relevance of our dedicated heuristic, we have tested 16 predefined heuristics from Choco on a small facade (400 × 100). We present the results for those ones that threw at least one solution. These strategies are: domOverWDeg which selects the variable with the smallest ratio $\frac{|d(x)|}{w(x)}$, where $|d(x)|$ denotes the domain size of a variable x and $w(x)$ its weighted degree; lexico_LB which chooses the first non-instantiated variable, and assigns it to its lower bound; lexico_Split which chooses the first non-instantiated variable, and removes the second half of its domain; maxReg_LB which chooses the non-instantiated variable with the largest difference between the two smallest values in its domain, and assigns it to its lower bound; minDom_LB which chooses the first non-instantiated variable with the smallest domain size, and assigns it to its lower bound and; minDom_MidValue which chooses the first non-instantiated variable with the smallest domain size, and assigns it to the value closest to the middle of its domain. The last entry is our own search heuristic. Recall that variables are ordered as b, o_1, o_2, l_1 and l_2.

Table 1. Heuristic comparison on a 400×100 (pixels) facade.

Strategy	First solution time (s)	Total time (s)	#nodes	#solutions
domOverWDeg	18.77	19.94	1412897	66
lexico_LB	**0.03**	0.22	2380	66
lexico_Split	**0.03**	**0.16**	**441**	66
maxReg_LB	**0.03**	0.22	2380	66
minDom_LB	0.74	19.96	1411183	66
minDom_MidValue	43.43	47.32	4755206	66
dedicated	**0.03**	0.85	10978	66

Table 2. Heuristic comparison on a 400×200 (pixels) facade with a $3-$minute time limit

Strategy	First solution time (s)	#nodes	#solutions
domOverWDeg	-	7286594	0
lexico_LB	-	5772501	0
lexico_Split	-	4966920	0
maxReg_LB	-	5490088	0
minDom_LB	-	11961712	0
minDom_MidValue	-	11157755	0
dedicated	**0.039**	**3499527**	**726**

Tables 1 and 2 respectively provide the results for a 400×100 and 400×200 instance. Although some predefined heuristics have a good performance on the first (small) instance, none of them scales. In fact, no predefined search heuristic finds a solution for a facade with size 400×200 in reasonable computational time whereas our dedicated heuristic already finds 726 different solutions in 180 seconds. Our heuristic clearly outperforms the others. More importantly, it enables to always output a solution fast, which is critical for the user.

7   Conclusions

This paper presents a solution to the facade-layout synthesis problem treated as a two-dimensional packing problem. Although there exists a great body of literature on two-dimensional packing, our industrial scenario includes three characteristics never considered simultaneously: Its deals with the allocation of an unfixed number of rectangular panels that must not overlap, frames over the facade must be overlapped by one and only one panel, and facades have specific areas to attach panels. Thus, as far as we know, no support system nor design system is well-suited for addressing such particularities.

We have used constraint satisfaction and constraint programming as underlying model and solving technique. Constraint programming technology is well-suited for addressing this industrial problem because on the one hand, an objective function is identified, namely minimize number of panels, and on the other hand, the building of a prototype using an open constraint programming environment is much easier, thanks to all the pre-defined constraints, search and provided abstractions. The modeling decisions was made by a four-person team whereas the development was carried out by a two-person team with knowledge on open constraint programming environments (e.g. Choco, Gecode, finite domain module of Mozart-Oz). The development was done in one month of work.

Considering that the number of panels is not know in advance we have used a variant of the DiffN constraint to handle optional rectangles by means of boolean variables. We have implemented a constraint for the mandatory frame overlapping and a dedicated search heuristic that takes advantage of the problem structure and thus is able to enumerate optimal solutions w.r.t. the number of used panels. Our proposed solutions have been implemented in the Choco solver and demonstrate the validity of our method. In particular, our model takes the benefit of both a greedy and a tree-search approach in order to output *several* good solutions very fast, which was the most critical requirement from the client. This highlights the importance of the great flexibility of Constraint Programming.

Future directions of this work include the definition of different heuristics that take into account aesthetic aspects, the inclusion of weight constraints over panels and supporting areas as well as variables for panel's thickness and isolation type that have an impact on the layout solution. In addition, the packing of different (convex) shapes should be stressed.

References

1. Barták, R.: Dynamic global constraints in backtracking based environments. Annals of Operations Research **118**(1–4), 101–119 (2003)
2. Beldiceanu, N., Carlsson, M., Poder, E., Sadek, R., Truchet, C.: A generic geometrical constraint kernel in space and time for handling polymorphic k-dimensional objects. In: Bessière, C. (ed.) CP 2007. LNCS, vol. 4741, pp. 180–194. Springer, Heidelberg (2007)
3. Beldiceanu, N., Carlsson, M., Demassey, S., Poder, E.: New filtering for the cumulative constraint in the context of non-overlapping rectangles. Annals of Operations Research **184**(1), 27–50 (2011)
4. The Energy Conservation Center: Energy Conservation Handbook. The Energy Conservation Center, Japan (2011)
5. Charman, P.: Solving space planning problems using constraint technology (1993)
6. U.S. Green Building Council: New Construction Reference Guide (2013)
7. Falcon, M., Fontanili, F.: Process modelling of industrialized thermal renovation of apartment buildings. In: eWork and eBusiness in Architecture, Engineering and Construction, pp. 363–368 (2010)
8. Imahori, S., Yagiura, M., Nagamochi, H.: Practical algorithms for two-dimensional packing. Chapter 36 of Handbook of Approximation Algorithms and Metaheuristics (Chapman & Hall/Crc Computer & Information Science Series) (2007)
9. Jelle, B.P.: Traditional, state-of-the-art and future thermal building insulation materials and solutions - properties, requirements and possibilities. Energy and Buildings **43**(10), 2549–2563 (2011)

10. Juan, Y.-K., Gao, P., Wang, J.: A hybrid decision support system for sustainable office building renovation and energy performance improvement. Energy and Buildings **42**(3), 290–297 (2010)
11. Laborie, P.: IBM ILOG CP optimizer for detailed scheduling illustrated on three problems. In: van Hoeve, W.-J., Hooker, J.N. (eds.) CPAIOR 2009. LNCS, vol. 5547, pp. 148–162. Springer, Heidelberg (2009)
12. Lee, K.J., Hyun, W.K., Lee, J.K., Kim, T.H.: Case- and constraint-based project planning for apartment construction. AI Magazine **19**(1), 13–24 (1998)
13. Robin, S.: Liggett. Automated facilities layout: past, present and future. Automation in Construction **9**(2), 197–215 (2000)
14. Montanari, U.: Networks of constraints: Fundamental properties and applications to picture processing. Information Sciences **7**, 95–132 (1974)
15. Olarte, C., Rueda, C., Valencia, F.D.: Models and emerging trends of concurrent constraint programming. Constraints **18**(4), 535–578 (2013)
16. Pérez-Lombard, L., Ortiz, J., Pout, C.: A review on buildings energy consumption information. Energy and Buildings **40**(3), 394–398 (2008)
17. Prud'homme, C., Fages, J.G.: An introduction to Choco 3.0, an open source java constraint programming library. In: International workshop on CP Solvers: Modeling, Applications, Integration, and Standardization, Uppsala, Sweden (2013)
18. Schulte, C., Tack, G., Lagerkvist, M.Z.: Modeling and programming with Gecode (2010)
19. Schutt, A., Feydy, T., Stuckey, P.J.: Scheduling optional tasks with explanation. In: Schulte, C. (ed.) CP 2013. LNCS, vol. 8124, pp. 628–644. Springer, Heidelberg (2013)
20. Smith, B.M.: Modelling for constraint programming (2005)
21. van Hoeve, W.-J., Hooker, J.N. (eds.): CPAIOR 2009. LNCS, vol. 5547. Springer, Heidelberg (2009)
22. van Hoeve, W.-J., Régin, J.-C.: Open constraints in a closed world. In: Beck, J.C., Smith, B.M. (eds.) CPAIOR 2006. LNCS, vol. 3990, pp. 244–257. Springer, Heidelberg (2006)
23. Vareilles, E., Barco Santa, A.F., Falcon, M., Aldanondo, M., Gaborit, P.: Configuration of high performance apartment buildings renovation: a constraint based approach. In: 2013 IEEE International Conference on Industrial Engineering and Engineering Management (IEEM), pp. 684–688, December 2013
24. Zawidzki, M., Tateyama, K., Nishikawa, I.: The constraints satisfaction problem approach in the design of an architectural functional layout. Engineering Optimization **43**(9), 943–966 (2011)

Power Capping in High Performance Computing Systems

Andrea Borghesi[1,2]($\boxtimes$), Francesca Collina[1,2], Michele Lombardi[1,2],
Michela Milano[1,2], and Luca Benini[1,2]

[1] DISI, University of Bologna, Bologna, Italy
[2] DEI, University of Bologna, Bologna, Italy
{andrea.borghesi3,michele.lombardi2,michela.milano,luca.benini}@unibo.it,
francesca.collina@gmail.com

Abstract. Power consumption is a key factor in modern ICT infrastructure, especially in the expanding world of High Performance Computing, Cloud Computing and Big Data. Such consumption is bound to become an even greater issue as supercomputers are envisioned to enter the Exascale by 2020, granted that they obtain an order of magnitude energy efficiency gain. An important component in many strategies devised to decrease energy usage is "power capping", i.e., the possibility to constrain the system power consumption within certain power budget. In this paper we propose two novel approaches for power capped workload dispatching and we demonstrate them on a real-life high-performance machine: the Eurora supercomputer hosted at CINECA computing center in Bologna. Power capping is a feature not included in the commercial Portable Batch System (PBS) dispatcher currently in use on Eurora. The first method is based on a heuristic technique while the second one relies on a hybrid strategy which combines a CP and a heuristic approach. Both systems are evaluated and compared on simulated job traces.

1 Introduction

Supercomputer peak performance is expected to reach the ExaFLOP level in 2018-2020 [14][15], however energy efficiency is a key challenge to be addressed to reach this milestone. Today's most powerful Supercomputer is Tianhe-2 which reaches 33.2 PetaFlops with 17.8 MWatts of power dissipation [13]. Exascale supercomputers built upon today's technology would led to an unsustainable power demand (hundreds of MWatts) while according to [9] an acceptable range for an Exascale supercomputer is 20MWatts; for this goal, current supercomputer systems must obtain significantly higher energy efficiency, with a limit of 50GFlops/W. Today's most efficient supercomputer achieves 5.2 GFlops/W, thus we still need to close an order of magnitude gap to fulfill the Exascale requirements.

Almost all the power consumed by HPC systems is converted into heat. In addition to the power strictly needed for the computation - which measures only the computational efficiency - the cooling infrastructure must be taken into

© Springer International Publishing Switzerland 2015
G. Pesant (Ed.): CP 2015, LNCS 9255, pp. 524–540, 2015.
DOI: 10.1007/978-3-319-23219-5_37

account, with its additional power consumption. The extra infrastructure needed for cooling down the HPC systems has been proved to be a decisively limiting factor for the energy performance [3]; a common approach taken to address this problem is the shift from air cooling to the more efficient liquid cooling [10].

Hardware heterogeneity as well as dynamic power management have started to be investigated to reduce the energy consumption [16][2]. These low-power techniques have been derived from the embedded system domain where they have proven their effectiveness [1]. However, a supercomputer is different from a mobile handset or a desktop machine. It has a different scale, it cannot be decoupled by the cooling infrastructures and its usage mode is peculiar: it is composed by a set of scientific computing applications which run on different datasets with a predicted end time [7]. Finally supercomputers are expensive (6 orders of magnitude more than an embedded device [19]) making it impossible for researchers to have them on their desk. These features have limited the development of ad-hoc power management solutions.

Current supercomputers cooling infrastructures are designed to withstand power consumption at the peak performance point. However, the typical supercomputer workload is far below the 100% resource utilization and also the jobs submitted by different users are subject to different computational requirements[29]. Hence, cooling infrastructures are often over-designed. To reduce overheads induced by cooling over-provisioning several works suggest to optimize job dispatching (resource allocation plus scheduling) exploiting non-uniformity in thermal and power evolutions [8][20][21][23]. Currently most of these works are based on simulations and model assumptions which unfortunately are not mature enough to be implemented on HPC systems in production, yet.

With the goal of increasing the energy efficiency, modern supercomputers adopt complex and hybrid cooling solutions which try to limit the active cooling (chiller, air conditioner) by allowing direct heat exchange with the ambient (Free-cooling). Authors in [12] show for the 2013's top GEEN500 supercomputer that the cooling costs increase four times when the ambient temperature moves from 10 C to 40 C. Moreover the authors show that for a given ambient temperature there is a well-defined maximum power budget which guarantees efficient cooling. With an ambient temperature of 10 C the power budget which maximizes the efficiency is 45KWatt while at 40 C is 15KWatt. Unfortunately, today's ITs infrastructure can control its power consumption only reducing the power and performance of each computing node. This approach is not suitable for HPC system where users require to execute their application in a portion of the machine with guaranteed performance. A solution commonly adopted in HPC systems is *power capping*[25][17][15][27][22], which means forcing a supercomputer not to consume more than a certain amount of power at any given time. In this paper we study a technique to achieve a power capping by acting on the number of job entering the system.

We propose two different methods to enforce power capping constraints on a real supercomputer: 1) a Priority Rules Based algorithm and 2) a novel hybrid approach which combines a CP and a heuristic technique.

2 System Description and Motivations for Using CP

The Eurora Supercomputer: As described in [6] Eurora has a heterogeneous architecture based on nodes (blades). The system has 64 nodes, each with 2 octa-core CPUs and 2 expansion cards configured to host an accelerator module: currently, 32 nodes host 2 powerful NVidia GPUs, while the remaining ones are equipped with 2 Intel MIC accelerators. Every node has 16GB of installed RAM memory. A few nodes (external to the rack) allow the interaction between Eurora and the outside world, in particular a login node connects Eurora to the users and runs the job dispatcher (PBS). A key element of the energy efficiency of the supercomputer is a hot liquid cooling system, i.e. the water inside the system can reach up to 50°C. This obviously allows to save a lot of energy, since no power is used to actively cool down the water; furthermore the heat in excess can be reused as an energy source for other activities.

The PBS Dispatcher: The tool currently used to manage the workload on Eurora system is PBS [28] (Portable Batch System), a proprietary job scheduler by Altair PBS Works which has the task of allocating computational activities, i.e. batch jobs, among available computing resources. The main components of PBS are a server (which manages the jobs) and several daemons running on the execution hosts (i.e. the 64 nodes of Eurora), which track the resource usage and answer to polling requests about the host state issued by the server component.

Jobs are submitted by the users into one of multiple queues, each one characterized by different access requirements and by a different estimated waiting time. Users submit their jobs by specifying 1) the number of required nodes; 2) the number of required cores per node; 3) the number of required GPUs and MICs per node (never both of them at the same time); 4) the amount of required memory per node; 5) the maximum execution time. All processes that exceed their maximum execution time are killed. The main available queues on the Eurora system are called *debug*, *parallel*, and *longpar*. PBS periodically selects a job to be executed considering the current state of the nodes - trying to find enough available resources to start the job. If there are not enough available resources the job is returned to its queue and PBS considers the following candidate. The choices are guided by priority values and hard-coded constraints defined by the Eurora administrators with the aim to have a good machine utilization and small waiting times. For more detailed information regarding the system see [7].

Why CP? In its current state, the PBS system works mostly as an on-line heuristic, incurring the risk to make poor resource assignments due to the lack of an overall plan; furthermore, it does not include a power capping feature yet.

Other than that, as we shown in our previous work the lack of an overall plan may lead to poor resource assignments. The task of obtaining a dispatching plan on Eurora can be naturally framed as a resource allocation and scheduling problem, for which CP has a long track of success stories. Nevertheless since the problem is very complex and we are constrained by very strict time limits (due to the real-time requirements of a supercomputer dispatcher) it is simply not possible to always find a optimal solutions. Therefore, we used CP in combination with multiple heuristic approaches in order to quickly find the resource allocation and schedule needed.

3   Problem Definition

We give now a more detailed definition of the problem. Each job i enters the system at a certain arrival time q_i, by being submitted to a specific queue (depending on the user choices and on the job characteristics). By analyzing existing execution traces coming from PBS, we have determined an estimated waiting time for each queue, which applies to each job it contains: we refer to this value as ewt_i.

When submitting the job, the user has to specify several pieces of information, including the maximum allowed execution time D_i, the maximum number of nodes to be used rn_i, and the required resources (cores, memory, GPUs, MICs). By convention, the PBS system considers each job as if it was divided into a set of exactly rn_i identical "job units", to be mapped each on a single node. Formally, let R be a set of indexes corresponding to the resource types (cores, memory, GPUs, MICs), and let the capacity of a node $k \in K$ for resource $r \in R$ be denoted as $cap_{k,r}$. Let $rq_{i,r}$ be the requirement of a unit of job i for resource r. The dispatching problem at time τ requires to assign a start time $s_i \geq \tau$ to each waiting job i and a node to each of its units. All the resource and power capacity limits should be respected, taking into account the presence of jobs already in execution. Once the problem is solved, only the jobs having $s_i = \tau$ are actually dispatched. The single activities have no deadline or release time (i.e. they do not have to end within or start after a certain date), nor the global makespan is constrained by any upper bound.

Along with the aforementioned resources in this problem we introduce an additional finite resource, the power. This will allow us to model and enforce the power capping constraint. The power capping level, i.e. the maximal amount of power consumed by the system at any given time, is specified by the user and represents the capacity of the fake power resource; the sum of all the power consumed by the running job plus the sum of power values related to idle resources must never exceed the given limit throughout the execution of the whole schedule. Another important thing to notice is that the power "resource" has not a linear behaviour in terms of the computational workload: the first activation of a core in a node causes greater power consumptions for the remaining cores in that node.

The goal is to reduce the waiting times, as a measure of the Quality of Service guaranteed to the supercomputer users.

4   The PRB Approach

First, we implemented a dispatcher which belongs to a class of scheduling techniques known in the literature as *Priority Rules Based* scheduling (PRB)[18]. The main idea underlying this approach is to order a set of tasks which need to be scheduled, constructing the ordered list by assigning priority for each task. Tasks are selected in the order of their priorities and each selected task is assigned to a node; even the resource are ordered and the ones with higher priority are preferred - if available. This is an heuristic technique and it is obviously not able to guarantee an optimal solution but has the great advantage of being extremely fast.

The jobs are ordered w.r.t to their expected wait times, with the "job demand" (job requirements multiplied by the job estimated duration) used to break ties. Therefore, jobs which are expected to wait less have higher priority, subsequently jobs with smaller requirements and shorter durations are preferred over heavier and longer ones. The mapper selects one job at time and maps it on a available node with sufficient resources. The nodes are ordered using two criteria: 1) at first, more energy efficient nodes are preferred (i.e. cores that operate at higher frequencies also consume more power) 2) in case of ties, we favour nodes based on their current load (nodes with fewer free resources are preferred[1]).

The PRB algorithm proceeds iteratively trying to dispatch all the activities that need to be run and terminates only when there are no more jobs to dispatch. We suppose that at time $t = 0$ all the resources are fully available, therefore the PRB algorithm starts by simply trying to fit as many activities as possible on the machine, respecting all resource constraints and considering both jobs and nodes in the order defined by the priority rules. Jobs that cannot start at time 0 are scheduled at the first available time slot. At each time-event the algorithm will try to allocate and start as many waiting jobs as possible and it will keep postponing those whose requirements cannot be met yet.

The algorithm considers and enforces constraints on all the resources of the system, including power.

Algorithm 1 shows the pseudo code of the PRB algorithm. Lines 1-6 initialize the algorithm; J is the set of activities to be scheduled and R is the set of nodes (ordered with the *orderByRules()* function which encapsulates the priority rules). Then the algorithm proceeds while there are still jobs to be scheduled; at every iteration we try to start as many jobs as possible (line 8). Each job unit is considered (line 10) and the availability of resources on every node is taken into account; the function *checkAvailability*(rn_i, r) (line 12) returns true if there are enough available resources on node r to map unit rn_i. If it is possible the unit is then mapped and the system usage is updated (*updateUsages*(rn_i, R), line 13), vice versa we register that at least a job unit could not be mapped (line 16). If all the units of a job have been mapped (line 17-21), then the job can actually start and is removed from the pool of jobs that still need to be scheduled. Conversely,

---

[1] This criterion should decrease the fragmentation of the system, trying to fit as many job as possible on the same node.

Algorithm 1. PRB algorithm

```
 1  time ← 0;
 2  startTimes ⟵ ∅;
 3  endTimes ⟵ ∅;
 4  runningJobs ⟵ ∅;
 5  orderByRules(R);
 6  orderByRules(J);
 7  while J ≠ ∅ do
 8      for j ∈ J do
 9          canBeMapped ← true;
10          for rnᵢ ∈ j do
11              for r ∈ R do
12                  if checkAvailability(rnᵢ, r) then
13                      updateUsages(rnᵢ, R);
14                      break;
15                  else
16                      canBeMapped ← false
17          if canBeMapped = true then
18              J ⟵ J − {j};
19              runningJobs = runningJobs ∪ {j};
20              startTimes(j) ← time;
21              endTimes(j) ← time + duration(j);
22          else
23              undoUpdates(j, R);
24      orderByRules(R);
25      orderByRules(J);
26      time ← min(endTimes);
```

if the job cannot start we must undo the possible changes made to the system usage we made in advance ($updateUsages(rn_i, R)$, line 23). Finally, after having dealt with all schedulable jobs we reorder the activities and nodes (the activity pool is changed and the nodes order depends on the usages), lines 24-25, and compute the closest time point where some used resource becomes free, following time-event, i.e. the minimal end time of the running activities (line 26).

It is important to note that since in this problem we have no deadline on the single activities nor a constraint on the global makespan, the PRB algorithm will always find a feasible solution, for example delaying the least important jobs until enough resources become available due to the completion of previously started tasks.

5   Hybrid Approach

As previously discussed in [7] the task of obtaining a proactive dispatching plan on Eurora can be naturally framed as a resource allocation and scheduling problem. Constraint Programming (CP) techniques have shown great success when dealing with this kind of problem, thanks to the expressive and flexible language used to model the problem and powerful algorithms to quickly find good quality solutions[5]. In this Section we describe the mixed approach we used to solve the resource allocation and scheduling problem under power capping constraints.

The key idea of our method is to decompose the allocation and scheduling problem in two stages: 1) obtain a schedule using a relaxed CP model of the problem 2) find a feasible mapping using a heuristic technique. Since we used a relaxed model in the first stage, the schedule obtained may contain some inconsistencies; these are fixed during the mapping phase, thus we eventually obtain a feasible solution, i.e. a feasible allocation and schedule for all the jobs. This two stages are repeated n times, where n has been empirically chosen after an exploratory analysis, keeping in mind the trade-off between the quality of the solution and the computational time required to find one. To make this interaction effective, we devised a feedback mechanism between the second and the first stage, i.e. from the infeasibilities found during the allocation phase we learn new constraints that will guide the search of new scheduling solutions at following iterations.

In this work we implemented the power capping requirements as an additional constraint: on top of the finite resources available in the system such as CPUs or memory, we treat the power as an additional resource with its own capacity (i.e. the user-specified power cap), which we cannot "over-consume" at any moment[11]. In this way, the power used in the active nodes (i.e. those on which a job is running) summed to the power consumed by the idle nodes will never exceed the given threshold.

The next sections will describe in more detail the two stages of the decomposed approach.

5.1 Scheduling Problem

The scheduling problem consists in deciding the start times of a set of activities $i \in I$ satisfying the finite resource constraints and the power capping constraint. Since all the job-units belonging to the same jobs must start at the same time, during the scheduling phase we can overlook the different units since we need only the start time for each job. Whereas in the actual problem the resources are split among several nodes, the relaxed version we use in our two-stages approach considers all the resources of the same type (cores, memory, GPUs, MICs) as a pool of resource with a capacity Cap_r^T which is the sum of all the single resource capacities, $Cap_r^T = \sum_{k \in K} cap_{k,r} \quad \forall r \in R$. As mentioned before the power is considered as an additional resource type of the system, so we have a set of indexes R' corresponding to the resource types (cores, memory, GPUs, MICs plus the power); the overall capacity Cap_{power}^T is equal to the user-defined power cap.

The CP model employs other relaxations: 1) only the power required by running jobs (active power) is considered, i.e. we use a lower bound of the total power consumed in the system, 2) we assume that the jobs always run on the nodes which require less power and 3) we overlook the non-linear behaviour of the power consumption. These relaxations may produce infeasible solutions, since a feasible schedule must take into account that the resource are actually split among heterogeneous nodes and consider also the idle nodes for the power capping. The following stage of our method will naturally fix these possible infeasibilities during the allocation phase.

We define the scheduling model using Conditional Intervals Variables (CVI)[24]. A CVI τ represents an interval of time: the start of the interval is referred to as $s(\tau)$ and its end as $e(\tau)$; the duration is $d(\tau)$. The interval may or may not be present, depending on the value of its existence expression $x(\tau)$ (if not present it does not affect the model). CVIs can be subject to several different constraints, among them the `cumulative` constraint[4] to model finite capacity resources.

$$\forall r \in R' \quad cumulative(\tau, req_r, Cap_r^T) \tag{1}$$

where τ is the vector with all the interval vars, where req_r are the job requirements for resource r - using past execution traces we learned a simple model to estimate the power consumed by a job based on its requirements. As mentioned in section 3 this model is not linear. The cumulative constraints in 1 enforce that at any given time, the sum of all job requirements will not exceed the available capacity (for every resource type).

With this model it would be easy to define several different goals, depending on the metric we optimize. Currently we use as objective function the *weighted queue time*, i.e. we want to minimize the sum of the waiting times of all the jobs, weighted on estimated waiting time for each job (greater weights to job which should not wait long):

$$\min \sum_{i \in I} \frac{\max ewt_i}{ewt_i}(s(\tau_i) - q_i) \tag{2}$$

To solve the scheduling model we implemented a custom search strategy derived from the Schedule Or Postpone strategy [26]. The criteria used to select an activity among all the available ones at each decision node follows the priority rules used in the heuristic algorithm, thus preferring jobs that can start first and whose resource demand is lower. This strategy proved to be very effective and able to rapidly find good solutions w.r.t. the objective function we are considering in this problem. With different goals we should change the search strategy accordingly (as with the priority rules).

5.2 Allocation Problem

The allocation problem consists in mapping each job unit on a node. Furthermore, in our approach the allocation stage is also in charge of fixing the infeasibilities generated at the previous stage. In order to solve this problem we developed an algorithm which falls in the PRB category. Typical PRB schedulers decide both mapping and start times, whereas in our hybrid approach we need only to allocate the jobs.

The behaviour of this algorithm (also referred as *mapper*) is very close to the PRB one described in Section 4 and in particular the rules used to order jobs and resources are identical. The key difference is that now we already know the start and end times of the activities (at least the possible ones, they may change if any infeasibility is detected). This algorithm proceeds by time-step: at each time event t it considers only the jobs that should start at time t according to the relaxed CP model described previously, while the simple PRB algorithm considers at each time-event all the activities that still need to be scheduled.

During this phase the power is also taken into account, again seen as a finite resource with capacity defined by the power cap; here we consider both active and idle nodes powers. If the job can be mapped somewhere in the system the start time t from the previous stage is kept, otherwise - if there are not enough resources available to satisfy the requirements - the previously computed start time is discarded and the job will become eligible to be scheduled at the next time-step t'. At the next time event t' all the jobs that should start are considered, plus the jobs that possibly have been postponed due to scarce resources at the previous time-step. Through this postponing we are fixing the infeasibilities inherited from the relaxed CP model.

Again, since in this problem we have no constraints on the total duration of a schedule, it is always possible to delay a job until the system will have enough available resources to run it, thus this method is guaranteed to find a feasible solution.

5.3   Interaction between the Stages

We designed a basic interaction mechanism between the two stages with the goal to find better quality solutions. The main idea is to exploit the information regarding the infeasibilities found during the allocation phase to lead the search of new solutions for the relaxed scheduling problems. In particular whenever we detect a resource over-usage at time tau which requires a job to be postponed during the second stage of our model we know that the set of job running at time τ is a *Conflict Set* (not minimal), i.e. not all activities in the set can run concurrently. A possible solution for a conflict set is for example to introduce precedence relationships among activities (thus eliminating jobs from the conflict set) until the infeasibility is resolved.

In our approach we use the conflict set detected in the mapping phase to generate a new set of constraints which impose that not all jobs in the set will run at the same time. We then introduce in the CP model a fake cumulative constraint for each conflict set. The jobs included in such cumulative constraint are those included in the conflict set, each of them with a "resource" demand of one; the capacity not to be exceeded is given by the conflict set size minus one. These cumulative constraints enforce that the jobs involved will not run at the same time. This mechanism does not guarantee yet that the new solution found by the CP scheduler will be feasible since the conflict sets we detect are not minimal, nevertheless it provides a way to help the CP model to produce solutions which will require less "fixing" by the mapper.

In conjunction with the additional cumulative constraint at each iteration we also cast a further constraint on the objective variable in order to force the new solution to improve in comparison to the previous one.

6   Added Value of CP

The dispatcher we realized is currently a prototype: it will eventually be deployed on the Eurora supercomputer, but this requires still considerable development and research effort (at the same time a previous version without power capping of this model[7] has already successfully been implemented). At this stage we are interested in investigating the kind of impact that introducing a power capping feature may have on the dispatcher behaviour and performance.

The dispatcher we realized can work in two different modes: *off-line*, i.e. the resource allocation and the schedule are computed before the actual execution, and *on-line*, i.e. allocation and scheduling decisions are taken at run-time, upon the arrival of new tasks in the system. Clearly the actual implementation on the supercomputer would require the on-line strategy since the workload is submitted by users and not statically decided a priori. At the same time we decided to use the off-line strategy to perform the experiments described in this paper since we wanted to test our approaches with reproducible conditions. We also needed our techniques to be stable in order to implement them on a production system and the off-line strategy allows us to better verify that.

The proposed methods were implemented using or-tools[2], Google's software suite for combinatorial optimization. We performed an evaluation of all our approaches on PBS execution traces collected from Eurora in a timespan of several months. From the whole set of traces we extracted different batches of jobs submitted at various times and we used them to generate several job instances of different size, i.e. the number of jobs per instance (a couple of hundreds of instances for each size). Since in this experimental evaluation we were concerned only in the off-line approach the real enter queue times were disregarded and in our instances we assume that all jobs enter the system at the same time. The main performance metric considered is the time spent by the jobs in the queues while waiting their execution to begin (ideally as low as possible).

6.1   Evaluation of Our Models

We decided to make our experiments using two artificial versions of the Eurora machine: A) a machine composed with 8 nodes and B) a machine with 32 nodes. In addition to the real traces (set *base*) we generated two more sets of instances, one which is composed by especially computationally intensive jobs, in terms of resource requested (*highLoad*), and one which presents jobs composed by

---

[2] https://developers.google.com/optimization/

an unusually high number of job-units (*manyUnits*). These additional groups
were generated using selected subsets of the original traces. From these sets we
randomly selected subsets of smaller instances with dimension of 25, 40, 50, 60,
80, 100, 150 and 200 jobs; for each size we used 50 different instances in our
tests. The instances of dimension 25 and 50 were run on the smaller machine A
while the remaining instances executed on machine B.

On each instance we ran the PRB algorithm (*PRB*), the hybrid approach
with no feedback iteration (*DEC_noFeedBack*) and the hybrid approach with
feedback (*DEC_feedBack*). We tested the hybrid approach both with and without the interaction between the two layers because the method without feedback
is much faster than the one with the interaction, therefore better suited to a real-
time application as a HPC dispatcher. The CP component of the decomposed
method has a timeout which forces the termination of the search phase. The
timeout is set to 5 seconds; if the solver finds no solution within the time limit,
the search is restarted with a increased timeout (we multiply by 2 the previous
timeout), until we reach a maximum value of 60 seconds - which is actually never
reached in our experiments. The *PRB* is extremely fast as finding a solution
takes only a fraction of second even with the larger instances; *DEC_noFeedBack*
requires up to 3-4 seconds with the larger instances (but usually a lower quality solution is found in less than a second) and the *DEC_noFeedBack* requires
significant larger times to compute a solution due to the multiple iterations, in
particular up to 15-20 seconds with the instances of 200 jobs.

Each experiment was repeated with different values of power capping to
explore how the bound on the power influences the behaviour of the developed
dispatcher.

At every run we measured the average weighted queue time of the solution,
that is the average time spent waiting by the jobs, weighted with the expected
wait time (given by the queue of the job). As a unit of measure we use the
Expected Wait Time, (EWT), i.e. the ratio between the real wait time and the
expected one. A EWT value of 1 tells us that a job has waited exactly as long
as it was expecting; values smaller than 1 indicate that the job started before
the expected start time and values larger than one that the job started later
than expected. To evaluate the performance of the different approaches we then
compute the ratio between the average weight queue time obtained by *PRB* and
by the two hybrid methods; finally we plot these ratios in the figures presented
in the rest of the section. Since the best average queue times are the lowest
ones, it is clear that if the value of the ratio goes below one the *PRB* approach
is performing better, while when the the value is above one then the hybrid
approaches are obtaining better results.

The following figures will show the ratios in the y-axis while the x-axis will
specify the power capping level; we only show significant power capping levels,
i.e. the one larger enough to allow a solution to the problem, hence the x-scale
range may vary.

Machine with 8 nodes: Figures 1, 2 and 3 show the results of the experiments
with the machine with 8 nodes; each figure corresponds respectively to the *base*

workload set of instances (both size 25 and 50 jobs), the *highLoad* case and finally the *manyUnits* case. The solid lines represent the ratios between the average queue times obtained by *PRB* and those obtained by *DEC_feedBack*; conversely, the dashed line is the ratio between *PRB* and *DEC_noFeedBack*. As we can see in Figure 1 with an average workload the hybrid approaches usually outperform the heuristic algorithm, markedly in the 25 jobs case and especially at tighter power capping. With less tight power capping levels usually the hybrid approaches and *PRB* offer more similar results; this is reasonable since when the power constraints are more relaxed the allocation and scheduling decisions are more straightforward and the more advanced reasoning offered by CP is less necessary.

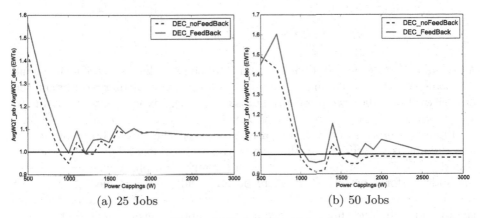

(a) 25 Jobs   (b) 50 Jobs

Fig. 1. 8 Node - *Base* Set

It is easy also to see that the hybrid method with the feedback mechanism always outperforms the feedback-less as we expected: the feedback-less solution has always inferior quality w.r.t. to those generated by the method with the interaction - the solution produced by *DEC_feedBack* is the same produced by *DEC_noFeedBack*. Our focus should be on the extent of the improvement guaranteed by the feedback mechanism in relation to the longer time required to reach a solution. For example, Fig.1 (corresponding to the original Eurora workloads) shows that the feedback method offers clear advantages, in particular with tighter power constraints (around 10%-15% gain over the feedback-less one), a fact that could justify its use despite the longer times required to reach a solution. Conversely Figure 2 reveals that the two hybrid approaches offer very similar performance if the workload is very intensive, leading us to prefer the faster *DEC_noFeedBack* in these circumstances in case of the implementation of the real dispatcher.

If we consider the workload characterized by an unusually high amount of job-units per job we can see slightly different results: as displayed in Figure 3, *DEC_feedBack* definitely outperforms the other two methods with tight power capping values, especially in the case of instances of 25 jobs. When the power constraints get more relaxed the three approaches offers almost the same results.

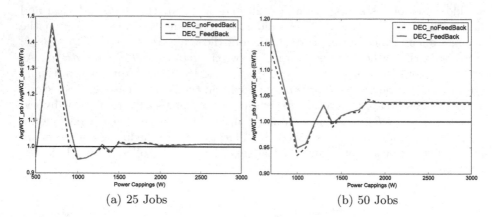

(a) 25 Jobs   (b) 50 Jobs

Fig. 2. 8 Node - *HighLoad* Set

Machine with 32 nodes: In Figures 4 and 5 we can see the results of some of the experiments done on the machine with 32 nodes (in particular we present the case size 40 and 100); we present only a subset of the experiments made due to space limitations, but the results not shown comply with those presented here.

Figure 4 shows again the comparison between the two hybrid approaches and the heuristic technique in the case of average workload. The pattern here is slightly different from the 8-nodes case: after an initial phase where the hybrid methods perform better, *PRB* offers better results for intermediate levels of power capping (around 3000W); after that we can see a new gain offered by the hybrid techniques until we reach a power capping level around 6000W (the power constraint relaxes), where the three approaches provide more or less the same results. It is evident again that the method with feedback outperforms the feedback-less one, especially with the larger instances.

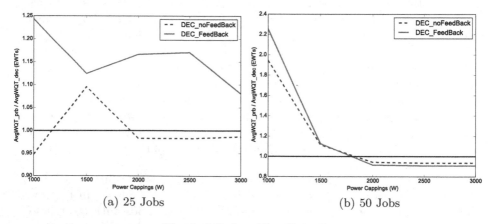

(a) 25 Jobs   (b) 50 Jobs

Fig. 3. 8 Node - *ManyUnits* Set

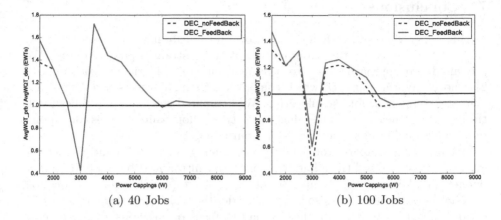

(a) 40 Jobs   (b) 100 Jobs

Fig. 4. 32 Node - *Base* Set

In Figure 5 we can see the results obtained with the computationally more intensive workload. In this case *PRB* performs generally better at lower power capping levels until the power constraint become less tight and the three methods produce similar outcomes again.

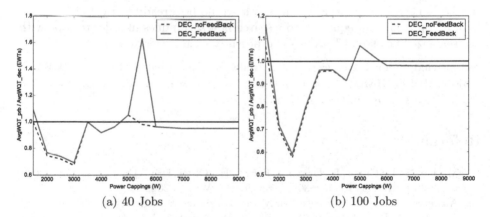

(a) 40 Jobs   (b) 100 Jobs

Fig. 5. 32 Node - *Highload* Set

We have performed a number of additional experiments, not shown here, and we could summarize that with the machine with 32 nodes with average workloads the hybrid approaches perform definitely better than the heuristic technique for most power capping levels except a small range of intermediate values (up to a 60% reduction of average queue times). On the contrary, if we force very intensive workloads we obtain better outcomes with *PRB*, even though with usually smaller, but still significant, differences (around 10%). We can also see that with these intensive workloads the difference between the two hybrid approaches is minimal and this suggest that the basic feedback mechanism needs to be refined.

7   Conclusions

In this paper we dealt with the allocation and scheduling problem on Eurora HCP system subject to the power consumption constraint - power capping. We presented two approaches to solve this problem: 1) a heuristic algorithm and 2) a novel, hybrid approach which combines a CP model for scheduling and a heuristic component for the allocation. We compared the two approaches using the average queue times as an evaluation metrics. Short waiting times correspond to a higher quality of service for the system users.

We tackled a complex problem due to the limited amount of multiple, heterogeneous, resources and the additional constraint introduced by the finite power budget. In addition to the complexity of the problem (scheduling and allocation are NP-Hard problems) we also had to consider the real-time nature of the application we wanted to develop, thus we had to focus on methods able to quickly produce good solutions. In this preliminary study we shown that the quality of the solution found by the different approaches varies with the levels of power capping considered.

As a long-term goal we plan to further develop the preliminary power capping feature presented here in order to integrate it within the actual dispatcher we already developed and implemented on Euora. In order to do that we will need to develop methods to allow our approach to operate quickly enough to match the frequency of job arrivals; the heuristic would be already fast enough but the hybrid approach will require us to research new techniques in order to cope with the real-sized version of Eurora - and possibly even larger systems.

Acknowledgement. This work was partially supported by the FP7 ERC Advance project MULTITHERMAN (g.a. 291125). We also want to thank CINECA and Eurotech for granting us the access to their systems.

References

1. Mudge, N.: In: Culler, P., Druschel, D.E. (eds.) OSDI. USENIX Association (2002). Operating Systems Review 36(Special Issue), Winter 2002
2. Auweter, A., et al.: A case study of energy aware scheduling on supermuc. In: Kunkel, J.M., Ludwig, T., Meuer, H.W. (eds.) ISC 2014. LNCS, vol. 8488, pp. 394–409. Springer, Heidelberg (2014)
3. Banerjee, A., Mukherjee, T., Varsamopoulos, G., Gupta, S.K.S.: Cooling-aware and thermal-aware workload placement for green hpc data centers. In: Green Computing Conference, pp. 245–256 (2010)
4. Baptiste, P., Laborie, P., Le Pape, C., Nuijten, W.: Constraint-based scheduling and planning. Foundations of Artificial Intelligence **2**, 761–799 (2006)
5. Baptiste, P., Le Pape, C., Nuijten, W.: Constraint-based scheduling. Kluwer Academic Publishers (2001)
6. Bartolini, A., Cacciari, M., Cavazzoni, C., Tecchiolli, G., Benini, L.: Unveiling eurora - thermal and power characterization of the most energy-efficient supercomputer in the world. In: Design, Automation Test in Europe Conference Exhibition (DATE), 2014, March 2014

7. Bartolini, A., Borghesi, A., Bridi, T., Lombardi, M., Milano, M.: Proactive work-load dispatching on the EURORA supercomputer. In: O'Sullivan, B. (ed.) CP 2014. LNCS, vol. 8656, pp. 765–780. Springer, Heidelberg (2014)
8. Bartolini, A., Cacciari, M., Tilli, A., Benini, L.: Thermal and energy management of high-performance multicores: Distributed and self-calibrating model-predictive controller. IEEE Trans. Parallel Distrib. Syst. **24**(1), 170–183 (2013)
9. Bergman, K., Borkar, S., Campbell, D., Carlson, W., Dally, W., Denneau, M., Franzon, P., Harrod, W., Hiller, J., Karp, S., Keckler, S., Klein, D., Lucas, R., Richards, M., Scarpelli, A., Scott, S., Snavely, A., Sterling, T., Williams, R.S., Yelick, K., Bergman, K., Borkar, S., Campbell, D., Carlson, W., Dally, W., Denneau, M., Franzon, P., Harrod, W., Hiller, J., Keckler, S., Klein, D., Kogge, P., Williams, R.S., Yelick, K.: Exascale computing study: Technology challenges in achieving exascale systems, September 2008
10. Chu, R.C., Simons, R.E., Ellsworth, M.J., Schmidt, R.R., Cozzolino, V.: Review of cooling technologies for computer products. IEEE Transactions on Device and Materials Reliability **4**(4), 568–585 (2004)
11. Collina, F.: Tecniche di workload dispatching sotto vincoli di potenza. Master's thesis, Alma Mater Studiorum Università di Bologna (2014)
12. Conficoni, C., Bartolini, A., Tilli, A., Tecchiolli, G., Benini, L.: Energy-aware cooling for hot-water cooled supercomputers. Proceedings of the 2015 Design. Automation & Test in Europe Conference & Exhibition, DATE 2015, pp. 1353–1358. EDA Consortium, San Jose (2015)
13. Dongarra, J.J.: Visit to the national university for defense technology changsha, china. Technical report, University of Tennessee, June 2013
14. Dongarra, J.J., Meuer, H.W., Strohmaier, E.: 29th top500 Supercomputer Sites. Technical report, Top500.org, November 1994
15. Fan, X., Weber, W.-D., Barroso, L.A.: Power provisioning for a warehouse-sized computer. In: ACM SIGARCH Computer Architecture News, vol. 35, pp. 13–23. ACM (2007)
16. Feng, W.-C., Cameron, K.: The Green500 List: Encouraging Sustainable Super-computing. IEEE Computer **40**(12) (2007)
17. Gandhi, A., Harchol-Balter, M., Das, R., Kephart, J.O., Lefurgy, C.: Power capping via forced idleness (2009)
18. Haupt, R.: A survey of priority rule-based scheduling. Operations-Research-Spektrum **11**(1), 3–16 (1989)
19. Kim, J.M., Chung, S.W., Seo, S.K.: Looking into heterogeneity: When simple is faster
20. Kim, J., Ruggiero, M., Atienza, D.: Free cooling-aware dynamic power management for green datacenters. In: 2012 International Conference on High Performance Computing and Simulation (HPCS), pp. 140–146, July 2012
21. Kim, J., Ruggiero, M., Atienza, D., Lederberger, M.: Correlation-aware virtual machine allocation for energy-efficient datacenters. In: Proceedings of the Conference on Design. Automation and Test in Europe, DATE 2013, pp. 1345–1350. EDA Consortium, San Jose (2013)
22. Kontorinis, V., Zhang, L.E., Aksanli, B., Sampson, J., Homayoun, H., Pettis, E., Tullsen, D.M., Rosing, T.S.: Managing distributed ups energy for effective power capping in data centers. In: 2012 39th Annual International Symposium on Computer Architecture (ISCA), pp. 488–499. IEEE (2012)

23. Kudithipudi, D., Qu, Q., Coskun, A.K.: Thermal management in many core systems. In: Khan, S.U., Koodziej, J., Li, J., Zomaya, A.Y. (eds.) Evolutionary Based Solutions for Green Computing. SCI, vol. 432, pp. 161–185. Springer, Heidelberg (2013)
24. Laborie, P., Rogerie, J.: Reasoning with conditional time-intervals. In: Proc. of FLAIRS, pp. 555–560 (2008)
25. Lefurgy, C., Wang, X., Ware, M.: Power capping: a prelude to power shifting. Cluster Computing 11(2), 183–195 (2008)
26. Pape, C.L., Couronné, P., Vergamini, D., Gosselin, V.: Time-versus-capacity compromises in project scheduling. In Proc. of the 13th Workshop of the UK Planning Special Interest Group, pp. 1–13 (1994)
27. Reda, S., Cochran, R., Coskun, A.K.: Adaptive power capping for servers with multithreaded workloads. IEEE Micro 32(5), 64–75 (2012)
28. Altair PBS Works: Pbs professional12.2 administrator's guide (2013). http://resources.altair.com/pbs/documentation/support/PBSProAdminGuide12.2.pdf
29. You, H., Zhang, H.: Comprehensive workload analysis and modeling of a petascale supercomputer. In: Cirne, W., Desai, N., Frachtenberg, E., Schwiegelshohn, U. (eds.) JSSPP 2012. Lecture Notes in Computer Science, vol. 7698, pp. 253–271. Springer, Heidelberg (2013)

A Constraint-Based Approach to the Differential Harvest Problem

Nicolas Briot[1]([✉]), Christian Bessiere[1], and Philippe Vismara[1,2]

[1] LIRMM, CNRS–University Montpellier, Montpellier, France
{briot,bessiere,vismara}@lirmm.fr
[2] MISTEA, Montpellier SupAgro, INRA, 2 Place Viala, 34060 Montpellier, France

Abstract. In this paper, we study the problem of differential harvest in precision viticulture. Some recent prototypes of grape harvesting machines are supplied with two hoppers and are able to sort two types of grape quality. Given estimated qualities and quantities on the different areas of the vineyard, the problem is to optimize the routing of the grape harvester under several constraints. The main constraints are the amount of first quality grapes to harvest and the capacity of the hoppers. We model the differential harvest problem as a constraint optimization problem. We present preliminary results on real data. We also compare our constraint model to an integer linear programming approach and discuss expressiveness and efficiency.

1 Introduction

In precision viticulture, many studies have proposed to define field quality zones [14]. They demonstrated the technical and economic value of a differential harvesting of these different zones. This interest justifies the recent development of prototypes of conventional grape harvesting machines able to sort two types of harvest quality, such as the EnoControlTM system prototype (*newHolland Agriculture*, PA, USA). These grape harvesting machines have two tanks, called hoppers, able to differentiate two types of grape quality, named A and B, according to the harvested zone.

Optimizing harvest consists on minimizing the working time of grape harvester. This time corresponds to travel time and emptying time of the machine. Ideally, this goal requires that both hoppers of the machine are full at each emptying. In the case of selective harvesting, the simultaneous filling of the two hoppers is combinatorial and complex. Indeed, the hopper that contains A grapes can fill up at a speed different from the hopper that contains B grapes, depending on the harvested zone. Other issues have to be considered. Top quality grapes should not be altered (mixed with lower quality grapes) when the harvester moves from one quality zone to another. Turning radius of the machine must also be taken into account.

This problem, called *Differential Harvest Problem*, has not been studied in the literature except in a preliminary study we published in [4]. A comparison with routing problems can be established. For instance, Bochtis *et al.* [1,2] show

© Springer International Publishing Switzerland 2015
G. Pesant (Ed.): CP 2015, LNCS 9255, pp. 541–556, 2015.
DOI: 10.1007/978-3-319-23219-5_38

various formulations for many related problems in agriculture, such as precision spraying, fertilizing in arable farming, etc. In [9], Kilby and Shaw give a Constraint Programming (CP) formulation for the *Vehicle Routing Problem* (VRP) and detail other exact methods and heuristics to solve the problem. *Balancing Bike Sharing System* (BBSS) is an example of real application solved with VRP formulation and CP approach. Di Gaspero *et al.* [7] show two novel CP models for BBSS and a Large Neighborhood Search approach is proposed.

In this paper, we explore how to solve the Differential Harvest Problem with a CP approach. We consider two models: a Step model we introduced in our previous work [4] and a new routing model that we call the Precedence model. We also compare these two models to an Integer Linear Programming approach.

The paper is organized as follows. Section 2 describes the problem of differential harvest in viticulture. Section 3 proposes the two CP models. We present preliminary results on real data in Section 4. We compare our CP model to an Integer Linear Programming approach and discuss expressiveness and efficiency in Section 5.

2   The Differential Harvest Problem

Precision viticulture is a new farming management approach that focuses on the intra-field variability in a vineyard. Thanks to high resolution satellite images and infield sensors, it is possible to predict the variability of quality and yield within a vine field. The Differential Harvest Problem (DHP) is defined on a vine field with two qualities of grapes. The first one will be denoted A grapes and corresponds to high quality grapes. These grapes will be used to produce high quality wine. The rest of the harvest will be put in the B-grapes class. Note that high quality grapes (A grapes) can be downgraded to the low quality B. But the converse is not true.

We consider a harvesting machine with two hoppers. We denote by $Cmax$ the maximum capacity of a hopper. With such a machine, two categories of grapes can be separated. Generally, one hopper (called A hopper) receives only A grapes whereas the other one (called $B\&A$ hopper) can contain either B grapes only or both A grapes and B grapes. It is important not to contaminate the grapes in A hopper with B grapes. When the hoppers are full, the machine can dump them into a bin located around the plot before continuing its work. Generally, the bin is immediately brought to the winery after each dump of the hoppers in order to preserve grapes quality. A second bin waits in stand-by to avoid delaying the harvester.

The harvesting machine takes some time between picking grapes and putting them into hoppers. This time (≈ 10 seconds), required to empty the conveyor belt, is called *latency*. When the harvesting machine passes over two zones, the quality of grapes harvested is not guaranteed during this time because latency may lead to mixing both qualities of grapes. Hence, when the machine leaves a zone of A grapes to enter a zone of B grapes or leaves a zone of B grapes to enter a zone of A grapes, the grapes must be considered B grapes. For the same row, the type of transitions may vary according to the direction that the row is harvested. Because of latency, the quantity of A grapes (resp. B grapes) that are collected within a given row can change with the direction. For instance, consider a row where the sequence of quality areas is A-B-A-B (see Figure 1). If the machine harvests the row from left to right, there is only one B-A transition. If the machine harvests the row in the opposite direction, two transitions B-A appear. Thus, quantities of A grapes and B grapes harvested will necessarily depend on the direction in which the rows are harvested.

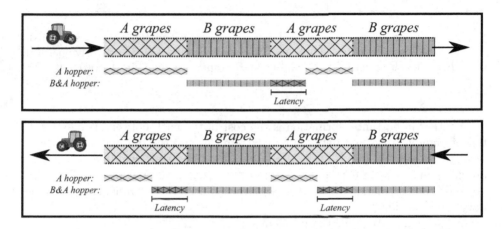

Fig. 1. Quantities of grapes depend on the direction of row harvest.

Consider $Rmin$, the desired minimum volume of A grapes harvested in the vine field. $Rmin$ is a threshold. In practice, $Rmin$ corresponds to shortfall of a volume of A grapes according to the objectives of the winery. If the vineyard contains more than $Rmin$ A grapes, the excess amount can be downgraded to B-grapes quality. As long as $Rmin$ has not been reached, A grapes are stored into the A hopper. Once $Rmin$ has been reached and after the hoppers have been emptied, A grapes and B grapes can be mixed in both hoppers. Regarding $Rmin$, there are three possibilities to fill the two hoppers. When the harvesting machine must differentiate qualities in hoppers and the A hopper is not full, A grapes are put in the A hopper and B grapes in the $B\&A$ hopper (see Figure 2.a). When A hopper is full, the machine can put A grapes in the $B\&A$ hopper (see Figure 2.b). In such a case, these A grapes are downgraded. Once $Rmin$ has been reached, the machine can mix grapes in both hoppers (see Figure 2.c).

The vine field, composed of n rows, is modelled by differentiating the two extremities of each row. A row $r \in \{0, \dots, n-1\}$ is represented by extremities $2r$ and $2r+1$. For each row, we denote $Q^A_{2r \to 2r+1}$ and $Q^B_{2r \to 2r+1}$ (resp. $Q^A_{2r+1 \to 2r}$ and $Q^B_{2r+1 \to 2r}$) the quantities of A grapes and B grapes that will be collected in row r with orientation $2r \to 2r+1$ (resp. $2r+1 \to 2r$). These quantities are computed according to the latency.

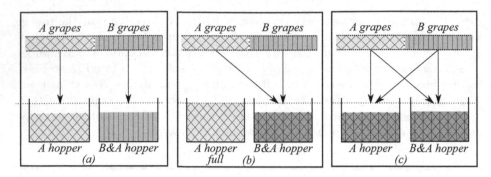

Fig. 2. Three possibilities to fill the hoppers. (a) A grapes and B grapes are separated. (b) When A hopper is full, A grapes and B grapes are mixed in the $B\&A$ hopper. (c) Once $Rmin$ has been reached, A grapes and B grapes are mixed in the two hoppers.

Another important information is the cost of the path between two extremities of different rows and between row extremities and the bin place. We denote $d(p,q) = d(q,p) \ \forall p, q \in \{0, 1, \ldots, 2n - 1\} \cup \{2n\}$ the time required to go from an extremity p to an extremity q (where $2n$ denotes the bin place). This cost depends on the distances between extremities and the turning radius of the harvesting machine.

We are ready to define the problem.

Definition 1 (Differential Harvest Problem). *Given a vine field described by a cost path matrix between row extremities (or the bin place) and an estimation of the quantity of A grapes and B grapes on each row according to the direction, given a harvesting machine with a hopper capacity of $Cmax$ and a latency delay, given a threshold of $Rmin$ A grapes to harvest, the Differential Harvest Problem consists in finding a sequence of extremities (that is, an order and orientation of the rows) that minimizes the time required to harvest the vine field and ensures at least $Rmin$ A grapes harvested.*

3 Constraint Programming for the DHP

The problem of differential harvest in precision viticulture is a recent problem. Its specification is not yet stable as it corresponds to a simplification of the real agronomic problem. In this context, other constraints will certainly appear when the first solutions will be applied. An example of probable extra constraint is shown in section 5.2. Constraint programming is well recognized for its flexibility and ease of maintenance. For these reasons, using constraint programming to model the DHP seems to be a good approach.

This section is devoted to present two CP models. First, the *Step model* consists in choosing which row to harvest step by step. Second, the *Precedence model* is devoted to find the best predecessor of each row.

Given a set of variables and a set of possible values for each variable, given a set of constraints, and given an optimization criterion, constraint optimization consists in finding an instantiation of variables that satisfies all constraints of the problem and minimizes the criterion.

3.1   The Step Model

In [4], we presented a preliminary CP model to solve the Differential Harvest problem. This model was dedicated to measure the gain of an optimized routing compared to the traditional approach. The traditional routing consists in systematically taking the second next row on the same side,[1] until the capacity of one hopper is reached. We showed in [4] that using a CP model to optimize the routing can reduce the harvest time of almost 40% compared to traditional harvest routing.

The *Step model* presented in [4] is illustrated in Figure 3. This model is based on three main kinds of variables that describe:

- the number of the row visited at each step of the route,
- the orientation of the visited row (harvest direction),
- the act of emptying the hoppers into the bin at the end of the visited row or not.

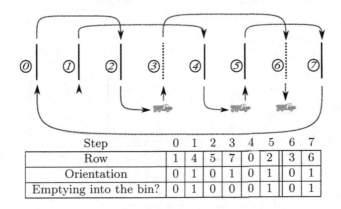

Step	0	1	2	3	4	5	6	7
Row	1	4	5	7	0	2	3	6
Orientation	0	1	0	1	0	1	0	1
Emptying into the bin?	0	1	0	0	0	1	0	1

Fig. 3. The Step model.

An *Alldifferent* constraint ensures that the machine passes exactly once in each row. Additional variables are used to represent the quantity of grapes harvested until each step. They are subject to a set of numeric constraints that control the capacities of the hoppers.

The first k steps of the route perform differential harvest. Before step k, B grapes cannot appear in A hoppers, i.e. harvest is separated. In step k, the

---

[1] Because of the turning radius of the machine, turning into the next row is generally longer than jumping to the second next row.

machine goes to the bin and must have harvested at least a total of $Rmin$ A grapes. After step k, A grapes and B grapes are mixed in both hoppers.

In the first version of this Step model we included k as a variable but experimental results were very bad. The combinatorial aspect of the problem decreases if we set k to a fixed value. Hence, each instance of the original problem is replaced by a set of sub-instances with different values of k. According to the capacity of the hoppers and the value of $Rmin$, few values of k must be considered. Since these instances are independent, they can be executed in parallel.

In section 4, we will see that this model is not effective. The next subsection gives another model for the DHP, based on models for routing problems.

3.2   The Precedence model

The *Precedence model* is based on variables that represent the previous row of a row. Our model shares similarities with the model presented in [9] for the Vehicle Routing Problem.

For a given instance of the Differential Harvest Problem, suppose that we have an upper bound λ on the number of times the hoppers have to be dumped into the bin. We note γ the number of dumps used in the differential part (beginning) of the harvest ($\gamma \leq \lambda \leq n$) where n is the number of rows in the vine field. We call $\mathcal{R} = \{0, \ldots, n-1\}$ the set of rows. $\mathcal{S} = \mathcal{R} \cup \{n, \ldots, n+\gamma-1, \ldots, n+\lambda-1\}$ denotes the set of *sites* (rows and hopper dumps into the bin).

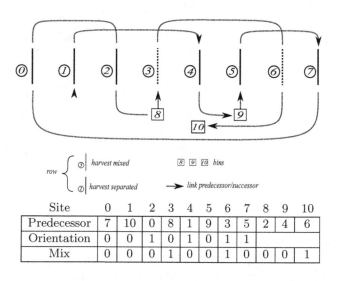

Site	0	1	2	3	4	5	6	7	8	9	10
Predecessor	7	10	0	8	1	9	3	5	2	4	6
Orientation	0	0	1	0	1	0	1	1			
Mix	0	0	0	1	0	0	1	0	0	0	1

Fig. 4. The Precedence model.

Variables. For each site $i \in \mathcal{S}$ we create:

- $\mathbf{P_i}, \mathbf{S_i}$: are integer variables that represent the direct predecessor (P_i) and direct successor (S_i) of site i. We have $D(P_i) = D(S_i) = \mathcal{S} \setminus \{i\}$;
- $\mathbf{Mix_i}$: is a Boolean variable that represents the harvest mode in site i. If $Mix_i = 0$ the A grapes are separated (differential harvest) otherwise they are mixed with B grapes. The domain of this variable is: $D(Mix_i) = \{0, 1\}$;
- $\mathbf{T_i}$: is an integer variable that represents the time to travel from P_i to site i, including the time to harvest row i. $D(T_i) = \mathbb{N}$;
- $\mathbf{U_i^A}$, $\mathbf{U_i^B}$ are integer variables that represent the quantity of A grapes and B grapes harvested up to site i since the last dump into the bin. The domains are: $D(U_i^A) = D(U_i^B) = \{0, \ldots, 2 \times Cmax\}$.

For each row $r \in \mathcal{R}$ we have:

- $\mathbf{Ori_r}$: is a Boolean variable that represents the orientation for row r (0 is the direction from odd to even extremities, as described in Section 2);
- $\mathbf{u_r^A}$ (resp. $\mathbf{u_r^B}$): represents the quantity of grapes of quality A (resp. B) harvested in row r according to the direction of harvest. $D(u_r^A) = D(u_r^B) = \mathbb{N}$;

Constraints. Predecessor variables form a permutation of rows and are subject to the *alldifferent* constraint (1).

$$AllDifferent(P_0, \ldots, P_{n+\lambda-1}) \tag{1}$$

Constraint (2) is a *channelling* constraint between predecessor and successor variables.

$$P_{S_i} = i \qquad\qquad S_{P_i} = i \qquad\qquad \forall i \in \mathcal{S} \tag{2}$$

Constraints (3) and (4) force the harvest mode to be differential or mixed according to the index of the corresponding hopper dump. Constraint (5) is an *element* constraint that communicates the harvest mode between successors:

$$Mix_i = 0 \qquad\qquad \forall i \in \{n, \ldots, n+\gamma-1\} \tag{3}$$
$$Mix_i = 1 \qquad\qquad \forall i \in \{n+\gamma, \ldots, n+\lambda-1\} \tag{4}$$
$$Mix_r = Mix_{S_r} \qquad\qquad \forall r \in \mathcal{R} \tag{5}$$

The following constraints give the quantities of grapes according to the orientation of the row. It can be implemented as a table constraint or as an *if ... then ... else* one. $\forall r \in \mathcal{R}$, we have :

$$u_r^A = Ori_r \times Q_{2r \rightarrow 2r+1}^A + (1 - Ori_r) \times Q_{2r+1 \rightarrow 2r}^A \tag{6}$$
$$u_r^B = Ori_r \times Q_{2r \rightarrow 2r+1}^B + (1 - Ori_r) \times Q_{2r+1 \rightarrow 2r}^B \tag{7}$$

Constraint (8) fixes quantities A and B for all sites representing dumps into the bin. Constraint (9) computes the quantities at the end of row i by adding the accumulated amount from predecessor P_i and the quantity in row i given by the precedent constraints.

$$U_i^\alpha = 0 \qquad\qquad \forall \alpha \in \{A, B\} \ \ \forall i \in \mathcal{S} \setminus \mathcal{R} \qquad (8)$$
$$U_i^\alpha = U_{P_i}^\alpha + u_i^\alpha \qquad\qquad \forall \alpha \in \{A, B\} \ \forall i \in \mathcal{R} \qquad (9)$$

Harvested quantities are limited by the capacity of hoppers. Variable U_i^A always have an upper bound of $2Cmax$ because A grapes can be put in the two hoppers. When variable $Mix_i = 0$, harvest is not mixed and quantity of B grapes is bound by $Cmax$ (10). When variable $Mix_i = 1$, A grapes and B grapes are mixed in the two hoppers. Constraint (11) checks that the total amount of harvested grapes is smaller than the capacity of the two hoppers:

$$U_i^B \leq (1 + Mix_i) \times Cmax \qquad\qquad \forall i \in \mathcal{R} \qquad (10)$$
$$U_i^A + U_i^B \leq 2 \times Cmax \qquad\qquad \forall i \in \mathcal{R} \qquad (11)$$

Constraint (12) requires to harvest at least $Rmin$ A grapes. Only the A grapes stored in A hopper must be considered. This quantity corresponds to the part of U_i^A which is smaller than the capacity of the A hopper. It concerns the differential harvest mode only, i.e. dumps from n to $n + \gamma - 1$.

$$\sum_{i=n}^{n+\gamma-1} min(U_{p_i}^A, Cmax) \geq Rmin \qquad (12)$$

It is possible to reformulate constraint (12) without the minimum operator. We add new variables called Ac_i $\forall i \in \{n, \ldots, n + \gamma - 1\}$ such that:

$$Ac_i \leq Cmax \qquad \text{that is: } D(Ac_i) = \{0, \ldots, Cmax\} \qquad (12\text{'a})$$
$$Ac_i \leq U_{P_i}^A \qquad \forall i \in \{n, \ldots, n + \gamma - 1\} \qquad (12\text{'b})$$
$$\sum_{i=n}^{n+\gamma-1} Ac_i \geq Rmin \qquad (12\text{'c})$$

Constraint (13) forces the exit from row P_i to be on the same side of the vine field as the entrance of row i. Hence, P_i and i have inverse orientations. This is the case in practice with traditional routing.

$$Ori_i = 1 - Ori_{P_i} \qquad\qquad \forall i \in \mathcal{R} \qquad (13)$$

Next constraints ((14a) and (14b)) require a unique cycle (subtour elimination) on predecessor variables and successor variables. Figure 5 is an example of filtering for the circuit constraint described in [5,12].

$$Circuit(1, (P_0, \ldots, P_{n+\lambda-1})) \qquad (14\text{a})$$
$$Circuit(1, (S_0, \ldots, S_{n+\lambda-1})) \qquad (14\text{b})$$

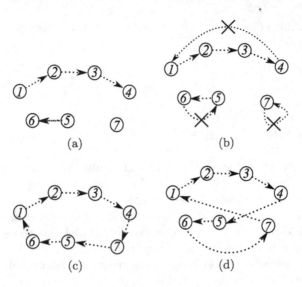

Fig. 5. An example of circuit constraint propagator on successor variables. Arrows represent the link (successor) between two variables (circles). (a) is a current instantiation. (b) filtering prohibits links between variables that represent the end and the start of a same path (i.e., the value that represents the start of a subpath is removed from the domain of variables that represent the end of this path). (c) and (d) represent two possible solutions in this example.

Constraints (15) force T_i to be equal to the time to travel from P_i to site i, including the time to harvest row i if it is the case. Function d gives the time to travel between two extremities of rows (or an extremity and the bin) as defined in Section 2. Note that for two consecutive dumps into the bin ($i \notin \mathcal{R}$ and $P_i \notin \mathcal{R}$) the travel time is equal to 0 (15b).

$$\text{if } P_r \in \mathcal{R} \text{ then } T_r = d(2P_r + Ori_{P_r}, 2r + 1 - Ori_r)$$
$$\text{else } T_r = d(2n, 2r + 1 - Ori_r) \qquad \forall r \in \mathcal{R} \qquad (15a)$$
$$\text{if } P_i \in \mathcal{R} \text{ then } T_i = d(2P_i + Ori_{P_i}, 2n) \text{ else } T_i = 0 \qquad \forall i \in \mathcal{S} \setminus \mathcal{R} \qquad (15b)$$

Constraint (16) is the objective function that minimizes the travel time of the grape harvester:

$$Minimize \sum_{i=0}^{n+\lambda-1} T_i \qquad (16)$$

Symmetry Breaking. The main variables of this model are predecessor variables. They take their values in a set of rows and dumps into the bin. For the set of dumps in the differential (or mixed) mode, there is a value symmetry caused by the possibility to dump indexes. To break this symmetry, we add an ordering constraint (17a) (resp. (17b)) on the variables that correspond to the same harvest mode.

$$P_i < P_{i+1} \qquad \forall i \in \{n, \ldots, n + \gamma - 2\} \qquad (17a)$$
$$P_i < P_{i+1} \qquad \forall i \in \{n + \gamma, \ldots, n + \lambda - 2\} \qquad (17b)$$

4   Experimental Results

In this section, we present some experimental results. We have implemented the two models in Choco [13] using the variable ordering strategy DomOverWDeg [3] limited to the P_i variables. All experiments were executed on a Linux server with an Intel(R) Xeon(R) CPU E5-2697 2.60GHz processor. It has 14 cores that makes it possible to solve several instances in parallel but the solver was not configured to exploit the multicore architecture.

Our models were tested on a real data from an experimental vineyard of INRA Pech-Rouge (Gruissan) located in southern France (see Figure 6). In this experiment we want to compute the optimal solution. To test the models on instances of increasing size, we have extracted subsets of continuous rows from the 24 rows vine field.

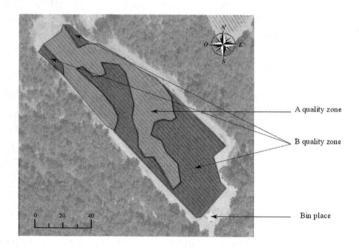

Fig. 6. A vine field with two qualities B grapes in red and A grapes in green.

Table 1 reports results with Step and Precedence models for 10, 12 and 14 rows. Each row in the table reports the average time on 12 instances of the given size generated from contiguous sequences of rows of the vine field. For each sequence, we experimented with two capacities of the hoppers (1000 and 2000) and two values for the desired threshold $Rmin$ (50% and 70% of the total amount of A grapes).

For the Step model, each instance of the Differential Harvest Problem is split into a set of more specific sub-instances of the constraint optimization problem. Each sub-instance is associated with a fixed value k of number of rows harvested on differential mode. For 10 (resp. 12) rows we varied k from 4 to 8 (resp. 6 to 10). Each set of sub-instances can be solved in parallel. So, Table 1 shows the maximum time needed to solve a sub-instance of a given set. The total time (for all sub-instances) corresponds to a single core architecture. Table 1 also gives the time for the sub-instance which has the optimal solution. With sequences of 14 rows, at least one sub-instance of each instance is not solved before the timeout (7200 sec).

Table 1. Comparison of the Step model and the Precedence model. All CPU times are averages on 12 instances and are given in seconds. Timeout at 7200 seconds.

| # rows | Step model | | | Precedence model |
	Parallel solving (maximum time)	Single processor (total time)	Time for the sub-instance with the optimal solution	Time
10	262	527	180	8
12	3414	8466	2749	118
14	timeout			8 instances solved

These results show that the Step model is clearly outperformed by the Precedence model. One disadvantage of the Step model is that it has to communicate the harvested quantities between steps. Hence, as long as the route is undefined, it is impossible to know if the capacity of hoppers is exceeded and if $Rmin$ is reached. This hinders propagation. There is another drawback of the Step model. It assumes a fixed value k for the step on which differential harvest stops (A grapes and B grapes are mixed from step k to step n). Finding an optimal solution requires to solve several instances of the problem with different values of k. This can be done in parallel as the problems are independent. But Table 1 shows that, even with an oracle predicting the value k of the sub-instance containing the optimal solution of the original instance, the Step model is outperformed by the Precedence model.

Table 2 shows results for instances of the Precedence model for sequences of 12, 14 and 16 rows. The first column shows the number of rows in the vine field. The second column gives the capacity of the hoppers ($Cmax$) and the third column gives the desired threshold $Rmin$. For each size of problem, we give the average time to solve it and the average number of nodes. Timeout is 7200 sec. For 16 rows with a $Cmax$ value of 1000 and for any greater number of rows, all instances exceeded the timeout.

Table 2. Precedence model on 12, 14 and 16 rows.

# Rows	Cmax	Rmin	Time (s)	# Nodes
12	1000	50%	227	663267
		70%	170	667112
	2000	50%	35	153300
		70%	41	175017
14	1000	50%	2/3 instances solved	
		70%	0/3 instances solved	
	2000	50%	52	181262
		70%	21	72468
16	2000	50%	856	2835199
		70%	106	318246

Table 3. Impact of different configurations for the Precedence model

	Precedence model		without symmetry breaking constraints		with min function in (12)		with minDomain variable ordering	
# Row	Time	# Nodes	Time	# Nodes	Time	# Nodes	Time	# Nodes
10	8	41811	17	95363	167	382345	101	1066550
12	118	441674	539	4017780	1042	4824446	776	3582405

These results indicate fairly high variability in CPU times. Instances with a small value of $Cmax$ seem to be more difficult to solve. When the threshold $Rmin$ is equal to 50% of A grapes, CPU time generally increases.

Table 3 shows results for different configurations of the Precedence model. All CPU times and node numbers are average on all instances of the same number of rows. The first two columns correspond to the complete Precedence model. The next columns give the results for variants of this model: without symmetry breaking (constraints (17a) and (17b)); with constraint (12) instead of constraints (12'a), (12'b) and (12'c); and with the default Choco variable ordering (minDomain) instead of the DomOverWDeg strategy.

Despite a small number of bins, adding constraints to eliminate symmetric solutions helps a lot to reduce the search effort. It is also the case when constraint (12) with the minimum function is replaced by constraints (12'a), (12'b) and (12'c), or when the default Choco variable ordering is replaced by DomOver-WDeg.

5   Discussion

Our experimental results show that the Precedence model is orders of magnitude faster than the Step model. For small instances (up to 16 rows), the Precedence model solves the Differential Havest Problem in a reasonable amount of time. For larger instances, the timeout of 2 hours is generally reached. This amount of time has to be compared to the time needed to collect data and to build the quality and yield map for the vine field. With additional improvements of the CP model, we can expect to solve larger real instances.

Our results confirm what other studies have reported on similar problems like the BBSS problem presented in [7]. In that paper, Di Gaspero *et al.* show that a Step formulation is outperformed by Routing (similar to Precedence) formulation for the optimal solution. In a second experiment, they show that large neighborhood search (LNS) is a good approach to decrease the time to solve, though the optimal solution can no longer be guaranteed. It could be interesting to explore that direction.

5.1   Comparison with an ILP Formulation

Contrary to our expectations, it was not difficult to express our model of the Differential Harvesting Problem as a set of linear constraints on integer variables. Such a problem can be solved with Integer Linear Programming (ILP).

So we have designed an ILP model using the two-index vehicle flow formulation of the Capacited Vehicle Routing Problem (CVRP) introduced by Laporte, Nobert, and Desrochers [10]. There is one Boolean variable $x_{t:i \rightarrow j}$, for each pair i, j of rows extremities and bin on tour t. Each tour corresponds to a single mode of harvest (differential or not). $x_{t:i \rightarrow j}$ is equal to 1 if the harvesting machine goes from i to j on tour t and 0 otherwise. Constraints on the $x_{t:i \rightarrow j}$ variables ensure that each extremity is visited exactly once on all routes. Quantities of grapes harvested in each tour are expressed by a summation over the traversed rows and similarly for travel cost (objective function). As in CVRP formulation, there are constraints on capacity of hoppers according to the tour t. Variables and linear constraints (12'a), (12'b), and (12'c) on threshold $Rmin$ can directly be added to the ILP formulation.

Unfortunately, the cycle constraint cannot be added so easily. The basic ILP approach to forbid sub-tours (cycles that do not pass through the bin) is to post constraints of the form $\sum_{i,j \in S} x_{t:i \rightarrow j} < |S|$ for any subset S of extremities of rows. The number of such constraints is exponential. Another formulation introduced by Miller, Tucker, and Zemlin ($MTZ - formulation$ [11]) makes the number of constraints polynomial but its linear relaxation generally produces a significantly weaker lower bound compared to the basic model [6]. Hence, a typical approach consists in adding the sub-tour elimination constraints incrementally, as follows: Step 1: find an optimal solution (without sub-tour constraints at the beginning). Step 2: If the solution does not contain sub-tours, it is an optimal solution of the problem; Otherwise, new constraints are added to forbid all the sub-tours in the current solution. Then, proceed to Step 1. In the worst case, this algorithm has to solve an exponential number of instances with finally an exponential number of constraints on sub-tours. Thus, each intermediate ILP loop gives relaxed solutions that contain sub-tours between rows. It is only in the last ILP step that the reported (optimal) solution does not contain sub-tours and can be exploited. As a consequence, if we interrupt the ILP solver at a given timeout limit, it has not yet produced any feasible solution, not even

Table 4. Comparison between CP model and ILP model for 12, 14, 16 and 24 rows, $Cmax = 1000$, $Rmin = 70\%$ and timeout of 7200s. In bold, optimal solutions.

# rows	upper-bound for harvesting time	CP model		ILP model	
		harvesting time	CPU time	harvesting time	CPU time
12	$+\infty$	**960**	556s	**960**	3s
	2671	**960**	496s		
14	$+\infty$	*1120*	timeout	**1112**	113s
	2436	*1116*	timeout		
16	$+\infty$	*1260*	timeout	-	timeout
	2694	*1260*	timeout		
24	$+\infty$	*1805*	timeout	-	timeout
	2059	*1800*	timeout		

suboptimal ones. This is an advantage of the CP model over the ILP model. Any of its solutions, at any time in the solving process, can be exploited.

We have implemented the ILP model with the Cplex solver [8]. Table 4 shows a comparison between the Precedence model and the ILP formulation using a single core. Tests are performed with and without an initial upper bound. This bound is an estimation of the harvest time using a manual routing based on a repetitive pattern. This upper bound does not improve the ILP results so we give only one value for the ILP model. Preliminary experiments show that the ILP formulation clearly outperforms the CP model for small instances. But for hard instances ($n \geq 16$ and $Cmax = 1000$), ILP search fails to obtain a solution before the timeout of 7200 seconds. On hard instances the CP model can find suboptimal solutions that significantly improve the manual routing solution whilst ILP provides no solution at all.

5.2 Complements to the Model

Another advantage of the CP model is its ability to integrate new constraints. The formulation of the problem presented in this paper is a quite simplified version of the Differential Harvest Problem and it is dedicated to evolve. For instance, consider the case where the harvesting machine finishes a tour on an extremity of a row opposite to the bin. If the row is in the middle of the vineyard, the shortest path to the bin passes through a row. But it is not possible to go to the bin by passing through a non-harvested row. This can be expressed by the following rule: the path to the bin, for an extremity which is on the side opposite to the bin, must pass through a row that precedes the extremity in the global route. Such a constraint is very difficult to add to the ILP formulation but can be easily implemented in the CP approach. The propagate procedure of the circuit constraint already computes the set of rows that precede the last step of any tour in order to detect subtours. So it is easy to find in this set what is the best row to go back to the bin. This can be done without changing the overall complexity of the propagate procedure.

6   Conclusion

In this paper, we have presented the Differential Harvest Problem in precision viticulture. We have proposed to use constraint programming to solve it. Two models have been presented, the Step model and the Precedence model. In the Step model, variables represent the row that is visited at a given time step and in which direction the row is traversed. In the Precedence model, variables connect a row to its predecessor and successor. The experiments we have performed to assess the behavior of these models show that the Precedence model is orders of magnitude faster than the Step model. We have also experimentally shown that an ILP formulation of the Differential Harvest Problem outperforms our CP approach on easy instances. However, such an ILP formulation requires an exponential space, and more importantly, fails to produce solutions on hard instances. All in all, our Precedence model seems to be a good approach. It allows to solve the problem on real data within reasonable time and it inherits the flexibility of CP models, that allows the addition of extra user-constraints in a simple way.

References

1. Bochtis, D.D., Sørensen, C.G.: The vehicle routing problem in field logistics: part i. Biosystems Engineering **104**, 447–457 (2009)
2. Bochtis, D.D., Sørensen, C.G.: The vehicle routing problem in field logistics: part ii. Biosystems Engineering **105**, 180–188 (2010)
3. Boussemart, F., Hemery, F., Lecoutre, C., Sais, L.: Boosting systematic search by weighting constraints. In: Proceedings of the 16th Eureopean Conference on Artificial Intelligence (ECAI 2004), Valencia, Spain, pp. 146–150 (2004)
4. Briot, N., Bessiere, C., Tisseyre, B., Vismara, P.: Integration of operational constraints to optimize differential harvest in viticulture. In: Proc. 10th European Conference on Precision Agriculture, July 2015 (to appear)
5. Caseau, Y., Laburthe, F.: Solving small TSPs with constraints. In: Naish, L. (ed.) Logic Programming, Proceedings of the Fourteenth International Conference on Logic Programming, July 8–11, 1997, pp. 316–330. MIT Press, Leuven, Belgium (1997)
6. Desrochers, M., Laporte, G.: Improvements and extensions to the miller-tucker-zemlin subtour elimination constraints. Operations Research Letters **10**(1), 27–36 (1991)
7. Di Gaspero, L., Rendl, A., Urli, T.: Constraint-based approaches for balancing bike sharing systems. In: Schulte, C. (ed.) CP 2013. LNCS, vol. 8124, pp. 758–773. Springer, Heidelberg (2013)
8. IBM ILOG. Cplex. http://www-01.ibm.com/software/integration/optimization/cplex-optimizer/ (accessed April 2015)
9. Kilby, P., Shaw, P.: Vehicle routing. In: Rossi, F., Beek, P.V., Walsh, T. (eds), Handbook of Constraint Programming, chapter 23, pp. 799–834. Elsevier (2006)
10. Laporte, G., Nobert, Y., Desrochers, M.: Optimal routing under capacity and distance restrictions. Operations Research **33**(5), 1050–1073 (1985)
11. Miller, C.E., Tucker, A.W., Zemlin, R.A.: Integer programming formulation of traveling salesman problems. Journal of the ACM (JACM) **7**(4), 326–329 (1960)

12. Pesant, G., Gendreau, M., Potvin, J.-Y., Rousseau, J.-M.: An exact constraint logic programming algorithm for the traveling salesman problem with time windows. Transportation Science **32**(1), 12–29 (1998)
13. Prud'homme, C., Fages, J.G., Lorca, X.: Choco3 Documentation. TASC, INRIA Rennes, LINA CNRS UMR 6241, COSLING S.A.S. (2014)
14. Tisseyre, B., Ojeda, H., Taylor, J.: New technologies and methodologies for site-specific viticulture. Journal International des Sciences de la Vigne et du Vin **41**, 63–76 (2007)

Constrained Minimum Sum of Squares Clustering by Constraint Programming

Thi-Bich-Hanh Dao$^{(\boxtimes)}$, Khanh-Chuong Duong, and Christel Vrain

University of Orléans, INSA Centre Val de Loire LIFO EA 4022,
F-45067 Orléans, France
{thi-bich-hanh.dao,khanh-chuong.duong,christel.vrain}@univ-orleans.fr

Abstract. The Within-Cluster Sum of Squares (WCSS) is the most used criterion in cluster analysis. Optimizing this criterion is proved to be NP-Hard and has been studied by different communities. On the other hand, Constrained Clustering allowing to integrate previous user knowledge in the clustering process has received much attention this last decade. As far as we know, there is a single approach that aims at finding the optimal solution for the WCSS criterion and that integrates different kinds of user constraints. This method is based on integer linear programming and column generation. In this paper, we propose a global optimization constraint for this criterion and develop a filtering algorithm. It is integrated in our Constraint Programming general and declarative framework for Constrained Clustering. Experiments on classic datasets show that our approach outperforms the exact approach based on integer linear programming and column generation.

1 Introduction

Cluster analysis is a Data Mining task that aims at partitioning a given set of objects into homogeneous and/or well-separated subsets, called classes or clusters. It is usually formulated as an optimization problem and different optimization criteria have been defined [18]. One of the most used criteria is minimizing the Within-Cluster Sum of Squares (WCSS) Euclidean distances from each object to the centroid of the cluster to which it belongs. The well-known k-means algorithm as well as numerous heuristic algorithms optimize it and find a local optimum [27]. Finding a global optimum for this criterion is a NP-Hard problem and even finding a good lower bound is difficult [1]. The best exact approach for clustering with this criterion is based on Integer Linear Programming (ILP) and column generation [2].

On the other hand, since this last decade, user-defined constraints have been integrated to clustering task to make it more accurate, leading to Constrained Clustering. User constraints usually make the clustering task harder. The extension to user constraints is done either by adapting classic algorithms to handle constraints or by modifying distances between objects to take into account constraints [4,9,30]. Recently an exact approach has been proposed, which aims at finding an optimal solution for the WCSS criterion satisfying all the user constraints [3]. This approach extends the method based on ILP and column

© Springer International Publishing Switzerland 2015
G. Pesant (Ed.): CP 2015, LNCS 9255, pp. 557–573, 2015.
DOI: 10.1007/978-3-319-23219-5_39

generation [2]. It integrates different kinds of user constraints but only handle the WCSS criterion.

Recently general and declarative frameworks for Data Mining have attracted more and more attention from Constraint Programming (CP) and Data Mining communities [3,11,16,28]. In our previous work [7,8] we have proposed a general framework based on CP for Constrained Clustering. Different from classic algorithms that are designed for one specific criterion or for some kinds of user constraints, the framework offers in a unified setting a choice among different optimization criteria and a combination of different kinds of user constraints. It allows to find a global optimal solution that satisfies all the constraints if one exists. Classic heuristic algorithms can quickly find a solution and can scale on very large datasets, however they do not guarantee the satisfaction of all the constraints or the quality of the solution. A declarative and exact approach allows a better understanding of the data, which is essential for small valuable datasets that may take years to collect. In order to make the CP approach efficient, it is important to invest in dedicated global constraints for the clustering tasks [8].

In this paper we propose a global optimization constraint *wcss* to represent the WCSS criterion. We propose a method based on dynamic programming to compute a lower bound and develop a filtering algorithm. The filtering algorithm filters not only the objective variable, but also the decision variables. The constraint extends our framework to the WCSS criterion. Experiments on classic datasets show that our approach based on CP outperforms the state-of-the-art approach based on ILP and column generation that handles user constraints [3].

Outline. Section 2 gives preliminaries on constrained clustering and reviews related work on existing approaches. Section 3 presents the framework based on CP and introduces the constraint *wcss*. Section 4 presents our contributions: the computation of a lower bound and the filtering algorithm for the constraint *wcss*. Section 5 is devoted to experiments and comparisons of our approach with existing approaches. Section 6 discusses perspectives and concludes.

2   Preliminaries

2.1   Constrained Clustering

A clustering task aims at partitioning a given set of objects into clusters, in such a way that the objects in the same cluster are similar, while being different from the objects belonging to other clusters. These requirements are usually expressed by an optimization criterion and the clustering task consists in finding a partition of objects that optimizes it. Let us consider a dataset of n objects $\mathcal{O}$ and a dissimilarity measure $d(o, o')$ between any two objects $o, o' \in \mathcal{O}$ (d is defined by the user). A partition Δ of objects into k classes $C_1, \ldots, C_k$ is such that: (1) $\forall c \in [1, k]^1$, $C_c \neq \emptyset$, (2) $\cup_c C_c = \mathcal{O}$ and (3) $\forall c \neq c'$, $C_c \cap C_{c'} = \emptyset$. The optimization criterion may be, among others:

---

[1] For integer value we use $[1, k]$ to denote the set $\{1, .., k\}$.

- Maximizing the minimal split between clusters, where the minimal split of a partition Δ is defined by: $S(\Delta) = \min_{c \neq c' \in [1,k]} \min_{o \in C_c, o' \in C_{c'}} d(o, o')$.
- Minimizing the maximal diameter of clusters, which is defined by: $D(\Delta) = \max_{c \in [1,k]} \max_{o, o' \in C_c} d(o, o')$.
- Minimizing the Within-Cluster Sum of Dissimilarities, which is defined by:

$$WCSD(\Delta) = \sum_{c \in [1,k]} \frac{1}{2} \sum_{o, o' \in C_c} d(o, o')$$

- Minimizing the Within-Cluster Sum of Squares:

$$WCSS(\Delta) = \sum_{c \in [1,k]} \sum_{o \in C_c} d(o, m_c)$$

where for each $c \in [1, k]$, m_c is the mean (centroid) of the cluster C_c and $d(o, m_c)$ is defined by the squared Euclidean distance $||o - m_c||^2$. When the dissimilarity between objects is defined by the squared Euclidean distance $d(o, o') = ||o - o'||^2$, we have [12, 18]:

$$WCSS(\Delta) = \sum_{c \in [1,k]} \frac{\frac{1}{2} \sum_{o, o' \in C_c} d(o, o')}{|C_c|}$$

All these criteria except the split criterion are NP-Hard (see e.g. [14] for the diameter criterion, [1] for WCSS, or concerning WCSD, the weighted max-cut problem, which is NP-Complete, is an instance with $k = 2$). Most of classic algorithms use heuristics and search for a local optimum [17]. For instance the k-means algorithm finds a local optimum for the WCSS criterion or FPF (Furthest Point First) [14] for the diameter criterion. Different optima may exist, some may be closer to the one expected by the user. In order to better model the task, user knowledge is integrated to clustering task. It is usually expressed by constraints, leading to Constrained Clustering. User constraints can be cluster-level, giving requirements on clusters, or instance-level, specifying requirements on pairs of objects. The last kind, introduced in [29], is the most used. An instance-level constraint on two objects can be a must-link or a cannot-link constraint, which states that the objects must be or cannot be in the same cluster, respectively.

Different kinds of cluster-level constraints exist. The minimal (maximal) capacity constraint requires that each cluster must have at least (at most, resp.) a given α (β, resp.) objects: $\forall c \in [1, k]$, $|C_c| \geq \alpha$ (resp. $\forall c \in [1, k]$, $|C_c| \leq \beta$). A diameter constraint sets an upper bound γ on the cluster diameter: $\forall c \in [1, k], \forall o, o' \in C_c$, $d(o, o') \leq \gamma$. A split constraint, named δ-constraint in [9], sets a lower bound δ on the split: $\forall c \neq c' \in [1, k]$, $\forall o \in C_c, o' \in C_{c'}, d(o, o') \geq \delta$. A density constraint, which extends ϵ-constraints in [9], requires that each object o must have in its neighborhood of radius ϵ at least m objects belonging to the same cluster as itself: $\forall c \in [1, k], \forall o \in C_c, \exists o_1, ..., o_m \in C_c, o_i \neq o \land d(o, o_i) \leq \epsilon$.

User-constraints can make the clustering task easier (e.g. must-link constraints) but usually they make the task harder, for instance the split criterion that is polynomial becomes NP-Hard with cannot-link constraints [10].

2.2 Related Work

We are interested in constrained clustering with the WCSS criterion. The clustering task with this criterion is NP-Hard [1]. The popular k-means algorithm finds a local optimum for this criterion without user constraints. This algorithm starts with a random partition and repeats two steps until a stable state: (1) compute the centroid of the clusters, (2) reassign the objects, where each one is assigned to the cluster whose centroid is the closest to the object. The found solution thus depends on the initial random partition. Numerous heuristic algorithms have been proposed for this criterion, see e.g. [27] for a survey. Exact methods are much less numerous than heuristic. Some of them are based on branch-and-bound [6,20] or on dynamic programming [19,23]. An algorithm based on Integer Linear Programming (ILP) and column generation is proposed in [22]. This algorithm is improved in [2] and to our knowledge, it is the most efficient exact approach. It can handle datasets up to 2300 objects with some heuristics [2], but it does not handle user constraints.

The COP-kmeans algorithm [30] extends the k-means algorithm to handle must-link and cannot-link constraints and tries to satisfy all of them. This algorithm changes k-means in step (2): it tries to assign each object to the cluster whose centroid is the closest, without violating the user constraints. If all the possible assignments violate a user constraint, the algorithm stops. This is a greedy and fast algorithm without backtrack, but it can fail to find a solution that satisfies all the user constraints, even when such a solution exists [10]. Another family of approaches tries to satisfy only most of the user constraints. The CVQE (Constrained Vector Quantization Error) algorithm [9] or an improved version LCVQE [24] extend the k-means algorithm to must-link and cannot-link constraints by modifying the objective function.

Recently, a general framework, which minimizes the WCSS criterion and which can handle different kinds of user constraints has been developed [3]. This framework extends the approach [2] based on ILP and column generation. It allows to find a global optimum and which satisfies all the user constraints. Due to the choice of variables in the linear program, the framework is however specified for this very criterion and does not handle other optimization criteria.

In our previous work, we have developed a general framework based on Constraint Programming for Constrained Clustering [7,8]. This framework offers a choice among different optimization criteria (diameter, split or WCSD) and integrates all popular kinds of user constraints. Moreover, the capacity of handling different optimization criteria and different kinds of user constraints allows the framework to be used in bi-criterion constrained clustering tasks [8]. In this paper we extend this framework to integrate the WCSS criterion.

3   Constraint Programming Model

We are given a collection of n points and a dissimilarity measure between pairs of points i, j, denoted by $d(i, j)$. Without loss of generality, let us suppose that points are indexed and named by their index (1 represents the first point).

In [8] we have presented a CP model for constrained clustering, which integrates the diameter, split and WCSD criteria. We present an extension of this model to the WCSS criterion.

In this model the number of clusters k is not set, but only bounded by k_{min} and k_{max} ($k_{min} \leq k \leq k_{max}$). The values k_{min} and k_{max} are set by the user. If the user needs a known number k of clusters, all he has to do is to set $k_{min} = k_{max} = k$. The clusters are identified by their index, which is a value from 1 to k for a partition of k clusters. In order to represent the assignment of points into clusters, we use integer value variables $G_1, \ldots, G_n$, each one having as domain the set of integers $[1, k_{max}]$. An assignment $G_i = c$ expresses that point i is assigned to the cluster number c. Let $\mathcal{G}$ be $[G_1, \ldots, G_n]$. To represent the WCSS criterion, we introduce a float value variable V, with $Dom(V) = [0, \infty)$.

3.1   Constraints

Any complete assignment of the variables in $\mathcal{G}$ defines a partition of points into clusters. In order to break symmetries between the partitions, we put the constraint $precede(\mathcal{G}, [1, \ldots, k_{max}])$ [21], which states that $G_1 = 1$ (the first point is in cluster 1) and $\forall c \in [2, k_{max}]$, if there exists $G_i = c$ with $i \in [2, n]$, then there must exists $j < i$ such that $G_j = c - 1$. Since the domain of the variables G_i is $[1, k_{max}]$, there are at most k_{max} clusters. The fact that there are at least k_{min} clusters means that all the values from 1 to k_{min} must appear in $\mathcal{G}$. We only need to require that k_{min} must appear in $\mathcal{G}$, since with the use of the constraint $precede$, if k_{min} is taken, then $k_{min} - 1, k_{min} - 2, \ldots, 1$ are also taken. This means $\#\{i \mid G_i = k_{min}\} \geq 1$ and can be expressed by the constraint: $atleast(1, \mathcal{G}, k_{min})$.

Instance-level user constraints can be easily integrated within this model. A must-link constraint on two points i, j is expressed by $G_i = G_j$ and a cannot-link constraint by $G_i \neq G_j$. All popular cluster-level constraints are also integrated. For instance, a minimal capacity α constraint is expressed by the fact that each point must be in a cluster with at least α points including itself. Therefore, for each $i \in [1, n]$, the value taken by G_i must appear at least α times in $\mathcal{G}$, i.e. $\#\{j \mid G_j = G_i\} \geq \alpha$. For each $i \in [1, n]$, we put: $atleast(\alpha, \mathcal{G}, G_i)$. This requirement allows to infer an upper bound on the number of clusters. Indeed, $G_i \leq \lfloor n/\alpha \rfloor$, for all $i \in [1, n]$. For other kinds of user constraints, such as maximal capacity, diameter or density constraints, we refer the reader to [8].

In order to express that V is the within-cluster sum of squares of the partition formed by the assignment of variables in $\mathcal{G}$, we develop a global optimization constraint $wcss(\mathcal{G}, V, d)$. The filtering algorithm for this constraint is presented in Section 4. The value of V is to be minimized.

3.2   Search Strategy

The branching is on the variables in $\mathcal{G}$. A mixed strategy is used with a branch-and-bound mechanism. A greedy search is used to find the first solution: at each branching, a variable G_i and a value $c \in Dom(G_i)$ such that the assignment $G_i = c$ increases V the least are chosen. The value of V at the first found solution gives an upper bound of the domain of V. After finding the first solution, the search strategy changes. In the new strategy, at each branching, for each unassigned variable G_i, for each value $c \in Dom(G_i)$ we compute the value a_{ic}, which is the added amount to V if point i is assigned to cluster c. For each unassigned variable G_i, let $a_i = \min_{c \in Dom(G_i)} a_{ic}$. The value a_i thus represents the minimal added amount to V when point i is assigned to a cluster. Since each point must be assigned to a cluster, at each branching, the variable G_i with the greatest value a_i is chosen, and for this variable, the value c with the smallest value a_{ic} is chosen. Two branches are then created, corresponding to two cases where $G_i = c$ and $G_i \neq c$. This strategy tends to detect failure more quickly or in case of success, to find a solution with the value of V as small as possible.

4 Filtering Algorithm for WCSS

We consider that the objects are in an Euclidean space and the dissimilarity measure is defined by the squared Euclidean distance. The sum of dissimilarities of a cluster C_c is defined by $WCSD(C_c) = \frac{1}{2} \sum_{o,o' \in C_c} d(o, o')$. The sum of squares of C_c is defined by $WCSS(C_c) = \frac{1}{|C_c|} WCSD(C_c)$. The WCSS of a partition $\Delta = \{C_1, \ldots, C_k\}$ is $WCSS(\Delta) = \sum_{c \in [1,k]} WCSS(C_c)$.

We introduce a new constraint $wcss(\mathcal{G}, V, d)$ and develop a filtering algorithm. This constraint maintains the relation that the float value variable V is the sum of squares of the clusters formed by the assignment of the decision variables of $\mathcal{G}$, given a dissimilarity measure d. Given a partial assignment of some variables in $\mathcal{G}$, we develop a lower bound computation for V and an algorithm to filter the domains of the variables. Since the variable V represents the objective function, this constraint is a global optimization constraint [13,26]. The filtering algorithm filters not only the domain of the objective variable V, but also the domain of decision variables in $\mathcal{G}$.

A partial assignment of variables of $\mathcal{G}$ represents a case where some points have been already assigned to a cluster and there are unassigned points. Let $k = \max\{c \mid c \in \bigcup_i Dom(G_i)\}$. The value k is the greatest cluster index among those remaining in all the domains $Dom(G_i)$ and thus it is the greatest possible number of clusters in the partition. Let $\mathcal{C}$ be the set of clusters $C_1, \ldots, C_k$. Some of these clusters can be empty, they correspond to indexes that remain in some non-singleton domains $Dom(G_i)$ but not in a singleton domain. For each cluster C_c, let n_c be the number of points already assigned to C_c ($n_c \geq 0$) and let $S_1(C_c)$ be the sum of dissimilarities of all the assigned points in C_c: $S_1(C_c) = \frac{1}{2} \sum_{i,j \in C_c} d(i,j)$. Let U be the set of unassigned points and let $q = |U|$.

4.1 Lower Bound Computation

We compute a lower bound of V considering all the possibilities for assigning all the points in U into the clusters $C_1, \ldots, C_k$. This is done in two steps:

1. For each $m \in [0, q]$ and $c \in [1, k]$, we compute a lower bound $\underline{V}(C_c, m)$ of $WCSS(C_c)$ considering all possible assignments of m points of U into C_c.

2. For each $m \in [0, q]$ and $c \in [2, k]$, we compute a lower bound $\underline{V}(C_1 \ldots C_c, m)$ of $WCSS(\{C_1, .., C_c\})$ considering all the possibilities for assigning any m points of U into the clusters $C_1, \ldots, C_c$.

Existing branch-and-bound approaches [6,20] are also based on the computation of a lower bound. However, these algorithms only make use of dissimilarities between the unassigned points. In our lower bound computation, we exploit not only dissimilarities between the unassigned points, but also the dissimilarities between unassigned points and assigned points. The computation is achieved by Algorithm 1. The two steps are detailed below.

Algorithm 1. Lower_bound()

---

    input : a partial assignment of $\mathcal{G}$, a set $U = \{i \mid G_i \text{ unassigned}\}$, $q = |U|$
    output: a lower bound of the sum of squares V

1 **foreach** $x \in U$ **do**
2   **for** $c \leftarrow 1$ **to** k **do**
3   **if** $c \notin Dom(G_x)$ **then** $s_2[x, c] \leftarrow \infty$;
4   **else** $s_2[x, c] \leftarrow 0$;

5   **foreach** $v \in [1, n]$ *such that* $|Dom(G_v)| = 1$ **do**
6   **if** $val(G_v) \in Dom(G_x)$ **then** $s_2[x, val(G_v)] = s_2[x, val(G_v)] + d(x, v)$;

7   sort $u \in U$ in the increasing order of $d(x, u)$
8     $s_3[x, 0] \leftarrow 0$
9   **for** $m \leftarrow 1$ **to** q **do**
10        $s_3[x, m] \leftarrow s_3[x, m - 1] + d(x, u_m)/2$

11 **for** $c \leftarrow 1$ **to** k **do**
12   **for** $m \leftarrow 0$ **to** q **do**
13   **foreach** $x \in U$ **do**
14             $s[x] = s_2(x, c) + s_3(x, m)$

15   sort the array s increasingly
16         $\underline{S_2}(C_c, m) \leftarrow \sum_{i=1}^{m} s[i]$
17   **if** $n_c + m = 0$ **then** $\underline{V}(C_c, m) \leftarrow 0$;
18   **else**
19             $\underline{V}(C_c, m) \leftarrow (S_1(C_c) + \underline{S_2}(C_c, m))/(n_c + m)$

20 **for** $c \leftarrow 2$ **to** k **do**
21     $\underline{V}(C_1..C_c, 0) \leftarrow \underline{V}(C_1..C_{c-1}, 0) + \underline{V}(C_c, 0)$
22   **for** $m \leftarrow 1$ **to** q **do**
23         $\underline{V}(C_1..C_c, m) \leftarrow \min_{i \in [0, m]}(\underline{V}(C_1..C_{c-1}, i) + \underline{V}(C_c, m - i))$

24 **return** $\underline{V}(C_1..C_k, q)$

Assignment of any m points of U into a cluster C_c. If we choose a subset $U' \subseteq U$ with $|U'| = m$ and we assign the points of U' into the cluster C_c, the sum of squares of C_c after the assignment will be:

$$V(C_c, U') = \frac{S_1(C_c) + S_2(C_c, U')}{n_c + m}$$

Here $S_1(C_c)$ is the sum of dissimilarities of the points already assigned in C_c and $S_2(C_c, U')$ is the sum of dissimilarities related to points in U'. The value of $S_1(C_c)$ is known. If the set U' is known the value $S_2(C_c, U')$ can be computed exactly by:

$$S_2(C_c, U') = \sum_{u \in U', v \in C_c} d(u, v) + \frac{1}{2} \sum_{u, v \in U'} d(u, v)$$

But $S_2(C_c, U')$ remains unknown while U' is not defined. However, for any subset U' of size m, we can compute a lower bound $\underline{S_2}(C_c, m)$ as follows. Each point $x \in U$, in case of assignment to the cluster C_c together with other $m - 1$ points of U, will contribute an amount $s(x, c, m) = s_2(x, c) + s_3(x, m)$, where:

- $s_2(x, c)$ represents the sum of dissimilarities between x and the points already in the cluster C_c. If $c \notin Dom(G_x)$ then $s_2(x, c) = +\infty$, since x cannot be assigned to C_c. Otherwise $s_2(x, c) = \sum_{v \in C_c} d(x, v)$. This value $s_2(x, c)$ is 0 if the cluster C_c is empty. It is computed by lines 2–6 in Algorithm 1.
- $s_3(x, m)$ represents a half of the sum of dissimilarities $d(x, z)$, for all the $m - 1$ other points z. These points z can be any points in U, however, if we order all the points $u \in U$ in an increasing order on $d(x, u)$ and we denote by u_i the i-th point in this order, we have a lower bound for $s_3(x, m)$ (lines 7–10 in Algorithm 1): $s_3(x, m) \geq \frac{1}{2} \sum_{i=1}^{m-1} d(x, u_i)$.

A lower bound $\underline{S_2}(C_c, m)$ is thus the sum of the m smallest contributions $s(x, c, m)$ (represented by $s[x]$ in Algorithm 1, for fixed values c and m), for all points $x \in U$. The lower bound $\underline{V}(C_c, m)$ is 0 if $n_c + m = 0$ or otherwise is computed by:

$$\underline{V}(C_c, m) = \frac{S_1(C_c) + \underline{S_2}(C_c, m)}{n_c + m} \tag{1}$$

This is computed for all $c \in [1, k]$ and $m \in [0, q]$ (lines 11–19 in Algorithm 1).

For an example, let us consider the partial assignment given in Figure 1 (A) where $k = 3$ and some points have been assigned into 3 clusters C_1 (square), C_2 (triangle) and C_3 (circle). The set of unassigned points U is $\{3, 4, 8, 9\}$. A lower bound $\underline{V}(C_3, 2)$ for the sum of squares of C_3 in case of assignment of any 2 points of U into C_3 is computed by Formula (1), with $n_c = 3$ and $m = 2$. In this formula, $S_1(C_3) = d(10, 11) + d(11, 12) + d(10, 12)$ and $\underline{S_2}(C_3, 2)$ is the sum of the 2 smallest contributions to C_3 among those of all the unassigned points. They are the contributions of points 4 and 9. Figure 1 (B) presents the dissimilarities used in the computation of $\underline{S_2}(C_3, 2)$. For the contribution of point 4, we make use of one ($= m - 1$) smallest dissimilarity from point 4 to the other unassigned points, which is $d(4, 3)$. Idem for point 9, where $d(9, 8)$ is used.

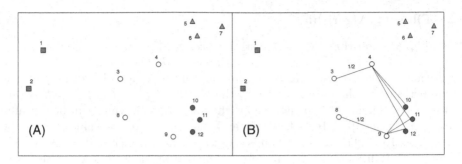

Fig. 1. Example: (A) partial assignment, (B) dissimilarities used in $\underline{S}_2(C_3, 2)$

Let us note that the contribution of each point is computed separately, in order to avoid combinatory cases. Therefore $d(4, 9)$ is not used, even though points 4 and 9 are assumed to be assigned to C_3.

Assignment of any m points of U into c clusters $C_1, .., C_c$. Any assignment of m points to c clusters is expressed by an assignment of some i points to the first $c - 1$ clusters and the remaining $m - i$ points to the last cluster. Reasoning only on the number of points to be assigned, we always have the following relation:

$$V(C_1..C_c, m) \geq \min_{i \in [0,m]} (V(C_1..C_{c-1}, i) + V(C_c, m - i))$$

A lower bound $\underline{V}(C_1..C_c, m)$ therefore can be defined by:

$$\underline{V}(C_1..C_c, m) = \min_{i \in [0,m]} (\underline{V}(C_1..C_{c-1}, i) + \underline{V}(C_c, m - i)) \qquad (2)$$

This is computed by a dynamic program for all $c \in [2, k]$ and $m \in [0, q]$ (lines 20–23 in Algorithm 1). Let us notice that with (2), for all $c \in [1, k]$ and $m \in [0, q]$:

$$\underline{V}(C_1..C_c, m) = \min_{m_1+..+m_c=m} (\underline{V}(C_1, m_1) + \cdots + \underline{V}(C_c, m_c)) \qquad (3)$$

Let us reconsider the example given in Figure 1. The value $\underline{V}(C_1..C_3, 4)$ computed by (2) corresponds to the case $\underline{V}(C_1..C_2, 1) + \underline{V}(C_3, 3)$, i.e when one point is assigned to the clusters C_1, C_2 and 3 points are assigned to C_3. The value $\underline{V}(C_1..C_2, 1)$ corresponds to the case $\underline{V}(C_1, 0) + \underline{V}(C_2, 1)$. The value $\underline{V}(C_2, 1)$ corresponds to the case where point 4 is assigned to cluster C_2 and $\underline{V}(C_3, 3)$ to the case where points 4, 8 and 9 are assigned to cluster C_3. We note that in this lower bound, point 4 is considered twice and point 3 is not considered.

Concerning the complexity, the complexity of the first loop (lines 1–10) is $O(q(k + n + q \log q + q)) = O(q^2 \log q + qn)$. The complexity of the second loop (lines 11–19) is $O(kq(q + q \log q)) = O(kq^2 \log q)$ and the complexity of the last loop (lines 20–23) is $O(kq^2)$. The complexity of Algorithm 1 is then $O(kq^2 \log q + qn)$. Let us notice that in clustering tasks, the number of clusters k is usually constant or much smaller than n.

4.2   Filtering Algorithm

The filtering algorithm for the constraint $wcss(\mathcal{G}, V, d)$ is presented in Algorithm 2, given a partial assignment of variables in $\mathcal{G}$. The value $\underline{V}(C_1..C_k, q)$ in Algorithm 1 represents a lower bound for the sum of squares V, for all possible assignments of all the points in U into the clusters $C_1, \ldots, C_k$. Let $[V.lb, V.ub)$ be the actual domain of V, where $V.lb$ is the lower bound, which can be initially 0, and $V.ub$ is the upper bound, which can be either $+\infty$ or the value of V in the previous solution. The upper bound is strict since in a branch-and-bound search the next solution must be strictly better than the previous solution. The lower bound $V.lb$ is then set to $\max(V.lb, \underline{V}(C_1..C_k, q))$. We present below the filtering of unassigned decision variables in $\mathcal{G}$.

For each value $c \in [1, k]$, for each unassigned variable G_i, if $c \in Dom(G_i)$, with the assumption of assigning point i into the cluster C_c, we compute a new lower bound of V. Let C'_c be the cluster $C_c \cup \{i\}$ and let $\mathcal{C}' = \{C_l \mid l \neq c\} \cup \{C'_c\}$. A new lower bound V' of V is the value $\underline{V}(\mathcal{C}', q - 1)$, since there still remain $q - 1$ points of $U \setminus \{i\}$ to be assigned to the k clusters. According to (2):

$$\underline{V}(\mathcal{C}', q - 1) = \min_{m \in [0, q-1]} (\underline{V}(\mathcal{C}' \setminus \{C'_c\}, m) + \underline{V}(C'_c, q - 1 - m))$$

For all $m \in [0, q-1]$, we revise the lower bounds $\underline{V}(\mathcal{C}' \setminus \{C'_c\}, m)$ and $\underline{V}(C'_c, m)$ by exploiting informations constructed by Algorithm 1. The revision will be detailed in the remainder of this subsection. The new lower bound V' is computed by line 8 of Algorithm 2. Therefore, since $Dom(V) = [V.lb, V.ub)$, if $V' \geq V.ub$, the variable G_i cannot take the value c. The value c is then removed from $Dom(G_i)$ (lines 9–10). The complexity of Algorithm 2 is the complexity of computing the lower bound $O(kq^2 \log q + qn)$ plus the complexity of the loop (lines 2–10) $O(kq^2)$. The overall complexity therefore is $O(kq^2 \log q + qn)$.

Algorithm 2. Filtering of $wcss(\mathcal{G}, V, d)$

---

input : a partial assignment of $\mathcal{G}$, a set $U = \{i \mid G_i \text{ unassigned}\}$, $q = |U|$

1   $V.lb \leftarrow \max(V.lb, \text{Lower\_bound}())$
2   **for** $c \leftarrow 1$ **to** k **do**
3   **for** $m \leftarrow 0$ **to** $q - 1$ **do**
4         $\underline{V}(\mathcal{C}' \setminus \{C'_c\}, m) \leftarrow \max_{m' \in [0, q-m]}(\underline{V}(\mathcal{C}, m + m') - \underline{V}(C_c, m'))$
5   **foreach** $i \in U$ *such that* $c \in Dom(G_i)$ **do**
6   **for** $m \leftarrow 0$ **to** $q - 1$ **do**
7            $\underline{V}(C'_c, m) \leftarrow ((n_c + m)\underline{V}(C_c, m) + s_2(i, c) + s_3(i, m))/(n_c + m + 1)$
8         $V' \leftarrow \min_{m \in [0, q-1]}(\underline{V}(\mathcal{C}' \setminus \{C'_c\}, m) + \underline{V}(C'_c, q - 1 - m))$
9   **if** $V' \geq V.ub$ **then**
10   remove c from $Dom(G_i)$

Computing $\underline{V}(C'_c, m)$. Let us recall that C'_c is the cluster C_c augmented by point i, and $\underline{V}(C'_c, m)$ is the lower bound of the sum of squares of C'_c after adding any m points of $U \backslash \{i\}$ into C'_c. According to (1):

$$\underline{V}(C'_c, m) = \frac{S_1(C'_c) + \underline{S_2}(C'_c, m)}{n_c + 1 + m}$$

We have $S_1(C'_c) = S_1(C_c) + s_2(i, c)$. The value of $\underline{S_2}(C'_c, m)$ can be revised from $\underline{S_2}(C_c, m)$ by:

$$\underline{S_2}(C'_c, m) = \underline{S_2}(C_c, m) + s_3(i, m)$$

According to (1), we have (line 7 of Algorithm 2):

$$\underline{V}(C'_c, m) = \frac{(n_c + m)\underline{V}(C_c, m) + s_2(i, c) + s_3(i, m)}{n_c + m + 1}$$

Computing $\underline{V}(C' \backslash \{C'_c\}, m)$. This value represents a lower bound of the sum of squares for any assignment of m points in $U \backslash \{i\}$ into the clusters different from C'_c. According to (3), for all $q' \in [m, q]$:

$$\underline{V}(C, q') = \min_{m + m' = q'} (\underline{V}(C \backslash \{C_c\}, m) + \underline{V}(C_c, m'))$$

so for all $q' \in [m, q]$ and with $m + m' = q'$, we have:

$$\underline{V}(C, m + m') \leq \underline{V}(C \backslash \{C_c\}, m) + \underline{V}(C_c, m')$$

which corresponds to:

$$\underline{V}(C \backslash \{C_c\}, m) \geq \underline{V}(C, m + m') - \underline{V}(C_c, m')$$

Since $m \leq q' \leq q$ and $m + m' = q'$, we have $0 \leq m' \leq q - m$. We then have:

$$\underline{V}(C \backslash \{C_c\}, m) \geq \max_{m' \in [0, q-m]} (\underline{V}(C, m + m') - \underline{V}(C_c, m'))$$

We also have:

$$\underline{V}(C' \backslash \{C'_c\}, m) \geq \underline{V}(C \backslash \{C_c\}, m)$$

since $C' \backslash \{C'_c\}$ and $C \backslash \{C_c\}$ denote the same set of clusters, $\underline{V}(C' \backslash \{C'_c\}, m)$ is computed for any m points of $U \backslash \{i\}$, while $\underline{V}(C \backslash \{C_c\}, m)$ is computed for any m points of U. Therefore we can exploit the columns computed by the dynamic program in Algorithm 1 to revise a new lower bound (line 4 in Algorithm 2):

$$\underline{V}(C' \backslash \{C'_c\}, m) = \max_{m' \in [0, q-m]} (\underline{V}(C, m + m') - \underline{V}(C_c, m'))$$

5   Experiments

Our model is implemented with Gecode library version 4.2.7[2], which supports float and integer variables. Experiments have been performed on a 3.0 GHz Core i7 Intel computer with 8 Gb memory under Ubuntu. All our programs are available at http://cp4clustering.com. We have considered the datasets Iris, Soybean and Wine from the UCI Machine Learning Repository[3]. The number of objects and the number of classes are respectively 150 and 3 for Iris, 47 and 4 for Soybean and 178 and 3 for Wine dataset. We compare our model with the approach proposed in [3], based on Integer Linear Programming and column generation and optimizing WCSS criterion with user constraints. Our approach is also compared to COP-kmeans [30] that extends k-means algorithm to integrate user constraints and to Repetitive Branch-and-Bound Algorithm (RBBA) [5], without user constraints, since this algorithm is not able to handle them.

In MiningZinc, a modeling language for constraint-based mining [15], it is shown that clustering with the WCSS criterion can be modeled[4]. The model can be translated to different backend solvers including Gecode. However, because of the intrinsic difficulty of the WCSS criterion, this example model cannot handle 14 points randomly selected from Iris dataset within 30 min, whereas our model takes 0.01s to solve them.

5.1   Optimizing WCSS in Presence of User Constraints

The most widespread constraints in clustering are must-link or cannot-link constraints, since they can be derived from partial previous knowledge (e.g. cluster labels known for a subset of objects). Therefore we choose these two kinds of constraints in the experiments. To generate user constraints, pairs of objects are randomly drawn and either a must-link or a cannot-link constraint is created depending on whether the objects belong to the same class or not. The process is repeated until the desired number for each kind of constraints is reached. For each number of constraints, five different constraint sets are generated for the tests. In each test, we compute the WCSS value, the Rand index of the solution compared to the ground truth partition and the total run-time. The Rand index [25] measures the similarity between two partitions, P and P^*. It is defined by $RI = (a + b)/(a + b + c + d)$, where a and b are the number of pairs of points for which P and P^* are in agreement (a is the number of pairs of points that are in the same class in P and in P^*, b is the number of pairs of points that are in different classes in P and in P^*), c and d are the number of pairs of points for which P and P^* disagree (same class in P but different classes in P^* and vice versa). This index varies between 0 and 1 and the better the partitions are in agreement, the closer to 1. Since experiments are performed on 5 sets of constraints, the mean value μ and the standard deviation σ are computed for

---

[2] http://www.gecode.org
[3] http://archive.ics.uci.edu/ml
[4] http://inductiveconstraints.eu/miningzinc/examples/kmeans.mzn

run-time and RI. A timeout is set to 30 min. A minus sign (-) in the tables means that the timeout has been reached without completing the search. Since the ground truth partition is used to generate user constraints, experiments are done with k equal to the ground truth number of classes for each dataset.

Table 1 gives results for our model (CP) and for the approach based on ILP and column generation (ILP) [3] for the Iris dataset with different numbers #c of must-link constraints. For both the execution time and the Rand index, the mean value of the five tests and the coefficient of variation (σ/μ) are reported. Since the two approaches are exact, they find partitions having the same WCSS value. It can be noticed that must-link constraints help to improve quality of the solution as well as to reduce the execution time for this dataset. Our approach can find a global optimal without user constraints, whereas ILP approach needs at least 100 must-link constraints to be able to prove the optimality. With more than 100 must-link constraints, our approach always takes less time to complete the search. Our approach is also more efficient to handle must-link and cannot-link constraints. Table 2 (left) reports mean execution time in seconds for the five tests and the coefficient of variation σ/μ. In each case, the same number #c of must-link and cannot-link constraints are added. We can see that when #c $\leq$ 75, our approach can complete the search within the timeout and in the other cases, it performs better than ILP. Concerning the Wine dataset, the two approaches cannot prove the optimality of the solution in less than 30 min, when there are less than 150 must-link constraints, as shown in Table 2 (right).

Our CP approach makes better use of cannot-link constraints, as shown in Table 3. This table reports the mean time in seconds and the percentage of tests for which each system completes the search within the timeout. The execution time varies a lot, depending on the constraints. If we consider the Iris database, Table 3 (left) shows that our model is able to find an optimal solution and to prove it for roughly 60 % cases, wheres ILP can solve no cases. If we consider the Wine dataset, Table 3 (right) shows that when 100 must-link and 100 cannot-link constraints are added, CP can solve all the cases, whereas ILP cannot solve them. When 125 must-link constraints and 125 cannot-link constraints are added, both approaches can solve all the cases, but our approach is less time-consuming.

Experiments with the Soybean dataset lead to the same observations. With a number of must-link constraints varying from 10 to 80, the mean run-times for

Table 1. Time (in seconds) and RI for Iris dataset with #c must-link constraints

#c	CP		ILP		RI	
	μ	σ/μ	μ	σ/μ	μ	σ/μ
0	888.99	0.83 %	-	-	0.879	0 %
50	332.06	78.96 %	-	-	0.940	1.66 %
100	7.09	40.18 %	62	74.24 %	0.978	1.68 %
150	0.31	36.39 %	0.45	44.55 %	0.989	0.66 %
200	0.07	24.83 %	0.11	48.03 %	0.992	0.66 %
250	0.05	10.63 %	0.06	35.56 %	0.996	0.70 %
300	0.04	9.01 %	0.04	19.04 %	0.998	0.35 %

Table 2. Time in seconds for Iris dataset with #c must-link and #c cannot-link constraints (left) and Wine dataset with #c must-link constraints (right)

#c	CP		ILP	
	μ	σ/μ	μ	σ/μ
25	969.33	51.98 %	-	-
50	43.85	46.67 %	-	-
75	4.97	150 %	-	-
100	0.41	49.8 %	107	72.35 %
125	0.09	52.07 %	4.4	95.85 %
150	0.06	22.6 %	0.8	50 %

#c	CP	ILP
150	6.84	12.98
200	0.11	0.32
250	0.08	0.11
300	0.08	0.06

Table 3. Iris dataset with #c cannot-link constraints (left) and Wine dataset with #c must-link and #c cannot-link constraints (right)

#c	CP		ILP	
	μ	solved	μ	solved
50	1146.86	20 %	-	0 %
100	719.53	80 %	-	0 %
150	404.77	60 %	-	0 %
200	1130.33	40 %	-	0 %
250	172.81	60 %	-	0 %
300	743.64	60 %	-	0 %

#c	CP		ILP	
	μ	solved	μ	solved
100	10.32	100 %	-	0 %
125	0.35	100 %	497.6	100 %
150	0.12	100 %	13.98	100 %

both CP and ILP approaches decrease from 0.3 s to 0.01 s. However, with different numbers of cannot-link constraints, CP always outperforms ILP approach. For instance, the mean time is 5.19 s (CP) vs. 278.60 s (ILP) with 20 cannot-link constraints, or 2.5 s (CP) vs. 126 s (ILP) with 80 cannot-link constraints.

5.2   Comparisons with COP-kmeans and RBBA

RBBA [5] is based on a repetitive branch-and-bound strategy and finds an exact solution for WCSS. This algorithm takes 0.53 s to find and prove the optimal solution for the Iris dataset and 35.56 s for the Wine dataset. But it does not handle user constraints. Concerning the quality of the lower bound, when most of the points are unassigned, the lower bound of RBBA is better than ours. However when more points are assigned to the clusters, our lower bound can be better than the lower bound of RBBA.

On the other hand, COP-kmeans algorithm [30] extends k-means algorithm to must-link and cannot-link constraints. This algorithm is based on a greedy strategy to find a solution that satisfies all the constraints. When there are only must-link constraints, COP-kmeans always finds a partition satisfying all the constraints, which is a local optimum of WCSS. Nevertheless, when considering also cannot-link constraints, the algorithm may fail to find a solution satisfying all the constraints, even when such a solution exists.

We perform the same tests, but for each set of constraints, COP-kmeans is run 1000 times and we report the number of times COP-kmeans has been able to find a partition. Figure 2 (left) shows the percentage of successes when cannot-link constraints are added. With the two datasets Iris and Wine, COP-kmeans fails to find a partition when 150 constraints are added. Our CP model always find a solution satisfying all the constraints. For Iris dataset, our model succeeds in proving the optimality for roughly 60 % cases (Table 3 left).

Figure 2 (right) gives the results when $\#c$ must-link and $\#c$ cannot-link constraints are added. For Wine dataset, COP-kmeans always fail to find a partition when $\#c = 75$, 125 or 150. Our CP approach finds a solution satisfying all the constraints in all the cases. It completes the search in all the cases for Iris dataset, as shown in Table 2 (left), and in all the cases where $\#c \geq 100$ for Wine dataset, as shown in Table 3 (right).

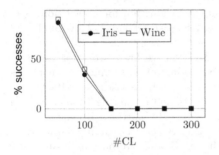 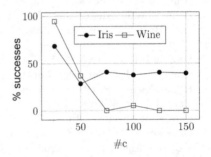

Fig. 2. COP-kmeans with cannot-link (left), with $\#c$ must-link and $\#c$ cannot-link constraints (right)

6 Conclusion

In this paper we address the well-known WCSS criterion in cluster analysis. We develop a new global optimization constraint $wcss$ and present a filtering algorithm, which filters not only the domain of the objective variable but also those of decision variables. This constraint integrated to our CP framework [7, 8] extends it to model constrained minimum sum of squares clustering tasks. Experiments on classic datasets show that our framework outperforms the state-of-the-art best exact approach, which is based on Integer Linear Programming and column generation [3].

Working on search strategies and on constraint propagation enables to improve substantially the efficiency of our CP model. We continue studying these aspects to make the framework able to deal with larger datasets. We are working on exploiting user constraints inside the filtering algorithm, either by using connected components or by modifying the dissimilarities according to the user constraints. We exploit the flexibility of the CP framework to offer a choice between exact or approximate solutions, by studying the use of approximate search strategies, such as local search methods.

References

1. Aloise, D., Deshpande, A., Hansen, P., Popat, P.: NP-hardness of Euclidean Sum-of-squares Clustering. Machine Learning **75**(2), 245–248 (2009)
2. Aloise, D., Hansen, P., Liberti, L.: An improved column generation algorithm for minimum sum-of-squares clustering. Mathematical Programming **131**(1–2), 195–220 (2012)
3. Babaki, B., Guns, T., Nijssen, S.: Constrained clustering using column generation. In: Proceedings of the 11th International Conference on Integration of AI and OR Techniques in Constraint Programming for Combinatorial Optimization Problems, pp. 438–454 (2014)
4. Bilenko, M., Basu, S., Mooney, R.J.: Integrating constraints and metric learning in semi-supervised clustering. In: Proceedings of the 21st International Conference on Machine Learning, pp. 11–18 (2004)
5. Brusco, M., Stahl, S.: Branch-and-Bound Applications in Combinatorial Data Analysis (Statistics and Computing). Springer, 1 edn. (2005)
6. Brusco, M.J.: An enhanced branch-and-bound algorithm for a partitioning problem. British Journal of Mathematical and Statistical Psychology **56**(1), 83–92 (2003)
7. Dao, T.B.H., Duong, K.C., Vrain, C.: A Declarative Framework for Constrained Clustering. In: Proceedings of the European Conference on Machine Learning and Principles and Practice of Knowledge Discovery in Databases, pp. 419–434 (2013)
8. Dao, T.B.H., Duong, K.C., Vrain, C.: Constrained clustering by constraint programming. Artificial Intelligence (2015). doi:10.1016/j.artint.2015.05.006
9. Davidson, I., Ravi, S.S.: Clustering with Constraints: Feasibility Issues and the k-Means Algorithm. In: Proceedings of the 5th SIAM International Conference on Data Mining, pp. 138–149 (2005)
10. Davidson, I., Ravi, S.S.: The Complexity of Non-hierarchical Clustering with Instance and Cluster Level Constraints. Data Mining Knowledge Discovery **14**(1), 25–61 (2007)
11. De Raedt, L., Guns, T., Nijssen, S.: Constraint Programming for Data Mining and Machine Learning. In: Proc. of the 24th AAAI Conference on Artificial Intelligence (2010)
12. Edwards, A.W.F., Cavalli-Sforza, L.L.: A method for cluster analysis. Biometrics **21**(2), 362–375 (1965)
13. Focacci, F., Lodi, A., Milano, M.: Cost-based domain filtering. In: Proceedings of the 5th International Conference on Principles and Practice of Constraint Programming, pp. 189–203 (1999)
14. Gonzalez, T.: Clustering to minimize the maximum intercluster distance. Theoretical Computer Science **38**, 293–306 (1985)
15. Guns, T., Dries, A., Tack, G., Nijssen, S., De Raedt, L.: Miningzinc: A modeling language for constraint-based mining. In: IJCAI (2013)
16. Guns, T., Nijssen, S., De Raedt, L.: k-Pattern set mining under constraints. IEEE Transactions on Knowledge and Data Engineering **25**(2), 402–418 (2013)
17. Han, J., Kamber, M., Pei, J.: Data Mining: Concepts and Techniques, 3rd edn. Morgan Kaufmann Publishers Inc., San Francisco, CA, USA (2011)
18. Hansen, P., Jaumard, B.: Cluster analysis and mathematical programming. Mathematical Programming **79**(1–3), 191–215 (1997)
19. Jensen, R.E.: A dynamic programming algorithm for cluster analysis. Journal of the Operations Research Society of America **7**, 1034–1057 (1969)

20. Koontz, W.L.G., Narendra, P.M., Fukunaga, K.: A branch and bound clustering algorithm. IEEE Trans. Comput. **24**(9), 908–915 (1975)
21. Law, Y.C., Lee, J.H.M.: Global constraints for integer and set value precedence. In: Wallace, M. (ed.) Proceedings of the 10th International Conference on Principles and Practice of Constraint Programming, pp. 362–376 (2004)
22. du Merle, O., Hansen, P., Jaumard, B., Mladenovic, N.: An interior point algorithm for minimum sum-of-squares clustering. SIAM Journal on Scientific Computing **21**(4), 1485–1505 (1999)
23. B.J. van Os, J.M.: Improving Dynamic Programming Strategies for Partitioning. Journal of Classification (2004)
24. Pelleg, D., Baras, D.: K-Means with Large and Noisy Constraint Sets. In: Kok, J.N., Koronacki, J., Lopez de Mantaras, R., Matwin, S., Mladenič, D., Skowron, A. (eds.) ECML 2007. LNCS (LNAI), vol. 4701, pp. 674–682. Springer, Heidelberg (2007)
25. Rand, W.M.: Objective Criteria for the Evaluation of Clustering Methods. Journal of the American Statistical Association **66**(336), 846–850 (1971)
26. Régin, J.C.: Arc consistency for global cardinality constraints with costs. In: Proceedings of the 5th International Conference on Principles and Practice of Constraint Programming, pp. 390–404 (1999)
27. Steinley, D.: k-means clustering: A half-century synthesis. British Journal of Mathematical and Statistical Psychology **59**(1), 1–34 (2006)
28. Ugarte Rojas, W., Boizumault, P., Loudni, S., Crémilleux, B., Lepailleur, A.: Mining (Soft-) Skypatterns Using Dynamic CSP. In: Simonis, H. (ed.) CPAIOR 2014. LNCS, vol. 8451, pp. 71–87. Springer, Heidelberg (2014)
29. Wagstaff, K., Cardie, C.: Clustering with instance-level constraints. In: Proceedings of the 17th International Conference on Machine Learning, pp. 1103–1110 (2000)
30. Wagstaff, K., Cardie, C., Rogers, S., Schrödl, S.: Constrained K-means Clustering with Background Knowledge. In: Proceedings of the 18th International Conference on Machine Learning, pp. 577–584 (2001)

A Constraint Programming Approach
for Non-preemptive Evacuation Scheduling

Caroline Even[1], Andreas Schutt[1,2]([✉]), and Pascal Van Hentenryck[1,3]

[1] NICTA, Optimisation Research Group, Melbourne, Australia
{caroline.even,andreas.schutt,pascal.hentenryck}@nicta.com.au
http://org.nicta.com.au
[2] The University of Melbourne, Parkville, Victoria3010 , Australia
[3] Australian National University, Canberra, Australia

Abstract. Large-scale controlled evacuations require emergency services to select evacuation routes, decide departure times, and mobilize resources to issue orders, all under strict time constraints. Existing algorithms almost always allow for preemptive evacuation schedules, which are less desirable in practice. This paper proposes, for the first time, a constraint-based scheduling model that optimizes the evacuation flow rate (number of vehicles sent at regular time intervals) and evacuation phasing of widely populated areas, while ensuring a non-preemptive evacuation for each residential zone. Two optimization objectives are considered: (1) to maximize the number of evacuees reaching safety and (2) to minimize the overall duration of the evacuation. Preliminary results on a set of real-world instances show that the approach can produce, within a few seconds, a non-preemptive evacuation schedule which is either optimal or at most 6% away of the optimal preemptive solution.

Keywords: Constraint-based evacuation scheduling · Non-preemptive scheduling · Phased evacuation · Simultaneous evacuation · Actionable plan · Real-world operational constraints · Network flow problem

1 Introduction

Evacuation planning is a critical part of the preparation and response to natural and man-made disasters. Evacuation planning assists evacuation agencies in mitigating the negative effects of a disaster, such as loss or harm to life, by providing them guidelines and operational evacuation procedures so that they can make informed decisions about whether, how and when to evacuate residents. In the case of controlled evacuations, evacuation agencies instruct each endangered resident to follow a specific evacuation route at a given departure time. To communicate this information in a timely fashion, evacuation planners must design plans which take into account operational constraints arising

NICTA is funded by the Australian Government through the Department of Communications and the Australian Research Council through the ICT Centre of Excellence Program.

G. Pesant (Ed.): CP 2015, LNCS 9255, pp. 574–591, 2015.
DOI: 10.1007/978-3-319-23219-5_40

in actual evacuations. In particular, two critical challenges are the deployment of enough resources to give precise and timely evacuation instructions to the endangered population and the compliance of the endangered population to the evacuation orders. In practice, the control of an evacuation is achieved through a mobilization process, during which mobilized resources are sent to each residential area in order to give instructions to endangered people. The number of mobilized resources determines the overall rate at which evacuees leave. Finally, to maximize the chances of compliance and success of a controlled evacuation, the evacuation and mobilization plans must be easy to deploy for evacuation agencies and should not leave, to the evacuees, uncontrolled alternative routes that would affect the evacuation negatively.

Surprisingly, local authorities still primarly rely on expert knowledge and simple heuristics to design and execute evacuation plans, and rarely integrate human behavioral models in the process. This is partly explained by the limited availability of approaches producing evacuation plans that follow the current practice. Apart from a few exceptions [2,5,6,9,11,13] existing evacuation approaches rely on free-flow models which assume that evacuees can be dynamically routed in the transportation network [3,10,14]. These free-flow models however violate a desirable operational constraint in actual evacuation plans, i.e., the fact that all evacuees in a given residential zone should preferably follow the same evacuation route.

Recently, a handful of studies considered evacuation plans where each residential area is assigned a single evacuation path. These studies define both a set of evacuation routes and a departure schedule. Huibregtse et al. [9] propose a two-stage algorithm that first generates a set of evacuation routes and feasible departure times, and then assigns a route and time to each evacuated area using an ant colony optimization algorithm. In subsequent work, the authors studied the robustness of the produced solution [7], and strategies to improve the compliance of evacuees [8]. Pillac et al. [13] first introduced the Conflict-based Path Generation (CPG) approach which was extended to contraflows by Even et al. [5]. CPG features a master problem which uses paths for each residential node to schedule the evacuation and a pricing problem which heuristically generates new paths addressing the conflicts in the evacuation schedule.

These evacuation algorithms however do not guarantee that evacuees will follow instructions. If the evacuation plan contains forks in the road, evacuees may decide to change their evacuation routes as the evacuation progresses. This issue is addressed in [1,6] which propose evacuation plans without forks. The resulting evacuation plan can be thought of as a forest of evacuation trees where each tree is rooted at a safe node (representing, say, an evacuation center) and with residential areas at the leaves. By closing roads or controlling intersections, these evacuation trees ensure the compliance of the evacuees and avoid congestions induced by drivers slowing down at a fork. Even et al. [6] produce such convergent evacuation plans by decomposing the evacuation problem in a tree-design problem and an evacuation scheduling problem. Andreas and Smith [1] developed a Benders decomposition algorithm that selects convergent evacuation routes that are robust to a set of disaster scenarios.

All of the approaches reviewed above allow preemption: The evacuation of a residential area can be interrupted and restarted arbitrarily. This is not desirable in practice, since such schedules will confuse both evacuees and emergency services, and will be hard to enforce. Disallowing preemption makes the evacuation plans harder to solve for these mathematical programming based approaches opposing to a constraint programming one. Non-preemptive schedules have been considered in [4,12] in rather different ways. In [4], a phased evacuation plan evacuates each area separately, guaranteeing that no vehicles from different areas travel on a same path at the same time. By definition, phased evacuation does not merge evacuation flows, which is motivated by empirical evidence that such merging can reduce the road network capacities. The algorithm in [12] is a column-generation approach for simultaneous evacuation, i.e., evacuations where multiple paths can share the same road segment at the same time. Each column represents the combination of a path, a departure time, and a response curve capturing the behavioral response of evacuees for each evacuation area. Here the flow rate of each evacuation area is restricted to pre-existing response curves, and columns are generated individually for each evacuation area. This column-generation approach requires a discretization of the evacuation horizon.

This paper proposes, for the first time, a constraint programming approach to generate non-preemptive evacuation schedules. It takes as input a set of evacuation routes, which are either chosen by emergency services or computed by an external algorithm. The constraint-based scheduling model associates a task with each residential area, uses decision variables for modeling the number of evacuees (i.e., the *flow*), the number of vehicles to be evacuated per time unit (i.e., the *flow rate*), and the starting time of the area evacuation; It also uses cumulative constraints to model the road capacities. In addition, the paper presents a decomposition scheme and dominance relationships that decrease the computational complexity by exploiting the problem structure. Contrary to [12], the constraint programming model uses a decision variable for the flow rate of each

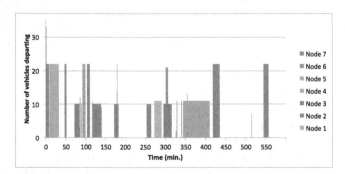

Fig. 1. Departure times and flows of 7 residential areas of the HN80 instance with the preemptive algorithm FSP from [6].

evacuation area (instead of a fixed set of values) and avoids discretizing time. In contrast to [4], the constraint programming model allows for simultaneous evacuation while satisfying practice-related constraints.

The constraint programming model was applied on a real-life evacuation case study for the Hawkesbury-Nepean region in New South Wales, Australia. This region is a massive flood plain protected from a catchment area (the blue mountains) by the Warragamba dam. A major spill would create damages that may reach billions of dollars and require the evacuation of about 80,000 people. Preliminary experimental results indicate that the constraint programming model can be used to generate non-preemptive schedules that are almost always within 5% of the optimal preemptive schedules generated in prior work. These results hold both for maximizing the number of evacuees for a given time horizon and for minimizing the clearance time (i.e., the earliest time when everyone is evacuated). These results are particularly interesting, given that the optimal preemptive solutions produce evacuation plans which are far from practical. Indeed, Fig. 1 shows the repartition of departure times for seven residential areas in the original HN80 instance using a preemptive schedule produced by the algorithm in [6]. Observe how departure times are widely distributed within the scheduling horizon, indicating that the plan makes heavy use of preemption and is virtually impossible to implement in practice. Finally, experimental results on the phased version of the algorithm indicate that phased evacuations are much less time effective, and should not be the preferred method for short-notice or no-notice evacuation.

The rest of this paper is organized as follows. Section 2 defines the problem. Section 3 presents the constraint programming model, including the core model, the decomposition scheme, the dominance relationships, and the search procedure. Section 4 presents the experimental results. Section 5 concludes the paper.

2   Problem Description

The Evacuation Planning Problem (EPP) was introduced by the authors in [11]. It is defined on a directed graph $\mathcal{G} = (\mathcal{N} = \mathcal{E} \cup \mathcal{T} \cup \mathcal{S}, \mathcal{A})$, where $\mathcal{E}$, $\mathcal{T}$, and $\mathcal{S}$ are the set of evacuation, transit, and safe nodes respectively, and $\mathcal{A}$ is the set of edges. The EPP is designed to respond to a disaster scenario, such as a flood, which may determine a time at which some edges become unavailable. Each evacuation node $k \in \mathcal{E}$ is characterized by a number of evacuees d_k, while each arc e is associated with a triple (t_e, u_e, b_e), where t_e is the travel time, u_e is the capacity, and b_e is the time at which the arc becomes unavailable. We denote by $e.tail$ (resp. $e.head$) the tail (resp. head) of an edge e. The problem is defined over a *scheduling horizon* $\mathcal{H}$, which depends on the disaster forecast and the time to mobilize resources. The objective is either (1) to maximize the total number of evacuees reaching a safe node (for a fixed horizon) or (2) to minimize the time at which the last evacuee reaches a safe zone (for a variable horizon). In the following, we assume that the evacuation is carried out using private vehicles, but the proposed approach could be adapted to other contexts, such as building evacuation.

This paper extends the EPP to the *non-preemptive* (simultaneous) evacuation planning problem (NEPP) and the *non-preemptive phased* evacuation planning problem (NPEPP). Both are designed to assist evacuation planners with the scheduling of fully controlled evacuations. Given a set of evacuation paths, the NEPP decides the start time, flow, and flow rate at which vehicles are evacuating each individual evacuation node, ensuring that the evacuation operates without interruption. The NEPP allows several evacuation nodes to use the same road segments at the same time. In contrast, the NPEPP guarantees that no two evacuation nodes use the same road segment at the same time. The practical interest of the NPEPP is to evacuate designated priority areas quickly and efficiently, by limiting the risk of any delay caused by slowdown or traffic accidents which may result from merging traffic.

Formally, an evacuation plan associates with each evacuation area $k \in \mathcal{E}$ exactly one *evacuation path* p_k which is used to route all residents in k to a same safe node. Let $\Omega_{p_k} = \bigcup_{k \in \mathcal{E}} p_k$ the *set of evacuation paths* for all evacuations nodes in $\mathcal{E}$. The characteristics of a path p_k are as follows. We denote by $\mathcal{A}_{p_k}$ (resp. $\mathcal{N}_{p_k}$) the *set of edges* (resp. *nodes*) of p_k and by $\mathcal{E}(e)$ the *set of evacuation areas* whose path contains edge e, i.e., $e \in \mathcal{A}_{p_k}$. The *travel time* $t_{k,n}$ between the evacuation area k and a node $n \in \mathcal{N}_{p_k}$ is equal to the sum of the path edges travel times separating k from n and t_{p_k} is the *total travel time* between the start and end of p_k. The *path capacity* u_{p_k} is the minimum edge capacity of p_k. The *last possible departure time* LASTDEP(p_k) along path p_k, i.e., the latest time at which a vehicle can depart on p_k without being blocked, is easily derived from all $t_{k,n}(n \in \mathcal{N}_{p_k})$ and the time b_e at which each path edge $e \in \mathcal{A}_{p_k}$ becomes unavailable. If none of the path edges $e \in \mathcal{A}_{p_k}$ are cut by the disaster then LASTDEP(p_k) $= \infty$; otherwise LASTDEP(p_k) $= \min_{e \in \mathcal{A}_{p_k}}(b_e - t_{k,e.head})$. Note that the latest path departure time only depends on the path characteristics and not on $\mathcal{H}$.

3   The Constraint Programming Model

The NEPP and NPEPP are two optimization problems whose objective is either to maximize the number of evacuees or to minimize the overall evacuation clearance time. The key contribution of this paper is to model them as constraint-based scheduling problems and to use CP to solve them. This modeling avoids time dicretization and makes it possible to design non-preemptive plans with variable flow rates. This section presents the constraint-based scheduling models, including their decision variables, their domains, and the constraints common to both problems. This section then presents the constraint-based scheduling models for NEPP and NPEPP.

3.1   Decision Variables

The models associate with each evacuation area $k \in \mathcal{E}$ the following decision variables: the total flow of vehicles evacuated FLOW_k (i.e., the number of vehicles evacuated from area k), the flow rate λ_{FLOW_k} representing the number of vehicles departing per unit of time, the evacuation start time START_k (i.e., the time at which the first vehicle is evacuated from area k), the evacuation end time END_k, and the total evacuation duration time DUR_k. The last three decision variables are encapsulated into a task variable TASK_k which links the evacuation start time, the evacuation end time, and the evacuation duration and ensures that $\text{START}_k + \text{DUR}_k = \text{END}_k$.

The decision variables range over natural numbers. The flow and flow rates can only be non-negative and integral since a number of vehicles is a whole entity. The models use a time step of one minute for flow rates and task variables which, from an operational standpoint, is a very fine level of granularity: Any time step of finer granularity would only be too complex to handle in practice. The domains of the decision variables are defined as follows. The flow variable is at most equal to the evacuation demand: $\text{FLOW}_k \in [0, d_k]$ where $[a, b] = \{v \in \mathbb{N} \mid a \le v \le b\}$. The flow-rate variable has an upper bound which is the minimum of the evacuation demand and the path capacity rounded down to the nearest integer, i.e., $\lambda_{\text{FLOW}_k} \in [1, \min(d_k, \lfloor u_{p_k} \rfloor)]$. The upper bounds for the evacuation start time and evacuation end time are the smallest of the scheduling horizon minus the path travel time, which is rounded up to the nearest integer, and the latest path departure time, i.e., $\text{START}_k \in [0, \min(\mathcal{H} - \lceil t_{p_k} \rceil, \lfloor \text{LASTDEP}(p_k) \rfloor)]$. The evacuation of an area k can last at most d_k minutes assuming the flow rate is set to one vehicle per minute: $\text{DUR}_k \in [0, d_k]$. Note that the lower bound for duration is zero in order to capture the possibility of not evacuating the area.

3.2   Constraints

The NEPP requires to schedule the flow of evacuees coming from each evacuation area k on their respective path p_k such that, at any instant t, the flow sent on all paths through the network does not exceed the network edges capacities. These flow constraints can be expressed in terms of cumulative constraints. Consider an edge e and the set $\mathcal{E}(e)$ of evacuation areas whose evacuation paths use e. For each evacuation area $k \in \mathcal{E}(e)$, the model introduces a new task TASK_k^e which is a view over task TASK_k satisfying:

$$\text{START}_k^e = \text{START}_k + t_{k,e.tail} \ , \quad \text{DUR}_k^e = \text{DUR}_k \ , \quad \text{END}_k^e = \text{END}_k + t_{k,e.tail} \ .$$

This new task variable accounts for the number of vehicles from evacuation area k traveling on edge e at any time during the scheduling horizon. Note that $t_{k,e.tail}$ is computed as the sum of the travel times on each edge, each rounded up to the next integer for consistency with the domains of the decision variables. While this approximation may slightly overestimates travel times, it also counterbalances possible slowdowns in real-life traffic, which are not taken into account in this model.

The constraint-based scheduling model for the NEPP introduces the following cumulative constraint for edge e:

$$\texttt{cumulative}(\{(\text{TASK}_k^e, \lambda_{\text{FLOW}_k}) \mid k \in \mathcal{E}(e)\}, u_e).$$

The constraint-based scheduling model for the NPEPP introduces a disjunctive constraint for edge e instead:

$$\texttt{disjunctive}(\{\text{TASK}_k^e \mid k \in \mathcal{E}(e)\}). \tag{1}$$

3.3   The Constraint-Based Scheduling Models

We are now in a position to present a constraint-based scheduling model for NEPP-MF:

$$\max \quad \text{OBJECTIVE} = \sum_{k \in \mathcal{E}} \text{FLOW}_k \tag{2}$$

$$\text{s.t.} \quad \text{FLOW}_k^{\text{UB}} = \text{DUR}_k \times \lambda_{\text{FLOW}_k} \qquad \forall k \in \mathcal{E} \tag{3}$$

$$\text{FLOW}_k = \min(\text{FLOW}_k^{\text{UB}}, d_k) \qquad \forall k \in \mathcal{E} \tag{4}$$

$$\texttt{cumulative}(\{(\text{TASK}_k^e, \lambda_{\text{FLOW}_k}) \mid k \in \mathcal{E}(e)\}, u_e) \qquad \forall e \in \mathcal{A} \tag{5}$$

The objective (2) maximizes the number of evacuated vehicles. Constraints (3) and (4) link the flow, flow rate, and evacuation duration together, by ensuring that the total flow for each area k is the minimum of the evacuation demand and the flow rate multiplied by the evacuation duration. They use an auxiliary variable $\text{FLOW}_k^{\text{UB}}$ denoting an upper bound on the number of vehicles evacuated from area k. Constraints (5) impose the capacity constraints.

The model NEPP-SAT is the satisfaction problem version of NEPP-MF where the objective (2) has been removed and the constraint

$$\text{FLOW}_k = d_k \quad \forall k \in \mathcal{E} \tag{6}$$

has been added to ensure that every vehicle is evacuated.

To minimize clearance time, i.e., to find the minimal scheduling horizon such that all vehicles are evacuated, it suffices to add the objective function to NEPP-SAT

$$\min \quad \text{OBJECTIVE} = \max_{k \in \mathcal{E}}(\text{END}_k + t_{p_k}) \tag{7}$$

and to relax the start and end time domains to $[0, \mathcal{H}^{\text{UB}}]$ where $\mathcal{H}^{\text{UB}}$ is an upper bound on the horizon required to evacuate all vehicles to a shelter. The resulting constraint-based scheduling model is denoted by NEPP-CT.

A constraint programming formulation NPEPP-MF of the non-preemptive phased evacuation planning problem can be obtained from NEPP-MF by replacing (5) with (1), which prevents the flows from two distinct origins to travel on the same edge at the same time. NPEPP-SAT, which is the satisfaction problem

version of NPEPP-MF, is obtained by removing the objective (2) and adding
the constraint (6). NPEPP-CT, which minimizes the evacuation clearance time,
is obtained from NPEPP-SAT by adding the objective (7). Note that since the
flow-rate bounds ensure that edge capacities are always respected in NPEPP, the
flow-rate variable can be directly set to its upper bound to maximize evacuation
efficiency. Hence, the following constraints are added to the NPEPP model:

$$\lambda_{\text{FLOW}_k} = \min(d_k, u_{p_k}) \quad \forall k \in \mathcal{E}. \tag{8}$$

3.4   Problem Decomposition

This section shows how the NEPP and NPEPP can be decomposed by understanding which paths compete for edges and/or how they relate to each other.
In the following we introduce the *path dependency relationship*.

Definition 1. *Two paths p_x and p_y are directly dependent, which is denoted
by $p_x \triangle p_y$, if and only if they share at least a common edge, i.e., $\mathcal{A}_{p_x} \cap \mathcal{A}_{p_y} \neq \emptyset$.*

Definition 2. *Two paths p_x and p_z are indirectly dependent if and only if
$\neg(p_x \triangle p_z)$ and there exists a sequence of directly dependent paths $p_{y_1}, \ldots, p_{y_n}$
such that $p_x \triangle p_{y_1}$, $p_{y_n} \triangle p_z$ and $p_{y_1} \triangle p_{y_2}, \ldots, p_{y_{n-1}} \triangle p_{y_n}$.*

Definition 3. *Two paths p_x and p_y are dependent, which is denoted by $p_x \top p_y$,
if they are either directly dependent or indirectly dependent. Conversely, paths p_x
and p_y are independent, which is denoted by $p_x \bot p_y$, if they are neither directly
dependent nor indirectly dependent.*

Obviously, the *path dependency* $\top$ forms an equivalence relationship, i.e., $\top$ is
reflexive, symmetric, and transitive.

The key idea behind the decomposition is to partition the evacuation areas
$\mathcal{E}$ into $\Upsilon = \{\mathcal{D}_0, \ldots, \mathcal{D}_n\}$ in such a way that any two paths, respectively from
the set of evacuation areas $\mathcal{D}_i$ and $\mathcal{D}_j$ $(0 \leq i < j \leq n)$, are independent. As a

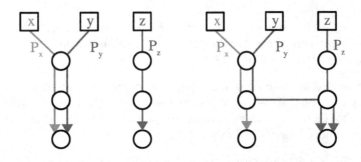

(a) Directly dependent paths   (b) Indirectly dependent paths

Fig. 2. Illustrating independent nodes and paths.

result, it is possible to solve the overall model by solving each subproblem $\mathcal{D}_i$ $(0 \leq i \leq n)$ independently and concurrently. Figure 2a illustrates two sets of evacuation nodes $\mathcal{D}_0 = \{x, y\}$ and $\mathcal{D}_1 = \{z\}$ where paths p_x and p_y are directly dependent and there are no indirectly dependent paths. Figure 2b illustrates a single set of evacuation nodes $\mathcal{D} = \{x, y, z\}$ where the set of paths $\{p_x, p_y\}$ and $\{p_y, p_z\}$ are directly dependent, while the set of paths $\{p_x, p_z\}$ are indirectly dependent. We now formalize these concepts.

Definition 4. *Let $\Omega^{\mathcal{D}_i}_{p_k}$ denote the paths of the set of nodes $\mathcal{D}_i$. Two sets $\mathcal{D}_0$ and $\mathcal{D}_1$ of evacuation areas are independent if and only if any two paths from $\Omega^{\mathcal{D}_0}_{p_k}$ and $\Omega^{\mathcal{D}_1}_{p_k}$ respectively are independent. They are dependent otherwise.*

Theorem 1. *Let $\Upsilon = \{\mathcal{D}_0, \dots, \mathcal{D}_n\}$ be a partition of $\mathcal{E}$ such that D_i and D_j $(0 \leq i < j \leq n)$ are independent. Then the NEPP and NPEPP can be solved by concatenating the solutions of their subproblems $\mathcal{D}_i$ $(0 \leq i \leq n)$.*

The partition $\Upsilon = \{\mathcal{D}_0, \dots, \mathcal{D}_n\}$ can be generated by an algorithm computing the strongly connected components of a graph. Let $\mathcal{G}_{\Omega_{p_k}}$ be the directed graph consisting of the edges and vertices of all paths $p_k \in \Omega_{p_k}$ and let $\mathcal{G}^u_{\Omega_{p_k}}$ be its undirected counterpart, i.e., the graph obtained after ignoring the direction of all edges in $\mathcal{G}_{\Omega_{p_k}}$. The strongly connected components of $\mathcal{G}^u_{\Omega_{p_k}}$ define the partition Υ.

3.5   Dominance Relationships

This section shows how to exploit dominance relationships to simplify the constraint-based scheduling models. The key idea is to recognize that the capacity constraints of some edges are always guaranteed to be satisfied after introducing constraints on other particular edges.

Definition 5. *Let $\mathcal{A}$ be a set of edges and $e, e' \in \mathcal{A}$, $e \neq e'$. Edge e dominates e', denoted by $e > e'$, if and only if*

- *For simultaneous evacuation, the capacity of e is less than or equal to the capacity of e': $u_e \leq u_{e'}$;*
- *The set of paths using e' is a subset of the set of paths using e: $\mathcal{E}(e') \subseteq \mathcal{E}(e)$;*
- *For non-convergent evacuation paths, the travel times for evacuation paths in $\mathcal{E}(e')$ between e and e' are the same.*

Note that two edges may be dominating each other. For this reason and without loss of generality, this paper breaks ties arbitrarily (e.g., by selecting the edge closer to a safe node as the dominating edge). Note also that the capacity condition is ignored for phased evacuation.

Theorem 2. *Let $\mathcal{A}^>$ the set of dominating edges in $\mathcal{A}$. We can safely substitute $\mathcal{A}^>$ for $\mathcal{A}$ in (5) in NEPP-MF such that the cumulative constraints are only stated for dominating edges. Similar results hold for NEPP-CT/SAT, and for the disjunctive constraints in NPEPP-MF/CT/SAT.*

3.6   Additional Constraints to a Real-World Evacuation Scheduling Problem

The flexibility of the constraint-based evacuation scheduling approach allows to easily include many constraints appearing in real-world evacuation scheduling. For example, each flow rate variable domain may be restricted to a subset of values only, in order to account for the number of door-knocking resources available to schedule the evacuation [12]. Other real-world evacuation constraints may restrict departure times in order to wait for a more accurate prediction of an upcoming disaster or to restrict evacuation end times to ensure that the last vehicle has left a certain amount of time before the disaster strikes the evacuation area.

3.7   Complexity of Phased Evacuations with Convergent Paths

When using convergent paths, phased evacuations may be solved in polynomial time.

Theorem 3. *Model NPEPP-MF can be solved in polynomial time for convergent paths if all evacuation paths share the same latest completion time at the safe node.*

Proof (Sketch). Using the decomposition method and the dominance criterion, each subproblem with at least two evacuation paths includes exactly one dominating edge e which is part of every evacuation path. An optimal schedule can then be obtained in two steps. The first step builds a preemptive schedule by a sweep over the time starting from the minimal earliest start time of the tasks on e and ending before the shared completion time. For each point in time, it schedules one eligible task (if existing) with the largest flow rate (ties are broken arbitrarily) where a task is eligible if the point in time is not after its earliest start time on e and it has not been fully scheduled before that time. Note if a task is almost fully scheduled except the last evacuation batch then this step considers the actual flow rate of the last batch instead, which may be smaller than the task flow rate. This preemptive schedule is optimal as for each point in time, the unscheduled eligible tasks do not have an (actual) greater flow rate than the scheduled ones. The second step converts this optimal preemptive schedule to a non-preemptive one by postponing tasks interrupting others until after the interrupted tasks are completed. This transformation does not change the flows and hence the non-preemptive schedule has the same objective value as the preemptive schedule.   □

3.8   The Search Procedure

The search procedure considers each *unassigned* task in turn and assigns its underlying variables. A task TASK_k is *unassigned* if the domain of any of its variables $\{\text{START}_k, \text{DUR}_k, \text{END}_k, \text{FLOW}_k, \lambda_{\text{FLOW}_k}\}$ has more than one value. The search procedure selects an unassigned task TASK_k and then branches on all its underlying variables until they all are assigned. Depending on the considered problem, the models use different heuristics to (1) find the next unassigned task and to (2) select the next task variable to branch on and the value to assign.

For NPEPP, the flow rate is directly set to the maximum value. The search strategy is determined by the problem objective as follows. If the objective maximizes the number of evacuees for a given scheduling horizon $\mathcal{H}$, the search is divided into two steps. The first step selects a task with an unfixed duration and the highest remaining actual flow rate. If the lower bound on duration of the task is at least two time units less than its maximal duration then a minimal duration of the maximal duration minus 1 is imposed and a maximal duration of the maximal duration minus 2 on backtracking. Otherwise the search assigns duration in decreasing order starting with the largest value in its domain. The second step selects tasks according to their earliest start time and assigns a start time in increasing order.[1] If the objective is to minimize the horizon $\mathcal{H}$ such that all vehicles are evacuated, then the search selects the next unassigned task with earliest start time by increasing order among all dominating edges, selecting the one with maximal flow rate to break ties, and to label the start time in increasing order.

For NEPP, the different search heuristics are as follows. For the choice (1), the strategy (1A) randomly selects an unassigned task, and performs geometric restarts when the number of backtracks equals twice the number of variables in the model, using a growth factor of 1.5. The strategy (1B) consists in selecting the next unassigned task in decreasing order of evacuation demand for the dominating edge with the greatest number of tasks. For the choice (2), the strategy (2A) first labels the flow rate in increasing order, then the task start time also in increasing order and, finally, the flow in decreasing order. The strategy (2B) first labels the flow rate in decreasing order, then the flow in decreasing order again and, finally, the start time in increasing order.

4   Experimental Results

This section reports experiments on a set of instances used in [6]. These instances are derived from a real-world case study: the evacuation of the Hawkesbury-Nepean (HN) floodplain. The HN evacuation graph contains 80 evacuated nodes, 5 safe nodes, 184 transit nodes, 580 edges and 38343 vehicles to evacuate. The experimental results also consider a class of instances HN80-Ix using the HN evacuation graph but a number of evacuees scaled by a factor $x \in \{1.1, 1.2, 1.4, 1.7, 2.0, 2.5, 3.0\}$ to model population growth. For simplicity, the experiments did not consider a flood scenario and assume that network edges are always available within the scheduling horizon $\mathcal{H}$. It is easy to generalize the results to various flood scenarios.

---

[1] Note that the search procedure does not assume convergent paths or restrictions on the latest arrival times at the safe node.

Table 1. The strongly connected components associated with each HN80-Ix instance.

Instance	#vehicles	#scc	scc details
HN80	38343	5	{22,9048}, {17,10169}, {14,6490}, {22,9534}, {5,3102}
HN80-I1.1	42183	4	{1,751}, {2,1281}, {40,23656}, {37,16495}
HN80-I1.2	46009	3	{2,1398}, {35,18057}, {43,26554}
HN80-I1.4	53677	5	{2,1631}, {28,16737}, {27,19225}, {4,3824}, {19,12260}
HN80-I1.7	65187	4	{22,16992}, {2,1980}, {42,33240}, {14,12975}
HN80-I2.0	76686	4	{15,13974}, {38,40612}, {2,2330}, {25,19770}
HN80-I2.5	95879	5	{32,36260}, {6,11214}, {16,17983}, {6,9324}, {20,21098}
HN80-I3.0	115029	5	{5,11574}, {12,14184}, {19,23403}, {7,13068}, {29,39651}

For each evacuation instance, a set of convergent evacuation paths was obtained from the TDFS approach [6]. The TDFS model is a MIP which is highly scalable as it aggregates edge capacities and abstracts time. The paths were obtained for the maximization of the number of evacuees within a scheduling horizon of 10 hours, allowing preemptive evacuation scheduling. Thus, for each instance, the set of evacuation paths can be thought of as forming a forest where each evacuation tree is rooted at a safe node and each leaf is an evacuated node. In this particular case, each evacuation tree is a strongly connected component. It is important to emphasize that paths are not necessarily the same for the different HN instances, nor necessarily optimal for non-preemptive scheduling, which explains some non-monotonic behavior in the results. Table 1 reports the evacuation paths, the number of vehicles to evacuate (#vehicles), the number of strongly connected components (#scc), the number of evacuated nodes, the number of vehicles per scc (scc details) for each HN80-Ix instance. Each strongly connected component is represented by a pair $\{x, y\}$ where x is the number of evacuated nodes and y the number of vehicles to evacuate.

The experimental results compare the flow scheduling results obtained with the NEPP and NPEPP approaches and the flow scheduling problem (FSP) formulation presented in [6]. The FSP is solved using a LP and it relaxes the non-preemptive constraints. Indeed, the flow leaving an evacuated node may be interrupted and restarted subsequently at any time $t \in \mathcal{H}$, possibly multiple times. Moreover, the flow rates in the FSP algorithm are not necessarily constant over time, giving substantial scheduling flexibility to the FSP but making it very difficult to implement in practice. Once again, the FSP comes in two versions. The objective of the core FSP is to maximize the number of vehicles reaching safety, while the objective of FSP-CT is to minimize the evacuation clearance time. In order to compare the FSP algorithm and the constraint programming approaches of this paper fairly, the FSP is solved with a time discretization of 1 minute. The experiments for the NEPP and NPEPP models use different search

Table 2. Percentage of Vehicles Evacuated with FSP, NEPP-MF/SAT, NPEPP-MF/SAT.

	FSP		NEPP-MF/SAT			NPEPP-MF/SAT	
Instance	CPU (s)	Perc. Evac.	CPU (s)	Perc. Evac.	Search	CPU (s)	Perc. Evac.
HN80	0.9	100.0%	0.2	100.0% (SAT)	{1B, 2B}	3.4	96.9%
HN80-I1.1	1.1	100.0%	1538.9	99.2%	{1A, 2B}	1295.9	58.4%
HN80-I1.2	1.0	100.0%	0.4	100.0% (SAT)	{1B, 2B}	1444.4s	57.7%
HN80-I1.4	1.3	100.0%	1347.5	99.3%	{1A, 2B}	307.0	73.0%
HN80-I1.7	1.8	100.0%	1374.9	97.8%	{1A, 2A}	0.3	59.0%
HN80-I2.0	2.0	97.9%	1770.1	93.1%	{1A, 2B}	5.9	52.8%
HN80-I2.5	1.8	82.2%	1664.1	79.2%	{1A, 2B}	0.1	51.5%
HN80-I3.0	1.4	69.2%	887.2	67.5%	{1A, 2B}	0.1	43.1%

heuristics and each experimental run was given a maximal runtime of 1800 seconds per strongly connected component. The results were obtained on 64-bit machines with 3.1GHz AMD 6-Core Opteron 4334 and 64Gb of RAM and the scheduling algorithms were implemented using the programming language JAVA 8 and the constraint solver Choco 3.3.0, except for NPEPP-MF where the search was implemented in ObjectiveCP.

Maximizing the Flow of Evacuees. Table 2 compares, for each HN80-Ix instance and a 10-hour scheduling horizon, the percentage of vehicles evacuated (Perc. Evac.) and the solving time in seconds (CPU (s)) with FSP, NEPP-MF/SAT and NPEPP-MF/SAT. All solutions found with FSP are optimal and are thus an upper bound on the number of vehicles that can be evacuated with NEPP and NPEPP. Prior to solving NEPP-MF (resp. NPEPP-MF), the algorithm attempts to solve NEPP-SAT (resp. NPEPP-SAT) with a 60s time limit and, when this is successful, the annotation (SAT) is added next to the percentage of vehicles evacuated. As we make use of decomposition and parallel computing, the reported CPU for NEPP/NPEPP is the latest of the time at which the best solution is found among all strongly connected components. The table reports the best results across the heuristics, i.e., the run where the most vehicles are evacuated ; for the random strategy, the best result is reported across 10 runs (note that the standard deviation for the objective value ranges between 0.4% and 1.1% only across all instances). The search strategy for the best run is shown in column (Search) as a combination {TaskVar, VarOrder} where TaskVar is the heuristic for choosing the next task variable and VarOrder is the heuristic for labeling the task variables.

The results highlight the fact that the constraint-based simultaneous scheduling model finds very high-quality results. On the first five instances, with population growth up to 70%, the solutions of NEPP-MF are within 2.2% of the

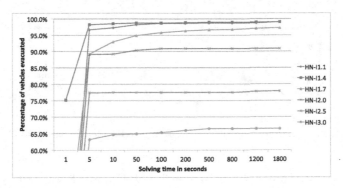

Fig. 3. Quality of solutions over Time for NEPP-MF.

preemptive bound. This is also the case for the largest instance. In the worst case, the constraint-based scheduling model is about 4.9% away from the preemptive bound. It is thus reasonable to conclude that the constraint-based algorithms may be of significant value to emergency services as they produce realistic plans for large-scale controlled evacuations. For NPEPP-MF, the solver found optimal solutions and proved optimality for all instances, except HN80-I1.1 and HN80-I1.2 for which the best found solution was within 0.1% of the optimal one.[2] The results indicate that a phased evacuation is much less effective in practice and decreases the number of evacuees reaching safety by up to 40% in many instances. Unless phased evacuations allow an evacuation of all endangered people, they are unlikely to be applied in practice, even if they guarantee the absence of traffic merging.

Figure 3 shows how the quality of solutions improves over time for all HN-Ix instances which are not completely evacuated, for a particular run. For all instances, a high-quality solution is found within 10 seconds, which makes the algorithm applicable to a wide variety of situations. When practical, giving the algorithm more time may still produce significant benefits: For instance, on HN-I1.7 the percentage of vehicles increases from 93.0% to 97.6% when the algorithm is given 1800s. Such improvements are significant in practice since they may be the difference between life and death.

Profile of the Evacuation Schedules. To demontrate the benefits of NEPP, it is useful to look at the evacuation profiles produced by the various algorithms. Recall that Fig. 1 displays a repartition of departure times for seven evacuated nodes in the original HN80 instance in the optimal solution produced by the FSP solver. *The key observation is that, for several residential areas, the departure times are widely distributed within the scheduling horizon, indicating that the FSP solution makes heavy use of preemption.* In the FSP solution, the number of vehicles departing at each time step is often equal to the path capacity. But

---

[2] In our experiments, the problem NPEPP satisfies the condition for Theorem 3. Thus, these instances can be solved almost instantly using the algorithm outlined in the proof of Theorem 3.

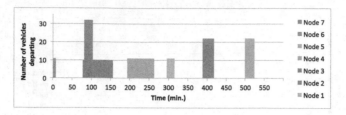

Fig. 4. Departure times and flows of 7 evacuated nodes for the NEPP solver.

Table 3. Evacuation clearance time (CT) with FSP-CT, NEPP-CT, and NPEPP-CT.

Instance	FSP-CT		NEPP-CT			NPEPP-CT	
	CPU (s)	CT (min)	CPU (s)	CT (min)	Search	CPU (s)	CT (min)
HN80	4.3	398	1370.9	409	{1B, 2A}	0.2	680
HN80-I1.1	7.5	582	280.3	616	{1A, 2A}	0.3	1716
HN80-I1.2	5.0	577	6.2	590	{1A, 2B}	0.3	1866
HN80-I1.4	6.1	587	1386.0	614	{1A, 2A}	0.4	1226
HN80-I1.7	7.3	583	1298.6	610	{1A, 2A}	0.3	2307
HN80-I2.0	4.0	625	1713.7	657	{1A, 2B}	0.4	2909
HN80-I2.5	8.4	1092	110.6	1133	{1A, 2B}	0.4	1884
HN80-I3.0	9.4	1232	212.6	1235	{1A, 2B}	0.3	2467

there are also some suprising combinations {evacuated node, time step}, such as {3, 50}, {3, 84} and {3, 85} where the flow rate is respectively 22, 3, and 12 for evacuation area 3. In summary, the FSP solution is unlikely to be the basis of a controlled evacuation: It is just too difficult to enforce such a complicated schedule. Figure 4 shows a repartition of departure times for the same nodes in the original HN80 instance using the NEPP. The evacuation profile for the departure times is extremely simple and can easily be the basis of a controlled evacuation. Its simplicity contrasts with the complexity of the FSP solution and demonstrates the value of the constraint programming approach promoted in this paper.

Minimizing the Clearance Time. Table 3 compares, for each HN80-Ix instance, the minimal clearance time in minutes (CT (min)) found with FSP-CT, NEPP-CT and NPEPP-CT. All solutions found with FSP-CT and NPEPP-CT are optimal for the given set of paths. Once again, solutions found by NEPP-CT are of high-quality and reasonably close to the preemptive lower bound produced by FSP-CT. In the worst case, the results are within 5.1% of the preemptive lower bound. The clearance times of the phased evacuations, which are optimal, are significantly larger than for the NEPP. Note again that paths are different between instances and are not necessarily optimal with respect to different scheduling horizons, which explain inconsistencies such as the horizon found for HN80-I1.4 being shorter than the horizon found for HN80-I1.2 with NPEPP-CT.

Table 4. Vehicles Evacuated with NEPP-MF with Flow Rates in $\{2, 6, 10, 15, 20\}$.

Instance	CPU (s)	Perc. Evac.	Search
HN80	0.4	100.0% (SAT)	{1B, 2B}
HN80-I1.1	1538.9	99.2%	{1A, 2B}
HN80-I1.2	0.9	100.0% (SAT)	{1B, 2B}
HN80-I1.4	986.0	99.5%	{1A, 2B}
HN80-I1.7	1289.5	97.1%	{1A, 2A}
HN80-I2.0	1614.3	91.0%	{1A, 2B}
HN80-I2.5	1784.9	77.0%	{1A, 2B}
HN80-I3.0	1558.7	65.6%	{1A, 2B}

Table 5. Comparison of FSP and NEPP problem sizes.

	FSP-10		FSP-15		NEPP-MF	
Instance	#cols	#rows	#cols	#rows	#vars	#ctrs
HN80	44651	145880	68651	218780	1958	2288

The Impact of the Flow Rates. The constraint-based scheduling models have the flow rates as decision variables, which increases the flexibility of the solutions. Table 4 studies the benefits of this flexibility and compares the general results with the case where the flow rates must be selected from a specific set, here $\{2, 6, 10, 15, 20\}$. This is similar to the approach proposed in [12], which uses a fixed set of response curves and their associated mobilization resources. Note that the column-generation algorithm in [12] does not produce convergent plans and discretizes time. The results seem to indicate that flexible flow rates sometimes bring benefits, especially for the larger instances where the benefits can reach 3.0% ; nonetheless the possible loss when using fixed rates is not substantial and may capture some practical situations.

Comparison of Model Sizes. One of the benefits of the constraint-based scheduling models is that they do not discretize time and hence are highly scalable in memory requirements. This is important for large-scale evacuations which may be scheduled over multiple days. Table 5 compares the FSP problem size for a scheduling horizon of 10 hours (FSP-10) and 15 hours (FSP-15) with the NEPP-MF problem size for the HN80 instance, when using 1 minute time steps. It reports the number of columns (#cols) and the number of rows (#rows) of the FSP model, as well as the number of variables (#vars) and the number of constraints (#ctrs) of the NEPP model. As can be seen, the number of variables and constraints grow quickly for the FSP model and are about 2 orders of magnitude larger than those in the NEPP-MF model which is time-independent.

5   Conclusion

This paper proposes, for the first time, several constraint-based models for controlled evacuations that produce practical and actionable evacuation schedules. These models address several limitations of existing methods, by ensuring non-preemptive scheduling and satisfying operational evacuation constraints over mobilization resources. The algorithms are scalable, involve no time discretization, and are capable of accommodating side constraints for specific disaster scenarios or operational evacuation modes. Moreover, the models have no restriction on the input set of evacuation paths, which can be convergent or not. Preliminary experiments show that high-quality solutions, with an objective value close to an optimal value of an optimal preemptive solution, can be obtained within a few seconds, and improve over time. Future work will focus on improving the propagation strength of the cumulative constraint for variable durations, flows, and flow rates, and on generalizing the algorithm for the joint evacuation planning and scheduling.

References

1. Andreas, A.K., Smith, J.C.: Decomposition algorithms for the design of a nonsimultaneous capacitated evacuation tree network. Networks **53**(2), 91–103 (2009)
2. Bish, D.R., Sherali, H.D.: Aggregate-level demand management in evacuation planning. European Journal of Operational Research **224**(1), 79–92 (2013)
3. Bretschneider, S., Kimms, A.: Pattern-based evacuation planning for urban areas. European Journal of Operational Research **216**(1), 57–69 (2012)
4. Cepolina, E.M.: Phased evacuation: An optimisation model which takes into account the capacity drop phenomenon in pedestrian flows. Fire Safety Journal **44**, 532–544 (2008)
5. Even, C., Pillac, V., Van Hentenryck, P.: Nicta evacuation planner: Actionable evacuation plans with contraflows. In: Proceedings of the 20th European Conference on Artificial Intelligence (ECAI 2014). Frontiers in Artificial Intelligence and Applications, vol. 263, pp. 1143–1148. IOS Press, Amsterdam (2014)
6. Even, C., Pillac, V., Van Hentenryck, P.: Convergent plans for large-scale evacuations. In: Proceedings of the 29th AAAI Conference on Artificial Intelligence (AAAI 2015) (in press 2015)
7. Huibregtse, O.L., Bliemer, M.C., Hoogendoorn, S.P.: Analysis of near-optimal evacuation instructions. Procedia Engineering 3, 189–203 (2010), 1st Conference on Evacuation Modeling and Management
8. Huibregtse, O.L., Hegyi, A., Hoogendoorn, S.: Blocking roads to increase the evacuation efficiency. Journal of Advanced Transportation **46**(3), 282–289 (2012)
9. Huibregtse, O.L., Hoogendoorn, S.P., Hegyi, A., Bliemer, M.C.J.: A method to optimize evacuation instructions. OR Spectrum **33**(3), 595–627 (2011)
10. Lim, G.J., Zangeneh, S., Baharnemati, M.R., Assavapokee, T.: A capacitated network flow optimization approach for short notice evacuation planning. European Journal of Operational Research **223**(1), 234–245 (2012)

11. Pillac, V., Even, C., Van Hentenryck, P.: A conflict-based path-generation heuristic for evacuation planning. Tech. Rep. VRL-7393, NICTA (2013), arXiv:1309.2693, submitted for publication
12. Pillac, V., Cebrian, M., Van Hentenryck, P.: A column-generation approach for joint mobilization and evacuation planning. In: International Conference on Integration of Artificial Intelligence and Operations Research Techniques in Constraint Programming for Combinatorial Optimization Problems (CPAIOR), Barcelona (May 2015)
13. Pillac, V., Van Hentenryck, P., Even, C.: A Path-Generation Matheuristic for Large Scale Evacuation Planning. In: Blesa, M.J., Blum, C., Voß, S. (eds.) HM 2014. LNCS, vol. 8457, pp. 71–84. Springer, Heidelberg (2014)
14. Richter, K.F., Shi, M., Gan, H.S., Winter, S.: Decentralized evacuation management. Transportation Research Part C: Emerging Technologies **31**, 1–17 (2013)

Solving Segment Routing Problems with Hybrid Constraint Programming Techniques

Renaud Hartert[✉], Pierre Schaus, Stefano Vissicchio, and Olivier Bonaventure

UCLouvain, ICTEAM, Place Sainte Barbe 2, 1348 Louvain-la-Neuve, Belgium
{renaud.hartert,pierre.schaus,stefano.vissicchio,
olivier.bonaventure}@uclouvain.be

Abstract. Segment routing is an emerging network technology that exploits the existence of several paths between a source and a destination to spread the traffic in a simple and elegant way. The major commercial network vendors already support segment routing, and several Internet actors are ready to use segment routing in their network. Unfortunately, by changing the way paths are computed, segment routing poses new optimization problems which cannot be addressed with previous research contributions. In this paper, we propose a new hybrid constraint programming framework to solve traffic engineering problems in segment routing. We introduce a new representation of path variables which can be seen as a lightweight relaxation of usual representations. We show how to define and implement fast propagators on these new variables while reducing the memory impact of classical traffic engineering models. The efficiency of our approach is confirmed by experiments on real and artificial networks of big Internet actors.

Keywords: Traffic engineering · Segment routing · Constraint programming · Large neighborhood search

1 Introduction

During the last decades, the Internet has quickly evolved from a small network mainly used to exchange emails to a large scale critical infrastructure responsible of significant services including social networks, video streaming, and cloud computing. Simultaneously, Internet Service Providers have faced increasing requirements in terms of quality of service to provide to their end-users, e.g., low delays and high bandwidth. For this reason, controlling the paths followed by traffic has become an increasingly critical challenge for network operators [22] – especially those managing large networks. Traffic Engineering – a field at the intersection of networking, mathematics, and operational research – aims at optimizing network traffic distribution. Among its main objectives, avoiding link overload is one of the most important as it leads to drop of network reliability, e.g., loss of packets and increasing delays [1]. New traffic engineering objectives recently emerged [29]. For instance, a network operator may want specific demands to

© Springer International Publishing Switzerland 2015
G. Pesant (Ed.): CP 2015, LNCS 9255, pp. 592–608, 2015.
DOI: 10.1007/978-3-319-23219-5_41

get through different sequences of network services, e.g., suspect traffic through a battery of firewalls and high-priority traffic via load-balancer and on low-delay paths [13].

Segment Routing (SR) [12] has been recently proposed to cope with those challenges. It is an emerging network architecture that provides enhanced packet forwarding capabilities while keeping a low configuration impact on networks. Segment Routing is both an evolution of MPLS (MultiProtocol Label Switching) [23] and of IPv6 [31]. Many actors of the network industry support segment routing and several internet service providers will implement segment routing to manage their networks [12,13,14,28]. All-in-one, segment routing seems to be a promising technology to solve traffic engineering problems.

The basic idea of Segment Routing is to prepend packets with a stack of labels, called segments, contained in a segment routing header. A segment represents an instruction. In this work, we focus on node segments that can be used to define paths in a weighted graph that represents the network topology. A node segment contains the unique label of the next router to reach. When such a router is reached, the current node segment is popped and the packet is sent to the router referenced by the next segment and so on. Note that segment routing exploits all the equal cost shortest-paths to reach a given destination. Such equal cost shortest-paths – called Equal-Cost Multi-Paths (ECMP) paths – are extremely frequent in network architectures and it is not rare to see as much as 128 different shortest-paths between a source and a destination within a single network [13]. Fig. 1 illustrates the use of two node segments to define a segment routing path from router s to router t. Note that the use of segments provides extreme flexibility in the path selection for different demands.

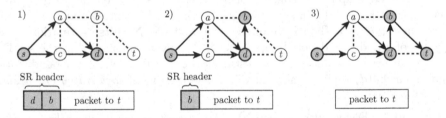

Fig. 1. A segment routing header with two segments prepended to a packet. First, the packet is sent to router d using the ECMP path from s to d (assuming unary link costs). Then, the packet is sent to the next label b following the ECMP path from d to b. Finally, all the segments have been processed and the original packet is sent to its destination t using the ECMP path from b to t.

Unfortunately, optimization approaches proposed in the past years do not consider the enhanced forwarding capabilities of segment routing [12]. Indeed, they all incur two major limitations. First, they typically focus on the basic problem of avoiding network congestion while not considering specific additional requirements [13,29]. Second, all past work assumes network technologies different from segment routing. For example, one of the most successful approaches also used in commercial traffic engineering tools is a tabu search algorithm [18] proposed by Fortz and Thorup in [15]. However, this method is based on the assumption that the network paths are computed as shortest paths in the network topology, hence it simply cannot be used to optimize segment routing networks. Similar considerations also apply to optimization frameworks used for more flexible protocols, like RSVP-TE [11,20].

In this work, we focus on the traffic placement problem using segment routing. Precisely, this problem consists in finding an SR-path for each demand such that the usage of each link is minimized while respecting a set of side constraints on the demands and the network (§2). We target large networks, like those of Internet Service Providers, hence both performance and scalability of the optimization techniques are crucial requirements, in addition to good quality of the solution. We note that classical constraint programming encoding of the problems generally lead to expensive memory consumption that cannot be sustained for large networks (several hundreds of nodes) (§3.1). We thus propose a data structure to encode the domain of our demands that is dedicated to this problem (§3.2 and §3.3). This encoding has the advantage of reducing the memory consumption of classical constraint programming from $\mathcal{O}(n^4)$ to $\mathcal{O}(n^3)$. We describe specific and light propagators for constraints defined on top of this new representation (§4). Our results are finally evaluated on synthetic and real topologies (§5 and §6). They highlight that constraint programming and large neighborhood search is a winning combination for segment routing traffic engineering.

2   The General Segment Routing Problem

Let us introduce some notations and definitions. A network is a strongly connected directed graph that consists of a set of nodes $\mathbf{N}$ (i.e. the routers) and a set of edges $\mathbf{E}$ (i.e. the links). An edge $e \in \mathbf{E}$ can be represented as the pair (u, v) where $u \in \mathbf{N}$ is the source of the edge and $v \in \mathbf{N}$ is its destination. The capacity of an edge is denoted $\texttt{capa}(e) \in \mathbb{N}$. For every demand d in the set $\mathbf{D}$, we have an origin $\texttt{src}(d) \in \mathbf{N}$, a destination $\texttt{dest}(d) \in \mathbf{N}$, and a bandwidth requirement $\texttt{bw}(d) \in \mathbb{N}$. We now introduce the notions of *forwarding graph* and *segment routing path*.

Definition 1 (Forwarding Graph). *A forwarding graph describes a flow between a pair of nodes in the network. Formally, a forwarding graph $FG(s,t)$ is a non-empty directed acyclic graph rooted in $s \in \mathbf{N}$ that converges towards $t \in \mathbf{N}$ and such that $s \neq t$.*

Definition 2 (Flow Function). *A forwarding graph $FG(s,t)$ is associated with a flow function $\texttt{flow}_{(s,t)}(e, b) \rightarrow \mathbb{N}$ that returns the amount of the bandwidth $b \in \mathbb{N}$ received at node s that is forwarded to node t through edge $e \in \mathbf{E}$ by the forwarding graph. Particularly, flow functions respect the equal spreading mechanism of ECMP paths. That is, each node of the forwarding graph $FG(s,t)$ splits its incoming traffic equally on all its outgoing edge contained in $FG(s,t)$. Flow functions thus respect the flow conservation constraints.*

Fig. 2 illustrates the flow function of some ECMP paths.

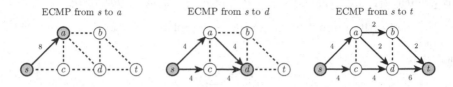

Fig. 2. Three different Equal-Cost Multi-Paths used to transfer 8 units of bandwidth on a network with unary link costs. Each router splits its incoming traffic equally on its outgoing edges. In this context, the value of $\mathtt{flow}_{(s,a)}((a,d),8)$ is 0, the value of $\mathtt{flow}_{(s,d)}((a,d),8)$ is 4, and the value of $\mathtt{flow}_{(s,t)}((a,d),8)$ is 2.

Definition 3 (Segment Routing Path). *A Segment Routing path (SR-path) from $s \in \mathbf{N}$ to $t \in \mathbf{N}$ is a non-empty sequence of forwarding graphs*

$$FG(s, v_1), FG(v_1, v_2), \ldots, FG(v_i, v_{i+1}), \ldots, FG(v_k, v_{k-1}), FG(v_k, t)$$

denoted $(s, v_1, \ldots, v_k, t)$ and such that the destination of a forwarding graph is the the source of its successor in the sequence. Also, the source of the first forwarding graph and the destination of the last forwarding graph respectively correspond to the source and the destination of the SR-path.

Segment routing paths are illustrated in Fig. 3.

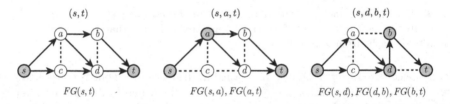

Fig. 3. Three different SR-paths based on the forwarding graphs of Fig. 2. An SR-path is represented by the sequence of nodes (top) corresponding to the extremities of its forwarding graphs (bottom).

We can now formalize the problem we want to solve. Let **FG** be a set of forwarding graphs on a network such that there is at most one forwarding graph for each pair of nodes $(u, v) \in \mathbf{N} \times \mathbf{N}$ with $u \neq v$. Let $SR(d)$ be the SR-path of demand $d \in \mathbf{D}$ using the forwarding graph in **FG**. The General Segment Routing Problem (GSRP) consists in finding a valid SR-path for each demand d such that the capacity of each edge is not exceeded

$$\forall c \subset \mathbf{E} : \sum_{d \in \mathbf{D}} \sum_{FG(i,j) \in SR(d)} \mathtt{flow}_{(i,j)}(e, \mathtt{bw}(d)) \leq \mathtt{capa}(e) \qquad (1)$$

and such that a set of constraints on the SR-paths and the network is respected.

Proposition 1. *The general segment routing problem is $\mathcal{NP}$-Hard.*

Proof. The Partition problem [6] is an $\mathcal{NP}$-complete problem that consists of n numbers $c_1, \ldots, c_n \in \mathbb{N}$. The question is whether there is a set $A \subseteq \{1, \ldots, n\}$ such that

$$\sum_{i \in A} c_i = \sum_{j \in \overline{A}} c_j$$

where $\overline{A}$ is the set of elements not contained in A. This problem can easily be reduced to the instance of the GSRP depicted in Fig. 4 with all edge capacities fixed to $\sum_{i=1}^{n} c_i/2$. First, consider n demands $d_1, \ldots, d_n$ from node s to node t such that $\texttt{bw}(d_i) = c_i$. Then, consider that forwarding graphs are defined such that there are only two possible SR-paths from node s to node t (see Fig. 4). Finding a valid SR-path for each demand amounts to find a solution to the Partition problem, i.e., demands having (s, A, t) as SR-path are part of the set A while the remaining demands are part of the set $\overline{A}$.   □

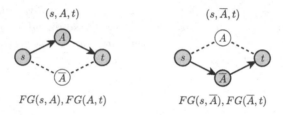

Fig. 4. The $\mathcal{NP}$-complete Partition problem can be encoded as an instance of the GSRP on this network. Forwarding graphs are defined such that there are only two possible SR-paths from node s to node t.

Practical considerations. Due to hardware limitations, segment routing headers usually contain no more than $k \in [4, 16]$ segments which limits the number of forwarding graphs to be contained in an SR-path to k. Furthermore, an SR-path should not include a loop, i.e., packets should never pass twice on the same link in the same direction.

3   Segment Routing Path Variables

This section is dedicated to the ad-hoc data structure we use to model the decision variables of the GSRP. Specialized domain representations are not new in constraint programming [34] and several data structures have already been proposed to represent abstractions such as intervals, sets, and graphs [8,10,16,17].

3.1   Shortcomings of Classical Path Representations

Observe that an SR-path from node s to node t can be seen as a simple path from s to t in a complete graph on $\mathbf{N}$ where each link (u, v) corresponds to the forwarding graph $FG(u, v)$. With this consideration in mind, let us motivate the need of an alternative representation by reviewing classical ways to model path variables, i.e., path-based, link-based, and node-based representations [39].

In *path-based* representations, a single variable is associated to each path. The domain of this variable thus contains all the possible paths to be assigned to the variable. This representation is usual in column generation frameworks [2]. In the context of the GSRP, complete path-based representations have an impracticable memory cost of $\Theta(|\mathbf{D}||\mathbf{N}|^k)$ where k is the maximal length of the SR-paths.

Link-based representations are surely the most common way to model path variables [25,42]. The idea is to associate a boolean variable $\mathbf{x}(d, e)$ to each demand $d \in \mathbf{D}$ and edge $e \in \mathbf{E}$. The boolean variable is set to true if edge e is part of the path, false otherwise. Hence, modeling SR-paths with link-based representations requires $\Theta(|\mathbf{D}||\mathbf{N}|^2)$ boolean variables.

Basically, a *node-based* representation models a path as a sequence of visited nodes [33]. This could be easily modeled using a discrete variable for each node that represents the successor of this node. The memory impact of such representation is $\Theta(|\mathbf{D}||\mathbf{N}|^2)$.

As mentioned above, memory and computational costs are of practical importance as we want to solve the GSRP on networks containing several hundreds of nodes with tens of thousands of demands. In this context, even link-based and node-based representations suffer from important shortcomings in both aspects.

3.2   Segment Routing Path Variables

Given the impossibility of using classic representations, we propose a new type of structured domain which we call *SR-path variable*. An SR-path variable represents a sequence of visited nodes[1] from the source of the path to its destination. The domain representation of this variable contains (see Fig. 5):

1. A prefix from the source to the destination that represents the sequence of already *visited* nodes ;
2. The set of possible nodes, called *candidates*, to append to the prefix of already visited nodes.

Basically, the domain of an SR-path variable is the set of all the possible extensions of the already visited nodes followed directly by a node contained in the set of candidates and finishing at its destination node. An SR-path variable can thus be seen as a relaxation of classic node-representations. An SR-path variable is assigned when the last visited node is the path destination. It is considered as invalid if it is unassigned and if its set of candidates is empty.

The particularity of SR-path variables is that the filtering of the domain is limited to the direct successor of the last visited node, i.e., the set of candidates

---

[1] Recall that an SR-path is a sequence of forwarding graphs denoted by a sequence of nodes (see Definition 3).

(see Fig. 5). Hence, no assumption can be made on the part of the sequence that follows the prefix of already visited nodes and the set of candidates. As a side effect, visiting a new node c automatically generates a new set of candidates as assumptions on the successor of c were impossible until now. It is then the responsibility of constraints to reduce this new set of candidates.

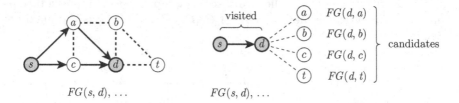

Fig. 5. A partial SR-path (left) and its corresponding SR-path variable (right). The SR-path variable maintains the partial sequence of visited nodes from the source s to the destination t and a set of candidates. Candidates represent the possible direct successors of the last visited node.

An SR-path variable S supports the following operations:

- $\mathtt{visited}(S)$: returns the sequence of visited nodes.
- $\mathtt{candidates}(S)$: returns the set of candidates.
- $\mathtt{visit}(S, c)$: appends c to the sequence of visited nodes.
- $\mathtt{remove}(S, c)$: removes c from the current set of candidates.
- $\mathtt{position}(S, c)$: returns the position of c in the sequence of visited nodes.
- $\mathtt{isAssigned}(S)$: returns true iff the path has reached its destination.
- $\mathtt{length}(S)$: returns an integer variable representing the length of the path.
- $\mathtt{src}(S), \mathtt{dest}(S), \mathtt{last}(S)$: returns the source node, the destination node, and the last visited node respectively.

3.3   Implementation

We propose to use array-based sparse-sets [5] to implement the internal structure of SR-path variables. Our data structure is similar to the one proposed to implement set variables in [8]. The sparse-set-based data structure relies on two arrays of $|\mathbf{N}|$ elements, $\mathtt{nodes}$ and $\mathtt{map}$, and two integers $\mathtt{V}$ and $\mathtt{R}$. The $\mathtt{V}$ first nodes in $\mathtt{nodes}$ correspond to the sequence of visited nodes. The nodes at positions $[\mathtt{V}, \ldots, \mathtt{R}[$ in $\mathtt{nodes}$ form the set of candidates. The $\mathtt{map}$ array maps each node to its position in $\mathtt{nodes}$. Fig. 6 illustrates this data structure and its corresponding partial SR-path.

The sparse-set-based implementation of SR-path variables offers several advantages. First, it is linear in the number of nodes since it only relies on two arrays and two integers. Also, it implicitly enforces the $\mathtt{AllDifferent}$ constraint on the sequence of visited nodes. Finally, it allows optimal computational complexity for all the operations presented in Table 1. Many of these operations

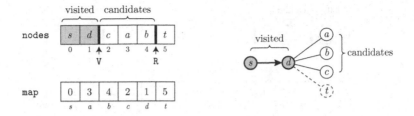

Fig. 6. An incremental sequence variable implemented in a sparse set. The sequence of visited nodes is s, d the set of candidates is $\{a, b, c\}$. Node t is not a valid candidate.

Table 1. A sparse-set-based implementation offers optimal time-complexities for several operations.

Iterate on the sequence of visited nodes	$\Theta(|visited|)$
Iterate on the set of candidates	$\Theta(|candidates|)$
Iterate on the set of removed candidates	$\Theta(|removed|)$
Returns the last visited node in the sequence	$\mathcal{O}(1)$
Test if a node has been visited, removed, or is still a candidate	$\mathcal{O}(1)$
Returns the position of a visited node in the sequence	$\mathcal{O}(1)$
Remove a candidate	$\mathcal{O}(1)$
Visit a candidate	$\mathcal{O}(1)$

can be implemented trivially by comparing node positions to the value of V and R. However, the `visit` and `remove` operations require more sophisticated manipulations of the data structure.

Visit a new node. Each time a node is visited, it swaps its position in **nodes** with the node at position V. Then, the value of V is incremented to append the visited node to the sequence of visited nodes. Finally, the value of R is set to $|\mathbf{N}|$ to restore the set of candidates to all the non-visited nodes. The `visit` operation is illustrated in Fig. 7. Observe that the sequence of visited nodes keeps its chronological order.

Remove a candidate. The remove operation is performed in a similar way as the visit operation. First, the removed node swaps its positions in **nodes** with the node at position R. Then, the value of R is decremented to exclude the removed node from the set of candidates. Fig. 8 illustrates this operation.

Backtracking is achieved in $\mathcal{O}(1)$. Indeed, we only need to trail the value of V to restore the previous sequence of visited nodes. Unfortunately, this efficient backtracking mechanism cannot restore the previous set of candidates. Nevertheless, this problem could be addressed by one of the following ways:

- The set of candidates could be recomputed on backtrack ;
- Changes in the set of candidates could be trailed during search ;
- Search could be restricted to visit all valid candidates with n-ary branching.

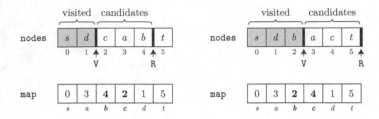

Fig. 7. State of the internal structure before (left) and after (right) visiting node b. First, node b exchanges its position with the node at position V in **nodes**, i.e., c. Then, V is incremented and R is set to $|\mathbf{N}|$.

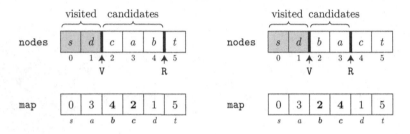

Fig. 8. State of the internal structure before (left) and after (right) removing node c. First, node c exchanges its position with the node at position R in **nodes**, i.e., b. Then, R is decremented.

We chose to apply the third solution as visiting a new node automatically restores the set of candidates to all the non-visited nodes (by updating the value of R to $|\mathbf{N}|$). This search mechanism thus allows us to keep backtracking in constant time since previous sets of candidates do not need to be recovered.

4   Constraints on SR-Path Variables

We model the GSRP by associating an SR-path variable to each demand in $\mathbf{D}$. These variables are the decisions variables of the problem. Also, each link of the network $e \in \mathbf{E}$ is associated with an integer variable $\texttt{load}(e)$ that represents the total load of this link, i.e., the total amount of traffic routed through e. We now present some constraints developed on top of the SR-path variables. These constraints are designed to meet requirements of network operators. Constraints on SR-path variables are awaken if one of both following events occurs in a variable of their scope:

- $\texttt{visitEvent}(S, a, b)$: tells the constraints that node b has been visited just after node a in the SR-path variable S ;
- $\texttt{removeEvent}(S, a, b)$: tells the constraints that node b is not a valid candidate to be visited after node a in the SR-path variable S.

We implemented a propagation algorithm specialized for these events in an AC5-like framework [41].

4.1 The Channeling Constraint

The role of the Channeling constraint is to ensure consistency between an SR-path variable S and the load of each link in the network. The Channeling constraint maintains the following property:

$$\forall c \in \mathtt{candidates}(S), \forall e \in FG(\mathtt{last}(S), c) :$$
$$\mathtt{load}(e) + \mathtt{flow}_{(\mathtt{last}(S),c)}(e, \mathtt{bw}(d)) \leq \mathtt{capa}(e) \tag{2}$$

The filtering of the Channeling constraint is enforced using two filtering procedures. The first filtering procedure is triggered each time a new candidate is visited by an SR-path variable to adjust the load variable of all the links traversed by the visited forwarding graph. Then, the second filtering procedure is called to reduce the set of candidates by removing forwarding graphs which cannot accommodate demand d due to the insufficient remaining bandwidth of their links. This second filtering procedure is also called each time the load variable of a link is updated.

The Channeling constraint relies intensively on forwarding graphs and flow functions. In our implementation, we chose to precompute these flow functions for fast propagation of the Channeling constraint. Such data structures may have an important impact of $\mathcal{O}(|\mathbf{N}|^2|\mathbf{E}|)$ in memory. Fortunately, this cost can be reduced to $\mathcal{O}(|\mathbf{N}||\mathbf{E}|)$ by exploiting the shortest-paths DAG[2] returned by Dijkstra's algorithm.[3] Indeed, we can test the presence of an edge (u, v) in a shortest-path from s to t using the transitive property of shortest-paths as follows:

$$\mathtt{dist}(s, u) + \mathtt{cost}(u, v) + \mathtt{dist}(v, t) = \mathtt{dist}(s, t). \tag{3}$$

This property allows one to use shortest-paths DAGs to recompute flow functions and forwarding graphs in $\mathcal{O}(|\mathbf{E}|)$ when required.

4.2 The Length Constraint

The Length constraint limits the number of forwarding graphs contained in an SR-path variable S:

$$|\mathtt{visited}(S)| = \mathtt{length}(S). \tag{4}$$

This constraint aims to respect real network hardware limitations in terms of maximum number of segments to be appended to packets [12]. We designed a simple filtering procedure for the Length constraint by comparing the number

---

[2] Shortest-paths DAGs can be easily computed by maintaining several shortest-paths to a given node in Dijkstra's algorithm.

[3] Recall that we make the assumption that forwarding graphs correspond to ECMP paths.

of visited nodes to the bounds of the length variable. Let $\min(\texttt{length}(S))$ and $\max(\texttt{length}(S))$ respectively be the lower bound and the upper bound of the length variable of S. Changes on the bounds of this variable trigger both following filtering procedures :

- If $|\texttt{visited}(S)| < \min(\texttt{length}(S)) - 1$, we know that the destination of the SR-path is not a valid candidate as visiting it next will result in a path that is smaller than the lower bound of its length variable ;
- If $|\texttt{visited}(S)| = \max(\texttt{length}(S)) - 1$, the destination of the SR-path must be visited next to respect the upper bound of its length variable.

The Length constraint is also awakened when a new node is visited. In this case, the filtering procedure first checks if the visited node is the destination of the path. If it is the case, the length variable is assigned to $|\texttt{visited}(S)|$. Otherwise, the lower bound of the length variable is updated to be strictly greater than $|\texttt{visited}(S)|$ as the path destination still has to be visited.

4.3   The ServiceChaining Constraint

Many operators have lately shown interest for *service chaining*, i.e., the ability to force a demand to pass through a sequence of services [29]. The aim of the ServiceChaining constraint is to force an SR-path variable S to traverse a particular sequence of nodes described by a sequence of services. Each service is traversed by visiting one node in a given set, i.e., the set of nodes providing the corresponding service in the network. Let $services = (service_1, \ldots, service_k)$ denote the sequences of services to be traversed. The filtering rule of the ServiceChaining constraint enforces the sequence of visited nodes to contain a non-necessarily contiguous subsequence of services:

$$\texttt{visited}(S) \quad = \quad \ldots, s_1, \ldots, s_2, \ldots, s_k, \ldots$$
$$\forall i \in \{1, \ldots, k\} : s_i \in service_i \tag{5}$$

The filtering of the ServiceChaining constraint uses similar procedures as those used by the Length constraint.

4.4   The DAG Constraint

The DAG constraint prevents cycles in the network. A propagator for this constraint has already been proposed in [10]. The idea behind this propagator is to forbid the addition of any link that will result in a cycle in the transitive closure of the set of edges visited by an SR-path variables. In the context of the GSRP, the filtering of the DAG constraint can be strengthened by taking in consideration that an SR-path is a sequence of forwarding graphs which are acyclic by definition. First, we know that the source (resp. destination) of an SR-path variable S has no incoming (resp. outgoing) traffic:

$$\nexists FG(u, v) \in S : \texttt{src}(S) \in \texttt{nodes}(FG(u, v)) \tag{6}$$

$$\nexists FG(u,v) \in S : \text{dest}(S) \in \text{nodes}(FG(u,v)) \qquad (7)$$

Additional filtering is achieved using the definition of SR-paths and the following proposition.

Proposition 2. *Let c be a node traversed by two forwarding graphs such that c is not one of the extremities of those forwarding graphs. Then, there is no acyclic SR-path that visits both forwarding graph.*

Proof. Let c be a node traversed by $FG(i,j)$ and $FG(u,v)$ such that $c \notin \{i,j,u,v\}$ and that $FG(i,j)$ and $FG(u,v)$ are part of the same SR-path. As SR-paths are defined as a sequence of forwarding graphs, we know that $FG(i,j)$ is traversed before $FG(u,v)$ or that $FG(u,v)$ is traversed before $FG(i,j)$. Let us consider that $FG(i,j)$ is traversed before $FG(u,v)$. In this case, we know that there is a path from node j to node u. According to the definition of forwarding graphs, we also know that there is a path from c to j and from u to c. Therefore, there is a path from c to c which is a cycle. The remaining part of the proof is done symmetrically.   □

The DAG constraint thus enforce the following redundant property:

$$\forall \{FG(u,v), FG(i,j)\} \subseteq S :$$
$$\text{nodes}(FG(u,v)) \cap \text{nodes}(FG(i,j)) \subseteq \{u,v\}. \qquad (8)$$

The filtering procedures of the DAG constraint are triggered each time an SR-path variable visits a new node. Note that the set of nodes traversed by the visited forwarding graph of an SR-path variable – necessary to implement these filtering procedures – can be maintained incrementally with a memory cost of $\mathcal{O}(|\mathbf{D}||\mathbf{N}|)$ or it can be recomputed when required with a time complexity of $\mathcal{O}(|\mathbf{N}|)$.

4.5 The MaxCost Constraint

Often, service level agreements imply that some network demands must be routed on paths with specifics characteristics, e.g., low delays. Such requirements can easily be enforced using the MaxCost constraint that ensures that the total cost of an SR-path does not exceed a maximum cost C. To achieve this, we associate a positive cost to each forwarding graph $FG(u,v)$. Let $\text{cost}(FG(u,v)) \in \mathbb{N}$ denote this cost and $\text{minCost}(s,t)$ denote the minimum cost of reaching node t from node s with an SR-path of unlimited length. Such minimum costs could be easily precomputed using shortest-paths algorithms and only require a space complexity of $\Theta(|\mathbf{N}|^2)$ to be stored. The filtering rule of the MaxCost constraint (based on the transitive property of shortest-paths) enforces:

$$\forall c \in \text{candidates}(S) :$$
$$\text{cost}(\text{visited}(S)) + \text{cost}(\text{last}(S), c) + \text{minCost}(c, \text{dest}(S)) \leq C \qquad (9)$$

where $\text{cost}(\text{visited}(S))$ is the total cost of already visited forwarding graphs.

Substantial additional filtering could be added by specializing the minCost function to also consider the length variable of an SR-path variable. This would require to pre-compute the all pair shortest-distance for all the k possible lengths of the path with, for instance, labeling algorithms [3,9].

5   Hybrid Optimization

Finding a first feasible solution of the GSRP can be a difficult task. In this context, it is often more efficient to relax the capacity constraint of each link and to minimize the load of the maximum loaded links until respecting the original capacity constraints.

Frameworks such as Large Neighborhood Search (LNS) [38] have been shown to be very efficient to optimize large scale industrial problems in many domains [4,21,26,27,32,37]. The idea behind LNS is to iteratively improve a best-so-far solution by relaxing some part of this solution. In the context of GSRP, this could be easily achieved by removing demands that seem inefficiently placed and to replace them in the network to improve the objective function. The selected set of removed demands defines the neighborhood to be explored by a branch-and-bound constraint programming search.

Instead of dealing with all the load variables with a unique aggregation function (e.g. the maximum load), we designed a specific hybrid optimization scheme to make the optimization more aggressive. At each iteration, we force a randomly chosen most loaded link to strictly decrease its load. The load of the remaining links are allowed to be degraded under the following conditions:

- The load of the most loaded links cannot increase;
- The load of the remaining links are allowed to increase but must remain under the maximum load, i.e., the set of the most loaded links cannot increase;
- The load of the non-saturated links must remain under their link capacity to not increase the number of saturated links.

This optimization framework can be seen as particular instantiation of the variable-objective large neighborhood search framework [36].

Neighborhoods to be explored by constraint programming are generated using the following procedure [38]:

1. Let $\mathbf{D}^{\max}$ be the set of demands routed through the most loaded links;
2. While fewer than k demands have been selected:
   (a) Let A be the array of the not yet selected demands in $\mathbf{D}^{\max}$ sorted by non-increasing bandwidth;
   (b) Generate a random number $r \in [0, 1[$;
   (c) Select the $\lfloor r^{\alpha}|A| \rfloor$ demand in array A.

The parameter $\alpha \in [1, \infty[$ is used to control the diversification of the selection procedure. If $\alpha = 1$ all demands have the same chance to be selected. If $\alpha = \infty$ the k demands with the highest bandwidth are always selected. The sub-problem generated by removing the selected demands is then explored by constraint programming with a CSPF-like (Constrained Shortest Path First) heuristic [7]:

1. Select the unassigned demand with the largest bandwidth;
2. Try to extend the SR-path variable of this demand as follows:
   (a) If the demand destination is a valid candidate, visit the destination to assign the SR-path variable;

(b) Otherwise, visit all the candidates sorted by non-decreasing order of their impact on the load of the most loaded links.
3. If the SR-path variable has no valid candidate, backtrack and reconsider the previous visit.

We call this heuristic an *assign-first* heuristic because it tends to assign SR-path variables as soon as possible.

6   Experiments and Results

We performed many experiments on real and synthetic topologies and demands. Real topologies are meant to assess the efficiency of our approach on real practical situations. Conversely, synthetic topologies were used to measure the behavior of our approach on complex networks. The data set is summarized in Table 2. Eight of these topologies are large real-world ISP networks and from the publicly available Rocket Fuel topologies [40]. Synthetic topologies were generated using the Delaunay's triangulation algorithm implemented in IGen [30]. Notice that these synthetic topologies are highly connected from a networking point of view, i.e., they have an average node degree of 5.72.

Whenever available – that is for topology RealF, RealG, and RealH – we used the real sets of demands measured by operators. In the other cases, we generated demands with similar characteristics as the ones in our real data sets as described in [35]. Finally, we proportionally scaled up all the demands until a maximum link usage of approximatively 120% was reached.[4] This allowed us to consider critical situations in our analysis.

We measured the efficiency of our approach by minimizing the maximum link load on each network presented in Table 2. Our experiments were performed on top of the open-source OscaR Solver [24] with a MacBook Pro configured with 2.6 GHz Intel CPU and 16GB of memory, using a 64-Bit HotSpot™ JVM 1.8.

Table 2. Topologies and results.

Topology	$|N|$	$|E|$	$|D|$	Before	Relaxation	Max load
RealA	79	294	6160	120%	68%	72%
RealB	87	322	7527	142%	72%	82%
RealC	104	302	10695	130%	**86%**	**86%**
RealD	141	418	19740	140%	**96%**	**96%**
RealE	156	718	23682	121%	76%	78%
RealF	161	656	25486	117%	65%	74%
RealG	198	744	29301	162%	37%	74%
RealH	315	1944	96057	124%	**76%**	**76%**
SynthA	50	276	2449	103%	69%	71%
SynthB	100	572	9817	100%	59%	75%
SynthC	200	1044	37881	135%	53%	84%

---

[4] In a solution with all demands routed through their ECMP path.

For each instance, we computed a linear programming lower bound on the maximum load by solving the linear multi-commodity flow problem on the same topology and demands with Gurobi 6.0 [19] (see column "Relaxation" of Table 2). We also show the initial maximum load of the network if no segment is used to redirect demands through the network (see column "Before" of Table 2). The results provided by our optimization framework after 60 seconds of computations are presented in the last column of Table 2. Note that the linear relaxation cannot be reached in real networks because it assumes that flows can be arbitrarily split, a function that does not exist in current routers. In practice, even if uneven splitting became possible in hardware, it would be impossible from an operational viewpoint to maintain different split ratio for each flow on each router.

As we see, our approach is able to find close to optimal solutions at the exception of instance RealG. Moreover, the solutions found by our approach are optimal for instances RealC, RealD, and RealH. This is due to the fact that these instances contain links that must be traversed by many demands. Fig. 9 illustrates the changes in link loads to highlight such bottleneck links.

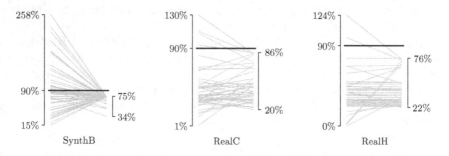

Fig. 9. Load of the 40 most loaded edges before (left) and after (right) optimization on a synthetic topology and two real ones. The load of many links have to be reduced to improve the maximum load on the artificial topologies while only a few have to be considered on the real ones. Real topologies contain more bottleneck links.

7   Conclusion

This paper presented an hybrid framework to solve segment routing problems with constraint programming and large neighborhood search. First, we introduced and formalized this new problem encountered by network operators. Then, we analyzed the shortcomings of classical constraint programming models to solve this problem on large networks. We thus proposed a new domain structure for path variable that we called *SR-path variable*. The particularity of this structure is that it sacrifices tight domain representation for low memory cost and fast domain operations. We explained how to implement common requirements of network operators in terms of constraint on this new variable. The efficiency of our approach was confirmed on large real-world and synthetic topologies.

References

1. Awduche, D., Chiu, A., Elwalid, A., Widjaja, I., Xiao, X.: Overview and principles of internet traffic engineering-rfc3272. IETF (2002)
2. Barnhart, C., Hane, C.A., Vance, P.H.: Using branch-and-price-and-cut to solve origin-destination integer multicommodity flow problems. Operations Research 48(2), 318–326 (2000)
3. Beasley, J.E., Christofides, N.: An algorithm for the resource constrained shortest path problem. Networks 19(4), 379–394 (1989)
4. Bent, R., Hentenryck, P.V.: A two-stage hybrid algorithm for pickup and delivery vehicle routing problems with time windows. Computers & Operations Research 33(4), 875–893 (2006)
5. Briggs, P., Torczon, L.: An efficient representation for sparse sets. ACM Letters on Programming Languages and Systems (LOPLAS) 2(1–4), 59–69 (1993)
6. Uslar, M., Specht, M., Rohjans, S., Trefke, J., Gonzalez, J.M.V.: Introduction. In: Uslar, M., Specht, M., Rohjans, S., Trefke, J., Vasquez Gonzalez, J.M. (eds.) The Common Information Model CIM. POWSYS, vol. 2, pp. 3–48. Springer, Heidelberg (2012)
7. Davie, B., Rekhter, Y.: MPLS: technology and applications. Morgan Kaufmann Publishers Inc. (2000)
8. le Clément de Saint-Marcq, V., Schaus, P., Solnon, C., Lecoutre, C.: Sparse-sets for domain implementation. In: CP workshop on Techniques for Implementing Constraint programming Systems (TRICS), pp. 1–10 (2013)
9. Desrochers, M., Soumis, F.: A generalized permanent labeling algorithm for the shortest path problem with time windows. INFOR Information Systems and Operational Research (1988)
10. Dooms, G., Deville, Y., Dupont, P.E.: CP(Graph): Introducing a Graph Computation Domain in Constraint Programming. In: van Beek, P. (ed.) CP 2005. LNCS, vol. 3709, pp. 211–225. Springer, Heidelberg (2005)
11. Figueiredo, G.B., da Fonseca, N.L.S., Monteiro, J.A.S.: A minimum interference routing algorithm. In: ICC, pp. 1942–1947 (2004)
12. Filsfils, C., et al.: Segment Routing Architecture. Internet draft, draft-filsfils-spring-segment-routing-00, work in progress (2014)
13. Filsfils, C., et al.: Segment Routing Use Cases. Internet draft, draft-filsfils-spring-segment-routing-use-cases-00, work in progress (2014)
14. Filsfils, C., et al.: Segment Routing with MPLS data plane. Internet draft, draft-filsfils-spring-segment-routing-mpls-01, work in progress (2014)
15. Fortz, B., Thorup, M.: Internet traffic engineering by optimizing OSPF weights. In: Proc. INFOCOM (March 2000)
16. Frei, C., Faltings, B.V.: Resource Allocation in Networks Using Abstraction and Constraint Satisfaction Techniques. In: Jaffar, J. (ed.) CP 1999. LNCS, vol. 1713, pp. 204–218. Springer, Heidelberg (1999)
17. Gervet, C.: Conjunto: Constraint logic programming with finite set domains. ILPS 94, 339–358 (1994)
18. Glover, F.: Tabu search: A tutorial. Interfaces 20(4), 74–94 (1990)
19. Inc., Gurobi Optimization. Gurobi optimizer reference manual (2015)
20. Kodialam, M., Lakshman, T.V.: Minimum interference routing with applications to mpls traffic engineering. In: Proceedings of the Nineteenth Annual Joint Conference of the IEEE Computer and Communications Societies, INFOCOM 2000, vol. 2, pp. 884–893. IEEE (2000)

21. Mairy, J.B., Deville, Y., Van Hentenryck, P.: Reinforced adaptive large neighborhood search. In: The Seventeenth International Conference on Principles and Practice of Constraint Programming (CP 2011), p. 55 (2011)
22. Nucci, A., Papagiannaki, K.: Design, Measurement and Management of Large-Scale IP Networks - Bridging the Gap between Theory and Practice. Cambridge University Press (2008)
23. RFC3031 Œ TF. Multiprotocol label switching architecture (2001)
24. OscaR Team. OscaR: Scala in OR (2012). https://bitbucket.org/oscarlib/oscar
25. Ouaja, W., Richards, B.: A hybrid multicommodity routing algorithm for traffic engineering. Networks **43**(3), 125–140 (2004)
26. Pacino, D., Van Hentenryck, P.: Large neighborhood search and adaptive randomized decompositions for flexible jobshop scheduling. In: Proceedings of the Twenty-Second International Joint Conference on Artificial Intelligence, vol. 3, pp. 1997–2002. AAAI Press (2011)
27. Shaw, P., Furnon, V.: Propagation Guided Large Neighborhood Search. In: Wallace, M. (ed.) CP 2004. LNCS, vol. 3258, pp. 468–481. Springer, Heidelberg (2004)
28. Previdi, S., et al.: IPv6 Segment Routing Header (SRH). Internet draft, draft-previdi-6man-segment-routing-header-00, work in progress (2014)
29. Quinn, P., Nadeau, T.: Service function chaining problem statement. draft-ietf-sfc-problem-statement-07 (work in progress) (2014)
30. Quoitin, B., Van den Schrieck, V., Francois, P., Bonaventure, O.: IGen: Generation of router-level Internet topologies through network design heuristics. In: ITC (2009)
31. Deering, S.: RFC2460 and R Hinden. Internet protocol
32. Ropke, S., Pisinger, D.: An adaptive large neighborhood search heuristic for the pickup and delivery problem with time windows. Transportation Science **40**(4), 455–472 (2006)
33. Giralt, L.R., Creemers, T., Tourouta, E., Colomer, J.R., et al.: A global constraint model for integrated routeing and scheduling on a transmission network (2001)
34. Rossi, F., Van Beek, P., Walsh, T.: Handbook of constraint programming. Elsevier Science (2006)
35. Roughan, M.: Simplifying the synthesis of internet traffic matrices. SIGCOMM Comput. Commun. Rev. **35**(5), 93–96 (2005)
36. Schaus, P.: Variable objective large neighborhood search (Submitted to CP13, 2013)
37. Schaus, P., Hartert, R.: Multi-Objective Large Neighborhood Search. In: Schulte, C. (ed.) CP 2013. LNCS, vol. 8124, pp. 611–627. Springer, Heidelberg (2013)
38. Shaw, P.: Using Constraint Programming and Local Search Methods to Solve Vehicle Routing Problems. In: Maher, M.J., Puget, J.-F. (eds.) CP 1998. LNCS, vol. 1520, pp. 417–431. Springer, Heidelberg (1998)
39. Simonis, H.: Constraint applications in networks. Handbook of Constraint Programming **2**, 875–903 (2006)
40. Spring, N., Mahajan, R., Wetherall, D., Anderson, T.: Measuring isp topologies with rocketfuel. IEEE/ACM Trans. Netw., 12(1) (2004)
41. Van Hentenryck, P., Deville, Y., Teng, C.-M.: A generic arc-consistency algorithm and its specializations. Artificial Intelligence **57**(2), 291–321 (1992)
42. Xia, Q., Simonis, H.: Primary/Secondary Path Generation Problem: Reformulation, Solutions and Comparisons. In: Lorenz, P., Dini, P. (eds.) ICN 2005. LNCS, vol. 3420, pp. 611–619. Springer, Heidelberg (2005)

Modeling Universal Instruction Selection

Gabriel Hjort Blindell[1]([✉]), Roberto Castañeda Lozano[1,2], Mats Carlsson[2],
and Christian Schulte[1,2]

[1] SCALE, School of ICT, KTH Royal Institute of Technology, Stockholm, Sweden
{ghb,cschulte}@kth.se
[2] SCALE, Swedish Institute of Computer Science, Kista, Sweden
{rcas,matsc}@sics.se

Abstract. Instruction selection implements a program under compilation by selecting processor instructions and has tremendous impact on the performance of the code generated by a compiler. This paper introduces a graph-based *universal* representation that unifies data and control flow for both programs and processor instructions. The representation is the essential prerequisite for a constraint model for instruction selection introduced in this paper. The model is demonstrated to be expressive in that it supports many processor features that are out of reach of state-of-the-art approaches, such as advanced branching instructions, multiple register banks, and SIMD instructions. The resulting model can be solved for small to medium size input programs and sophisticated processor instructions and is competitive with LLVM in code quality. Model and representation are significant due to their expressiveness and their potential to be combined with models for other code generation tasks.

1 Introduction

Instruction selection implements an input program under compilation by selecting instructions from a given processor. It is a crucial part of code generation in a compiler and has been actively researched for over four decades (see [23] for a recent survey). It is typically decomposed into identifying the applicable instructions and selecting a combination of applicable instructions to meet the semantics of the input program. Combinations differ in efficiency and hence selecting one is naturally an optimization problem. Finding an efficient combination is crucial as efficiency might differ by up to two orders of magnitude [35].

Common approaches use graph-based techniques that operate on the dataflow graph of a program (nodes represent operations, edges describe data flow). However, state-of-the-art approaches are restricted to trees or DAGs to avoid NP-hard methods for general graphs. This restriction is severe: ad-hoc routines are needed for handling control flow; many instructions of modern processors, such as DSPs (digital signal processors), cannot be handled; and the scope of instruction selection is typically *local* to tiny parts of the input program and hence by design forsakes crucial optimization opportunities.

© Springer International Publishing Switzerland 2015
G. Pesant (Ed.): CP 2015, LNCS 9255, pp. 609–626, 2015.
DOI: 10.1007/978-3-319-23219-5_42

This paper introduces a *universal* representation based on general graphs. It is universal as it simultaneously captures both data and control flow for both programs and processor instructions. By that it overcomes the restrictions of current approaches. In particular, the universal representation enables a simple treatment of *global code motion*, which lets the selection of instructions to be *global* for an entire function. The representation is compatible with state-of-the-art compilers; the paper uses LLVM [26] as compiler infrastructure.

However, the very reason of the proposed approach is that instruction selection can be expressed as a constraint model. Due to the expressiveness of the universal representation, the introduced model accurately reflects the interaction between control and data flow of both programs and processor instructions. The paper presents the model in detail and discusses how it supports processor features that are out of reach of state-of-the-art approaches, such as advanced branching instructions, multiple register banks, and SIMD instructions.

The paper shows that the described approach is feasible. The resulting model can be solved for small to medium size input programs and challenging SIMD processor instructions and is competitive with LLVM in code quality.

Model and representation are significant for two reasons. First, they are considerably more powerful than the state of the art for instruction selection and can capture common features in modern processors. Crucially, instruction selection with a universal representation is only feasible with an approach as expressive as a constraint model. Second, the paper's approach will be essential to integrate instruction selection with register allocation and instruction scheduling as the other code generation tasks that we have explored in previous work [10,11]. It is only the combination of all three interdependent tasks that will enable generating optimal code for modern processors.

Paper outline. Section 2 introduces graph-based instruction selection and Sect. 3 introduces the representations that enable universal instruction selection. The corresponding constraint model is introduced in Sect. 4. Section 5 experimentally evaluates the paper's approach followed by a discussion of related work in Sect. 6. Section 7 concludes the paper.

2   Graph-Based Instruction Selection

The most common approach for instruction selection is to apply graph-based methods. As is common, the unit of compilation is a single *program function*, which consists of a set of basic blocks. A *basic block*, or just *block*, is a sequence of *computations* (like an addition or memory load) and typically ends with a *control procedure* (like a jump or function return). Each block has a single entry point and a single exit point for execution. A program function has exactly one block as entry point, called *entry block*.

For a program function a *data-flow graph* called *program graph* is constructed, where each node represents a computation and each edge indicates that one computation uses the value produced by another computation. As a data-flow graph does not incorporate control-flow information, a single program function typically results in several program graphs (at least one per block). A *local* instruction selector selects instructions for program graphs of a single block, whereas a *global* instruction selector does so for an entire function.

For each instruction of a given processor data-flow graphs called *pattern graphs* are also constructed. The set of all pattern graphs for a processor constitute a *pattern set*. Thus, the problem of identifying the applicable instructions is reduced to finding all instances where a pattern graph from the pattern set is subgraph isomorphic to the program graph. Such an instance is called a *match*.

A set of matches *covers* a program graph if each node in the program graph appears in exactly one match. By assigning a cost to each match that corresponds to the cost of the selected instruction, the problem of finding the best combination of instructions is thus reduced to finding a cover with least cost. It is well known that the subgraph isomorphism problem and the graph covering problem are NP-complete in general [8,20], but can be solved optimally in linear time if the program and pattern graphs are trees [2].

Program trees are typically not expressive enough and hence modern compilers commonly use pattern trees and program DAGs, which are then covered using greedy heuristics. This, however, suffers from several limitations. First, pattern trees significantly limit the range of instructions that can be handled. For example, ad-hoc routines are required to handle even simple branch instructions as pattern trees cannot include control procedures. Second, program DAGs exclude handling of sophisticated instructions of modern processors that span multiple blocks. Third, more efficient code can be generated if certain computations can be moved across blocks, provided the program semantics is kept.

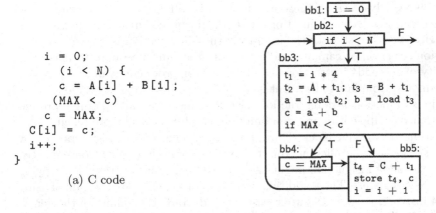

(a) C code

Fig. 1. An example program that computes the saturated sums of two arrays, where A, B, and C are integer arrays of equal lengths and stored in memory, and N and MAX are integer constants representing the array length and the upper limit, respectively. An int value is assumed to be 4 bytes.

A Motivating Example. Assume the C program shown in Fig. 1a, which computes the saturated sums of two arrays. Saturation arithmetic "clamps" the computed value such that it stays within a specified range and does not wrap around in case of overflow, a property commonly desired in many digital signal processing applications. For simplicity the program in Fig. 1a only clamps on the upper bound, but the following argumentation can easily be extended to programs that clamp both bounds. Most compilers do not operate directly on source code but on the *internal representation (IR)* of a program, shown in Fig. 1b, which can be viewed as a high-level assembly language. Programs written in this form are typically portrayed as a *control-flow graph* where each node represents a block and each edge represents a potential jump from one block to another. Most modern compilers, however, employ a slightly different representation that will be discussed in Sect. 3.

Assume further that this program will be executed on a processor whose instruction set includes the following instructions: satadd computes the saturated sum of two integer values; repeat iteratively executes an instruction sequence a given number of times; and add4 can compute up to four ordinary integer sums simultaneously (a so-called *vector* or *SIMD (Single-Input Multiple-Data)* instruction). Clearly, this program would benefit from selecting the satadd instruction to compute the value c and from selecting the repeat instruction to implement the control of the loop consisting of blocks bb2 through bb5. What might be less obvious, however, is the opportunity to select the add4 instruction to compute values t_2 and t_3 together with t_4 and i . Since these additions all reside inside the loop and are independent from one another, they can be computed in parallel provided they can be performed in the same block. This notion of moving computations across blocks is referred to as *global code motion*.

But taking advantage of these instructions is difficult. First, describing the saturated sum involves computations and control procedures that span several blocks. However, the state of the art in instruction selection is limited to local instruction selection or cannot handle control procedures. Second, to maximize the utility of the add4 instruction the additions for producing values t_4 and i must be moved from bb5 to bb3. But it is not known how to perform global code motion in conjunction with instruction selection, and hence all existing representations that describe entire program functions inhibit moves by pre-placing each computation to a specific block. Third, for many processors the registers used by vector instructions are different from those of other instructions. Consequently, selecting the add4 instruction might necessitate further instructions for copying data between registers. These additional instructions can negate the gain of using the add4 instruction or, in the worst case, even degrade the quality of the generated code. Making judicious use of such instructions therefore requires that the instruction selector is aware of this overhead. Fourth, since the program graph must be covered exactly, the computation of the value i cannot be implemented by both the add4 and repeat instructions. Instruction selection must therefore evaluate which of the two will result in the most efficient code, which depends on their relative costs and the restrictions imposed by the processor.

```
bb1:  int i₁ = 0;
bb2:  int i₂ = φ(i₁:bb1, i₃:bb5);
      if (i₂ < N) goto bb3; else goto end;
bb3:  int c₁ = A[i₂] + B[i₂];
      if (MAX < c₁) goto bb4; else goto bb5;
bb4:  int c₂ = MAX;
bb5:  int c₃ = φ(c₁:bb3, c₂:bb4);
      C[i₂] = c₃; int i₃ = i₂ + 1;
      goto bb2;
```

Fig. 2. The C program from Fig. 1a in SSA form.

3   Representations for Universal Instruction Selection

This section introduces representations for both programs and instructions and how they can be used to express covering and global code motion.

Program Representation. The key idea is to combine both data flow and control flow in the very same representation. We start by modifying control-flow graphs such that the nodes representing blocks no longer contain any computations. We refer to these as *block nodes*. In addition, the control procedures previously found within the blocks now appear as separate *control nodes*, where each control node has exactly one inbound control-flow edge indicating to which block the procedure belongs. Consequently, the control-flow edges that previously originated from the block nodes now originate from the control nodes. An example will be provided shortly.

To capture the data flow of entire program functions as a data-flow graph we use the *SSA (Static Single Assignment) graph* constructed from programs in *SSA form*. SSA is a state-of-the-art program representation where each program variable must be defined only once [14]. When the definition depends on the control flow, SSA uses so-called φ-*functions* to disambiguate such cases by taking a value and the block from where it originates as arguments. Fig. 2 shows the C program from Fig. 1a in SSA form, and the corresponding control-flow and SSA graphs are shown in Fig. 3a and Fig. 3b, respectively. Originally the SSA graph only consists of nodes representing computations – called *computation nodes* – but we extend it such that each value is represented by a *value node*. Also note that copying of values is not represented as separate computation nodes in the SSA graph. Therefore, in Fig. 3b the program variables i_1 and c_2 have been replaced by 0 and MAX , respectively.

To unify the control-flow graph and the SSA graph, we first add data-flow edges from value nodes to the control nodes that make use of the corresponding values. For the control-flow graph shown in Fig. 3a this entails the c.br nodes, which represent conditional jumps. Note that the SSA graph does not indicate in which block a given computation should be performed. Although we want to keep such pre-placements to a minimum, having *no* pre-placements permits moves that violate the program semantics. For example, in Fig. 1b the assignment c = MAX must be performed in block bb4 as its execution depends on whether

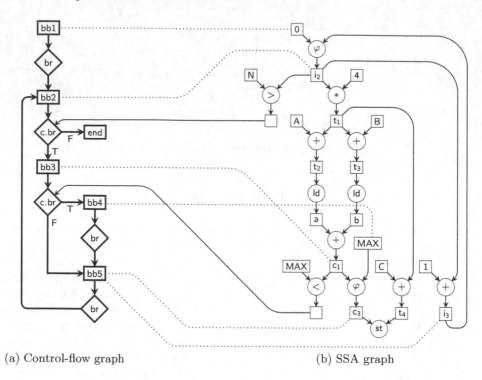

(a) Control-flow graph   (b) SSA graph

Fig. 3. The program graph constructed from the program shown in Fig. 2. Thick-lined diamonds, boxes, and arrows represent control nodes, block nodes, and control-flow edges, respectively. Thin-lined circles, boxes, and arrows represent computation nodes, value nodes, and data-flow edges, respectively. Dotted lines represent definition edges.

$MAX < c$ holds. Fortunately, these cases can be detected whenever a φ-function appears in the program. The SSA-based program in Fig. 2, for example, has in block bb2 a statement $i_2 = \varphi(i_1:\text{bb1}, i_3:\text{bb5})$. Thus, the program variable i_2 is assigned either the value i_1 or the value i_3 depending on whether the jump to bb2 was made from block bb1 or block bb5. These conditions can be ensured to hold in the generated code by requiring that (i) due to the arguments to the φ-function, the values i_1 and i_3 must be computed in blocks bb1 and bb5, respectively; and (ii) due to the location of the φ-function in the program, the value i_2 must be assigned its value in block bb2. We encode these constraints into the program graph by introducing *definition edges* to signify that a certain value must be produced in a specific block. Hence, in case of the example above three definition edges are added: one from value node 0 to block node bb1, one from value node i_3 to block node bb5, and another from value node i_2 to block node bb2. Such edges are also added for the statement $c_3 = \varphi(c_1:\text{bb3}, c_2:\text{bb4})$, which results in the program graph shown in Fig. 3.

Instruction Representation. The procedure for constructing pattern graphs is almost identical to constructing program graphs. The only exception is that the control-flow graph of a pattern graph could be empty, which is the case for instructions whose result does not depend on any control flow. For example, Fig. 4 shows the pattern graph of satadd (introduced in Sect. 2), which has a substantial control-flow graph. In comparison, the pattern graph of a regular add instruction would only comprise an SSA graph,

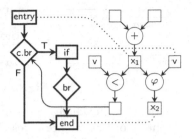

Fig. 4. satadd's pattern graph.

consisting of one computation node and three value nodes. Like with the program graph, it is assumed that all pattern graphs with a non-empty control-flow part have exactly one block node representing the instruction's entry point.

Since the program graph now consists of several kinds of nodes, we need to refine the notion of coverage. A program graph is *covered* by a set of matches if each *operation* in the program graph appears in exactly one match from the set, where an operation is either a computation or a control node. Likewise, a match *covers* the operations in the program graph corresponding to the operations in the pattern graph. Consequently, matches are allowed to partially overlap on the block and value nodes in the program graph. This property is deliberate as it enables several useful features, which will be seen shortly.

But not all covers of a given program graph yield valid solutions to the global code motion problem. For example, assume a cover of Fig. 3 where a match corresponding to the add4 instruction has been selected to cover the computation nodes that produce values t_2, t_3, and i_3. Because t_2 and t_3 are data-dependent on value i_2, which must be produced in block bb2 due to a definition edge, these values cannot be produced earlier than in bb2. Likewise, because value c_1 must be produced in block bb3 and is data-dependent on t_2 and t_3, these values cannot be produced later than in bb3. At the same time, i_3 must be produced in block bb5. Hence no single instruction that computes t_2, t_3, and i_3 can be placed in a block such that all conditions imposed by the program graph are fulfilled.

We use the above observation to formalize the global code motion problem as follows. If a *datum* refers to a particular value node in the program graph, a match m *defines* respectively *uses* a datum d if there exists an inbound respectively outbound data-flow edge to d in the pattern graph of m. Hence a datum can be both defined and used by the same match. We also refer to the data used but not defined by a match as its *input data* and to the data defined but not used as its *output data*. Next, a block b in the program graph *dominates* another block b' if every control-flow path from the program function's entry block to b' goes through b. By definition, a block always dominates itself. Using these notions, we define a placement of selected matches to blocks to be a solution to the global code motion problem if each datum d in the program graph is defined in some block b such that b dominates every block wherein d is used. Note that

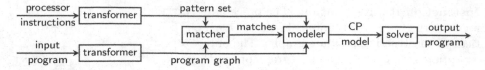

Fig. 5. Overview of our approach.

this definition excludes control procedures because moving such operations to another block rarely preserves the semantics of the program.

Some instructions impact both the covering and global code motion problems simultaneously. Assume for example a match in Fig. 3 of the pattern graph from Fig. 4, which corresponds to the satadd instruction. A match *spans* across the blocks in the program graph corresponding to the blocks appearing in the pattern graph. Hence, the match above spans across blocks bb3, bb4, and bb5. Of these, we note that the control procedures involving bb4 are all covered by this match. Consequently, the computations performed within this block must all be implemented by the satadd instruction. The match therefore *consumes* bb4, meaning no other matches can be placed in this block. Consequently, a universal instruction selector must take both the covering problem and the global code motion problem into account when selecting the matches.

4   A Constraint Model for Universal Instruction Selection

This section introduces a constraint model for universal instruction selection. An overview of the approach is shown in Fig. 5.

Match Identification. For a given program graph G and pattern set P, the matches are identified by finding all instances M where a pattern graph in P is subgraph isomorphic to G. Hence, separating identification from selection of matches is an optimality-preserving decomposition. We use an implementation of VF2 [13] to solve the subgraph isomorphism problem.

Depending on the pattern graph, certain matches may lead to cyclic data dependencies and must therefore be excluded from the set of identified matches. An example is a match of add4 in Fig. 3 defining t_2 or t_3 together with c_1. Such matches can be detected by finding all connected components in the match and checking whether a path exists between two components in the program graph.

Global Instruction Selection. The set of variables $\mathbf{sel}(m) \in \{0, 1\}$ models whether a match $m \in M$ is selected, where M denotes the set of identified matches. As explained in Sect. 3, each operation $o \in O$, where O denotes the set of operations in the program graph, must be covered by exactly one of the selected matches. Hence, if $\mathsf{covers}(m) \subseteq O$ denotes the set of operations covered by match m, then the condition can be expressed as:

$$\sum_{\substack{m \in M \text{ s.t.} \\ o \in \mathsf{covers}(m)}} \mathbf{sel}(m) = 1 \quad \forall o \in O \tag{1}$$

Global Code Motion. The set of variables $\mathbf{place}(m) \in B \cup \{b_{\text{null}}\}$ models in which block a match m is placed. B denotes the set of blocks in the program graph, and b_{null} denotes an additional block (not part of the program graph) in which non-selected matches are placed. In other words:

$$\mathbf{sel}(m) \Leftrightarrow \mathbf{place}(m) \neq b_{\text{null}} \quad \forall m \in M \tag{2}$$

where $\mathbf{sel}(m)$ abbreviates $\mathbf{sel}(m) = 1$.

Control procedures cannot be placed in another block than originally indicated in the program graph. Let $\mathsf{entry}(m) \in B$ denote the entry block of match m if the pattern graph of m has such a node, otherwise $\mathsf{entry}(m) \notin B$. This condition can then be expressed as:

$$\mathbf{sel}(m) \Rightarrow \mathbf{place}(m) = \mathsf{entry}(m) \quad \forall m \in M, \mathsf{entry}(m) \in B \tag{3}$$

The set of variables $\mathbf{def}(d) \in B$ models in which block a datum $d \in D$ is defined, where D denotes the set of data in the program graph. As explained in Sect. 3, each datum d must be defined in some block b such that b dominates every block wherein d is used. In addition, the conditions imposed by the definition edges must be maintained. Let $\mathsf{dom}(b) \subseteq B$ denote the set of blocks that dominate block b, where it is assumed that $\mathsf{dom}(b_{\text{null}}) = B$. Also let $\mathsf{uses}(m) \subseteq D$ denote the set of data used by match m, and let DE denotes the set of definition edges in the program graph. The conditions above can then be expressed as:

$$\mathbf{def}(d) \in \mathsf{dom}(\mathbf{place}(m)) \quad \forall m \in M, \forall d \in \mathsf{uses}(m) \tag{4}$$

$$\mathbf{def}(d) = b \quad \forall \{d, b\} \in DE \tag{5}$$

The $\mathbf{def}(\cdot)$ variables must be connected to the $\mathbf{place}(\cdot)$ variables. Intuitively, if a selected match m is placed in block b and defines a datum d, then d should also be defined in b. However, a direct encoding of this condition leads to an over-constrained model. Assume again a match in Fig. 3 of the pattern graph from Fig. 4, which thus defines the values c_1 and c_3. Due to the definition edges incurred by the φ-node, c_1 and c_3 must be defined in blocks bb3 and bb5, respectively. But if $\mathbf{def}(d) = \mathbf{place}(m)$ is enforced for every datum d defined by a match m, then the match above will never become eligible for selection because c_1 and c_3 must be defined in different blocks whereas a match can only be placed in a single block. In such cases it is sufficient to require that a datum is defined in any of the blocks spanned by the match. Let $\mathsf{spans}(m) \subseteq B$ denote the set of blocks spanned by match m and $\mathsf{defines}(m) \subseteq D$ denote the set of data defined by m. Then the condition can be relaxed by assigning $\mathbf{def}(d)$ to $\mathbf{place}(m)$ when $\mathsf{spans}(m) = \varnothing$, otherwise to any of the blocks in $\mathsf{spans}(m)$. Both these conditions can be combined into a single constraint:

$$\mathbf{sel}(m) \Rightarrow \mathbf{def}(d) \in \{\mathbf{place}(m)\} \cup \mathsf{spans}(m) \quad \forall m \in M, \forall d \in \mathsf{defines}(m) \tag{6}$$

Finally, matches cannot be placed in blocks that are consumed by some selected match. It $\mathsf{consumes}(m) \subseteq B$ denotes this set for a match m, then this condition can be expressed as:

$$\mathbf{sel}(m) \Rightarrow \mathbf{place}(m') \neq b \quad \forall m, m' \in M, \forall b \in \mathsf{consumes}(m) \tag{7}$$

Data Copying. Instructions typically impose requirements on the values that they operate on, for example that its input and output data must be located in particular registers. Combinations of such instructions may therefore require additional copy instructions to be selected in order to fulfill these requirements.

The set of variables $\mathbf{loc}(d) \in L \cup \{l_{\text{null}}\}$ models in which location a datum d is available. L denotes the set of possible locations, and l_{null} denotes the location of a value computed by an instruction that can only be accessed by this very instruction. For example, the address computed by a memory load instruction with a sophisticated addressing mode cannot be reused by other instructions. Thus, if $\mathsf{stores}(m, d) \subset L$ denotes the locations for a datum d permitted by a match m – where an empty set means no restrictions are imposed – then such conditions can be imposed as:

$$\mathbf{sel}(m) \Rightarrow d \in \mathsf{stores}(m, d) \quad \forall m \in M, \forall d \in D \text{ s.t. } \mathsf{stores}(m, d) \neq \varnothing \qquad (8)$$

This alone, however, may cause no solutions to be found for processors where there is little overlap between the locations – like in the case of multiple register banks – that can be used by the instructions. This problem is addressed using a method called *copy extension*. Prior to identifying the matches, the program graph is expanded such that every data usage is preceded by a special operation called a *copy*, represented by *copy nodes*. This expansion is done by inserting a copy node and value node along each data-flow edge that represents a use of data, as shown in the figure to the right. The definition edges are also moved such that they remain on the data adjacent to the φ-nodes. The same expansion is also applied to the pattern graphs except for the data-flow edges that represent use of input data.

Consequently, between the output datum of one match that is the input datum to another match, there will be a copy node that is not covered by either match. It is therefore assumed that a special *null-copy pattern*, consisting of a single copy node and two value nodes, is always included in the pattern set. A match derived from the null-copy pattern has zero cost but requires, if selected, that both data are available in the same location. This means that if the null-copy pattern can be used to cover some copy node, then the two matches connected by that copy node both use the same locations for its data. Hence there is no need for an additional instruction to implement this copy. If the locations are not compatible, however, then a match deriving the null-copy pattern cannot be selected as that would violate the requirement that the locations must be the same. An actual instruction is therefore necessary, and the restrictions on the $\mathbf{loc}(\cdot)$ variables ensure that the correct copy instruction is selected.

Fall-Through Branching. Branch instructions often impose constraints on the distance that can be jumped from the instruction to a particular block. For example, conditional branch instructions typically require that the FALSE block be located directly after the branch instruction, which is known as a *fall-through*.

This condition may also be imposed by instructions that span multiple blocks. Consequently, the order in which the blocks appear in the generated code is interconnected with the selection of certain matches.

The set of variables $\mathbf{succ}(b) \in B \cup \{b_{null}\}$ models the successor of block b. A valid block order then corresponds to a cycle formed by the $\mathbf{succ}(\cdot)$ variables. Using the global circuit constraint [27], this condition can be expressed as:

$$\text{circuit} \left(\cup_{b \in B \cup \{b_{null}\}} \{\mathbf{succ}(b)\} \right) \tag{9}$$

$$\mathbf{succ}(b_{null}) = b_{entry} \tag{10}$$

Hence, if the instruction corresponding to a match m performs a fall-through to a block b, then this condition can be expressed as $\mathbf{sel}(m) \Rightarrow \mathbf{succ}(\mathbf{place}(m)) = b$. This approach can readily be extended to incorporate generic jump distances as required by some processors.

Objective Function. The objective function depends on the desired characteristics of the generated code. For example, if the compiler should maximize the performance of the program under compilation, then the execution time of each block should be minimized. In addition, the execution time should be weighted with the relative execution frequency of each block. Thus, if $\mathsf{freq}(b) \in \mathbb{N}_1$ denotes the relative execution frequency of block b, then the objective function to be minimized is:

$$\sum_{b \in B} \mathsf{freq}(b) \times \sum_{\substack{m \in M \text{ s.t.} \\ \mathbf{place}(m)=b}} \mathsf{cost}(m) \tag{11}$$

where $\mathsf{cost}(m) \in \mathbb{N}_0$ is the relative time it takes to execute the instruction corresponding to a match m. Note that non-selected matches are always placed in the b_{null} block, which is not part of the B set. If the compiler should optimize for code size, then the weight $\mathsf{freq}(b)$ is dropped from Eq. 11 and $\mathsf{cost}(m)$ is redefined as the size of the instruction corresponding to match m.

Implied Constraints. Similarly to Eq. 1, the set of data must be defined by the set of selected matches:

$$\sum_{\substack{m \in M \text{ s.t.} \\ d \in \mathsf{defines}(m)}} \mathbf{sel}(m) = 1 \quad \forall d \in D \tag{12}$$

For every block b in which a datum is defined, there must exist a selected match that either is placed in b or spans b:

$$\mathbf{def}(d) = b \Rightarrow \mathbf{sel}(m) \wedge b \in \{\mathbf{place}(m)\} \cup \mathsf{spans}(m) \\ \forall b \in B, \forall d \in D, \exists m \in M \tag{13}$$

If two matches impose conflicting requirements on input or output data locations, or impose conflicting fall-through requirements, then at most one of these matches may be selected.

Dominance Constraints. By analyzing the constraints on the $loc(\cdot)$ variables, one can identify subsets S of values such that any solution with $loc(d) = v$ and $v \in S$ can be replaced by an equivalent solution with $loc(d) = \max(S)$, for any $d \in D$. Consequently, all values in $S \setminus \{\max(S)\}$ can be a priori removed from the domains of all $loc(\cdot)$ variables.

Suppose that there are two mutually exclusive matches m and m' with $cost(m) \leq cost(m')$ and the constraints imposed by m are compatible with and no stricter than the constraints imposed by m'. Then any solution that selects m' can be replaced by a solution of less or equal cost that selects m. Consequently, $sel(m') = 0$ can be set a priori for all such m'. In case m and m' have identical cost and impose identical constraints, a lexicographic ordering rule is used.

Branching Strategy. Our branching strategy only concerns those $sel(\cdot)$ variables of matches not corresponding to copy instructions. Let M' denote this set of matches. We branch on $\{sel(m) \mid m \in M'\}$ ordered by non-increasing $| covers(m)|$, trying $sel(m) = 1$ before $sel(m) = 0$. The intuition behind this is to eagerly cover the operations. The branching on the remaining decision variables is left to the discretion of the solver (see Sect. 5 for details).

Model Limitations. A constant value that has been loaded into a register for one use cannot be reused for other uses. Consequently, the number of selected matches may be higher than necessary. This problem is similar to spilling reused temporaries in [11] and can be addressed by adapting the idea of *alternative temporaries* introduced in [10].

For some processors, conditional jumps can be removed by predicating instructions with a Boolean variable that determines whether the instruction shall be executed [3]. For example, assume that the statement c = MAX in Fig. 1b is implemented using a copy instruction. If this instruction can be predicated with the Boolean value MAX < c, then the conditional jump to block bb4 becomes superfluous. This is known as *if-conversion*. Such instructions can be described using two pattern graphs: one representing the predicated version, and another representing the non-predicated version. But because every operation must be covered by exactly one match, the predicated version can only be selected if the match implements all computations in the conditionally executed block.

5   Experimental Evaluation

The model is implemented in MiniZinc [31] and solved with CPX [1] 1.0.2, which supports FlatZinc 1.6. The experiments are run on a Linux machine with an Intel Core i7-2620M 2.70 GHz processor and 4 GB main memory using a single thread.

We use all functions (16 in total) from MediaBench [28] that have more than 5 LLVM IR instructions and do not contain function calls or memory computations (due to limitations in the toolchain but not in the constraint model). The size of their corresponding program graphs ranges between 34 and 203 nodes. The functions are compiled and optimized into LLVM IR files using LLVM 3.4 [26] with the -O3 flag (optimizes execution time). These files serve as input to our

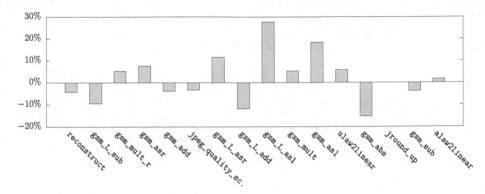

Fig. 6. Estimated execution speedup over LLVM for simple instructions.

toolchain. However, as LLVM currently lacks a method re-entering its backend after instruction selection we cannot yet execute the code generated by our approach. Instead the execution time is estimated using LLVM's cost model.

In the following, the full runtimes comprising matching, flattening the MiniZinc model to FlatZinc, and solving are summarized. Thus the time for producing the LLVM IR files and transforming them into program graphs is not included. The time spent on matching is negligible (less than 1% of the full runtime), while the time spent on flattening is considerable (more than 84% of the full runtime). The solving time is measured until proof of optimality. All runtimes are averaged over 10 runs, for which the coefficient of variation is less than 6%.

Simple Instructions. For proof of concept, the processor used is MIPS32 [33] since it is easy to implement and extend with additional instructions. The pattern graphs are manually derived from the LLVM instruction set description (to enable comparison), however the process could be easily automated as patterns are already described as trees in LLVM. As greedy heuristics already generate code of sufficient quality – in many cases even optimal – for MIPS32, we should not, in general, expect to see any dramatic execution speedup.

The shortest full runtime for the benchmarks is 0.3 seconds, the longest is 83.2 seconds, the average is 27.4 seconds, and the median is 10.5 seconds.

Fig. 6 shows the estimated execution speedup of our approach compared to LLVM. The geometric mean speedup is 1.4%; in average our approach is slightly better than LLVM. As the figure reveals, our approach has the potential to generate better code than the state of the art even for simple and regular architectures such as MIPS32, mainly due to its ability of moving computations to blocks with lesser execution frequency. The cases where our approach generates code that is worse than LLVM are due to the model limitations described in Sect. 4. Some of these cases are aggravated by the fact that LLVM's instruction selector is capable of, where appropriate, combining chains of binary φ-functions into a single φ-function in order to reduce the number of branch operations. This feature has yet to be implemented in our toolchain.

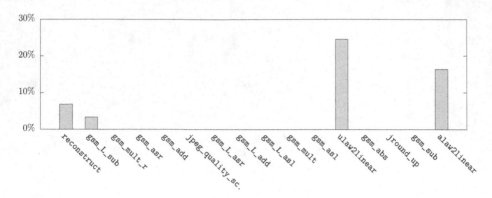

Fig. 7. Estimated execution speedup by adding SIMD instructions to MIPS32.

An interesting aspect is that due to the branching described in Sect. 4 solving yields a very good first solution, which provides a tight upper bound. We conjecture that this together with the lazy clause learning in the CPX solver is the reason why we can prove optimality for all benchmarks.

SIMD Instructions. In the following experiments we extend the MIPS32 architecture with SIMD instructions for addition, shifting, and Boolean logic operations as motivated by potential matches in the considered benchmark functions. The SIMD instructions use the same registers as the simple instructions.

The shortest full runtime for the benchmarks is 0.3 seconds, the longest is 146.8 seconds, the average is 44.2 seconds, and the median is 10.5 seconds.

Fig. 7 shows the estimated execution speedup for MIPS32 together with SIMD instructions. The geometric mean speedup compared to our approach for basic MIPS32 is 3%. The best cases correspond to functions that are computation-intensive (e.g. `ulaw2linear`) as more data parallelism can be exploited by the SIMD instructions. The worst cases (no improvement) correspond to control-intensive functions (e.g. `gsm_L_asr`), where there are either no matches for SIMD instructions or matches that are not profitable. In the latter case, the solver often discards SIMD matches that would require moving operations to *hot* blocks which are estimated to be executed often. A SIMD instruction in a hot block would be more costly than multiple primitive operations in the *colder* blocks where they originally reside. Our approach is unique in that it reflects this trade-off accurately. A traditional approach – vectorization first, then instruction selection – would greedily select the SIMD instruction and generate worse code. If a certain primitive operation must already be executed in a hot block then the solver will bring in operations of the same type from colder blocks to form a *speculative* SIMD instruction (this is the case e.g. for `alaw2linear`).

Hence our approach can exploit sophisticated instructions (such as SIMD) and has the potential to improve over traditional approaches since it accurately reflects the trade-offs in instruction selection. We also expect to see further benefits when integrated with register allocation and instruction scheduling.

6   Related Work

Several linear time, optimal algorithms exist for local instruction selection on tree-based program and pattern graphs [2,22,32]. These have subsequently been extended to DAG-based program graphs [17,18,25], but at the loss of optimality. Together with instruction scheduling or register allocation, the problem has also been approached using integer programming (IP) [7,21,36] and constraint programming (CP) [6,19,30]. Far fewer methods exist for global instruction selection, which so far only has been approached as a partitioned Boolean quadratic problem [9,15,16]. Common among these techniques is that they are restricted to pattern trees or pattern DAGs.

Many methods exist for selecting vector instructions separately from instruction selection, but attempts have been made at combining these two tasks using IP [29,34] and CP [4]. Of these, however, only [34] takes the cost of data copying into account, and none is global.

Global code motion has been solved both in isolation [12] as well as in integration with register allocation [5,24]. Both [5] and [12] use a program representation that is similar to ours, but where the data are not explicitly represented as nodes. To the best of our knowledge, no previous attempt has been made in combining global code motion with instruction selection.

7   Conclusions and Future Work

This paper introduces a *universal* graph-based representation for programs and instructions that unifies control and data flow. From the representations a new constraint model for instruction selection is derived. The paper shows that the approach is more expressive and generates code of similar quality compared to LLVM as a state-of-the-art compiler. The constraint model is robust for small to medium-sized functions as well as expressive processor instructions and is competitive with LLVM in code quality.

One line of future work is to extend the model to address the limitations discussed in Sect. 4. Additional improvements include exploring more implied and dominance constraints and pre-solving techniques to increase the model's robustness. We intend to explore both larger input programs as well as more processor architectures with a more robust model.

Instruction selection is but one task in code generation. We will explore in detail how this paper's model can be integrated with a constraint model for register allocation and instruction scheduling introduced in [10,11].

Acknowledgments. This research has been partially funded by LM Ericsson AB and the Swedish Research Council (VR 621-2011-6229). The authors are grateful for helpful discussions with Frej Drejhammar and Peter van Beek and for constructive feedback from the anonymous reviewers.

References

1. Opturion CPX user's guide: Version 1.0.2. Tech. rep., Opturion Pty Ltd (2013)
2. Aho, A.V., Ganapathi, M., Tjiang, S.W.K.: Code Generation Using Tree Matching and Dynamic Programming. Transactions on Programming Languages and Systems 11(4), 491–516 (1989)
3. Allen, J.R., Kennedy, K., Porterfield, C., Warren, J.: Conversion of Control Dependence to Data Dependence. In: ACM SIGACT-SIGPLAN Symposium on Principles of Programming Languages, pp. 177–189 (1983)
4. Arslan, M.A., Kuchcinski, K.: Instruction Selection and Scheduling for DSP Kernels on Custom Architectures. In: EUROMICRO Conference on Digital System Design (2013)
5. Barany, G., Krall, A.: Optimal and Heuristic Global Code Motion for Minimal Spilling. In: Jhala, R., De Bosschere, K. (eds.) Compiler Construction. LNCS, vol. 7791, pp. 21–40. Springer, Heidelberg (2013)
6. Bashford, S., Leupers, R.: Constraint Driven Code Selection for Fixed-Point DSPs. In: ACM/IEEE Design Automation Conference, pp. 817–822 (1999)
7. Bednarski, A., Kessler, C.W.: Optimal Integrated VLIW Code Generation with Integer Linear Programming. In: Nagel, W.E., Walter, W.V., Lehner, W. (eds.) Euro-Par 2006. LNCS, vol. 4128, pp. 461–472. Springer, Heidelberg (2006)
8. Bruno, J., Sethi, R.: Code Generation for a One-Register Machine. Journal of the ACM 23(3), 502–510 (1976)
9. Buchwald, S., Zwinkau, A.: Instruction Selection by Graph Transformation. In: International Conference on Compilers, Architectures and Synthesis for Embedded Systems, pp. 31–40 (2010)
10. Castañeda Lozano, R., Carlsson, M., Blindell, G.H., Schulte, C.: Combinatorial spill code optimization and ultimate coalescing. In: Kulkarni, P. (ed.) Languages, Compilers, Tools and Theory for Embedded Systems, pp. 23–32. ACM Press, Edinburgh, UK (2014)
11. Lozano, R.C., Carlsson, M., Drejhammar, F., Schulte, C.: Constraint-Based Register Allocation and Instruction Scheduling. In: Milano, M. (ed.) CP 2012. LNCS, vol. 7514, pp. 750–766. Springer, Heidelberg (2012)
12. Click, C.: Global Code Motion/Global Value Numbering. In: ACM SIGPLAN 1995 Conference on Programming Language Design and Implementation, pp. 246–257 (1995)
13. Cordella, L.P., Foggia, P., Sansone, C., Vento, M.: A (Sub)Graph Isomorphism Algorithm for Matching Large Graphs. IEEE Transactions on Pattern Analysis and Machine Intelligence 26(10), 1367–1372 (2004)
14. Cytron, R., Ferrante, J., Rosen, B.K., Wegman, M.N., Zadeck, F.K.: Efficiently Computing Static Single Assignment Form and the Control Dependence Graph. ACM TOPLAS 13(4), 451–490 (1991)
15. Ebner, D., Brandner, F., Scholz, B., Krall, A., Wiedermann, P., Kadlec, A.: Generalized Instruction Selection Using SSA-Graphs. In: ACM SIGPLAN-SIGBED Conference on Languages, Compilers, and Tools for Embedded Systems, pp. 31–40 (2008)
16. Eckstein, E., König, O., Scholz, B.: Code Instruction Selection Based on SSA-Graphs. In: Anshelevich, E. (ed.) SCOPES 2003. LNCS, vol. 2826, pp. 49–65. Springer, Heidelberg (2003)

17. Ertl, M.A.: Optimal Code Selection in DAGs. In: ACM SIGPLAN-SIGACT Symposium on Principles of Programming Languages, pp. 242–249 (1999)
18. Ertl, M.A., Casey, K., Gregg, D.: Fast and Flexible Instruction Selection with On-Demand Tree-Parsing Automata. In: ACM SIGPLAN Conference on Programming Language Design and Implementation, pp. 52–60 (2006)
19. Floch, A., Wolinski, C., Kuchcinski, K.: Combined Scheduling and Instruction Selection for Processors with Reconfigurable Cell Fabric. In: International Conference on Application-Specific Systems, Architectures and Processors, pp. 167–174 (2010)
20. Garey, M., Johnson, D.: Computers and Intractability: A Guide to the Theory of NP-Completeness. Freeman (1979)
21. Gebotys, C.H.: An Efficient Model for DSP Code Generation: Performance, Code Size, Estimated Energy. In: International Symposium on System Synthesis, pp. 41–47 (1997)
22. Glanville, R.S., Graham, S.L.: A New Method for Compiler Code Generation. In: ACM SIGACT-SIGPLAN Symposium on Principles of Programming Languages, pp. 231–254 (1978)
23. Hjort Blindell, G.: Survey on Instruction Selection: An Extensive and Modern Literature Study. Tech. Rep. KTH/ICT/ECS/R-13/17-SE, KTH Royal Institute of Technology, Sweden (October 2013)
24. Johnson, N., Mycroft, A.: Combined Code Motion and Register Allocation Using the Value State Dependence Graph. In: International Conference of Compiler Construction, pp. 1–16 (2003)
25. Koes, D.R., Goldstein, S.C.: Near-Optimal Instruction Selection on DAGs. In: IEEE/ACM International Symposium on Code Generation and Optimization, pp. 45–54 (2008)
26. Lattner, C., Adve, V.: LLVM: A compilation framework for lifelong program analysis & transformation. In: IEEE/ACM International Symposium on Code Generation and Optimization (2004)
27. Laurière, J.L.: A Language and a Program for Stating and Solving Combinatorial Problems. Artificial Intelligence 10(1), 29–127 (1978)
28. Lee, C., Potkonjak, M., Mangione-Smith, W.H.: MediaBench: A tool for evaluating and synthesizing multimedia and communications systems. In: IEEE MICRO-30, pp. 330–335 (1997)
29. Leupers, R.: Code Selection for Media Processors with SIMD Instructions. In: Conference on Design, Automation and Test in Europe, pp. 4–8 (2000)
30. Martin, K., Wolinski, C., Kuchcinski, K., Floch, A., Charot, F.: Constraint-Driven Instructions Selection and Application Scheduling in the DURASE System. In: International Conference on Application-Specific Systems, Architectures and Processors, pp. 145–152 (2009)
31. Nethercote, N., Stuckey, P.J., Becket, R., Brand, S., Duck, G.J., Tack, G.R.: MiniZinc: Towards a Standard CP Modelling Language. In: Bessière, C. (ed.) CP 2007. LNCS, vol. 4741, pp. 529–543. Springer, Heidelberg (2007)
32. Pelegrí-Llopart, E., Graham, S.L.: Optimal Code Generation for Expression Trees: An Application of BURS Theory. In: ACM SIGPLAN-SIGACT Symposium on Principles of Programming Languages, pp. 294–308 (1988)

33. Sweetman, D.: See MIPS Run, Second Edition. Morgan Kaufmann (2006)
34. Tanaka, H., Kobayashi, S., Takeuchi, Y., Sakanushi, K., Imai, M.: A Code Selection Method for SIMD Processors with PACK Instructions. In: International Workshop on Software and Compilers for Embedded Systems, pp. 66–80
35. Živojnović, V., Martínez Velarde, J., Schläger, C., Meyr, H.: DSPstone: A DSP-Oriented Benchmarking Methodology. In: Conference on Signal Processing Applications and Technology, pp. 715–720 (1994)
36. Wilson, T., Grewal, G., Halley, B., Banerji, D.: An Integrated Approach to Retargetable Code Generation. In: International Symposium on High-Level Synthesis, pp. 70–75 (1994)

Optimizing the Cloud Service Experience Using Constraint Programming

Serdar Kadioglu[(✉)], Mike Colena, Steven Huberman, and Claire Bagley

Oracle Corporation, Burlington, MA 01803, USA
{serdar.kadioglu,mike.colena,steven.huberman,
claire.bagley}@oracle.com

Abstract. This paper shows how to model and solve an important application of the well-known assignment problem that emerges as part of workforce management, particularly in cloud based customer service center optimization. The problem consists of matching a set of highly skilled agents to a number of incoming requests with specialized requirements. The problem manifests itself in a fast-paced online setting, where the complete set of incoming requests is not known apriori, turning this into a challenging problem where rapid response time and quality of assignments are crucial for success and customer satisfaction. The key contribution of this paper lies in the combination of a high-level constraint model with customizable search that can take into account various objective criteria. The result is an efficient and flexible solution that excels in dynamic environments with complex, conflicting and often changing requirements. The constraint programming approach handles hundreds of incoming requests in real-time while ensuring high-quality agent assignments.

1 Introduction

Introduced in early 1950s, the Assignment Problem is one of the fundamental combinatorial optimization problems. It deals with the question of how to assign agents (resources, jobs, etc.) to requests (tasks, machines, etc.) in the "best" possible way on a one-to-one basis. Equivalently, the problem consists of finding a maximum (or minimum) weighted matching in a weighted bipartite graph [1]. While the definition of the best depends on the particular application, it is usually a value associated with each agent-request assignment representing time, cost, profit, sales and etc.

The assignment problem enjoys numerous practical applications such as the assignment of doctors/nurses to hospitals in the residency problem, students to schools in the college admission problem, human organs to recipients in the transplant matching problem, and aircrafts to gates as part of airline scheduling problem (see [22] for a detailed survey). Recognizing the importance of the work in this area, A. Roth and L. Shapley were awarded the Nobel Prize in Economic Sciences in 2012, for their contributions to the theory of stable matchings.

© Springer International Publishing Switzerland 2015
G. Pesant (Ed.): CP 2015, LNCS 9255, pp. 627–637, 2015.
DOI: 10.1007/978-3-319-23219-5_43

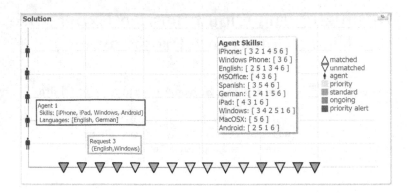

Fig. 1. An Example of the Agent Matching Problem.

In terms of solution methods, the assignment problem can be solved with the simplex algorithm as it is a special case of linear programming. However, there exist more efficient algorithms for this special structure. The prominent solution method is the Hungarian algorithm (due to Hungarian mathematicians D. Kőnig and J. Egerváry [17]). The algorithm runs in cubic time in the number of agents (tasks) in the worst-case and is strongly polynomial [18], i.e., there exists no algorithm for this structure with better time complexity. Other solutions include the shortest augmenting path algorithms (see e.g., [13]), or flow-based approaches, which have similar worst-case complexity.

Moving beyond the assignment problem to a more general setting, the next section describes a number of important dimensions that extends the classical definition. While these extensions bring the problem closer to real-life applications, it also becomes much harder to solve using a specialized approach such as the Hungarian algorithm. To that end, this paper introduces a better solution based on Constraint Programming (CP).

2   Service Center Optimization: Agent Matching

An important application of the assignment problem emerges as part of workforce management, particularly in cloud based customer service center optimization [12]. Conceptually, the problem consists of assigning a set of skilled agents to a number of incoming requests with specialized requirements in a continuous, 24/7, fast-paced setting.

Definition 1 (The Agent Matching Problem (AMP)). *Given a set of Agents, A, and a set of Requests, R, the goal is to assign every request $r \in R$ to an agent $a \in A$ such that: (i) the language of the agent matches the required language of the request, (ii) preferably, the skill set of the agent contains the skill demanded by the request, and (iii) each agent has a limit on the number of requests that they can work on simultaneously; therefore, capacity constraints must be respected at all times.*

It is important to note that not all agents possess the same skill level. Given a particular skill, there is a natural ordering among agents based on their expertise and experience. Similarly, not every request is equally important. Depending on the service level agreement, some requests may have priority over others. In fact, there might even be multiple priority levels. The ultimate goal is to assign the most skillful agent to the most important request while increasing agent utilization and decreasing the wait time. This strategy is referred to as the *best agent matching*.

Figure 1 presents an example of the AMP with 5 agents, shown in blue icons, and 15 requests, shown in either green or yellow triangles depending on whether they are standard or priority requests respectively. This is the initial state of the problem before any matching has taken place. Each agent possesses a number of technical skills and speaks one or more languages (see e.g., Agent 1). Similarly, each request demands a certain technical skill and a language (see e.g., Request 3). Agents are ordered with respect to their skills based on their experience level as shown in the Agent Skills list.

3   Why CP?

The AMP is a generalization of the assignment problem and differs from the classical definition in a number important ways. Unlike the original definition, the number of tasks and requests might not align, leading to an unbalanced matching problem. Secondly, this is an online problem as opposed to its one-time-only offline counterpart, meaning that, the complete set of incoming requests cannot be determined apriori. Consequently, the number of available agents changes dynamically as the system proceeds and requests become active (or inactive) through time. The agents are distinguishable in their skill set. Not every agent can handle every request, i.e., there exists forbidden pairs of assignments. Moreover, agents are not bound to perform a single task as in the classical assignment problem; they can work on multiple requests simultaneously. In addition, there is no single objective to optimize; it is often the case that multiple conflicting objectives are present. For example, while minimizing the number of agents is important to reduce the operating cost, employing more agents reduces the wait time, which in turn improves the customer satisfaction.

Such real-life requirements complicate the efficient algorithms that are designed for the original setting. The Hungarian method, and similarly flow-based approaches, have to generate a weighted combination of different objectives and cost values for possible assignments. These values are not only hard to predict but also need to be updated each time a constraint or preferences change. Both the objective weighting schema and cost values highly influence the actual solutions found. In practice, even the exact numeric values become important for the stability and the convergence of the algorithm. Finally, introducing artificial high cost values to forbid infeasible agent-request assignments or copies of agents (or tasks) are prohibitive for the scalability of the approach as they have direct impact on the asymptotic complexity of the algorithm.

These considerations turn AMP into a challenging problem where rapid response time and quality of assignments are crucial for success and customer satisfaction. Poor decisions can lead to under-staffing, poor customer service, and loss of revenue. The nature of the problem demands a fast real-time solution that can run iteratively in such a dynamic environment, and at the same time, take into account various objective criteria. This paper proposes a solution in this direction using constraint programming.

4   Added Value of CP

Expressiveness: As shown in Section 5.1, the CP model is high-level and offers a succinct representation of the problem without decomposing it, e.g., into Boolean assignment variables. There is almost no gap between the application and implementation, which is highly desired in a production environment and reduces the time to solution.

Flexibility: The plug-and-play nature of the constraint model makes it easy to add/remove restrictions to the problem. Customizable search heuristics, an example of which is given in Section 5.2, allow expressing various preferences, even conflicting ones.

Maintainability: There is a clear separation between the model and the search. Each component can be developed on its own and can be modified without changing the other parts. This is a strong property in an industrial setting. Essentially, the same abstract model can be utilized to solve the problem while different customers implement different preferences.

Efficiency: Powerful constraint engines are capable of handling hundreds of requests per second while assigning the best possible agents. As shown in Section 7, this amounts to handling millions of requests an hour, which satisfies beyond the current needs of service centers at peak hours. In addition, performing the best agent selection incurs almost no additional cost on the runtime.

5   Solution Approach

In principle, a constraint model consists of three elements: a set of *variables* each of which can take a value from a set of possible values, called *domains*, and a set of restrictions on the possible combinations of values referred to as *constraints*. [27].

The AMP lends itself naturally to a high-level model thanks to the declarative nature of constraint programming. First, a high-level constraint model that can capture the combinatorial structure of the problem is presented. Then, this model is combined with a customizable greedy search heuristic to find a good solution quickly.

5.1   The Constraint Model

The model uses two sets of decision variables; one that corresponds to the requests, and one that accounts for the number requests assigned to an agent.

Request Variables: There is a one-to-one mapping between these variables and requests. Each request is treated as an integer decision variable with a domain that consists of only the agents that speak the required language. Notice how language compatibility is implicitly guaranteed within this formalism via the domains. There is no need to generate artificial high cost values to forbid infeasible agent-request assignments; those pairs simply do not exist. This reduces the size of the problem considerably.

Given that some requests might not have an available agent at a particular time, a *wildcard* agent, that is capable of serving any request, is added as the last value to each domain. This way, the model is guaranteed to have a feasible solution while an assignment to the wildcard agent indicates an unmatched request.

Cardinality Variables: In order to count the number of times an agent is assigned to a request, a cardinality variable is created for each agent with a domain from 0 (no assignment) to the amount of current work capacity of the agent. The domain for the wildcard agent is set from 0 to the number of requests. The CP formalism allows agents to have different capacities with ease depending on their experience level. Similarly, the upper bound in each domain implicitly enforces the maximum capacity of an agent.

Global Cardinality Constraint: The request and cardinality variables are linked together via the (extended) global cardinality constraint [4,21,25]:

$$gcc(requestVars, \{Agents, wildCardAgent\}, cardVars)$$

5.2   The Search Heuristic

Variable Selection: Sorting the request variables, first, by their priority, and then, by their wait time yields a natural variable ordering heuristic. Without loss of generality, the variable heuristic can be extended into more levels of priorities or other sorting schemas.

Value Selection: Given a request variable, the next step is to find an agent to assign to it. Agents can be selected as directed by a "best" matching heuristic. Several interesting alternatives exist; an exemplar value selection procedure is outlined in Algorithm 1. The algorithm treats priority requests specially; if it is possible, priority requests are assigned to the most skillful agent for the task (line 6), otherwise, they are still matched with an agent with the same language (line 10).

```
Input: request variable, var
Output: request-agent assignment, σ(var ← val)
 1: val := −1
 2: if var is a priority request then
 3:     agentsSortedBySkill = a best-first ordering of all agents wrt the desired skill
        of var
 4:     for all v ∈ agentsSortedBySkill do
 5:         if v ∈ the domain of var then
 6:             val := v; break
 7:         end if
 8:     end for
 9:     if val == wildcard then
10:         val := the first agent in the domain of var
11:     end if
12: else
13:     val := the first agent in the domain of var
14: end if
15: return σ(var ← val)
```

Algorithm 1: Value Selection: Best Agent Matching Heuristic

Figure 2 illustrates the assignments found when the incoming requests in Figure 1 are matched to the agents without (on the left) and with (on the right) the best agent matching heuristic. When the heuristic is in play, notice how a priority request, namely Request 10, is assigned to Agent 4, who is the most skilled representative on technical issues with iPads (see Agent Skills list in Figure 1). Each agent works at capacity, i.e. two simultaneous requests, and agent utilization is maximized. No priority request is waiting for service since the heuristic tries to serve priority requests first. On the other hand, when the heuristic is not active, two priority requests are placed on the waitlist, as shown in red alerting triangles. Also, while Request 10 is still served in this case, it is served by Agent 5, who speaks the same language, but not technically savvy on iPad issues at all. This sharp contrast between the two solutions clearly reveals the impact of customized search heuristics for improving the quality of the assignments.

Again, without loss of generality, the value selection heuristic can be modified to serve other purposes. For example, among the most skillful agents, the agent with the least current load can be favored to balance the work among agents, and so on.

5.3   A Note on the Filtering Power of GCC

It is important to note that customized search heuristics, such as the one described in Algorithm 1, might assign values from the inner range of the domains instead of the bounds. This leads to creating holes in the domains. Therefore, such heuristics should be used with care when employing bounds consistency

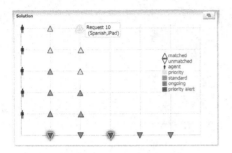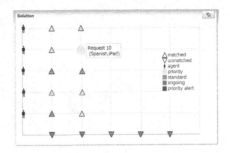

Fig. 2. The Solution Quality. Solutions found for the AMP instance in Figure 1 when the Best Agent Matching heuristic is not active (on the left) and is active (on the right).

filtering on the cardinality variables (e.g., as in [15,24]). In the case of AMP, we recognized the following behavior:

Remark 1. When the maximum cardinality of an agent is less than the number of requests demanding the same skill, for which, this agent is the most skilled, the best agent matching heuristic, in combination with a bounds consistent gcc operator, might lead to thrashing.

To enable stronger filtering of cardinality variables, it is beneficial to consider all the bounded variables and update their associated cardinalities. This improvement is similar to the Assigned Var Removal optimization described in the empirical gcc survey of [20].

6   Related Work

The existing approaches for solving the variants of the assignment problem broadly fall into two main categories: (i) ad-hoc, heuristic implementations, and (ii) generic methods. As opposed to the approaches in the first category, which produce very specific solutions, lack modeling support, and hard to adapt changes, our approach belongs to the second category, where, traditionally, Operational Research and Artificial Intelligence methods are employed. In particular, constraint (logic) programming techniques were successfully used for assignment problems, such as stand or counter allocation in airports, and berth allocation to container ships (see e.g., [6,7,10,23]). CP is also applied to the special case of personnel assignment as it is well-suited for changing work rules and labor regulations, such as rostering nurses in hospitals, scheduling technicians, and generating work plans to cover different skills (see e.g., [5,8,11]).

The work presented in [3,19,26] uses CP to assign highly-skilled professionals and employs similar decision variables. However, it focuses on one-to-one matching, whereas, in our context, multiple assignments are allowed. Also, our solution is geared toward a rapid operational setting whereas other approaches

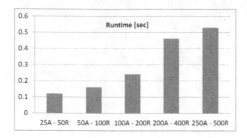

Fig. 3. Scalability Experiments. The effect of increasing the number of agents, A, and the number of requests, R.

seek long-term strategic or tactical planning. As in the work of [28], our approach also avoids the costly optimality search and constructs a good solution instead. Finally, it is possible to take into account not only the preferences of requests but also the agents as shown in [2].

7   Numerical Results

The constraint model and the search heuristic are implemented using an in-house constraint programming solver. All experiments were run on a Dell laptop with Intel Core i5 CPU @2.5 GHz and 8.00 Gb RAM.

Benchmarks: Experiments were run on instances of varying complexity and size based on the information gathered from our collaboration with other internal groups. Each data set consists of 50 synthetic instances whereby the largest instance deals with 250 agents and 500 request. Half of the requests selected uniformly at random as priority, and are treated specially. Regarding the capacity constraint, each agent is allowed to handle up to four requests simultaneously.

The goal of the experiments is to analyze the runtime and scalability of the approach. Notice that, in real applications, there is a dynamic nature of the problem: ongoing chat sessions remain active in the system for some unknown period, and as agents become available, the system continues to serve the previously unmatched requests as well as the new incoming requests that arrived in the meantime. In that sense, the following experiments stress test the *initial state* of the problem.

Impact of Problem Size: In an online setting such as the AMP, the response time is crucial, therefore, only a restricted runtime is allowed. As shown in Figure 3, the CP approach spends minimal time, around half of a second on average for the largest data set, while performing the best agent matching for 50% of the requests. If the proposed solution is deployed in a service center, this would amount to dealing with more than 2 million requests per hour which satisfies the business requirements of typical cloud based service centers under heavy loads.

Next, the impact of increasing the number of agents and the number of requests in isolation is studied. While the former leads to larger domains, the latter corresponds to handling more decision variables. The tests revealed that the complexity of the CP model depends more on the number of requests than the number of agents.

Comparison with the Hungarian Algorithm: To test the CP approach with the best-known algorithm, an in-house, specialized implementation of the Hungarian method is considered. Experiments showed that the CP approach was more than an order of magnitude faster on average for the same data sets. It should be noted that the Hungarian algorithm solves an optimization problem, albeit a fictitious one generated by imaginary agent-request assignment costs, while the CP approach finds a greedy solution empowered with a constraint model. The modification of the problem instances to fit into the right format (such as, creating dummy nodes and copies of agents, generating artificial values to express preferences and to forbid incompatible pairs) turned out to be prohibitive in practice, especially for an online setting where multiple assignment problems are solved successively. The same issue applies to flow-based approaches as well. Moreover, this not only requires pre-processing the instances, but also, post-processing to map the solution found back, which further complicates the code and maintenance in the long run.

Impact of Best Agent Matching: Finally, the overhead of the special treatment for priority requests is studied. The percentage of priority requests is adjusted from none (0%) to all (100%) using increments of size 5. The number of agents is fixed to 200 and the number of requests is fixed to 400.

The experiments shown in Figure 4 reveals an interesting property of these generated instances. The AMP problem goes through a transition when the $20(\pm3)\%$ of requests are marked as priority, which might be attributed to an easy-hard-less–hard pattern [9,14,16]. The behavior is due to the opposing forces between the best agent matching heuristic, which favors most skilled agents, versus the cardinality constraints, which allows only up to four simultaneous requests. The combination stretches the cardinality constraint for validation. Across all priority levels, the impact of the heuristic remains marginal. In a

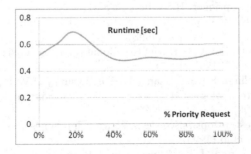

Fig. 4. The Effect of the Best Agent Matching Heuristic.

sense, within the CP formalism, preference based search, i.e., the considered best agent matching heuristic, comes at almost no additional cost, which makes the approach more attractive.

8   Conclusion

Optimizing a service center involves many decisions one of which deals with assigning agents in the best possible way. A good long-term hiring policy, skill-based routing, and the right mix of employee contract types are essential for good workforce management. This paper focused on runtime efficiency and assignment quality in a real-time environment. The expressive modeling language of constraint programming enabled a compact and flexible representation of this combinatorial problem. This model provided an efficient solution that can handle hundreds of requests per second, and more than 2 million per hour. The flexibility of variable and value selection allowed customizable search that can account for multiple objectives. The model and the search remained completely independent from each other allowing practitioners to develop and modify one without altering the other. Overall, the proposed constraint programming solution, stands out as an attractive approach to be considered for industrial applications of the service center optimization problem.

References

1. Ahuja, R.K., Magnanti, T.L., Orlin, J.B., Reddy M.R.: Applications of Network Optimization. In: Network Models. Handbooks in Operations Research and Management Science, vol. 7. North-Holland (1995)
2. Alsheddy, A., Tsang, E.P.K.: Empowerment scheduling for a field workforce. J. Scheduling 14(6), 639–654 (2011)
3. Boni, O., Fournier, F., Mashkif, N., Naveh, Y., Sela, A., Shani, U., Lando, Z., Modai, A.: Applying Constraint Programming to Incorporate Engineering Methodologies into the Design Process of Complex Systems. In: Proceedings of the Twenty-Fourth Conference on Innovative Applications of Artificial Intelligence (2012)
4. Bourdais, S., Galinier, P., Pesant, G.: HIBISCUS: A Constraint Programming Application to Staff Scheduling in Health Care. In: Rossi, F. (ed.) CP 2003. LNCS, vol. 2833, pp. 153–167. Springer, Heidelberg (2003)
5. Chan, P., Heus, K., Veil, G.: Nurse scheduling with global constraints in CHIP: Gymnaste. In: Proc. PACT 1998 (1998)
6. Chow, K.P., Perrett, M.: Airport counter allocation using constraint logic programming. In: Proc. PACT 1997 (1997)
7. Chun, A.H.W., Chan, S.H.C., Tsang, F.M.F., Yeung, D.W.M.: Stand allocation with constraint technologies at Chek Lap Kok international airport. In: Proc. PACLP 1999 (1999)
8. Collignon, C.: Gestion optimisée de ressources humaines pour l'Audiovisuel. In: Proc. CHIP users club (1996)
9. Crawford, J.M., Auton, L.D.: Experimental Results on the Crossover Point in Random 3sat. Artificial Intelligence 81, 31–57 (1996)

10. Dincbas, M., Simonis, H.: APACHE - a constraint based, automated stand alloca-
    tion system. In: Proc. of Advanced Software technology in Air Transport (ASTAIR
    1991) (1991)
11. Dubos, A., Du Jeu, A.: Application EPPER planification des agents roulants. In:
    Proc. CHIP users club (1996)
12. Durbin, S.D., Warner, D., Richter, J.N., Gedeon, Z.: RightNow eService Center:
    Internet customer service using a self-learning knowledge base. In: Proceedings of
    the Thirteenth Annual Conference of Innovative Applications of Artificial Intelli-
    gence (IAAI 2002), pp. 815–821 (2002)
13. Fredman, M.L., Tarjan, R.E.: Fibonacci Heaps and Their Uses in Improved Net-
    work Optimization Algorithms. J. ACM **34**(3), 596–615 (1987)
14. Hogg, T., Huberman, B.A., Williams, C.P.: Phase Transitions and the Search Prob-
    lem. Artif. Intell. **81**(1–2), 1–15 (1996)
15. Katriel, I., Thiel, S.: Complete Bound Consistency for the Global Cardinality Con-
    straint. Constraints **10**(3), 191–217 (2005)
16. Kirkpatrick, S., Selman, B.: Critical Behavior in the Satisfiability of Random
    Boolean Expressions. Science **264**(5163), 1297–1301 (1994)
17. Kuhn, H.W.: The Hungarian method for the assignment problem. Naval Research
    Logistics Quarterly **2**(1–2), 83–97 (1955)
18. Munkres, J.: Algorithms for the Assignment and Transportation Problems. Journal
    of the Society for Industrial and Applied Mathematics **5**(1), 32–38 (1957)
19. Naveh, Y., Richter, Y., Altshuler, Y., Gresh, D.L., Connors, D.P.: Workforce opti-
    mization: Identification and assignment of professional workers using constraint
    programming. IBM Journal of Research and Development **51**(3/4), 263–280 (2007)
20. Nightingale, P.: The extended global cardinality constraint: An empirical survey.
    Artif. Intell. **175**(2), 586–614 (2011)
21. Oplobedu, A., Marcovitch, J., Tourbier, Y.: CHARME: Un langage industriel de
    programmation par contraintes, illustré par une application chez Renault. In: Pro-
    ceedings of the Ninth International Workshop on Expert Systems and their Appli-
    cations: General Conferencehnical, vol. 1, pp. 55–70 (1989)
22. Pentico, D.W.: Assignment problems: A golden anniversary survey. European Jour-
    nal of Operational Research **176**(2), 774–793 (2007)
23. Perrett, M.: Using constraint logic programming techniques in container port plan-
    nings. ICL Technical Journal, 537–545 (1991)
24. Quimper, C.-G., Golynski, A., Lopez-Ortiz, A., van Beek, P.: An Efficient Bounds
    Consistency Algorithm for the Global Cardinality Constraint. Constraints **10**(2),
    115–135 (2005)
25. Régin, J.C., Gomes, C.P.: The Cardinality Matrix Constraint. In: 10th Interna-
    tional Conference on Principles and Practice of Constraint Programming, CP 2004,
    pp. 572–587 (2004)
26. Richter, Y., Naveh, Y., Gresh, D.L., Connors, D.P.: Optimatch: applying con-
    straint programming to workforce management of highly skilled employees. Int. J.
    of Services Operations and Informatics **3**(3/4), 258–270 (2008)
27. Rossi, F., van Beek, P., Walsh, T.: Handbook of Constraint Programming. Elsevier
    (2006)
28. Yang, R.: Solving a Workforce Management Problem with Constraint Program-
    ming. Technical Report, University of Bristol, Bristol (1996)

Find Your Way Back: Mobility Profile Mining with Constraints

Lars Kotthoff[1]([⊠]), Mirco Nanni[2], Riccardo Guidotti[2], and Barry O'Sullivan[3]

[1] University of British Columbia, Vancouver, Canada
larsko@cs.ubc.ca
[2] KDDLab – ISTI-CNR and CS Department, University of Pisa, Pisa, Italy
{mirco.nanni,riccardo.guidotti}@isti.cnr.it
[3] Insight Centre for Data Analytics, University College Cork, Cork, Ireland
barry.osullivan@insight-centre.org

Abstract. Mobility profile mining is a data mining task that can be formulated as clustering over movement trajectory data. The main challenge is to separate the signal from the noise, i.e. one-off trips. We show that standard data mining approaches suffer the important drawback that they cannot take the symmetry of non-noise trajectories into account. That is, if a trajectory has a symmetric equivalent that covers the same trip in the reverse direction, it should become more likely that neither of them is labelled as noise. We present a constraint model that takes this knowledge into account to produce better clusters. We show the efficacy of our approach on real-world data that was previously processed using standard data mining techniques.

1 Introduction

Clustering is one of the fundamental tasks in data mining whose general aim is to discover structure hidden in the data, and whose means consist in identifying the homogeneous groups of objects (the clusters) that emerge from the data [7]. Typically, this translates into putting together objects that are similar to each other, and keep separated as much as possible those that are different. The applications of this basic task are many and varied, ranging from customer segmentation for business intelligence to the analysis of images, time series, web search results and much more.

This paper focuses on the application domain of mobility data mining, which is concerned with the study of trajectories of moving objects and other kinds of mobility-related data [6]. Trajectories are sequences of time-stamped locations (typically longitude-latitude pairs) that describe the movements of an object. A most relevant example, which will also be the reference context for this work, is the sequence of GPS traces of cars collected by on-board devices, such as a satellite navigation system or an anti-theft platform [5].

In this context, clustering can be used to identify popular paths followed by several moving objects, such as common routes adopted by car drivers in a city. A different perspective to the problem was proposed in recent years in [8]: the

© Springer International Publishing Switzerland 2015
G. Pesant (Ed.): CP 2015, LNCS 9255, pp. 638–653, 2015.
DOI: 10.1007/978-3-319-23219-5_44

objective of the analysis is not directly the collective mobility of a population, but instead passes through an intermediate phase where the mobility of the single individual is studied first. Clustering the trajectories of an individual has the main purpose of highlighting which movements repeat consistently in his/her life, and therefore can be considered an important and representative part of the user's mobility. In the realm of urban traffic and transportation engineering such movements constitute the *systematic* component of the mobility of a person – and then of a city, as a consequence.

Identifying such *mobility profiles* is an important current research direction that has many applications, for example in urban planning [4]. If trips that individuals make frequently in their cars can be identified, the schedule and routes of the public transport system could be adjusted to provide people with an incentive to leave the car at home and do something for the environment.

In this paper, we study the problem from a constraint programming point of view. The trajectories that form part of mobility profiles have to conform to certain restrictions that are hard to incorporate in traditional data mining algorithms, but easy to express as constraints. We present a constraint programming model and demonstrate its benefits over previous approaches.

There are a number of approaches that model data mining problems with constraints (e.g. [1,2,9]), but to the best of our knowledge, this is the first time that this particular problem is studied in this context.

2   Drawbacks of the Pure Data Mining Approach

Existing data mining approaches to extract mobility profiles are usually customised algorithms that have been developed specifically for this purpose. The approach proposed in [8] (schematically depicted in Figure 1) consists of two ingredients: a trajectory distance function that encapsulates the context-specific notion of *similar trajectories* and a clustering schema that takes care of grouping similar trajectories together (center of Figure 1). It also ensures that the resulting groups have a certain minimum size, e.g. the cluster composed of the single dashed trajectory is removed (center of Figure 1). Once clusters have been formed, a representative trajectory for each of them is simply extracted by selecting the most central element of the cluster, i.e. the one that minimizes the sum of distances from the others (see the selected trajectories on the right of Figure 1).

We impose the minimum size constraint because we aim to identify the trips that are made regularly and are not interested in trips that occur regularly, but infrequently (such as a trip to a holiday home). The clustering procedure yields a list of trip groups, each of them essentially representing a mobility routine followed by the user. The set of routines of a user form the mobility profile of that user.

A very typical (and expected) result is that an individual's mobility contains at least two routines, one for the home-to-work systematic trips and one for the symmetric work-to-home return journey. Indeed, symmetric routines (the

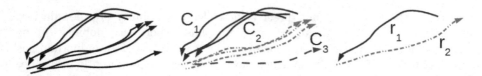

Fig. 1. The three steps of profile extraction: identify the trajectories (single trips) from the GPS data, discover clusters by grouping trajectories that are close, refine the clusters to remove the noise and extract representative routines.

home-work-home cycle being the most typical example) appear very naturally in real life, and they are a very important component of an individual's mobility. Therefore, while it is simply *important* to discover all the routines hidden in an individual's mobility data, it is *crucial* to discover symmetric routines if they exist.

From the viewpoint of a traffic management application for example, symmetric routines represent simple and regular mobility cycles that might neatly fit a public transportation offer. Asymmetric routines on the other hand are generally a symptom of a more complex daily mobility, for instance involving bring-and-get activities interleaved within a home-work-home cycle, which are much more difficult to serve by a public transportation service.

These observations lead to conclude that it would be advisable to equip the routine extraction procedure with mechanisms that boost the discovery of symmetric routines. Unfortunately, the plain clustering-based approach provided so far in [8] follows a local perspective, focused on guaranteeing properties of each single cluster – namely, mutual similarities of its members and a certain minimum size of the cluster – without considering the relation between different clusters or trajectories in different clusters.

Therefore, symmetric routines, which correspond to separate clusters even though they are connected by this particular property, are treated in isolation. What we would need to stimulate the discovery of symmetric routines, is a way to link such separate clusters, and change the cluster construction criteria accordingly. The purpose of this paper is to explore this way by exploiting a constraint formulation of the clustering problem, which can then be easily enriched and customized to include the application-specific requirements we mentioned above.

Setting the Minimum Cluster Size

Being able to take the symmetry of trajectories into account provides an additional benefit when clustering. The minimum size of a cluster is currently enforced by providing a static value to the algorithm that applies to all clusters. This value needs to be chosen carefully – if it is too high, no clusters at all may be identified and if it is too low, too many clusters will be identified. In particular, the value should be set such that trajectories that are noise as far as a mobility profile is concerned are not identified as clusters.

In practice, symmetric trajectories are often part of clusters of different sizes. There are many reasons for this. Consider for example the home-to-work routine, where an individual drives straight from home to work every day, but on the way back goes to the gym on certain days and shopping on others. Therefore, the size of the home-to-work cluster will be larger than that of the symmetric work-to-home cluster.

If the threshold for the minimum size of a cluster is set higher than the size of the work-to-home cluster, it will be categorized as noise even though there is a symmetric equivalent that supports it. Indeed, there could be cases like this with symmetric clusters of different sizes where no fixed value for the minimum size threshold will ensure that all symmetric clusters are identified while the noise trajectories are not.

In a constraint model, this can be taken into account easily. Instead of considering only the size of a cluster to justify its existence, we can explicitly model symmetric clusters as supports as well.

3 Constraint Model

We present the constraint model that takes all the considerations mentioned above into account. The constraints encode when two trajectories should be in the same cluster, when they should be in different clusters, when they are noise and the conditions that apply when there are symmetric trajectories.

For the sake of simplicity, trajectories are represented by a start and an end point. For our purposes, the route taken to get from the start to the end point is secondary, as we are interested in the extraction of mobility profiles that do not contain this information. For example, the way from home to work could be different from the way from work to home because of traffic conditions, one-way streets, or similar, but both are still part of the same routine. If the user stops, e.g. to do some shopping, the trajectory is split into two parts with two separate start and end points.

The model itself is general enough to accommodate routes represented by more than two points as well, but would likely reduce the number of symmetric trajectories as pairs would look less similar because of additional (irrelevant) information. Overall, the model would become larger and more complex.

Individual trajectories $x \in X$ are modelled as variables whose domains consist of the clusters they can be assigned to. That is, each trajectory can be assigned a cluster ID, including a special value which denotes that the trajectory is noise. We use $dist(x, y)$ to denote the distance (both spatial and temporal) between two trajectories x and y. Note that for the temporal distance, only the time of day is relevant, not the date. The reason for this is that we want to cluster trips that occur regularly, e.g. the daily commute to work. The distance of two trajectories is defined as the spatial and temporal distance between their start points and their end points, so $dist(x, y) < A$ is short hand for $dist(x_{start}, y_{start}) < A \wedge dist(x_{end}, y_{end}) < A$. T denotes the distance above which two trajectories are considered to be not *close* and R the distance above which they are considered *far*, $T < R$. If a trajectory is noise and should not be part of any cluster, it is denoted $x = noise$.

Our formulation consists of two complementary parts. The first part is a set of constraints that specify under which conditions two trajectories should belong to different clusters or be labelled as noise. The second one is a multi-objective optimization function that requires to cluster as many trajectories as possible and to fit them into as few clusters as possible. This requires the identified clusters to be as large as possible.

Our optimization function can be formulated as follows, where we use $\|X\|$ to denote the cardinality of the set X:

$$\begin{aligned} &\text{minimize } \|unique(\{x \mid x \in X\})\| &&\text{(no. distinct clusters)} \\ &\text{minimize } \|\{x \mid x \in X, x = noise\}\| &&\text{(no. noise trajectories)} \end{aligned} \quad (1)$$

Our constraint model includes three kinds of constraints. First we specify when two trajectories should *not* be in the same cluster:

$$\forall x, y \in X : dist(x, y) > R \rightarrow x \neq y \lor x = noise \lor y = noise \quad (2)$$

If two trajectories are far apart, they should be in different clusters or at least one of them should be noise. In our application, we are trying to identify mobility patterns which are defined by clusters of trajectories that are all close together. We therefore want to avoid clustering trajectories that are close to a set of intermediate trajectories, but far apart themselves.

Symmetric trajectories x and y are denoted as $symm(x, y)$ and defined as follows:

$$\begin{aligned} \forall x, y \in X : symm(x, y) \equiv\ &dist(x_{start}, y_{end}) \leq T \land \\ &dist(x_{end}, y_{start}) \leq T \end{aligned} \quad (3)$$

We now introduce the symmetry constraints that represent the main advantage of the constraint model over existing data mining algorithms. We denote the threshold for the minimum number of trajectories in a cluster S. This is a fixed value specified by the user and potentially prevents identifying clusters that are small, but supported by a symmetric cluster. We therefore introduce the threshold size for a pair of symmetric clusters S', where $S' > S$. If a trajectory x is symmetric to another trajectory y, they should belong to different clusters if they are not noise and the sum of the sizes of the clusters they belong to must be greater than S'.

$$\begin{aligned} \forall x, y \in X : symm(x, y) &\land x \neq noise \land y \neq noise \\ \land\ (\|\{z \mid z \in X, z = x\}\| &+ \|\{z \mid z \in X, z = y\}\| > S') \rightarrow x \neq y \end{aligned} \quad (4)$$

This set of constraints encodes most of the background knowledge on symmetric trajectories. We both take symmetry into account explicitly and mitigate the drawbacks of having a fixed minimum cluster size. The fixed threshold S is the only means of separating signal from noise in the data mining application. In the constraint model, we can relax this requirement in the presence of symmetric clusters – if a symmetric pattern is identified, we require the sum of the cluster sizes to be greater than the threshold for a pair of symmetric clusters, which is greater than the threshold for a single cluster.

If, on the other hand, there is no trajectory that is symmetric to x, then x is either noise or the size of the cluster that it is assigned to is greater than threshold S for the size of a single cluster.

$$\forall x \in X : \|\{y \mid y \in X, symm(x,y)\}\| = 0 \rightarrow$$
$$x = noise \vee (x \neq noise \wedge \|\{z \mid z \in X, z = x\}\| > S) \tag{5}$$

These constraints describe the requirements we place on trajectories to be part of clusters as well as considering the relation between clusters by taking the symmetry into account. We do not rely on any special constraints or esoteric constructs – our model is generic and can be implemented and solved with almost any constraint solver. In contrast, the data mining algorithm requires a specialised implementation that, while being able to leverage existing work, needs considerable efforts.

4   Model Implementation

We implemented the constraint model described above in the Minion constraint solver [3]. Minion is a general-purpose constraint solver. Our choice of solver was motivated by nothing but the authors' familiarity with this particular solver and its free availability.

4.1   Minion Model

Our Minion model is an almost direct translation of the model presented in the previous section. For each trajectory, we add one variable whose domain values represent the clusters the trajectory can be assigned to. We also add the dual representation where each cluster is represented by an array of Booleans, one for each trajectory. If a particular trajectory is in a particular cluster, the corresponding Boolean value is set to 1, else 0. We also add the channelling constraints between the one-variable-per-trajectory and the one-array-per-cluster representations. We require the dual representation to be able to post constraints on cluster sizes.

The noise trajectories are assigned to a special cluster that is represented by the value 0 in the domains of the trajectory variables. Modelling noise this way allows us to treat it in the same way as other clusters.

The Minion input language is flat and does not support quantifications over the variables. We therefore instantiate the free variables in Equations 2 to 5 and add constraints for each grounding. Note that the left hand side (before the implication) in all of these equations except 4 can be computed statically before solving the problem as it only depends on the trajectory data. We add constraints corresponding to the right hand side of the implications to the Minion model only if the left hand side is true. This preprocessing step helps to reduce the size of the Minion model, which is potentially very large because of the nested quantifications.

The implication on the right hand side of Equation 4 is modelled with the help of auxiliary variables that represent the left hand side of the implication being true. We need auxiliary variables here because Minion does not support implications between constraints, but only between a constraint and a variable. All variables and constraints are treated the same way by Minion and arc consistency is enforced during search. We use Minion's default parameters.

The thresholds T, R, S and S' are specified by the user. In the experiments section we will see the effect of these parameters on the results.

4.2   Optimising the Clustering

The optimisation objective of mobility profile mining is to cluster as many trajectories as possible, subject to the constraints, while minimising the overall number of clusters. We cannot easily express this multi-objective optimisation in our Minion model, but it turns out that we do not have to. We instead solve the simpler satisfaction problem that is defined as follows. Given a number of clusters, we assign values to trajectory variables in a descending manner. That is, we start with the highest cluster and assign a trajectory to the noise cluster (value 0) only if all other possibilities have been exhausted. This ensures that as many trajectories as possible are clustered as non-noise. In Minion, we simply specify a descending value ordering for the search.

We now minimise the number of clusters as follows. We start by specifying only two clusters in addition to the noise cluster. If the constraint satisfaction problem (CSP) has a solution, we have found our clustering. If there are more symmetric trajectories that need to be in different non-noise clusters however, this first CSP will have no solution. In this case, we increase the number of clusters by one, generate a new CSP and try to solve it. This way, we identify the solution with minimum number of clusters.

In practice, the number of clusters is usually small. The number of routine trips an individual makes that we have observed empirically was fewer than five in most cases, such that our method is likely to find the solution after only a few iterations.

4.3   Scalability

The main reason why data mining and machine learning applications use specialised approximate algorithms instead of complete optimisation approaches such as constraint programming is the scalability of such methods. Approximate methods are almost always orders of magnitude faster than corresponding complete methods while delivering solutions of similar quality.

While scalability is a concern for our application as well, it is not an issue in practice, due to the high modularity of the problem: indeed, each user can be analyzed separately, making the complexity of the problem linear in the number of users. The critical factor is, instead, the number of trajectories of a single user. However, the vast majority of individuals have a relatively low number of trips (in the dataset used for our experiments, which covers a time span of 5 weeks, most users have fewer than 100 trajectories), and therefore the constraint models can be solved in reasonable time. Our application is not time-critical, but puts the main emphasis on the quality of the solutions. We are happy to spend more time solving the problem as long as we get a better quality result.

The main factor that contributes to the size of the CSP apart from the number of trajectories is the number of clusters. Each additional cluster increases the size of the domains of the trajectory variables, requires an additional array of Boolean values and additional constraints to account for the possibility of a trajectory being assigned to that cluster. The iterative approach for setting the number of clusters described above keeps the size of the CSP as small as possible while not compromising on the quality of the solution.

We have found that this approach has a very significant impact on the efficiency with which we can solve this clustering problem. Often, increasing the number of clusters beyond the optimal number increases the solve time of the model by orders of magnitude.

5 Experimental Evaluation

We evaluated the quality of the clusters that the constraint model finds on real-world trajectories collected in Italy. We will show how our constraint model improves on the existing solution for extracting mobility profiles, thus evaluating the impact of taking symmetric clusters into account on the results. We briefly describe the dataset used in the experiments, then summarize the results, and finally show an example of profiles found on a single user with the different approaches.

5.1 Dataset

Our dataset contains the trajectories of the cars of 150 users living around the city of Pisa, Italy, collected over a period of 5 weeks (Figure 2). We cannot make the actual data available because of privacy reasons; an anonymized version is available at http://www.cs.ubc.ca/~larsko/downloads/cp2015_anonymized_dataset.tar.gz.

The users were selected in such a way to have a good representation of different mobility complexities, i.e. different numbers of trajectories. In our stratified sample, the number of trajectories per user roughly follows a uniform distribution that ranges from 1 to 100 (see Figure 3).

Experiments were carried out adopting a spatio-temporal distance $dist(x, y)$ between two points that is computed as the standard Euclidean distance between

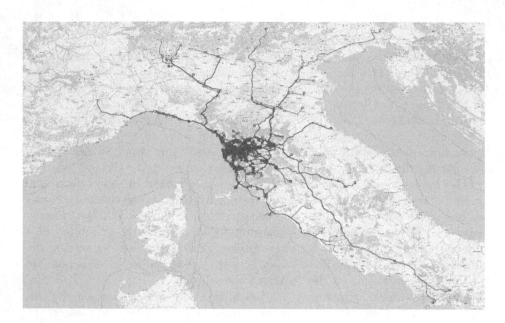

Fig. 2. Dataset of trajectories used in the experiments, containing 150 users centered around Pisa, Italy.

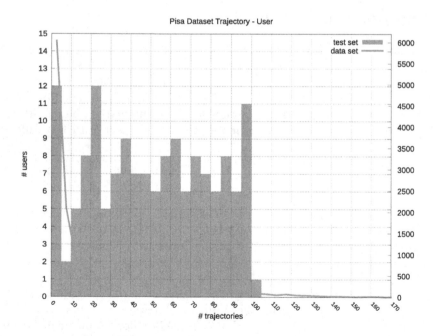

Fig. 3. Distribution of number of trajectories per user used in the experiments (blue bars) vs. distribution of original dataset (green line).

the spatial components of x and y if their temporal gap is less than a threshold $\tau = 1hr$, and ∞ otherwise. Therefore, we used the following ranges of values for parameters S (the minimum number of trajectories in a cluster) and T (the distance threshold between trajectories, expressed in meters): $S \in \{3, 4, 5, 6\}$, $T \in \{100, 200, 300, 400, 500\}$. Moreover, the parameters S' (the threshold for the combined size of two symmetric clusters) and R (the maximum distance between trajectories in a cluster) are set as functions of S and T, respectively: $S' = 1.5 \cdot S$, $R = 3 \cdot T$. The coefficients used to compute S' and R have been chosen empirically through a set of preliminary experiments over the dataset. Different datasets may exhibit different properties and a recalibration of these values may be required.

Computation over a medium-low-end desktop machine (i3 CPU at 3GHz, with 4GB RAM, running Ubuntu 12) took an overall time below 15 minutes. In particular, single users with 20 or less trajectories (which are the most common in the original dataset) are dealt with in less than one second, while those with 50 to 100 trajectories (less common, yet over-represented in our sample set, to evaluate their impact) took up to 15 seconds each.

5.2   Evaluation of Results

The main advantage of our constraint model over previous approaches is that it allows to take symmetric clusters into account. Without the constraints that do this, we should achieve results that are equivalent to the data mining solution proposed in [8]. However, the constraint model should find an optimal solution instead of a heuristic one.

In our first set of experiments, we compare both approaches without taking symmetric clusters into account to get a baseline for the performance of our new approach. This also allows to evaluate the impact of taking the symmetric clusters into account later.

Figures 4 and 5 show a comparison of the two approaches for the different parameter values (S and T) in terms of the number of non-noise trajectories (Figure 4) and of number of clusters found (Figure 5).

Both the number of non-noise trajectories and clusters increases as we relax the constraints, i.e. allow fewer trajectories that are further apart in a cluster. This is true for both the data mining and the CP approach. In general, both the number of non-noise trajectories and clusters is larger for the CP approach, in some cases significantly so. For example, the number of clusters for $S = 3$ that the CP approach finds is almost twice the number the data mining approach finds. As S is increased, the difference becomes smaller. The reason for this phenomenon is that the data mining approach partitions the set of trajectories into clusters through a heuristics that considers the distribution of data only locally, and therefore can create small clusters that are later discarded, as shown in the example of Figure 1. This kind of situations are avoided by the CP approach, which instead considers the assignment of trajectories to clusters from a global viewpoint.

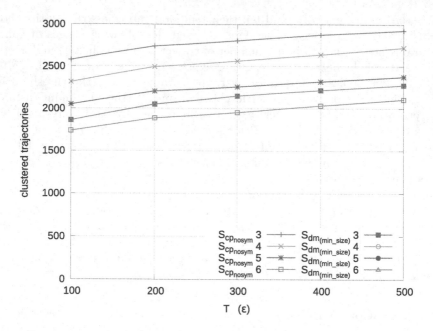

Fig. 4. Number of clustered trajectories found by the data mining approach (dm) and the constraint model without symmetry constraints (CP_{nosym}) for different distance threshold and cluster size values T and S.

In our second set of experiments, we focus on evaluating the impact of taking symmetric clusters into account. We adopt two measures for comparing the complete constraint model with all constraints to the one that does not take symmetric clusters into account as above.

- the *trajectory coverage* of method A over method B, defined as:

$$tr\_cov(A, B) = \frac{\|\{x \mid x \in X_A, x \neq noise\} \cap \{x \mid x \in X_B, x \neq noise\}\|}{\|\{x \mid x \in X_B, x \neq noise\}\|} \quad (6)$$

where X_A (resp. X_B) represents the overall set of clustered trajectories obtained with method A (resp. B). When A completely covers B, $tr\_cov(A, B) = 1$.

- the *cluster coverage* of method A over method B, defined as:

$$cl\_cov(A, B) = \frac{\sum_{u \in U} (C_u \cdot cl\_cov_u(A, B))}{\sum_{u \in U} C_u} \quad (7)$$

where U is the set of users and C_u the number of clusters found by A for user u. Function $cl\_cov_u(A, B)$, in turn, is computed over each user u as the average trajectory coverage of each cluster of u found by A w.r.t. those found by B. The matching between clusters is performed by taking the best match,

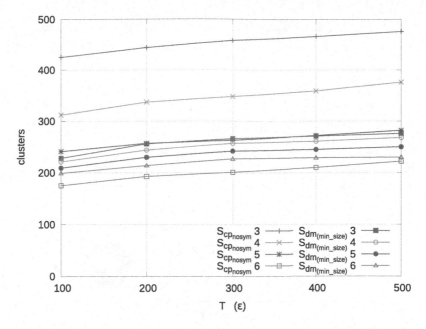

Fig. 5. Number of clusters found by the data mining approach (dm) and the constraint model without symmetry constraints (CP_{nosym}) for different distance threshold and cluster size values T and S.

i.e. maximizing the *trajectory coverage* limited to the two clusters compared. Clusters left unpaired are assigned to an empty set. When A completely covers B, $cl\_cov(A, B) = 1$.

While the trajectory coverage compares the overall ability to recognize systematic trips, the cluster coverage looks at how much each single cluster is discovered by the other method. Figures 6 and 7 report the results obtained on our dataset. Lines labelled with $S_{CP_{sym}}$ represent the coverages of the model with symmetry constraint (CP_{sym}) over that without symmetry (CP_{nosym}), and $S_{CP_{nosym}}$ the opposite.

The plots clearly show that the model that takes symmetric trajectories into account completely covers the other one, both in global terms of trajectory coverage (Figure 6) and in terms of (the more precise) cluster coverage (Figure 7), since the corresponding values are always very close to 1. The model that does not take symmetric trajectories into account on the other hand loses at least 50% of trajectories and 70% of cluster coverage. These results are consistent across the range of values for S and T that we consider here, with little variation.

Our results show that considering the symmetry of clusters has a large impact on results. We are able to cluster a lot more trajectories that were mistakenly identified as noise before, allowing to recover several *backward* trips that would otherwise not be recognized as routines.

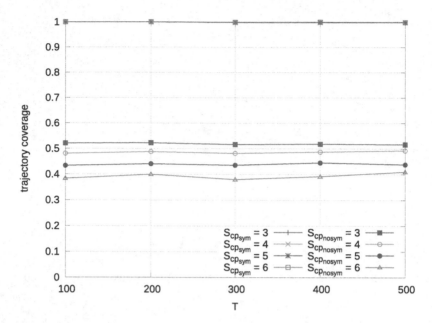

Fig. 6. Trajectory coverage for constraint models with symmetry constraint (CP_{sym}) and without (CP_{sym}) for different distance threshold and cluster size values T and S.

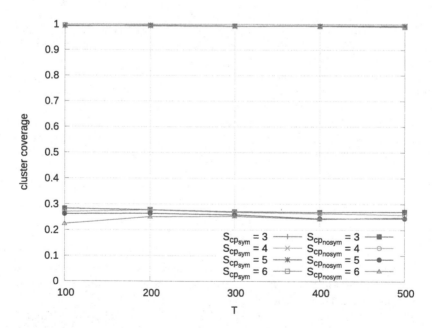

Fig. 7. Cluster coverage for constraint models with symmetry constraint (CP_{sym}) and without (CP_{sym}) for different distance threshold and cluster size values T and S.

5.3   Qualitative Evaluation

The quality of the clustering is more difficult to assess, as there are no auto-mated means to determine whether a set of clusters "makes sense". In practice, clusterings are evaluated through assessment by a human expert. We obviously cannot do this for all of the users we consider in this paper and consider the clusters for a single user as an example here.

While not all users have symmetric trajectories, the following example demonstrates the motivation of our approach and the case that is difficult to solve for specialised data mining methods. A clustering example that shows the benefit of the constraint model is presented in Figure 8. The image was obtained by using $T = 100$ and $S = 6$.

The red and the blue trajectories represent symmetric parts of the same routine. However, there are fewer red than blue trajectories; in particular the number of red trajectories is smaller than the threshold S. If the symmetry is not taken into account, only the blue trajectories are clustered, whereas the red ones are discarded as noise. The CP approach that does consider symmetry identifies the red trajectories correctly as a non-noise cluster.

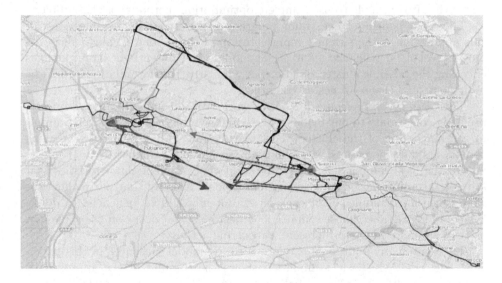

Fig. 8. Trajectory clusters found on a sample user (best seen in colors).

For the eastward journey, our user mainly follows the blue routine, but for the return journey he/she often changes path, either for traffic reasons or for personal needs, and the red trajectories represent the most frequent choice. The figure clearly shows the benefit of being able to take symmetric trajectories into account. Indeed, there are only three trajectories in the red cluster, which has been discarded as noise by the other method.

Lowering S to 3 would have clustered the red trajectories as non-noise, but at the same time erroneously identified a number of additional clusters that are in fact noise for our purposes. Therefore, the model that does not consider symmetric trajectories (as well as the specialised data mining algorithm discussed above) could not have identified the same set of clusters for this user.

6   Conclusion and Future Work

We have presented a constraint programming formulation of the problem of mining mobility profiles from GPS trajectories. This is an important research direction with many real-world applications that, to date, has been the domain of specialised data mining algorithms. We have described the drawbacks of these specialised algorithms and how they are alleviated with constraint programming.

The implementation of our model and its experimental evaluation on real-world data showed that the model stands up to its promise and delivers clusterings of a higher quality than those that the previously-used specialised data mining algorithms achieve. In particular, we are able to identify a much larger number of non-noise trajectories and clusters. The experiments further demonstrate that it is feasible to use complete optimisation methods for this problem.

The main benefit of constraint programming over similar technologies for this application is the succinct representation of the model. In particular SAT models would be much larger to accommodate the potentially large domains encoding the possible cluster assignments. Similarly, ILP formulations would be more complex because of nested constructs.

There are several avenues for future work. We did not add any symmetry-breaking constraints to our model even though symmetries occur – for example in the numbering of the clusters. Formulating and adding symmetry-breaking constraints may increase the scalability of our approach further.

On the data mining side, there are additional problem characteristics that are not taken into account in our current model. The day that particular trips were taken for example is not considered, but it would be beneficial to do so. We could for instance boost symmetric trajectories that occur on the same day, as they are more likely to represent a recurring activity.

There are many additional extensions to the model presented in this paper. Even our relatively basic formulation shows the promise of the approach by delivering results of a higher quality than specialised approaches while being easier to maintain and modify.

Acknowledgments. This work was supported by the European Commission as part of the "ICON - Inductive Constraint Programming" project (contract number FP7-284715). The Insight Centre for Data Analytics is supported by Science Foundation Ireland under grant number SFI/12/RC/2289. We thank the anonymous reviewers for their helpful feedback.

References

1. Davidson, I., Ravi, S.S., Shamis, L.: A SAT-based Framework for Efficient Constrained Clustering. In: SIAM International Conference on Data Mining, pp. 94–105 (2010)
2. De Raedt, L., Guns, T., Nijssen, S.: Constraint programming for data mining and machine learning. In: AAAI, pp. 1671–1675 (2010)
3. Gent, I.P., Jefferson, C.A., Miguel, I.: MINION: a fast, scalable, constraint solver. In: ECAI. pp. 98–102. IOS Press (2006)
4. Giannotti, F., Nanni, M., Pedreschi, D., Pinelli, F.: Trajectory pattern analysis for urban traffic. In: Proceedings of the Second International Workshop on Computational Transportation Science, IWCTS 2009, pp. 43–47 (2009)
5. Giannotti, F., Nanni, M., Pedreschi, D., Pinelli, F., Renso, C., Rinzivillo, S., Trasarti, R.: Unveiling the complexity of human mobility by querying and mining massive trajectory data. The VLDB Journal **20**(5), 695–719 (2011). http://dx.doi.org/10.1007/s00778-011-0244-8
6. Giannotti, F., Pedreschi, D. (eds.): Mobility, Data Mining and Privacy - Geographic Knowledge Discovery. Springer (2008)
7. Tan, P.N., Steinbach, M., Kumar, V.: Introduction to Data Mining. Addison-Wesley (2005)
8. Trasarti, R., Pinelli, F., Nanni, M., Giannotti, F.: Mining mobility user profiles for car pooling. In: KDD, pp. 1190–1198 (2011)
9. Wagstaff, K., Cardie, C.: Clustering with Instance-level Constraints. In: Seventeenth International Conference on Machine Learning, ICML 2000, pp. 1103–1110. Morgan Kaufmann Publishers Inc., San Francisco, CA, USA (2000)

Joint Vehicle and Crew Routing and Scheduling

Edward Lam[1,2]($\boxtimes$), Pascal Van Hentenryck[1,3], and Philip Kilby[1,3]

[1] NICTA, Eveleigh, NSW 2015, Australia
edward.lam@nicta.com.au
[2] University of Melbourne, Parkville, VIC 3010, Australia
[3] Australian National University, Acton, ACT 2601, Australia

Abstract. Traditional vehicle routing problems implicitly assume only one crew operates a vehicle for the entirety of its journey. However, this assumption is violated in many applications arising in humanitarian and military logistics. This paper considers a Joint Vehicle and Crew Routing and Scheduling Problem, in which crews are able to interchange vehicles, resulting in space and time interdependencies between vehicle routes and crew routes. It proposes a constraint programming model that overlays crew routing constraints over a standard vehicle routing problem. The constraint programming model uses a novel optimization constraint that detects infeasibility and bounds crew objectives. Experimental results demonstrate significant benefits of using constraint programming over mixed integer programming and a vehicle-then-crew sequential approach.

1 Introduction

A vehicle routing problem (VRP) aims to design routes for a fleet of vehicles that minimize some cost measure, and perhaps, while adhering to side constraints, such as time windows, or pickup and delivery constraints. In practice, however, VRPs are not solved in isolation; they typically arise in a sequential optimization process that first optimizes vehicle routes and then crew schedules given the vehicle routes [33]. This sequential methodology has the advantage of reducing the computational complexity. However, by assigning vehicle routes first, this approach may lead to suboptimal, or even infeasible, crew schedules since decisions in the first stage ignore crew constraints and objectives. Hence it is imperative to simultaneously consider vehicle and crew constraints and objectives, particularly in industries in which crew costs outstrip vehicle costs.

This paper considers a Joint Vehicle and Crew Routing and Scheduling Problem (JVCRSP) motivated by applications in humanitarian and military logistics. In these settings, vehicles (e.g., airplanes) travel long routes, and serve a variety of pickup and delivery requests under various side constraints. Vehicles must be operated by crews who have limitations on their operating times. Crews are also able to interchange vehicles at different locations and to travel as passengers before and after their operating times. JVCRSPs are extremely challenging computationally since vehicle and crew routes are now interdependent. In addition, vehicle interchanges add an additional time complexity, since two vehicles must

G. Pesant (Ed.): CP 2015, LNCS 9255, pp. 654–670, 2015.
DOI: 10.1007/978-3-319-23219-5_45

be synchronized to allow for the exchange to proceed. It is thus necessary to decide whether vehicles wait, and the waiting duration, since both vehicles must be present at the same location for an interchange to take place.

This paper proposes a constraint programming formulation of the JVCRSP that jointly optimizes vehicle and crew routing in the hope of remedying some limitations of a multi-stage model. The formulation overlays crew routing constraints over a traditional vehicle routing problem, and also adds a number of synchronization constraints to link vehicles and crews. In addition, the formulation includes a novel optimization constraint that checks feasibility and bounds crew costs early in the search while the focus is on vehicle routing. The constraint programming model is solved using a large neighborhood search (LNS) that explores both vehicle and crew neighborhoods.

The constraint program is compared to a mixed integer program and a two-stage method that separates the vehicle routing and crew scheduling problems. Experimental results on instances with up to 96 requests and different cost functions indicate that (1) the joint optimization of vehicle and crew routing produces considerable benefits over the sequential method, (2) the constraint program scales significantly better than the mixed integer program, and (3) vehicle interchanges are critical in obtaining high-quality solutions. These preliminary findings indicate that it is now in the realm of optimization technology to jointly optimize vehicle and crew routing and scheduling, and that constraint programming has a key role to play in this promising direction.

The rest of this paper is organized as follows. Section 2 describes the Joint Vehicle and Crew Routing and Scheduling Problem and some high-level modeling decisions. Section 3 outlines the two-stage methodology, while Section 4 describes the mixed integer formulation of the JVCRSP. Section 5 details the constraint programming formulation, two novel global constraints, and the search procedure. Section 6 reports the experimental results. Section 7 reviews related work and Section 8 concludes this paper.

2   Problem Description

The traditional vehicle routing problem with pickup and delivery and time windows (VRPPDTW) [30] consists of a fleet of vehicles, stationed at a common depot, that services pickup and delivery requests before returning to the depot. Every pickup request has one associated delivery request, with an equal but negated demand, that must be served after the pickup and by the same vehicle. In addition, every request has a time window, within which service must begin. Vehicles can arrive earlier than the time window, but must wait until the time window opens before starting service. Given a route, fixed travel times dictate the arrival and departure times of requests along the route.

The JVCRSP considered in this paper overlays, on top of the VRPPDTW, crew routing constraints that explicitly capture the movements of crews. The JVCRSP groups requests by location, and defines travel times between locations, thereby removing the one-to-one mapping between requests and locations

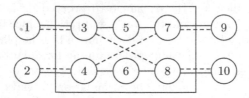

Fig. 1. A location at which vehicle interchange occurs. Nodes 3 to 8 belong to the same location. A vehicle (solid line) and a crew (dotted line) travel from node 1 to node 3, and another vehicle and crew travel from node 2 to node 4. The two vehicles service nodes 3 to 8, while the crews travel to nodes 7 and 8 to depart on a different vehicle. The movements into nodes 3 and 4, and out of nodes 7 and 8 must be synchronized between the vehicles and crews, but movements within the location are independent.

found in many routing problems. Vehicles require a crew when moving between two locations, but require no crew when servicing requests within a location. Similarly, crews require a vehicle to move between two locations, but can move independently of vehicles within a location. Crews are restricted to at most one driving shift, which has a maximum duration. No distance or time limitations are placed on crews before and after the driving shift.

The JVCRSP allows crews to switch vehicles at any given location. This idea is illustrated in Fig. 1. Vehicle interchange occurs when a crew moves from a node that is serviced by one vehicle to a node that is serviced by another vehicle, provided that the departure of the second vehicle is after the arrival of the first. Hence vehicle interchange requires explicit modelling of arrival and departure times. The JVCRSP retains the time windows on service start times and permits early arrival, but also allows vehicles to wait after service to depart at a later time. This functionality is necessary to facilitate vehicle interchange.

The objective function of the JVCRSP minimizes a weighted sum of the number of vehicles and crews used, and the total vehicle and crew travel distances.

3 A Two-Stage Method for the JVCRSP

One of the goals of this paper is to assess the benefits of jointly routing and scheduling vehicles and crews. A baseline to determine the potential of joint optimization is a two-stage approach that separates these two aspects by first solving a vehicle routing problem and then a crew pairing problem. The first stage can be solved with any appropriate technology, and this paper uses a large neighborhood search over a constraint programming model (e.g., [29,21,1]). Arrival and departure times are fixed at their earliest possible starting times to produce vehicle schedules. The crew pairing problem is modeled with a set covering problem typical in the airline industry [33]. It is solved using column generation with a resource-constrained shortest path pricing problem, and then a mixed integer programming solver to find integer solutions once the column generation process has converged.

4   A Mixed Integer Program for the JVCRSP

This section sketches a mixed integer programming (MIP) formulation for the JVCRSP. The model consists of a vehicle component and a crew component. The vehicle component is the traditional multi-commodity flow model of the VRPPDTW, which is omitted for familiarity and space reasons. The crew component contains similar routing constraints, but omits travel time and request cover constraints, since these are already included in the vehicle component. The vehicle and crew components are linked via synchronization constraints, which are key aspects of the formulation.

Table 1 lists the relevant inputs and decision variables. The usual vehicle flow variables $veh_{v,i,j}$ are duplicated for crews; the variable $crew_{c,i,j}$ denotes whether crew c traverses the edge (i,j). Additionally, the $driver_{c,i,j}$ variable indicates whether crew c drives the vehicle that traverses the arc (i,j), and is used to synchronize the vehicles and crews in space.

Figure 2 describes the key synchronization constraints that coordinate the vehicles, crews, and drivers in the MIP model. Equations (1) and (2) state that when moving from one location to another, a crew requires a vehicle and a vehicle requires a driver. Equation (3) restricts the driver along an arc to be one of the crews that traverses that arc. Equations (4) and (5) link the arrival, service, and departure times. Equation (6) constrains the start and end nodes to a common arrival and departure time. Equations (7) to (9) synchronize time at request, start, and end nodes, respectively. For every crew c that drives from a request i to any request j, Eq. (10) bounds the start of driving to be no later than the departure time at i. Similarly, Eq. (11) bounds the end of driving. Equation (12) restricts the driving duration of each crew.

5   A Constraint Program for the JVCRSP

This section introduces the constraint programming (CP) formulation for the JVCRSP. It describes the instance data, decision variables, and constraints of the model, and the specifications of two global constraints, and the search procedure.

The data and decision variables are described in Table 2. Like the MIP model, the CP model is divided into a vehicle component and a crew component.

The vehicle component, described in Figure 3, is based on the standard constraint programming modeling for vehicle routing (e.g., [21,28]). Successor variables succ($\cdot$) capture the sequence of requests along vehicle routes. By artificially joining the end of one route to the start of another (Eqs. (13) and (14)), the successor variables then describe a Hamiltonian cycle, which enables the use of the CIRCUIT global constraint for subtour elimination (Eq. (15)). Equations (16) and (17) track a vehicle along its route, and Eq. (18) determines if a vehicle is used. Equations (19) and (20) order the arrival, service, and departure times at each request. Equation (21) restricts the start and end nodes to one common arrival/departure time. Equation (22) enforces travel times. Symmetry-breaking, load, and pickup and delivery constraints are omitted due to space limitations.

Table 1. Instance data and decision variables of the MIP model.

Name	Description
$T > 0$	Time horizon.
$\mathcal{T} = [0, T]$	Time interval.
$\mathcal{V}$	Set of vehicles.
$\mathcal{C}$	Set of crews.
$\overline{T} \in \mathcal{T}$	Maximum driving duration of a crew.
s	Start node.
e	End end.
$\mathcal{R}$	Set of all requests.
$\mathcal{N} = \mathcal{R} \cup \{s, e\}$	Set of all nodes, including the start and end nodes.
$\mathcal{A}$	Arcs to be traversed by the vehicles and crews, excluding the arc (s, e) representing an unused vehicle/crew.
$\mathcal{L}$	Set of locations, including one depot location.
$l_i \in \mathcal{L}$	Location of $i \in \mathcal{N}$.
$a_i \in \mathcal{T}$	Earliest start of service at $i \in \mathcal{N}$.
$b_i \in \mathcal{T}$	Latest start of service at $i \in \mathcal{N}$.
$t_i \in \mathcal{T}$	Service duration of $i \in \mathcal{N}$.
$\text{veh}_{v,i,j} \in \{0, 1\}$	1 if vehicle $v \in \mathcal{V}$ traverses $(i, j) \in \mathcal{A} \cup \{(s, e)\}$.
$\text{crew}_{c,i,j} \in \{0, 1\}$	1 if crew $c \in \mathcal{C}$ traverses $(i, j) \in \mathcal{A} \cup \{(s, e)\}$.
$\text{driver}_{c,i,j} \in \{0, 1\}$	1 if crew $c \in \mathcal{C}$ drives on $(i, j) \in \mathcal{A}$, $l_i \neq l_j$.
$\text{arr}_{v,i} \in \mathcal{T}$	Arrival time of vehicle $v \in \mathcal{V}$ at $i \in \mathcal{N}$.
$\text{serv}_{v,i} \in [a_i, b_i]$	Start of service by vehicle $v \in \mathcal{V}$ at $i \in \mathcal{N}$.
$\text{dep}_{v,i} \in \mathcal{T}$	Departure time of vehicle $v \in \mathcal{V}$ at $i \in \mathcal{N}$.
$\text{crewTime}_{c,i} \in \mathcal{T}$	Arrival/departure time of crew $c \in \mathcal{C}$ at $i \in \mathcal{N}$.
$\text{driveStart}_c \in \mathcal{T}$	Start time of driving for crew $c \in \mathcal{C}$.
$\text{driveEnd}_c \in \mathcal{T}$	End time of driving for crew $c \in \mathcal{C}$.
$\text{driveDur}_c \in [0, \overline{T}]$	Driving duration of crew $c \in \mathcal{C}$.

$$\text{crew}_{c,i,j} \leq \sum_{v \in \mathcal{V}} \text{veh}_{v,i,j}, \qquad\qquad \forall c \in \mathcal{C}, (i,j) \in \mathcal{A}, l_i \neq l_j, \quad (1)$$

$$\sum_{v \in \mathcal{V}} \text{veh}_{v,i,j} = \sum_{c \in \mathcal{C}} \text{driver}_{c,i,j}, \qquad\qquad \forall (i,j) \in \mathcal{A}, l_i \neq l_j, \quad (2)$$

$$\text{driver}_{c,i,j} \leq \text{crew}_{c,i,j}, \qquad\qquad \forall c \in \mathcal{C}, (i,j) \in \mathcal{A}, l_i \neq l_j, \quad (3)$$

$$\text{arr}_{v,i} \leq \text{serv}_{v,i}, \qquad\qquad \forall v \in \mathcal{V}, i \in \mathcal{R}, \quad (4)$$

$$\text{serv}_{v,i} + t_i \leq \text{dep}_{v,i}, \qquad\qquad \forall v \in \mathcal{V}, i \in \mathcal{R}, \quad (5)$$

$$\text{arr}_{v,i} = \text{serv}_{v,i} = \text{dep}_{v,i}, \qquad\qquad \forall v \in \mathcal{V}, i \in \{s, e\}, \quad (6)$$

$$\text{arr}_{v,i} \leq \text{crewTime}_{c,i} \leq \text{dep}_{v,i}, \qquad\qquad \forall v \in \mathcal{V}, c \in \mathcal{C}, i \in \mathcal{R}, \quad (7)$$

$$\text{crewTime}_{c,s} - \text{dep}_{v,s} \leq M_1(2 - \text{veh}_{v,s,j} - \text{crew}_{c,s,j}), \ \forall v \in \mathcal{V}, c \in \mathcal{C}, j : (s,j) \in \mathcal{A}, \quad (8)$$

$$\text{arr}_{v,e} - \text{crewTime}_{c,e} \leq M_2(2 - \text{veh}_{v,i,e} - \text{crew}_{c,i,e}), \quad \forall v \in \mathcal{V}, c \in \mathcal{C}, i : (i,e) \in \mathcal{A}, \quad (9)$$

$$\text{driveStart}_c - \text{crewTime}_{c,i} \leq M_3 \left(1 - \sum_{j:(i,j) \in \mathcal{A}} \text{driver}_{c,i,j} \right), \qquad \forall c \in \mathcal{C}, i \in \mathcal{R} \cup \{s\}, \quad (10)$$

$$\text{crewTime}_{c,i} - \text{driveEnd}_c \leq M_4 \left(1 - \sum_{h:(h,i) \in \mathcal{A}} \text{driver}_{c,h,i} \right), \qquad \forall c \in \mathcal{C}, i \in \mathcal{R} \cup \{e\}, \quad (11)$$

$$\text{driveDur}_c = \text{driveEnd}_c - \text{driveStart}_c, \qquad\qquad \forall c \in \mathcal{C}. \quad (12)$$

Fig. 2. Core constraints of the MIP formulation.

Table 2. Instance data and the decision variables of the CP model.

Name	Description
$T \in \{1, \ldots, \infty\}$	Time horizon.
$\mathcal{T} = \{0, \ldots, T\}$	Set of time values.
$V \in \{1, \ldots, \infty\}$	Number of vehicles.
$\mathcal{V} = \{1, \ldots, V\}$	Set of vehicles.
$\mathcal{C} = \{1, \ldots, |\mathcal{C}|\}$	Set of crews.
$\mathcal{C}_0 = \mathcal{C} \cup \{0\}$	Set of crews, including a 0 value indicating no crew.
$\overline{T} \in \mathcal{T}$	Maximum driving duration of a crew.
$R \in \{1, \ldots, \infty\}$	Total number of requests.
$\mathcal{R} = \{1, \ldots, R\}$	Set of all requests.
$\mathcal{S} = \{R + 1, \ldots, R + V\}$	Set of vehicle start nodes.
$\mathcal{E} = \{R + V + 1, \ldots, R + 2V\}$	Set of vehicle end nodes.
$\mathcal{N} = \mathcal{R} \cup \mathcal{S} \cup \mathcal{E}$	Set of all nodes.
$\mathcal{N}_0 = \mathcal{N} \cup \{0\}$	Set of all nodes, including a dummy node 0.
$s(v) = R + v$	Start node of vehicle $v \in \mathcal{V}$.
$e(v) = R + V + v$	End node of vehicle $v \in \mathcal{V}$.
$\mathcal{L}$	Set of locations, including one depot location.
$l(i) \in \mathcal{L}$	Location of $i \in \mathcal{N}$.
$d(i, j) \in \mathcal{T}$	Distance and travel time from $i \in \mathcal{N}$ to $j \in \mathcal{N}$.
$a(i) \in \mathcal{T}$	Earliest start of service at $i \in \mathcal{N}$.
$b(i) \in \mathcal{T}$	Latest start of service at $i \in \mathcal{N}$.
$t(i) \in \mathcal{T}$	Service duration of $i \in \mathcal{N}$.
w_1	Cost of using one vehicle.
w_2	Cost of using one crew.
w_3	Cost of one unit of vehicle distance.
w_4	Cost of one unit of crew distance.
$\text{vehUsed}(v) \in \{0, 1\}$	Indicates if vehicle $v \in \mathcal{V}$ is used.
$\text{succ}(i) \in \mathcal{N}$	Successor node of $i \in \mathcal{N}$.
$\text{veh}(i) \in \mathcal{V}$	Vehicle that visits $i \in \mathcal{N}$.
$\text{arr}(i) \in \mathcal{T}$	Arrival time at $i \in \mathcal{N}$.
$\text{serv}(i) \in \{a(i), \ldots, b(i)\}$	Start of service at $i \in \mathcal{N}$.
$\text{dep}(i) \in \mathcal{T}$	Departure time at $i \in \mathcal{N}$.
$\text{crewUsed}(c) \in \{0, 1\}$	Indicates if crew $c \in \mathcal{C}$ is used.
$\text{crewDist}(c) \in \{0, \ldots, \infty\}$	Distance traveled by crew $c \in \mathcal{C}$.
$\text{crewSucc}(c, i) \in \mathcal{N}_0$	Successor of $i \in \mathcal{N}_0$ for crew $c \in \mathcal{C}_0$.
$\text{crewTime}(i) \in \mathcal{T}$	Arrival/departure time of every crew at $i \in \mathcal{N}$.
$\text{driveStart}(c) \in \mathcal{T}$	Start time of driving for crew $c \in \mathcal{C}_0$.
$\text{driveEnd}(c) \in \mathcal{T}$	End time of driving for crew $c \in \mathcal{C}_0$.
$\text{driveDur}(c) \in \{0, \ldots, \overline{T}\}$	Driving duration of crew $c \in \mathcal{C}$.
$\text{driver}(i) \in \mathcal{C}_0$	Driver of vehicle $\text{veh}(i)$ from $i \in \mathcal{R} \cup \mathcal{S}$ to $\text{succ}(i)$, with 0 indicating no driver.

The crew component, depicted in Fig. 4, overlays the vehicle component to upgrade the VRP into a JVCRSP. Crew successor variables $\text{crewSucc}(c, \cdot)$ model a path for every crew c from any start node to any end node via a sequence of arcs that cover the vehicle movements. To allow crews to start and end on any vehicle, the model includes a dummy node 0; crews start and end at this dummy node, whose successor must be a start node and whose predecessor must be an end node. Equations (23) to (27) formalize this idea. The modeling uses the SUBCIRCUIT constraint [14] to enforce subtour elimination, since a crew does not visit all nodes. The SUBCIRCUIT constraint differs from the CIRCUIT constraint seen in the vehicle component by allowing some nodes to be excluded from the Hamiltonian cycle. The successor of an excluded node is the node itself.

Every node i has an associated $\text{driver}(i)$ variable that denotes the driver of vehicle $\text{veh}(i)$ when it moves from i to its successor $\text{succ}(i)$ at a different location,

$$\text{succ}(e(v)) = s(v+1), \hspace{3cm} \forall v \in \{1, \ldots, V-1\}, \quad (13)$$

$$\text{succ}(e(V)) = s(1), \hspace{6cm} (14)$$

$$\text{CIRCUIT}(\text{succ}(\cdot)), \hspace{6cm} (15)$$

$$\text{veh}(s(v)) = \text{veh}(e(v)) = v, \hspace{4cm} \forall v \in \mathcal{V}, \quad (16)$$

$$\text{veh}(\text{succ}(i)) = \text{veh}(i), \hspace{3.5cm} \forall i \in \mathcal{R} \cup \mathcal{S}, \quad (17)$$

$$\text{vehUsed}(v) \leftrightarrow \text{succ}(s(v)) \neq e(v), \hspace{3cm} \forall v \in \mathcal{V}, \quad (18)$$

$$\text{arr}(i) \leq \text{serv}(i), \hspace{5cm} \forall i \in \mathcal{R}, \quad (19)$$

$$\text{serv}(i) + t(i) \leq \text{dep}(i), \hspace{4cm} \forall i \in \mathcal{R}, \quad (20)$$

$$\text{arr}(i) = \text{serv}(i) = \text{dep}(i), \hspace{3cm} \forall i \in \mathcal{S} \cup \mathcal{E}, \quad (21)$$

$$\text{dep}(i) + d(i, \text{succ}(i)) = \text{arr}(\text{succ}(i)), \hspace{2cm} \forall i \in \mathcal{R} \cup \mathcal{S}. \quad (22)$$

Fig. 3. Key constraints of the vehicle component of the CP model.

$$\text{crewSucc}(c, 0) \in \mathcal{S} \cup \{0\}, \hspace{4cm} \forall c \in \mathcal{C}, \quad (23)$$

$$\text{crewSucc}(c, i) \in \mathcal{R} \cup \{i\}, \hspace{3.5cm} \forall c \in \mathcal{C}, i \in \mathcal{S}, \quad (24)$$

$$\text{crewSucc}(c, i) \in \mathcal{R} \cup \mathcal{E}, \hspace{3.5cm} \forall c \in \mathcal{C}, i \in \mathcal{R}, \quad (25)$$

$$\text{crewSucc}(c, i) \in \{0, i\}, \hspace{3.5cm} \forall c \in \mathcal{C}, i \in \mathcal{E}, \quad (26)$$

$$\text{SUBCIRCUIT}(\text{crewSucc}(c, \cdot)), \hspace{4cm} \forall c \in \mathcal{C}, \quad (27)$$

$$l(\text{succ}(i)) = l(i) \leftrightarrow \text{driver}(i) = 0, \hspace{2.5cm} \forall i \in \mathcal{R} \cup \mathcal{S}, \quad (28)$$

$$\text{crewSucc}(\text{driver}(i), i) = \text{succ}(i), \hspace{2.5cm} \forall i \in \mathcal{R} \cup \mathcal{S}, \quad (29)$$

$$\text{crewUsed}(c) \leftrightarrow \bigvee_{i \in \mathcal{N}_0} \text{crewSucc}(c, i) \neq i \leftrightarrow \bigvee_{i \in \mathcal{R} \cup \mathcal{S}} \text{driver}(i) = c, \hspace{1cm} \forall c \in \mathcal{C}, \quad (30)$$

$$\text{crewUsed}(c) \geq \text{crewUsed}(c+1), \hspace{2cm} \forall c \in \{1, \ldots, C-1\}, \quad (31)$$

$$l(\text{crewSucc}(c, i)) = l(i) \vee \text{crewSucc}(c, i) = \text{succ}(i), \hspace{1cm} \forall c \in \mathcal{C}, i \in \mathcal{R} \cup \mathcal{S}, \quad (32)$$

$$\text{CREWSHORTCUT}(\text{succ}(\cdot), \text{crewSucc}(\cdot, \cdot), \text{arr}(\cdot), \text{dep}(\cdot), \mathcal{C}, \mathcal{R}, \mathcal{S}, \mathcal{E}, l(\cdot)), \hspace{1cm} (33)$$

$$\text{arr}(i) \leq \text{crewTime}(i) \leq \text{dep}(i), \hspace{4cm} \forall i \in \mathcal{N}, \quad (34)$$

$$\text{crewTime}(i) \leq \text{crewTime}(\text{crewSucc}(c, i)), \hspace{1.5cm} \forall c \in \mathcal{C}, i \in \mathcal{R} \cup \mathcal{S}, \quad (35)$$

$$\text{CREWBOUND}(\text{crewDist}(c), \text{crewSucc}(c, \cdot), \text{crewTime}(\cdot), \mathcal{R}, \mathcal{S}, \mathcal{E}, d(\cdot, \cdot)), \hspace{0.5cm} \forall c \in \mathcal{C}, \quad (36)$$

$$\text{driveStart}(\text{driver}(i)) \leq \text{dep}(i), \hspace{3cm} \forall i \in \mathcal{R} \cup \mathcal{S}, \quad (37)$$

$$\text{driveEnd}(\text{driver}(i)) \geq \text{arr}(\text{succ}(i)), \hspace{2.5cm} \forall i \in \mathcal{R} \cup \mathcal{S}, \quad (38)$$

$$\text{driveDur}(c) = \text{driveEnd}(c) - \text{driveStart}(c), \hspace{3cm} \forall c \in \mathcal{C}. \quad (39)$$

Fig. 4. Crew component of the CP model.

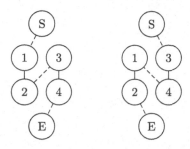

Fig. 5. A partial solution obtained at some stage during search. A crew is known to traverse the arcs $(1, 2)$ and $(3, 4)$, but it is not yet known whether the start node S connects (directly or via a path) to node 1 or node 3, and whether node 4 or node 2 connects to the end node. It is also not known whether such a path exists.

and takes the value 0 when the successor is at the same location, indicating that no driver is necessary (Eq. (28)). Equation (29) requires drivers to move with their vehicles. Equation (30) determines whether a crew is used. Equation (31) is a symmetry-breaking constraint. Equation (32) is a space synchronization constraint that requires crews to move to a node at their current location, or move with a vehicle (to a different location). The CREWSHORTCUT global constraint in Eq. (33) enforces the crew shortcut property, and is detailed in Section 5.2. Equation (34) synchronizes time between vehicles and crews. The time window in this constraint allows a crew to switch vehicle by moving to another node at the same location, provided that this vehicle departs later than this crew. Equation (35) forces crews to only move forward in time. Equation (36) is an optimization constraint, described in Section 5.1, that bounds crew distances and checks whether crews can return to the depot. Equations (37) to (39) determine the driving time of each crew.

The objective function (Eq. (40)) minimizes a weighted sum of the vehicle and crew counts, and the total vehicle and crew travel distances.

$$\min w_1 \sum_{v \in \mathcal{V}} \text{vehUsed}(v) + w_2 \sum_{c \in \mathcal{C}} \text{crewUsed}(c) +$$

$$w_3 \sum_{i \in \mathcal{R} \cup \mathcal{S}} d(i, \text{succ}(i)) + w_4 \sum_{c \in \mathcal{C}} \text{crewDist}(c). \quad (40)$$

5.1   Feasibility and Bounding of Crew Routes

Since crew routing is superimposed on a vehicle routing problem, it is important to determine during the search whether each crew has a feasible route and to compute a lower bound on the crew distance. Consider Fig. 5, which illustrates a partial solution. Here, the crew has been allocated to travel on some arcs but it is not known whether there exists a path from a start node to an end node through these arcs, and how long this path is, if it exists.

To detect infeasibility and to compute lower bounds to crew distances, an optimization constraint can be used [13,10,12,11]. It needs to find a shortest

$$\min \sum_{i \in \mathcal{R} \cup \mathcal{S}} \sum_{j \in \mathrm{cSucc}(i)} d(i,j) \cdot x_{i,j}, \tag{41}$$

$$\sum_{i \in \mathcal{S}} \sum_{j:(i,j) \in \mathcal{B}} x_{i,j} = 1, \tag{42}$$

$$\sum_{i \in \mathcal{E}} \sum_{h:(h,i) \in \mathcal{B}} x_{h,i} = 1, \tag{43}$$

$$\sum_{h:(h,i) \in \mathcal{B}} x_{h,i} = \sum_{j:(i,j) \in \mathcal{B}} x_{i,j}, \qquad \forall i \in \mathcal{R}, \tag{44}$$

$$\sum_{j:(i,j) \in \mathcal{B}} x_{i,j} \le 1, \qquad \forall i \in \mathcal{R} \cup \mathcal{S}, \tag{45}$$

$$t_i + d(i,j) - t_j \le M(1 - x_{i,j}), \qquad (i,j) \in \mathcal{B}, \tag{46}$$

$$x_{i,j} \in [0,1], \qquad (i,j) \in \mathcal{B}, \tag{47}$$

$$t_i \in [\min(\mathrm{cTime}(i)), \max(\mathrm{cTime}(i))], \qquad i \in \mathcal{N}. \tag{48}$$

Fig. 6. The linear programming relaxation for optimization constraint CREWBOUND.

path through specified nodes, which is NP-hard [24,20,9,32]. Instead, the CREWBOUND constraint solves a linear relaxation, which is presented in Fig. 6.

The set $\mathcal{B}$ represents the current set of arcs that can be traversed by the crew. It is defined as

$$\mathcal{B} = \{(i,j) : i \in \mathcal{R} \cup \mathcal{S}, j \in \mathrm{cSucc}(i), i \ne j\},$$

where the set $\mathrm{cSucc}(i)$ represents the domain of $\mathrm{crewSucc}(c,i)$ for every node $i \in \mathcal{R} \cup \mathcal{S}$ and the appropriate crew c. The set $\mathrm{cTime}(i)$ represents the domain of $\mathrm{crewTime}(i)$ for every node $i \in \mathcal{N}$.

The $x_{i,j}$ variables indicate whether the crew traverses the arc $(i,j) \in \mathcal{B}$, and the t_i variables represent the arrival time to node i.

The objective function minimizes the total distance (Eq. (41)). (Equations (42) to (44) are the flow constraints, which ensure the existence of a path from a start node to an end node. Equation (45) is a redundant constraint that strengthens the linear program. Equation (46) is the time-based subtour elimination constraint, where M is a big-M constant. Equations (47) and (48) restrict the domains of the $x_{i,j}$ and t_i variables respectively.

5.2   Symmetry-Breaking within Locations

The constraint program also uses the global constraint CREWSHORTCUT to remove symmetries in the crew routes. The intuition is best explained using Fig. 1. Recall that, in this example, one crew moves from node 3 to 8 and the other crew moves from node 4 to 7. Both crews thus shortcut the intermediate nodes, which are only visited by vehicles. The global constraint CREWSHORTCUT ensures that crews never visit intermediate nodes, but move directly from incoming to outgoing nodes at a location. This pruning eliminates symmetric solutions in which crews visit various sequences of intermediate nodes within a location.

Let $\text{pred}(i) \in \mathcal{N}$ be the predecessor of a node $i \in \mathcal{N}$. The constraint implements simple propagation rules, such as,

$$l(\text{pred}(i)) = l(i) = l(\text{succ}(i)) \leftrightarrow \bigwedge_{c \in \mathcal{C}} \text{crewSucc}(c, i) = i, \qquad \forall i \in \mathcal{R}, \quad (49)$$

which ensures that crews never visit a node whose predecessor and successor are at the same location as the node itself.

5.3   The Search Procedure

This section briefly describes the labeling procedure and the LNS procedure.

The Labeling Procedure. The labeling procedure first assigns vehicle routes, and then crew routes.

To determine vehicle routes, the search procedure considers each vehicle in turn and assigns the successor variables from the start node to the end node. The value selection strategy first chooses nodes at the same location ordered by earliest service time, and then considers nodes at other locations, also ordered by earliest service time.

To assign crew routes, the labeling procedure begins by ordering all nodes $i \in \mathcal{R} \cup \mathcal{S}$ that have an unlabeled driver(i) variable by earliest departure time. The first of these nodes is assigned a driver, and then a path from this node to any end node is constructed for this crew by labeling both the crew successor and the driver variables. This process is repeated until all driver($\cdot$) variables are labeled. The value selection heuristic for crew successor variables favors the node that is visited next by the current vehicle of the crew, provided that the crew can drive from this node; otherwise, the successor is chosen randomly.

The LNS Procedure. The LNS procedure randomly explores two neighborhoods:

- **Request Neighborhood:** This neighborhood relaxes a collection of pickup and delivery requests for insertion, in addition to all crew routes;
- **Crew Neighborhood:** This neighborhood fixes all vehicle routes, and relaxes a number of crew routes.

6   Experimental Results

This section compares the solution quality and features of the MIP and CP models, and the two-stage procedure.

Table 3. The number of locations and requests in each of the ten JVCRSP instances.

	A	B	C	D	E	F	G	H	I	J
$|\mathcal{L}|$	7	7	7	11	11	12	11	11	5	6
$|\mathcal{R}|$	12	16	20	40	46	46	40	80	54	96

The Instances. The instances are generated to capture the essence of applications in humanitarian and military logistics. These problems typically feature fewer locations than traditional vehicle routing applications, but comprise multiple requests at each location. Both the locations and the time windows are generated randomly. Table 3 describes the size of the instances. The maximum number of vehicles and crews are not limited in the instances.

The Algorithms. As mentioned previously, the two-stage method uses a constraint programming model for its first step. The model is solved using a large neighborhood search with a time limit of one minute, and the first step takes the best VRP solution out of ten runs. The crew pairing problem is solved optimally using the output of the first stage. All instances are solved to optimality in under 0.02 seconds. Hence the two-stage method requires a total of ten minutes. Obviously, the VRP runs can be performed in parallel.

The MIP model is solved using Gurobi 6.0.0 on an eight-core CPU for twelve hours with default parameters. The linear programming relaxation is rather weak and the dual bounds are not informative, which is why they are not reported here. An alternative MIP model that folds Eqs. (8) and (9) into Eq. (7) by modifying the network is also tested, but is inferior to the one presented.

The CP model is implemented in Objective-CP [31], and initialized with the same VRP solution as in the two-stage approach. This allows us to assess the benefits of the joint optimization reliably, since the second stage and the CP model are seeded with the same vehicle routes. The LNS procedure over the constraint model is then performed ten times for five minutes each. The model essentially requires a total of sixty minutes, which includes ten minutes to produce the initial VRP solution. Of course, all these computations can be executed in parallel in six minutes of wall clock time.

Solution Quality. Tables 4 to 6 depict the experimental results for various vehicle and crew costs. The tables only report the CPU time for the MIP model, since the two-stage and CP models have been given short execution times, as explained above. The experimental results highlight a few important findings:

1. Jointly optimizing vehicle and crew routing and scheduling achieves significant benefits. On the above instances, the average benefit of the CP model over the two-stage approach is 12%. When crew costs are higher, the benefits can exceed 45%.
2. The benefits obviously increase when the crew costs increase. However, even when vehicle costs are five times as large as crew costs, the benefits are still substantial and exceed 5% in more than half of the instances.

Table 4. Solutions found by the three models with equal vehicle and crew costs ($w_1 = w_2 = 1000$, $w_3 = w_4 = 1$). The objective value columns separate vehicle and crew costs in parenthesis. Time (in seconds) is provided for the MIP model. The last column lists the percentage improvement of the CP model compared against the two-stage method.

	Two-Stage	MIP		CP	
	Objective Value	Objective Value	Time	Objective Value	%
A	8328 (2086 + 6242)	8318 (2086 + 6232)	5	8318 (2086 + 6232)	-0.1%
B	7198 (2065 + 5133)	6154 (2063 + 4091)	27132	6167 (2068 + 4099)	-16.7%
C	8216 (2079 + 6137)	6192 (2088 + 4104)	37254	6200 (2079 + 4121)	-32.5%
D	14879 (3298 + 11581)	-	-	13855 (3302 + 10553)	-7.4%
E	18105 (2275 + 15830)	-	-	17181 (2278 + 14903)	-5.4%
F	26334 (4406 + 21928)	-	-	24570 (4412 + 20158)	-7.2%
G	14110 (3351 + 10759)	-	-	13074 (3355 + 9719)	-7.9%
H	25617 (6585 + 19032)	-	-	24975 (6587 + 18388)	-2.6%
I	8894 (2316 + 6578)	-	-	7795 (2309 + 5486)	-14.1%
J	16932 (3710 + 13222)	-	-	13877 (3710 + 10167)	-22.0%

Table 5. Best solutions from the three models with higher crew costs ($w_1 = 1000$, $w_2 = 5000$, $w_3 = w_4 = 1$).

	Two-Stage	MIP		CP	
	Objective Value	Objective Value	Time	Objective Value	%
A	32328 (2086 + 30242)	32318 (2086 + 30232)	5	32318 (2086 + 30232)	0.0%
B	27198 (2065 + 25133)	22154 (2056 + 20098)	8352	22167 (2068 + 20099)	-22.7%
C	32216 (2079 + 30137)	28198 (3086 + 25112)	4869	22200 (2079 + 20121)	-45.1%
D	58879 (3298 + 55581)	-	-	53905 (3317 + 50588)	-9.2%
E	78105 (2275 + 75830)	-	-	73175 (2280 + 70895)	-6.7%
F	110334 (4406 + 105928)	-	-	100502 (4413 + 96089)	-9.8%
G	54110 (3351 + 50759)	-	-	43996 (3353 + 40643)	-23.0%
H	97617 (6585 + 91032)	-	-	92928 (6585 + 86343)	-5.0%
I	32894 (2316 + 30578)	-	-	27817 (2326 + 25491)	-18.3%
J	64932 (3710 + 61222)	-	-	49930 (3710 + 46220)	-30.0%

Table 6. Best solutions from the three models with higher vehicle costs ($w_1 = 5000$, $w_2 = 1000$, $w_3 = w_4 = 1$).

	Two-Stage	MIP		CP	
	Objective Value	Objective Value	Time	Objective Value	%
A	16328 (10086 + 6242)	16318 (10086 + 6232)	7	16318 (10086 + 6232)	-0.1%
B	15198 (10065 + 5133)	14154 (10056 + 4098)	2327	14167 (10068 + 4099)	-7.3%
C	16216 (10079 + 6137)	14200 (10091 + 4109)	32135	14200 (10079 + 4121)	-14.2%
D	26879 (15298 + 11581)	-	-	24946 (15303 + 9643)	-7.7%
E	26105 (10275 + 15830)	-	-	25172 (10278 + 14894)	-3.7%
F	42334 (20406 + 21928)	-	-	40597 (20412 + 20185)	-4.3%
G	26110 (15351 + 10759)	-	-	24008 (15353 + 8655)	-8.8%
H	49617 (30585 + 19032)	-	-	48911 (30585 + 18326)	-1.4%
I	16894 (10316 + 6578)	-	-	15800 (10326 + 5474)	-6.9%
J	28932 (15710 + 13222)	-	-	25930 (15710 + 10220)	-11.6%

3. On the smaller instances, the MIP marginally outperforms the CP model when run over a significantly longer period of time.
4. The MIP model contains numerous big-M synchronization constraints that weaken its linear relaxation, making it difficult to solve. It can only prove optimality on the trivial instance A, and it also fails to scale to large instances, contrary to the CP model, which quickly finds high-quality solutions.

The Impact of Vehicle Interchanges. One of the main advantages of jointly optimizing vehicle and crew routing and scheduling is the additional flexibility for vehicle interchanges. For example, instance C in Tables 4 to 6 highlight some of the benefits of this flexibility. The CP model is able to delay vehicles to favor interchanges, thereby decreasing crew costs significantly.

Table 7 quantifies vehicle interchanges for the instances with equal vehicle and crew costs. It reports the total number of crews required in the solution, the number of crews participating in vehicle interchanges, the total number of vehicle interchanges, and the average number of interchanges per crew that switches vehicle. These results indicate that the constraint programming model allows more vehicle interchanges on average. For the larger instances, each crew travels on almost three vehicles on average. The total number of vehicle interchanges is roughly the same for the CP model and the two-stage approach, but the CP model produces solutions with fewer crews.

The Impact of the Joint Vehicle and Crew Routing and Scheduling. It is also interesting to notice in Tables (4) to (6) that the CP model often prefers vehicle routes whose costs are higher than those of the two-stage approach. This is the case despite the fact that the CP search is initialized with the first-stage VRP solution. This indicates that the CP model trades off crews costs for vehicle costs. Running a two-stage version of the CP model, in which the vehicle routes

Table 7. Statistics on vehicle interchange in the solutions to the runs with equal vehicle and crew costs. For both the two-stage and the CP models, this table provides the number of crews used in the solution, the number of crews that participates in vehicle interchange, the total number of interchanges, and the average number of interchanges for each crew that undertakes interchange.

	Two-Stage				CP			
	Crews Used	Intchng Crews	Vehicle Intchngs	Average Intchngs	Crews Used	Intchng Crews	Vehicle Intchngs	Average Intchngs
A	6	2	2	1.0	6	3	3	1.0
B	5	2	2	1.0	4	3	3	1.0
C	6	3	3	1.0	4	4	4	1.0
D	11	8	12	1.5	10	7	7	1.0
E	15	9	14	1.6	14	12	21	1.8
F	21	20	35	1.8	19	19	36	1.9
G	10	8	9	1.1	9	8	11	1.4
H	18	17	28	1.6	17	16	29	1.8
I	6	5	5	1.0	5	3	5	1.7
J	12	11	15	1.4	9	8	15	1.9

(but not the vehicle schedules) are fixed, also confirms these benefits. The CP model significantly outperforms this two-stage CP approach on a number of instances. These results indicate that both the flexibility in interchanges and the joint optimization are key benefits of the CP model.

7   Related Work

JVCRSPs have not attracted very much interest in the literature at this point, probably due to their inherent complexity [5]. This contradicts with the numerous studies of VRPs and related problems (e.g., [30,23,2]). This section reviews the most relevant work.

A JVCRSP is considered in [22], in which vehicles transport teams to service customers, and are able to move without any team on board. The problem features three types of tasks that must be serviced in order, and all customers have one task of each type. Each team can only service one compatible type of task. The MIP formulation has variables indexed by five dimensions, and is intractable. The paper also develops a simple iterative local search algorithm embedded within a particle swarm metaheuristic. This approach is very specific and cannot accommodate side constraints easily.

A JVCRSP modeling a mail distribution application is solved using a two-stage heuristic column generation approach in [19]. This model features multiple depots at which vehicle routes begin and end. In the first stage, trips that begin and end at these depots are computed. The second stage takes a scheduling approach and assigns vehicles and crews to the trips. Vehicle interchange can only occur at the depots at the start or end of a trip. In addition, the model features a 24-hour cyclic time period, and variables indexed by discretized time blocks. Since the solution method is based on two stages, it cannot jointly optimize vehicles and crews. Observe also that the CP formulation in our paper does not require time-indexed variables, and allows for vehicle interchange at all locations.

Another JVCRSP considers European legislation and includes relay stations where drivers are allowed to rest and interchange vehicles [8]. Upon reaching a relay station, the vehicle must wait a fixed amount of time. Vehicle interchange can only occur if other drivers arrive at this relay station during this time interval. The problem also provides a shuttle service, separate to the fleet of vehicles, that can be used to move drivers between relay stations. The problem is solved using a two-stage LNS method. In the first stage, a VRP is solved and the resulting routes form the customers of a VRP in the second stage, in which the crews perform the role of vehicles. Observe that this approach also cannot jointly optimize vehicle and crew routing. The model features several fixed parameters, such as the duration during which a vehicle waits at a relay station, and a search procedure that only explores a limited number of nearby relay stations. Both these restrictions reduce the search space, but negatively impact vehicle interchange, leading the authors to conclude that allowing for vehicle interchange does not significantly improve the objective value. Our model proves the contrary since it does not fix the duration that a crew waits, and can explore all locations for vehicle interchange.

JVCRSPs belong in the class of problems known as the VRPs with multiple synchronization constraints [5]. Synchronization is a feature present in certain VRPs, in which decisions about one object (e.g., vehicle, route, request) imply actions that may or must be taken on other objects. A complex VRP with synchronization is the VRP with Trailers and Transshipments (VRPTT) [7,4]. It features two vehicle classes: lorries which can move independently, and trailers which must be towed by an accompanying lorry. All lorries begin at a single depot with or without a trailer. A lorry can detach its trailer at transshipment locations in order to visit customers who are unable to accommodate a trailer (e.g., due to size). Lorries can transfer load into and/or attach with any trailer at any transshipment location. A lorry that has detached its trailer can also return to the depot without reattaching a trailer, leaving its trailer behind at a transshipment location to be collected by another lorry at a future time. Several sophisticated MIP formulations are presented, which are solved using branch-and-cut on instances with up to eight customers, eight transshipment locations and eight vehicles. The VRPTT can be reformulated into a JVCRSP by casting crews as lorries with zero capacity, and vehicles as trailers [6].

The VRPTT is most closely related to our JVCRSP model, since lorry routes and trailer routes are jointly computed, and the search space is not artificially limited. However, one major difference is that the VRPTT incorporates load synchronization, which the JVCRSP does not consider.

Finally, observe that vehicle and crew *scheduling* problems (VCSPs) are thoroughly studied (e.g., [18,27,17,15,16,3,25,26]). VCSPs aim to assign vehicles and crews to a predetermined set of trips, with each trip consisting of a fixed route, and usually fixed arrival and departure times. Trips in VCSPs correspond to parts of a route in JVCRSPs, which are not available a priori, but instead, must be computed during search, thereby increasing the computational challenges.

8   Conclusion

Motivated by applications arising in humanitarian and military logistics, this paper studied a Joint Vehicle and Crew Routing and Scheduling Problem, in which crews are able to interchange vehicles, resulting in interdependent vehicle routes and crew routes. The paper proposed a constraint programming model that overlays crew routing constraints over a standard vehicle routing problem. The model uses a novel optimization constraint to detect infeasibility and to bound crew objectives, and a symmetry-breaking global constraint. The model is compared to a sequential method and a mixed integer programming model. Experimental results demonstrate significant benefits of the constraint programming model, which reduces costs by 12% on average compared to the two-stage method, and scales significantly better than the mixed integer programming model. The benefits of the constraint programming model are influenced by both the additional flexibility in vehicle interchanges and in trading crew costs for vehicle costs. Future work will investigate more sophisticated search techniques and global constraints to scale to larger instances.

References

1. Bent, R., Van Hentenryck, P.: A two-stage hybrid local search for the vehicle routing problem with time windows. Transportation Science **38**(4), 515–530 (2004)
2. Berbeglia, G., Cordeau, J.F., Gribkovskaia, I., Laporte, G.: Static pickup and delivery problems: a classification scheme and survey. TOP **15**(1), 1–31 (2007)
3. Cordeau, J.F., Stojković, G., Soumis, F., Desrosiers, J.: Benders decomposition for simultaneous aircraft routing and crew scheduling. Transportation Science **35**(4), 375–388 (2001)
4. Drexl, M.: On some generalized routing problems. Ph.D. thesis, RWTH Aachen University, Aachen (2007)
5. Drexl, M.: Synchronization in vehicle routing–a survey of VRPs with multiple synchronization constraints. Transportation Science **46**(3), 297–316 (2012)
6. Drexl, M.: Applications of the vehicle routing problem with trailers and transshipments. European Journal of Operational Research **227**(2), 275–283 (2013)
7. Drexl, M.: Branch-and-cut algorithms for the vehicle routing problem with trailers and transshipments. Networks **63**(1), 119–133 (2014)
8. Drexl, M., Rieck, J., Sigl, T., Press, B.: Simultaneous vehicle and crew routing and scheduling for partial- and full-load long-distance road transport. BuR - Business Research **6**(2), 242–264 (2013)
9. Dreyfus, S.E.: An appraisal of some shortest-path algorithms. Operations Research **17**(3), 395–412 (1969)
10. Focacci, F., Lodi, A., Milano, M.: Embedding relaxations in global constraints for solving TSP and TSPTW. Annals of Mathematics and Artificial Intelligence **34**(4), 291–311 (2002)
11. Focacci, F., Lodi, A., Milano, M.: Optimization-oriented global constraints. Constraints **7**(3–4), 351–365 (2002)
12. Focacci, F., Lodi, A., Milano, M.: Exploiting relaxations in CP. In: Milano, M. (ed.) Constraint and Integer Programming, Operations Research/Computer Science Interfaces Series, vol. 27, pp. 137–167. Springer, US (2004)
13. Focacci, F., Lodi, A., Milano, M., Vigo, D.: Solving TSP through the integration of OR and CP techniques. Electronic Notes in Discrete Mathematics **1**, 13–25 (1999)
14. Francis, K.G., Stuckey, P.J.: Explaining circuit propagation. Constraints **19**(1), 1–29 (2014)
15. Freling, R., Huisman, D., Wagelmans, A.: Applying an integrated approach to vehicle and crew scheduling in practice. In: Voß, S., Daduna, J. (eds.) Computer-Aided Scheduling of Public Transport. Lecture Notes in Economics and Mathematical Systems, vol. 505, pp. 73–90. Springer, Berlin Heidelberg (2001)
16. Freling, R., Huisman, D., Wagelmans, A.: Models and algorithms for integration of vehicle and crew scheduling. Journal of Scheduling **6**(1), 63–85 (2003)
17. Freling, R., Wagelmans, A., Paixão, J.: An overview of models and techniques for integrating vehicle and crew scheduling. In: Wilson, N. (ed.) Computer-Aided Transit Scheduling. Lecture Notes in Economics and Mathematical Systems, vol. 471, pp. 441–460. Springer, Berlin Heidelberg (1999)
18. Haase, K., Desaulniers, G., Desrosiers, J.: Simultaneous vehicle and crew scheduling in urban mass transit systems. Transportation Science **35**(3), 286–303 (2001)
19. Hollis, B., Forbes, M., Douglas, B.: Vehicle routing and crew scheduling for metropolitan mail distribution at Australia Post. European Journal of Operational Research **173**(1), 133–150 (2006)

20. Ibaraki, T.: Algorithms for obtaining shortest paths visiting specified nodes. SIAM Review **15**(2), 309–317 (1973)
21. Kilby, P., Prosser, P., Shaw, P.: A comparison of traditional and constraint-based heuristic methods on vehicle routing problems with side constraints. Constraints **5**(4), 389–414 (2000)
22. Kim, B.I., Koo, J., Park, J.: The combined manpower-vehicle routing problem for multi-staged services. Expert Systems with Applications **37**(12), 8424–8431 (2010)
23. Laporte, G.: What you should know about the vehicle routing problem. Naval Research Logistics (NRL) **54**(8), 811–819 (2007)
24. Laporte, G., Mercure, H., Norbert, Y.: Optimal tour planning with specified nodes. RAIRO - Operations Research - Recherche Opérationnelle **18**(3), 203–210 (1984)
25. Mercier, A., Cordeau, J.F., Soumis, F.: A computational study of benders decomposition for the integrated aircraft routing and crew scheduling problem. Computers & Operations Research **32**(6), 1451–1476 (2005)
26. Mercier, A., Soumis, F.: An integrated aircraft routing, crew scheduling and flight retiming model. Computers & Operations Research **34**(8), 2251–2265 (2007)
27. Mesquita, M., Paias, A.: Set partitioning/covering-based approaches for the integrated vehicle and crew scheduling problem. Computers & Operations Research **35**(5), 1562–1575 (2008), part Special Issue: Algorithms and Computational Methods in Feasibility and Infeasibility
28. Rousseau, L.M., Gendreau, M., Pesant, G.: Using constraint-based operators to solve the vehicle routing problem with time windows. Journal of Heuristics **8**(1), 43–58 (2002)
29. Shaw, P.: Using Constraint Programming and Local Search Methods to Solve Vehicle Routing Problems. In: Maher, M.J., Puget, J.-F. (eds.) CP 1998. LNCS, vol. 1520, pp. 417–431. Springer, Heidelberg (1998)
30. Toth, P., Vigo, D.: The Vehicle Routing Problem. Society for Industrial and Applied Mathematics (2002)
31. Van Hentenryck, P., Michel, L.: The Objective-CP Optimization System. In: Schulte, C. (ed.) CP 2013. LNCS, vol. 8124, pp. 8–29. Springer, Heidelberg (2013)
32. Volgenant, T., Jonker, R.: On some generalizations of the travelling-salesman problem. The Journal of the Operational Research Society **38**(11), 1073–1079 (1987)
33. Yu, G.: Operations Research in the Airline Industry. International Series in Operations Research & Management Science: 9, Springer, US (1998)

Constructing Sailing Match Race Schedules: Round-Robin Pairing Lists

Craig Macdonald, Ciaran McCreesh, Alice Miller, and Patrick Prosser[✉]

University of Glasgow, Glasgow, Scotland
patrick.prosser@glasgow.ac.uk

Abstract. We present a constraint programming solution to the problem of generating round-robin schedules for sailing match races. Our schedules satisfy the criteria published by the International Sailing Federation (ISAF) governing body for match race pairing lists. We show that some published ISAF schedules are in fact illegal, and present corresponding legal instances and schedules that have not previously been published. Our schedules can be downloaded as blanks, then populated with actual competitors and used in match racing competitions.

1 Introduction

This work describes a round-robin competition format that arises from sailing, known as match racing. Competitors, i.e. *skippers*, compete against each other in a series of matches taking place in rounds, known as *flights*. A match is composed of two skippers, with each skipper in a boat on their own. Skippers in a match set off together, one on the port side, the other starboard, and first home wins that match. Criteria published by the International Sailing Federation (ISAF) dictate what makes a legal match race schedule and what should be optimized. This is a rich set of criteria, 13 in all, and as far as we know, all published schedules have been produced by hand. This is a daunting task. A close inspection of these schedules reveals that most are illegal, violating many of the match race scheduling criteria, many schedules are missing, and some that are legal are far from the ISAF definition of optimality.

This paper describes the scheduling problem. We present a constraint programming solution and document how this model was constructed incrementally. We believe this illustrates how adaptable constraint programming is: a robust model can be developed incrementally to address all of the required criteria. Our schedules are then constructed in a sequence of stages, not dissimilar to the hybrid and decomposition based approaches by Lombardi and Milano [3]. Some ISAF published schedules are presented along with some of our own, some of these being new. These schedules are essentially blank forms, made available to event organizers, which can be filled in with names of actual competitors, i.e. these schedules are reusable.

The paper is organized as follows: we introduce the 13 match racing criteria, then explain the constraint models and four stages of schedule construction, and then schedules are presented. We finish with discussion and a conclusion.

© Springer International Publishing Switzerland 2015
G. Pesant (Ed.): CP 2015, LNCS 9255, pp. 671–686, 2015.
DOI: 10.1007/978-3-319-23219-5_46

2   Problem Definition: Round-Robin Pairing Lists

The guidelines for running match racing events [2] places a number of criteria on how the competing skippers are scheduled into flights and matches, known as pairing lists. Several of these criteria are applicable when the number of skippers exceeds the number of available boats, in which case skippers have to change in and out of boats, and are allowed time in the schedule to check and fine-tune the boat they change into. Typically, matches set off at 5 minute intervals with each match taking about 20 minutes. Therefore, if we have 10 skippers and 10 boats we have 45 matches, 9 flights each of 5 matches, and if each flight takes about 50 minutes, eight or nine flights a day is a reasonable achievement for most events [2]. Consequently, the number of boats and skippers involved is typically small.

The criteria for match racing schedules (pairing lists) are given in ISAF "International Umpires' and Match Racing Manual" [2], section M.2 "Recommended Criteria for Round Robin Pairing Lists", and are detailed below.

Principal Criteria in Order of Priority:

1. Each skipper sails against each other skipper once.
2*. When skippers have an even number of matches, they have the same number of port and starboard assignments.
3*. When skippers have an odd number of matches, the first half of the skippers will have one more starboard assignment.
4. No skipper in the last match of a flight should be in the first match of the next flight.
5. No skipper should have more than two consecutive port or starboard assignments.
6. Each skipper should be assigned to match 1, match 2, etc. in a flight as equally as possible.
7. In flights with five or more matches, no skipper should be in the next-to-last match in a flight and then in the first match of the next flight.
8. If possible, a skipper should be starboard when meeting the nearest lowest ranked skipper (i.e. #1 will be starboard against #2, #3 will be starboard against #4).
9. Close-ranked skippers meet in the last flight.
10. Minimize the number of boat changes.
11. Skippers in the last match of a flight do not change boats.
12. Skippers in new boats do not sail in the first match of the next flight.
13. Skippers have a reasonable sequence of matches and blanks.

Note that criteria 10, 11 and 12 only apply when there are fewer boats than skippers; 11 and 12 override 6 when changes are required, and 13 applies when there are more boats than skippers.

We have rephrased criteria 2 and 3 (denoted *) to clarify an error in the manual [2]. Note that the order of matches within a flight is significant and is constrained, as is the order of flights within a schedule and the position of skippers within a match. This permits fair schedules that provide sufficient changeover time for skippers changing boats, etc.

Criterion 6 allows us to measure the *balance* of a schedule. Perfect balance would be when a skipper has as many matches in first position in flights as in second position in flights, and so on. For example, if we had 9 skippers, each skipper would have 8 matches and could be in the first, second, third or fourth match of a flight. In perfect balance, each of the 8 skippers would appear twice in each position. Balance is one of our optimization criteria, and we do this unconditionally, even when there are more skippers than boats.

Criterion 13 uses the term *blanks* where conventionally we use the term *bye*. That is, a blank is a bye and a bye is a flight where a skipper does not compete (and is ashore).

Criterion 12 discusses boat changes: if in flight i this skipper is not in a match (i.e. it is a bye for this skipper) but is in a match in flight $i + 1$ then he has to get into a boat, rearrange all his kit and set the boat to his preference before racing commences, and this is a boat change.

Next, we note that skippers can be ranked prior to an event, based on their performance in previous events[1], which is used to seed the skippers. The ordering of skippers within a match signifies their starting position (port or starboard), as skippers allocated the starboard starting position gain a small competitive advantage in that match—criterion 8 accomplishes the seedings.

Criterion 10 is ambiguous: this could be interpreted on a schedule basis: i.e. to minimize the overall number of changes in the entire pairing list schedule; or alternatively as well as a fair, but minimal number of changes across all skippers. In our work, we minimize the overall number of changes.

There are some conflicts inherent in the criteria. Take criteria 4 and 11, assume we have 4 boats, and in flight i the last match is the pair of skippers (x, y). Criterion 11 dictates that skippers x and y must appear in the next flight, $i + 1$, and criterion 4 that neither can be first in the next flight. This forces skippers x and y to compete in flight $i + 1$ as the last match, violating criterion 1. Therefore, although not explicitly stated, it is not possible to satisfy every criteria with fewer than 6 boats.

3   The Constraint Models

Our approach produces schedules in four stages. The first stage produces a schedule that respects criteria 1, 4, 11 and 12, and minimizes criteria 10 (boat changes). The second stage constructs a schedule that respects criteria 1, 4, 11, 12 (again), has criterion 10 as a hard constraint generated from first stage, and minimizes criterion 6 (balance). This results in a multi-objective schedule that

---

[1] http://www.sailing.org/rankings/match/

Flight	Matches		
0	(0,1)	(2,3)	(4,5)
1	(0,2)	(4,6)	(1,5)
2	(2,6)	(0,5)	(1,3)
3	(5,6)	(0,3)	(1,4)
4	(3,5)	(1,6)	(2,4)
5	(3,6)	(0,4)	(2,5)
6	(3,4)	(1,2)	(0,6)

Table 1. A round-robin schedule with 7 skippers, 6 boats and 7 flights. Skippers are numbered 0 to 6. Note that the order of skippers within a flight is significant, as is position within a match (port or starboard). This schedule is the result of stages 1 and 2, and has yet to be renumbered and oriented (stages 3 and 4).

attempts to minimize boat changes *and* balance. The third stage satisfies criterion 9, and is a simple translation of the schedule produced in stage 2. The final stage orients skippers within matches to address criteria 2, 3, 5 and 8.

We now describe the constraint models and processes used in each stage. In the text below we assume there are n skippers and b boats, with $m = b/2$ matches in a flight. In total there are $t = n(n-1)/2$ matches and $f = \lceil n(n-1)/m \rceil$ flights. We use the schedule in Table 1, for 7 skippers and 6 boats, to help illustrate the constraint models.

3.1 Stage 1: Minimizing Boat Changes

Modeling skippers: The first thing we model, in the first half of Figure 1, is a skipper. Each skipper σ has both a temporal view of their schedule ("who, if anyone, am I racing in the match at time t?"), and a state view ("where am I racing within flight f?"). The temporal view is via the array *timeSlot* defined in (V1). If variable $timeSlot[i] = k$ and $k \geq 0$ then this skipper is in a match with skipper k at time i.

Variable , $[i]$ (V2) gives the state of this skipper in flight i, corresponding to time slots $timeSlot[m \cdot i]$ to $timeSlot[m(i + 1) - 1]$. The cases in (C1) state that a skipper can be in a match in the first time slot in the flight, or in the middle of the flight[2], or in the last time slot of the flight. Alternatively, if all time slots in a flight are equal to -1 then the skipper is in a bye (i.e. not in a match in this flight), and if all time slots in a flight are equal to -2 then the skipper has finished all matches.

We must then ensure that each skipper σ is in a match with all other skippers $\{0, \ldots, n - 1\} \setminus \{\sigma\}$. This is (C2), which is enforced by imposing a global cardinality constraint [5] on the array *timeSlot*.

State transitions: The , variables are then used to impose match race criterion 4 (if last in flight i then not first in flight $i + 1$), criterion 11 (if last in flight i then not in a bye in flight $i + 1$) and criterion 12 (if flight i is a bye then flight $i + 1$ is not first). These criteria are imposed in (C3) by the deterministic finite automaton (DFA) shown in Figure 2 using Pesant's *regular* constraint [4]. The

---

[2] i.e. not first and not last.

A copy of these variables and constraints is created for each skipper σ:

$$\forall \tau \in \{0 \dots t{-}1\} : timeSlot[\tau] \in \{-2 \dots t{-}1\} \setminus \{\sigma\} \tag{V1}$$

$$\forall i \in \{0 \dots f{-}1\} : state[i] \in \{\textsc{First}, \textsc{Mid}, \textsc{Last}, \textsc{Bye}, \textsc{End}\} \tag{V2}$$

$$\forall i \in \{0 \dots f{-}1\} : \tag{C1}$$

$state[i] = \textsc{First} \Leftrightarrow timeSlot[m \cdot i] \geq 0$

$state[i] = \textsc{Mid} \quad \Leftrightarrow \exists j \in \{m \cdot i + 1 \dots m \cdot (i+1) - 2\} : timeSlot[j] \geq 0$

$state[i] = \textsc{Last} \Leftrightarrow timeSlot[m \cdot (i+1) - 1] \geq 0$

$state[i] = \textsc{Bye} \quad \Leftrightarrow \forall j \in \{m \cdot i \dots m \cdot (i+1) - 1\} : timeSlot[j] = -1$

$state[i] = \textsc{End} \quad \Leftrightarrow \forall j \in \{m \cdot i \dots m \cdot (i+1) - 1\} : timeSlot[j] = -2$

$$\text{eachOccursExactlyOnce}(timeSlot, \{0, \dots, n{-}1\} \setminus \{\sigma\}) \tag{C2}$$

$$\text{regular}(state, \text{Figure 2}) \tag{C3}$$

$$\forall i \in \{0 \dots f{-}2\} : change[i] \in \{0,1\} \tag{V3}$$

$$totalChanges \in \mathbb{N} \tag{V4}$$

$$\forall i \in \{0 \dots f{-}2\} : change[i] = 1 \Leftrightarrow state[i] = \textsc{Bye} \wedge state[i+1] \neq \textsc{Bye} \tag{C4}$$

$$totalChanges = \sum change \tag{C5}$$

Match and temporal perspectives:

$$\forall i \in \{0 \dots n{-}2\} : \forall j \in \{i{+}1 \dots n{-}1\} :$$

$$match[i,j] \in \{0 \dots t{-}1\} \tag{V5}$$

$$match[j,i] \equiv match[i,j]$$

$$\forall k \in \{0 \dots t{-}1\} :$$

$$match[i,j] = k \Leftrightarrow \sigma[i].timeSlot[k] = j \wedge \sigma[j].timeSlot[k] = i \tag{C6}$$

$$\forall i \in \{0 \dots n{-}2\} : \forall j \in \{i{+}1 \dots n{-}1\} :$$

$$modMatch[i,j] \in \{0 \dots f{-}1\} \tag{V6}$$

$$modMatch[j,i] \equiv modMatch[i,j]$$

$$modMatch[i,j] = match[i,j]/m \tag{C7}$$

$$\forall i \in \{0 \dots n{-}1\} : \text{allDifferent}(modMatch[i]) \tag{C8}$$

$$\forall \tau \in \{0 \dots t{-}1\} :$$

$$time[\tau] \in \{(0,1) \dots (n{-}2, n{-}1)\} \tag{V7}$$

$$\forall i \in \{0 \dots n{-}2\} : \forall j \in \{i{+}1 \dots n{-}1\} : time[\tau] = (i,j) \Leftrightarrow match[i,j] = \tau \tag{C9}$$

$$totalBoatChanges = \sum \sigma.totalChanges \tag{V8}$$

$$\text{minimise}(totalBoatChanges) \tag{C10}$$

Fig. 1. Our constraint model, from a skipper perspective (top) and a match and temporal perspective (below). The number of skippers is n, and m is the number of matches in a flight. The number of flights is $f = \lceil n(n-1)/m \rceil$, and there are $t = n(n-1)/2$ matches (and time slots) in total. We define $\mathbb{N}$ to include zero.

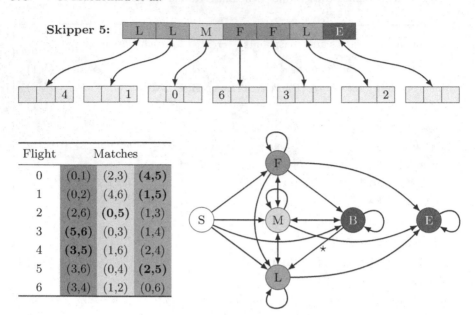

Fig. 2. A pictorial representation of a skipper (skipper 5) with multicoloured state and grey timeslots. The schedule for skipper 5 is in bold and the DFA for criteria 4, 11 and 12 is drawn with state START in white, FIRST in blue, MID in yellow, LAST in green, BYE in pink and END in red. The edge marked ⋆ is explained in the text.

arcs in Figure 2 represent possible transitions from one state to another. The DFA is encoded as a set of transition objects $\langle q_i, \iota, q_j \rangle$, where a transition is made from state q_i to state q_j when encountering input ι. In addition we specify the accepting states, which are all states except the start state (white) and the bye state (pink). This constraint also ensures that if a skipper has finished all matches in flight i (i.e. is in state END) then that skipper will also have finished all matches in flight $i + 1$ (and so on). Note that when we have as many boats as skippers, the directed edge (BYE, LAST) (marked ⋆) becomes bidirectional.

Figure 2 also shows the skipper-oriented variables and constraints corresponding to skipper 5 in the schedule shown in Table 1. The *state* array is presented as coloured boxes, and below that is the *timeSlot* array. The arrows represent the channeling constraints (C1). The *state* array is coloured with green representing LAST, yellow for MID, blue for FIRST, red for LAST and (not shown) pink for BYE. The schedule of Table 1 is reproduced with the matches for skipper 5 emboldened.

Boat changes: A boat change occurs when a skipper has been in a BYE state in flight i and then is in a match in the next flight $i + 1$. This is encoded using the array of zero/one variables (V3), and the constraint (C4), and accumulated into *totalChanges* (V4, C5).

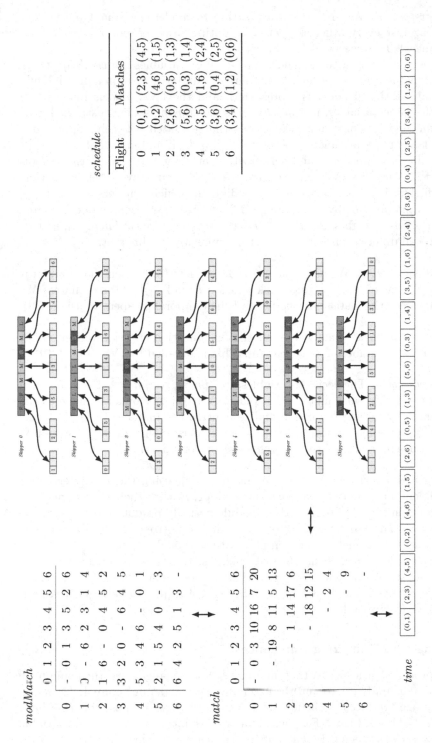

Fig. 3. A pictorial representation of the entire model of the schedule for 7 skippers and 6 boats. The schedule is reproduced on the right. In the centre we have the 7 skippers, on left the *modMatch* and *match* arrays, and along the bottom the *time* array. The arrows signify channeling between parts of the model.

Match perspective: We are now in a position to model criterion 1 (each skipper sails against every other skipper) and optimization criterion 10 (minimize boat changes). In the second half of Figure 1 we present a match perspective of the schedule, using a two dimensional array of variables *match* (V5), where $match[i,j]$ is the time slot in which skippers $\sigma[i]$ and $\sigma[j]$ meet in a match. Only the half above the diagonal is represented, and the lower half is made up of exactly the same variables, i.e. $match[i,j]$ is *exactly* the same constrained variable as $match[j,i]$. Constraint (C6) states that a match between skippers $\sigma[i]$ and $\sigma[j]$ takes place at time k (i.e. $match[i,j] = k$) if and only if skipper $\sigma[i]$'s k^{th} time slot is skipper $\sigma[j]$ and conversely that skipper $\sigma[j]$'s k^{th} time slot is skipper $\sigma[i]$. Variables (V6) and constraints (C7) then convert this from time slots to flights, i.e. $modMatch[i,j]$ is the flight in which skippers $\sigma[i]$ and $\sigma[j]$ meet for a match. Finally, constraint (C8) ensures that each skipper's match occurs in different flights. Also, since $modMatch[i,j] \equiv modMatch[j,i]$, setting rows to be all different also ensures that columns are all different.

Temporal perspective: We now take a temporal view (V7), such that $time[\tau]$ is a pair (i,j), stating that at time τ skippers $\sigma[i]$ and $\sigma[j]$ are in a match. We channel between the match perspective and the temporal perspective using (C9).

Optimization criteria: Finally we have the optimization criterion (criterion 10) to minimize the number of boat changes. This is handled by (V8) and (C10).

Decision variables: The decision variables are $time[0]$ to $time[t-1]$, i.e. for each time slot we decide what match takes place. A symmetry break is made at the top of search by forcing the first round to contain the matches $(0,1),(2,3),...,$ i.e. $\forall i \in \{0...m-1\} : match[2i, 2i+1] = i$. (This is independent of criterion 9, which will be enforced in a subsequent stage by renumbering.)

Figure 3 gives a pictorial view of the entire model, using the 7 skipper and 6 boat problem. On the right we have the 7 skippers with their states and time slots. Again, on the right we give the schedule actually produced. On the left we have the *modMatch* and *match* arrays, and at the bottom the *time* array. The arrows show the channeling between parts of the model.

The schedule presented has 6 boat changes: there are no boat changes in flight 0 (first flight), in flight 1 skipper 6 has a boat change, in flight 2 skipper 3 has a boat change, in flight 3 skipper 4, in flight 4 skipper 6, in flight 5 skipper 0, and flight 6 skipper 1.

3.2   Stage 2: Minimizing Imbalances

Having produced a schedule that minimizes boat changes, we then minimize imbalance (i.e. apply criterion 6) by extending the model in Figure 4. Assuming the first stage produces a schedule with β boat changes we post this as a hard constraint (C11) in stage 2. For each skipper σ we introduce a zero/one variable $position[i,j]$ (V9) which is 1 if and only if skipper σ is in a match in time $m \cdot j + i$,

The objective value from stage 1 is used as a hard constraint:

$$totalBoatChanges \leq \beta \qquad \text{(C11)}$$

A copy of these variables and constraints is created for each skipper σ:

$\forall i \in \{0 \ldots m-1\} : \forall j \in \{0 \ldots f-1\} :$

$$position[i,j] \in \{0,1\} \qquad \text{(V9)}$$
$$position[i,j] = 1 \Leftrightarrow timeSlot[m.j + i] \geq 0 \qquad \text{(C12)}$$

$\forall i \in \{0 \ldots m-1\} :$

$$imbalance[i] \in \mathbb{N} \qquad \text{(V10)}$$

$$imbalance[i] = \left| \frac{n-1}{m} - \sum_{j=0}^{f-1} position[i,j] \right| \qquad \text{(C13)}$$

$$maxImbalance \in \mathbb{N} \qquad \text{(V11)}$$
$$maxImbalance = \text{maximum}(imbalance) \qquad \text{(C14)}$$

We minimize the maximum imbalance over all skippers:

$$\forall i \in \{0 \ldots n-1\} : imbalance[i] \equiv \sigma[i].maxImbalance \qquad \text{(V12)}$$
$$maxImbalance \in \mathbb{N} \qquad \text{(V13)}$$
$$maxImbalance = \text{maximum}(imbalance) \qquad \text{(C15)}$$
$$\text{minimise}(maxImbalance) \qquad \text{(C16)}$$

Fig. 4. Additions to the constraint model in Figure 1 for stage 2. The constant β is the minimum number of boat changes found in stage 1.

i.e. in the ith match of the jth flight (C12). Imbalance (V10) is then computed for each position, as the absolute amount over or under the expected presence in a given position (C13). Finally, we compute the maximum imbalance over all positions for this skipper (V11, C14).

We then capture the maximum imbalance over all skippers (V12, V13 and C15) and minimize this (C16). As before, the decision variables are $time[0]$ to $time[t-1]$ and the same symmetry breaking is imposed at top of search.

3.3   Stage 3: Renaming Skippers

Stage 3 then takes as input the schedule produced in stage 2 and renames skippers in the last round of the last flight to be $(0, 1)$, satisfying criterion 9. This is a simple linear process. The schedule produced for 7 skippers and 6 boats is shown in Table 2.

Flight	Matches		
0	(0,6)	(2,3)	(4,5)
1	(0,2)	(4,1)	(6,5)
2	(2,1)	(0,5)	(6,3)
3	(5,1)	(0,3)	(6,4)
4	(3,5)	(6,1)	(2,4)
5	(3,1)	(0,4)	(2,5)
6	(3,4)	(6,2)	(0,1)

Table 2. A round-robin schedule with 7 skippers, 6 boats and 7 flights, which is the result of stages 1, 2, and 3. It has 6 boat changes and imbalance 1. An example of imbalance is skipper $\sigma[0]$ with 2 matches in position 0 (first), 3 in position 1 and 1 (mid) in position 2 (last). Perfect balance would be 2 matches in each position, only achieved by $\sigma[2]$.

3.4 Stage 4: Orienting Matches

The final stage, stage 4, orients skippers within matches, such that they are either port (red) or starboard (green). This is done to satisfy criteria 2, 3, 5 and 8. A sketch of the constraint model is given with a supporting diagram, Figure 5. A two dimensional array of constrained integer variables $orient[i, j] \in \{0, 1\}$ has $orient[i, j] = 0$ if and only if skipper $\sigma[i]$ is on port side in his j^{th} match (starboard otherwise). For example, taking the schedule in Table 2, skipper $\sigma[2]$ meets skippers in the order 3,0,1,4,5,6; skipper $\sigma[3]$ meets skippers in the order 2,6,0,5,1,4 and $\sigma[5]$ meets skippers in the order 4,6,0,1,3,2. Therefore $orient[2, 0] = 1$ and $orient[3, 0] = 0$ (criterion 8), $orient[2, 4] \neq orient[5, 5]$ and $orient[3, 3] \neq orient[5, 4]$. Summation constraints on each skipper enforce criteria 2 and 3 (equal number of port and starboard matches). Criterion 8 is enforced by posting constraints such that in a match $(\sigma[i], \sigma[j])$ where $|i - j| = 1$, the higher indexed skipper (lower ranked) is on the port side. A *regular* constraint is posted for criterion 5 (restricting sequences of port and starboard assignments).

In Figure 5, port is red and starboard green. The sequence of zero/one variables for each skipper is shown on the left. Bottom right is the DFA, and top right is the final schedule produced from stages 1 to 4. The schedule has 6 boat changes and a maximum imbalance of 1. The schedule presented here is new, not appearing in the example schedules provided in the manual [2].

4   Sample Schedules

We present four pairs of schedules. The first pair is for 8 skippers and 6 boats, one published by ISAF, the other generated by us. Next, we present two new schedules, not appearing in the ISAF manual, for 9 skippers and 6 boats and 9 skippers with 8 boats, then 10 skippers and 6 boats, again both published by ISAF and generated by us. Finally, we show what is a relatively easy problem, 8 skippers and 8 boats. For each of our schedules (with the exception of the latter "easy-8" problem) we used 1 day CPU time for the first stage (minimizing boat changes) then a further 2 days to minimize imbalance. We were unable to prove optimality in this time, therefore some of these schedules might be improved. The processor used was an Intel (R) Xeon (R) E5-2660 at 2.20GHz. Note, stages 3 and 4 (renaming and orientation) completed in less than a second.

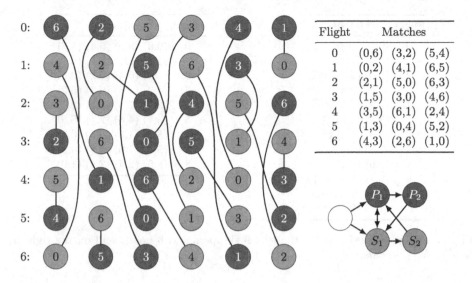

Flight	Matches		
0	(0,6)	(3,2)	(5,4)
1	(0,2)	(4,1)	(6,5)
2	(2,1)	(5,0)	(6,3)
3	(1,5)	(3,0)	(4,6)
4	(3,5)	(6,1)	(2,4)
5	(1,3)	(0,4)	(5,2)
6	(4,3)	(2,6)	(1,0)

Fig. 5. A pictorial representation of the orientation process, stage 4. Top right is the actual oriented schedule. Below that is the DFA to satisfy criterion 5. On the left is the zero/one variables, a row representing a skipper's sequence of competitors. An edge between variables corresponds to a matched pair, that must take different values.

Flight	Matches		
0	(5,2)	(4,3)	(1,6)
1	(4,2)	(6,5)	(3,1)
2	(6,4)	(2,1)	(3,5)
3	(6,2)	(5,0)	(1,7)
4	(5,1)	(0,6)	(2,7)
5	(7,5)	(2,0)	(6,3)
6	(0,7)	(4,1)	(3,2)
7	(7,4)	(0,3)	
8	(4,0)	(7,3)	
9	(5,4)	(7,6)	(1,0)

Flight	Matches		
0	(2,7)	(3,0)	(5,4)
1	(0,2)	(4,3)	(7,5)
2	(0,4)	(7,6)	(5,3)
3	(4,6)	(5,0)	(3,1)
4	(6,5)	(3,2)	(1,4)
5	(6,3)	(4,7)	(2,1)
6	(7,3)	(1,6)	(4,2)
7	(7,1)	(2,5)	(6,0)
8	(5,1)	(0,7)	(6,2)
9	(1,0)		

Fig. 6. Schedules for 8 skippers and 6 boats. On the left, the published ISAF schedule (illegal), and on the right is our schedule.

8 skippers and 6 Boats: We first analyze the ISAF schedule, Figure 6 on the left. Criterion 4 (skippers in last match in a flight cannot be first in next flight) is violated on 3 occasions (skipper 1 in flight 3, skipper 7 in flight 4, skipper 0 in flight 7). Criterion 5 (no more than two consecutive port or starboard assignments) is violated for skipper 6 (flights 1 to 3) and skipper 7 (flights 7 to 9). Criterion 6 (imbalance) is violated for skippers 6 and 8. Criterion 12 (new boats do not sail first) is violated in flight 9 for skipper 5. Finally, the schedule has 8 boat changes. Our schedule, Figure 6 on the right, respects all criteria and has 6 boat changes and a maximum imbalance of 1.

Flight	Matches				Flight	Matches			
0	(0,7)	(3,2)	(5,4)		0	(0,6)	(3,2)	(5,4)	(1,7)
1	(0,2)	(7,4)	(5,3)		1	(2,0)	(6,3)	(4,1)	(8,7)
2	(4,0)	(7,3)	(2,5)		2	(6,4)	(0,5)	(7,2)	(3,8)
3	(3,0)	(2,4)	(6,5)		3	(4,7)	(1,8)	(0,3)	(5,2)
4	(4,3)	(5,1)	(8,6)		4	(3,1)	(7,5)	(8,0)	(2,6)
5	(3,1)	(6,2)	(4,8)		5	(3,7)	(4,8)	(2,1)	(6,5)
6	(6,3)	(1,4)	(8,2)		6	(8,2)	(1,6)	(5,3)	(0,4)
7	(4,6)	(2,7)	(1,8)		7	(5,1)	(7,0)	(6,8)	(4,3)
8	(7,6)	(0,8)	(2,1)		8	(8,5)	(2,4)	(7,6)	(1,0)
9	(8,7)	(5,0)	(6,1)						
10	(8,5)	(1,7)	(0,6)						
11	(7,5)	(3,8)	(1,0)						

Fig. 7. Two new schedules. On the left, 9 skippers and 6 boats, and on the right 9 skippers and 8 boats.

9 skippers: Two new schedules are presented in Figure 7, on the left for 6 boats and on the right for 8 boats. Neither of these schedules appear in the ISAF Umpires' Manual. Both schedules respect all criteria. For 6 boats there are 8 boat changes (no skipper making more than 1 boat change) and a maximum imbalance of 2. For 8 boats there are again 8 boat changes, no skipper with more than 1 boat change, and each skipper 4 times on starboard side and 4 on port side.

10 skippers and 6 boats: Figure 8 shows the ISAF schedule on the left for 10 skippers and 6 boats and on the right, our schedule. The ISAF schedule violates criterion 5 (no more than 2 consecutive port or starboard assignments) ten times (twice for skippers 0, 1 and 6 and once for skippers 3, 7, 8 and 9). Criterion 12 (new boats do not sail first) is violated 7 times (in flight 3 for skippers 6 and 2, flight 5 for skipper 8, flight 6 for skipper 7, flight 8 for skipper 1, flight 9 for skipper 4 and flight 12 for skipper 4 again). There are 22 boat changes, with the minimum changes for a skipper being 1 and the maximum 3. Our schedule, right of Figure 8, satisfies all criteria, has 12 boat changes with the minimum changes for a skipper being 0 and the maximum 2, and a maximum imbalance of 1.

8 skippers and 8 boats (easy-8): Figure 9 shows the ISAF schedule on the right for 8 skippers and 8 boats and on the right, our schedule. The ISAF schedule violates criterion 4 (last match in a flight cannot be first in next flight) for skipper 5 in flights 1 and 2. Also, criterion 5 (no skipper should have more than 2 consecutive port or starboard assignments) is violated for skippers 0, 1, 3, 4 and 6. Furthermore, skipper 1 appears in the last match of four flights resulting in significant imbalance. Our schedule, on the right, satisfies all criteria, is optimal (maximum imbalance of 1) and took less than 10 seconds in total (all stages) to produce.

Flight	Matches		
0	(8,3)	(1,7)	(0,9)
1	(7,3)	(0,8)	(1,9)
2	(0,7)	(9,3)	(1,8)
3	(6,2)	(5,0)	(7,4)
4	(5,2)	(6,4)	(3,0)
5	(8,6)	(2,9)	(4,1)
6	(7,2)	(4,8)	(9,6)
7	(2,4)	(0,6)	(7,5)
8	(6,1)	(5,9)	(8,2)
9	(9,4)	(8,7)	(6,5)
10	(9,7)	(6,3)	(8,5)
11	(7,6)	(3,1)	(9,8)
12	(4,3)	(5,1)	(2,0)
13	(3,5)	(4,0)	(2,1)
14	(5,4)	(3,2)	(1,0)

Flight	Matches		
0	(4,7)	(3,2)	(0,5)
1	(2,4)	(7,0)	(3,5)
2	(0,4)	(7,3)	(5,2)
3	(4,3)	(0,2)	(5,7)
4	(3,0)	(6,5)	(2,7)
5	(6,3)	(7,1)	(8,2)
6	(7,6)	(2,1)	(9,8)
7	(2,6)	(8,7)	(4,9)
8	(6,8)	(1,4)	(7,9)
9	(8,1)	(5,4)	(9,6)
10	(5,1)	(9,2)	(4,6)
11	(1,9)	(6,0)	(8,4)
12	(0,9)	(3,8)	(6,1)
13	(8,0)	(9,5)	(1,3)
14	(5,8)	(9,3)	(1,0)

Fig. 8. Schedules for 10 skippers and 6 boats. On the left, the ISAF published schedule (illegal), and on the right, our schedule.

Flight	Matches			
0	(5,2)	(4,3)	(1,6)	(0,7)
1	(2,4)	(0,6)	(1,7)	(5,3)
2	(0,5)	(7,2)	(6,3)	(4,1)
3	(7,3)	(5,1)	(4,0)	(6,2)
4	(3,0)	(4,7)	(6,5)	(2,1)
5	(6,4)	(7,5)	(2,0)	(3,1)
6	(7,6)	(5,4)	(3,2)	(1,0)

Flight	Matches			
0	(4,3)	(2,0)	(1,5)	(7,6)
1	(3,2)	(0,4)	(6,1)	(5,7)
2	(4,1)	(7,3)	(5,2)	(0,6)
3	(1,7)	(6,5)	(3,0)	(2,4)
4	(0,5)	(4,6)	(7,2)	(3,1)
5	(7,0)	(2,1)	(6,3)	(5,4)
6	(6,2)	(5,3)	(4,7)	(1,0)

Fig. 9. Schedules for 8 skippers and 8 boats. On the left, the published ISAF schedule (illegal), and on the right, our schedule.

5   Discussion

It was anticipated that this problem would be easy to model. This naïvety was due to the use of the term "round-robin" for these schedules, leading us to believe that we could use some of the models already in the literature [1,6,7]. This naïvety slowly dissolved as we addressed more and more of the ISAF criteria. A number of models we produced were scrapped due to misunderstanding about the actual problem, i.e. a communication problem between the authors. Eventually this was resolved by presenting the majority of the criteria as a deterministic finite automaton. This became our Rosetta Stone, with the surprising benefit that it not only improved the team's communications, it was also a constraint (the *regular* constraint).

The staged approach was taken cautiously. We expected that isolating the orientation of schedules as a post-process might leave us with a hard or insoluble problem. So far, every schedule we have produced has been oriented with very little search, typically taking less than a second.

We used the Choco constraint programming toolkit and one of our goals was to use only toolkit constraints, i.e. we wanted to see how far we could go without implementing our own specialized constraints. We are pleased to say that we did not have to do anything out of the box, although we did feel that we had used just about every constraint there was.

There are many redundant variables in our model. One of our first valid models was essentially that shown in Figure 2, and did not use *modMatch* and *match* variables. Performance was hopeless, struggling to produce a 7 skipper 6 boat schedule in days. The *modMatch* and *match* variables were added. This improved domain filtering, and a flattened *match* array was used as decision variables. That is, the decisions where "when do we schedule this match?". At this point we had not yet produced a 9 skipper 6 boat schedule and we suspected that the combined criteria might exclude such a schedule. A non-constraint model was developed, primarily to attempt to prove that there was no schedule for 9 skippers with 6 boats. This program used a backtracking search with decision variables being positions within flights, i.e. time slots being assigned to matches. At the top of search, the matches in the first flight were set, and symmetry breaking was used to reduce the number of legal second flights. A solution was found in seconds! With this experience we added in the *time* variables, channeled these into the existing model and used these as decision variables, i.e. the question now was "what match will we do now?". We also anchored the matches in the first flight. With these changes we began to produce schedules in acceptable time. The model was then built upon, incrementally adding criteria. This took a

Fig. 10. Scene from an imaginary dinghy race

surprisingly short amount of time, sometimes minutes of coding to add in a new feature. The model was then enhanced so that it would address optimization rather than just satisfaction, again a trivial programming task.

We have only reported a handful of our schedules, however there are missing schedules i.e. unpublished and not yet produced by us. Examples of these are 10 skippers and 8 boats, 11 skippers with fewer than 10 boats, 12 skippers and 8 boats, 13 skippers and 8 boats and 14 skippers with fewer than 10 boats. None of these schedules have been published by the ISAF, although we expect that they can be produced by selectively violating some of constraints. We have also not yet encoded criterion 7, addressing the situation of 10 or more boats. To do this we will have to modify the DFA used by the *regular* constraint in the first two stages. In addition, we have yet to address the format of *two-group* round-robin schedules as also found in the manual [2].

6   Conclusion

We have produced new and better match race schedules. These schedules can be used by anyone who competes under the criteria published by ISAF. Our schedules can be downloaded as blank schedules and then populated with the names of the actual competing skippers.

So, why did we use constraint programming? The answer is obvious: we are constraint programmers, or to borrow Mark Twain's words "To a man with a hammer, everything looks like a nail". But actually, constraint programming has been a good way to go. From an engineering perspective, it has allowed us to prototype solutions quickly and to build solutions incrementally. There is also an advantage that we might exploit in the future: we can now investigate the effect different criteria have on our ability to produce schedules and how relaxation of those affect optimization criteria. That is, the constraint program might be used to design the next set of criteria for match race schedules, or allow the event organizer to decide which criteria to sacrifice to ensure a faster schedule with fewer boat changes.

References

1. Henz, M., Müller, T., Thiel, S.: Global constraints for round robin tournament scheduling. European Journal of Operational Research **153**(1), 92–101 (2004)
2. International Sailing Federation. ISAF International Umpires' and Match Racing Manual (2012)
3. Lombardi, M., Milano, M.: Optimal methods for resource allocation and scheduling: a cross-disciplinary survey. Constraints **17**(1), 51–85 (2012)
4. Pesant, G.: A regular language membership constraint for finite sequences of variables. In: Proceedings of the 10th International Conference on Principles and Practice of Constraint Programming, CP 2004, Toronto, Canada, September 27-October 1, pp. 482–495 (2004)

5. Quimper, C.-G., van Beek, P., López-Ortiz, A., Golynski, A., Sadjad, S.B.S.: An Efficient Bounds Consistency Algorithm for the Global Cardinality Constraint. In: Rossi, F. (ed.) CP 2003. LNCS, vol. 2833, pp. 600–614. Springer, Heidelberg (2003)
6. Rasmussen, R.V., Trick, M.A.: Round robin scheduling - a survey. European Journal of Operational Research **188**(3), 617–636 (2008)
7. Trick, M.A.: Integer and Constraint Programming Approaches for Round-Robin Tournament Scheduling. In: Burke, E.K., De Causmaecker, P. (eds.) PATAT 2002. LNCS, vol. 2740, pp. 63–77. Springer, Heidelberg (2003)

Design and Evaluation of a Constraint-Based Energy Saving and Scheduling Recommender System

Seán Óg Murphy[1,2(✉)], Óscar Manzano[1,2], and Kenneth N. Brown[1,2]

[1] International Energy Research Centre, Cork, Ireland
[2] Insight Centre for Data Analytics, Department of Computer Science,
University College Cork, Cork, Ireland
{seanog.murphy,oscar.manzano,ken.brown}@insight-centre.org
http://www.insight-centre.org

Abstract. Development of low-cost and inter-operable home sensing products in recent years has motivated the development of consumer-level energy and home monitoring software solutions to exploit these new streams of data available to end-users. In particular, this opens up the home energy space as an area of high potential for the use of consumer-level energy optimisation with home-owners actively engaged with data about their energy use behaviour. We describe the development of a tablet-based home energy cost saving and appliance scheduling system which calculates behaviour change suggestions that save occupants on their energy bills while avoiding disruption to their regular routines. This system uses a Constraint Satisfaction Problem Solver to compute savings based on real-world sensor data, and to generate revised schedules in a user-friendly format, operating within a limited computing environment and achieving fast computation times.

1   Introduction

In-home networked sensing has, in recent years, reached a level of cost, reliability and protocol maturity that enables consumer-level home sensing to a degree not possible before. The increasing variety and availability of Zigbee-compatible wireless sensing allows for rich sensor coverage of homes with minimal impact or installation, providing rich data streams from around the home, monitoring use of space, appliance use, temperature and more. Providing these data streams in an easy to understand format is of substantial benefit to home-owners to help understand inefficient heating of spaces, electricity efficiency and overall energy costs [24].

Furthermore, developments in energy infrastructure and generation have lead to a greater variety of energy production sources and tariff schemes, leading to a greater incentive (both in terms of cost and in environmental impact) for home-owners to take pay attention to their energy consumption over different time periods. Some regions use variable tariffs based on the time of day [25], and

© Springer International Publishing Switzerland 2015
G. Pesant (Ed.): CP 2015, LNCS 9255, pp. 687–703, 2015.
DOI: 10.1007/978-3-319-23219-5_47

some schemes feature time-variable peak-demand limits with punitive charges incurred where total concurrent energy use exceeds a limit at a particular time of the week [1].

In this work we describe a Constraint-based system that recommends adjustments to energy use patterns to provide targetted cost savings in the home, based on historical and current observation of appliance use patterns in that home. The data is obtained from a deployed sensor network and energy monitoring system. The adjusted appliance use pattern accepted by the home-owner then forms the basis of a appliance use scheduling application, which recommends a set of activation times that correspond to observed historical behaviour while respecting limits of overall capacity and variable demand pricing schemes. This system operates on a low-power Android tablet device, as part of a Reporting Tool app which provides robust, easy-to-understand feedback to home-owners from the data streams provided by the home sensor deployment.

2   AUTHENTIC Smart Home Project

AUTHENTIC is a multi-institutional initiative to design and deliver a Home Area Network (HAN) infrastructure capable of supporting opportunistic decision making for effective energy management within the home.

The Authentic HAN consists of a collection of Zigbee wireless sensors communicating over a multi-hop wireless network with a low-power gateway Linux PC. This hub stores the sensor readings in a MySQL database. The readings are retrievable through a REST (Representational State Transfer) interface [19] in the form of JSON (Javascript Object Notation) Objects [6] which are transformed and can be displayed to the users via a reporting tool (discussed below). The first phase of deployments included smartplug sensors attached to power sockets which report appliance activations and energy consumption, passive infrared motion detectors reporting space occupancy, temperature, light and humidity sensors, and contact sensors reporting when doors and windows are opened or closed (Figure 1). Future phases include wireless smart meters for measuring total electricity, gas and oil consumption, sensors for monitoring the use and settings of heating, ventilation, air conditioning and water heating, and sensors and meters for monitoring the available of energy from renewable sources.

In order to assist the users to monitor and change their energy consumption, and to track their energy bills, a Reporting Tool app has been developed for Android tablets. The app provides direct and instant feedback, converting complex sensed data into customisable human-readable reports; sample screenshots are shown in Figure 2. Responding to user queries for specified time periods, the app can create visualisation of, for example, total energy consumption, consumption by room, consumption by appliance, room occupancy patterns, correlations between room occupancy and temperature or appliance use, and comparisons between different appliances and different time periods. The app includes a facility for real-time alerts when specified or anomalous patterns occur (e.g. windows opening in a heated room, or instantaneous consumption above

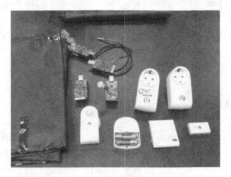

Fig. 1. AUTHENTIC sensors (movement, light, humidty, contact and smartplug sensors)

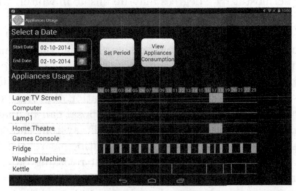

(a) Appliance Use History

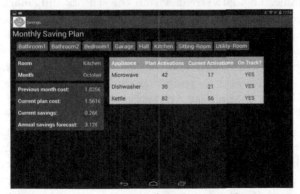

(b) Plan Recommendation (number of appliance activations)

Fig. 2. Android reporting tool

a threshold). Finally, the app provides facilities for guiding the user to change their behaviour. At a basic level, the app allows the user to set annual energy goals, and to track progress towards those goals. More sophisticated feedback, including advice on how to achieve specified reductions, is the main topic of this paper, and is discussed in the remaining sections.

3 Requirements

Users need guidance to help change their energy consumption patterns in accordance with their own goals and preferences. We consider two high level goals: reducing cost, and reducing total energy consumption[1]. The first steps are understanding how much energy is consumed by each activity and what cost is incurred by those activities at different times. Reducing consumption itself is easy – appliances can simply be switched off – but the difficulty is in balancing the reduction in consumption with the users' preferences for safe and comfortable living. Lowering a heating thermostat by 2 degrees may be unsafe in winter, while switching off an entertainment console may be impractical.

The aim is to propose changes to a user's behaviour which achieve their energy goals, while still being acceptable for their health, comfort and enjoyment. Further, those proposed changes should be personal, tailored to the users in a specific home. Thirdly, the interaction load imposed on the users must be manageable: systems which require extensive elicitation of preferences and utility functions under different scenarios before any proposal can be made are simply impractical in most domestic settings. Finally, the system should respect the privacy of the users, and should be capable of operating entirely within the home, on devices with limited computing power, without releasing data or relying on data obtained from other users outside the home.

Our intention is to guide the users based on the observed behaviour of those users in their home. The data streams from the sensors and smart plugs provide a history of appliance use, from which we can extract daily and weekly patterns. Given a goal (e.g. reduce the energy bill by 10%), we will then search for minimal changes to those patterns which satisfy the goal. Our proposals will be on two levels. The first level will recommend the total number (or duration) of activations for each appliance over a specified period, and if differential tariffs are in use, the total number of appliance activations in each price band.

The second level will propose a schedule of use for each day, ensuring the total simultaneous load is within any threshold, again representing a minimal change to observed behaviour patterns. Thus in each case an initial proposal is tailored to each home, without requiring any interactive preference elicitation apart from setting the initial high level goal. In each case, the users should then be able to interact with the system, imposing new constraints or objectives if the proposal is not satisfactory, and requesting a new recommendation. Further, the system will then monitor energy consumption within the home, alerting the

---

[1] Additional goals of reducing carbon footprint and reducing energy consumption from non-renewable sources are being developed.

user when targets are not going to be met, and offering recomputation of the
schedules or activation levels.

Our problem thus involves multiple levels of decision support and optimisation, over constrained variables. The problems range from relatively simple linear
assignment problems to constrained scheduling, and in each case the problems
may be extended by additional user constraints. Thus we choose to use constraint
programming as the single technology for all problem variants because of the
dedicated support for modelling cumulative scheduling problems and because
of its flexibility in adding what ultimately could be arbitrary side constraints
and preferences. The Authentic system also imposes some technical limitations,
which influence our choice of solver. The Reporting Tool app (Section 2) runs
under Android, and is limited to JDK 1.6 for library compatibility issues. Thus
we model and solve the decision and optimisation problems using the Choco
2.1.5 Java library, as it is also JDK1.6 compatible. This places a restriction on
the variable types and constraints we use.

4   Constraint-Based Recommender System

4.1   System Overview

The overall system is structured as follows. The Authentic HAN produces a
database of sensor readings for the home. This data is transformed into appliance activation reports. Information is presented to the user, who specifies an
energy or cost goal for the next period. From the appliance reports, the system
constructs a constraint optimisation model, and proposes a high-level energy
consumption plan to the user. It iterates with the user until an acceptable plan
is agreed. From the high-level plan, the system then creates a scheduling problem
and proposes a schedule to the user that respects the plan and requires a minimal
change to previous behaviour. Again, the user iterates with the system until an
acceptable schedule is agreed. The user is in control of all activations, and the
system monitors compliance, alerts the user when targets are to be missed, and
propose new plans or schedules in response. We describe the constraint-based
modules in Sections 4 and 5, and evaluate performance in Section 6.

4.2   Data Preprocessing (Data Analysis Module)

To pre-process the sensor data, we developed a Data Analysis module which
interfaces with the Authentic RESTful API [2] [18] to retrieve home sensor data
which it discretises into forms more amenable to analysis and optimisation.

Sensor readings and events are retrieved in the form of JSON Objects which
are then parsed by the module. By parsing the sensor information, data manipu-
lation can be performed to convert this information into alternative forms more
suitable for optimisation problems or efficiency analysis. In general, the facility
is provided to convert Occupancy, Temperature, Light, Humidity and Appliance
Activation information into arrays of samples (at a user-configurable sampling

rate). This allows for the like-for-like comparison of readings between arbitrary timepoints and aribtrary sensing types (e.g. plotting Occupancy vs Television, Light vs Lamp, or Humidity vs Dryer).

An additional function provided by this module is the generation of "ActivationTuples", data objects representing the usage pattern for appliances in the home which we use as input for the Appliance Use Optimisation Module (Section 4.3).

ActivationTuples contain the following information:

- Appliance Name
- Average Consumption Per Activation (Watt-Hours)
- Number of Activations per Tariff (in the case of Day/Night or other variable tariffs)
- Power Draw Profile (time series, Watts)

Activations are considered from the time an appliance sensor reports instantaneous demand above a minimum threshold until the instantaneous demand returns to below that threshold. The average consumption is based on the watt-hour (WH) consumption per activation period. In the case of appliances with variable activation profiles (for example, different washing machine settings), we can subdivide the appliance into seperate sub-appliances, one for each setting or pattern. The Power Draw Profile represents the instantaneous power demand in watts over a time series with configurable sampling rate (e.g. 5 minutes). The Power Draw Profile is used with the Scheduling Module to ensure that the total concurrent power load in a home or on a circuit at a given time can be accounted for in the constraint satisfaction problem. As the reporting interval of the smart-plug sensors is generally 5 minutes, higher resolution sampling rates are not required.

4.3   Appliance Use Optimisation Module (Solver)

The aim of the Appliance Use Optimisation module ("Solver") is to propose high-level patterns of appliance use which will achieve desired cost or energy savings while minimising disruption to the household. These patterns are simply counts of each appliance use in each price tariff. The module retrieves usage patterns from the analysis module in the form of activation tuples, and uses these to create a constraint satisfaction or optimisation problem. In this work we consider three tariffs (high, medium, low) available each day, although the solver is flexible and any number of price bands can be represented. The solver then searches for a modified pattern of use, by moving activations to lower price bands or reducing the number of activations, with a constraint on total cost. To avoid disruption, and to promote acceptance of the plan, we include an objective to minimise the deviation between the historical activation use pattern and the proposed pattern, represented as the sum of the squared differences of the appliance use counts). The constraint model is shown in Tables 1 and 2.

Table 1. Appliance Use Optimisation variables

Variable Name	Description
	Constants
$InputCost$	Cost of the historical period
$Target$	Reduction in cost required (e.g. 0.9 of the previous cost)
H, L, M	Price per unit(wH) of electricity at High, Medium and Low tariffs
$A_i InputAct$	Original Number of activations of Appliance i in input
$A_i Cons$	Average Consumption of Appliance i (wH) per activation
$A_i InputH$	Original number of activations at High tariff for Appliance i in input
$A_i InputM$	Original number of activations at Medium tariff for Appliance i in input
$A_i InputL$	Original number of activations at Low tariff for Appliance i in input
	Variables
$A_i HAct$	Number of activations of Appliance i at High tariff times
$A_i MAct$	Number of activations of Appliance i at Medium tariff times
$A_i LAct$	Number of activations of Appliance i at Low tariff times
	Auxilliary Variables
$A_i HDiff$	The difference between historical High use and the new plan for Appliance i
$A_i MDiff$	The difference between historical Medium use and the new plan for Appliance i
$A_i LDiff$	The difference between historical Low use and the new plan for Appliance i
$A_i HCost$	Total Cost for Appliance i at High tariff
$A_i MCost$	Total Cost for Appliance i at Medium tariff
$A_i LCost$	Total Cost for Appliance i at Low tariff
$A_i TotalCost$	Total cost for Appliance i
$A_i TotalDiff$	Sum of the squared differences for Appliance i
$TotalCost$	Total cost of the new plan
TotalDiff	Objective variable. Sum of squared differences
$A_i Act$	Total Number of activations of Appliance i

The only input required from the user for the first solution is to specify the required energy or cost reduction, and thus the interaction load on the user is minimal. However, it is likely that the proposed changes may not satisfy the user's requirements. Thus after the first schedule is proposed, the user is invited to specify additional constraints. The main type of user constraint is a domain restriction, specifying for a single appliance the acceptable range of activations in any price band or in total. Additional constraints, including for example, that the total number of activations for one appliance must be not less than the number for another can easily be handled by the solver, but are not yet implemented in the interface. The intention is to regard these user constraints as critiques of the

Table 2. Appliance Use Optimisation constraints

Constraints	Description
$A_iHCost = A_iHAct * H$	Set the cost at H for Appliance i
$A_iMCost = A_iHAct * M$	Set the cost at M for Appliance i
$A_iLCost = A_iHAct * L$	Set the cost at L for Appliance i
$A_iHDiff = (A_iHAct - A_iInputH)^2$	Set the Squared Difference in H.
$A_iMDiff = (A_iMAct - A_iInputM)^2$	Set the cost at Squared Difference in M
$A_iLDiff = (A_iLAct - A_iInputL)^2$	Set the cost at Squared Difference in L
$A_iTotalDiff = A_iHDiff + A_iMDiff + A_iLDiff$	Set the sum Difference for Appliance i
$A_iTotalCost = A_iHCost + A_iMCost + A_iLCost$	Set the total cost for Appliance i
$A_iAct = A_iHAct + A_iMAct + A_iLAct$	Set the total number of activations for Appliance i
$TotalDiff = \sum_{i=1}^{n}(A_iTotalDiff)$	Set the total sum of Differences
$TotalCost = \sum_{i=1}^{n}(A_iTotalCost)$	Set the total Cost
$TotalCost \leq InputCost * Target$	Ensure the TotalCost is below the target price
$Minimise(TotalDiff)$	Objective is to minimise the sum of squared differences

proposed solutions, and thus the user is not required to specify any preference that is already satisfied. However, the added user constraints are stored, and will be applied to future solving instances (and the user is invited to remove any constraints that no longer apply).

5   Scheduling Module

The output from the optimisation module is a high level plan specifying the number of activations of each appliance in each price band over a specified period. The user is in control, and will choose when to activate an appliance. However, as discussed above, the system will monitor the activations and alert the user if the observed activations are not on track to meet the goals at the end of the period. In cases where there are multiple price bands during a day, or where there are peak power thresholds which invove a higher price, it may be difficult for the user to balance their appliance use appropriately. Thus, we also provide a task scheduler, which recommends start times for appliance use over a number of days in order to respect the cost constraints. As with the high level plan, there is an objective to find appliance start times which are as close as possible to the observed historical patterns for the user. The system allows users to modify aspects of the schedule, by adding or modifying tasks and constraints. The user is in control, but as before, the system will monitor appliance use compared to the schedule and alert the user when thresholds are likely to be breached, and offer the option of rescheduling.

The first aim is to generate a schedule which respects the cost constraints. The number of activations for each appliance is obtained from the high-level plan, and the required number of task instances are generated. In order to handle the time granularity for different constraints, each task instance is then broken up into multiple sub-tasks each of fixed duration, and sequencing constraints are applied between the subtasks. Disjunctive constraints[8] are imposed on multiple instance of the same appliance, to ensure that they cannot be scheduled to operate simultaneously. The high level plan also specifies the number of activations to take place under each tariff, and so for each task instance, we designate its highest permissible tariff and impose constraints preventing that task running in a higher tariff's time period. Note that this is a permissive approach, allowing the solver to find schedules with a lower cost than expected.

The next constraint to be considered is the peak power threshold. We handle this by imposing a cumulative constraint [3] over all tasks. We associate heights with each appliance task relative to their power demand (and in cases where the power demand varies over the activation cycle, the heights vary across the different sub-tasks). The cumulative constraint height parameter is set to to the maximum power threshold, and in the case of variable maximum demand schemes we create dummy tasks with appropriate heights to represent temporary lower thresholds over appropriate time periods.

To ensure each schedule mimics prior behaviour, we analyse historic usage patterns and infer a time series of the frequency with which an appliance was active at a given time. The aim is then to minimise the abnormality of appliance activation times; that is, to avoid recommending appliance use at times with very low historic use frequency. To achieve this, we divide each day of the week into timeslots based on the sub-task time granularity (in this work, 5 minutes). For each appliance activation in the historical data, we increment a count in each timeslot that matches the day and time (e.g. for a 15-minute activation we might increment Tuesday 4:00. Tuesday 4:05 and Tuesday 4:10). Having done this for several weeks worth of historical data, we now have a time series representing the frequency any appliance was in use at any given time during the 7-day week. We invert this time series (where F = the highest frequency in the series, all entries E are changed to F-E, Figure 5) which we use to produce a set of dummy constant Task Variables (each of height equal to the appropriate inverted frequency) for use with another cumulative constraint.

The peak value for the cumulative constraint is then a variable, with the aim being to minimise that peak, subject to the lower bound equal to the maximum height of the inverse frequencies. For each appliance activation, the height of all of its sub-tasks are scaled to be the gap between the highest and lowest frequencies for that appliance, and the sub-tasks are added to the cumulative constraint (Figure 5). Thus, if we schedule a task so that the full duration of

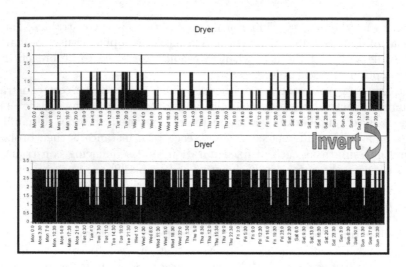

Fig. 3. Frequency Time Series and inversion to produce Masking Gradient

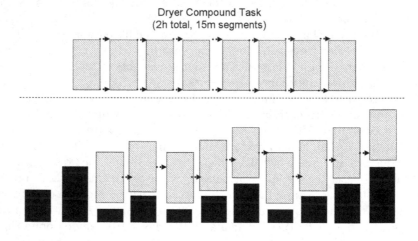

Fig. 4. Scheduling appliances using Masking Gradient

a task (i.e. each of its sub-task components) is at a time of highest frequency, the cumulative height variable is set to the lower bound; if the appliance task is scheduled so that it is active at the time of lowest historic frequency, then the cumulative height is at its maximum. All appliances are then normalised, and the optimisation objective is to minimise the sum of the appliance cumulative heights; that is, we minimise the sum of the 'abnormalities' of the activation times over each appliance use.

The initial solution to the scheduling can now be generated and displayed to the user. Although it may seem counter-intuitive to schedule uses of appliances like televisions or games consoles, it is important to account for their expected use, in order to manage the use of other appliances around that time. Thus we learn when those devices are most often used, and build that expected usage into the schedule. This initial solution does not require any interaction with the user to elicit preferences or objectives. We then allow the user to interact with the scheduler to improve the schedule, if necessary. The user can add new tasks or delete existing tasks for a particular day, can extend the duration of a task, can fix the start time of a task (by sliding the task position on the displayed schedule), and can add limited temporal constraints between tasks (for example, a specific dryer task must start within 30 minutes of the completion of specific washing task, or that the shower and washing machine cannot be active at the same time), and then can request a rescheduling. For new tasks, the same historical frequency pattern is applied; for user task time changes, the start time is fixed, and the frequency mask is adjusted to that no penalty is applied.

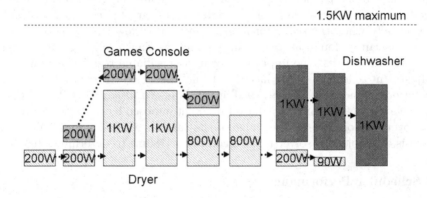

Fig. 5. Scheduling three appliances accounting for concurrent power demand

6   Performance Evaluation

A series of experiments were performed to determine the performance of the optimisation module on an Android tablet (Samsung Galaxy Tab 4 , Android version 4.4.2). Three months of data from a real-world home deployment formed the basis for the historical appliance use in these experiments, and the computation time taken for the Savings module was typically approximately 900ms and the Scheduler completed its task in approximately 17 seconds (using 30 minute time slots). These timings are well within the limits for acceptable response time, and demonstrate the success of the constraint programming formulation of the problems. As there is the facility for users to influence the savings problem through the addition of constraints on the number of appliance activations at different tariffs (i.e. the user can specify more limited ranges for the number of activations of appliances during particular tariffs, or to specify limits on the amount of adjustment made), we investigated the impact of the introduction of preferences on the running time and solvability of the savings problem. We also investigate the impact of larger and smaller time-slots on scheduler performance, with and without some user constraints applied.

6.1   Solver Performance

To investigate performance on the extreme end of user interaction with the appliance savings module, we performed experiments where we gradually restrict the scope of adjustment available to the solver. From savings targets of 10% through to 60%, we performed the savings routine while gradually restricting the range of acceptable values (or "freedom") for appliance activations at each tariff. For instance, at "40%" freedom, the solver is free to adjust the number of activations for appliances at any particular tariff band to within 40% of the historical values. As the freedom is reduced, solutions at high savings targets become unavailable, and eventually when the scope for adjustment is very low (low freedom value) no solutions can be found at any savings target, as shown in Figure 6. In these results, we observe that the imposition of user-specifiable restrictions on scope for adjustment to appliance use has little-to-no impact on the computation time (all solutions found took between 900 and 1150ms to discover), with solutions unavailable where the limits on value adjustment imposed by preferences prevent the discovery of solutions in the range of the target savings amount. Where no solution is available, the solver takes a consistent 780ms to determine this.

6.2   Scheduler Performance

To evaluate the performance of the scheduling module, we took the output of the activations saving module (without preferences) as the basis for scheduling the next week of appliance activations. In this evaluation, we schedule using a range of time-slot resolution values, from 5 minutes up to 60 minutes, and observe the computation time. As the users can introduce custom constraints to the scheduler, we re-perform the scheduling preocedure at each time-slot resolution level with the addition of an Inter-Appliance Disjunction constraint between two appliances (two particular appliances cannot operate simultaneously), and the further addition of a temporal link between two appliances (wherever one appliance ends, a particular other appliance should start shortly afterwards). The results of these experiments are shown in Figure 7. We observe that the performance with the addition of user-specified constraints remains approximately the same as without ("Default"), and that while computation time for small time-slot resolution is extreme (up to 400 seconds), with timeslots of 20 minutes or larger the computation time is much more reasonable (from around 30 seconds computation time at 20-minute slots down to 5 seconds at 60 minute slots).

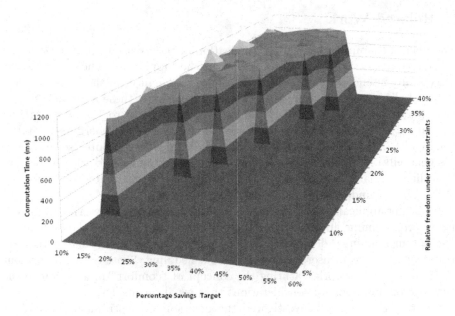

Fig. 6. Savings module performance under restricted scope

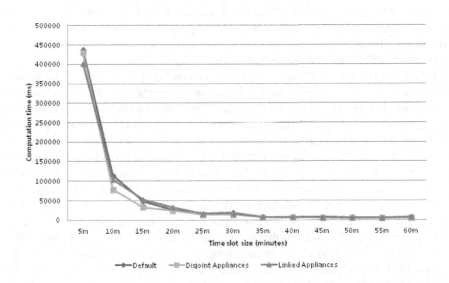

Fig. 7. Scheduler Results

7   Related Work

A number of examples of work in optimisation of energy use to balance demand and reduce costs exist intended to operate at the supply side. These approaches are typically intended to integrate with "smart" appliances in the home to allow for load balancing across the so-called "smart grid". As these approaches typically operate centrally at the utility, Mixed Integer Linear Programming (MILP) is commonly used to compute power use schedules based on energy availability [4,5,13,14,22,23]. MILP approaches take some time to compute (several minutes generally), and as they are solving over several homes and with respect to the utility company's energy supply requirements, there is limited scope for the home-owner to understand their personal energy use.

While not applicable to operating on tablets, Lim et al [17] describes a mixed-integer programming approach to air conditioning scheduling based on scheduled meetings in buildings, demonstrating the integration of energy-based requirements with real-world user requirements in scheduling. Scott et al [21] present a MILP-based framework featuring the concept of "comfort" in addition to the consumption and demand considerations that we also consider.

Felfernig & Burke[12] investigated the position of constraint-based models as the basis of Recommender Systems, noting that limited attention or understanding of complex recommender systems on the part of the user is a major consideration in the design of recommender systems. In our work, we use a simple-to-understand tablet-based interface to allow users to enter a dialogue with the system, introducing requirements which are interpreted as constraints, but without requiring particular expertise on the part of the user. Like our work, Ha et al [15] proposed a constraint-based approach for demand-side appliance scheduling, though this work was demonstrated using randomly generate synthetic data and compution time was in the order of minutes.

He et al [16] consider the case where electricity pricing is highly variable in time and the active use of energy by homes impacts the pricing, a potential future direction for energy production. They developed a market model combined with optimisation of community electricity use (using both constraint programming and mixed integer programming) for demand management and cost saving. Tushar et al [26] described a game-theory-based approach to demand-side energy management, modeling the interaction between home-owners and utilities through the trade and consumption of variable energy supplies and costs.

Darby et al [11] and Carroll [9] note that end-user perception has a significant role to play in motivating changes in energy use behaviours, with greater engagement and savings achieved simply through making more information available to the homeowner as found in smart-metering, from variable tariff trials in Ireland [10]. Bu et al [7] presented an MILP-based adaptive home energy scheduling approach with an initial discretisation step similar to the role of data pre-processing model in our work, though using synthetic test data rather than real-world data streams, and with relatively high computation time and resources compared to our approach. De Sá Ferreira et al [20] investigate highly variable tariff design at the utility level, generating optimised tariff pricing on a day-to-day basis.

Highly variable tariffs generated through such a scheme could function well combined with the energy saving and scheduling described in our work, allowing for dynamic schedule generation on a daily basis while respecting the home-owner's typical behaviour patterns.

8   Conclusions and Future Work

We presented a constraint-based energy saving recommender system which uses real-world appliance energy use data to generate recommended behaviour changes to achieve energy saving goals in the home. The system is computationally efficient and operates on a low powered tablet. Integrating user preferences allows for a continuous, interactive optimisation process with users introducing additional requirements as constraints in the problem, and through periodic recalculation of solutions when the sensor-observed energy use behaviour differs from the suggested adjustments. This aspect allows for users to interact with the model and solutions without any special knowledge, and could lead to greater understanding on the part of the home-owner as to their energy use habits and their impact on costs and emissions.

In future we will expand the deployments and models, incorporating add room and water heating sensing, and the capacity to remotely actuate appliances attached to smart-plugs. Remote actuation would allow for the optimised schedules to be automatically implemented in the home without requiring the users to physically attend the devices, and would substantially ease the burden on the occupant to conform to the scheduled plans. As the AUTHENTIC project expands, further home deployments and long-term feedback from users will motivate expanding the range of user-specified constraints and further investigation into the performance of the scheduler in these scenarios. We will expand on the peak-load model for cases where temporally-restricted renewable energy or home-energy-storage is a feature. In the case that a home has a large energy storage solution (e.g. an electric car battery), a limited capacity of cheap energy could be stored overnight and used to power appliances the next day during higher tarriff times. Similarly, solar panels could augment the energy supply depending on the weather, and this could motivate opportunistic appliance scheduling in reaction to volatile cheap energy availability.

Acknowledgments. This paper is funded by Enterprise Ireland through grant TC20121002D. We are grateful to the IERC and our academic and industry partners in Authentic for the design and development of the Authentic HAN, and to the anonymous occupants of our pilot homes for trialling the Authentic HAN, providing their data, and testing the Reporting Tool and Recommender System.

References

1. Application of delivery prices, orion group energy. http://www.oriongroup.co.nz/downloads/ApplicationOfDeliveryPrices.pdf (issued February 2, 2015, accessed April 21, 2015)

2. Authentic restful api. http://authentic.ucd.ie/documentation (accessed: December 1, 2014)
3. Aggoun, A., Beldiceanu, N.: Extending chip in order to solve complex scheduling and placement problems. Mathematical and Computer Modelling **17**(7), 57–73 (1993)
4. Agnetis, A., de Pascale, G., Detti, P., Vicino, A.: Load scheduling for household energy consumption optimization. IEEE Transactions on Smart Grid **4**(4), 2364–2373 (2013)
5. Bajada, J., Fox, M., Long, D.: Load modelling and simulation of household electricity consumption for the evaluation of demand-side management strategies. In: 2013 4th IEEE/PES Innovative Smart Grid Technologies Europe (ISGT EUROPE), pp. 1–5. IEEE (2013)
6. Bray, T.: The javascript object notation (json) data interchange format (2014)
7. Bu, H., Nygard, K.: Adaptive scheduling of smart home appliances using fuzzy goal programming. In: The Sixth International Conference on Adaptive and Self-Adaptive Systems and Applications, ADAPTIVE 2014, pp. 129–135 (2014)
8. Carlier, J.: The one-machine sequencing problem. European Journal of Operational Research **11**(1), 42–47 (1982)
9. Carroll, J., Lyons, S., Denny, E.: Reducing household electricity demand through smart metering: The role of improved information about energy saving. Energy Economics **45**, 234–243 (2014)
10. CER11080a: Electricity smart metering customer behaviour trials (cbt findings report) (2011)
11. Darby, S.J., McKenna, E.: Social implications of residential demand response in cool temperate climates. Energy Policy **49**, 759–769 (2012)
12. Felfernig, A., Burke, R.: Constraint-based recommender systems: technologies and research issues. In: Proceedings of the 10th International Conference on Electronic Commerce, p. 3. ACM (2008)
13. Georgievski, I.: Planning for coordination of devices in energy-smart environments. In: Proceedings of the Twenty-Third International Conference on Automated Planning and Scheduling (ICAPS 2013) (2013)
14. Ha, D.L., Joumaa, H., Ploix, S., Jacomino, M.: An optimal approach for electrical management problem in dwellings. Energy and Buildings **45**, 1–14 (2012)
15. Ha, L.D., Ploix, S., Zamai, E., Jacomino, M.: Tabu search for the optimization of household energy consumption. In: 2006 IEEE International Conference on Information Reuse and Integration, pp. 86–92. IEEE (2006)
16. He, S., Liebman, A., Rendl, A., Wallace, M., Wilson, C.: Modelling rtp-based residential load scheduling for demand response in smart grids. In: International Workshop on Constraint Modelling and Reformulation (Carlos Ansotegui 08 September 2014 to 08 September 2014), pp. 36–51 (2014)
17. Lim, B.P., van den Briel, M., Thiébaux, S., Bent, R., Backhaus, S.: Large Neighborhood Search for Energy Aware Meeting Scheduling in Smart Buildings. In: Michel, L. (ed.) CPAIOR 2015. LNCS, vol. 9075, pp. 240–254. Springer, Heidelberg (2015)
18. O'Sullivan, T., Muldoon, C., Xu, L., O'Grady, M., O'Hare, G.M.: Deployment of an autonomic home energy management system. In: The 18th IEEE International Symposium on Consumer Electronics (ISCE 2014), pp. 1–2. IEEE (2014)
19. Richardson, L., Ruby, S.: RESTful web services. O'Reilly Media, Inc. (2008)
20. de Sá Ferreira, R., Barroso, L.A., Rochinha Lino, P., Carvalho, M.M., Valenzuela, P.: Time-of-use tariff design under uncertainty in price-elasticities of electricity demand: A stochastic optimization approach. IEEE Transactions on Smart Grid **4**(4), 2285–2295 (2013)

21. Scott, P., Thiébaux, S., van den Briel, M., Van Hentenryck, P.: Residential Demand Response under Uncertainty. In: Schulte, C. (ed.) CP 2013. LNCS, vol. 8124, pp. 645–660. Springer, Heidelberg (2013)
22. Sou, K.C., Kordel, M., Wu, J., Sandberg, H., Johansson, K.H.: Energy and co 2 efficient scheduling of smart home appliances. In: 2013 European Control Conference (ECC), pp. 4051–4058. IEEE (2013)
23. Sou, K.C., Weimer, J., Sandberg, H., Johansson, K.H.: Scheduling smart home appliances using mixed integer linear programming. In: 2011 50th IEEE Conference on Decision and Control and European Control Conference (CDC-ECC), pp. 5144–5149. IEEE (2011)
24. Sweeney, J.C., Kresling, J., Webb, D., Soutar, G.N., Mazzarol, T.: Energy saving behaviours: Development of a practice-based model. Energy Policy **61**, 371–381 (2013)
25. Torriti, J.: Price-based demand side management: Assessing the impacts of time-of-use tariffs on residential electricity demand and peak shifting in northern italy. Energy **44**(1), 576–583 (2012)
26. Tushar, W., Zhang, J.A., Smith, D.B., Thiebaux, S., Poor, H.V.: Prioritizing consumers in smart grid: Energy management using game theory. In: 2013 IEEE International Conference on Communications (ICC), pp. 4239–4243. IEEE (2013)

Scheduling Running Modes of Satellite Instruments Using Constraint-Based Local Search

Cédric Pralet[1]([⊠]), Solange Lemai-Chenevier[2], and Jean Jaubert[2]

[1] ONERA – The French Aerospace Lab, F-31055 Toulouse, France
cedric.pralet@onera.fr
[2] CNES Toulouse, Toulouse, France
{solange.lemai-chenevier,jean.jaubert}@cnes.fr

Abstract. In this paper, we consider a problem involved in the management of Earth observation satellites. The input of the problem is a time-stamped observation plan and the output is a schedule of transitions between running modes of instruments available on board. These transitions must be scheduled so as to effectively realize the acquisitions requested, while satisfying several constraints, including duration constraints, non-overlapping constraints, and resource constraints over the thermal consumption of instruments. Criteria such as the minimization of the number of times instruments are switched off are also considered, for long-term reliability issues. The scheduling problem obtained needs to be solved several times per day, and the requirement is to produce good quality solutions in a few seconds. To produce such solutions, we propose a specific constraint-based local search procedure. Experiments are performed on realistic scenarios involving hundreds of observations, and the approach is compared with other techniques.

1 Problem Description

The study described in this paper is performed in relation with a future space program involving an observation satellite orbiting around the Earth at a low-altitude orbit (several hundreds of kilometers). The task of this satellite is to realize acquisitions over specific Earth areas using three optical instruments $I1$, $I2$, $I3$ available on board, and to download collected data towards ground reception stations. For such a satellite, each orbit takes about one hour and a half, and different areas are overflied during successive orbits due to the rotation of the Earth on itself. The set of areas selected for observation is defined by a ground mission center, which produces so-called *acquisition plans* for the satellite. Each acquisition in these plans is defined by its start and end times, by a set of instruments to be used, and by *instrument configurations*, which correspond to specific settings regarding instrument signal processing.

In addition to acquisition plans, the ground mission center must also provide the satellite with low-level plans specifying in more detail the operations

© Springer International Publishing Switzerland 2015
G. Pesant (Ed.): CP 2015, LNCS 9255, pp. 704–719, 2015.
DOI: 10.1007/978-3-319-23219-5_48

to execute in order to make instruments ready for acquisition when requested. These low-level plans must define, for each instrument, the successive transitions to make between several possible running modes: the (O)ff mode, the (S)tandby mode, the (H)eat mode, and the (A)cquisition mode. The possible transitions between these modes are described in Fig. 1a. To effectively realize an acquisition a over time interval $[t_1, t_2]$, each instrument required for a must be in the acquisition mode over $[t_1, t_2]$. For this, low-level plans must specify when instruments must perform a transition from the off mode to the standby mode, when the right instrument configuration must be loaded, when a transition to the heat mode must be performed to warm up the instrument, and when a transition to the acquisition mode must be triggered to have the instrument ready at t_1. Between two acquisitions, the instrument may come back to the heat, standby, or off mode depending on the time and resources available. Each transition will then result in a sequence of even more basic commands executed on board.

In this paper, we consider this problem of scheduling on the ground the transitions between instrument modes, from a time-stamped acquisition plan given as an input. In this problem, several kinds of constraints must be taken into account, including (1) *duration constraints*, enforcing a minimum duration spent in each mode and a fixed duration for each mode transition; (2) *non-overlapping constraints* over transitions; more precisely, transitions over instruments I1 and I2 cannot overlap except when they correspond to the same transition; the reason for this is that instruments I1 and I2 are managed on board by the same monitor, which is able to send a common sequence of basic commands to both I1 and

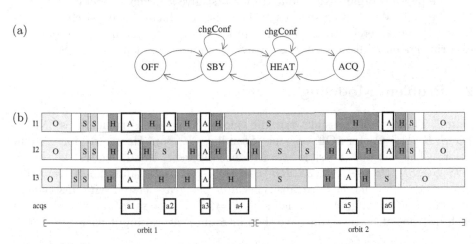

Fig. 1. (a) Mode automaton associated with each instrument; chgConf stands for a change of the configuration of the instrument; (b) example of a schedule defining mode transitions for instruments I1, I2, I3, given an acquisition plan involving 6 acquisitions a_1 to a_6 over 2 orbits; transitions correspond to empty rectangles between modes; transitions between the same two modes correspond to changes in instrument configuration; an acquisition such as a_1 uses all instruments, while a_4 uses only I2.

I2, but not two distinct command sequences in parallel; similarly, for instrument monitoring reasons, it is not possible to perform a transition between off and standby over I3 in parallel to a transition between off and standby over I1 or I2; moreover, no transition must be performed during acquisitions to avoid disturbances; (3) *resource constraints*, imposing that the total thermal consumption induced by the activities performed on each instrument over each orbit cannot exceed a maximum limit. In terms of optimization criterion, the main objective is to limit the number of times instruments are switched off, for long-term reliability issues. For a given number of times instruments are switched off, a secondary objective is to minimize the total thermal consumption of the instruments (better resource usage). In the end, the goal is to produce schedules such as the one provided in Fig. 1b.

This mode transition scheduling activity is the last step of the complete mission programming process which may be done several times a day, depending on the station opportunities to upload a new plan to the satellite. It is required to be solved in less than a few seconds. Constraint Programming (CP) comes into play because the realization of consistent mode transition schedules becomes more and more challenging for space engineers, due to the increasing complexity of satellites. One expected benefit in using CP is a better exploration of the search space, at least better than with hard-coded decision rules. Another advantage is that in first design phases as the one we are in, the declarative nature of CP is more flexible in case of changes in the satellite specifications, and it allows to easily assess the impact of new design choices.

The paper is organized as follows. We first give a formal model of the problem (Sect. 2). We then present a specific constraint-based local search encoding (Sect. 3) and a dedicated local search procedure (Sect. 4). Finally, we describe experiments on realistic scenarios involving hundreds of acquisitions (Sect. 5).

2 Problem Modeling

Data. The set of possible running modes of the instruments is $\mathcal{M} = \{O, S, H, A\}$: O for Off, S for Standby, H for Heat, A for Acquisition. The set of possible transitions between these modes is $\mathcal{T} = \{OS, SO, SH, HS, HA, AH, SS, HH\}$. In this set, mm' stands for a transition from mode m to mode m', transitions mm being associated with instrument configuration loading. We also define as $\mathcal{S} = \mathcal{M} \cup \mathcal{T}$ the set of possible states of each instrument at each time. We then introduce several input data:

- a set of contiguous orbits $\mathcal{O}$; each orbit $o \in \mathcal{O}$ has a start time $\mathbf{TsOrb}_o$ and an end time $\mathbf{TeOrb}_o$; the planning horizon considered is $[\mathbf{Ts}, \mathbf{Te}]$ with $\mathbf{Ts}$ the start time of the first orbit and $\mathbf{Te}$ the end time of the last orbit;
- a set of instruments $\mathcal{I}$; for each instrument $i \in \mathcal{I}$ and each mode $m \in \mathcal{M}$, $\mathbf{DuMin}_{i,m}$ denotes a minimum duration for i in mode m, and for each transition of type $t \in \mathcal{T}$, $\mathbf{Du}_{i,t}$ denotes the fixed duration of transition t for instrument i; a thermal consumption rate $\mathbf{ThRate}_{i,s}$ is associated with

each state $s \in \mathcal{S}$ of instrument i; last, **ThMax**$_i$ denotes a maximal thermal consumption allowed for instrument i over each orbit;

- a set of acquisitions $\mathcal{A}$; each acquisition $a \in \mathcal{A}$ has a fixed start time **TsAcq**$_a$ and a fixed duration **DuAcq**$_a$; for each instrument $i \in \mathcal{I}$, **Conf**$_{a,i}$ denotes the configuration required by acquisition a on instrument i (value nil when instrument i is not used in the realization of a); without loss of generality, we assume that each instrument is used at least once in the acquisition plan;
- a set of incompatibilities **Inc** between mode transitions; each element in **Inc** is a 4-tuple $(i, i', t, t') \in \mathcal{I}^2 \times \mathcal{T}^2$ specifying that transition t on instrument i cannot be performed in parallel to transition t' on instrument i'; for instance, $(I1, I2, SH, OS) \in$ **Inc** because transition SH over $I1$ cannot be performed in parallel to transition OS over $I2$.

For each instrument $i \in \mathcal{I}$, several additional data can be derived, such as the number of acquisitions **Nacqs**$_i$ for which i is used. For every $u \in [1..$**Nacqs**$_i]$, **Acq**$_{i,u} \in \mathcal{A}$ denotes the acquisition associated with the uth use of instrument i.

Allowed State Sequences. In order to produce plans such as the one given in Fig. 1b, it is necessary to determine, for each instrument, the sequence of its contiguous states over planning horizon [**Ts**, **Te**], as well as the exact start and end times associated with these states. As the desired state sequences must contain the acquisitions, the main question is actually how to handle each instrument outside these acquisitions, that is (1) from the beginning of the planning horizon, where instruments are switched off, to the first acquisition, (2) between two successive acquisitions which use the same instrument configuration, (3) between two successive acquisitions which use distinct configurations, and (4) from the last acquisition to the end of the horizon, where instruments must all be switched off for safety reasons. These four types of global transitions between successive uses of instruments must be composed of state sequences that are consistent with the automaton given in Fig. 1a.

To limit the search space, one key point is to avoid considering state sequences such as [**A**,AH,H,HS,S,**SH**,**H**,**HS**,S,SH,H,HA,**A**], in which the instrument is warmed up and cooled down (part **SH**,**H**,**HS**) in the middle of two acquisitions. To do this, we define beforehand a set of allowed state sequences for each global transition between instrument uses. The decompositions allowed are provided in Fig. 2. It can be seen that from the beginning of the horizon to the first acquisition, there are only two relevant state sequences, performing the configuration loading operation in the standby mode and heat mode respectively. For global transitions between acquisitions which use the same instrument configuration, there are four relevant decompositions: one which comes back to the heat mode between the acquisitions, one which comes back to the standby mode, and two which come back to the off mode and differ in the place where the configuration is loaded (the loaded configuration is lost when using the off mode). Similarly, there are six possible decompositions for global transitions between acquisitions which use distinct configurations. For the last global transition, there is a unique possible decomposition.

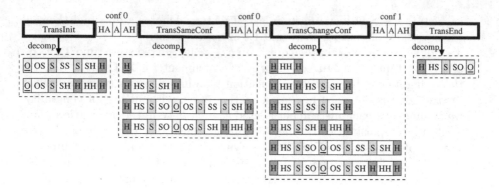

Fig. 2. Global transitions between instrument uses and their possible decompositions

Decision Variables. In the following, we define a *state interval* as a temporal interval over which a unique state is active. Formally, a state interval *itv* is defined by a boolean presence variable $pres(itv)$, by an integer start time variable $start(itv)$, by an integer end time variable $end(itv)$, by a fixed state $\mathbf{S}(itv) \in \mathcal{S}$, and by a fixed instrument $\mathbf{I}(itv)$ over which it is placed. In case of acquisition intervals, $\mathbf{I}(itv)$ contains all instruments required for the acquisition. We consider integer dates because the equipments which send mode transition commands to the instruments work with a discrete time clock. We also consider *global transition intervals gitv*, which are defined by an integer start time variable $start(gitv)$, by an integer end time variable $end(gitv)$, and by a fixed type among $\{TransInit, TransSameConf, TransChangeConf, TransEnd\}$.

Then, to describe schedules at a global level (upper part in Fig. 2), we introduce the following elements:

- for each acquisition $a \in \mathcal{A}$, one state interval $\mathbf{itvAcq}_a$ of state A is introduced for representing the temporal interval during which a is realized;
- for each instrument i and each use $u \in [1..\mathbf{Nacqs}_i]$ of this instrument, two mandatory state intervals $\mathbf{itvHA}_{i,u}$ and $\mathbf{itvAH}_{i,u}$ of respective states HA and AH are introduced for representing the transitions between the heat mode and the acquisition mode just before and just after acquisition $\mathbf{Acq}_{i,u}$;
- for each instrument i and each use $u \in [1..\mathbf{Nacqs}_i + 1]$ of this instrument, one global transition interval $\mathbf{itvTrans}_{i,u}$ is defined for representing the global transition performed before the uth use of the instrument (use $\mathbf{Nacqs}_i + 1$ corresponds to the last global transition before the end of the horizon);
- for each instrument i, if $[a_1, \ldots, a_n]$ denotes the sequence of acquisitions performed by i, the sequence of intervals associated with i is defined by:
$$\mathbf{Seq}_i : [\mathbf{itvTrans}_{i,1}, \mathbf{itvHA}_{i,1}, \mathbf{itvAcq}_{a_1}, \mathbf{itvAH}_{i,1},$$
$$\ldots$$
$$\mathbf{itvTrans}_{i,n}, \mathbf{itvHA}_{i,n}, \mathbf{itvAcq}_{a_n}, \mathbf{itvAH}_{i,n}, \mathbf{itvTrans}_{i,n+1}].$$

To describe schedules at a finer level (lower part in Fig. 2), we introduce the following elements:

- for each instrument i and each use $u \in [1..\mathbf{Nacqs}_i + 1]$ of i, one integer decision variable $\mathbf{decomp}_{i,u} \in [1..\mathbf{Ndec}_{i,u}]$ is introduced for representing the index of the decomposition chosen for realizing global transition $\mathbf{itvTrans}_{i,u}$, among those given in Fig. 2; parameter $\mathbf{Ndec}_{i,u}$ stands for the number of possible decompositions of $\mathbf{itvTrans}_{i,u}$;
- for each instrument i, for each use $u \in [1..\mathbf{Nacqs}_i + 1]$ of i, and for each decomposition $d \in [1..\mathbf{Ndec}_{i,u}]$ of $\mathbf{itvTrans}_{i,u}$, one sequence of state intervals $\mathbf{Dec}_{i,u,d} = [itv_1, \ldots, itv_k]$ is introduced; the latter is determined following the decomposition schemes provided in Fig. 2; state intervals in sequence $\mathbf{Dec}_{i,u,d}$ are optional (they will be present only in case $\mathbf{decomp}_{i,u} = d$).

The set of state intervals defined in the model is then $Itv = \{\mathbf{itvAcq}_a \,|\, a \in \mathcal{A}\} \cup \{\mathbf{itvHA}_{i,u} \,|\, i \in \mathcal{I}, u \in [1..\mathbf{Nacqs}_i]\} \cup \{\mathbf{itvAH}_{i,u} \,|\, i \in \mathcal{I}, u \in [1..\mathbf{Nacqs}_i]\} \cup \{itv \in \mathbf{Dec}_{i,u,d} \,|\, i \in \mathcal{I}, u \in [1..\mathbf{Nacqs}_i + 1], d \in [1..\mathbf{Ndec}_{i,u}]\}$. Each interval itv in this set has a minimum start time $Min(itv)$ and a maximum end time $Max(itv)$, both obtained from the known dates of acquisitions and from the minimum durations of state intervals. The set of intervals in Itv which may overlap orbit o and which concern instrument i is denoted by $ItvOrb_{i,o}$.

Constraints. We now express the set of constraints holding over the integer variables, intervals, and sequences of intervals previously introduced. These constraints are given in Eq. 1 to 15. Each constraint introduced is active only when the intervals over which it holds are present. Constraint 1 expresses that all intervals involved in sequence $\mathbf{Seq}_i$ must be present. Constraints 2-3 express that the start and end times of this sequence must coincide with the start and end times of the planning horizon. Constraint 4 imposes that successive intervals belonging to sequence $\mathbf{Seq}_i$ must be contiguous (no period during which the state of the instrument is undefined). Constraint 5 expresses that for a given global transition, only state intervals involved in the decomposition chosen for this transition are present. Constraints 6-7 impose that the start and end times of the sequence of intervals associated with a decomposition must coincide with the start and end times of the global transition from which it is derived. Constraint 8 specifies that successive intervals belonging to sequence $\mathbf{Dec}_{i,u,d}$ must be contiguous (no period during which the state of the instrument is undefined). Constraints 9-10 define a minimum duration for state intervals labeled by a mode and a fixed duration for state intervals labeled by a transition. Constraints 11-12 express that acquisition intervals must start at the required start time and have the required duration. Constraint 13 imposes that no transition must occur during acquisitions. Constraint 14 expresses that incompatible transitions must not overlap. Constraint 15 imposes a maximum thermal consumption over each orbit of the satellite for each instrument. In this constraint, the left term of the inequality takes into account only the part of intervals which overlaps the orbit considered.

$\forall i \in \mathcal{I},$

$$\forall itv \in \mathbf{Seq}_i,\ pres(itv) = 1 \tag{1}$$

$$start(first(\mathbf{Seq}_i)) = \mathbf{Ts} \tag{2}$$

$$end(last(\mathbf{Seq}_i)) = \mathbf{Te} \tag{3}$$

$$\forall itv \in \mathbf{Seq}_i \mid itv \neq last(\mathbf{Seq}_i),\ end(itv) = start(next(itv, \mathbf{Seq}_i)) \tag{4}$$

$\forall i \in \mathcal{I}, \forall u \in [1..\mathbf{Nacqs}_i + 1], \forall d \in [1..\mathbf{Ndec}_{i,u}],$

$$\forall itv \in \mathbf{Dec}_{i,u,d},\ pres(itv) = (\mathbf{decomp}_{i,u} = d) \tag{5}$$

$$start(first(\mathbf{Dec}_{i,u,d})) = start(\mathbf{itvTrans}_{i,u}) \tag{6}$$

$$end(last(\mathbf{Dec}_{i,u,d})) = end(\mathbf{itvTrans}_{i,u}) \tag{7}$$

$$\forall itv \in \mathbf{Dec}_{i,u,d} \mid itv \neq last(\mathbf{Dec}_{i,u,d}),\ end(itv) = start(next(itv, \mathbf{Dec}_{i,u,d})) \tag{8}$$

$$\forall itv \in Itv\ s.t.\ \mathbf{S}(itv) \in \mathcal{M},\ duration(itv) \geq \mathbf{DuMin}_{\mathbf{I}(itv),\mathbf{S}(itv)} \tag{9}$$

$$\forall itv \in Itv\ s.t.\ \mathbf{S}(itv) \in \mathcal{T},\ duration(itv) = \mathbf{Du}_{\mathbf{I}(itv),\mathbf{S}(itv)} \tag{10}$$

$\forall a \in \mathcal{A},$

$$start(\mathbf{itvAcq}_a) = \mathbf{TsAcq}_a \tag{11}$$

$$duration(\mathbf{itvAcq}_a) = \mathbf{DuAcq}_a \tag{12}$$

$$\forall itv \in Itv \mid \mathbf{S}(itv) \in \mathcal{T},\ noOverlap(itv, \mathbf{itvAcq}_a) \tag{13}$$

$$\forall itv, itv' \in Itv\ s.t.\ (\mathbf{I}(itv), \mathbf{I}(itv'), \mathbf{S}(itv), \mathbf{S}(itv')) \in \mathbf{Inc},\ noOverlap(itv, itv') \tag{14}$$

$$\forall i \in \mathcal{I}, \forall o \in \mathcal{O},\ \sum_{itv \in ItvOrb_{i,o}} pres(itv) \cdot d(itv) \cdot \mathbf{ThRate}_{i,\mathbf{S}(itv)} \leq \mathbf{ThMax}_i \tag{15}$$

$$\text{with } d(itv) = 0 \text{ if } (end(itv) \leq \mathbf{TsOrb}_o) \vee (start(itv) \geq \mathbf{TeOrb}_o)$$
$$\min(end(itv), \mathbf{TeOrb}_o) - \max(start(itv), \mathbf{TsOrb}_o)) \text{ otherwise}$$

Optimization Criteria. The first objective is to minimize the number of times instruments are switched off, for long-term reliability issues:

$$\text{minimize } card\{itv \in Itv \mid (\mathbf{S}(itv) = O) \wedge (pres(itv) = 1)\} \tag{16}$$

The second (and less important) criterion is to minimize the total thermal consumption over all instruments:

$$\text{minimize } \sum_{itv \in Itv} pres(itv) \cdot (end(itv) - start(itv)) \cdot \mathbf{ThRate}_{\mathbf{I}(itv),\mathbf{S}(itv)} \tag{17}$$

The two criteria defined are antagonistic because for instance the addition of off periods into the plan reduces the total thermal consumption. Criteria imposing a fair sharing between instruments could also be considered, as well as a minimization of the number of off intervals for a particular instrument.

Problem Analysis. The decision problem obtained is a constraint-based scheduling problem involving temporal constraints, resource constraints, and optional tasks [1]. It can be related with standard Job Shop Scheduling Problems (JSSP [2]). In our case, there would be one job per instrument, and the (optional) operations associated with each job would be the set of state intervals associated with each instrument. However, there are several differences with JSSP. For instance, some operations are shared between jobs, namely the acquisition tasks that require using several instruments simultaneously. Also, successive operations must be contiguous. Contiguity is important for ensuring that the state of the instrument is defined at every time and for getting the right thermal consumption. This consumption depends on the duration of state intervals, which is not fixed for state intervals associated with modes.

Another significant feature of the problem is the presence of global transition tasks which can be decomposed in several possible ways. Such a hierarchical aspect is also found in the framework of Hierarchical Task Networks (HTN [3]) developed in the planning community. In this framework, some abstract tasks are specified and the goal is to decompose these tasks into basic sequences of actions, through a set of available decomposition methods. In CP, the possible decompositions of global transitions could be expressed using an *alternative constraint* [4]. In another direction, the set of allowed sequences of state intervals could be represented using a *regular constraint* [5].

Last, the criteria considered are not standard criteria such as the makespan or the tardiness. On this point, it is worth noting that some transitions should be scheduled as early as possible to reduce the resource consumption, whereas others should be scheduled as late as possible for the same reason (more details on these points later in the paper).

3   Specific Problem Encoding

To obtain an efficient approach, several difficulties must be overcome. We present these difficulties and we show how they are handled.

3.1   Problem Complexity Versus Allowed Running Times

The problem obtained could be modeled and solved using CP solvers such as IBM ILOG CpOptimizer, which contains efficient scheduling techniques. It could also be represented using Mixed Integer Programming [6], which would be particularly adapted to deal with linear expressions related to thermal aspects. However, we must be able to solve in a few seconds large instances, which can involve several hundreds of acquisitions generating several thousands or tens of thousands candidate state intervals.

To overcome this difficulty, we use Constraint-Based Local Search (CBLS [7]), which has the capacity to quickly produce good quality solutions for large-size problems. As in standard constraint programming, CBLS models are defined by decision variables, constraints, and criteria. One distinctive feature is that in CBLS, all decision variables are assigned when searching for a solution, *i.e.* the approach manipulates complete variable assignments. The search space is then explored by performing local moves which reassign some decision variables, and it is explored more freely than in tree search with backtracking. One specificity of CBLS models is that they manipulate so-called *invariants*. The latter are one-way constraints $x \leftarrow exp$ where x is a variable (or a set of variables) and exp is a functional expression of other variables of the problem, such as $x \leftarrow sum(i \in [1..N])\, y_i$. During local moves, these invariants are efficiently maintained thanks to specific procedures that incrementally reevaluate the output of invariants (left part) in case of changes in their inputs (right part).

For tackling the application, we use the InCELL CBLS engine [8]. The latter offers several generic building blocks, including a large catalog of CBLS invariants together with incremental reevaluation techniques, and generic elements for defining specific local search procedures. In InCELL, a temporal interval *itv* is represented by a presence variable $pres(itv)$, a start time-point $start(itv)$, and an end time-point $end(itv)$. For incrementally managing temporal constraints over intervals, InCELL uses the *Simple Temporal Network* invariant [9]. The latter takes as input an STN [10], which is a set of simple temporal constraints over time-points, and it maintains as an output the temporal consistency of this STN as well as, for each time-point t, two variables $earliestTime(t)$ and $latestTime(t)$ giving its earliest and latest temporal position in a consistent schedule respectively. In the following, given an interval *itv*, variables $earliestTime(start(itv))$, $latestTime(start(itv))$, $earliestTime(end(itv))$, and $latestTime(end(itv))$ are denoted more succinctly as $est(itv)$, $lst(itv)$, $eet(itv)$, and $let(itv)$.

3.2   Specific Manipulation of Topological Orders

To implement a specific local search procedure, we need to control the main decisions involved in the problem. These decisions first include the decompositions chosen for global transitions, which are directly represented by variables $\mathbf{decomp}_{i,u}$. They also include the ordering between state intervals which must not overlap. To represent such decisions, we explicitly manipulate in the CBLS model a *topological ordering* **topoOrder** over intervals involved in non-overlapping constraints (Constraints 13-14). Technically speaking, this topological ordering is a sequence which expresses that if two intervals *itv*, *itv'* must not overlap and if *itv* is placed before *itv'* in the sequence, denoted by $before(itv, itv', \mathbf{topoOrder}) = 1$, then $end(itv) \leq start(itv')$ must hold. Constraints $noOverlap(itv, itv')$ appearing in Eq. 13-14 are then reformulated as:

$$(before(itv, itv', \mathbf{topoOrder}) = 1) \rightarrow (end(itv) \leq start(itv')) \qquad (18)$$

$$(before(itv', itv, \mathbf{topoOrder}) = 1) \rightarrow (end(itv') \leq start(itv)) \qquad (19)$$

In the local search algorithm defined, some of the local moves are performed by updating directly the topological ordering of conflicting intervals. To obtain an efficient approach, it is crucial to avoid considering local moves which create precedence cycles between intervals, where for instance an interval itv_1 is requested to be placed both before and after another interval itv_2. For avoiding this, as in Job Shop Scheduling, we exploit a *mandatory precedence graph*. The latter captures all mandatory precedences over transitions and acquisitions, including: (1) precedences between successive transition intervals involved in the same sequence $\mathbf{Dec}_{i,u,d}$ (sequence associated with the dth possible decomposition of the global transition realized before the uth use of instrument i), and (2) precedences between the first (resp. last) transition intervals in $\mathbf{Dec}_{i,u,d}$ and the acquisition that precedes (resp. follows) $\mathbf{Dec}_{i,u,d}$. From this precedence graph, when a transition interval itv must be moved just after another interval itv' in topological order $\mathbf{topoOrder}$, all mandatory successors of itv placed before itv' in the order are also moved just after itv, to avoid the creation of a precedence cycle.

Also, for each instrument $i \in \mathcal{I}$ and each use of this instrument $u \in [1..\mathbf{Nacqs}_i]$, we merge intervals $\mathbf{itvHA}_{i,u}$ and $\mathbf{itvAcq}_{\mathbf{Acq}_{i,u}}$ in the topological order. Indeed, as transitions and acquisitions cannot overlap, a transition interval itv which is in conflict with $\mathbf{itvHA}_{i,u}$ either ends before $\mathbf{itvHA}_{i,u}$, or starts after $\mathbf{itvAcq}_{\mathbf{Acq}_{i,u}}$, hence there is no need to dissociate $\mathbf{itvHA}_{i,u}$ and $\mathbf{itvAcq}_{\mathbf{Acq}_{i,u}}$ in the order. For the same reason, intervals $\mathbf{itvAH}_{i,u}$ and $\mathbf{itvAcq}_{\mathbf{Acq}_{i,u}}$ are also merged.

3.3   Specific Encoding of the Thermal Resource Constraint

Once a decomposition is selected for each global transition ($\mathbf{decomp}_{i,u}$ variables) and once an ordering between conflicting intervals is chosen ($\mathbf{topoOrder}$ variable), the temporal consistency of the plan can be determined as well as the value of the first criterion (number of times instruments are switched off). The only remaining freedom concerns the choice of precise start and end times of all intervals in the model. There is usually a huge freedom in the choice of these precise times, because we must deal with a large possible set of time steps. To avoid loosing time exploring all precise time values and to detect thermal inconsistency earlier, there is a need to build an approximation of the impact of decisions made so far on the thermal resource constraint given in Eq. 15 and on the thermal consumption criterion given in Eq. 17.

To do this, we again use a specificity of the problem which is that in each decomposition $\mathbf{Dec}_{i,u,d}$, it is possible to identify one particular interval whose thermal consumption is the lowest, and therefore one particular interval which should be enlarged as much as possible. In Fig. 2, for each possible decomposition, this particular interval to enlarge is represented with an underlined label. From this observation, we build an approximation of the thermal constraint (Constraint 15) as follows: in Eq. 15, (1) if itv is an interval to enlarge, $start(itv)$ and $end(itv)$ are replaced by $est(itv)$ and $let(itv)$ respectively (start as early as possible, end as late as possible); (2) if itv is placed on the left of the interval to enlarge in a decomposition, then $start(itv)$ and $end(itv)$ are replaced by $est(itv)$ and $eet(itv)$ respectively (start and end as early as possible); (3) if itv is placed on the right of the interval to enlarge in a decomposition, then

$start(itv)$ and $end(itv)$ are replaced by $lst(itv)$ and $let(itv)$ respectively (start and end as late as possible). The schedule obtained with these dates may not be feasible since it mixes earliest dates and latest dates, but the thermal consumption obtained over each orbit for each instrument can be shown to be a lower bound on the real consumption.[1] The same rewriting is used for building a lower approximation of the thermal criterion (Eq. 17).

3.4   Specific Management of Some Hard Constraints

As almost always in local search, there is a choice between handling constraints as hard constraints, to be always satisfied by the current assignment, or as constraints returning a constraint violation degree. Here, we choose to make soft the constraints enforcing due dates (Constraints 3 and 11) and the thermal resource constraint (Constraint 15), and we keep the other constraints as hard. Such a choice is made because finding a first plan satisfying Constraints 3, 11, 15 is not straightforward, therefore we need to temporarily explore plans which violate these constraints to progressively get a first consistent solution. On the other side, for all other constraints, e.g. for all constraints enforcing contiguity between successive activities, it is easy to produce a first consistent schedule.

More precisely, we transform Constraint 11 $start(\mathbf{itvAcq}_a) = \mathbf{TsAcq}_a$ into one permanent constraint (Eq. 20), one constraint whose presence can be controlled based on an additional variable named $\mathbf{lock}_a$ (Eq. 21), and one constraint violation degree called $\mathbf{delay}_a$, defined as a CBLS invariant (Eq. 22). The latter represents the gap between the earliest start time of a and its required start time. Variable $\mathbf{lock}_a$ serves us to explore either consistent schedules or temporarily inconsistent schedules when needed. All $\mathbf{lock}$ variables will be set to 1 at the end of the search, meaning that all acquisitions will be realized on time.

$$start(\mathbf{itvAcq}_a) \geq \mathbf{TsAcq}_a \tag{20}$$

$$(\mathbf{lock}_a = 1) \rightarrow (start(\mathbf{itvAcq}_a) \leq \mathbf{TsAcq}_a) \tag{21}$$

$$\mathbf{delay}_a \leftarrow est(\mathbf{itvAcq}_a) - \mathbf{TsAcq}_a \tag{22}$$

The same transformation is used for Constraint 3, with a $\mathbf{lock}_{|\mathcal{A}|+1}$ variable and with a violation degree $\mathbf{delay}_{|\mathcal{A}|+1} \leftarrow \max_{i \in \mathcal{I}}(eet(last(\mathbf{Seq}_i)) - \mathbf{Te})$. When all $\mathbf{lock}$ variables are set to 0, provided that there is no precedence cycle, there always exists a schedule which satisfies all active temporal constraints, essentially because there is no maximum duration for off, standby, and heat intervals.

Last, given an instrument i and an orbit o, the violation degree $\mathbf{thv}_{i,o}$ associated with Constraint 15 is the maximum between 0 and the difference between the thermal consumption obtained and the maximum thermal consumption allowed:

---

[1] Without going into further details, this result holds under the assumption that the thermal consumption rate of the SS (resp. HH) state is equal to the thermal consumption of the S (resp. H) state, which is true in practice.

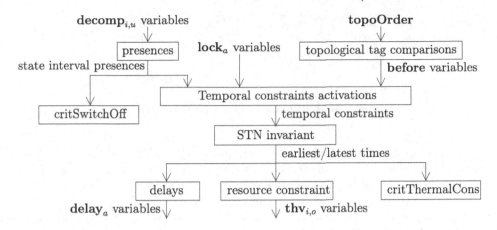

Fig. 3. Global view of the CBLS model

$$\mathbf{thv}_{i,o} \leftarrow \max(0, \sum_{itv \in ItvOrb_{i,o}} pres(itv) \cdot d(itv) \cdot \mathbf{ThRate}_{i,\mathbf{S}(itv)} - \mathbf{ThMax}_i)\,(23)$$

Fig. 3 gives a global view of the structure of the resulting CBLS model. The figure does not represent all individual invariants of the model.

4   A Multi-phase Local Search Algorithm

We now describe how we actually solve the problem, using a specific local search procedure. In the latter, the first step is to perform some preprocessing to restrict the set of feasible decompositions of global transitions. This step is able to automatically detect that for temporal reasons, the decomposition using the heat mode is the only feasible one between successive acquisitions placed very close to each other. It is also able to detect that for thermal reasons, only decompositions which come back to the off mode are feasible between successive acquisitions placed very far from each other.

After this preprocessing step, the main difficulty is that the problem involves antagonistic aspects, because for instance the best option regarding the on/off criterion is to never come back to the off mode between two successive acquisitions, whereas it is the worst option from a thermal point of view. This is why the resolution process is divided into several search phases (see Fig. 4).

Phase 1: First Heuristic Schedule. In the first phase, we build an initial schedule from simple decision rules. For decomposing global transitions, we systematically choose the option whose total minimum duration is the lowest. For choosing a topological order between intervals, we use a dispatching rule called the *best date heuristic*. In this heuristic, we compute, for each global transition and for each of its possible decompositions *dec*, the optimal start time of each transition

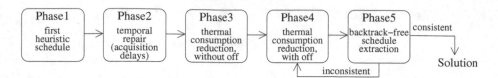

Fig. 4. Multi-phase local search algorithm

interval *itv* involved in *dec* when non-overlapping constraints are discarded. For intervals *itv* placed after the interval to enlarge in *dec*, this optimal start time corresponds to the latest start time, denoted by *lstAlone(itv)*, and for the others it corresponds to the earliest start time, denoted by *estAlone(itv)*. To get a first schedule, transitions with the smallest best dates are inserted first into the topological ordering. During this phase, all **lock** variables are set to 0.

Phase 2: Temporal Repair. The schedule obtained after Phase 1 may violate temporal constraints on the realization dates of acquisitions. To repair such constraint violations, we consider acquisitions chronologically. When considering an acquisition a, we try to enforce the realization date of a as long as the constraint violation degree associated with it is not equal to 0 (Eq. 22). To do this, we build the critical path explaining the late date obtained for a, we randomly select on this path a critical arc $itv \rightarrow itv'$ relating two intervals associated with distinct instruments, and we try to place *itv* just after *itv'* in the topological order. This local move is accepted iff it strictly reduces the delay of acquisition a. When all arcs in the critical path have been considered without improving the delay of a, the algorithm returns an inconsistency, meaning that the acquisition plan is too tight and should be revised. Otherwise, once the delay of acquisition a is equal to 0, variable $\mathbf{lock}_a$ is set to 1 before considering the next acquisition, meaning that future moves are not allowed to delay the realization of a. After the traversal of the acquisition plan, all acquisitions meet their due date. The same principle is used when the last state interval ends after the end of the planning horizon.

Phase 3: Resource Optimization. During the third phase, we consider the satisfaction of the thermal consumption constraint, by working on possible decompositions of global transitions. More specifically, we traverse the schedule chronologically, from the first acquisition to the last. At each step, we consider an acquisition a and we lock the realization dates of all acquisitions except for a ($\mathbf{lock}_a = 0$). For each instrument involved in the realization of a, we try to change the decomposition of the global transition leading to a. Only decompositions which do not use additional switch off operations are considered, and decompositions which induce lower thermal consumption are considered first. For assessing the impact of changing the current decomposition used *dec* for another decomposition *dec'*, we remove from the topological order all transition intervals belonging to *dec*, and we insert into it all transition intervals belonging to *dec'*. To do this, we use an insertion heuristic which works as follows: the transition intervals in *dec'* which must be scheduled as early as possible (resp. as late as possible) are inserted from the first to the last (resp. from the last to the first); the insertion of an interval itv is made just before the first (resp. after the last) interval itv' in the topological order such that $estAlone(itv) < start(itv')$ (resp. $lstAlone(itv) > start(itv')$) and such that the insertion does not lead to an inconsistency. Possible delays over the realization time of the acquisition considered are also repaired following the mechanism described in Phase 2. If delays cannot be repaired, decomposition dec' is rejected and we go on with the next possible decomposition. Otherwise, dec' is accepted as the new decomposition.

Phase 4: Resource Repair. After the third phase, the schedule may still be inconsistent from a thermal point of view. In this case, it is repaired by testing decompositions which add new off intervals during orbits where the thermal resource constraint is violated. To do this, the plan is traversed chronologically, orbit by orbit, and we work on an orbit until the thermal constraint of the CBLS model is satisfied. For introducing off intervals, we consider first global transitions which have the longest duration, because inserting off intervals during these global transitions is more likely to better reduce the total thermal consumption. For testing each decomposition, the same mechanism as in Phase 3 is used.

Phase 5: Solution Extraction. In the fifth phase, we use a backtrack-free procedure for determining good start and end times for intervals. This procedure traverses the topological ordering and considers at each step the next interval itv in the order. It sets the start time of itv to its latest consistent value when itv appears on the right of the interval to enlarge in a decomposition, and to its earliest consistent value otherwise. After setting the start time of itv, the procedure propagates the effect of this decision in the temporal network before going on to the next step. The schedule obtained at the end of the traversal of the topological ordering is temporally consistent, essentially because of the *decomposability* property of Simple Temporal Networks [10]. If the schedule obtained satisfies thermal constraints over each orbit, it is returned as a solution and the algorithm stops. Otherwise, we come back to fourth phase to insert additional off intervals. If no more off interval can be added, an inconsistency is returned.

5   Experiments

Three approaches are compared: (1) the local search algorithm described previously, built on top of the InCELL CBLS library [8], (2) a MIP approach based on IBM ILOG CPLEX 12.5; in the MIP model, boolean variables $before(itv, itv')$ are introduced for every pair of intervals itv, itv' which must not overlap; (3) a CP approach based on IBM ILOG CpOptimizer 12.5; in the CP model, CpOptimizer constraints such as the *alternative constraint* or the *noOverlap* constraint are used. From the definition of the formal model (Eq. 1-17) and from the definition of data structures for representing the instances, obtaining a first implementation with CpOptimizer took us less than one day, obtaining a first implementation with CPLEX took us approximately one day, and obtaining the local search

Table 1. Results obtained on an Intel i5-520 1.2GHz, 4GBRAM for 8 realistic instances, with a max. CPU time set to 2 minutes; *#orbits* gives the number of orbits which contain acquisitions; *#offs* stands for the number of times instruments are switched off and *conso* for the total thermal consumption; '*' indicates that the solution is optimal

	#acqs	#orbits	CpOptimizer			CPLEX			Specific CBLS		
			time (s)	#offs	conso	time (s)	#offs	conso	time (s)	#offs	conso
sat1	6	1	120	6	1048	0.11	6*	1046.0*	0.43	6	1074
sat2	11	2	-	-	-	14.01	7*	774.38*	0.61	7	785.87
sat3	37	1	-	-	-	1.41	8*	2922.30*	0.95	8	2933.80
sat4	42	1	-	-	-	55.35	13*	3395.49*	0.95	13	3416.27
sat5	47	1	-	-	-	3.29	8*	2895.54*	1.10	8	2903.03
sat6	82	2	-	-	-	120	20	6413.07	1.76	19	6569.49
sat7	129	3	-	-	-	-	-	-	2.25	17	8768.01
sat8	233	10	-	-	-	-	-	-	3.86	39	17270.62

approach took us approximately one day for implementing the CBLS model and one week for specifying and testing specific local search procedures.

Experiments were performed over realistic instances provided by the French space agency, ranging from small instances involving 6 acquisitions to large instances involving more than 200 acquisitions (see Table 1). To give an idea of the problem sizes, the largest instance (sat8) leads to 61515 variables and 176842 constraints in the CpOptimizer model, 103428 rows and 62434 columns in the CPLEX model, and 831074 variables and 258224 invariants in the InCELL CBLS model. Globally, it can first be observed that the attempt made with CpOptimizer is not efficient. We believe that the behavior observed is due to the fact that the sum constraint involved in the computation of the total thermal consumption does not allow to prune earliest/latest dates of activities that much because of its duration-dependent nature. Another point is that the domain of temporal variables may contain millions of values. As for the MIP approach, it performs very well on small and medium instances, for which it is able to find optimal solutions.[2] However, for the largest instances, the MIP approach does not find any solution in less than two minutes. Last, the CBLS approach scales well and delivers good quality solutions in a few seconds, which is compatible with the application needs. On the largest instance (sat8), the distribution of computation times among local search phases is as follows: CBLS model creation: 2.5s, preprocessing step: 0.69s, Phase 1: 0.4s, Phase 2: 0.02s, Phase 3: 0.20s, Phase 4: 0.03s, Phase 5: 0.02s. Moreover, there is no come back to Phase 4 after Phase 5.

6   Conclusion

This paper presents a specific CBLS approach for tackling a scheduling problem from the space domain. This problem raises several questions for CP. Indeed,

---

[2] Without further details, we relax the requirement of having integer dates, essentially because an integer solution can almost often be directly extracted from the solution found by the MIP.

whereas scheduling is known to be one of the most successful application area of CP, one lesson learned is that it can be difficult for standard CP solvers to deal with scheduling problems where the duration of some tasks is not fixed initially and where the duration of these tasks has an impact on resource consumption. Also, we believe that alternative choices between task decompositions make the propagation of the resource constraint even harder. On the opposite, it is easy to enlarge the right state intervals for the specific local search procedure or for the MIP approach. Due to the good performances of MIP on small and medium instances, it would probably be useful either to run CBLS and MIP in parallel, or to use the MIP model for realizing Large Neighborhood Search (LNS [11]). In another direction, the treatment of this application will also make us add new elements in our CBLS solver, for representing abstract tasks decomposable into basic activities following a set of available decomposition methods.

References

1. Baptiste, P., Pape, C.L., Nuijten, W.: Constraint-based Scheduling: Applying Constraint Programming to Scheduling Problems. Kluwer Academic Publishers (2001)
2. Pinedo, M.: Scheduling: Theory, Algorithms, and Systems. Springer (2012)
3. Erol, K., Hendler, J., Nau, D.S.: HTN Planning: Complexity and Expressivity. In: Proc. of the 12th National Conference on Artificial Intelligence (AAAI 1994), pp. 1123–1128 (1994)
4. Laborie, P., Rogerie, J.: Reasoning with Conditional Time-Intervals. In: FLAIRS Conference, pp. 555–560 (2008)
5. Pesant, G.: A Regular Language Membership Constraint for Finite Sequences of Variables. In: Proc. of the 10th International Conference on Principles and Practice of Constraint Programming (CP 2004), pp. 482–495 (2004)
6. Nemhauser, G., Wolsey, L.: Integer and Combinatorial Optimization. John Wiley & Sons (1988)
7. Hentenryck, P.V., Michel, L.: Constraint-Based Local Search. The MIT Press (2005)
8. Pralet, C., Verfaillie, G.: Dynamic Online Planning and Scheduling Using a Static Invariant-based Evaluation Model. In: Proc. of the 23th International Conference on Automated Planning and Scheduling (ICAPS 2013) (2013)
9. Pralet, C., Verfaillie, G.: Time-dependent Simple Temporal Networks: Properties and Algorithms. RAIRO Operations Research (2013)
10. Dechter, R., Meiri, I., Pearl, J.: Temporal Constraint Networks. Artificial Intelligence **49**, 61–95 (1991)
11. Shaw, P.: Using Constraint Programming and Local Search Methods to Solve Vehicle Routing Problems. In: Proc. of the 4th International Conference on Principles and Practice of Constraint Programming (CP 1998), 417–431 (1998)

Abstracts of Papers Fast Tracked
to *Constraints* Journal

Using Finite Transducers for Describing and Synthesising Structural Time-Series Constraints

Nicolas Beldiceanu[1], Mats Carlsson[2], Rémi Douence[1], and Helmut Simonis[3]

[1] TASC-ASCOLA (CNRS/INRIA), Mines Nantes, FR – 44307 Nantes, France
FirstName.LastName@mines-nantes.fr
[2] SICS, P.O. Box 1263, SE – 164 29 Kista, Sweden
Mats.Carlsson@sics.se
[3] Insight Centre for Data Analytics, University College Cork, Ireland
Helmut.Simonis@insight-centre.org

Abstract. We describe a large family of constraints for structural time series by means of function composition. These constraints are on aggregations of features of patterns that occur in a time series, such as the number of its peaks, or the range of its steepest ascent. The patterns and features are usually linked to physical properties of the time series generator, which are important to capture in a constraint model of the system, i.e. a conjunction of constraints that produces similar time series. We formalise the patterns using finite transducers, whose output alphabet corresponds to semantic values that precisely describe the steps for identifying the occurrences of a pattern. Based on that description, we automatically synthesise automata with accumulators, which can be used as constraint propagators, as well as constraint checkers. We show the correctness of the synthesis method using a parametrised formalisation. The description scheme not only unifies the structure of the existing 30 time-series constraints in the Global Constraint Catalogue, but also leads to over 600 new constraints, with more than 100,000 lines of synthesised code. Examples of initial use-cases based on the new constraints are given.[1]

Acknowledgements. The first author was partially supported by the Gaspard-Monge program that initially motivated this research. The first and third authors were partially supported by the European H2020 FETPROACT-2014 project "GRACeFUL". The last author was partially supported by a senior researcher chair offered by the Région Pays de la Loire. He was also supported by EU FET grant ICON (project number 284715).

---

[1] A full description of this work will appear in Constraints 21(1).

G. Pesant (Ed.): CP 2015, LNCS 9255, p. 723, 2015.
DOI: 10.1007/978-3-319-23219-5

Projection, Consistency, and George Boole

J.N. Hooker

Carnegie Mellon University, Pittsburgh, USA
jh38@andrew.cmu.edu

Abstract. Although George Boole is best known for his work in symbolic logic, he made a strikingly original contribution to the logic of probabilities. He formulated the probabilistic inference problem as an optimization problem we now call linear programming. He solved the problem with a projection method we now call Fourier-Motzkin elimination. In a single stroke, he linked the concepts of projection, optimization, and logical inference.

Boole's insight extends to constraint programming as well, because consistency maintenance is a form of projection. Projection is, in fact, the root idea that unites the other three concepts. Optimization is projection of the feasible set onto a variable that represents the objective value. Inference can be understood as projection onto a desired subset of variables. Consistency maintenance is a form of inference that is likewise equivalent to projection.

We propose to exploit this unifying vision in constraint programming by addressing consistency maintenance explicitly as a projection problem. Existing types of consistency are already forms of projection, but viewing them in this light suggests a particularly simple type of consistency that has apparently not seen application. We call it J-consistency, which is achieved by projecting the problem's solution set onto a subset J of variables.

After a review of Boole's contribution, we first present a projection method for inference in propositional logic, based on logic-based Benders decomposition. We show how conflict clauses generated during solution of the satisfiability problem can deliver Benders cuts that describe the projection.

We next observe that domain consistency is a particularly simple form of projection, while k-consistency is a less obvious form. We then indicate how achieving J-consistency can reduce backtracking when the solver propagates through a richer structure than a domain store, such as a relaxed decision diagram. We show how to compute projections for a few popular global constraints, including among, sequence, regular, and all-different constraints. These results suggest that achieving J-consistency could have practical application in a solver.

We conclude by proposing a generalization of bounds consistency as projection of the convex hull of the solution set. This implies that computing generalized bounds consistency is related to the the project of identifying valid cutting planes, which has long been pursued in mathematical programming.

Full paper to be published in *Constraints* 21(1).

On Computing Minimal Independent Support and Its Applications to Sampling and Counting (Extended Abstract)

Alexander Ivrii[1], Sharad Malik[2], Kuldeep S. Meel[3], and Moshe Y. Vardi[3]

[1] IBM Research Lab, Haifa, Israel
[2] Princeton University
[3] Department of Computer Science, Rice University

Abstract. Constrained sampling and counting are two fundamental problems arising in domains ranging from artificial intelligence and security, to hardware and software testing. Recent approaches to approximate solutions for these problems rely on employing combinatorial (SAT/SMT) solvers and universal hash functions that are typically encoded as XOR constraints of length $n/2$ for an input formula with n variables. It has been observed that lower density XORs are easy to reason in practice and the runtime performance of solvers greatly improves with the decrease in the length of XOR-constraints. Therefore, recent research effort has focused on reduction of length of XOR constraints. Consequently, a notion of *Independent Support* was proposed, and it was shown that constructing XORs over independent support (if known) can lead to a significant reduction in the length of XOR constraints without losing the theoretical guarantees of sampling and counting algorithms [1].

In this paper, we present the first algorithmic procedure (and a corresponding tool, called MIS) to determine minimal independent support for a given CNF formula by employing a reduction to group minimal unsatisfiable subsets (GMUS). Extensive experiments demonstrate that MIS scales to large formulas with tens of thousands of variables, and the minimal independent support computed by MIS is typically of 1/10 to 1/100th size of the support of the formulas. By utilizing minimal independent supports computed by MIS, we provide new tight theoretical bounds on the size of XOR constraints for constrained counting and sampling – in some cases, even better than previously observed empirical bounds. Furthermore, the universal hash functions constructed from independent supports computed by MIS provide one to two orders of magnitude performance improvement in state-of-the-art constrained sampling and counting tools, while still retaining theoretical guarantees.

References

1. S. Chakraborty, K. S. Meel, and M. Y. Vardi. Balancing scalability and uniformity in SAT witness generator. In *Proc. of DAC*, pages 1–6, 2014.

Authors names are ordered alphabetically by last name and does not indicate contribution.

The full version appears in Constraints 21(1).

© Springer International Publishing Switzerland 2015
G. Pesant (Ed.): CP 2015, LNCS 9255, p. 725, 2015.
DOI: 10.1007/978-3-319-23219-5

General Game Playing with Stochastic CSP

Frédéric Koriche, Sylvain Lagrue, Éric Piette, and Sébastien Tabary

Université Lille-Nord de France
CRIL - CNRS UMR 8188 Artois, F-62307 Lens, France
{koriche,lagrue,epiette,tabary}@cril.fr
http://www.cril.univ-artois.fr/~koriche,lagrue,epiette,tabary

Abstract. [1] The aim of *General Game Playing* (GGP) is to devise game playing algorithms which are not dedicated to a particular strategic game, but are general enough to effectively play a wide variety of games. A tournament is held every year by AAAI, in which artificial game players are supplied the rules of arbitrary new games and, without human intervention, have to play these game optimally. Games rules are described in a declarative representation language, called GDL for *Game Description Language*. The lastest version of this language is expressive enough to describe finite multi-player games with uncertain and incomplete information. GGP algorithms include, among others, answer set programming methods, automated construction of evaluation functions, and Monte-Carlo methods such as Upper Confidence bounds for Trees (UCT). Beyond its play value, GGP offers a rigorous setting for modeling and analyzing sequential decision-making algorithms in multi-agent environments.

By providing a declarative approach for representing and solving combinatorial problems, Constraint Programming appears as a promising technology to address the GGP challenge. Currently, several constraint-based formalisms have already been proposed to model and solve games; they include notably *Quantified* CSP, *Strategic* CSP and *Constraint Games*. Most of these formalisms are, however, restricted to deterministic, perfect information games: during each round of the game, players have full access to the current state and their actions have deterministic effects. This paper focuses on *stochastic* games, with chance events, using the framework of stochastic constraint networks.

More precisely, we study a fragment of the Stochastic Constraint Satisfaction Problem (SCSP), that captures GDL games with uncertain (but complete) information. Interestingly, the SCSP for this class of games can be decomposed into a sequence of μSCSPs (a.k.a one-stage stochastic constraint satisfaction problems). Based on this decomposition, we propose a sequential decision-making algorithm, MAC–UCB, that combines the MAC algorithm (*Maintaining Arc Consistency*) for solving each μSCSP, and the multi-armed bandits *Upper Confidence Bound* (UCB) method for approximating the expected utility of strategies. We show that in practice MAC–UCB significantly outperforms (the multi-player version of) UCT, which

---

[1] This paper has been published in Constraints 21(1), the Journal Fast Track issue of CP'15.

G. Pesant (Ed.): CP 2015, LNCS 9255, pp. 726–727, 2015.
DOI: 10.1007/978-3-319-23219-5

is the state-of-the-art GGP algorithm for stochastic games. MAC–UCB also domi-nates FC–UCB, a variant where MAC is replaced with the Forward Checking (FC) algorithm. Such conclusions are drawn from comparing the performance of these algorithms, using extensive experiments (about $1,350,000$ face-offs) over a wide range of GDL games.

Visual Search Tree Profiling

Maxim Shishmarev[1] Christopher Mears[1],
Guido Tack[2], and Maria Garcia de la Banda[2]

[1] Faculty of IT, Monash University, Australia
[2] National ICT Australia (NICTA) Victoria
{maxim.shishmarev,chris.mears,guido.tack,maria.garciadelabanda}@monash.edu

Finding the best combination of model, solver and search strategy for solving a given combinatorial problem is a challenging, iterative process with three classical steps: (1) *observing* the behaviour of the program; (2) developing a *hypothesis* about why certain unwanted behaviour occurs; (3) *modifying* the program to test the hypothesis by observing a *change in behaviour*.

Performance profiling tools are designed to support and speed up this process in a variety of ways. For constraint programs based on tree search, the existing profiling tools mostly focus on visualising the search tree and related information or gathering crude aggregate measures. While these tools can provide very useful information, they are also limited by requiring close coupling between the profiling tool and the solver or search strategy. This makes it difficult for the profiling tool to integrate new solvers and search strategies, or to compare different solvers. Also, they often leave the user to explore the search tree in a relatively unguided way, which is particularly problematic for large trees. Finally, and most importantly, almost none of these tools support the *comparison* of different executions of the same problem, where one or more aspects of the model, solver, or search strategy have been modified.

We present an architecture and tool set specifically designed to support programmers during the performance profiling process. In particular, the profiling tool enables programmers to *extract* the information they need to build hypotheses as to why the search is behaving in a particular way. Then it helps to *validate* their hypotheses by identifying and visualising the effect of changes in the program (be they changes in the model, solver, or search strategy). The former is achieved by providing different views and analyses of the execution as well as efficient navigation tools to help users focus their attention on the parts of the execution that might be worth modifying. The latter is achieved by providing two tools: one that can *replay* searches using a different model or solver, so that the user can isolate the effect of a change; and a second tool that can visually "merge" the common parts of two executions and allows users to explore the parts that differ.

The architecture is designed to easily accommodate new solvers and search strategies, and it allows for solvers to be easily integrated without imposing an unreasonable overhead on the program's execution. The modularity of the design supports the combination of many different visualisation and analysis components. Our implementation is available at https://github.com/cp-profiler.

This paper is published in *Constraints* 21(1).

© Springer International Publishing Switzerland 2015
G. Pesant (Ed.): CP 2015, LNCS 9255, p. 728, 2015.
DOI: 10.1007/978-3-319-23219-5

Long-Haul Fleet Mix and Routing Optimisation with Constraint Programming and Large Neighbourhood Search

Philip Kilby and Tommaso Urli

Optimisation Research Group, NICTA Canberra Research Lab, 7 London Circuit,
Canberra, 2601 ACT, Australia
{philip.kilby,tommaso.urli}@nicta.com.au

Abstract. We present an original approach to compute efficient mid-term fleet configurations, at the request of a Queensland-based long-haul trucking carrier. Our approach considers one year's worth of demand data, and employs a constraint programming (CP) model and an adaptive large neighbourhood search (LNS) scheme to solve the underlying multi-day multi-commodity split delivery capacitated vehicle routing problem. A Pareto-based pre-processing step allows to tackle the large amount of data by extracting a set of representative days from the full horizon. Our solver is able to provide the decision maker with a set of Pareto-equivalent fleet setups trading off fleet efficiency against the likelihood of requiring on-hire vehicles and drivers. Moreover, the same solver can be used to solve the daily loading and routing problem. We carry out an extensive experimental analysis based on real-world data, and we discuss the properties of the generated fleets. We show that our approach is a sound methodology to provide decision support for the mid- and short-term decisions of a long-haul carrier[1].

---

[1] An extended version of this paper, with the title *Fleet Design Optimisation From Historical Data Using Constraint Programming and Large Neighbourhood Search*, will be published in Constraints 21(1).

© Springer International Publishing Switzerland 2015
G. Pesant (Ed.): CP 2015, LNCS 9255, p. 729, 2015.
DOI: 10.1007/978-3-319-23219-5

Abstracts of Published Journal Track Papers

On the Reification of Global Constraints
(Abstract)

Nicolas Beldiceanu[1], Mats Carlsson[2], Pierre Flener[3], and Justin Pearson[3]

[1] TASC (CNRS/INRIA), Mines Nantes, FR – 44307 Nantes, France
[2] SICS, P.O. Box 1263, SE – 164 29 Kista, Sweden
[3] Uppsala University, Dept of Information Technology, SE – 751 05 Uppsala, Sweden

Conventional wisdom has it that many global constraints cannot be easily reified, i.e., augmented with a 0-1 variable reflecting whether the constraint is satisfied (value 1) or not (value 0). Reified constraints are useful for expressing propositional formulas over constraints, and for expressing that a number of constraints hold (e.g., a cardinality operator). The negation of constraints is important for inferring constraints from negative examples and for proving the equivalence of models. Many early CP solvers, such as CHIP, GNU Prolog, Ilog Solver, and SICStus Prolog, provide reification for arithmetic constraints. However, when global constraints started to get introduced (e.g., ALLDIFFERENT and CUMULATIVE), reification was not available for them. It was believed that reification could only be obtained by modifying the filtering algorithms attached to each constraint. We present a reification method that works in all CP solvers.

A global constraint $G(\mathcal{A})$ can be defined by restrictions R on its arguments $\mathcal{A}$, say on their bounds, and by a condition C on $\mathcal{A}$, i.e., $G(\mathcal{A}) \equiv R(\mathcal{A}) \wedge C(\mathcal{A})$. We define the *reified version* of $G(\mathcal{A})$ as $R(\mathcal{A}) \wedge (C(\mathcal{A}) \Leftrightarrow b)$, where b is a 0-1 variable reflecting whether $G(\mathcal{A})$ holds or not. We require the negation of $G(\mathcal{A})$ to satisfy the same restrictions $R(\mathcal{A})$. It turns out that the condition $C(\mathcal{A})$ can often be reformulated as a conjunction $F(\mathcal{A}, \mathcal{V}) \wedge N(\mathcal{A}, \mathcal{V})$ of constraints, where $\mathcal{A}$ and $\mathcal{V}$ are disjoint sets of variables, $R(\mathcal{A})$ implies that $F(\mathcal{A}, \mathcal{V})$ is a total-function constraint uniquely determining $\mathcal{V}$ from any $\mathcal{A}$, and $N(\mathcal{A}, \mathcal{V})$ is a Boolean combination of linear arithmetic (in)equalities and 0-1 variables; we assume such Boolean combinations are already reifiable, as is the case in all CP solvers we are aware of. The reified version of $G(\mathcal{A})$ is then defined as $R(\mathcal{A}) \wedge F(\mathcal{A}, \mathcal{V}) \wedge (N(\mathcal{A}, \mathcal{V}) \Leftrightarrow b)$. For example, ALLDIFFERENT($\langle v_1, \dots, v_n \rangle$) is reified by SORT($\langle v_1, \dots, v_n \rangle, \langle w_1, \dots, w_n \rangle) \wedge ((w_1 < w_2 \wedge \dots \wedge w_{n-1} < w_n) \Leftrightarrow b)$, where $\langle v_1, \dots, v_n \rangle$ uniquely determine $\langle w_1, \dots, w_n \rangle$ in the total-function constraint SORT($\langle v_1, \dots, v_n \rangle, \langle w_1, \dots, w_n \rangle$).

Surprisingly, this insight that many constraints can naturally be defined by a *determine and test* scheme, where the *determine* part is associated to total-function constraints that determine additional variables, and the *test* part to a reifiable constraint on these variables, allows us to reify at least 313 of the 381 (i.e., 82%) constraints of the Global Constraint Catalogue. Most of the constraints not covered are graph constraints involving set variables.

Beldiceanu, N., Carlsson, M., Flener, P., Pearson, J.: On the reification of global constraints. Constraints 18(1), 1–6 (January 2013), http://dx.doi.org/10.1007/s10601-012-9132-0, full details at http://soda.swedish-ict.se/5194

© Springer International Publishing Switzerland 2015
G. Pesant (Ed.): CP 2015, LNCS 9255, p. 733, 2015.
DOI: 10.1007/978-3-319-23219-5

MDD Propagation for Sequence Constraints

David Bergman[1], Andre A. Cire[2], and Willem-Jan van Hoeve[3]

[1] School of Business, University of Connecticut
[2] Department of Management, University of Toronto Scarborough
[3] Tepper School of Business, Carnegie Mellon University
david.bergman@business.uconn.edu, acire@utsc.utoronto.ca,
vanhoeve@andrew.cmu.edu

We study MDD propagation for the well-known SEQUENCE constraint, which finds applications in, e.g., car sequencing and employee scheduling problems. It is defined as a specific conjunction of AMONG constraints, where an AMONG constraint restricts the occurrence of a set of values for a sequence of variables to be within a lower and upper bound. It is known that conventional domain consistency can be established for SEQUENCE in polynomial time, and that MDD consistency can be established for AMONG constraints in polynomial time. However, it remained an open question whether or not MDD consistency for SEQUENCE can be established in polynomial time as well.

In this work, we answer that question negatively and our first contribution is showing that establishing MDD consistency on the SEQUENCE constraint is NP-hard. This is an important result from the perspective of MDD-based constraint programming. Namely, of all global constraints, the SEQUENCE constraint has perhaps the most suitable combinatorial structure for an MDD approach; it has a prescribed variable ordering, it combines sub-constraints on contiguous variables, and existing approaches can handle this constraint fully by using bounds reasoning only. As our second contribution, we show that establishing MDD consistency on the SEQUENCE constraint is fixed parameter tractable with respect to the lengths of the sub-sequences (the AMONG constraints), provided that the MDD follows the order of the SEQUENCE constraint. The proof is constructive, and follows from a generic algorithm to filter one MDD with another. The third contribution is a partial MDD propagation algorithm for SEQUENCE, that does not necessarily establish MDD consistency. It relies on the decomposition of SEQUENCE into 'cumulative sums', and a new extension of MDD filtering to the information that is stored at its nodes.

Our last contribution is an experimental evaluation of our proposed partial MDD propagation algorithm. We compare its performance with existing domain propagators for SEQUENCE, and with the currently best known MDD approach that uses the natural decomposition of SEQUENCE into AMONG constraints. Our experiments demonstrate that our algorithm can reduce the search tree size, and solving time, by several orders of magnitude in both cases. Our results thus provide further evidence for the power of MDD propagation in the context of constraint programming.

This is a summary of the paper "D. Bergman, A. A. Cire, and W.-J. van Hoeve. MDD Propagation for Sequence Constraints. *JAIR*, Volume 50, pages 697-722, 2014".

© Springer International Publishing Switzerland 2015
G. Pesant (Ed.): CP 2015, LNCS 9255, p. 734, 2015.
DOI: 10.1007/978-3-319-23219-5

Discrete Optimization with Decision Diagrams

David Bergman[1], Andre A. Cire[2], Willem-Jan van Hoeve[3], and John Hooker[3]

[1] School of Business, University of Connecticut
[2] Department of Management, University of Toronto Scarborough
[3] Tepper School of Business, Carnegie Mellon University
david.bergman@business.uconn.edu, acire@utsc.utoronto.ca,
vanhoeve@andrew.cmu.edu, jh38@andrew.cmu.edu

Decision diagrams (DDs) are compact graphical representations of Boolean functions originally introduced for applications in circuit design and formal verification. Recently, DDs have also been used for a variety of purposes in optimization and operations research. These include facet enumeration in integer programming, maximum flow computation in large-scale networks, solution counting in combinatorics, and learning in genetic programming.

In this paper we provide a number of contributions for the application of decision diagrams to *discrete optimization* problems. First, we propose a novel *branch-and-bound* algorithm for discrete optimization in which binary decision diagrams (BDDs) play the role of the traditional linear programming relaxation. In particular, *relaxed BDD representations* of the problem provide bounds and guidance for branching, while *restricted BDDs* supply a primal heuristic. We show how to compile relaxed and restricted BDDs given a dynamic programming (DP) model of the problem. This compilation process is based on augmenting the DP model with problem-specific *relaxation* and *restriction* operators, which are applied on the state transition graph of the model to derived relaxed and restricted BDDs, respectively. Finally, our novel search scheme branches within relaxed BDDs rather than on variable values, eliminating symmetry during search.

We report computational results of our branch-and-bound algorithm on a previously introduced DP model of the maximum weighted independent set problem, as well as on new DP formulations for the maximum cut problem and the maximum 2-SAT problem. The results are compared with those obtained from generic state-of-the-art solvers, in particular an integer programming and a semidefinite programming solver. We show that a rudimentary BDD-based solver is already competitive with or superior to existing technology. Specific to the maximum cut problem, we tested the BDD-based solver on a classical benchmark set and identified tighter relaxation bounds than have ever been found by any technique, nearly closing the entire optimality gap on four large-scale instances.

This is a summary of the paper "D. Bergman, A. A. Cire, W.-J. van Hoeve, and J. Hooker. Discrete Optimization with Decision Diagrams. *INFORMS Journal on Computing*, forthcoming 2015.

© Springer International Publishing Switzerland 2015
G. Pesant (Ed.): CP 2015, LNCS 9255, p. 735, 2015.
DOI: 10.1007/978-3-319-23219-5

A Hybrid Approach Combining Local Search and Constraint Programming for a Large Scale Energy Management Problem

Haris Gavranović[1] and Mirsad Buljubašić[2]

[1] International University of Sarajevo, Bosnia and Herzegovina
haris.gavranovic@gmail.com
[2] Ecole des Mines d'Ales, LGI2P Research Center, Nimes, France
mirsad.buljubasic@mines-ales.fr

In this work we take the perspective of a large utility company, tackling their problems in modeling and planning production assets, i.e., a multitude of power plants. The goal is to fulfill the respective demand of energy over a time horizon of several years, while minimizing a total operation cost.

The proposed problem consists of modeling the production assets and finding an optimal outage schedule that includes two mutually dependent and related subproblems:

1. Determining the schedule of plant outages. This schedule must satisfy a certain number of constraints in order to comply with limitations on resources, which are necessary to perform refueling and maintenance operations.
2. Determining an optimal production plan to satisfy demand. Production must also satisfy some technical constraints.

The work presents a heuristic approach combining constraint satisfaction, local search and a constructive optimization algorithm for a large-scale energy management and maintenance scheduling problem. The methodology shows how to successfully combine and orchestrate different types of algorithms and to produce competitive results. The numerical results obtained on the close-to-real EDF instances testify about the quality of the approach. In particular, the method achieves 3 out of 15 possible best results.

It is possible to use different approaches, such as local search, constraint programming or constructive procedure, to determine a feasible outage schedule. Constraint programming with a performant solver emerged to be the most reliable, and probably the most elegant approach, with very good numerical results.

The set of additional, artificial, constraints are added to the model to improve the diversity and the feasibility of the remaining production assignment problem. Finally, the feasible solution found in this way is then improved using a series of local improvements.

References

1. H. Gavranović, and M. Buljubaić, A Hybrid Approach Combining Local Search and Constraint Programming for a Large Scale Energy Management Problem, RAIRO - Operations Research, Volume 47,Issue 04,2013,pp 481-500,DOI 10.1051/ro/2013053,

© Springer International Publishing Switzerland 2015
G. Pesant (Ed.): CP 2015, LNCS 9255, p. 736, 2015.
DOI: 10.1007/978-3-319-23219-5

Representing and Solving Finite-Domain Constraint Problems using Systems of Polynomials (Extended Abstract)

Chris Jefferson[1], Peter Jeavons[2], Martin J. Green[3], and M.R.C. van Dongen[4]

[1] Department of Computer Science, University of St Andrews, UK
[2] Department of Computer Science, University of Oxford, UK
[3] Department of Computer Science, Royal Holloway, University of London, UK
[4] Department of Computer Science, University College Cork, Ireland

In this paper [1] we investigate the use of a system of multivariate polynomials to represent the restrictions imposed by a collection of constraints. A system of polynomials is said to allow a particular combination of values for a set of variables if the simultaneous assignment of those values to the variables makes all of the polynomials in the system evaluate to zero.

The use of systems of polynomials has been considered a number of times in the constraints literature, but is typically used to represent constraints on continuous variables. Here we focus on the use of polynomials to represent finite domain constraints. One advantage of representing such constraints using polynomials is that they can then be treated in a uniform way along with continuous constraints, allowing the development of very general constraint-solving techniques. Systems of polynomials can be processed by standard computer algebra packages such as Mathematica and REDUCE, so our approach helps to unify constraint programming with other forms of mathematical programming.

Systems of polynomials have been widely studied, and a number of general techniques have been developed, including algorithms that generate an equivalent system with certain desirable properties, called a Gröbner Basis. Given a Gröbner Basis, it is possible to obtain the solutions to a system of polynomials very easily (or determine that it has no solutions). A Gröbner Basis provides a convenient representation for the whole set of solutions which can be used to answer a wide range of questions about them, such as the correlations between individual variables, or the total number of solutions.

In general, the complexity of computing a Gröbner Basis for a system of polynomials is doubly-exponential in the size of the system. However, we observe that with the systems we use for constraints over finite domains this complexity is only singly exponential, and so comparable with other search techniques for constraints. Our main contributions are a family of polynomial-time algorithms, related to the construction of Gröbner Bases, which simulate the effect of the local-consistency algorithms used in constraint programming. Finally, we discuss the use of adaptive consistency techniques for systems of polynomials.

References

1. C. Jefferson, P. Jeavons, M. J. Green, and M. R. C. van Dongen. Representing and solving finite-domain constraint problems using systems of polynomials. *Annals of Mathematics and Artificial Intelligence*, 67(3-4):359–382, 2013.

© Springer International Publishing Switzerland 2015
G. Pesant (Ed.): CP 2015, LNCS 9255, p. 737, 2015.
DOI: 10.1007/978-3-319-23219-5

A Quadratic Extended Edge-Finding Filtering Algorithm for Cumulative Resource Constraints

Roger Kameugne[1,2], Laure Pauline Fotso[3], and Joseph Scott[4]

[1] Univ. of Maroua, Dept. of Mathematics, Cameroon
rkameugne@yahoo.fr
[2] Univ. of Yaoundé I, Dept. of Mathematics, Yaoundé, Cameroon
[3] Univ of Yaoundé I, Dept. of Computer Sciences, Cameroon
laurepfotso@yahoo.com
[4] Uppsala Univ., Dept. of Information Technology, Sweden
joseph.scott@it.uu.se

Extended edge-finding is a well-known filtering rule used in constraint-based scheduling problems for the propagation of constraints over both disjunctive and cumulative resources. A cumulative resource is renewable resource of fixed capacity, which may execute multiple tasks simultaneously so long as the resource requirement of tasks scheduled together never exceeds the capacity. In constraint-based scheduling, a single cumulative resource is typically modelled by means of the global constraint CUMULATIVE, which enforces the resource constraint for a set of tasks. While complete filtering for CUMULATIVE is NP-complete there are known polynomial time algorithms for solving specific relaxations of the constraint, including timetabling, edge-finding variants (i.e., standard, extended, and timetable), not-first/not-last, and energetic reasoning. These relaxations may lead to different pruning, so CUMULATIVE is commonly implemented as a conjunction of relaxations, yielding stronger fixpoints and smaller search trees.

Extended edge-finding is similar to the better known edge-finding algorithm: both reduce the range of possible start (resp. end) times of tasks through the deduction of new ordering relations. For a task $i \in T$, an extended edge-finder searches for a set of tasks that *must* end before the end (resp. start before the start) of i. Based on this newly detected precedence, the earliest start time (resp. latest completion time) of i is updated. The best previously known complexity for a cumulative extended edge-finding algorithm was $\mathcal{O}(kn^2)$, where n is the number of tasks on the resource, and k is the number of distinct capacity requirements.

We propose an $\mathcal{O}(n^2)$ extended edge-finding algorithm for cumulative resources. Similar to previous work on edge-finding, this new algorithm uses the notion of *minimum slack* to detect when extended edge-finding may strengthen a domain. The algorithm is more efficacious on domains already at the fix point of edge-finding. Experimental evaluation on a standard benchmark suite confirms that our algorithm outperforms previous extended edge-finding algorithms. We find

This is a summary of the paper: R. Kameugne, L.P. Fotso, and J. Scott. A Quadratic Extended Edge-Finding Filtering Algorithm for Cumulative Resource Constraints. International Journal of Planning and Scheduling 1(4):264–284, 2013.

© Springer International Publishing Switzerland 2015
G. Pesant (Ed.): CP 2015, LNCS 9255, pp. 738–739, 2015.
DOI: 10.1007/978-3-319-23219-5

that the conjunction of edge-finding and extended edge-finding leads to a slight decrease in efficiency in most cases when compared to edge-finding alone; however, on some instances the stronger pruning of the combined algorithm results in substantial performance gains. Furthermore, we show that our method is competitive with the current state-of-the-art in edge-finding based algorithms.

Achieving Domain Consistency and Counting Solutions for Dispersion Constraints

Gilles Pesant

École Polytechnique de Montréal, Montreal, Canada
CIRRELT, Université de Montréal, Montreal, Canada
gilles.pesant@polymtl.ca

Many combinatorial problems require of their solutions that they achieve a certain balance of given features. For example in assembly line balancing the workload of line operators must be balanced. In rostering we usually talk of fairness instead of balance, because of the human factor — here we want a fair distribution of the workload or of certain features such as weekends off and night shifts. In the Balanced Academic Curriculum Problem, courses are assigned to semesters so as to balance the academic load between semesters. Because of additional constraints (prerequisite courses, minimum and maximum number of courses per semester) and a varying number of credits per course, reaching perfect balance is generally impossible. A common way of encouraging balance at the modeling level is to set reasonable bounds on each load, tolerating a certain deviation from the ideal value, but it has the disadvantage of putting on an equal footing solutions with quite different balance. Another way is to minimize some combination of the individual deviations from the ideal value, but if other criteria to optimize are already present one must come up with suitable weights for the different terms of the objective function. Yet another way is to bound the sum of individual deviations.

In Constraint Programming the `spread` and `deviation` constraints have been proposed to express balance among a set of variables by constraining their mean and their overall deviation from the mean. Currently the only practical filtering algorithms known for these constraints achieve bounds consistency. In [1] we propose a more general constraint, `dispersion`, which includes both previous constraints as special cases and other variants of interest. We improve filtering by presenting an efficient domain consistency algorithm and also extend it to count solutions so that it can be used in counting-based search, a generic and effective family of branching heuristics that free the user from having to write problem-specific search heuristics.

We provide empirical evidence that these can lead to significant practical improvements in combinatorial problem solving. In particular we improve the state of the art on the Nurse to Patient Assignment Problem and on the Balanced Academic Curriculum Problem.

References

1. Gilles Pesant. Achieving Domain Consistency and Counting Solutions for Dispersion Constraints. *INFORMS Journal on Computing*, forthcoming. (Also available as https://www.cirrelt.ca/DocumentsTravail/CIRRELT-2015-08.pdf)

© Springer International Publishing Switzerland 2015
G. Pesant (Ed.): CP 2015, LNCS 9255, p. 740, 2015.
DOI: 10.1007/978-3-319-23219-5

meSAT: Multiple Encodings of CSP to SAT

Mirko Stojadinović and Filip Marić

Faculty of Mathematics, University of Belgrade, Serbia
{mirkos,filip}@matf.bg.ac.rs

Abstract. One approach for solving Constraint Satisfaction Problems (CSP) (and related Constraint Optimization Problems (COP)) involving integer and Boolean variables is reduction to propositional satisfiability problem (SAT). A number of encodings (e.g., direct, log, support, order, compact-order) for this purpose exist as well as specific encodings for some constraints that are often encountered (e.g., cardinality constraints, global constraints). However, there is no single encoding that performs well on all classes of problems and there is a need for a system that supports multiple encodings. In our paper *meSAT: Multiple encodings of CSP to SAT* we describe a system that translates specifications of finite linear CSP problems into SAT instances using several well-known encodings, and their combinations. The system also provides choice of using different SAT encodings of cardinality constraints. The experiments show that efficiency of direct, support and order encodings can be improved by their combinations, namely direct-support and direct-order encodings. Description of efficient encodings of global constraints within direct encoding and the correctness proof of a variant of order encoding are also provided. We present a methodology for selecting a suitable encoding based on simple syntactic features of the input CSP instance. The selection method is a modification of approach ArgoSmArT-kNN that uses k-nearest neighbors algorithm and was originally developed for SAT. Thorough evaluation has been performed on large publicly available corpora and our encoding selection method improves upon the efficiency of existing encodings and state-of-the-art tools used in comparison. We also show that if families of syntactically similar instances can be formed, then the selection can be trained only on very easy instances thus significantly reducing the training time. The idea is based on the observation that very often the solver that is efficient in solving easier instances of some problem will be also efficient in solving harder instances of the very same problem. The obtained results are very good, despite using short training time.

This is extended abstract of the paper: M. Stojadinović, F. Marić, mesat: Multiple encodings of csp to sat *Constraints*, Volume 19(4), pp. 380-403, (2014.)

This work was partially supported by the Serbian Ministry of Science grant 174021.

© Springer International Publishing Switzerland 2015
G. Pesant (Ed.): CP 2015, LNCS 9255, p. 741, 2015.
DOI: 10.1007/978-3-319-23219-5

Constraint programming for LNG ship scheduling and inventory management (Abstract)

Willem-Jan van Hoeve

Tepper School of Business, Carnegie Mellon University
5000 Forbes Avenue, Pittsburgh, PA 15213, USA
vanhoeve@andrew.cmu.edu

This work considers a central operational issue in the liquefied natural gas (LNG) industry: designing schedules for the ships to deliver LNG from the production (liquefaction) terminals to the demand (regasification) terminals. Operationally, a delivery schedule for each customer in the LNG supply chain is negotiated on an annual basis. The supply and demand requirements of the sellers and purchasers are accommodated through the development of this annual delivery schedule, which is a detailed schedule of LNG loadings, deliveries and ship movements for the planning year. Developing and updating this schedule is a complex problem because it combines ship routing with inventory management at the terminals.

A specific application of our work is to support supply chain design analysis for new projects. Typically, new LNG projects require building not only the natural gas production and treatment facilities, but also constructing the ships and the production-side terminals, berths and storage tanks. Goel et al. (*Journal of Heuristics*, 2012) first introduced an LNG inventory routing problem for this purpose. They proposed a mixed integer programming (MIP) model as well as a MIP-based heuristic for finding good solutions to this problem. Yet, both the MIP model and the MIP heuristic suffered from scalability issues for larger instances.

We propose two constraint programming (CP) models, both of which are based on a disjunctive scheduling representation. Visits of ships to facilities are modeled as optional activities, and distances are represented as sequence-dependent setup times for the disjunctive resource that represents each ship's schedule. In addition, we model the inventory at each location as a cumulative resource. Even if the number of (optional) activities is proportional to the number of time points, our CP models are still much smaller than the corresponding MIP models and much more scalable. We also propose an iterative search heuristic to generate good feasible solutions for these models. Computational results on a set of large-scale test instances demonstrate that our approach can find better solutions than existing approaches based on mixed integer programming, while being 4 to 10 times faster on average.

This is a summary of the paper "V. Goel, M. Slusky, W.-J. van Hoeve, K. C. Furman, and Y. Shao. Constraint Programming for LNG Ship Scheduling and Inventory Management. *European Journal of Operational Research* 241(3): 662-673, 2015.".

G. Pesant (Ed.): CP 2015, LNCS 9255, p. 742, 2015.
DOI: 10.1007/978-3-319-23219-5

Revisiting the Limits of MAP Inference by MWSS on Perfect Graphs

Adrian Weller

University of Cambridge
aw665@cam.ac.uk

We examine Boolean binary weighted constraint satisfaction problems without hard constraints, and explore the limits of a promising recent approach that is able to solve the problem exactly in polynomial time in a range of cases.

Specifically, we are interested in the problem of finding a configuration of variables $x = (x_1, \ldots, x_n) \in \{0,1\}^n$ that maximizes a score function, defined by unary and pairwise real terms $f(x) = \sum_{i=1}^{n} \psi_i(x_i) + \sum_{(i,j) \in E} \psi_{ij}(x_i, x_j)$. In the machine learning community, this is typically known as *MAP* (or *MPE*) *inference*, yielding a configuration of variables with maximum probability.

[2] showed that this problem (for discrete models of any arity) may be reduced to finding a maximum weight stable set (MWSS) in a derived weighted graph called an *NMRF* (closely related to the microstructure complement of the dual representation [1,3,4]), and introduced the technique of *pruning* certain nodes. Different *reparameterizations* can lead to different pruned NMRFs. A *perfect* pruned NMRF allows the initial problem to be solved in polynomial time.

Here we extend [5] to consider all possible reparameterizations of Boolean binary models. This yields a simple, exact characterization of the new, enlarged set of such models tractable with this approach. We also show how these may be efficiently identified, thus settling the power of the approach on this class.

The main result is that the method works for all valid cost functions if and only if each *block* of the original model is *almost balanced*.[1] This hybrid condition combines restrictions on both the structure and the types of pairwise constraints/score terms (no restrictions on unary terms).

References

1. M. Cooper and S. Živný. Hybrid tractability of valued constraint problems. *Artificial Intelligence*, 175(9):1555–1569, 2011.
2. T. Jebara. MAP estimation, message passing, and perfect graphs. In *Uncertainty in Artificial Intelligence*, 2009.
3. P. Jégou. Decomposition of domains based on the micro-structure of finite constraint-satisfaction problems. In *AAAI*, pages 731–736, 1993.
4. J. Larrosa and R. Dechter. On the dual representation of non-binary semiring-based CSPs. In *CP2000 workshop on soft constraints*, 2000.
5. A. Weller and T. Jebara. On MAP inference by MWSS on perfect graphs. In *Uncertainty in Artificial Intelligence (UAI)*, 2013.

This is a summary of the paper "A. Weller. Revisiting the Limits of MAP Inference by MWSS on Perfect Graphs. *JMLR*, Volume 38, pages 1061-1069, 2015".

[1] A block is a maximal 2-connected subgraph. A balanced block contains no frustrated cycle. An almost balanced block may be rendered balanced by removing one node.

G. Pesant (Ed.): CP 2015, LNCS 9255, p. 743, 2015.
DOI: 10.1007/978-3-319-23219-5

Author Index

Printed in the United States
By Bookmasters